David Hoyle's latest work on this ISO 9000 handbook once again "cracks the code", encrypted by the ISO TC 176 authors and offers, actionable, pragmatic advice for users and quality practitioners around the world on how to understand and effect the ISO 9000 family of documents. With so many competing publications on the subject, this 7th edition should be the essential, go-to handbook for quality professionals seeking to understand and benefit from an ISO 9001 modelled quality system.

Sidney Vianna, *Management System Professional, DNV GL, USA*

A must-have for anyone tasked with facilitating the creation and delivery of value to stakeholders, David Hoyle's staggeringly powerful latest edition of Quality Systems Handbook expertly and thoroughly demystifies the updated ISO 9000 Standards. An indispensable tool for managing quality and successful certification.

John Colebrook, *Director, Enhanced Operating Systems Ltd, New Zealand*

A valuable resource that explores salient concepts arising from the ISO 9001:2015 standard in an unambiguous manner. David Hoyle using his inimitable step by step approach demonstrates the applicability of quality management standard to any organisational context. This book is definitely recommended reading for both academics and business leaders seeking to gain an in depth understanding of quality management systems.

Dr Lowellyne James, *Lecturer in Quality Management, Robert Gordon University, IEMA Certificate in Sustainability Strategy Programme Leader, Scotland*

To be up-to-date with the latest in *ISO 9000 Quality Systems* (ISO9000:2015), this is the definitive guide to refer to. David Hoyle's 7th edition stays true to its purpose of ensuring interpreting *the Standard*, comprehensible.

Christopher Seow, *Visiting Lecturer "Six Sigma for Managers" Cass Business School, City, University of London, UK*

David Hoyle pulls no punches in this very comprehensive seventh edition of his ISO 9000 Handbook. Whether you are a scholar, business executive, quality manager, consultant or a Certification Body, this book is for you. He goes out of his way to change the misconception of what the ISO 9001:2015 standard is really all about. He meticulously explains that it is aimed at improved performance and not just a set of requirements for compliance.

David convinces the reader that Quality should be managed as an integral part of every business.

Paul Harding, *Managing Director, South African Quality Institute, South Africa*

Yet again David Hoyle has produced another excellent book that explains the new standard and its requirements in straightforward terms. It is the book we have all been waiting for that will help us implement the new Standard and will become the new bible for quality management Systems. If you follow the advice in Hoyle's book you will not have any issues when the auditor arrives at your door. It is suitable for all people involved in the standard from company directors who need to know more about their responsibilities to experienced and busy Quality Managers who need to implement and inform others about the changes.

Rhian Newton, *HSEQ Manager, Morgan Advanced Materials, UK*

The book does a great job of showing how a quality system built to the principles of ISO 9000 and conforming to the new ISO 9001:2015 requirements can be an integral part of leadership's strategic approach to running the business versus letting a quality management system operate in a silo.

The book explores and explains the ISO9001:2015 requirements in a way that is easy to follow yet at the same time deep and meaningful especially for those new concepts and requirements such as leadership, context of the organization and managing risk. This feels like a book I'll keep referring to for many years to come.

Richard Allan, *Director, Quality Assurance, Kimberly-Clark Corporation, UK*

ISO 9000 Quality Systems Handbook

Completely revised to align with ISO 9001:2015, this handbook has been the bible for users of ISO 9001 since 1994, helping organizations get certified and increase the quality of their outputs.

Whether you are an experienced professional, a novice, or a quality management student or researcher, this is a crucial addition to your bookshelf. The various ways in which requirements are interpreted and applied are discussed using published definitions, reasoned arguments and practical examples. Packed with insights into how the standard has been used, misused and misunderstood, *ISO 9000 Quality Systems Handbook* will help you to decide if ISO 9001 certification is right for your company and will gently guide you through the terminology, requirements and implementation of practices to enhance performance.

Matched to the revised structure of the 2015 standard, with clause numbers included for ease of reference, the book also includes:

*   Graphics and text boxes to illustrate concepts, and points of contention;
*   Explanations between the differences of the 2008 and 2015 versions of ISO 9001
*   Examples of misconceptions, inconsistencies and other anomalies
*   Solutions provided for manufacturing and service sectors.

This new edition includes substantially more guidance for students, instructors and managers in the service sector, as well as those working with small businesses.

Don't waste time trying to achieve certification without this tried and trusted guide to improving your business – let David Hoyle lead you towards a better way of thinking about quality and its management and see the difference it can make to your processes and profits!

David Hoyle as a manager, consultant, author and mentor has been helping individuals and organizations across the world understand and apply ISO 9001 effectively since its inception in 1987. He has held senior positions in quality management with British Aerospace and Ferranti International and worked with such companies as General Motors, the UK Civil Aviation Authority and Bell Atlantic on their quality improvement programmes. Although neither a member of ISO nor BSI technical committees, he has been a member of the CQI for over 40 years and through his work with them built a network of like-minded professionals including members of ISO and BSI technical committees, which gave him privileged access to many reports, presentations and early drafts and thus gained an insight into the thinking behind ISO 9001:2015.

ISO 9000 Quality Systems Handbook

Increasing the Quality of an Organization's Outputs

Seventh Edition

David Hoyle

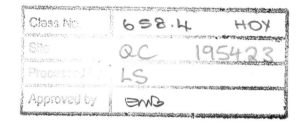

Routledge
Taylor & Francis Group

LONDON AND NEW YORK

Learning Resources Centre

Seventh edition published 2018
by Routledge
2 Park Square, Milton Park, Abingdon, Oxon OX14 4RN

and by Routledge
711 Third Avenue, New York, NY 10017

Routledge is an imprint of the Taylor & Francis Group, an informa business

© 2018 David Hoyle

First edition published by Butterworth-Heinemann 1994

Sixth edition published by Butterworth-Heinemann 2009 and by Routledge 2013

British Library Cataloguing-in-Publication Data
A catalogue record for this book is available from the British Library

Library of Congress Cataloging-in-Publication Data
Names: Hoyle, David, author.
Title: ISO 9000 quality systems handbook : using the standards as a framework
    for business improvement / David Hoyle.
Description: Seventh edition. | Abingdonm, Oxon ; New York, NY : Routledge,
    2017. | Includes bibliographical references and index.
Identifiers: LCCN 2016056514 (print) | LCCN 2016057166 (ebook) | ISBN
    9781138188631 (hardback : alk. paper) | ISBN 9781138188648 (pbk. : alk.
    paper) | ISBN 9781315642192 (ebook) | ISBN 9781315642192 (eBook)
Subjects: LCSH: ISO 9000 Series Standards—Handbooks, manuals, etc. |
    Quality control—Auditing—Handbooks, manuals, etc.
Classification: LCC TS156.6 .H69 2017 (print) | LCC TS156.6 (ebook) |
    DDC 658.5/620218—dc23
LC record available at https://lccn.loc.gov/2016056514

ISBN: 978-1-138-18863-1 (hbk)
ISBN: 978-1-138-18864-8 (pbk)
ISBN: 978-1-315-64219-2 (ebk)

Typeset in Times New Roman
by Apex CoVantage, LLC

Visit the companion website: http://www.routledge.com/cw/hoyle

Printed and bound in Great Britain by
TJ International Ltd, Padstow, Cornwall

Dedicated to my very dear friend and colleague, Tony Brown 1945–2016

Contents

Preface to the seventh edition   xiii

PART 1
Introduction   1

  1 Putting ISO 9001 in context   3

  2 Comparison between 2008 and 2015 editions   19

  3 How the 2015 version has changed misconceptions   37

    Key messages from Part 1   48

PART 2
Anatomy and use of the standards   51

  4 The ISO 9000 family of standards   53

  5 A practical guide to using these standards   61

    Key messages from Part 2   84

PART 3
Terminology   87

  6 Quality   91

  7 Requirements   102

  8 Management system   108

  9 Process and the process approach   126

 10 Risk and opportunity   148

 11 Interested parties and stakeholders   159

    Key messages from Part 3   168

PART 4

Context of the organization   171

12 Understanding the organization and its context   175

13 Understanding the needs and expectations of interested parties   188

14 Scope of the quality management system   199

15 Quality management system   210

16 Processes needed for the QMS   219

    Key messages from Part 4   251

PART 5

Leadership   255

17 Leadership and commitment   257

18 Customer focus   283

19 Policy   289

20 Organizational roles, responsibilities and authorities   305

    Key messages from Part 5   320

PART 6

Planning   323

21 Actions to address risks and opportunities   325

22 Quality objectives and planning to achieve them   347

23 Planning of changes   365

    Key messages from Part 6   373

PART 7

Support   375

24 People   379

25 Infrastructure   387

26 Environment for the operation of processes   396

27 Monitoring and measuring resources   409

28 Organizational knowledge   428

29 Competence   439

30 Awareness   457

31 Communication   464

32 Documented information   478

    Key messages from Part 7   514

PART 8
Operation   519

33 Operational planning and control   521

34 Customer communication   537

35 Requirements for products and services   551

36 Review of requirements for products and services   562

37 Design and development planning   575

38 Design and development inputs   597

39 Design and development controls   608

40 Design and development outputs   621

41 Design and development changes   630

42 Control of externally provided processes, products and services   637

43 Evaluation, selection and monitoring of external providers   649

44 Information for external providers   661

45 Control of production and service provision   671

46 Identification and traceability   686

47 Property belonging to external providers   692

48 Preservation of process outputs   698

49 Control of changes   704

50 Release, delivery and post-delivery of products and services   708

51 Control of nonconforming outputs   719

    Key messages from Part 8   730

PART 9
Performance evaluation 733

52 Monitoring, measurement, analysis and evaluation 735

53 Customer satisfaction 747

54 Analysis and evaluation 756

55 Internal audit 771

56 Management review 791

Key messages from Part 9 804

PART 10
Improvement 807

57 Determining and selecting opportunities for improvement 811

58 Nonconformity and corrective action 818

59 Continual improvement of the QMS 839

Key messages from Part 10 846

Appendix A: Common acronyms 848
Appendix B: Glossary of terms 850
Bibliography 861
Index 868

Preface to the seventh edition

Purpose and intended readership

This book is aimed primarily at professionals, researchers and students seeking to understand the concepts and requirements in the ISO 9000 family of standards in terms of what they mean, why they are necessary and how they might be addressed and conformity demonstrated. I have made no assumptions about my readers' prior knowledge or experience. In recognition that those seeking guidance on ISO 9001 may be experienced professionals but also people at various levels in an organization with no prior knowledge or experience of ISO 9001 or managing quality, the book is intended to meet the needs of both groups. Explanations may therefore appear laboured to some readers but provide new insight to others.

There are over 300 requirements in ISO 9001:2015, and the explanation of each of these forms the major portion of the book. The book is therefore intended as a source of reference for those using the standard to obtain or maintain ISO 9001 certification and as a tool to improve the quality of an organization's outputs and seeking meaning, justification and/or solutions.

Reason for the new edition

It is now 23 years since the first edition of the *Quality System Handbook*, and each subsequent edition has either built on experience gained in developing management systems or has followed revisions in the standard. This new edition is prompted by the major revision of ISO 9001 in 2015. Unlike the 2008 version, which contained no changes in requirements, the 2015 version is a complete revision, first to bring it in conformity with the new structure and common text of all new and future revisions of management system standards and second, to reflect the changes in the trading environment and user needs and expectations. The new structure and common text were introduced in 2014 through ISO/IEC directives that provide the rules to be followed by ISO committees. As ISO standards are reviewed every five years, it was inevitable that any future revision would have to take account of applicable ISO/IEC directives. However, if you look closely at the standard, you'll see that most of the requirements remain but are expressed in a different form. Less than a quarter are entirely new, but it will be the 20% of changes that give users 80% of the problems with the new version.

This seventh edition of the Handbook is a complete revision, but I have included many sections from the sixth edition where relevant. Although of comparable size to the sixth edition, a desire to maintain compatibility between the parts of the Handbook and the main sections of the standard has led to there being 10 parts as opposed to 8 in the sixth edition and 59 chapters as opposed to 40 in the sixth edition. System assessment, certification and continuing development that was in Part 8 of the sixth edition has been moved to the companion website, as have some of the appendices.

Correlation with ISO 9001 structure

The standard was not written for authors to write books about it or to guide users in developing a quality management system; if it had, the order in which the requirements are presented would have been different. Therefore, with very few exceptions, I have chosen to explain the requirements of the standard in the sequence in which they are presented in the standard and have added clause numbers to the headings to make it user friendly. Unlike previous editions, the Handbook now addresses all clauses of the standard, including those not containing specific requirements, as shown in the following table.

Table 0.1 Correlation with ISO 9001 structure

ISO 9001 Section	Handbook	
	Parts	*Chapters*
0.1 General	1, 4 & 7	3, 12 & 32
0.2 Quality management principles	1	5
0.3 Process approach	1	9
0.3.1 General	1	9
0.3.2 Plan-Do-Check-Act cycle	1	5
0.3.3 Risk-based thinking	3	10
0.4 Relationship with other management system standards	1	2
1 Scope	1	1
2 Normative references	2	4
3 Terms and definitions	3	6–11
4 Context of the organization	4	12–16
5 Leadership	5	17–20
6 Planning	6	21–23
7 Support	7	24–32
8 Operation	8	33–51
9 Performance evaluation	9	52–56
10 Improvement	10	57–59
A.1 Structure and terminology	1	2
A.2 Products and services	1	2
A.3 Understanding the needs and expectations of interested parties	4	13
A.4 Risk-based thinking	3 & 6	10 & 21
A.5 Applicability	1	1
A.6 Documented information	7	32
A.7 Organizational knowledge	7	28
A.8 Control of externally provided processes, products and services	8	42

Preparation for the seventh edition

Being neither a member of ISO nor BSI technical committees, this independence has allowed me the freedom to comment on international standards without bias. However, I have been a member of the Chartered Quality Institute (CQI) and its predecessor, the Institute of Quality Assurance (IQA), for over 40 years and through my work with them have built a network of like-minded professionals, some of whom are members of ISO and BSI technical committees.

I was involved with the IQA's Standards Development Group (SDG) during the 1994 revision of ISO 9001 and in 2008 was invited to re-join the SDG by its then chair, Tony Brown, who was at that time a member of BSI TC QS/1 and the UK representative on TC 176/SC1 and TC 69. In January 2010, I was invited to lead a project on behalf of the CQI to influence the next revision of the ISO 9000 family of standards and specifically ISO 9000 and ISO 9001. The project team included representatives from academia, industry, BSI Technical Committees (TC), consultancy and auditing, so it was a good mix of people to work with. We were privileged to have access to many TC reports, presentations and early drafts and thus gain an insight into their thinking. Following a member survey, we received many hundreds of e-mails, which helped shape our views and prepare a position paper on the next revision of the ISO 9000 family of standards for the CQI. The paper was published in May 2011 and was well received by BSI and TC 176. The team also provided extensive comments on ISO Guide 83 (the forerunner to Annex SL), ISO 9000 and ISO 9001 during the development process. I captured feedback from several presentations I gave at CQI branch events, lengthy discussions I had with members of TC 176, including Dr Nigel Croft (chair of TC 176/SC2), and I engaged with contributors on LinkedIn up to and following the release of the final versions of the two standards in September 2015.

In conducting research for this edition, I realized that the paradigm that has influenced the previous versions of ISO 9001, including the 2008 version, was that of scientific management, a belief that prescribing better ways of doing things and training people in those better ways will produce better quality outputs. There is no doubt that the application of scientific management does bring about great improvement in output quality, but it's predicated on treating the organization as a machine in which the parts have no choice. In reality, an organization is far more complex. For one thing, it is multi-minded, meaning that everyone in it has a choice. Organizations are also influenced by external factors, and ignoring these factors will result in its current performance being unsustainable.

Our understanding of how organizations function has evolved since World War II, and ISO management system standards have not kept pace. With each revision of ISO 9001, a few concepts were changed or added that moved it beyond managing an organization as if it were a machine to managing an organization as if it were a system of interdependent parts, but it's become a hybrid of systems theory and scientific management. During the development of this edition, I therefore encountered many issues that arose from ambiguities and inconsistencies in ISO 9000 and ISO 9001 that were not resolved by ISO TS 9002, the guide to the application of ISO 9001. Many of these arose first, from the definitions in ISO 9000 and the use of these terms in ISO 9001 and second, from the way requirements were expressed. At one stage I thought I'd not complete it at all as there were so many issues I had not resolved. But by ignoring some ISO definitions and putting my own interpretation on phrases like "establish, implement and maintain a quality management system", I think I have resolved these issues. Although many people may not choose to read the standards closely enough to appreciate these inconsistencies, it is important that they are highlighted and explanations provided.

One influential change in ISO 9001 has been the separation of product from service, and because the service sector is so diverse, I have used the example of a fast food outlet to illustrate how many of the requirements can be addressed. A fast food outlet has the advantage of being a provider of both products and services, and it's one most readers will have experienced at one time or another so I reckoned they will relate to it.

On several occasions, I tried testing opinions using LinkedIn groups, but what I have learnt is that it is easy to be deluded into believing there is understanding, despite the fact there appears to be a consensus. Time and again I find we don't all attribute the same meaning to the words we use. The meaning is strongly influenced by how we see things, what we call a paradigm, and I have addressed this and other communication issues in the book. The views I express mainly result from research and deduction rather than conjecture, and I have tried to present explanations that are faithful to the intent of the standard.

How to use the book

The contents list shows the parts and chapters of the book, and the chapter headings mirror the major clause headings of ISO 9001 from Chapter 12 onwards. A contents list is included in the introduction to each chapter, and the section headings are consistent with the subclauses or subject of the requirements within a clause. This should make it relatively easy to navigate the book.

I realize it's a hefty tome which may appear daunting to those unfamiliar with this work; therefore, for those readers who want a quick summary of each of the 10 parts, I have included a small number of pages at the end of each part that contain a few key messages from each chapter.

Although the book may be used as a source of reference where readers may look up a clause to find out what a requirement means, why it's necessary or how to address it or audit it, it is strongly recommended that the chapters in Parts 1, 2 and 3 be studied first. The reason for this is that without an understanding of the concepts, principles, terminology, recent changes, common misconceptions and different approaches, an unprepared reader can easily misinterpret what is written, both in the standard and in this book. The ways in which the term quality is commonly used may create fewer problems because its use in ISO 9001 is limited to specific concepts, but differences in the ways the terms risk, system, process, procedure and interested party are used may result in requirements being interpreted far differently from that which is intended.

In previous editions, the requirements of the standard have been paraphrased, but that has required the prior approval of BSI as the UK copyright holder. It is believed that this is no longer necessary, as judicious phrasing of chapter and section headings together with cross-references to clause numbers can achieve the same objective, but it is obviously desirable for users of this Handbook to have copies of ISO 9001:2015 and ISO 9000:2015 to hand.

Extensive use of cross-referencing is made throughout the book to avoid repetition. At the end of each chapter is a bibliography listing all the cited references in the chapter, and these are also recommended for further reading. As only chapters are numbered, cross-referencing to sections is by the ISO 9001 clause number that appears in the section heading. Also as tables, figures and text boxes are identified with the chapter number, this facilitates cross-referencing to specific locations within a chapter or section. Although tables and figures are referenced within the text, this is not always the case for text boxes.

As the Handbook addresses the requirements of ISO 9001, I decided to use the language of ISO 9001 to avoid any confusion. In the sixth edition I referred to a management system rather than a quality management system, but I have found that this could create ambiguities because it's clear from the requirements of ISO 9001 that they only apply to part of the management system. It may be a large part of the management system, but nonetheless implying otherwise would be confusing to say the least.

Companion website

A companion website will be created which contains checklists, examples of forms, case studies, PowerPoint presentations and other pedagogical features as they become available. See bottom of page vi for the url

Acknowledgements

I can't take credit for all the ideas expressed in this book. Some are my own, and some are from the people with whom I have had discussions, but many come from management philosophies, theories and techniques that have been pertinent to the field of quality management over the last 100 years. I have included over 150 citations from literature and over 24 citations from related ISO documents. I am indebted to Chris Cox, Chris Paris, Iain Moore, Janette Large, John Broomfield, John Colebrook, Mustafa Ghaleiw, Nigel Croft, Olec Kovalevsky, Paul Harding, Peter Fraser, Rhian Newton, Richard Allan, Sidney Vianna, Winston Edwards and the late Tony Brown with whom I have corresponded over the last eight years or so and who helped form my views on various aspects of ISO 9001 and ISO 9000. They each brought a different perspective to the discussions and helped reveal to me interpretations I hadn't considered. It was good to be able to share ideas with a group of professionals I knew and could trust.

Tony Brown was my ardent companion on this journey, always providing support and encouragement until his untimely illness and death dealt a devastating blow and cut short our almost daily Skype calls. Winston Edwards has helped enormously with the chapter on management systems, and Chris Cox from TC 176/SC1 was invaluable for his knowledge about ISO definitions and their development. I would also like to convey special thanks to Peter Fraser, Rhian Newton and Richard Allan who provided useful input and helped scrutinize several chapters of the manuscript, providing numerous suggestions for improvement. I am also indebted to my former business partner, John Thompson, who many years ago changed the way I think about so many different aspects of management and influenced the views on process management and the service industry presented in this book. There are also countless others with whom I have engaged in one way or another and shared views and experiences that in some way will have influenced something I have written in this book. From Taylor & Francis, I would like to thank Amy Laurens, my commissioning editor, and Nicola Cupit and Laura Hussey, editorial assistants, who arranged the independent reviews and have provided excellent support in the preparation of the manuscript. I also pay tribute to Tina Cottone for her project leadership skills, patience and cooperation in editing the first proof and bringing it up to T&F's production standards. Finally, I would like to thank my wife, Angela, who has been so patient these past 18 months while I laboured away working on the manuscript, often working into the early hours so as not to lose an idea. It is my wish that all readers will learn something from these pages that will not only enable them to increase the quality of their organization's outputs, but also provide a source of inspiration in the years to come.

David Hoyle
Monmouth
November 2016
E-mail: hoyle@transition-support.com

Part 1

Introduction

Introduction to Part 1

Consider these two scenarios:

A. The organization already exists; it is delivering services and providing products and services to customers, but getting quite a few complaints so it's becoming difficult to compete on quality. It is making a profit but not enough to invest for the future because the managers spend a lot of time firefighting instead of improving the efficiency and effectiveness of their processes. It's doing what it can to comply with regulations but occasionally breaches employment laws and environmental legislation as it strives to balance competing objectives. No matter how many times problems are fixed, similar problems seem to arise again elsewhere, and often quick fixes lead to bigger problems much later.

B. The organization already exists; it is delivering services and providing products and services to customers and mostly receiving compliments, but competition is tough. It is making enough profit to invest for the future because its managers are proactive, putting a lot of effort into ensuring risks to success are mitigated. Thus, this organization doesn't need to spend much time firefighting and can instead pursue opportunities for improving the efficiency and effectiveness of its processes. By striving to satisfy customers in a way that meets the needs of the other stakeholders, it has found it can balance competing objectives and has not had compliance issues of any significance. When problems arise, managers tend not to go for the quick fix, but spend time ensuring that actions to prevent their recurrence won't have adverse consequences later.

These scenarios represent situations where both organizations are likely to obtain benefits from adopting ISO 9001:2015. In scenario A, the organization will gain a competitive advantage from a significant improvement in its performance and demonstrable capability, and in scenario B, the organization will gain a competitive advantage by demonstrable capability. Most organizations are likely to be positioned between these two extremes and will therefore benefit to varying degrees from adopting ISO 9001:2015. Creating a competitive advantage in quality should be a goal of top management, and perhaps the most widely recognized tool for doing this is ISO 9001. The standard can be used in ways that make your organization less competitive, which is why it is so important that you digest Part 1 of this book first before deciding on your course of action.

In Chapter 1 we put the ISO 9000 family of standards in context. We examine its purpose, scope, content and application and the process by which it was developed. We include some statistics on its use and summarize the expected outcomes of accredited certification.

In Chapter 2 we compare the 2008 and 2015 editions, highlighting the significant changes, the rationale for the change in structure, the new requirements and the withdrawal of some requirements that were introduced in the first version nearly 30 years ago.

In Chapter 3 we examine the many misconceptions about the ISO 9000 family of standards that have grown since its inception. There are many views about the value of ISO 9001 certification, some positive and some negative. It has certainly spawned an industry that has not delivered as much as it promised, and even with the release of the 2015 version there is still much to be done to improve the standards, improve the image, improve the associated infrastructure and improve organizational effectiveness.

1 Putting ISO 9001 in context

Introduction

There shall be standard measures of wine, ale, and corn (the London quarter), throughout the kingdom. There shall also be a standard width of dyed cloth, russet, and haberject,[1] namely two ells within the selvedges.[2] Weights are to be standardised similarly.

(Magna Carta, 1215)

In the 800th anniversary year of Magna Carta, the International Organization for Standardization publishes a major revision to its most popular standard, ISO 9001. As will be understood from the quotation, standards have been used for centuries – in fact standards for quality have been traced as far back as 11th century BCE in China's Western Zhou Dynasty, but the notion of quality systems emerged after World War II when the industrial practices were still largely based on scientific management as defined by Frederick Winslow Taylor at the turn of the 20th century.

Box 1.1   Steve Jobs on quality

Customers don't form their opinions on quality from marketing, they don't form their opinions on quality from who won the Deming Award or who won the Baldridge Award. They form their opinions on quality from their own experience with the products or the services.

(Jobs, 1990)

Many people have started their journey towards ISO 9001 certification by reading the standard and trying to understand the requirements. They get so far and then call for help, but they often haven't learnt enough to ask the right questions. The helper might assume that the person already knows they are looking at ISO 9001 and therefore may not spend the necessary time for them to understand what it is all about, what pitfalls may lie ahead and whether, indeed, they need to make this journey at all. This should become clear when you read this chapter.

When you encounter ISO 9001 for the first time, it may be in a conversation, on the Internet, in a leaflet or brochure from your local chamber of commerce or, as many have done, from a customer. If you are a busy manager you could be forgiven for either putting it out of your mind or getting someone else to look into it. But you know that as a manager you are either maintaining the status quo or changing it, and if you stay with the status quo for too long, your organization will go into decline. So, you need to know whether there is an issue with product or service quality and:

*   What the issue is?
*   Why it's an issue?
*   What it's costing you?
*   What you should do about it?
*   What the impact of it will be?
*   How much it will cost to improve performance so that these types of issues don't recur?
*   Where the resources are going to come from?
*   When you need to act?
*   What the alternatives are and their relative costs?
*   What the consequences are of doing nothing?

In this chapter, we put ISO 9001 in context by explaining:

*   What ISO 9001 is intended to do, making the link between ISO 9000 and the fundamental basis for trade
*   The process by which international standards are developed and the roles of the various groups involved
*   Reasons for using ISO 9001 – the interdependent duo of capability and confidence
*   The scope of ISO 9001 and what it means
*   Applicability of ISO 9001 – why it's difficult to rule anything out
*   Design or assessment standard – the users' choice
*   ISO 9001 and the free movement of goods and services
*   Popularity of ISO 9001 certification – some facts, figures and trends
*   What accredited certification means and doesn't mean

What ISO 9001 is intended to do

Box 1.2   ISO 9000 in a nutshell

The standards were created to facilitate international trade.

Organizations use ISO 9001 to demonstrate their capability and in so doing give their customers confidence that they will satisfy their needs and expectations and are committed to continual improvement.

Customers use ISO 9001 to obtain an assurance of product and service quality that they can't get simply by examining them.

All other standards in the ISO 9000 family address particular aspects of quality management.

Since the dawn of civilization, the survival of communities has depended on trade. As communities grow, they become more dependent on others providing goods and services they are unable to provide from their own resources. Trade continues to this day on the strength of the customer–supplier relationship. The relationship survives through trust and confidence at each stage in the supply chain. A reputation for delivering a product or a service to an agreed specification, at an agreed price, on an agreed date is hard to win, and organizations will protect their reputation against external threats at all costs. But reputations are often damaged, not necessarily by those outside, but by those inside the organization and by other parties in the supply chain. Broken promises, whatever the cause, harm reputation, and promises are broken when an organization does not do what it has committed to do. This can arise either because the organization accepted a commitment it did not have the capability to meet or it had the capability but failed to manage it effectively.

This is what the ISO 9001 is all about. It is a set of criteria that, when satisfied by an organization, enable it to demonstrate its capability and in so doing give their customers confidence that they will meet their needs and expectations. Customers use it to obtain an assurance of product and service quality that they can't get simply by examining them. It can be applied to all organizations regardless of type, size and product or service provided. When applied correctly these standards will help organizations develop the capability to create and retain satisfied customers in a manner that satisfies all the other stakeholders. They are not product or service standards – there are no requirements for specific products or services – they contain criteria that apply to the management of an organization in satisfying customer needs and expectations in a way that satisfies the needs and expectations of other stakeholders.

Box 1.3   ISO standards

ISO develops only those standards that are required by the market. This work is carried out by experts coming from the industrial, technical and business sectors which have asked for the standard, and which subsequently put them to use.

(ISO, 2009)

ISO standards are voluntary and are based on international consensus among the experts in the field. ISO is a non-governmental organization, and it has no power to enforce the implementation of the standards it develops. It is a network of the national standards bodies from 162 countries, and its aim is to facilitate the international coordination and unification of industrial standards.

Most internationally agreed standards apply to specific types of products and services with the aim of ensuring interchangeability, compatibility, interoperability, safety, efficiency and reduction in variation. Mutual recognition of standards between trading organizations and countries increases confidence and decreases the effort spent in verifying that suppliers have shipped acceptable products or delivered acceptable services.

The ISO 9000 family of standards is just one small group of standards among over 19,500 internationally agreed standards and other types of normative documents in ISO's portfolio that are instrumental in facilitating international trade.

The standards in the ISO 9000 family provide a vehicle for consolidating and communicating concepts in the field of quality management. It is not their purpose to fuel the certification, consulting, training and publishing industries. The primary users of the standards are intended to be organizations acting as either customers or suppliers.

You don't need to use any of the standards in the ISO 9000 family to develop the capability of satisfying your stakeholders; there are other models, but none are more widely used.

Overview of the ISO standards development process

To understand why the 2015 version of ISO 9001 is the way it is, an appreciation of the standards development processes is necessary.

The International Organization of Standardization (ISO) is a network of national standards bodies (NSB). Each member represents ISO in its country. Individuals or companies cannot become ISO members. ISO standards are developed by groups of experts which form technical committees (TCs). Each TC deals with a different subject and is made up of representatives of industry, non-governmental organizations (NGOs), governments and other stakeholders, who are put forward by ISO's members. Being nominated as an expert doesn't mean a person is academically qualified or is more knowledgeable than anyone else in a subject. They must be able to demonstrate expertise to their peers on the TC in some area of the committee's work and be available to attend ISO committee meetings wherever they are convened. Not every subject matter expert can commit to this level of participation, as it can be quite time consuming. The experts are not employees of ISO, and therefore ISO's role is as a facilitator rather than a developer of standards, as it doesn't fund their development.

The process stages are as follows:

1   New standard is proposed to TC. This is the ***Proposal stage***, and if the proposal is accepted it moves to stage 2.
2   Working group of experts (WG) start discussion to prepare a working draft. This is the ***Preparatory stage*** which, when complete, the working draft moves to stage 3.
3   First working draft (WD) is shared with the TC and with ISO CS (Central Secretariat). If the committee uses the ***Committee stage***, a committee draft (CD) is circulated to the members of the committee, who then comment and vote. If consensus is reached within the TC, it moves to stage 4.
4   Draft international standard (DIS) is prepared and shared with all ISO national members, who have three months to comment. This is the ***Enquiry stage***, and if consensus is reached it moves to stage 5.
5   Final draft (FDIS) is sent to all ISO members. This is the ***Approval stage***, and the standard is approved if a two-thirds majority of the participating members (P-members) of the TC/SC is in favour and not more than one-quarter of the total number of votes cast are negative. Only editorial corrections are made to the final text.
6   The ISO international standard is published.

As can be seen, there are two stages where the decision to proceed is based on consensus rather than unanimity. ISO standards are reviewed every five years to determine whether they should be revised, withdrawn or confirmed extant for a further five years, and those that are to be revised pass through the aforementioned process. Further details are on the ISO website (ISO-SD, 2016).

Reasons for using ISO 9001

Trading organizations need to achieve sustained success in a complex, demanding and ever-changing environment. This depends on their capability to:

a) anticipate and/or identify the needs and expectations of their customers and other stakeholders;
b) convert the anticipated or identified needs and expectations of customers into products and services that will satisfy all stakeholders;
c) attract customers to the organization;
d) supply the products and services that meet customer requirements and deliver the expected benefits;
e) operate in a manner that satisfies the needs of the other stakeholders.

Many organizations develop their own ways of working and strive to satisfy their customers in the best way they know how. We will explain this further in more detail, but in simple terms the management system enables the organization to do (a)–(e) and includes both a technical capability and a people capability. Many organizations develop the technical capability but not the people capability and are thus forever struggling to do what they say they will do.

In choosing the best solution for them, they can either go through a process of trial and error, select from the vast body of knowledge on management or utilize one or more management models available that combine proven principles and concepts to develop the organization's capability. ISO 9001 represents one of these models. Others are Business Excellence Model, Six Sigma and Business Process Management (BPM).

Box 1.4   Food for thought

Many ISO 9001–registered organizations fail to satisfy their customers, but this is largely their own fault – they simply don't do what they say they will do, and even those that do may not be listening to their customers. If your management is not prepared to change its values, it will always have problems with quality.

Having given the organization the capability to do (a)–(e) above, in many business-to-business relationships, organizations are able to give their customers confidence in their capability without becoming registered to ISO 9001. In some market sectors, there is a requirement to demonstrate capability through independently regulated conformity assessment procedures before products and services are purchased. In such cases the organization has no option but to seek ISO 9001 certification if it wishes to retain business from that particular customer or market sector. However, conformity with ISO 9001 may not be a USP (unique selling point) in the markets that some organizations operate, as it no longer confers special status. It may already be perceived as a given without certification being expected or mandated.

Scope of ISO 9001

The scope of ISO 9001 is expressed in clause 1, and it's worth pulling this apart to explain the concepts it contains because it's an essential part of the standard that is often overlooked.

Requirements for a quality management system

ISO 9001 is not a quality management system (QMS); it contains requirement for a QMS. These are requirements that the industry representatives of national standards bodies believe will adversely affect the quality of products and services were they not to be met. A QMS is that part of an organization's management system that creates and retains customers by understanding their needs and designing and providing products and services that satisfy those needs (see Chapter 8 for further explanation).

Need to demonstrate its ability

Levels of confidence

It is clearly stated that the standard is for use by organizations that need to demonstrate their ability, but where would that need come from? From its inception ISO 9001 has been a business-to-business standard. Customers need confidence that their suppliers can meet their quality, cost and delivery requirements and have a choice as to how they acquire this confidence. They can select their suppliers:

a)   purely based on past performance, reputation or recommendation. (This option is often selected for general services, inexpensive or non-critical products coupled with some basic receipt or service completion checks.);

b)   by assessing the capability of potential suppliers themselves. (This option is often selected for bespoke services and products where quality verification by the purchaser is possible.);

c)   based on an assessment of capability performed by a third party. (This option is often selected for professional services and complex or critical products where their quality cannot be verified by external examination of the output alone.).

Most customers select their suppliers using option (a) or (b), but there will be cases where these options are not appropriate because there is no evidence for using option (a) or resources are not available to use option (b), or it is uneconomic to travel halfway around the world when an accredited certification body is on the spot to do the same job. Whether the certification body would do this job as well as the customer is the subject of much debate and is not an issue addressed by ISO 9001. ISO 9001 was developed for use in situations (b) and (c), enabling customers to impose common QMS requirements on their suppliers and either assess those suppliers themselves or use third-party audit as a means of obtaining confidence that their requirements will be met.

Overview of certification process

An organization wishing to do business with a customer and assure them they can meet their requirements submits to a second-party audit performed by their customer or a third-party audit performed by an accredited certification body independent of both customer and supplier. An audit is performed against the requirements of ISO 9001, and if no major nonconformities are found, a certificate is awarded. This certificate provides evidence that the organization has demonstrated its ability to meet certain requirements. Customers are now able to acquire the confidence they require, simply by establishing whether a supplier holds a current ISO 9001 certificate covering the types of products and services they are seeking. However, the credibility of the certificate rests on the competence of the auditor and the integrity of the certification body, neither of which are guaranteed.

If an organization's customers are not demanding ISO 9001 certification, the use of ISO 9001 is optional (i.e. there is no need to demonstrate its ability). However, many organizations wish to use ISO 9001 to create a competitive advantage in their market and may perceive there are tangible benefits from obtaining ISO 9001 certification. In such cases the need arises from top management rather than the customer. It is important to recognize that there is no requirement in the ISO 9000 family of standards for certification. Only where customers are imposing ISO 9001 in purchase orders and contracts would it be necessary to obtain ISO 9001 certification.

Meeting customer requirements

It is often believed that the customer requirements referred to in ISO 9001 are those specified by the customer verbally or in writing, but this is untrue. ISO 9000:2015 defines a requirement as a "need or expectation that is stated, generally implied or obligatory". It therefore includes specified requirements as in those defined in a contract, order or specification, but also standards that a reasonable person would regard as expected such as safety, reliability and maintainability. These are discussed further in Chapter 35.

It is also often believed that the customer referred to in ISO 9001 is the person or organization that purchases the organization's products or services, but again this in untrue. ISO 9000:2015 defines a customer as a "person or organization that could or does receive a product or a service that is intended for or required by this person or organization". It therefore includes the consumer, client, end user, retailer, beneficiary and purchaser. The customer is therefore anyone who may use your products and services however they may have come by them. But a person may have acquired your organization's products, and regardless of how long after they were put on the market, if they were intended for their use, they are classed as customers for the purposes of ISO 9001.

Meeting applicable statutory and regulatory requirements

Box 1.5   What are statutory and regulatory requirements?

Statutory requirements are obligatory requirements specified by a legislative body (persons who make, amend or repeal laws).

Regulatory requirements are obligatory requirements specified by an authority mandated by a legislative body (e.g. an agent of a national government).

(ISO 9000:2015)

ISO 9001 is not requiring your organization to meet all statutory and regulatory requirements. The word *applicable* in the scope statement means those statutory and regulatory requirements that are applicable to the product or service being provided. These requirements differ depending on the sector of the population, country, market and industry sector (see also Chapter 18).

If there is a law prohibiting the sale of products containing certain substances, that statutory requirement applies to the products offered for sale. It does not prohibit such substances being used in manufacturing processes providing they don't contaminate the product supplied. If there is a law granting maternity leave to employees, this applies to the organization but not the product or service being supplied. However, if there is a law governing hygiene in places where food is consumed, it applies to the organization and not the product offered, but will apply to any service offered where food is consumed as part of the service.

Consistently provide products and services

The ability the organization needs to demonstrate is an ability to "consistently" provide products and services of a certain standard. The key concept here is "consistent provision", which is important because it has more than one meaning. The word consistent means "Remaining in the same state or condition" (*Oxford English Dictionary*, 2013) but is it a specific product or service that should remain consistent or the meeting of customer and applicable statutory and regulatory requirements that should remain consistent?

The concept of consistent provision was introduced in the 2000 version of the standard. Demonstrating an ability to consistently provide products and services that meet customer and applicable statutory and regulatory requirements, which now includes their needs and expectations, therefore means not only meeting those requirements in every product and service that is provided to every customer, but also possessing the ability to anticipate future requirements of customers and offer products and services that will meet their requirements. Put another way, "consistently provide" can mean that every Ford Mondeo we provide meets the specification for that model of car. But it can also mean every time we provide a car, it will meet the customer's needs and expectations regardless of its specification.

This extension in scope was not obvious in the 2000 and 2008 versions, but in the 2015 version it is expressed through requirements in clause 8.2.2 and 8.2.3, where it addresses the determination and review of requirements related to products and services *to be offered* to customers and where it required product and service improvements to address *future needs and expectations* in clause 10.1a). The phrase "to be offered" implies an intention and therefore no longer does the standard only apply after a customer has expressed an interest in the organization's products and services but applies before a customer is even aware of the organization's products and services. Why the change? Well, some organizations are reactive, reacting to what customers' demand of them and aiming to satisfy their needs and expectations. Then there are others that are proactive, aiming to create a need and expectation and offering products and services the customer had not even dreamed of. The organization seeks opportunities to create new customers and through its marketing builds customer expectations. The products and services it eventually provides need to consistently meet these expectations, and this is expressed through the requirement in clause 8.2.2d where it states "the organization shall ensure that it can meet the claims for the products and services it offers."

Enhancing customer satisfaction

The standard has referred to meeting customer requirements so it may appear tauto-logical to place "meet customer requirements" in clause 1a)" and "enhance customer satisfaction" in clause 1b), but one may satisfy customers by meeting most of their requirements and enhance customer satisfaction by meeting all their requirements, which, as we have stated, means meeting stated, generally implied or obligatory needs and expectations.

Applicability of ISO 9001

There are several possible situations where the requirements of ISO 9001 could be deemed applicable:

A. After an organization has received a contract or order for specific products and services
B. When a customer has indicated an intention to place a contract or order for specific products and services
C. When a customer has expressed an interest in the organization's capability
D. When an organization seeks to create a new market for existing products and services
E. When an organization seeks opportunities for developing new products and services in its chosen market

ISO 9001 was introduced to facilitate national and international trade between orga-nizations and was therefore a tool of business-to-business relationships. The premise on which the first edition of ISO 9001 was based was that customers would be seeking suppliers that were able to demonstrate they could provide products and services that satisfied their requirements. Until the 2015 edition, ISO 9001 applied to situations A, B and C. In these situations, the product or service being offered already existed, with one exception; in situation C the customer may be attracted to the supplier because of its potential capability to design a product or service to their performance specification. In situations D and E, the requirements for a product or service have yet to be determined and there is no specific customer – only an unsatisfied need or potential want has been identified, and the organization may not yet have developed the necessary capability to satisfy it. In the 2015 revision, the applicability of ISO 9001 was extended to include situations D and E.

   The requirements specified in the standard are complementary to requirements for products and services. In fact there are no product or service requirements in the standard – all the requirements apply to the organization, but which parts of an organization? As will be seen so far, the focus of ISO 9001 is on customers; therefore, it does not apply to other stakeholders except in so far as they affect the ability of the organization to consistently provide products and services that satisfy its customers. The standard does not apply to management of the environment, occupational health and safety, finances, business risks and any other factor, provided those factors do not positively or negatively affect the ability of the organization to consistently provide products and services that satisfy its customers. However, it should not be assumed these factors are outside scope altogether. One property of a system is that everything is connected to everything else, so it is difficult to rule anything out.

Used for design and assessment purposes

When reading ISO 9001 it is easy to be confused over its purpose. The scope in clause 1 implies it's used to demonstrate ability, but when readers reach clause 4.4 they'll find that it requires a QMS to be established, thus in effect implying that the standard is used to design a QMS. When ISO 9001 is invoked in contracts, this requirement could be interpreted as requiring a QMS to be established specifically for the contract. In fact, it is used for both design and assessment purposes, but there are only 5 requirements for the organization to demonstrate something that would be indicative of a standard used for assessment purposes, and there are 50 requirements for something to be either determined or established, which is indicative of a standard used for design purposes. Although ISO 9001 can be used to design a QMS, it was intended to be used in contractual situations by customers seeking confidence in their supplier's capability.

The quest for confidence through regulated standards evolved in the defence industry. Defence quality assurance standards were based on the principle that when contractors can substantiate by objective evidence that they have systems in place to maintain control over the design, development and manufacturing operations and have performed inspection which demonstrates the acceptability of products and services, the customer can be assured that the products and services will be or are what they are claimed to be and will be, are being and have been produced under controlled conditions.

ISO 9001 and the free movement of goods and services

With the formation of the European Union (EU) in 1993 there was a need to remove barriers to the free movement of goods across the Union. One part of this was to harmonize standards. At the time, each country had its own standards for testing products and for controlling the processes by which they were conceived, developed and produced. This led to a lack of confidence and, consequently, to the buying organizations undertaking their own product testing and, in addition, assessment of the seller's quality management systems. The Council of the European Union has therefore adopted a common framework for marketing products. This broad package of provisions is intended to remove obstacles to the free circulation of products and represents a major boost for trade in goods between EU member states. ISO 9001 is perceived by the EU Council as ensuring health and safety requirements are met because ISO 9001 now requires organizations to demonstrate that they have the ability to consistently provide a product that meets customer and applicable statutory and regulatory requirements. Such requirements would be specified in EU directives. Where conformity with these directives can be verified by inspection or test of the end product, ISO 9001 is not a requirement for those seeking to supply within and into the EU. Where conformity cannot be ensured without control over design and/or production processes, conformity with ISO 9001 needs to be assessed by a *"notified body"*.

Since the formation of the EU, several other "common markets" have been formed throughout the world adopting many of the founding principles of the EU. These common markets are currently as identified here:

1 European Union Single Market (EU) of 28 countries
2 Sistema de la Integración Centroamericana (SICA) of 8 countries
3 Caribbean Single Market and Economy (CSME) of 12 countries
4 Eurasian Economic Space of 5 countries located primarily in northern Eurasia
5 Southern Common Market (Mercosur) of 6 countries in South America
6 Gulf Cooperation Council (GCC) that includes all Arab states of the Persian Gulf except for Iraq and Yemen
7 Association of Southeast Asian Nations (ASEAN) Economic Community (AEC) that includes 27 countries
8 The proposed East African Community (EAC) that includes six countries in the African Great Lakes region in eastern Africa

Although regulations may vary, as a general principle, if your organization is planning to export products into any one of these markets and conformity with that country's health and safety regulations cannot be verified by inspection or test of the product alone, conformity with ISO 9001 may be required to be demonstrated.

Popularity of ISO 9001 certification

ISO 9001 has gained in popularity since 1987 when the UK led the field holding the highest number of ISO 9001, 9002 and 9003 certificates. Since then certification in the UK has declined from a peak of 66,760 in 2001 to a low of 35,517 by 2008 and rising to 40,161 in 2015, putting the UK into fifth place. The latest year for which there are published figures is 2015 (up to 31 December 2014), when 33 of 195 countries (17%) held 90% of the total number of certificates issued, as detailed Table 1.1.

As quality system standards for automotive and medical devices (ISO/TS 16949 and ISO 13485) include all requirements from ISO 9001, certification to these standards can be added to the numbers of ISO 9001 certificates. The numbers only include data from certification bodies that are accredited by members of the International Accreditation Forum (IAF). The data are for numbers of certificates issued and not for the number of organizations to which certificates are issued, which may be less as some organizations register each location. In the sixth edition of this Handbook, data were used from the 2008 ISO Survey, and comparing the ranking reveals how certifications have increased and declined over the intervening period.

Table 1.1 17% of countries possessing 90% of ISO 9001 certificates in 2015

Rank	Country	ISO 9001	ISO/TS 16949	ISO 13485	Total	% of Total	Rank 2008
	Total	1,033,936	62,944	26,255	1,123,135		
1	China	292,559	25,498	1,961	320,018	28.49%	1
2	Italy	132,870	1,345	2,635	136,850	12.18%	2
3	Germany	52,995	3,473	2,508	58,976	5.25%	5
4	Japan	47,101	1,482	1,064	49,647	4.42%	4
5	United States of America	33,103	4,345	5,231	42,679	3.80%	8
6	United Kingdom	40,161	629	1,651	42,441	3.78%	6
7	India	36,305	4,992	439	41,736	3.72%	7
8	Spain	32,730	952	379	34,061	3.03%	3
9	France	27,844	1,012	1,400	30,256	2.69%	10
10	Romania	20,524	326	95	20,945	1.86%	19
11	Brazil	17,529	1,229	107	18,865	1.68%	14
12	Korea, Republic of	11,992	5,089	1,042	18,123	1.61%	9
13	Australia	13,636	79	138	13,853	1.23%	22
14	Switzerland	12,218	121	1,164	13,503	1.20%	15
15	Malaysia	11,963	529	366	12,858	1.14%	26
16	Colombia	12,324	44	39	12,407	1.10%	26
17	Czech Republic	10,648	737	269	11,654	1.04%	20
18	Poland	10,681	621	292	11,594	1.03%	17
19	Taipei, Chinese	8,766	1,323	750	10,839	0.97%	18
20	Netherlands	10,381	134	251	10,766	0.96%	12
21	Thailand	8,688	1,468	130	10,286	0.92%	29
22	Israel	9,085	30	582	9,697	0.86%	28
23	Russian Federation	9,084	329	125	9,538	0.85%	11
24	Turkey	8,538	906	88	9,532	0.85%	13
25	Mexico	7,418	1,441	136	8,995	0.80%	30
26	Indonesia	8,613	301	34	8,948	0.80%	31
27	Portugal	7,498	191	68	7,757	0.69%	34
28	Canada	6,417	530	606	7,553	0.67%	24
29	Argentina	7,112	249	39	7,400	0.66%	23
30	Greece	6,187	2	116	6,305	0.56%	27
31	Hungary	5,789	381	92	6,262	0.56%	21
32	Singapore	5,786	81	202	6,069	0.54%	36
33	Slovakia	5,683	290	77	6,050	0.54%	41
	Others					9.50%	

The 23-year trend since records began in 1993 is shown in Figure 1.1 for the following management systems certifications. Data are only available between the dates indicated here:

- ISO 9001 Quality management systems   (1993–2015)
- ISO/TS 16949 ISO 9001 for automotive production and relevant
  service part organizations   (2004–2015)

- ISO 13485 — Medical Devices – Quality Management Systems   (2004–2015)
- ISO 22000 — Food Safety Management Systems   (2007–2015)
- ISO 22301 — Business Continuity Management Systems   (2014–2015)
- ISO/IEC 20000–1 — Information Technology – Service Management   (2015)

The temporary arrest in growth during the transition from ISO 9001:1994 to ISO 9001:2000 is clearly evident, and it is predicted that there will be another arrest after the 2015 transition. The continued growth is likely to be due to the European Union directives and the expansion in globalization.

A major change in the 2015 revision was to make ISO 9001 appeal more to the service sector. As can been seen in Table 1.2, classifying the industry codes as either product or service shows that in 2015 roughly 75% of the certifications were of organizations supplying services with products, and 25% were of organizations providing a service without a product – that is approximately 199,281 certificates, far more if code 29 separated service organizations that did not provide a product. However, more and more organizations that predominately provide products are seeing themselves as service providers offering service solutions where the product is one component.

What accredited certification to ISO 9001 means

The IAF and the ISO support the following concise statement of outcomes that are to be expected as a result of accredited certification to ISO 9001. The intent is to promote

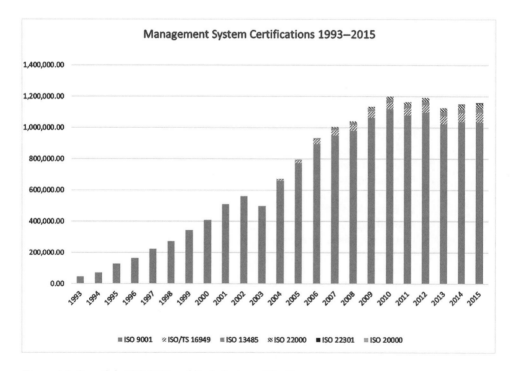

Figure 1.1 Growth in ISO 9001 and its derivate certifications

Table 1.2 ISO 9001 certificates by industry sector in 2015

EA Code	ISO 9001:2008 & 2015 CERTIFICATES	2015		
17	Basic metal and fabricated metal products	104,652	13.18%	p
19	Electrical and optical equipment	75,260	9.48%	p
28	Construction	67,354	8.48%	p
29	Wholesale and retail trade; repairs of motor vehicles and personal and household goods	66,975	8.44%	p
18	Machinery and equipment	56,413	7.11%	p
35	Other services	50,696	6.39%	s
14	Rubber and plastic products	41,101	5.18%	p
34	Engineering services	36,346	4.58%	s
12	Chemicals, chemical products and fibres	29,744	3.75%	p
33	Information technology	29,161	3.67%	p
31	Transport, storage and communication	27,053	3.41%	s
3	Food products, beverages and tobacco	26,602	3.35%	p
38	Health and social work	22,342	2.81%	s
37	Education	16,657	2.10%	s
32	Financial intermediation, real estate, rental	15,621	1.97%	s
4	Textiles and textile products	12081	1.52%	p
16	Concrete, cement, lime, plaster, etc.	11234	1.41%	p
22	Other transport equipment	10,972	1.38%	p
23	Manufacturing not elsewhere classified	10,558	1.33%	p
15	Non-metallic mineral products	10,441	1.32%	p
39	Other social services	10,017	1.26%	s
7	Pulp, paper and paper products	8,156	1.03%	p
9	Printing companies	7,500	0.94%	p
36	Public administration	6,580	0.83%	s
6	Wood and wood products	5,312	0.67%	p
30	Hotels and restaurants	4,340	0.55%	s
25	Electricity supply	4,249	0.54%	s
1	Agriculture, fishing	4,236	0.53%	p
2	Mining and quarrying	3,535	0.45%	p
13	Pharmaceuticals	3,532	0.44%	p
24	Recycling	3,432	0.43%	s
27	Water supply	1,948	0.25%	s
20	Shipbuilding	1,930	0.24%	p
5	Leather and leather products	1,908	0.24%	p
26	Gas supply	1,818	0.23%	p
21	Aerospace	1,783	0.22%	p
10	Manufacture of coke and refined petroleum products	1,445	0.18%	p
11	Nuclear fuel	569	0.07%	p
8	Publishing companies	409	0.05%	p
	Total	793,962	100.00%	

Note:

p = product

s = service

a common focus throughout the entire conformity assessment chain to achieve these expected outcomes and thereby enhance the value and relevance of accredited certification. To achieve conforming products and services, the accredited certification process is expected to provide confidence that the organization has a quality management system that conforms to the applicable requirements of ISO 9001. It is to be expected that the organization:

a)   has established a quality management system that is suitable for its products, services and processes and is appropriate for its certification scope;
b)   analyses and understands customer needs and expectations, as well as the relevant statutory and regulatory requirements related to its products and services;
c)   ensures that product and service characteristics have been specified in order to meet customer and statutory/regulatory requirements;
d)   has determined and is managing the processes needed to achieve the expected outcomes (conforming products, services and enhanced customer satisfaction);
e)   has ensured the availability of resources necessary to support the operation and monitoring of these processes;
f)   monitors and controls the defined product and service characteristics;
g)   aims to prevent nonconformities and has systematic improvement processes in place to:

  i   correct any nonconformities that do occur (including product and service nonconformities that are detected after delivery);
  ii   analyse the cause of nonconformities and take corrective action to avoid their recurrence;
  iii   address customer complaints.

h)   has implemented an effective internal audit and management review process;
i)   is monitoring, measuring and continually improving the effectiveness of its quality management system.

What accredited certification to ISO 9001 does not mean

It is important to recognize that ISO 9001 defines the requirements for an organization's quality management system, not for its products and services. Accredited certification to ISO 9001 should provide confidence in the organization's ability to "consistently provide products and services that meet customer and applicable statutory and regulatory requirements". It does not necessarily ensure that the organization will always achieve 100% product and service conformity, though this should, of course, be a permanent goal.

ISO 9001 accredited certification does not imply that the organization is providing a superior product or service, or that the product or service itself is certified as meeting the requirements of an ISO (or any other) standard or specification.

Summary

In this chapter, we discovered that ISO 9001 is intended to enable organizations to demonstrate their capability so as to give confidence to customers they'll get what they expected. We examined the way ISO standards are developed and revised, noting that they are consensus based and not the unanimous verdict of ISO members. We gave five reasons for using ISO 9001 and then explained the concepts that are enshrined in what is probably the most important section of the standard: the scope statement in clause 1 of the standard. We addressed its use in the free movement of goods and services, and from the statistics provided we found that 17% of the countries hold 90% of the certificates, with China having the most. Although in most developed countries the service economy is dominant, only 25% of ISO 9001 certificates are held by service organizations. Finally, we summarized what ISO 9001 certification means to an organization.

Notes

1 Haberject was a kind of cloth made in very early times in England, said to be a cloth of a mixed colour and have been worn chiefly by monks.
2 A selfage is an edge produced on woven fabric during manufacture that prevents it from unraveling.

Bibliography

ISO. (2009). Selection and use of the ISO 9000 Family of Standards. Retrieved from International Organization for Standardization: www.iso.org/files/live/sites/isoorg/files/archive/pdf/en/iso_9000_selection_and_use-2009.pdf.

ISO 9000:2015. (2015). *Quality Management Systems – Fundamentals and Vocabulary*. Geneva: ISO.

ISO-SD. (2016, September 15). How Does ISO Develop Standards? Retrieved from ISO: www.iso.org/iso/home/standards_development.htm.

Jobs, S. (1990). Steve Jobs on Joseph Juran and Quality. (ASQ, Interviewer). Retrieved from www.youtube.com/watch?v=XbkMcvnNq3g&feature=youtu.be.

Juran Foundation, Inc. (1995). *A History of Managing for Quality*. Milwaukee: ASQC Quality Press.

OED. (2013). Retrieved from Oxford English Dictionary: oed.com

2 Comparison between 2008 and 2015 editions

> Progress is impossible without change, and those who cannot change their minds cannot change anything.
>
> George Bernard Shaw (1856–1950)

Introduction

The last major revision of requirements was in 2000 when the structure was changed and the eight quality management principles were introduced. One of the principles, the process approach, had a significant impact on the way requirements were expressed. The intermediate revision in 2008 corrected errors and clarified normative and informative text, so there has not been a real update for 15 years, and therefore a major revision was due as the world had changed quite a lot.

To better understand current and future customer needs for ISO 9001 and ISO 9004 and to ensure their relevance into the future, an online survey was launched in October 2010. The results were used in the planning process and in establishing the strategic direction for the future of quality management systems standards. The survey provided an opportunity to sound out users as to whether there was a desire for different sets of requirements depending on the level of risk, as there are no grades of certification to ISO 9001: it's all or nothing. Just over half of 7,400 respondents wanted the standard to remain a single standard where all requirements remain equally mandatory.

In the full report, the major conclusion was that the survey responses indicate that "ISO 9001 is a good document that is relevant for the future, with some enhancement." Comments made by respondents also suggest major changes are not required, but improvements could be made, and special attention should be given to ensure the correct application of the standard (ISO/TC 176/SC 2/N1017, 2011). An online summary of results from the survey proclaims customer satisfaction to be the primary reason for seeking ISO 9001 certification and the most important benefit from implementing the QMS standard (Jarvis & MacNee, 2011).

ISO released a PowerPoint presentation (see ISO/TC176/SC2/N1282, 2015) in which they provided five reasons why the standard needed to change:

- To adapt to a changing world
- To reflect the increasingly complex environments in which organizations operate
- To provide a consistent foundation for the future
- To ensure the new standard reflects the needs of all relevant interested parties
- To ensure alignment with other management system standards

In this chapter, we provide:

- An overview of the changes
- The intent and impact of Annex SL: how it came about, who it's aimed at, what it aims to achieve and what impact it has
- A summary of the more significant changes and, in particular, the movement of requirements, the new requirements, the modified requirements and the requirements that have been withdrawn
- A summary of the changes in terminology and their impact

In addition to these summaries, changes to specific requirements are highlighted in this Handbook under the relevant heading.

Overview of the changes

The committee view

Two of the most important objectives in the revision of the ISO 9000 series of standards stated in (ISO/TC 176/SC2/N1276, 2015) were:

a) to develop a simplified set of standards that will be equally applicable to small as well as medium and large organizations;
b) for the amount and detail of documentation required to be more relevant to the desired results of the organization's process activities.

There are changes in structure, changes in terminology, changes in requirements and changes in emphasis, but the intent, objective, the stated scope and applicability of ISO 9001 have not changed.

TC 176, the committee responsible for ISO 9001, state that the main differences in content between the old and new versions are:

- The adoption of the high-level structure as set out in Annex SL of ISO directives part 1
- An explicit requirement for risk-based thinking to support and improve the understanding and application of the process approach
- Fewer prescriptive requirements
- More flexibility regarding documentation
- Improved applicability for services
- A requirement to define the boundaries of the QMS
- Increased emphasis on organizational context
- Increased leadership requirements
- Greater emphasis on achieving desired process results to improve customer satisfaction

The statistics

An analysis of the differences between the 1994, 2008 and 2015 versions is shown in Table 2.1 in terms of some key parameters. The elements are the major headings, and the clauses are the numbered sections containing requirements. The "shalls" are requirements

Table 2.1 How the versions differ at a simplistic level

Parameter	1994	2008	2015
Elements	20	5	10
Clauses	59	51	51
Shalls	138 (204)	136 (262)	132 (301)
Procedures	26	6	0
Records	20	21	17
Documents	8	10	9

Table 2.2 The magnitude of change

	Parameter	No	%	% excluding (E)
A	No Change in Intent of Requirement	92	26	28
B	Modified Requirement	89	25	27
C	No Change in Requirement	84	24	26
D	New Requirement	62	18	19
E	Requirement Withdrawn	24	7	
		351	100.00	

containing the word *shall*, and the numbers in parentheses is the total number of requirements when every "and" or a comma or a list is taken into account. As the 2015 version does not separate documents and records as in previous versions, these figures are obtained by whether there is a requirement for documented information to be retained or maintained.

Another analysis by the magnitude of changes is shown in Table 2.2 and shows some interesting statistics. At the level of individual requirements there are only 62 new requirements which, if we exclude the requirements withdrawn, represents 19% of all requirements, but the numbers hide their significance. However, this means that roughly 80% of the requirements remain more or less the same. Some are expressed in the same way (C), others are expressed differently but have the same intent (A) and others are modified in some way that changes the scope or applicability of the requirement, but not so much that it's an entirely new requirement. These statistics have been produced by the author for this edition of the Handbook. They have not been obtained or derived from any official source, and other people may interpret the changes differently. It has been done simply to represent the scale of the change. Even with the major proportion of requirements remaining similar to those in the 2008 version, the new requirements and changes to the definitions of terms creates a different context, and therefore it cannot be assumed that the requirements will have the same meaning.

Consideration replaces prescription

The degree of prescription has been reduced with the 2015 version. There are many clauses requiring the organization to give consideration to various factors, and this means that they have either taken such matters into account (i.e. the factors have influenced the decisions taken) or the factors have been discounted on the basis that they are deemed not relevant in the particular situation. In reaching such conclusions, those giving consideration are intended to apply risk-based thinking and thereby will be weighing up the potential benefits and harms of exercising one choice of action over another.

The intent and impact of Annex SL

As of 2015, all management system standards were required to conform to Annex SL of ISO directives (ISO/IEC, 2015) which provide the complete set of procedural rules to be followed by ISO committees. The intent is to harmonize common text and terminology so that they are easy to use and are compatible with each other.

When ISO 9001 was revised in 1994, it was still the only international management system standard. BS 7750 on Environmental Management Systems, which had been published in 1992, did not develop into the international standard ISO 14001 until 1996. BS 7799 on Information Security did not develop into the international standard ISO/IEC 27001 until 2005, when ISO 22000 on Food Safety Management systems also appeared. It became apparent that there were elements common to each of these standards, and so in 2006 BSI published Publically Available Specification (PAS) 99 to assist organizations in adopting an integrated approach to management systems to simplify the implementation of multiple system standards and any associated conformity assessment.

By 2009 the ISO Technical Management Board (TMB) were receiving requests to develop other management system standards, and it was already under pressure to standardize these standards so that users could take an integrated approach to their implementation. Consequently, in 2010 work started on formulating a high-level structure and identical text for management system standards, together with common management system terms and core definitions, which resulted in ISO Guide 83 in 2011. The aim of the guide was that all ISO management system "requirements" standards were aligned and the compatibility of the standards enhanced. It was envisaged that individual management systems standards would add additional *discipline-specific* requirements as needed. Although this guide was intended only for writers of standards, BSI revised PAS 99 based on ISO Guide 83 in 2012 to enable users of ISO 9001, ISO 14001, etc., to develop an integrated approach to these standards. ISO Guide 83, which was for use by standards developers, was eventually released as Annex SL. The full document is available from the ISO website. By October 2015 the number of management system standards has grown from 1 in 1987 to 28 in October 2015, as will be seen in Table 2.3.

In aligning the standards, it became apparent that there were topics common to all management systems and topics unique to specific applications. The TMB chose to refer to these applications as *disciplines*, a term used to indicate specific subject(s) to which a management system standard refers (e.g. energy, quality, records, environment, etc.). Common text, terms and definitions were developed, and additional discipline-specific text is not supposed to affect harmonization or contradict or undermine the intent of the high-level structure, identical core text, common terms and core definitions. If due to exceptional circumstances the high-level structure or any of the identical core text, common terms and core definitions cannot be applied in the management system standard, then the technical committee concerned submits their explanation to the TMB for review.

Table 2.3 Management system standards, October 2015

#	Reference	Title
1.	ISO 9001:2015	Quality management systems – Requirements
2.	ISO 14001:2015	Environmental management systems – Requirements with guidance for use
3.	ISO 15378:2011	Primary packaging materials for medicinal products – Particular requirements for the application of ISO 9001:2008, with reference to Good Manufacturing Practice (GMP)
4.	ISO 18091:2014	Quality management systems – Guidelines for the application of ISO 9001:2008 in local government
5.	ISO 18788:2015	Management system for private security operations – Requirements with guidance
6.	ISO 19600:2014	Compliance management systems – Guidelines
7.	ISO 20121:2012	Event sustainability management systems – Requirements with guidance for use
8.	ISO 21001*	Quality management systems – Requirements for the application of ISO 9001:2008 educational organizations
9.	ISO 21101:2014	Adventure tourism – Safety management systems – Requirements
10.	ISO 22000:2005**	Food safety management systems – Requirements for any organization in the food chain
11.	ISO 22301:2012	Societal security – Business continuity management systems – Requirements
12.	ISO 24526*	Water efficiency management systems – Requirements with guidance for use
13.	ISO 30301:2011	Information and documentation – Management systems for records – Requirements
14.	ISO 34001.3*	Security management system – Fraud countermeasures and controls
15.	ISO 37001:2016	Anti-bribery management systems
16.	ISO 37101:2016	Sustainable development and resilience of communities – Management systems – General principles and requirements
17.	ISO 39001:2012	Road traffic safety (RTS) management systems – Requirements with guidance for use
18.	ISO 41001*	Facilities management – Integrated management system – requirements [with guidance for use]
19.	ISO 45001*	Occupational health and safety management systems – Requirements
20.	ISO 50001:2011	Energy management systems – Requirements with guidance for use
21.	ISO 55001:2014	Asset management – Management systems – Requirements
22.	ISO 19443*	Quality management systems – Specific requirements for the application of ISO 9001 and IAEA GS-R requirements by organizations in the supply chain of the nuclear Energy sector
23.	ISO/IEC 27001:2013	Information technology – Security techniques – Information security management systems – Requirements
24.	ISO/IEC 27010:2012	Information technology – Security techniques – Information security management for inter-sector and inter-organizational communications
25.	ISO/IEC 80079–34:2011	Explosive atmospheres – Part 34: Application of quality systems for equipment manufacture
26.	ISO/IEC 90003:2014	Software engineering – Guidelines for the application of ISO 9001:2008 to computer software
27.	ISO/TS 17582:2014	Quality management systems – Particular requirements for the application of ISO 9001:2008 for electoral organizations at all levels of government
28.	ISO 13485:2016	Medical devices – Quality management systems – Requirements for regulatory purposes

*Under development **Under revision

For current status see https://www.iso.org/management-system-standards-list.html

Annex SL is a framework with which all new and future revisions of existing management systems standards are expected to conform, so they will all share a common structure and common text. It will therefore reduce conflict arising from similar requirements being expressed in different ways in management system standards.

ISO's Joint Technical Coordination Group (JTCG) on Management System Standards (MSSs) produced a guide containing answers to some frequently asked questions about Annex SL, and among these the following are of interest, but for further information see (ISO/TMB/JTCG N 359, 2013):

- Annex SL defines a set of interdependent requirements that function as a whole, often referred to as a *systems approach.*
- It defines what has to be achieved, not how it should be achieved.
- It specifies requirements. There is no inherent assumption of sequence or order in which they are to be implemented by an organization. There is no inherent demand that all activities in a specific clause must be done before activities in another clause are started.
- It does not dictate or imply a specific model for how to achieve the requirements.
- It was written with the aim of avoiding repeating words and using plain English.
- It deliberately separated clause 4.1 from 4.2 because of a wish to address interested parties separately and specifically.
- It used cross-referencing to show linkage.
- It deliberately used bullets to avoid presenting an inherent assumption of sequence or order. If standards writers want to, they can use a), b), etc., instead of the bullet as a symbol.
- It developed definitions with the aim of finding words that explained the concept behind the term in its most general approach. The process used the ISO requirement for development of definitions in ISO 704:2009 Terminology work – Principles and methods. In discipline-specific standards, it is possible to add notes to explain and complete the sense; however, it should be understood that notes to terminology are normative according to ISO Directives and cannot contain requirements.

Movement of requirements

As Annex SL applies a new structure to ISO 9001, requirements that remain in the new version have been moved under new headings as shown in Figure 2.1. It's not possible to show details at this scale, but the diagrams provide an overall impression of the extent of the change.

Preventive action

There is, however, one move that needs some explanation, and that is the clause 8.5.3 on preventive action. This is what JTCG Guide N359 has to say about the move:

> The high-level structure and identical text does not include a clause giving specific requirements for "preventive action". This is because one of the key purposes of a formal management system is to act as a preventive tool. Consequently, a MSS requires an assessment of the organization's "external and internal issues that are relevant to its purpose and that affect its ability to achieve the intended outcome(s)" in clause 4.1, and

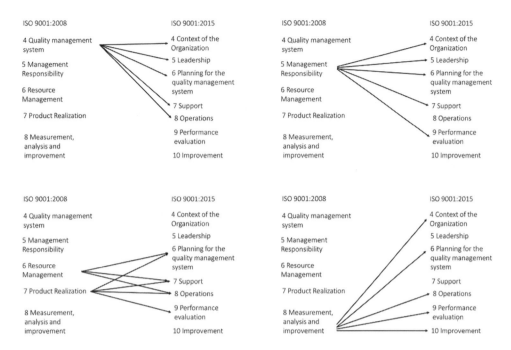

Figure 2.1 Movement of requirements

to "determine the risks and opportunities that need to be addressed to: assure the XXX management system can achieve its intended outcome(s); prevent, or reduce, undesired effects; achieve continual improvement." in clause 6.1. These two sets of requirements are considered to cover the concept of "preventive action", and also to take a wider view that looks at risks and opportunities.

(ISO/TMB/JTCG N 359, 2013)

As will be evident in Chapter 21 where we address risk, we reveal that the way the requirements are expressed in clause 6.1 does not cover the concept of preventive action as it was addressed in the 2008 version.

New requirements

Understanding the context of the organization

This is a new clause in ISO 9001 from Annex SL, which has been introduced in recognition that the management system does not function as a separate entity but is influenced by the environment in which it functions. To conform to ISO 9001:2015 organizations will now need to:

- determine, monitor and review issues relevant to the organization's purpose and direction that affect the ability of the QMS to achieve its intended results;

- determine, monitor and review relevant interested parties and their needs and expectations that are relevant to the organization's purpose and direction;
- take account of the identified issues and requirements of the interested parties when establishing or developing the QMS.

Leadership

This clause previously carried the heading "Management Commitment". The change to "Leadership" reflects the purpose of demonstrating commitment and aligns better with the second quality management principle (see Chapter 5). To conform to ISO 9001:2015 the organization's top management will now need to:

- take accountability for the effectiveness of the QMS;
- integrate the QMS requirements into the organization's business processes;
- promote the use of the process approach and risk-based thinking;
- communicate the importance of effective quality management;
- engage, direct and support persons to contribute to the effectiveness of the QMS.

QMS planning

This clause contains modified versions of the requirements of the 2008 version with a few significant additions. To conform to ISO 9001:2015 organizations will now need to:

- identify and address risks and opportunities to give assurance that the QMS can achieve its intended result(s);
- monitor achievement of quality objectives;
- carry out changes to the QMS in a planned manner and consider the purpose of the changes and their potential consequences.

Knowledge management

This is a new clause in ISO 9001 which has not been introduced through Annex SL. It recognizes that knowledge is a resource and vital to success, and therefore it's important to ensure there is a sufficient supply of it. To conform to ISO 9001:2015 organizations will now need to:

- determine and maintain the knowledge necessary for the operation of its processes and to achieve conformity of products and services;
- determine how to acquire or access any necessary additional knowledge and required updates.

Human resources

The requirements relating to human resources are addressed under the headings of Competence, Awareness and the Environment for the operation of processes. Many of the requirements of the 2008 version remain in one form or another, but to conform to ISO 9001:2015, personnel under the control of the organization will now need to be aware of the implications of not conforming with the QMS requirements.

Documentation control

Apart from requirements that have been withdrawn and changes in terminology, the remaining documentation requirements are very like those in the 2008 version with two minor exceptions. To conform to ISO 9001:2015 organizations will now need to:

- ensure documentation is in the appropriate format;
- ensure documentation retained as evidence of conformity is protected from unintended alterations.

Operations planning

This clause replaces the clause on planning for *Product Realization* and retains many of the requirements. To conform to ISO 9001:2015 organizations will now need to:

- ensure that planned changes to the processes needed to meet the requirements for the provision of products and services are controlled;
- review the consequences of unintended changes, taking action to mitigate any adverse effects, as necessary.

Customer communication

These requirements are like those in the 2008 version except that to conform to ISO 9001:2015 organizations will now need to establish specific requirements for contingency actions.

Requirements related to products and services

In previous versions of ISO 9001 the product realization requirements applied to work that follows from a customer enquiry as expressed by the requirement in ISO 9001:2008 clause 7.2.1: "The organization shall determine requirements specified by the customer." The requirements did not apply to the research and development that would go into the creation of products and services before they were offered to customers. This changes in the 2015 version by bringing marketing into the scope of the standard, and to conform to ISO 9001:2015 organizations will now need to:

- define requirements for the products and services to be offered to customers;
- ensure that the organization can meet the claims for the products and services it offers.

Design and development

In previous versions of ISO 9001 the clauses on design and development addressed planning, interfaces, inputs, outputs and changes and separated controls into reviews, verification and validation. In the 2015 version, all the requirements addressing controls are under the heading "*Design and Development Controls*", which is ironic because the complete section was headed "*Design Control*" back in 1987. Some new requirements codify what has been common practice in design engineering for some time; therefore, to conform to ISO 9001:2015 organizations will now need to:

- consider the nature, duration and complexity of the design and development activities when determining design and development stages and controls;
- consider the need for involvement of customers and users in the design and development process;
- consider the necessary requirements for subsequent provision of products and services;
- consider standards or codes of practice that the organization has committed to implement;
- determine the potential consequences of failure due to the nature of the products and services;
- ensure that actions are taken on problems determined during the reviews or verification and validation activities;
- retain documented information of design and development activities.

External provision

What was classed as purchasing and outsourcing in the 2008 version is now referred to as external provision so that products, services and processes supplied by external providers are brought together under one heading. Most of the requirements in the 2008 version have been transferred across, but to conform to ISO 9001:2015 organizations will now need to:

- consider the potential impact of the externally provided processes, products and services on the organization's ability to consistently meet customer and applicable statutory and regulatory requirements;
- consider the effectiveness of the controls applied by the external provider before their selection;
- communicate its requirements for release of products and services to external providers.

Production and service provision

This section contains all the requirements of section 7.5 of ISO 9001:2008. To conform to ISO 9001:2015 organizations will now need to:

- implement actions to prevent human error;
- consider the nature, use and intended lifetime of the products and services when determining the extent of post-delivery activities;
- consider customer feedback when determining the extent of post-delivery activities;
- consider statutory and regulatory requirements when determining the extent of post-delivery activities;
- review and control unplanned changes essential for production or service provision;
- retain documented information describing the results of change reviews and identifying the personnel authorizing the change.

Control of nonconforming outputs

Control of nonconformity was included under section 8 of the 2008 version of ISO 9001 on *Measurement, analysis and improvement*, but clearly it's none of these and therefore was in the wrong section. However, there are now two clauses that address nonconformity: clause 8.7 and 10.2. The difference between these two is that clause 8.7 addresses outputs from production and service provision processes, whereas clause 10.2 addresses nonconforming systems, products and services. To conform to ISO 9001:2015 organizations will now need to inform the customer of nonconforming process outputs detected after delivery of products, during or after the provision of services.

Analysis and evaluation

The requirements on analysis and evaluation are almost the same as those in the 2008 version, except that to conform to ISO 9001:2015 organizations will now need to demonstrate that planning has been successfully implemented.

Management review

Management review was previously located under clause 5 on *Management Responsibility*, but as it performs an evaluation function, it is has now been located under section 9 on *Performance Evaluation*. To conform to ISO 9001:2015 organizations will now need to:

* consider information on monitoring and measurement results;
* consider information on issues concerning external providers and other relevant interested parties;
* consider the adequacy of resources required for maintaining an effective QMS;
* consider the effectiveness of actions taken to address risks and opportunities.

Improvement

In the 2008 version, improvement was addressed in clause 8.5 and included three sections: *Continual improvement, corrective action and preventive action*. The preventive action requirements have been incorporated into clause 6.1 of the 2015 version under the heading *"Actions to address risks and opportunities"*. The scope of the requirements for improvement in the 2008 version could be interpreted as only applying to the QMS and not the products and services the organization provides, but this anomaly has now been corrected. To conform to ISO 9001:2015 on improvement organizations will now need to:

* determine and select opportunities for improvement that enhance customer satisfaction;
* improve processes to prevent nonconformities;
* improve products and services to meet known and predicted requirements.

Nonconformity and corrective action

To conform to ISO 9001:2015 organizations will now need to deal with the consequences when a nonconformity occurs, including those arising from complaints.

Requirements withdrawn

Documenting the QMS

From its inception, ISO 9001 has required the QMS to be documented, but this requirement has now been removed. However, whether the organization chooses to document its QMS or any part of it is now at the discretion of its management. There are exceptions where specific activities and results are to be documented (see Table 32.1), but in general the only documentation required is that which the organization determines as being necessary for the effectiveness of the quality management system. More on this topic can be found in Chapter 15 and Chapter 32.

Quality manual

The requirement for a quality manual was introduced in the 1987 version, but it was only required when stated as a customer requirement. In the 1994 version, it became a general requirement and was required to cover the requirements of the standard. It was this requirement that created the mould from which all quality manuals were to be cast; it was this requirement that expressed the purpose of the quality manual. The vast majority of manuals paraphrased the requirements of the standard and included the quality policy, quality objectives, quality procedures and other associated instructions, forms and records as is portrayed by the pyramid documentation structure. The guidance provided in ISO 10013:1995 perpetuated the approach whereby the manual responded to each requirement of ISO 9001, which is demonstrated in Annex C of that standard showing an example section of a quality manual.

  With the publication of the 2000 version, the requirements for a quality manual were changed, requiring it to describe the scope of the QMS and the interaction between the processes of the QMS and to include or reference the procedures. The term *quality manual* was defined in ISO 9000:2005 as "a document specifying the quality management system of an organization". The requirement remained in the 2008 version, but ISO 10013 was converted into a Technical Report and became ISO TR 10013:2001. Although the example section of a quality manual was withdrawn, the standard still included the pyramid documentation structure; however, the role of the documented procedures was changed from describing the activities of individual functional units needed to implement the quality system elements to describing the interrelated processes and activities required to implement the quality management system.

  In the 2015 version, there is no reference to a quality manual, and the former requirement is replaced with a requirement for the QMS to include "documented information determined by the organization as being necessary for the effectiveness of the quality management system". It is therefore at the organization's discretion whether it needs a document it wishes to label as a quality manual. The term *quality manual* remains in ISO 9000:2015, but its definition has changed. It is now defined as a "specification for the quality management system of an organization", where a specification is defined as a "document stating requirements". So, a quality manual no longer describes how the organization works, but specifies the characteristics or requirements of the QMS.

  Is a quality manual necessary? To address this question, we need to examine the purpose of the quality manual. There are three basic types:

A. Manuals that specify requirements for the organization's QMS. These may or may not mirror the clauses of the standard.
B. Manuals that explain to interested parties how the organization satisfies the requirements of ISO 9001. These mirror the clauses of the standard.
C. Manuals that explain how the quality of the organization's products and services is managed. These may describe the network of management and operational processes and the flow of information or product as it passes into, through and out of them.

Type A manuals are a vehicle for management to communicate their policies, and they might be structured around the clauses of ISO 9001 or some other sequence. They are used by process and procedure developers and internal auditors.

Type B manuals are of use to customers, external auditors, internal auditors, personnel preparing tenders and negotiating contracts and those training for such roles. Some may be slender volumes, and others may contain procedures and other information more suitable for a type C manual.

Type C manuals may contain confidential information and therefore may not be made available to outside bodies. They may be used internally as a means of:

*   communicating the organization's mission, vision, values, policies and objectives and how they are realized through the business processes;
*   showing how the system has been designed;
*   showing how work flows into, through and out of the organization among the departments and the linkages between them;
*   showing who does what;
*   training new people;
*   analysing potential improvements.

There will be a need to do all of this in one form or another; therefore, the removal of the requirement from ISO 9001 does not remove the need for the information. In deciding whether to retain the quality manual, relabel it, revise it or scrap it, give due consideration to the points mentioned here.

Documented procedures

Requirements for documented procedures were introduced in the 1987 version and continued through to the 2008 version. To begin with, 30 documented procedures were required which had been reduced to 6 in the 2008 version. In the 1987 version, requirements were to be implemented through procedures, and in the 1994 revision a distinction was made between procedures and documented procedures as not all procedures were intended to be documented. In the 2000 revision, a general requirement was introduced for QMS documentation to include "documents needed by the organization to ensure the effective planning, operation and control of its processes", which appeared to contradict an additional requirement for six documented procedures. In the 2015 version, this contradiction has been removed, as no procedures are required.

As with the quality manual, we need to examine the purpose of procedures and documented procedures in particular. A procedure is defined in ISO 9000:2015 as a "specified way to carry out an activity or a process", where the word *specified* means "stated explicitly, clearly and definitely" (ISO Glossary, 2016) and therefore any activity that is required to be carried out in a certain way is a candidate for a procedure. If it is necessary to document the procedure because it is too complicated for it to become a habit, or to expect people to memorize it or too risky to rely on memory, a documented procedure may be necessary. Therefore, the removal of the requirement from ISO 9001 for documented procedures does not remove the need for them. This topic is also addressed in Chapter 32.

Management representative

A requirement for the appointment of a management representative was introduced into the 1987 version and remained until its removal in the 2015 version. Because the term *management representative* was introduced in North Atlantic Treaty Organization (NATO) standards in the 1960s, it has always been expressed in the singular. The reason for its removal in part was necessitated by an ISO/IEC Directive that stipulates that standards be performance based as opposed to design prescriptive. This rule comes about as a result of the adoption of the World Trade Organization Technical Barriers to Trade Agreement (WTO/TBT), which placed an obligation on ISO to ensure that the international standards it develops, adopts and publishes are globally relevant. The stipulation mentioned earlier is one of several that, if not met by an international standard, opens it to being challenged as creating a barrier to free trade (ISO/IEC, 2015). It is likely that the way the requirement for a management representative was expressed in the 2008 version of ISO 9001 was deemed prescriptive because it disallowed a company from sharing the duties between two or more individuals, and therefore the change was made to bring ISO 9001 into compliance with international trade rules.

Completion of design validation

Design validation was introduced into the 1994 version, although design qualification and test was addressed in the 1987 version. The requirement for design validation to be completed prior to the delivery or implementation of the product wherever practicable was introduced in the 2000 version. In the committee draft of the 2015 version there was a requirement for transfer from development to production or service provision to take place when actions outstanding or arising from development have been completed. However, this requirement did not appear in the final draft, and so any constraint on when design validation is to take place is at the organization's discretion. Should problems arise after the release of product that can be traced back to a premature release of its design, the two requirements where nonconformity may be cited are clauses 8.3.4d) and e) (see also Chapter 39).

Changes in terminology

New terms and definitions for old terms

Eighty-four terms were defined in ISO 9000:2005, and in ISO 9000:2015 there are 146. Of the 62 new terms, only 24 are used in ISO 9001, and these are as follows:

Activity*	Engagement	Measurement*	Policy*
Complaint*	External provider	Monitoring*	Regulatory requirement*
Context of the organization	Feedback*	Objective*	Risk
Data*	Improvement*	Output*	Statutory requirement*
Determination*	Innovation	Outsource*	
Documented information	Involvement	Performance*	

*Terms used in ISO 9001:2008 but not defined in ISO 9000:2005.

Documented information

Data, documentation and records are now frequently processed electronically. Therefore, the new term *documented information* has been created to describe and take account of this situation. The term subsumes the previous concepts of documentation, documents, documented procedures and records (ISO/TMB/JTCG N 359, 2013).

Simplification

Some common words have been introduced into the definitions to simplify their construction and ensure consistency across multiple management system standards.

Object

The word *object* has been introduced into definitions of the terms *capability*, *design and development*, *quality* and *review* in place of a list of objects (systems, products, services, process, etc.) or, as in the case of the term *quality*, inserting the term *object* where it was absent previously. This particular change is significant because it allows the term *quality* to be used for the quality of anything and therefore not restrict its use to the unstated, but often assumed, *product or service*.

Objective

The word *objective* has been defined separately, which simplifies the definition of the term *quality objective* but within ISO 9001 no other type of objective is referred to other than a quality objective.

Policy

The word *policy* has been defined separately, which simplifies the definition of the term *quality policy* but within ISO 9001 no other type of policy is referred to other than a quality policy.

New definitions for old terms

Auditor

The term *auditor* has been simplified by removing the unnecessary statement "with the demonstrated personal attributes and competence" because the standard requires persons to be competent.

Competence

The term *competence* has been redefined to signify that someone who is competent must be able to achieve intended results, and this will have significant consequences as is discussed in Chapter 29.

Continual improvement

The term *continual improvement* has been redefined so that it relates to enhancing performance rather than fulfilling requirements, and this, too, will have significant consequences as is addressed in Chapter 59.

Corrective action

The term *corrective action* has been redefined emphasizing that a corrective action is an action to prevent recurrence. In fact, the way the term was used in ISO 9001:2008 clause 8.5.2 was clearer than the definition in ISO 9000:2005.

Customer satisfaction

The term *customer satisfaction* has been modified, replacing the word *requirements* with *expectations*. This removes an implied limitation that ISO 9001 is only concerned in meeting requirements specified by customers (see Chapter 54).

Customer

The term *customer* has been redefined so as to be applicable to people or organizations that could receive a product or a service, thereby bringing marketing into the QMS.

Interested party

The term *interested party* has been modified, removing the condition of the party being interested in an organization's success. This term was not used in ISO 9001:2008.

Management system

The term *management system* has been modified by including the definition of the term *system* (reversing the practice of simplification) and including *processes* but in a way that makes the definition ambiguous (see Chapter 8).

Organization

The term *organization* has been modified to bring in the concept that it exists to achieve objectives. This is not a significant change, but it does rule out any gathering of people from qualifying as an organization in the context of ISO 9001.

Process

The term *process* has been changed by removing the concept of transformation of inputs, recognizing that work can qualify as a process by simply using inputs. It does not have to transform them (see also Chapter 9).

Product

The term *product* has been redefined so that it now only applies to the organization's outputs and not the outputs from internal processes, as was previously the case. This clarifies the applicability of some requirements and should remove confusion, for example, a nonconforming product is no longer a nonconforming output from an internal process but a nonconforming deliverable to a customer. In redefining the term *product*, it also now only applies to an output that can be produced without any transaction taking place between the organization and the customer. Therefore, an order is received and a tangible output is provided such as hardware, software, information, etc. In the 2008 version, the term *product* included service, which caused some confusion, so now the two are separated.

Quality management system

The term *quality management system* has been modified to recognize that it is part of a management system and not a separate management system. This could be a significant change, depending on the prevailing beliefs. *Part of* does not mean it's a subsystem (see Chapter 8).

Summary

ISO 9001:2015 looks very different from ISO 9001:2008 primarily because of the new structure imposed on all management systems standards by ISO Directives. Perhaps the most significant change is the change in approach from a prescriptive standard defining what an organization has to do to demonstrate it can provide conforming products and services, to one that places that responsibility firmly in the hands of top management. The introduction of understanding the context of the organization and the identification of risks as a prerequisite to establishing the QMS is a new approach which should make the QMS more suitable for its purpose. Although most requirements in the 2008 version have been carried over, the introduction of risk-based thinking allows management to decide the level at which requirements apply to deliver products and services that enhance customer satisfaction. The removal of requirements for a quality manual, documented procedures and a management representative will create confusion among some users before they realize the advantages. The new terminology may not even be noticed but may create confusion when it is, but overall the changes will help organizations improve their performance. A video of the potential impact of the changes is available on the companion website.

Bibliography

ISO/IEC. (2015). *ISO/IEC Directives, Part 1 — Consolidated ISO Supplement – Procedures Specific to ISO*. Geneva: ISO General Secretariat.
ISO/TC 176/SC 2/N1017. (2011). *ISO 9000 User Survey Report*. London: BSI Standards.

ISO/TC176/SC2/N1276. (2015). Guidance on the requirements for Documented Information. Retrieved from ISO/TC 176 Home Page: http://isotc.iso.org/livelink/livelink/fetch/2000/2122/-8835176/-8835848/8835872/8835883/Documented_Information.docx.

ISO/TC176/SC2/N1282. (2015). ISO 9001 Summary of changes. Retrieved from ISO/TC 176 Home Page: http://isotc.iso.org/livelink/livelink/fetch/2000/2122/-8835176/-8835848/8835872/8835883/ISO9001Revision.pptx

ISO Glossary. (2016). Guidance on selected words used in the ISO 9000 family of standards. Retrieved from www.iso.org/iso/03_terminology_used_in_iso_9000_family.pdf

ISO/TMB/JTCG N 359. (2013, December 03). JTCG Frequently Asked Questions in Support of Annex SL. Retrieved from Annex SL Guidance documents: http://isotc.iso.org/livelink/livelink?func=ll&objId=16347818&objAction=browse&viewType=1.

Jarvis, A., & MacNee, C. (2011). Improved Customer Satisfaction – Key Result of ISO 9000 User Survey. Retrieved from ISO: www.iso.org/iso/home/news_index/news_archive/news.htm?refid=Ref1543

3 How the 2015 version has changed misconceptions

Introduction

In the sixth edition of this Handbook was a chapter with the title "A flawed approach", which addressed some 15 criticisms and misconceptions about ISO 9001 using, in some cases, arguments made by John Seddon in his book *The Case against ISO 9000*. Seddon's criticism was based on his observations on the use of ISO 9001 prior to publication of the 2000 revision. However, several criticisms remained valid and, in some cases, will always be made against ISO 9001 because of the fundamental principle on which it is based: that standards used in a regulatory framework improve organizational effectiveness. Whether they do or don't improve organizational effectiveness rather depends on how they are used. If an organization seeks to make the pursuit of quality its first priority and, having consulted ISO 9001 identifies opportunities for improving its performance, ISO 9001 has proved to be a useful tool. But if an organization is coerced by its customers into using ISO 9001 to build a system of documents that enable it to become certificated, ISO 9001 has proved to be a means to an end that has neither changed the organization's attitude towards quality nor added value. There are, of course, organizations that sit somewhere between these two extremes for which ISO 9001 does bring benefits but also some unwelcome bureaucracy.

In this chapter, we will address the following criticisms and misconceptions:

- Whether the standard remains a requirement for doing business
- The extent to which certification is perceived as the goal
- Whether the standard remains a prescription for conformance
- Whether it is designed for auditors or for the business
- The extent to which variation, a key factor in managing quality, has been neglected
- Whether organizations have management systems which function independently of one another
- The notion that the standard only applies to administrative functions of professional services
- Whether the "document what you do" approach has been banished for good
- Whether the new version will sideline those auditors who insisted on documentation for the *one in a million event*
- Whether it is management or customers that lead organizations towards using ISO 9001
- Whether the purpose is to conform to the standard or enhance customer satisfaction
- Whether effectiveness is measured by conformity or by alignment of results
- Whether the focus remains on conformity to procedures

A requirement for doing business

Several ISO standards – mainly those concerned with quality, health, safety or the environment – have been adopted in some countries as part of their regulatory framework or are referred to in legislation for which they serve as the technical basis. However, such adoptions are sovereign decisions by the regulatory authorities or governments of the countries concerned. ISO itself does not regulate or legislate. Although voluntary, ISO standards may become a market requirement, as has happened in the case of ISO 9001, and this has led to the perception that ISO 9001 is a requirement for doing business.

ISO 9001 was designed for use by customers to gain an assurance of product and service quality. It replaced a multitude of customer-specific requirements which suppliers had to meet and thus made it easier for them to bid for work. Coupled with the certification scheme, it enabled suppliers to demonstrate that they had the ability to consistently meet customer requirements and thus reduce multiple assessments and reduce costs.

It is not that this approach to quality assurance is flawed, for it goes back centuries to when traders joined guilds to prove their competence and keep charlatans out of their market. What is flawed is the approach of using ISO 9001 for situations where it is simply inappropriate. ISO 9001 was originally designed for situations where there was a contractual relationship between customer and supplier, but in the 1994 version that provision was removed and the standard is now only applicable where an organization needs to demonstrate its ability to consistently provide products and services that meet customer and legal requirements, but many customers have invoked it in contracts, regardless of the need.

ISO 9001 is now being used through the supply chain as a means of passing customer requirements down the line and saving the purchaser from having to assess for themselves the capability of suppliers, and this has led to certification becoming the goal.

There has been no change in ISO 9001 that affects this situation.

Making certification the goal

Many organizations have been driven to seek ISO 9001 certification by pressure from customers rather than as an incentive to improve business performance and have therefore sought the quickest route. One of the problems with this is that it creates an opportunity for "people to cheat, they do what they need to do to avoid the feared consequence of not being registered" (Seddon, 2000). Seddon calls this coercion and argues that it does not foster learning which is, of course, true and is an unfortunate consequence of any separate inspection regime. It is therefore not surprising that some organizations will play the game to win at all costs.

The flaw in the approach was that customers were led to believe that imposing ISO 9001 would improve product and service quality. To top it all, the organizations themselves believed that by getting the certificate they have somehow, overnight, become a champion of quality. Putting the badge on the wall made them feel "World Class" but in reality, not very much had changed. Whether the CEO was committed to quality before ISO 9001 was adopted, it often did not change the way of thinking.

To achieve anything in our society we inevitably have to impose rules and regulations – what the critics regard as command and control – but unfortunately, any progress we make masks the disadvantages of this strategy. There is a need for regulations to keep sharks out of the bathing area, but if the regulations prevent bathing, we defeat the objective, as did many of the customers that imposed ISO 9001.

Although ISO 9001:2015 is less prescriptive and does place responsibility for the effectiveness of the management system on top management rather than their representative, there is nothing in the new version that will stop certification from being perceived as a goal.

Conformance or performance

The flaw in the certification process prior to the adoption of the 2015 version is that when the standard was used as the acceptance criteria, it was so prescriptive that it was easy to find nonconformity. The standard is now more performance based, where the intended results of the QMS are derived from stakeholder needs and expectations, and therefore this together with risk-based thinking should mark a decline in the prescriptive approach taken with previous versions of the standard and may make the certificate worth having and easier to determine capability.

Designed for auditors, not for the business

Invariably, ISO 9001 is implemented incorrectly. It is an assessment standard but has been used as a design standard, resulting in new systems of documentation that exist for the benefit of auditors and not the business. By focusing only on the assurance requirements as interpreted by external auditors (see later), the management systems have been designed to pass the scrutiny of the third-party auditors rather than the scrutiny of top management. In some cases, the standard has been used wisely by looking at what it requires that is not done and assessing the benefits of change, but this is quite rare.

In being less prescriptive, the new version makes its use more difficult for auditors. The organization now determines the scope of the QMS, what it considers important, how it will address risks and opportunities and what documentation it needs. There are now requirements that address human factors, so the auditor is therefore faced with revealing evidence that the decisions taken by top management have not impaired the organization's ability to consistently provide products and services that meet customer and applicable legal requirements and enhance customer satisfaction. So, in summary, the new version looks as though it's been designed for businesses rather than for auditors.

Neglecting variation

In 2014 the author carried out a review of standards pertinent to quality management systems and extracted statements that refer to the words *variation*, *variability* and *variable* regardless of context. It was found that none of the primary specification quality system standards in use over the last 50 years include the words *variation*, *variability* or *variable* among the requirements. Ford's Q101 standard stands out as an exception, with the concept of variation included in both requirements and guidance provisions. Strangely, this did not follow through to its successor, ISO TS 16949, but would have been conveyed through customer-specific requirements.

ISO 9000:2015, which purports to address fundamental concepts, is devoid of the words *variation*, *variability* and *variable*. The concept of variation appears to have been sidelined, confined to guides and isolated to sector-specific standards, although variation itself is ubiquitous and as relevant to the service and public sectors as it is to the manufacturing sector. Those who fail to manage variation do so at their peril, but ISO 9001:2015 introduces the concept of risk, which is defined therein as the "effect of uncertainty". Variability is one of the four types of uncertainty; therefore, it may be argued that the concept of variation is already embodied in ISO 9001:2015 through use of the word *risk*.

Multiple management systems

ISO 9001 requires organizations to establish a quality management system as a means of ensuring that customer requirements are met. The misconception here is that many organizations failed to appreciate that they already had a management system, and because the language used in ISO 9001 was not consistent with the language of their business, many people did not see the connection between how results were achieved and what the standard required. So instead of mapping the requirements of ISO 9001 onto the business, they started to create a paper system that responded to the requirements of ISO 9001, thus separating this *system* from the business as shown in Figure 3.1.

The danger is that as more and more management system standards emerge, more and more management systems will be created, separating more parts from the business.

In 1987 ISO 9001 was the only international standard specifying requirements for a management system, but by 2015 there were 28 such standards, as identified in Table 2.3. From a systems perspective, this could be interpreted as nothing more than an increase in the number of systems of interest, where a system is a model constructed for the purposes of studying a particular situation. However, from a user's perspective, it has resulted in multiple management systems, each being developed in response to an ISO standard. Each of the standards required there to be a policy for XXX that provided a framework

Figure 3.1 Separate management systems

for objectives for XXX that were achieved by means of an XXX management system owned by an XXX manager as illustrated in Figure 3.2. Although in 2015 ISO had not published standards for all the management systems shown in the diagram, it remains a real possibility, notwithstanding the introduction of Annex SL, which itself is based on the premise that an organization may have multiple management systems which function independently of one another.

Annex SL encourages standards writers and users to believe that organizations establish management systems in response to management system standards. An organization has one purpose and thus one system for accomplishing that purpose. Different aspects of this system can be assessed to examine particular capabilities, and it is these aspects that should be expressed in "management system standards".

For the purpose of studying capability or for external certification, it may be convenient to examine the organization from the perspective of particular objectives and the organization's capability of achieving those objectives. These perspectives might be financial, product or service quality, social responsibility, environment, occupational health and safety, information security, etc. This may result in these different perspectives being labelled financial management system, quality management system,

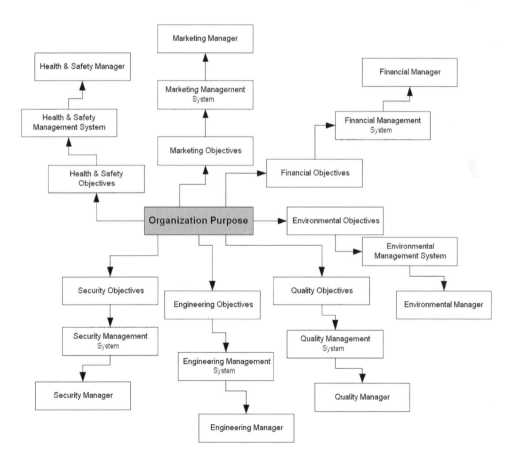

Figure 3.2 Multiple systems heading in different directions led by disparate teams

environmental management system, etc., but these perspectives are of the whole system – they are not distinct subsystems because they each contain common elements; see Chapter 8 for further explanation.

Although the new version does perpetuate the notion of a separate management system, it now requires the QMS requirements to be integrated into the organization's business processes, so there is an attempt to break from the past and ensure the way quality is managed is integral to the business and not separate from it.

Misunderstanding in professional services

There has also been a perception in the service industries that quality management systems conforming to ISO 9001 only deal with the procedural aspects of a service and not the professional aspects. For instance, in a medical practice, the ISO 9001 quality management system is often used only for processing patients and not for the medical treatment. In legal practices, the quality management system again has been focused only on the administrative aspects and not on the legal issues. The argument for this is that there are professional bodies that deal with the professional side of the business. In other words, the quality management system only addresses the non-technical issues, leaving the profession to address the technical issues. This is not quality management. The quality of the service depends on both the technical and non-technical aspects of the service. Patients who are given the wrong advice would remain dissatisfied even if their papers were in order or even if they were given courteous attention and advised promptly. To achieve quality, one must consider both the product and the service. A faulty product delivered on time, within budget and with a smile remains a faulty product!

Although the new version permits the organization to determine the scope of the QMS and, in theory, separate it from technical work, the requirements clearly place the organization under an obligation to ensure conformity of its products and services and the enhancement of customer satisfaction and to ensure persons doing work under its control that affects the performance and effectiveness of QMS are competent. This draws the work of the professional into the QMS.

The "document what you do" approach

An approach to ISO 9001 that found favour was that of "Document what you do, do what you document and prove it." It sounded so simple, and it appeared to match the expectations of third-party auditors who often asked questions such as:

- What do you do?
- In which procedure is it documented?
- Can you show me evidence of conformity with this procedure?

This approach was described by Jack Small of IBM (Small, 1997). Although Small explains the approach slightly differently, the explanatory statements he made were often overlooked by those who adopted the approach:

- Say what you do (i.e. establish appropriate quality controls and systems).
- Do what you say (i.e. ensure that everyone involved follows the established processes).
- Show me (i.e. demonstrate compliance of your quality system to an external auditor).

Box 3.1   SAY – DO

When Admiral Rickover stepped down as head of U.S. Navy nuclear power programs in the early 1980s he addressed a joint session of Congress in which he mentioned a trend he had observed in the Navy that concerned him greatly. This consisted of a leader devising a plan to address a problem and then just simply not executing it. He called it SAY – DO, meaning "SAYing" that something would be done, but not actually "DOing" it. This was a flaw in the implementation of ISO 9001 a decade or so after Rickover's observations as firms issued their quality policies and procedures and then failed to implement them.

It may appear as though this was the tenet of ISO 9001 prior to the 2000 version, but was not in fact what the standard required. The standard actually required the organization "to establish, document and maintain a quality system as a means of ensuring that product or service conforms to specified requirements". It was not recognized then that it needed far more than documented procedures to ensure that a product or service conforms to specified requirements. In subsequent versions, it requires the organization "to establish, document, implement and maintain a quality management system to ensure that customer requirements are determined and are met with the aim of enhancing customer satisfaction". Therefore, if after documenting what you do and doing what you documented and proving it to third-party auditors, your quality management system failed to:

a)   ensure that products and services conform to specified requirements;
b)   ensure customer requirements were determined;
c)   ensure customer requirements were met;
d)   enhance customer satisfaction;

. . . then clearly your quality management system should be deemed ineffective.

By *documenting what you do* you overlook the possibility that what you are doing is not consistent with the requirements. You may, in fact, be doing things that result in delivery of nonconforming products or services and that result in customer complaints; therefore, why would you want to document these? This approach also tends to focus only on tangible activities and overlooks the way people think, the informal network that makes things happen and the values that shape behaviour and lead to action, and so the result of *documenting what you do* creates an imperfect representation of how the organization is managed.

By proving only that you do what you have documented, you overlook the objectives of the system and the results it is delivering. If you test products before shipment and document you do this, then demonstrate that you are testing products before shipment, *you have documented what you do, you have done what you documented and you have proven it*. However, if the people doing these tests are not customer focused, they might skip some tests to avoid the tedium and go home early. If all you have is a record that the test had been performed or a tick in the appropriate box, you would be none the wiser. To be confident that what was done was what was supposed to be done, you need confidence in the people. This requires a different approach and could be why in ISO 9001:2015 there is greater emphasis on human factors by:

- addressing risks to prevent or reduce undesired effects;
- managing the social and psychological factors of the process environment;
- ensuring persons are aware of the implications of not conforming with the quality QMS requirements;
- taking action to prevent human error.

Documentation for the one in a million event

The persistence of the auditors to require documentation led to situations where documentation only existed in case something went wrong – in case someone was knocked down by a bus. The flaw in this approach is that although the unexpected can result in disaster for an organization, it needs to be based on a risk assessment. There was often no assessment of the risks or the consequences. This could have been avoided simply by asking the question "so what?" So there are no written instructions for someone to take over the job, but even if there were, would it guarantee there were no hiccups? Would it ensure product or service quality? Often the new person sees improvements that the previous person missed or deliberately chose not to make – often the written instructions are of no use without training, and often the written instructions are of no value whatsoever because they were written by people who were not doing the job. Requiring documented instructions for every activity would be sensible if what we were creating was a computer program because the instructions were needed to make the computer function as intended. People don't need written instructions to make them function; a management system is not a computer program. Those people who have been brought into the organization to accomplish an objective will seize the opportunity and begin to work without waiting for written instructions.

ISO 9001:2015 now requires risks to be identified and addressed, and this requirement can be used to great effect. It can be used to reduce the amount of documentation maintained by the organization and even justify not applying requirements in certain situations.

Management-led or customer-led approach

From clause 1 of ISO 9001 it can be deduced that the standard is to be used to assess the organization's ability to meet customer, statutory and regulatory requirements applicable to the product or service. It is not designed to be used as a design specification for management systems. In the Guide to Selection and Use (ISO, 2016) it states that

> it is highly recommended that you use ISO 9000 to become familiar with the basic concepts and the language used before you adopt ISO 9001 to achieve a first level of performance. The practices described in ISO 9004 may then be implemented to make your quality management system more effective and efficient in achieving your business goals and objectives.

(At the time of going to press, ISO 9004 was undergoing revision.)

The flaw in this approach is that management systems are being established to meet the requirements of ISO 9001 at the demand of customers or the market, and therefore this is a market- or customer-led approach. It may not result in outcomes which will satisfy all stakeholders. In such a documented system, there are likely to be no processes beyond those specified in ISO 9001 and within those processes no activities that could not be traced to a requirement in ISO 9001. Invariably, users go no further and do not embrace ISO 9004. Had ISO 9004 been promoted and used as a system design requirement, the management system would be designed to enable the organization to deliver outcomes that satisfied all stakeholders. ISO 9001 could then be used to assess the organization's ability to meet customer requirements and, if necessary, ISO 14001 could be used to assess the organization's ability to meet environmental requirements and so on for health, safety, security, etc.

A misunderstood purpose

It was believed that by operating in accordance with documented procedures, errors would be reduced and consistency of output would be ensured. If you find the best way of achieving a result, put in place measures to prevent variation, document it and train others to apply it, it follows that the results produced should be consistently good. The flaw in this argument is that you can't build a system from a set of procedures as though a management system is just a pile of paper. If it were a pile of paper, it wouldn't do very much on its own; there has to be some energizing force for the system to ensure customer requirements are met.

In the third edition of ISO 9001 in 2000, only six documented procedures were required and the emphasis placed upon processes. Some organizations going through the transition from previous versions only produced six documented procedures and converted the other procedures into work instructions or renamed them as processes, which largely missed the point. The tragedy was that certification body auditors accepted this approach. They misunderstood the difference between procedures and processes and continued to prescribe activities as though by doing so they were describing processes. Those starting afresh were not so constrained and had the opportunity to take a process approach, but invariably this has resulted in procedures presented as flow charts instead of text, thus again exhibiting a misunderstanding between procedures and processes.

Application of ISO 9001 can result in organizations writing procedures and following them, regardless of losing sight of the objective. This claim is hard to refute, and it is true not only for ISO 9001–registered organizations but also for any organization that places adherence to procedures above achievement of objectives. In ISO 9001:2015 there are no requirements for specific procedures and more emphasis is placed on ensuring products and services meet requirements, but the standard still misses the point. Look at the first requirement in clause 4.4 where it requires the organization to "establish, implement and maintain a quality management system and continually improve its effectiveness in accordance with the requirements of this International Standard". This emphasizes that the purpose of the system is to meet the requirements of ISO 9001 and not to provide products and services that satisfy the organization's stakeholders. A simple amendment would have changed the focus considerably. If we look at clause 9.2.2 on internal audit, we see that it requires the organization to conduct internal audits to determine whether the quality management system conforms to the requirements of ISO 9001; again, a misunderstanding of purpose. There is no requirement to audit the system to establish how effective it is in enabling the organization to satisfy its customers. Again, a simple amendment would make internal audits add value instead of a being a box-ticking exercise.

Box 3.2   Applying the process approach

If auditors apply the process approach, they would first look at what results were being achieved and whether they were consistent with the intent of ISO 9001 and then discover what processes were delivering these results, and only after doing this, establish whether these processes complied with stated policies, procedures and standards.

Measure of effectiveness

ISO 9001 requires top management to review the management system at defined intervals to ensure its continuing suitability, adequacy and effectiveness, where effectiveness was deemed to be the extent to which the system was implemented. The flaw in this approach is that it led to quality being thought of as conformity with procedures. This preoccupation with documentation alienated upper management so that internal auditors had great difficulty in getting commitment from managers to undertake corrective action. Where auditors do discover serious breaches of company policy or non-adherence to procedure, the managers might commit to take action, but when most of the audit findings focus on what they might regard as trivia, the auditor loses the confidence of management. The management reviews were fuelled by customer complaints and nonconformities from audits and product and process inspections, which when resolved maintained the status quo but did not measure the effectiveness of the system to achieve the organization's objectives. But as the system was not considered to be how the organization achieved its results, it was not surprising that these totally inadequate management reviews continued in the name of keeping the *badge on the wall*. Had top management understood that the system was simply how the organization functioned and that reviewing the system was synonymous with reviewing the effectiveness of the organization in meeting its goals, they might have held a different perception of management reviews and committed more time and energy into making them effective.

ISO 9001:2015 now ties the QMS to the organization's purpose and strategic direction. It requires that the issues affecting the ability of the QMS to achieve its intended results and the needs and expectations of relevant stakeholders be considered when determining the scope of the QMS. It therefore defines its success criteria rather better than conformity with the standard, as was previously the case.

Measuring conformity with procedures

ISO 9001 has indeed encouraged the notion that following the correct procedures was all that was needed to provide a quality product or service. When people are subjected to external controls, they will be inclined to pay attention only to those things which are affected by the controls. There is a tendency for people to *do what you count and not what counts*. This approach was one of the factors that led to the death of a baby in Haringey, North London, UK, in August 2007 as social workers stuck by the rules and their supervisor defended their position. The belief that following the correct procedures produces quality may not be the case in top management, but in a large organization, managers at lower levels are often judged on their ability to play the game, stick to the rules and adhere to the policy and procedures. Under an authoritarian management style, people don't step out of line for fear of losing their jobs. It is true in most organizations and particularly within those where targets are set for every conceivable variable.

Many quality managers feel obliged to take external auditors seriously because their boss would not be pleased to receive reports of nonconformity. Instead of appointing a person with a wealth of experience in quality management who might expect a salary appropriate to their experience, organizations sometimes chose for their quality manager a less quali-fied person who was at an immediate disadvantage with the third-party auditor. Sometimes they select a person with many years of experience with a certification body. This can have the desired effect of facing like with like, but a third-party auditor might not have sufficient experience in developing quality management systems. It is more important for a quality manager to understand the factors on which the achievement of quality depends and know how to influence them rather than understand the requirements of ISO 9001 because such a person will be able to prevent the organization from being adversely influenced by an external auditor.

ISO 9001:2008 did not eliminate this perception because it required the organization to implement the measurement processes needed to ensure conformity of the quality management system. But in ISO 9001:2015 it now requires the measurement needed to ensure valid results which is focused on performance rather than conformance and should lead to there being more focus on results and less on following procedures.

Summary

When ISO 9001 was first published in 1987 it consolidated what was at that time com-mon practices within the Western manufacturing industry for managing product quality. These had evolved from scientific management introduced at the turn of the 20th century and were therefore predicated on a *prescriptive approach* to management. This created many misconceptions of how work should be managed to increase the quality of its out-puts, and the language used in ISO 9001 reinforced this approach. This changed with the 2000 version of ISO 9000, but remnants of the approach remained, and the 2015 version has addressed some of these as has been shown in this chapter. Whether those people encountering ISO 9001 in 2015 for the first time will have the misconceptions addressed in this chapter or an entirely different set is impossible to say, but it's likely that wherever standards are used in a regulatory framework to improve organizational effectiveness, misconceptions will prevail.

Bibliography

ISO. (2016). *Selection and Use of the ISO 9000 Family of Standards*. Retrieved from: https://www.iso.org/files/live/sites/isoorg/files/archive/pdf/en/selection_and_use_of_iso_9000_family_of_standards_2016_en.pdf.

Seddon, J. (2000). *The Case against ISO 9000*. Oxford, UK: Oak Tree Press.

Small, J. E. (1997). *ISO 9000 for Executives*. Sunnyvale, CA: Lanchester Press, Inc.

Key messages from Part 1

Chapter 1   Putting ISO 9001 in context

1   Trade depends on the strength of the customer–supplier relationship, and this relationship survives through trust and confidence at each stage in the supply chain.

2   The intent of ISO 9001 is to enable an organization to demonstrate its capability and in so doing give their customers confidence that they will meet their needs and expectations and consequently build trust.

3   Standards for quality have been traced as far back as 11th century BCE, and those for quality systems came in just after WWII. The ISO 9000 family of standards is just one small group of standards among over 19,500 internationally agreed upon standards that are instrumental in facilitating international trade.

4   The primary users of these standards are intended to be organizations acting as either customers or suppliers.

5   ISO standards are developed by industry experts and not by employees of ISO.

6   ISO standards receive public scrutiny prior to their approval, which is based on a consensus of two-thirds of the participating members.

7   ISO 9001 is not a quality management system, nor does it define requirements for an organization's products and services, but is a set of requirements for a quality management system.

8   ISO 9001 applies to organizations that need to demonstrate their ability to meet customer requirements, either because it's mandated by contract or because it provides a competitive advantage or because organizations prefer to use internationally recognized standards.

9   Customer requirements are intended to include requirements of consumers, clients, end users, retailers, beneficiaries and purchasers.

10   Conformity with ISO 9001 implies that every time a product or service is provided by the organization, it will meet the customer's needs and expectations, regardless of its specification because it has anticipated and taken account of those needs and expectations.

11   Conformity with ISO 9001 also implies that should an organization satisfy its customers by meeting most of their requirements, it can be relied on to undertake continual improvement to meet all their requirements and thereby enhance customer satisfaction.

12   ISO 9001 can apply after an organization has received an order for specific products and services, but it can also apply in situations when the requirements for a product or service have yet to be determined and there is only an unsatisfied need or potential want.

13   ISO 9001 is intended to be used in contractual situations by customers seeking confidence in their supplier's capability, and therefore it is an assessment standard rather than a design standard.

14 ISO 9001 certification may be mandated in the directives of common markets where the free movement of goods and services is a condition of entry.

15 ISO 9001 certification worldwide peaked in 2010 at 1.2 million with an average fall of 6% over the last five years.

Chapter 2   Comparison between 2008 and 2015 editions

16 The primary objectives of the revision to ISO 9000 and ISO 9001 were to develop a simplified set of standards that will be applicable to all types of organizations and for the documentation required to be more relevant to the desired results of the organization.

17 Roughly 80% of the requirements of the 2008 version are embedded in the 2015 version.

18 New requirements include those for:

    a. determining issues that affect the ability of the QMS to achieve its intended results;
    b. holding top management to account for the effectiveness of the QMS;
    c. integrating QMS requirements into the organization's business processes;
    d. promotion of risk-based thinking and the process approach;
    e. identifying and addressing risks and opportunities;
    f. determining the knowledge necessary for the operation of the organization's processes;
    g. planning of changes;
    h. considering the needs and expectations of relevant interested parties;
    i. implementing actions to prevent human error.

19 Requirements withdrawn include those for:

    a. documenting the QMS;
    b. documented procedures;
    c. quality manual and management representative.

20 New concepts include:

    a. *documented information* to subsume the previous concepts of documentation, documents, documented procedures and records;
    b. *competence* has been redefined to signify that someone who is competent must be able to achieve intended results;
    c. *product* is now referred to as product and service and used only to designate the organization's outputs;
    d. *external providers* replace suppliers.

Chapter 3   How the 2015 version has changed misconceptions

21 One of the principles upon which ISO 9001 is based is that standards used in a regulatory framework improve organizational effectiveness.

22 ISO 9001:2015 is far less prescriptive than previous versions and this flexibility may change perceptions that it has been designed more for businesses than for auditors.

23 Although variation is ubiquitous, it's not been mentioned in the standard since its inception; however, by introducing risk and hence uncertainty, variability has indirectly been introduced.

24   Although the standard perpetuates the notion of separate management systems, it now necessitates the QMS requirements to be integrated into the organization's business processes, thereby ensuring the way quality is managed is integral to the business and not separate from it.

25   The introduction of risks and opportunities can be used to reduce the amount of documentation maintained by the organization and even justify not applying requirements in certain situations.

26   ISO 9001:2015 now ties the QMS to the organization's purpose and strategic direction, thereby making its effectiveness judged by performance rather than conformance, and this should lead to there being more focus on results and less on following procedures.

Part 2

Anatomy and use of the standards

Introduction to Part 2

ISO 9000 is a specific standard, but is also a general term for what has become the ISO 9000 phenomenon, meaning not just the single standard, but the infrastructure that has grown around ISO 9001 certification. It might therefore be a surprise to learn that there are 18 standards in the ISO 9000 family and although a detailed knowledge of all of them is not essential to meet the intent of ISO 9001, Chapter 4 provides an overall appreciation of the range of standards, together with the basic requirements, and Chapter 5 provides a practical guide for using these standards which, if adopted, will enable organizations to obtain greater benefit from the ISO 9000 family of standards.

A historical perspective

The story of ISO 9001 is a story of standards, methods and regulation. From when the Egyptians built the pyramids in around 3000 BCE, through China's Western Zhou Dynasty from 1600 BCE and into the modern era, there is evidence to show that:

• standards are an ancient concept that has survived several millennia;
• a means of verifying compliance often follows the setting of standards;
• the formalizing of working practices is centuries old and seen as a means to consistently meet standards;
• market regulation (relative to the standard of goods and services) has been around for centuries for the protection of both craftsmen and traders.

Some milestones in the evolution of standards and practices in the management of quality follow:

1100   Guild system in Europe monopolized trade by setting standards for the quality of goods and the integrity of trading practices. Some outside merchants were prohibited altogether from participating in a particular trade.
1300   Edward I, King of England, brought in a statute that no gold or silver be sold until tested by the "Gardiens of the Craft" and struck with the leopard's head first known as the king's mark.
1776   Adam Smith, a Scottish social philosopher, wrote *Inquiry into the nature and causes of the Wealth of Nations* which influenced the way work was organized for the next 200 years.

1911   Fredrick Winslow Taylor developed his principles of scientific management.

1913   The Aeronautical Inspection Department of the UK War Office was formed to regulate the quality of aircraft production.

1940   Regulations requiring the approval of firms supplying materiel to the Ministry of Aircraft Production and the certification of supplies (Form 649).

1956   10CFR 50 Appendix B Quality Assurance Criteria for Nuclear Power Plants and Fuel Reprocessing Plants.

1959   Mil-Q-9858, Quality Program Requirements.

1967   BS 9000–1 General requirements for a system for electronic components of assessed quality.

1968   AQAP 1, NATO Quality Control Requirement for Industry.

1972   BS 4891, A guide to quality assurance.

1973   Def Stan 05–21, Quality Control Requirements for Industry.

1974   CSA CAN3 Z299 Quality Assurance Program

1974   BS 5179, Guide to the Operation and Evaluation of Quality Assurance Systems.

1979   BS 5750, Quality Systems.

1987   ISO 9000, Quality Management Systems – First series.

1994   ISO 9000, Second series.

2000   ISO 9000, Third series.

2015   ISO 9000, Fourth series.

4 The ISO 9000 family of standards

> The standards provide guidance and tools for companies and organizations who want to ensure that their products and services consistently meet customer's requirements, and that quality is consistently improved.
>
> International Organization of Standardization (2016)

Introduction

In this chapter, we provide an overall appreciation of the range of standards and in particular:

- The identity of the standards that make up the ISO 9000 family
- The characteristics of the core standards in the ISO 9000 family
- The relationship between the standards to illustrate how they are used
- The basis and purpose of the requirements
- A summary of the management requirements
- A summary of the assurance requirements

The ISO 9000 family of standards

All generic international quality management and quality assurance standards are the responsibility of ISO technical committee (ISO/TC) 176. They include standards commonly referred to as the ISO 9000 family. Related standards that are sector specific are the responsibility of other ISO technical committees (see Table 4.1). In this table TS means Technical Specification and TR means Technical Report. In October 2016 IATF 16949 was published by the International Automotive Task Force (IATF) and superseded and replaced ISO/TS 16949. The IATF will take over administration of the standard but remains a participant on TC 176.

It should be noted that there are now three requirement standards in the ISO 9000 family: ISO 9001, ISO 10012 and ISO/TS 17582.

The purpose of these generic standards is to assist organizations in operating effective quality management systems, thereby facilitating international trade. It does this to facilitate mutual understanding in national and international trade and help organizations achieve sustained success. This notion of sustained success is brought out in the title of ISO 9004, showing clearly the broad intent. However, as most organizations are driven towards the ISO 9000 family to gain certification to ISO 9001 rather than to use ISO 9004, its purpose and intent are often overlooked. Although ISO 9001 specifies requirements to be met by the organization, it does not dictate how these requirements

Table 4.1 Standards in the ISO 9000 family

International standard	Title	What it's used for
ISO 9000:2015	Quality management systems – Fundamentals and vocabulary	Used to understand the concepts and terms in ISO 9001
ISO 9001:2015	Quality management systems – Requirements	Used to assess the ability of a QMS to consistently provide conforming products and services and enhance customer satisfaction
ISO 9004:2009	Managing for the sustained success of an organization – A quality management approach	Used to design a management system that will deliver outputs that satisfy the needs of all of an organization's stakeholders
ISO 10001:2007	Quality management – Customer satisfaction – Guidelines for codes of conduct for organizations	Used to guide development of a code expressing the organization's policies for maintaining customer satisfaction in advertising, sales, delivery and post-delivery. Applies in Chapters 34 and 50.
ISO 10002:2014	Quality management – Customer satisfaction – Guidelines for complaints handling in organizations	Used to guide development of a complaints handling process. Applies in Chapters 34 and 58.
ISO 10003:2007	Quality management – Customer satisfaction – Guidelines for dispute resolution external to organizations	Used to guide development of a complaints handling process. Applies in Chapters 34 and 58.
ISO 10004:2012	Quality management – Customer satisfaction – Guidelines for monitoring and measuring	Used to guide development of a customer satisfaction monitoring process. Applies in Chapter 53.
ISO 10005:2005	Quality management systems – Guidelines for quality plans	Used to guide development of quality plans for specific projects. Applies in Chapter 22.
ISO 10006:2003	Quality management systems – Guidelines for quality management in projects	Used to guide the management of projects. Applies in Chapter 33.
ISO 10007:2003	Quality management systems – Guidelines for configuration management	Used to guide development of processes for managing the physical and functional characteristics of products and services during their development, production and use. Applies in Chapters 37 and 41.
ISO 10008:2013	Quality management – Customer satisfaction – Guidelines for business-to-consumer electronic commerce transactions	Used to guide development of sales processes that use e-commerce. Applies in Chapter 34.
ISO 10012:2003	Measurement management systems – Requirements for measurement processes and measuring equipment	Used to assess whether measurement management systems are fit for purpose. Applies in Chapters 27 and 52.
ISO/TR 10013:2001	Guidelines for quality management system documentation	Used to guide preparation of QMS documentation. Applies in Chapter 32.

International standard	Title	What it's used for
ISO 10014:2006	Quality management – Guidelines for realizing financial and economic benefits	Used to guide the application of quality management principles. Applies in Chapter 5.
ISO 10015:1999	Quality management – Guidelines for training	Used to guide competence development. Applies in Chapter 29.
ISO/TR 10017:2003	Guidance on statistical techniques for ISO 9001:2000	Used to guide the selection and application of statistical techniques in production and service provision. Applies in Chapters 45 and 52.
ISO 10018:2012	Quality management – Guidelines on people involvement and competence	Used to guide people engagement in the development, operation and improvement of the QMS. Applies in Chapter 29
ISO 10019:2005	Guidelines for the selection of quality management system consultants and use of their services	Used to guide QMS certification planning.
ISO/TS 17582:2014	Quality management systems – Particular requirements for the application of ISO 9001:2008 for electoral organizations at all levels of government	Used to assess ability of a QMS to consistently provide conforming products and services and enhance customer satisfaction for electoral organizations at all levels of government.
ISO 18091:2014	Quality management systems – Guidelines for the application of ISO 9001:2008 in local government	Used to guide development and operation of a QMS conforming to ISO 9001 in local government.
ISO 19011:2011	Guidelines for auditing management systems	Used to guide the planning, conduct and reporting of management system audits. Applies in Chapter 55

should be met; that is entirely up to the organization's own management. It therefore leaves significant scope for use by different organizations operating in different markets and cultures.

The associated certification schemes (which are not a requirement of any of the standards in the ISO 9000 family) were launched to reduce costs of customer-sponsored audits performed to verify the capability of their suppliers. The schemes were born out of a reticence of customers to trade with organizations that had no credentials in the marketplace.

The core standards in the ISO 9000 family

The core standards are a subset of the family of ISO/TC 176 standards. Together they form a coherent set of quality management system standards facilitating mutual understanding in national and international trade. Use of these standards is addressed later in this chapter, but it is important that each is put in the correct context (see Figure 4.1).

At the core is the organization sitting in an environment in which it desires sustained success. To reach this state the fundamental concepts and vocabulary as expressed in ISO 9000:2015 must be understood; then, if necessary, the organization demonstrates that it has the capability of satisfying customers through assessment against ISO 9001 conducted in accordance with ISO 19011, and finally using ISO 9004, all parts of the management system are managed, producing results that satisfy the needs and expectations of all stakeholders, thereby delivering sustained success. Each of these standards has a different purpose, intent, scope and applicability as indicated in Table 4.2.

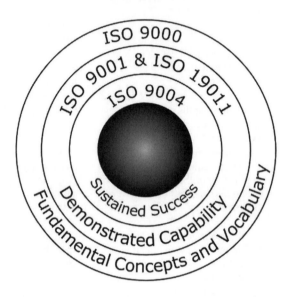

Figure 4.1 ISO 9000 core standards relationship

Table 4.2 Core standards in the ISO 9000 family

Attribute	*ISO 9000*	*ISO 9001*	*ISO 9004*	*ISO 19011*
Purpose	To facilitate common understanding of the concepts and language used in the family of standards	To provide an equitable basis for assessing the capability of organizations to meet customer and applicable regulatory requirements	To assist organizations in achieving sustained success in a complex, demanding and ever-changing environment	To assist organizations in achieving greater consistency and effectiveness in auditing practices
Intent	For use in conjunction with ISO 9001 and ISO 9004. It is invoked in ISO 9001 and therefore forms part of the requirements	This standard is a prescriptive assessment standard used for obtaining an assurance of quality and therefore for contractual and certification purposes only	This standard is a descriptive standard and therefore for guidance only and not intended for certification, regulatory or contractual use	For use in internal and external auditing of management systems

Attribute	ISO 9000	ISO 9001	ISO 9004	ISO 19011
What it covers	Defines the principles and fundamental concepts and terms used in the ISO 9000 family	Defines the requirements of a quality management system, the purpose of which is to enable the organization to continually satisfy their customers	Describes how organizations can achieve sustained success by applying the quality management principles	Provides guidance on the principles of auditing, managing audit programmes, conducting management system audits and guidance on the competence of management system auditors
What it applies to	Applies to all terms used in the ISO 9000 family of standards	Applies where an organization needs to demonstrate its ability to provide products and services that meet customer and regulatory requirements and aims to enhance customer satisfaction	Applies to any organization, regardless of size, type and activity seeking sustained success	Applies to all organizations needing to conduct internal or external audits of quality and/ or environmental management systems or to manage an audit programme
Facts and figures	131 definitions; no requirements	10 sections; 51 clauses; 301 requirements	8 sections; 64 clauses; no requirements	7 sections; 45 clauses; no requirements
Comment	The context and interpretation of the requirements will not be understood without an appreciation of the concepts that underpin them. Also without an understanding of the terms, the standards are prone to misinterpretation.	In theory, if suppliers satisfy ISO 9001, only conforming products or services are provided. This would reduce the need for customers to verify the product on receipt. However, ISO 9001 does not define everything an organization needs to do to satisfy its customers.	There are significant benefits in using the standard as a basis for assessing current capability. There is no doubt that if an organization were to follow the guidance given in ISO 9004, it would have no problem in demonstrating it had an effective management system	ISO 19011 expands the requirements of ISO 9001 clause 9.2 on internal auditing. The guidance is equally applicable to any type of management system

The terms and definitions in ISO 9000 are invoked in ISO 9001 and thus form part of the requirements and will be the basis on which an auditor can judge the acceptability of something, for example, whether an action qualifies as a corrective action or whether a party qualifies as an interested party, as these and another 140 terms are defined in ISO 9000:2015.

ISO 9004 is referenced in ISO 9001 as a guide to improvement beyond the requirements of ISO 9001.

The requirements

The basis for the requirements

The requirements of ISO 9001 are claimed to have been based on set of seven quality management principles (see Chapter 5). However, they have not been derived from them, as the requirements have evolved in parallel over many decades, and the principles were only defined for the 2000 version. ISO 9001 contains over 300 requirements spread over seven sections, but the way the requirements are grouped creates some anomalies and some ambiguities which are addressed in later chapters.

Purpose of requirements

The purpose of these requirements is to enable an organization to provide an assurance of product and service quality. They are not intended to be for developing a quality management system. If we ask of every requirement, "would confidence in the quality of the product or service be diminished if this requirement was not met?" we should, in principle, get an affirmative response, but this might not always be the case. Validation of the 2015 version sought to solicit answers to the following four questions from the committee members responsible for its development (TC 176):

- Is the clause easy to understand?
- Is the clause easy to use or apply?
- Is the clause easy to translate from English to your language?
- Is the clause easy to audit?

There was a change in direction in 2000 when the ISO 9000 family changed its focus from procedures to processes, and another change in 2015 when understanding the organization's context became the driving force in QMS development. These changes are illustrated Figure 4.2. ISO 9001:2008 clearly positions the system of managed processes as the means for generating conforming products or services with the intent that these create satisfied customers. ISO 9001:2015 goes further and creates a cycle of sustained performance, driven from understanding the organization's context through an extended system of managed processes to produce results (consistently conforming products and services) that satisfy all stakeholders.

The basic management requirements

Further on in this book we comment on the structure of ISO 9001 and the 300+ requirements in more detail, but we can condense these into the following eight management requirements:

1   Purpose – review the organization's purpose and the needs and expectations of the stakeholders relative to this purpose. This is addressed by clause 4.2.

Figure 4.2 The changing purpose in ISO 9001

2 Context – scan the environment to determine the factors affecting the ability of the organization to fulfill its purpose, decide priorities for action and set the strategic direction. This is addressed by clause 4.1.

3 Policy – define the overall intentions, principles and guiding values related to quality commensurate with the organization's purpose and strategic direction. This is addressed by clause 5.2.

4 Planning – establish objectives, measures and targets for fulfilling the organization's purpose and its policies; assess risks and develop plans and processes; and determine the resources needed for achieving the objectives that take due account of these risks. This is addressed by clauses 4.4, 6.1, 6.2 and 7.1.

5 Implementation – resource, operate and manage the plans and processes to deliver outputs that achieve the planned results. This is addressed by clauses 7.1 to 7.5 and clause 8.

6 Measurement – monitor, measure and audit processes; the achievement of objectives; and adherence to policies and the satisfaction of stakeholders. This is addressed by clauses 8, 9.1.1, 9.1.2 and 9.2.

7 Review – analyse and evaluate the results of measurement, determine performance against objectives and determine changes needed in policies, objectives, measures, targets and processes for the continuing suitability, adequacy and effectiveness of the system. This is addressed by clauses 9.1.3 and 9.3.

8 Improvement – undertake action to bring about improvement by better control, better utilization of resources and better understanding of stakeholder needs. This might include innovation and learning. This is addressed by clauses 6.3 and 10.

The basic assurance requirements

We can also condense the requirements of ISO 9001 into five assurance requirements:

1   The organization shall demonstrate its commitment to the achievement of quality. This is addressed by clause 5.2.
2   The organization shall demonstrate that it has effective policies for creating an environment that will motivate its personnel into satisfying the needs and expectations of its customers and applicable statutory and regulatory requirements. This is addressed by clause 5.1.
3   The organization shall demonstrate that it has effectively translated the needs and expectations of its customers and applicable statutory and regulatory requirements into measurable and attainable objectives. This is addressed by clause 6.2.
4   The organization shall demonstrate that it has a network of processes for enabling the organization to meet these objectives in the most efficient way. This is addressed by clauses 7.1 and 8.1.
5   The organization shall demonstrate that it is achieving these objectives as measured, that they are being achieved in the best way and that they remain consistent with the needs and expectations of its stakeholders. This is addressed by clause 9.

Summary

In this chapter, we focused on the ISO 9000 family of standards, explaining the composition of the family, their relationships, intent and applicability. There are only three requirements standards in the family, but the most commonly used standard is ISO 9001. There are over 300 requirements in ISO 9001 spread over 7 sections, and so it's not easy to grasp the full intent simply by scanning through the pages. We therefore examined the purpose of the requirements and summarized them in two sets, a set of eight management requirements and a set of five assurance requirements, and cross-referred to the relevant clauses of the standard to act as a simple guide.

5   A practical guide to using these standards

Introduction

There are three ways of using these standards:

*   As a source of information on good practices that can be consulted to identify opportunities for improvement in business performance
*   As set of requirements and recommendations that are implemented by the organization
*   As criteria for assessing the capability of a management system or any of its component parts

In this chapter, we address the pros and cons of either consulting, implementing or applying management system standards and in particular:

*   The usefulness of the standards
*   What you should do before, during and after consulting these standards
*   What you should do before you change anything
*   The misnomer of implementing ISO 9001
*   How to go about applying these standards from the point of view of the organization and a customer or third party
*   How the right level of attention becomes a critical success factor
*   The relevance and use of the quality management principles
*   The relevance and use of PDCA in developing a quality management system

Usefulness of the standards

The standards capture what may be regarded as good practice in a particular field. The information has been vetted by those deemed to be experts by ISO member states, and therefore one can defer to any of these standards as a legitimate authority in the absence of anything more appropriate. They are, however, but one of several sources of authoritative information. With this caveat in mind, these standards can be useful in:

*   Forming ideas
*   Settling arguments
*   Clarifying terminology, concepts and principles
*   Identifying the right things to do
*   Identifying the conditions for ensuring things are done right

Before consulting the standards

Before consulting any of the standards, either a need for improvement in performance or a need for demonstration of capability should have been identified and agreed with senior management.

Ideally the objectives for change and a strategy for change should also have been established to indicate the direction and the means of getting there. This will place these standards in the correct context. Consulting the standards before doing this will prejudice the strategy and may result in compliance, with the standard becoming the objective, thereby changing perceptions as to the motivator for change.

The need for improvement might arise from:

- a performance analysis showing a declining market share or significant number of customer complaints either with the product or the associated services;
- a competitor analysis showing that productivity needs to be increased to compete on price and delivery;
- a market analysis showing a demand for confidence that operations are being managed effectively;
- an analysis of the environment that identifies opportunities for creating new markets, products or services.

If the organization is currently satisfying its stakeholders but lacks a means of demonstrating its capability to customers or regulators that demand it, certification to ISO 9001 may provide a satisfactory solution, but it is not the only solution unless given no option by the customer.

While consulting the standards

There is no doubt that ISO 9001 is the top-selling international standard of all time, but other standards in the family have not had similar success, which creates a major problem with the use of these standards.

When consulting these standards, bear in mind the following:

- They reflect the collective wisdom of various organizations and *experts* that participate in the development of national and international standards.
- They have been produced by different committees and therefore as a group of standards will contain inconsistencies, ambiguities and even conflicting statements.
- They do not yet fit together as a system with all prescriptions and descriptions aligned to an overarching purpose and set of principles.
- Compromises often must be made for the standards to be a consensus of at least 75% of the voters in the ISO community.
- What you read is not necessarily the latest thinking on a topic or the result of the latest research primarily because of the review cycle (often five years) being so protracted.
- The standards reflect practices that are well proven and possibly now outdated in some quarters but have stood the test of time and are used universally.
- Common terms may be given an uncommon meaning, but terminology is by no means consistent across this class of standards thus making their use more difficult (management system, correction and corrective action being typical examples).

- Some phrases might appear rather unusual and this is to preserve meaning when translated into other languages (use of the term interested party in place of stakeholder is one example) and as a result creates ambiguities.
- Requirements are not necessarily placed in their true relationship and context due to the constraints of the medium by which the requirements are conveyed. As a result, users and auditors often treat requirements in isolation when in fact they are all interrelated.

Although there is the opportunity for changing these standards, there may not be any desire for change because of the various vested interests. If organizations have based their approach on one or more of these standards, they will be reluctant to sponsor any change that might result in additional costs, regardless of the benefits. These organizations might be willing to institute the changes informally rather than to have them imposed through an externally assessed standard.

Box 5.1   Axiom

1   Understand the intent.
2   Understand the impact.
3   Understand how to make it happen.
4   Understand how to make it a habit.

When a family of standards is embraced, studied and applied intelligently, there can be enormous benefits from its use. However, standards of this type can lend themselves to misuse by spreading the information so widely across several documents and by not translating the concepts into requirements with a clarity that removes any ambiguity.

The most important factor is that whatever the statement in these standards, it is necessary to understand the intent (i.e. what it is designed to achieve). There is simply no point in following advice unless you fully understand the consequences (i.e. what the impact will be) and have a good idea of what you might have to do to make it happen and to sustain the benefits it will bring. Sustained levels of performance will only arise when the new practices become ingrained in the culture and become habitual. This makes it imperative that you do not limit your reading to ISO 9001 alone but also include the guidance standards and other relevant literature.

After consulting the standards

Having consulted the standards, you need to:

- Put your findings in context, as not everything you read will be applicable in your organization.
- Assess the impact (benefits, drawbacks) on the organization of applicable provisions.
- Validate your findings with other sources (books, articles, peers etc.).

If it seems like what is expressed in the standards accords with good practice and offers practical benefits, then by all means follow the advice given.

Before you change anything

At some stage after you have obtained your ISO 9001 certificate, people will ask you about the benefits it has brought to the organization. Unless you capture the state of the organization and its performance beforehand, you can only provide a subjective opinion. It is therefore highly advisable to record a series of benchmarks that you can use later to determine how far you have progressed. A simple model is provided with the self-assessment in relation to the quality management principles further on in this chapter, but you also need measurements against your key performance indicators such as:

- Time to market (time it takes to get a new product or service into the market)
- Customer satisfaction (customer perception of your organization and its products and services)
- Conformity (measure of conformity, e.g. ratio of the number of products returned to those shipped or system availability if you are a service provider such as phone company, energy supplier, etc.)
- Supplier relationships (supplier perception of your organization and the way you deal with them)
- On-time delivery
- Processing delays (the impact of shortages, bottlenecks, down time)
- Employee satisfaction (employee perception of your organization and the way you attend to their needs)
- External failure costs (costs of correcting failures after product or service delivery)
- Internal failure costs (costs of correcting errors detected before product delivery or during service delivery)
- Appraisal costs (costs of detecting errors)

Remember to use the same measurement process after certification; otherwise, the results will be invalidated.

Box 5.2   Applying or implementing ISO 9001?

Do we implement or apply ISO 9001? The two words do have different meanings, and therefore their use produces different results. It's the difference between putting into effect (implement) and bringing something to bear on another (apply).

Implementing ISO 9001 results in companies setting up separate ISO 9001 quality management systems, whereas application of ISO 9001 results in using the standard to see where improvements to existing policies and practices can be improved. You can't apply a coat of paint to a surface that does not exist or a principle to a situation that does not exist.

The word *application* is used in the title of ISO TS 9002, and the word *apply* is used in clause 4.3 of ISO 9001.

Implementing management system standards

Some ways in which these standards have been promoted have not helped their cause because they have been perceived as addressing issues separate from the business of managing the organization. Invariably organizations are being told to implement ISO 9001 or some other standard, but implementation is often not the best approach to take. Hence, in response, some organizations have set up new systems of documentation that run in parallel to the operating systems in place. Regrettably, certification has followed implementation, and it is certification that has driven the rate of adoption rather than a quest for economic performance.

When we implement something, we put it into effect, we fulfil an obligation. In fact, many organizations have implemented these standards because they have put it into effect and fulfilled an obligation to do as required and recommended by the standard.

Box 5.3   Stop what you are doing

The biggest mistake many make is in following the ritual; document what you do, do what you document and prove it, and continue to pursue activities and behaviours that adversely affect performance. This approach is like taking medicine but continuing the lifestyle that prompted the medication in the first place.

Implementation implies we pick up the standard and do what it requires. As the standards don't tell us to stop doing those things that adversely affect performance, these things continue. If the culture is not conducive to the pursuit of quality, these things will not only continue but also make any implementation of standards ineffective. Doing as the standards require will not necessarily result in improved performance. A far better way is to consult the standards (as described earlier), establish a system that enables the organization to fulfil its goals (as described in Chapter 15) and then assess the system by applying the standards as described next.

Applying management system standards

By the organization

If you apply these standards instead of implementing them, you design a system that enables you to achieve your goals and then use ISO 9001 to assess whether this system conforms to the requirements. The guides may help you consider various options, even find the right things to do, but it is your system, your organization, so only you know what is relevant.

In applying these standards, you should not create a separate system but look at the organization as if it were a system and look for alignment with the requirements and recommendations of the various standards. Only change the organization's processes to bring about an improvement in its performance, utilization of resources or alignment with stakeholder needs and expectations. Where there is no alignment:

a)   Verify that the requirement is really applicable in your circumstances.
b)   Change the organization's processes only if it will yield a business benefit.

Changing a process simply to meet the requirements of a standard is absurd; there must be a real benefit to the organization. If you can't conceive any benefit, take advice from experts who should be able to explain what benefits your organization will get from a change. ISO 9001:2015 required risk-based thinking, and therefore it's expected that organizations will use its resources wisely.

If the organization is seeking to develop a QMS that conforms to ISO 9001, it will be tempting to label the project "ISO 9001 implementation" or similar and the project leader "ISO 9001 project manager". Try and avoid such labels, as they detract from what the organization is trying to do. The goal is not to meet ISO 9001. That's like saying my object in going to university is to get a degree when it should be to get a university education. The goal is to give the organization the capability to satisfy its customers so labels such as quality improvement or customer first send out more appropriate signals.

By the customer or a third party (conformity assessment)

Customers and third-party certification bodies use the assessment standards such as ISO 9001 and ISO 14001 to determine the capability of other organizations to satisfy certain requirements (customer, environment, security, etc.). This is called conformity assessment, which refers to a variety of processes whereby goods and/or services are determined to meet voluntary or mandatory standards or specifications. Conformity assessment is therefore limited to the scope of the standard being used, and thus (unlike the excellence model or the self-assessment criteria in ISO 9004) it is not intended to grade organizations on their capability. An organization either conforms or it doesn't conform.

ISO 9001 was primarily intended for situations where customers and suppliers were in a contractual relationship. It was not intended for use where there were no contractual relationships. It was therefore surprising that schools, hospitals, local authorities and many other organizations not having a contractual relationship with their "customer" would seek ISO 9001 certification. Even in contractual situations, demonstration of capability is often only necessary when the customer cannot verify the quality of the products or services after delivery. The customer may not have any way of knowing that the product or service meets the agreed requirements until it is put into service by which time it is costly in time, resource and reputation to make corrections. In cases where the customer has the capability to verify conformity, the time and effort required are an added burden, and their elimination helps reduce costs to the end user.

In many cases, using ISO 9001 as a contractual requirement is like using a sledgehammer to crack a nut – it was totally unnecessary and much simpler models should have been used.

Level of attention to quality

In the first section of the Introduction to ISO 9001 there is a statement that might appear progressive, but depending on how it is interpreted, it could be regressive. The statement is: "The adoption of a quality management system should be a strategic decision of an organization." The very idea of adopting a QMS implies it's not a system but a set of principles or methods – but more on that in Chapter 8.

What would top management be doing if they adopted a QMS? Would they be agreeing to:

- implement the requirements of ISO 9001 and subject the organization to periodic third-party audit as evidence of commitment to quality?
- document the approach they take for the management of product or service quality and subsequently do what they have documented?
- manage the organization as a system that delivers stakeholder satisfaction?

It all comes down to their understanding of the word *quality* and this is what will determine the level of attention to quality.

Although the decision to make the management of quality a strategic issue will be an executive decision, the attention it is given at each level in the organization will have a bearing on the degree of success attained.

There are three primary organization levels: the enterprise level, the business level and the operations level. Between each level there are barriers. At the enterprise level, the executive management responds to the *voice* of the stakeholders, and on one level is concerned with profit, return on capital employed, market share, etc., and on another level with care of the environment, its people and the community. At the business level, the managers are concerned with products and services and so respond to the *voice* of the customer. At the operational level, the middle managers, supervisors, operators, etc., focus on processes that produce products and services and so respond to the *voice* of the processes carried out within their own function.

In reality, these levels overlap, particularly in small organizations. The chief executive officer (CEO) of a small company will be involved at all three levels, whereas in the large multinational, the CEO spends all the time at the enterprise level, barely touching the business level except when major deals with potential customers are being negotiated. Once the contract is won, the CEO of the multinational may confine his or her involvement to monitoring performance through metrics and goals.

Quality should be a strategic issue that involves the owners because it delivers fiscal performance. Low quality will ultimately cause a decline in fiscal performance.

The typical focus for a quality management system is at the operations level. ISO 9001 is perceived as an initiative for work process improvement. The documentation is often developed at the work process level and focused on functions. Much of the effort is focused on the processes within the functions rather than across the functions and only involves the business level at the customer interface, as illustrated in Table 5.1. For the application of ISO 9001 to be successful, quality must be a strategic issue, with every function of the organization embraced by the management system that is focused on satisfying the needs of all stakeholders.

Table 5.1 Attention levels

Organizational Level	Principle Process Focus	Basic Team Structure	Performance Issue Focus	Typical Quality System Focus	Ideal Quality System Focus
Enterprise	Strategic	Cross-Business	Ownership	Market	Strategic
Business	Business	Cross-Functional	Customer	Administrative	Business Process
Operations	Work	Departmental	Process	Task Process	Work Process

In conversations about quality it is not unusual to find that the focus shifts from product quality, to process quality, to organization quality without there being any signal that the shift has taken place. A useful definition of quality is fitness for purpose, so when we say the QMS only addresses those aspects of the organization which affect the quality of the goods and services provided to customers, we are focusing on product and service quality. But if we focus on process quality, we bring in more factors that affect the ability of the process to fulfil its purpose because a process serves the needs of the organization as well as the needs of customers and therefore needs to be efficient as well as effective. If we focus on the organization as a whole, we bring in even more factors that affect the ability of the organization to fulfil its purpose because an organization serves the needs of all the stakeholders as well as the needs of customers. So, if an activity or a process is required for the organization, then it must contribute – directly or indirectly – to the overall quality of the organization.

The relevance and use of the quality management principles

The management principles

If we ask ourselves on what does the achievement of quality depend, we will find that it rather depends upon our point of view.

In Edwards Deming's seminars, he suggested that if you ask people to answer *Yes* or *No* to the question "Do you believe in quality?" no one would answer *No*. They would also know what to do to achieve it, and he cited several examples which have been put into the cause-and-effect diagram shown in Figure 5.1 (Neave, 1990). It is the causes below the labels on the ends of each line that are the determinants. The labels simply categorize the causes. Deming regarded these factors as all wrong. Either singularly or all together they will not achieve quality. They all require money or learning a new skill, and as Lloyd Dobyns (Deming's collaborator on the video library) says "They allow management to duck the issue." However,

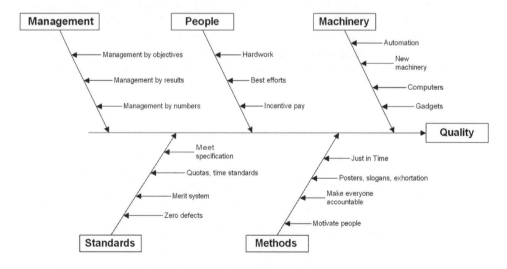

Figure 5.1 Inappropriate determinants of quality

he tells us that the fact that they won't work does not mean each of them is wrong. Once the processes are predictable and the system is stable, a technique such as Just-in-Time is a smart thing to do.

We need principles to help us determine the right things to do and understand why we do what we do. The more prescription we have, the more we get immersed in the detail and lose sight of our objectives – our purpose – our reason for doing what we do. Once we have lost sight of our purpose, our actions and decisions follow the mood of the moment. They are swayed by the political climate or fear of reprisals.

Since the dawn of the industrial revolution, when man came out of the fields into the factories, management became a subject for analysis and synthesis in an attempt to discover some analytical framework upon which to build managerial excellence. Many management principles emerged, and indeed they continue to emerge in an attempt to help managers deal with the challenges of management more effectively. Over the last 20 years several principles have been developed that appear to represent the factors upon which the achievement of quality depends:

1   Understanding customer needs and expectations (i.e. a customer focus)
2   Creating a unity of purpose and a quality culture (i.e. leadership)
3   Developing and motivating the people (i.e. engagement of people)
4   Managing processes effectively (i.e. the process approach)
5   Understanding the complex relationship between cause and effect (i.e. the systems approach)
6   Continually seeking better ways of doing things (i.e. continual improvement)
7   Basing decisions on facts (i.e. evidence-based decision-making)
8   Realizing that you need others to succeed (i.e. relationship management)

These eight factors represent the causes of quality as shown in the cause-and-effect diagram in Figure 5.2. In this diagram, it is the causes below the labels on the ends of each line that are a few of the factors influencing quality either positively or negatively. A failure either to understand the nature of any one of these factors or to manage them effectively will

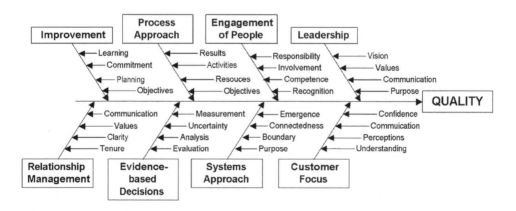

Figure 5.2 Quality management principles

invariably lead to a quality failure, the consequences of which may be disastrous for the individual, the customer, the organization, the country and the planet.

In the ISO/TC 176 guide: Quality management principles, it is stated that "quality management principles are a set of fundamental beliefs, norms, rules and values that are accepted as true and can be used as a basis for quality management" (ISO, 2015). It is a pity that this definition includes the word *rule* because principles are not rules (see Box 5.4), but guides to action, implying flexibility and judgement as to their appropriateness.

Box 5.4   Principles or rules

A principle is a fundamental law, truth or assumption that is verifiable. Management principles are a guide to action; they are not rules. "No entry to unauthorized personnel" is a rule that is meant to be obeyed without deviation, whereas a principle is flexible; it does not require rigid obedience. A principle may not be useful under all conditions, and a violation of a principle under certain conditions may not invalidate the principle for all conditions.

A violation of a principle results in consequences, usually by making operations more inefficient or less effective, but that may be a price worth paying under certain circumstances. In order to make this judgement, managers need a full understanding of the consequence of ignoring the principles.

All the requirements of ISO 9001:2015 are related to one or more of these principles. These principles provide the reasons for the requirements and are thus very important. Each of these is addressed later. In the 2015 revision, the systems approach principle was removed and the process approach principle revised to include the phrase "interrelated processes that function as a coherent system". This change does not fully recognize the principle involved, that the behaviour of a system is inherent in its structure, that it's the interconnectedness among the elements that produces its behaviour. The process approach treats work as a process, a series of linear actions, whereas the systems approach looks for the interactions, the causal loops (see Figure 8.7). For this reason, the systems approach principle has been retained in this Handbook. See also Chapter 9.

Customer focus

This principle was expressed in ISO 9000:2005 as follows:

> Organizations depend on their customers and therefore should understand current and future customer needs, meet customer requirements and strive to exceed customer expectations.

This was changed in 2015 to "the primary focus of quality management is to meet customer requirements and to strive to exceed customer expectations", which is more like a statement of purpose than of a principle. Perhaps a better way of expressing this as a principle would be as follows: *When an organization manages in a way that increases the quality of its outputs, it will more likely meet customer requirements and exceed customer expectations.*

An organization applying the customer focus principle would be one in which people:

- understood customer needs and expectations;
- met customer requirements in a way that met the needs and expectations of all other stakeholders;
- communicated these needs and expectations throughout the organization;
- have the knowledge, skills and resources required to satisfy the organization's customers;
- measured customer satisfaction and acted on results;
- understood and managed customer relationships;
- could relate their actions and objectives directly to customer needs and expectations;
- were sensitive to customer preferences and acted in a way that put the customer first.

Leadership

This principle is expressed in (ISO, 2015) as follows:

> Leaders at all levels establish unity of purpose and direction and create conditions in which people are engaged in achieving the organization's quality objectives.

An organization applying the leadership principle would be one in which leaders are:

- establishing and communicating a clear vision of the organization's future;
- establishing shared values and ethical role models at all levels of the organization;
- being proactive and leading by example;
- understanding and responding to changes in the external environment;
- considering the needs of all stakeholders;
- building trust and eliminating fear;
- providing people with the required resources and freedom to act with responsibility and accountability;
- promoting open and honest communication;
- educating, training and coaching people;
- setting challenging goals and targets aligned to the organization's mission and vision;
- communicating and implementing a strategy to achieve these goals and targets;
- using performance measures that encourage behaviour consistent with these goals and targets.

Engagement of people

This principle is expressed in (ISO, 2015) as follows:

> Competent, empowered and engaged people at all levels throughout the organization are essential to enhance its capability to create and deliver value.

Previously the principle focused on involvement which is a passive concept, whereas engaged employees move in the same direction and at the same pace as their leaders.

An organization applying the engagement of people principle would be one in which people:

- feel personally and emotionally connected to the organization;
- feel pride in recommending it as a good place to work to other people;
- get more than just a wage or salary from working there and are attached to the intrinsic rewards they gain from being with the organization, and
- feel a close attachment to the values, ethics and actions embodied by the organization.

Process approach

This principle is expressed in (ISO, 2015) as follows:

> Consistent and predictable results are achieved more effectively and efficiently when activities are understood and managed as interrelated processes that function as a coherent system.

This is the result of merging two principles that were previously referred to as the process approach and the systems approach. An organization applying the process approach principle would be one in which people:

- know the objectives they must achieve and the process that will enable them to produce outputs that achieve these objectives;
- know what measures will indicate whether the objectives have been achieved;
- have clear responsibility, authority and accountability for the results;
- perform only those activities that are necessary to achieve these objectives and deliver these outputs;
- assess risks before taking action and act in a way that mitigates the impact of the risks;
- know what resources, information and competences are required to achieve the objectives;
- know whether the process is achieving its objectives as measured;
- understand how the outputs of one process affect the outputs of other processes;
- find better ways of achieving the process objectives and of improving process efficiency;
- regularly confirm that the objectives and targets they are aiming for remain relevant to the needs of the organization.

Systems approach to management

This principle is expressed as follows:

> An approach to managing an organization that recognizes its performance results from the interaction of interrelated elements and cannot be predicted by analysing each element taken separately.

An organization applying the system approach principle would be one in which people:

- are able to visualize the organization as a system of interdependent elements;
- recognize that there is no "final" model and that their model is a mental construct;
- understand their own mental models and experiment with using different models;
- constantly examine and make transparent their mental models, habits, values and assumptions;
- whatever problem they are experiencing is related to larger forces and interactions;
- understand that time delays and the chain effects of actions often mask the connection between cause and effect;
- understand that reality is a potential with multiple outcomes – it does not consist of simple cause-and-effect relationships;
- don't lose sight of the whole even when dealing with the detail.

Improvement

This principle is expressed in (ISO, 2015) as follows:

Successful organizations have an ongoing focus on improvement.

An organization applying the improvement principle would be one in which people are:

- improving products, services, processes and systems – an objective for every individual in the organization;
- applying the basic improvement concepts of incremental improvement and breakthrough improvement;
- using periodic assessments against established criteria of excellence to identify areas for potential improvement;
- continually improving the efficiency and effectiveness of all processes;
- promoting prevention-based activities;
- providing every member of the organization with appropriate education and training on the methods and tools of continual improvement;
- establishing measures and goals to guide and track improvements;
- recognizing improvements.

Evidence-based decision-making

This principle is expressed in (ISO, 2015) as follows:

Decisions based on the analysis and evaluation of data and information are more likely to produce desired results.

An organization applying the factual approach principle would be one in which people are:

- defining performance measures that relate to the quality characteristics required for the process, product or service being measured;

- taking measurements and collecting data and information relevant to the product, process or service objective;
- ensuring that the data and information are sufficiently accurate, reliable and accessible;
- analysing the data and information using valid methods;
- understanding the value of appropriate statistical techniques;
- making decisions and taking action based on the results of logical analysis balanced with experience and intuition.

Relationship management

This principle is expressed in (ISO, 2015) as follows:

> For sustained success, an organization manages its relationships with interested parties, such as suppliers.

An organization applying the relationship management principle would be one in which people:

- understand which interested parties the organization depends on for its success and which may threaten its success;
- jointly establish clear understanding of the needs and expectations of those parties on which the organization depends for its success;
- endeavour to mitigate the effect of those parties that threaten its success;
- establish relationships that balance short-term gains with long-term considerations for the organization and society at large;
- create clear and open communications;
- initiate joint development and improvement of products, services and processes where economically viable;
- sharing information and future plans where it's beneficial to the organization;
- recognizing improvements and achievements with those interested parties that have contributed to the organization's success.

Using the principles

Validating process design

The principles can be used in validating the design of processes, validating decisions and auditing systems and processes. You look at a process and ask:

- Where is the customer focus in this process?
- Where in this process are there leadership, guiding policies, measurable objectives and the environment that motivate the workforce to achieve these objectives?
- Where in this process is the engagement of people in the design of the process, the making of decisions, the monitoring and measurement of performance and the improvement of performance?
- Where is the process approach to the accomplishment of these objectives?

- Where in the management of these processes is recognition of the consequences of actions and decisions, the optimization of performance and the elimination of bottle-necks and delays?
- Where in the process are decisions based on fact?
- Where is there continual improvement in performance, efficiency and effectiveness of this process?
- Where is there a mutually beneficial relationship with stakeholders in this process?

Identifying issues, risks and opportunities

Another use of the principles is as a tool for identifying issues, risks and opportunities. Clause 4.1 requires we determine external and internal issues that affect our ability to achieve the intended results of the QMS, and clause 6.1 requires that we consider the issues referred to in clause 4.1 and determine the risks and opportunities that need to be addressed. We could therefore ask, what internal and external issues could potentially affect:

What internal and external issues could potentially affect:

(a) the ability of the organization to meet customer requirements and exceed customer expectations?
(b) conditions in which people are engaged in achieving the organization's quality objectives?
(c) the competence, empowerment and engagement of people at all levels throughout the organization and their ability to enhance the organization's capability to create and deliver value?
(d) the ability of the organization to manage activities as interrelated processes that function as a coherent system?
(e) the ability of the organization to focus on improvement?
(f) the ability of the organization to base decisions on the analysis and evaluation of data and information?

The resultant factors can then be assessed to establish the extent to which an increase or decrease impedes or facilitates performance. For example, we might identify management style as a key factor in people engagement and conclude that the recent appointment of an autocratic manager puts employee morale at risk.

Determining system maturity

Also, you can review the actual measures used for assessing leadership, customer relationships, personnel, processes, systems, decisions, performance and stakeholder relationships for alignment with the principles. Table 5.2 is a maturity grid to test where your organization is relative to the seven quality management principles. An organization fully committed to quality would be at maturity level III. Simply score your organization against the criteria, placing your score at the most appropriate level. For example, if you believe that there is no proactive process for understanding customer needs, you would place a 1 in level I. There is a range (e.g. 4–6) to allow for "not sure, might be, some but not all" responses, etc.

Table 5.2 Self-assessment using the quality management principles

PRINCIPLE	MATURITY LEVEL					
	Level I (1–3)		Level II (4–6)		Level III (7–10)	
1. **Customer focused** Understanding customer needs and expectations	No proactive process for understanding customer needs		A proactive process exists but not in the QMS		The process is fully integrated into the QMS	
2. *Leadership* Creating a unity of purpose and a quality culture	No clearly defined and communicated organization purpose, values and objectives		We know where we are going, but we are not all pulling in the same direction		Everyone understands the organization's purpose and objectives and is motivated and supported to achieve them	
3. *Engagement of people* Developing and motivating the people	People are just another resource to be used to achieve our results		We involve everyone in decisions that affect them		We value our people and achieve our results through teamwork	
4. *Process approach* Managing processes effectively	We have a set of random task-based procedures that are independent of the business objectives		We have departmental processes that serve departmental goals		We design our processes to achieve objectives derived from the organization's objectives and continually measure, review and improve their performance	
5. *Improvement* Continually seeking better ways of doing things	Improvement is perceived as correcting mistakes only		Improvement is perceived as responding to problems		Improvement is perceived as proactively seeking opportunities to improve performance in everything we do	
6. *Evidence-based decision-making* Basing decisions on facts	We don't use any data generated by the QMS to make business decisions		We mainly use audit data, customer complaints and nonconformity data as inputs to decision-making		We base our decisions on process performance data generated by the management system	
7. **Relationship management** Realizing that we need others to succeed	We treat our suppliers as adversaries and keep them at arm's length		We work with our suppliers and employees to improve our overall performance		We involve our suppliers, employees, customers and other stakeholders in our future strategy	
Total						

The relevance and use of PDCA

Historical perspective

Plan-Do-Check-Act (PDCA) has its origins in the scientific method that has evolved over 400 years. Shewart turned the linear specification, production and inspection process, which corresponded to the scientific process of acquiring knowledge, into a circular path. This, he advocated, would represent the idealized case where no evidence is found on inspection to indicate a need to change the specification no matter how many times the three steps are repeated. This is shown pictorially in Figure 5.3 (Shewhart, 1939; republished 1986).

There is no Act stage in this cycle where inspection finds no errors outside specification limits. Were errors to be found, the cycle is not formed. Moen and Norman explain how Deming built on Shewhart's ideas and introduced them in his lectures in Japan in the 1950s and from this how the Japanese modified the Deming cycle to create PDCA (Moen & Norman, 2010). The PDCA concept as explained in ISO 9001:2015 is not quite what was described by Masaaki Imai in his explanation of the PDCA cycle (Imai, 1986), and according to Moen and Norman, Deming commenting on the PDCA cycle said, "What you propose is not the Deming cycle." It appears that Deming disliked the word *check* because he understood it to mean "hold back" and not what he advocated for this stage in the cycle which was "study and learn". Neave suggests that the Inspection stage of the Shewart cycle may be divided into two steps, Observation and Analysis, rather than incorporating the Act stage, and both the Shewart cycle and Deming's cycle is drawn in a way that makes clear the sequence of steps may be repeated (Neave, 1990). The cycle advocated by Deming was Plan-Do-Study-Act and is often illustrated as in Figure 5.4. Deming labelled the figure as "A flow diagram for learning and for improvement of a process or a product" (Deming, 1994).

It is important to understand the intent beneath the labels PDCA because they can so often be taken too literally, and therefore some further explanation is necessary.

Relationship of PDCA to ISO 9001

PDCA is a methodology on which the structure of ISO 9001 is based, and this is evident from Figure 5.2 in the standard where it shows:

- Plan – clause 6: Planning
- Do – clauses 7 and 8: Support and Operation
- Check – clause 9: Performance Evaluation
- Act – clause 10: Improvement

This means that the requirements in these clauses are broadly in line with the steps of PDCA. This leaves out clause 4, which is shown as relating to inputs to planning, and clause 5, which relates to leadership. There is no requirement to use the PDCA methodology. It is more like a way of thinking.

Purpose

Before using any method it's necessary to understand its purpose. PDCA as originally designed is, as stated by Imai, a series of activities for improvement (Imai, 1986). Deming

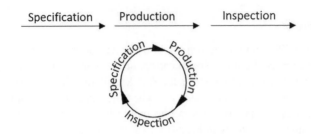

Figure 5.3 Shewhart cycle

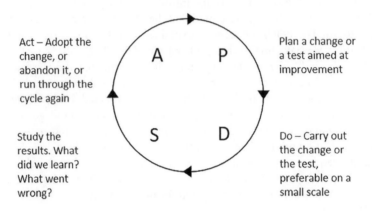

Figure 5.4 The PDCA cycle as advocated by Deming

stressed the importance of the interaction among the stages, and Interaction here is not simply a flow of information from one stage to another but the effect that information has on changing the quality of the output. In this respect PDCA is a systematic approach to problem-solving. PDCA is not a systemic approach because it does not start with identifying the whole of which the problem is a part; it only treats one problem at a time (for more on systematic vs. systemic see Box 8.1).

Imai also introduces the concept of SDCA where the first stage is Standardize. This, he explains, is a cycle that is employed before PDCA which is used to raise standards. He argues that before PDCA is employed, it is essential that the current standards be stabilized because only when a process is stable can we move on to upgrade current standards. This emphasizes once again that PDCA is a method of improvement, not a method of control.

Plan – what do we want to happen?

Planning begins with a study of the current situation. Pascal Dennis, a former manager with Toyota Canada, introduces a precursor to PDCA he refers to as GTS – Grasp The Situation, which entails asking questions such as:

- What is actually happening?
- What should be happening?
- What must be happening?
- What is the ideal condition?

Dennis puts GTS at the centre of the PDCA cycle, as shown in Figure 5.5, because it supports each phase (Dennis, 2006).

The data gathered are analysed to identify a problem, the possible causes and provide a compelling reason for change and the predicted benefits that will result. The reason is expressed as an objective, and the benefits translated into measures for judging whether the objective has been achieved. A plan or method is then formulated to achieve these objectives (i.e. a plan for bringing about the desired result). The plan not only identifies what is to be done and in what sequence and when, but who is to be involved and what resources will be required, in other words, the schedule, processes, resources and responsibilities.

In ISO 9001 clause 0.3.1, the brevity with which the planning stage is expressed limits it to establishing objectives of a system and its processes and resources, but when we examine clauses 4 and 6 we are presented with a much wider range of actions that address all the elements identified earlier except the *problem and its cause* is expressed as *internal and external issues* in clause 4.1.

If we take the word *plan* literally it means scheme, programme or method worked out beforehand for the accomplishment of an objective (ISO Glossary, 2016) and therefore the setting of the objective precedes the formulation of the plan. Also, objectives should not be set before assessing the current situation, which is why some critics of the PDCA methodology prefer to put A for Assess first. Without knowledge of the current situation, there's no basis for knowing whether a change has brought about an improvement. This observation resulted in a development of Deming's PDCA cycle in the form of three questions (Moen & Norman, 2010):

- What are we trying to accomplish?
- How will we know that a change is an improvement?
- What change can we make that will result in improvement?

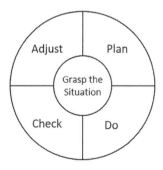

Figure 5.5 Grasp the situation at the centre of PDCA

It is therefore imperative that the first stage in the cycle does not focus only on the word *Plan* and that it is approached with these questions in mind.

Do – make it happen

This stage is generally expressed as "implement the plan" but it hides an important concept, that PDCA is a fractal (i.e. a repeating pattern that displays at every scale) and therefore there will be planning or preparation, doing, checking (examining, inspecting or testing) and correcting for many of the activities defined in the plan. In preparing to implement a plan to change standards of performance, training and education may be necessary.

Implementation of the plan is not complete until all of these actions are complete. Deming suggested the changes were implemented on a small scale to test whether the desired level of improvement results. It is also important to stick to the plan and keep in mind the objective. The plan or solution is one which has been predicted to work, so if what was implemented was not what was planned, the results may lead to the conclusion that the solution was a success or a failure when in reality it was neither.

Check or study – what did we learn?

This stage commences after work on implementation of the change is complete and is intended to establish whether the predicted results were achieved. Implementation needs to run for a long enough period that data can be gathered and studied to establish whether the change had the desired effect (i.e. that the predicted benefits have been realized). Checking is not only about comparing data to predictions and evaluating the benefits but about learning, which is why Deming used the word *Study*. In observing the results of a planned change some things may have gone right and other things gone wrong. From this we can learn but only if we discover why things worked out the way they did. This is important to know so that we may not only avoid the failures but repeat the successes in the future.

Act – what should we do now?

This stage commences when we have gathered sufficient information on which to make a decision as to whether:

- the change should be abandoned; it didn't work and there's no prospect of it ever working with current technologies or financial constraints;
- the cycle should be repeated using the knowledge gained to change the plan or method;
- the change should be adopted and rolled out in the processes to which it applies.

ISO 9001 describes this stage as taking actions necessary to improve performance, implying as in the earlier statement that sometimes the cycle must be repeated, abandoned or adopted.

It is this last statement that contains the purpose of PDCA – that it's a tool for improvement but that is not made clear before the methodology is described, and this misconception sometimes results in PDCA being used in circumstances that are not appropriate, as Imai explains – that SDCA is used for stabilizing standards and PDCA is used for changing standards.

Using PDCA in developing a QMS

PDCA is used in situations where the problem and the objectives can be defined and agreed by those engaged in its resolution. It is also only used for improvement. It is not a method of control, but unfortunately by using it as a framework for ISO 9001 it gives the impression that it can be used for both control and improvement. Although the letters *PDCA* can be used in both ways, the deeds beneath the labels for each use are not the same as explained earlier.

After assimilating the information presented so far in this chapter, it will now be apparent that PDCA can be used as a framework for undertaking a change in the way a part of the organization functions such as when developing a QMS. However, as everything is connected to everything else, a systems approach is needed to ensure changes in parts of the organization also bring about a change in its overall performance. This was a major problem with many Total Quality Management (TQM) programmes of the 1980s when a narrow focus on using quality improvement produced short-term benefits that were not sustained. Russell Ackoff remarks that "Until managers take into account the systemic nature of their organizations most of their efforts to improve their performance are doomed to failure" (Ackoff, 1994). This point is addressed in Chapter 8.

When organizations encounter ISO 9001 and either as an internal initiative or under pressure from customers they choose to develop a formal QMS, PDCA in the way it's described in this chapter can help in the following way:

Plan

*   Study the current situation and get a clear understanding of current performance, the issues and opportunities.
*   Model the organization from the perspective of how it creates and retains customers to a level where a comparison with ISO 9001 can be made.
*   Review the model against the requirements of ISO 9001 and identify the differences in both deeds and behaviours.[1]
*   Determine what needs to change, the benefits of making changes to the way quality is managed and revise the model accordingly.
*   Define the objectives of the change and formulate a plan for achieving them that includes training and the reorientation of attitudes and beliefs, etc.
*   Get agreement to the plan from top management and those who will be affected by it.

Do

*   Prepare to implement the agreed changes, acquire the resources, train the people, etc.
*   Implement the plan preferably in stages.
*   Verify that the plan is being implemented as intended.

Check

- Gather data on the effect of the changes in the process affected and in the performance of the organization as a whole.
- Study the results and determine whether the predicted benefits have been obtained and establish what lessons have been learnt.
- Where the changes did bring improvement in a particular process, check it didn't adversely affect interfacing processes or overall performance of the organization.

Act

- Where the changes didn't bring overall improvement, review the objective and the plan, find better solutions and repeat the cycle.
- Where the changes did bring the predicted benefits, use the knowledge gained to roll out the change in policies and practices throughout the organization.

These are the basic steps, and more detail can be found throughout this Handbook.

Summary

At a practical level, the ISO 9000 family of standards is very useful, but it's unwise to treat the standards as a prescription for what has to be done without considerable preparedness. We outlined the necessary steps that should be taken before, during and after consulting these standards, the dangers of simply doing what the standards require and the importance of getting the attention of top management and understanding what quality and its management mean to them. As the quality management principles form the foundation on which the family of standards is based, we explained the principles and what the organization or its people would be doing if the principles were being applied. We also showed how the principles can be used to validate process design, identify risks and determine system maturity. Finally, we addressed another foundation on which the ISO 9000 family is based, and that is the PDCA methodology. We took a historical perspective so as to appreciate its original intent and reveal misconceptions, and we followed this by explaining the intent behind the stages of Plan, Do, Check and Act when being used to develop a QMS.

Note

1 The differences are more than a list of additional things to do. They affect the way people think, and that influences their behaviour. Undertaking new actions but retaining the same way of thinking won't lead to improvement.

Bibliography

Ackoff, R. L. (1994). Beyond Continuous Improvement. Retrieved from www.youtube.com/watch?v=OqEeIG8aPPk..

Deming, E. W. (1994). *The New Economics for Industry, Government and Education* – Second Edition. Cambridge, MA: The MIT Press.

Dennis, P. (2006). *Getting the Right Things Done – A Leaders Guide to Planning and Execution*. Cambridge, MA, USA: The Lean Enterprise Institute.

Imai, M. (1986). *KAIZEN: The Key to Japan's Competitive Success*. Singapore: McGraw-Hill.

ISO. (2015). *Quality Management Principles*. Geneva: ISO Central Secretariat. Retrieved from: https://www.iso.org/files/live/sites/isoorg/files/archive/pdf/en/pub100080.pdf

ISO Glossary. (2016). Guidance on selected words used in the ISO 9000 family of standards. Retrieved from www.iso.org/iso/03_terminology_used_in_iso_9000_family.pdf

Moen, R. D., & Norman, C. L. (2010, November). Circling Back – Clearing Up Myths about the Deming Cycle and Seeing How It Keeps Evolving. *Quality Progress*, 22–28.

Neave, H. R. (1990). *The Deming Dimension*. Knoxville: SPC Press.

Shewhart, W. A. (1939 (Republished 1986)). *Statistical Methods from the Viewpoint of Quality Control*. New York: Dover Publications.

Key messages from Part 2

Chapter 4 The ISO 9000 family of standards

1 Market regulation (relative to the standard of goods and services) has been around for centuries for the protection of both craftsmen and traders.
2 The first quality system standard was the American military standard, M-Q-9858 in 1959.
3 ISO 9001 derives its pedigree from BS 5750, a British standard that was itself derived from military standards.
4 There are 18 standards in the ISO 9000 family, 3 of which are requirement standards, namely: ISO 9001; ISO 10012, which applies to measurement labs; and ISO/TS 17582, which applies to electoral organizations.
5 Although ISO 9001 specifies requirements to be met by the organization, it does not dictate how these requirements should be met.
6 Certification is not a requirement of any of the standards in the ISO 9000 family.
7 ISO 9001:2015 goes further than the 2008 version and creates a cycle of sustained performance, driven from understanding the organization's context through an extended system of managed processes to produce products and services that satisfy the needs and expectations of all stakeholders.
8 There are over 300 requirements in ISO 9001.

Chapter 5 A practical guide to using these standards

9 The standards capture what may be regarded as good practice in a particular field.
10 Before consulting any of the standards, either a need for improvement in performance or a need for demonstration of capability should have been identified and agreed with the senior management.
11 These standards reflect the collective wisdom of various experts but they will contain inconsistencies, they won't necessarily reflect the latest thinking, common terms may have an uncommon meaning and requirements cannot be treated in isolation as they are all interrelated.
12 Organizations that have based their approach on one or more of these standards will be reluctant to sponsor any change that might result in additional costs, regardless of the benefits.
13 Whatever the statement in these standards, it is necessary to understand its intent (i.e. what it is designed to achieve). There is simply no point in following advice unless the consequences are fully understood.

14 Do not limit your reading to ISO 9001 alone but also include the guidance standards and other relevant literature. Remember that ISO 9000 is indispensable for the application of ISO 9001.

15 Before you change anything, record a series of benchmarks that you can use later to determine how far you have progressed.

16 Implementation implies we pick up the standard and do what it requires. As the standards don't tell us to stop doing those things that adversely affect performance, these things continue. If the culture is not conducive for the pursuit of quality, these things will not only continue but also make any implementation of standards ineffective.

17 A far better way is to consult the standards, establish a system that enables the organization to fulfil its goals and then use ISO 9001 to assess whether this system conforms to the requirements.

18 Changing a process simply to meet the requirements of a standard is absurd; there must be a real benefit to the organization.

19 The goal should not be to meet ISO 9001 but to give the organization the capability to satisfy its customers.

20 The process by which customers and certification bodies determine the capability of organizations to satisfy certain requirements is called conformity assessment. It is limited to the scope of the standard being used and thus is not intended to grade organizations on their capability. An organization either conforms or it doesn't conform.

21 ISO 9001 was primarily intended for situations where customers and suppliers were in a contractual relationship. It was not intended for use where there were no contractual relationships.

22 Demonstration of capability is often only necessary when the customer cannot verify the quality of the products or services during or after delivery.

23 Quality should be a strategic issue that involves the owners because it delivers economic performance. Low quality will ultimately cause a decline in economic performance.

24 If we ask ourselves on what does the achievement of quality depend, we will find that it rather depends upon our point of view.

25 No one will admit they don't believe in quality, but they will often disagree about how quality can be achieved.

26 We need principles to help us determine the right things to do and understand why we do what we do. The more prescription we have, the more we get immersed in the detail and lose sight of our objectives – our purpose – our reason for doing what we do.

27 The quality management principles can be used to validate process design; identify issues, risks and opportunities; and determine system maturity.

28 Plan-Do-Check-Act (PDCA) has its origins in the scientific method that has evolved over 400 years.

29 Deming disliked the word *check* because he understood it to mean "hold back" and not what he advocated for this stage in the cycle, which was *study and learn*.

30 PDCA as originally designed is a series of activities for improvement.

Part 3

Terminology

Introduction to Part 3

If you wish to converse with me, define your terms.

François-Marie Arouet aka Voltaire (1694–1778)

All our work, our whole life is a matter of semantics, because words are the tools with which we work, the material out of which laws are made, out of which the Constitution was written. Everything depends on our understanding of them.

Felix Frankfurter (American jurist, 1882–1965)

Box P3.1   The trouble with words

The trouble with words is that no sooner do we hear them then we distort them, tinting with pigments of prejudice, shading them with our opinions, ending them to fit the slots of our experience. Yet the words we use, and the meanings we assign to them, influence our interpretations of events and so, ultimately our decisions and our actions.

(Price, 1984)

Section 3 of ISO 9001 includes the statement "for the purposes of this document, the terms and definitions given in ISO 9000 apply." What this means is that the definitions in ISO 9000 form part of the requirements of ISO 9001. For example, the term *top management* is used several times in ISO 9001, and it is defined in ISO 9000 as "a person or group of people who directs and controls an organization at the highest level". We can therefore substitute this definition for the term where it is used in a requirement. For example, in clause 5.1.1 the requirement commences "Top management shall demonstrate leadership and commitment with respect to the quality management system . . ." By substitution this becomes "The person or group of people who directs and controls an organization at the highest level shall demonstrate leadership and commitment with respect to the quality management system . . ." The meaning of the requirement has not changed by the substitution. This is what is meant by "definitions given in ISO 9000 apply." Therefore, for a proper understanding of the requirements, the definitions in ISO 9000 are indispensable for the application of this document as indeed it states in ISO 9001 clause 2.

Box P3.2   Handling misunderstanding

It is easy to be deluded into believing there is an understanding when two people use the same words. If you have a disagreement you first need to establish what actions and deeds the other person is talking about. Once these are understood, communication can proceed whether or not there is agreement on the meaning of the words.

(Juran, 1974)

However, it is important to also understand that the terms and definitions in ISO 9000 are not intended to substitute terms and definitions that are used within the standard user's organization. The terms and definitions in ISO 9000 are not only intended for interpreting the requirements in ISO 9001, but also for use in interpreting other standards, primarily those for which ISO TC 176 is responsible and for many other standards in which those terms are used.

The International Organization for Standardization provides open access to a database that can be searched for terms and definitions (see Figure P3.1). The URL is www.iso.org/obp/ui/.

A concern that many people have with definitions is they appear to conflict with dictionary definitions and common usage. With this in mind, the chapters in this part of the Handbook address a few terms that have particular significance in ISO 9001:2015. It's necessary to acquire an appreciation of the differences between common usage of these terms and their usage in ISO 9001 because any differences will not be quickly assimilated by readers of the standard and therefore the dictionary definitions are compared with the ISO definitions.

The relationship between the terms *product, service, output, customers* and *external providers* has changed three times in the development of ISO 9001, as illustrated in Figure P3.2. The term *product* is now only used for intended outputs that enter or exit the organization. It

Welcome to the Online Browsing Platform (OBP)

Access the most up to date content in ISO standards, graphical symbols, codes or terms and definitions. Preview content before you buy, search within documents and easily navigate between standards.

● All ○ Standards ○ Collections ○ Publications ○ Graphical symbols ○ Terms & Definitions ○ Country codes

| | English ▾ | Q |

More options [+]

Need help getting started? Check our **Quick start** guide here!

Figure P3.1 ISO Online browsing platform screen

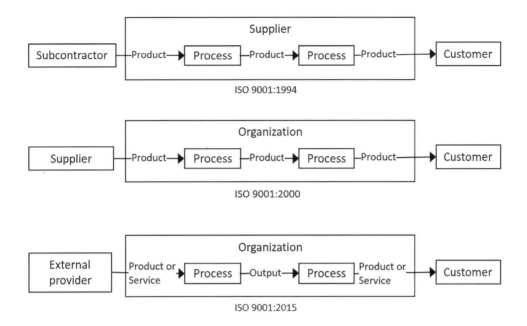

Figure P3.2 Evolution of terms in ISO 9001 and their relationships

will be seen that in the 2015 edition, *service* is being identified separately. See also "Changes in terminology" in Chapter 2.

When debates arise over the meaning of words, some people dismiss them as "only semantics" but Deming understood the importance of having operational definitions and devoted a whole chapter to operational definitions in *Out of the Crisis*. Why? Because as Deming remarks: "The only communicable meaning of any word, prescription, instruction, specification, measure, attribute, regulation, law system, edict is the record of what happens on application of a specified operation or test" (Deming, 1982).

In this part of the Handbook the following six significant terms are explained to impart a depth of understanding that is necessary to fully appreciate the requirements of ISO 9001:2015:

* Quality (see Chapter 6)
* Requirements (see Chapter 7)
* Management system (see Chapter 8)
* Process and the process approach (see Chapter 9)
* Interested parties and stakeholders (see Chapter 10)

Bibliography

Deming, W. E. (1982). *Out of the Crisis*. Cambridge, MA: The MIT Press.
Juran, J. M. (1974). *Quality Control Handbook* – Third Edition. New York: McGraw-Hill.
Price, F. (1984). *Right First Time*. Aldershot, England: Wildwood House.

6   Quality

Quality is not an act, it is a habit.

Aristotle (384–322 BCE)

Introduction

As the achievement of product and service quality is the raison d'être for the ISO 9000 family, no reader should pass by this chapter without getting an insight into the meaning of the term *quality* as used in ISO 9001.

If you are thinking about ISO 9001, you can't get past the front page of the standard without noticing the title in which the first word is *quality* and the second word is *management*. Therefore, it would be unwise to go further without a clear understanding of what quality is and how the achievement of quality is managed in the context of ISO 9001 because the term may be used differently in other contexts. It is also vital that managers have a unified understanding of quality to build a coherent strategy for its achievement.

However, in discussions in which the word *quality* is used, people will differ in their viewpoint either because the word *quality* has more than one meaning or they have different perceptions of what the word means or because they are drawing conclusions from different premises or concepts. Some of the people are perhaps thinking that quality means goodness or perfection or that quality means adherence to procedure, following the rules, etc., or that fewer defects mean higher costs or that quality means high class and is expensive. Others might be thinking that controlling quality means rigid systems, inspectors in white coats or that if they push production, quality suffers, or that quality management is what the quality department does – all of which may be true or flawed depending on the context.

You may consult ISO 9000, which is invoked in ISO 9001, to gain some appreciation of the concepts and the terms used, but this is a rather clinical treatment that does not allow for the wide variation in their application and usage in the real world, and this is what this chapter aims to provide.

In this chapter, we examine:

* The origin of the word *quality*
* The different ways in which the word *quality* is used, drawing on dictionary definitions
* The ISO 9000 definition of the term *quality* and how and why it has changed over the last 28 years

- Other ways in which the term *quality* is used
- Some problems we encounter in its usages within quality management
- The way the term *quality* is used in ISO 9001:2015 and the phrases used to avoid misconceptions

Dictionary definitions

In the *Oxford English Dictionary*, there are 21 pages devoted to the word *quality* addressing several ways in which it is used. It is used primarily as a noun, but also as an adjective. When it's used as an adjective it implies something of a high standard or of excellence (e.g. quality wines, quality cars, quality newspapers). The word *quality* is not used as an adjective in ISO 9001.

As a noun, there are two primary uses, one of which is regarding a person, but this use of the word is not present in ISO 9001. The other primary use of the word is regarding a thing, and this is the sense in which the word is used in ISO 9001. However, ignoring rare uses of the word, it still leaves us with several from which to choose:

a)   An attribute, property; a special feature or characteristic. E.g. "Her art, although brilliant, had a quality pale and luminous, as delicate as a white rose-petal."

(From Charles Chaplin's *My Autobiography*)

b)   The standard or nature of something as measured against other things of a similar kind; the degree of excellence possessed by a thing; e.g. "The landlady called it French cooking, by which she meant that the poor quality of the materials was disguised by ill-made sauces."

(From W. Somerset Maugham's *Of Human Bondage* xciv. 494)

c)   Without article and regarding excellence, superiority; e.g. "Quality of colour means purity or truth of hue."

(From R. Tyrwhitt's *Our Sketching Club* 255)

d)   With article, referring to a particular class, kind, or grade of something, as determined by its character, especially its excellence; e.g. "Industrial concerns need a quality of water from which all minerals are removed."

(*The Times* 4 Oct 1901)

Usage c) is perhaps where perfection might be perceived, but if we are referring to the word *quality* in the context of quality management then it is usage b) that most users of ISO 9001 would be familiar with. As one can see this is not about perfection but a degree of excellence. Quality is relative in this use of the term.

Operational definitions

We are likely to know what quality is when we see or experience it. We are also more likely to ponder the real meaning of the word when we buy something that fails to do what we originally bought it to do. We thus judge quality by making comparisons, based on our own experiences, but defining it in terms that convey the same meaning to others can be difficult. There are dictionary definitions that express how the word *quality* is used but they don't help when we try to take action. When we set out to provide a quality product or service, formulate a strategy for quality, produce a quality policy, control the quality of something or are faced with an angry customer, we need to know what quality means so that we involve the right people and judge whether the action to be taken is appropriate.

There are several definitions in use, each of which is valid when used in a certain context. These are summarized next and then addressed in more detail in the sections that follow.

a) Freedom from deficiencies or defects (Juran) – The meaning used by those making a product or delivering a service.
b) Conformity to requirements (Crosby) – The meaning used by those designing a product or a service or assessing conformity.
c) Fitness for use (Juran) – The meaning used by those accepting a product or service.
d) Fitness for purpose (Sales and Supply of Goods Act 1994) – The meaning used by those selling and purchasing goods.
e) The degree to which a set of inherent characteristics of an object fulfils requirements (ISO 9000:2015) – The meaning used by those managing or assessing the achievement of quality.
f) Sustained satisfaction (Deming) – The meaning used by those in upper management using quality for competitive advantage.

It therefore becomes important to establish the context of a statement in which the term *quality* is used; for example, it would be wrong to say that quality doesn't mean freedom from defects but if the context is a discussion on corporate strategy, it would be foolish to limit one's imagination to that meaning of the word when the purpose of the discussion is to devise a means of gaining a competitive advantage. Even if your products and services were totally free of deficiencies, you would not gain a competitive advantage if your products or services lacked the latest features or were not innovative.

There are other ways in which we think of quality. Masaaki Imai in his book on KAIZEN writes that "when speaking of quality one tends to think first of product quality" (Imai, 1986), and this is indeed the most common context for quality. But Imai goes on to write "when discussed in the context of KAIZEN strategy the foremost concern is with the quality of people."

Then there is the quality of life, the quality of management, the quality of education, etc., and in all these cases we are invoking a definition of quality that leans more towards the degree of excellence that is expressed in the OED. It is helpful to remember that dictionaries record common usage and implied meanings, not legally correct definitions or definitions resulting from the deliberations of a team of experts. The latter two meanings noted earlier are embodied in the more formal definitions that follow.

Freedom from defects or deficiencies

The idea that quality means freedom from defects or deficiencies is based on the premise that the fewer the errors, the better the quality, so a product with zero defects is a product of superior quality. A defect is nonconformity with a specified requirement. Therefore, if the requirement has been agreed with the customer, a defect-free product should satisfy the customer. However, at the level where decisions on nonconformity are made, the requirement is likely to be the supplier's own specification and might not address all product characteristics necessary to reflect customer needs; therefore, a defect-free product or service might not be the one with characteristics that satisfy customers.

Juran contrasts two definitions of quality: that of freedom from deficiencies and product features which meet customer requirements (Juran, 1992). He observes from a manager's perspective that:

- Product features affect sales, so higher quality in this sense usually costs more.
- Product deficiencies affect costs, so higher quality in this sense usually costs less.

In the eyes of the customer, they see only one kind of quality. The product or service must satisfy their needs and expectations, and this means that it should possess all the necessary features and be free of deficiencies. It would be foolish to simply focus on reducing defects as a quality strategy because, as Deming remarked, reducing defects does not keep the plant open. Innovation is necessary to create new products or service features to maintain customer loyalty (Deming, 1994)

Conformity to requirements or specification

The idea that quality means conformance or conformity to the requirements is based on the premise that if a product or service conforms to all the requirements for that product or service, it is a quality product or service. This was the view of the American quality guru, Philip B. Crosby. It became one of his four absolutes of quality (Crosby, 1986). This approach depends on the customer or the supplier defining all characteristics that are essential for the product or service to be fit for its use under all conditions it will be used. However, it removes the subjectivity associated with words like goodness, perfection and excellence and eliminates opinions and feelings. It means that no one is in any doubt as to what must be achieved.

The implication with this definition is that should a product or service not conform to the specified requirements, it will be rejected and deemed poor quality when it might well satisfy the customer. It led Rolls Royce in the 1980s to declare its quality policy as "Meet the requirements or cause them to be changed" in order to prevent products being rejected for trivial reasons. There was and still is a tendency with this definition to pursue ever more detailed requirements in an attempt to capture every nuance of customer needs by defining what is and what is not acceptable. Where customer requirements are very detailed, it means that the simplest decision on fitness for use must be deferred to the customer rather than being made locally. However, the specification is often an imperfect definition of what a customer needs. Some needs can be difficult to express clearly, and by not conforming, it doesn't mean that the product or service may be unsatisfactory to the customer. In the food industry, the quest for conformity to such specifications has led to 30% of food going to waste. Major supermarkets, in meeting consumer expectations, will often reject entire crops of perfectly edible fruit and vegetables at the farm because they do not meet exacting marketing standards for their physical characteristics, such as size and appearance (IMechE, 2015).

Conformance to the requirements can be an appropriate definition at the operational level where customer needs have been translated into requirements to levels where acceptance decisions are made. Crosby was credited with a 25% reduction in the overall rejection rate and a 30% reduction in scrap costs (Wikipedia (1), 2016), so understanding quality as conformity to the requirements can bring significant benefits for the supplier and the customer.

It is also possible that a product that conforms to requirements may be unfit for use. It all depends on whose requirements are being met. Companies often define their own requirement as a substitute for conducting in-depth market research and misread the market. On the other hand, if the standards are well in excess of what the customer requires, the price may well be much higher than what customers are prepared to pay – there probably isn't a market for a gold-plated mousetrap, except as an ornament perhaps!

The conformance to requirements definition relies on there being specified requirements with which to conform. The definition does not recognize potential requirements or future needs or wants, so as a strategy it is rooted in the present.

Fitness for use

The idea that quality means fitness for use is based on the premise that an organization will retain satisfied customers only if it offers for sale products or services that respond to the needs of the user in terms of price, delivery and fitness for use. Juran defined fitness for use as "the extent to which the product or service successfully serves the purpose of the user during usage" (not just at the point of sale) and rather than invent a word for this concept settled on the word *quality* as being acceptable for this purpose (Juran, 1974). It is interesting to note that Juran did not sit down and ponder on what the word *quality* meant. He had identified a concept, then looked around for a label he could use that would adequately convey his intended meaning. It is only in the ensuing decades that the word *quality* has been abused and misused.

Juran later recognized that the fitness for use definition did not provide the depth for managers to take action and conceived of two branches: product features that meet customer needs and freedom from deficiencies. Nonetheless, as a strategy this definition is also rooted in the present and does not consider the future needs of customers.

In societies where business is conducted face to face in the marketplace, there is little need for specifications. Each party knows what makes a product or a service fit for its use, and if there are issues, the provider will be quickly informed by the receiver. However, where the creation and provision of products and services involve many individuals, groups and organizations that are widely dispersed geographically, only a few will understand how their contribution affects the goal of fitness for use and so specifications are provided as a substitute (Juran, 1974).

Fitness for purpose

The UK Sales and Supply of Goods Act 1994, Chapter 35, makes provision as to the terms to be implied in certain agreements for the transfer of property and other transactions. (An extract from this act is contained in Box 6.1). This definition for quality appears to be based on the premise that quality is a standard that a reasonable person would regard as satisfactory, taking account of any description of the goods, the price (if relevant) and all the other relevant circumstances. The only notion excluded is that of delighting customers, but that is where some organizations develop a competitive advantage.

Box 6.1   Extract from UK Sale and Supply of Good Act 1994, Chapter 35, Section 1

Where the seller sells goods in the course of a business, there is an implied term that the goods supplied under the contract are of satisfactory quality. For the purposes of this Act, goods are of satisfactory quality if they meet the standard that a reasonable person would regard as satisfactory, taking account of any description of the goods, the price (if relevant) and all the other relevant circumstances. For the purposes of this Act, the quality of goods includes their state and condition and the following (among others) are in appropriate cases aspects of the quality of goods:

- fitness for all the purposes for which goods of the kind in question are commonly supplied,
- appearance and finish,
- freedom from minor defects,
- safety, and
- durability

Sustained satisfaction

Deming wrote that a product or service possesses quality if it helps somebody and enjoys a good and sustainable market (Deming, 1994). If organizations produce products and services that satisfy their customers, and a satisfied customer is deemed as one who does not complain, then the customer may choose a competitor's product or service next time, not because of dissatisfaction with the previous organization's products or services but because a more innovative product or service came on to the market. Even happy customers and loyal customers will switch to suppliers offering innovative products. This does not arise from meeting present customer needs and expectations; it arises from not recognizing that markets change.

Before the age of mobile phones customers were not hammering on the door of the telephone companies demanding mobile phones; before we had video recorders that could pause live TV we were watching, we were not demanding digital video recorders with hard drives; these innovations arose because the designers looked for better and different solutions that would make life easier for their customers. The innovations do not have to involve high technology. It has now become commonplace in the UK for restaurants to provide chocolate mints after a meal. For a while it delighted customers, as they were not expecting it, but once it became the norm, its power to delight has diminished and so the restaurant trade must look to other innovations to keep the customers coming through the door. In business-to-business relationships a quality service is not simply satisfying customers, but enabling your customers to be more successful with their business by using your services. At Lockheed Martin, they say that the core purpose of their corporation is to achieve mission success, which they define by saying that "mission success is when we make our customers successful".

Sustained satisfaction therefore takes the meaning of quality beyond the present and attempts to secure the future.

Satisfactory and unsatisfactory quality

The definition of quality in ISO 9000:2015 contains the notion of degree, implying that quality is not an absolute but a variable. This concept of degree is present in the generally accepted definition of quality in the *Oxford English Dictionary* and is also implied in the UK Sales and Supply of Goods Act through the phrase *satisfactory quality*. The concept of degree is illustrated in Figure 6.1. The diagram expresses several truths:

- Needs, requirements and expectations are constantly changing.
- Performance needs to be constantly changing to keep pace with the needs.
- Quality is the difference between the standard stated, implied or required and the standard reached.
- Satisfactory quality is where the standard reached is within the range of acceptability defined by the required standard.
- Superior quality is where the standard reached is above the standard required.
- Inferior quality is where the standard reached is below the standard required.

We need to express our relative satisfaction with products and services and therefore use subjective terms. When a product or service satisfies our needs, we are likely to say it is of good quality or satisfactory quality, and likewise when we are dissatisfied we say the product or service is of poor quality or of inferior quality. When the product or service exceeds our needs, we will probably say it is of high quality or superior quality, and likewise if it falls well below our expectations we say it is of low or unsatisfactory quality.

Products or services that do not possess the right features and characteristics, either by design or by construction, are products or services of poor quality. Those that fail to give customer satisfaction by being uneconomic to use are also products or services of poor quality, regardless of their conformance to specifications. Often people might claim that a product or service is of good quality but of poor design, or that a product or service is

Figure 6.1 The meaning of quality

of good quality but it has a high maintenance cost. A product or service may not need to possess defects for it to be regarded as poor quality; for instance, it may not possess the features that we would expect, such as access for maintenance. These are design features that give a product or service its saleability. Products and services that conform to customer requirements are considered to be products or services of acceptable quality. If an otherwise acceptable product has a blemish, is it now unacceptable? Perhaps not because it may still be far superior to other competing products in those features and have characteristics that are acceptable.

For companies supplying products and services, a more precise means of measuring quality is needed. To the supplier, a quality product or service is the one that meets in full the perceived customer requirements. To the customer, a quality product or service is one that meets in full the stated customer requirements, and it is the supplier's responsibility to ensure that the perceived and stated requirements are within the range of acceptability (see Box 6.2).

Box 6.2   Who decides quality?

The decision as to whether something is of satisfactory quality rests with the receiver, not the producer. Producers offer what they perceive will satisfy the needs and expectations of those who will receive their outputs, but only recipients can judge whether their needs and expectations have been satisfied.

> In the final analysis it is the customers who set the standards for quality and they do this by deciding which products to purchase and whom to buy them from.
>
> (Imai, 1986)

Satisfaction and dissatisfaction are not necessarily opposites as observed by Juran (Juran, 1992) and Deming (Deming, 1994). Many products have conformed to requirements and were fit for use and free of defects when produced but no longer satisfy customers because their target market has changed – magnetic tape recorders, carburettors, carbon paper and valve radios are a few examples. They did satisfy large numbers of customers at one time but have been replaced by devices offering different functionality and greater satisfaction. Therefore, when judging the quality of a product or service, you need to be sure you are judging competing alternatives.

Internationally agreed definitions

Since 1986 the term *quality* has been defined in International Standards, and its definition changed as can be seen in Table 6.1.

When the definition mentioned requirements in 2000 rather than needs, it appears it was reaching back to an era where conformity to requirements was the accepted norm. However, we can remove the implied limitation by combining the definition of the terms *quality*, *object* and *requirement* in ISO 9000:2015, and therefore quality can be expressed as "the degree to which a set of inherent characteristics of anything perceivable or conceivable fulfils a need or expectation that is stated, generally implied or obligatory".

Table 6.1 Evolution of international definitions of quality

Definition	Source	Remarks
Totality of characteristics of a product or service that bear on its ability to satisfy stated or implied needs	ISO 8402:1986 clause 3.1	Products or service was a limiting factor
Totality of characteristics of an entity that bear on its ability to satisfy stated or implied needs	ISO 8402:1994 clause 2.1	The limiting factor was removed by use of the term *entity*
The degree to which a set of inherent characteristics fulfils requirements	ISO 9000:2000 clause 3.1.1 ISO 9000:2005 clause 3.1.1	The word *entity* has been removed, possibly because it didn't translate well
Degree to which a set of inherent characteristics of an object fulfils requirements	ISO 9000:2015 clause 3.6.2	The word *entity* has now been reintroduced and replaced by the word *object* which is defined as "anything perceivable or conceivable"

This implies that quality is relative to what something should be and what it is. The something may be a product, service, a document, piece of information, any output from a process or any action or decision. It should therefore not be assumed that when we use the term *quality* we are only referring to products and services or to the requirements of customers. This implies that when we talk of anything using the word *quality* we are referring to the extent or degree to which a need or expectation is met. It also implies that all the principles, methodologies, tools and techniques in the field of quality management serve one purpose: that of enabling organizations to close the gap between the standard reached and the standard required and, if desirable, exceed it. In this context, problems with performance, environmental, safety, security and health problems become *quality problems* because an expectation or a requirement for an object has not been met. If the expectation had been met, there would be no problem. Further elaboration of this concept may be found on the companion website.

The definition appears to be rooted in the present because it makes no acknowledgement as to whether the *needs* are present needs or future needs, but if we imagine that customers expect continual improvement including innovation, then the definition is sound.

Attainment of levels of quality

The definitions we have examined all have their place. None of them is entirely incorrect – they can all work but they suggest that there are levels of attainment with respect to quality as shown in Box 6.3.

If we perceive quality as freedom from deficiencies or defects, we are limiting our understanding of quality to the current and local requirements. We will lose customers if the local requirements don't align with the customer requirements. We will also reduce costs with this mind-set, but we will only retain customers for as long as our products and services are valued.

If we perceive quality as conformity with customer requirements, we recognize that a conforming product or service is one that is free of deficiencies and meets all local and customer requirements. We are, however, limiting our understanding of quality to the current customer requirement and not future needs. With this mind-set, we will reduce costs and retain more customers, but again only as long as our products or services are valued.

Box 6.3   Attainment levels of quality

1   Freedom from deficiencies which requires better controls and results in lower costs but does not necessarily retain satisfied customers.
2   Conformity with customer requirements which requires capable processes and results in lower costs but does not necessarily retain satisfied customers.
3   Satisfying customer needs and expectations which requires innovation as well as capable processes and results in lower operating costs and higher development costs, but in return creates and retains satisfied customers and leads to sustained success.

If we perceive quality as satisfying customer needs and expectations, we recognize that a quality product or service is the one that is free of deficiencies, conforms to customer requirements and satisfies customer needs and expectations. We are not limiting our understanding of quality to current requirements and thus take in future needs and expectations. For example, the customer may not have a requirement to pause live TV but once you make him aware that this is now available, it becomes a customer need and after a month or two, he finds he can't live without it, and any other supplier that cannot offer this feature is not even considered. With this mind-set, we will reduce production costs and increase research and development costs, but the bonus is that we will also create and retain more customers if we can continue to innovate.

Use of the term *quality* in ISO 9001

The word *quality* is used some 171 times in ISO 9001:2015 but 109 of these are within the compound term *quality management system*. The other uses are in the following compound terms:

- Quality policy
- Quality objectives
- Quality management principles – only used in the foreword, introduction and bibliography
- Quality plans – only used in Annex A.6 and the bibliography
- Quality manual – only used in Annexes A.1 and A.6

If we go back to the 1987 version, we'll find that the word *quality* was used very differently (e.g. commitment to quality, work affecting quality, product quality problems and quality activities).

So why the change? One of the issues was that in the 1980s, adding the word *quality* was often interpreted as pertaining to the Quality Department and therefore quality activities were perceived as being activities of the Quality Department. The quality system was even perceived as a system imposed by the Quality Department to control quality. There were exceptions in phrases such as "where the absence of such instructions would adversely affect quality", which is using the term as fitness for use, and the phrase "quality of the product" would mean "the degree to which the product met requirements".

There was no change in this situation in the 1994 version, but the 2000 version brought significant changes. All references to the word *quality* were removed except as labels for policy, objectives and the management system. This practice has been adopted in the 2015 version by using such phrases as *conformity to customer* and *applicable statutory and regulatory requirements* and the *degree of customer satisfaction*.

As the word *requirement* in ISO 9001 means "need or expectation that is stated, generally implied or obligatory" a conforming product or service is one that satisfies the needs or expectations that are stated, generally implied or obligatory. It is therefore important that judgements about product or service quality are informed by the recipients stated, generally implied or obligatory needs or expectations and not limited to specified requirements which are used as a substitute (see "Fitness for use" earlier).

Summary

The word *quality* is used in so many ways that in a specification such as ISO 9001, it's necessary to avoid misinterpretation. However, it's such a common word that it's difficult for people to put aside the many ways in which the term is defined in literature on quality management. We addressed five popular definitions, their advantages and disadvantages and emphasized the importance of context, because the meanings vary depending on the context in which the term is used. We revealed how ISO definitions have changed since the inception of ISO 9001 and examined the way the term is used in ISO 9001. The frequency in which the term *quality* is used has rapidly declined in ISO 9001 to avoid the word being associated with the work of a quality department, and in its place phrases such as *conformity to customer requirements* are now used.

Bibliography

Crosby, P. B. (1986). *Quality without Tears*. New York: McGraw-Hill.

Deming, E. W. (1994). *The New Economics for Industry, Government and Education* – Second Edition. Cambridge, MA: The MIT Press.

Imai, M. (1986). *KAIZEN: The Key to Japanese Competitive Success*. New York: McGraw-Hill.

IMechE. (2015, November 2). Global Food – Waste Not, Want Not. Retrieved from Institution of Mechanical Engineers: www.imeche.org/knowledge/themes/environment/global-food

Juran, J. M. (1974). *Quality Control Handbook* – Third Edition. New York: McGraw-Hill.

Juran, J. M. (1992). *Juran on Quality by Design*. New York: The Free Press, Division of Macmillan Inc.

Wikipedia (1). (2016, September). Philip B Crosby. Retrieved from Wikipedia: https://en.wikipedia.org/wiki/Philip_B._Crosby

7   Requirements

Introduction

Box 7.1   Requirement

A need or expectation that is stated, generally implied or obligatory.

Generally implied means that it is custom or common practice for the organization and interested parties, that the need or expectation under consideration is implied.

A specified requirement is one that is stated, for example, in documented information.

(ISO 9000:2015)

Throughout ISO 9001 mention is made of customer requirements and yet the heading to clause 4.2 of the standard is "Understanding the needs and expectation of interested parties", and indeed one of those interested parties is the customer. So why not assign the title "Understanding the requirements of interested parties", particularly when beneath the title is a requirement to determine the requirements of interested parties? The answer lies in the ISO definition of the term *requirement* (see Box 7.1).

Organizations are created to achieve a goal, mission or objective but they will only do so if they satisfy the requirements of their stakeholders. Their customers, as one of the stakeholders, will be satisfied only if they receive products and services that meet their needs and expectations (i.e. their requirements). They will retain their customers if they continue to delight them with superior service and convert wants into needs. But they also must satisfy their customers without harming the interests of the other stakeholders, which means giving investors, employees, suppliers and society what they want in return for their contributions to the organization.

Many of these requirements must be discovered by the organization as they won't all be stated in contracts, orders, regulations and statutes. In addition, the organization has an obligation to determine the intent behind these requirements so that they may provide products and services that are fit for purpose.

This creates a language of demands that is expressed using a variety of terms, each signifying something different, but collectively we can refer to these as requirements.

In this chapter, we examine the various ways in which requirements are expressed as needs, wants, expectations, desires, preferences, intent, demands and constraints.

Needs

Needs are essential for life, to maintain certain standards, or essential for products and services, to fulfil the purpose for which they have been acquired. For example, a car needs a steering wheel, and the wheel needs to withstand the loads put upon it, but it does not need to be clad in leather and hand stitched for it to fulfil its purpose.

Everyone's needs will be different, and therefore instead of every product and service being different and being prohibitively expensive, we must accept compromises and live with products and services that in some ways will exceed what we need and in other ways will not quite match our needs. To overcome the diversity of needs, customers define requirements, often selecting existing products and services because they appear to satisfy their need but might not have been specifically designed to do so.

Wants

By focusing on benefits resulting from products and services, needs can be converted into wants such that a need for food may be converted into a want for a particular brand of chocolate. Sometimes the want is not essential but the higher up the hierarchy of needs we go, the more a want becomes essential to maintain our social standing or esteem or to realize our personal goals.

In growing their business, organizations create a demand for their products and services, but far from the demand arising from a want that is essential to maintain our social standing, it is based on an image created for us by media advertising. We don't need spring vegetables in the winter but because industry has created the organization to supply them, a demand is created that becomes an expectation. Spring vegetables have been available in the winter now for so long that we expect them to be available in the shops and will go elsewhere if they are not. But they are not essential for survival, to safety, to esteem or to realize our potential, and their consumption may in fact harm our health because we are no longer absorbing the right chemicals to help us survive the cold winters. We might want it, even need it, but it does us harm and regrettably, there are plenty of organizations ready to supply us products or services that will harm us.

Expectations

Expectations are implied needs or requirements. They have not been requested because we take them for granted – we regard them to be understood within our particular society as the accepted norm. They may be things to which we are accustomed, based on fashion, style, trends or previous experience. One therefore expects sales staff to be polite and courteous, electronic products to be safe and reliable, policemen to be honest, coffee and soup to be hot, etc. One would like businessmen to be honest but in some markets, we have come to expect them to be unethical, corruptible and dishonest. As expectations are also born out of experience, after frequent poor service from a train operator, our expectations are that the next time we use that train operator we will once again be disappointed. We would therefore be delighted if, through some well-focused quality initiative, the train operator exceeded our expectations on our next journey and was on time.

Specified requirements

Specified requirements are stated needs, expectations and wants but often we don't fully realize what we need until after we have stated our requirements. For example, now that we own a smartphone we discover we need it to be waterproof as it may slip from our hand when taking photographs. A costlier example is with software projects where customers keep on changing the specified requirements after the architecture has been established. Our specified requirements at the moment of sale therefore may or may not express all our needs. Requirements may also go beyond needs and include characteristics that are nice to have but not essential. They may also encompass rules and regulations that exist to protect society, prevent harm, prevent fraud and other undesirable situations.

Anything can be expressed as a requirement, whether it is essential or whether the circumstances it aims to prevent might ever occur or the standards invoked might apply.

Specified requirements are often an imprecise expression of needs, wants and expectations. Some customers believe that they must define every characteristic; otherwise, there is a chance that the product or service will be unsatisfactory. For this reason, parameters may be assigned tolerances that are arbitrary simply to provide a basis for acceptance/rejection. It does not follow that a product that fails to meet the specified requirement will not be fit for use. It simply provides a basis for the customer to use judgement on the failures. The difficulty arises when the producer has no idea of the conditions under which the product will be used. For example, a power supply may be used in domestic, commercial, military or even in equipping a spacecraft. Variations acceptable in domestic equipment might not be acceptable in military equipment but the economics favour selection for use rather than a custom design which would be far costlier.

Box 7.2   References to statutory and regulatory requirements

The phrase *statutory and regulatory requirements* is used on 11 occasions in the standard (clauses 1 to 10.3). There are three requirements to determine statutory and regulatory requirements (4.4, 8.2.2 and 8.3.3), one to review them (8.2.3), another to ensure they are defined and met (5.1.2a) and one to ensure that focus on meeting them is maintained (5.1.2c).

Legal requirements are defined by the legislature, not the organization, so the meaning of the word *defined* in this context means that out of all the thousands of legal requirements that exist which ones apply to the products and services the organization intends to provide.

Desires and preferences

When an organization scans the environment to determine why people buy one product or service over another competing product or service, they are attempting to reveal customer preferences and opportunities for developing new products and services.

Customers express their requirements in different ways, but as we have seen earlier, these may go beyond what is essential and may include mandatory regulations as well as things that are nice to have – what we can refer to as desires. Sometimes a customer will distinguish between those characteristics that are essential and those that are desirable by using the word *should*. Desired requirements might also be expressed as preferences, for example, a customer might prefer milk in glass bottles rather than plastic bottles because of the perception that glass is more environmentally friendly than plastic. The determination of stakeholder preferences is an important factor in decision-making at the strategic level and in product and service development.

Intent

Behind every want, need, requirement, expectation or desire will be an intent: what the customer is trying to accomplish as a result or the reason for the requirement. In many cases clarifying the intent is not necessary because the requirements express what amounts to common sense or industry practice and norms. But sometimes requirements are expressed in terms that clarify the intent. A good example can be taken from ISO 9001 where in clause 8.7.1 it states: "The organization shall ensure that outputs that do not conform to their requirements are identified and controlled to prevent their unintended use or delivery." The phrase *to prevent their unintended use or delivery* signifies the intent of the requirement but not all requirements are as explicit as this. For example, in clause 5.3 of ISO 9001 it states: "Top management shall ensure that the responsibilities and authorities for relevant roles are assigned, communicated and understood within the organization."

There is no expression of intent in this requirement; that is, it does not clarify why responsibilities and authority need to be assigned, communicated and understood, although in this example it might appear obvious. However, in the 1994 version of the standard it also required responsibilities and authority to be documented without stating why. Knowing the intent of a requirement is very important when trying to convince someone to change their behaviour.

Demands and constraints

Requirements become demands at the point when they are imposed on an organization through contract, order, regulation or statute. Until then they simply don't apply.

From the outside looking in, all demands imposed on the organization are requirements, but from the inside looking out these requirements appear as two distinct categories: one category addresses the objective of the required product or service, which we can refer to as product or service requirements, and another category addresses the conditions that affect the way in which the required product or service is produced and provided, which we can refer to as constraints. Whereas customer requirements may be translated into product or service requirements and constraints, the other stakeholder requirements are only translated into constraints because if the customer requirements were to be removed, there would be no activity upon which to apply the constraint. In other words, requirements and constraints are not mutually exclusive. If we do not have any oil platforms, the safety regulations governing personnel working on oil platforms cannot apply to us.

It might be argued that in theory the customer is always right; therefore, even if the customer makes a demand that cannot be satisfied without compromising corporate values, the organization has no option but to satisfy that demand. However, in reality, organizations can choose not to accept demands that compromise their values or the constraints of other stakeholders, particularly those concerning the environment, health, safety and national security. ISO 9001 clause 4.2 provides a way of avoiding requirements by requiring that those that are relevant to the QMS to be determined (see Chapter 13).

The product or service may be able to fulfil the requirements without satisfying the constraints. However, in not satisfying the constraints the stakeholders could censure the organization and stop it from continuing production; for example, if manufacture of the product
causes illegal pollution, the environmental regulatory authority will sanction or close the
production facility. If the organization treats its employees unfairly and against recognized
codes of practices or employment legislation, regulators and employees could litigate against
the organization and thus damage its reputation. In other words, *product and service requirements* define the true focus for the organization, and *constraints* define parameters that influence the way the organization meets those requirements.

It could be said that fulfilling product and service requirements is the only true objective
because all other demands generate constraints on the way the objective is to be met (what
John Bryson refers to as mandates). If the objective was to supply freeze-dried coffee to
supermarkets, then generating a net profit of 15% using raw materials sourced only under
fair trade agreements processed without using ozone-depleting chemicals are all constraints
and not objectives. In practice objectives tend to be set based upon both requirements and
constraints, which often lead to the relationships between requirements and constraints
being confused, and consequently the focus on the true objective being lost or forgotten (see
also Chapter 16).

The language used differs depending upon whether we are addressing demands, requirements, constraints, objectives, wants, intents or desires; the direction they are coming from;
and how they are responded to.

In stated or implied requirements:	*We respond by declaring that:*
• demands are placed;	• demands have been met,
• requirements are defined;	• requirements have been fulfilled,
• constraints are imposed;	• constraints have been satisfied,
• objectives are established;	• objectives have been achieved,
• intentions are stated or declared;	• intentions have been honoured,
• desires and preferences are expressed;	• desires and preferences have been addressed,

If the direction or response is not clearly understood, what might be perceived and labelled
as one of these turns out to be another and as a consequence inappropriate influence and priority are applied.

When the same label is used for two different types of requirements, it can result in some
people or departments prioritizing actions inappropriately and both types of requirements
being pursued independently of each other. If it helps to bring about improvement in performance by labelling constraints as objectives, this is not a bad thing, provided that people
understand they are not trading off customer satisfaction when doing so.

Summary

The word *requirement* is used many times in ISO 9001 but it is not often realized that it
has a special meaning. It's often believed that a requirement is a condition that is specified,
when in fact, the way the term is used in ISO 9001 is as "a need or expectation that is stated,

customarily implied or obligatory". This will have significant implications in the customer–supplier relationship as it affects the meaning of nonconformity. A requirement does not need to be documented for there to be a requirement that has not been met. We addressed several different ways in which requirements are expressed to provide some insight into what's involved in establishing customer requirements. We distinguished between requirements that relate to the product or service to be provided and requirements that constrain how those products and services are to be provided, emphasizing the danger of losing sight of the true objective if we treat both as objectives to be achieved independently of each other.

Bibliography

ISO 9000:2015. (2015). *Quality Management Systems – Fundamentals and Vocabulary*. Geneva: ISO.

8 Management system

Introduction

In the last 50 years or so organizations have been encouraged to establish management systems to bring about predictability in performance and instil customers and regulators with confidence that they are doing the right things right. Much of the encouragement has come from national and international standards developed to provide benchmarks against which these systems may be judged.

The world is an immensely complicated and sometimes chaotic place in which countries, organizations and people interact in an environment that is constantly changing. When managers of organizations are confronted with large complex problems, they often break them down into more manageable parts and arrange to have each part solved separately and not necessarily at the same time or under the same management. The result of these separate efforts is then presented as a solution to the original problem, but as Ackoff wisely remarked many years ago, "We can be sure that the sum of the best solutions obtained from the parts taken separately is not the best solution to the whole" (Ackoff, 1999). As organizations became more complex, we have become more aware of the dangers of a silo mentality in which, as Sherwood remarks, a fix "here" simply shifts the problem to "there" and organizational myopia in which a fix "now" gives rise to a much bigger problem "later" (Sherwood, 2002). The reason why this happens is because of the connectedness between actions and events. Over the last 60 years or so new theories have emerged which help us understand complexity by looking at situations as wholes using the concept of a bounded system of linked components. This is the essence of the systems approach to management.

We also look on organizations differently depending on the way we see reality. As Morgan explains, "All theories of organization and management are based on implicit images or metaphors that lead us to see, understand, and manage organizations in distinctive yet partial ways" (Morgan, 1997, p. 4). This leads us to view organizations as if they were machines, living organisms, brains, cultures, political systems and other metaphors. The problem is that all of these are partially true. Some parts of the organization function like machines, whereas other parts function like a social system in which the strongest, the most devious and the most manipulating of people set the direction and force others to follow. But as Morgan explains, "Metaphor is inherently paradoxical. It can create powerful insights that also become distortions, as the *way of seeing* created through a metaphor becomes a way of *not seeing*" (Morgan, 1997, p. 5, original emphasis), and that is the dilemma with ISO 9001.

The standard requires we establish a *quality management system* but doesn't provide an operational definition that removes doubt as to what type of system it is. As we will all see the organization differently, what we perceive to be the quality management system will depend on the way we *see* the organization.

In this chapter, we:

- Explain the two basic uses of the word *system* and choose an operational definition
- Examine the different ways in which the terms *management system* and *quality management system* are explained in ISO documents and their implications
- Examine different type of systems and consider whether a QMS could be an abstract, physical or human activity system
- Explain why systems are mental models and not objects that exist in the real world
- Examine the nature of systems, including their purpose, key characteristics and their structure from different perspectives
- Summarize the points made and construct an operational definition for a QMS

What is a management system?

The word *management* in the compound term *management system* is used as an adjective to describe the noun *system.* It is therefore necessary that we have an understanding of the word *system* before moving onto the term *management system.*

Definition of a system

The word *system* is used in an increasing variety of ways to express ideas in different contexts. From the *Oxford English Dictionary* and the *Merriam-Webster Dictionary* it appears there are two basic uses of the word: (a) a connected group of objects forming a complex whole and (b) an orderly way of doing something, and it is this distinction that creates communication problems because when the term *system* is used, it may not be clear whether it is being used in sense (a) or sense (b).

Box 8.1   Systemic vs. systematic

Systematic means doing things in an orderly way, following a method, whereas systemic is not at all about the way things are done but about the interconnectedness amongst entities and the effect that has on the whole.

When using the word *system* in sense (a) the emphasis is on the systemic way entities interact to produce results, but when using it in sense (b), the emphasis is on systematically applying codified methods of work to produce the desired results; for example, a gardener uses a systemic weed killer to kill every part of the plant because all its parts are interconnected, and he proceeds to apply the weed killer systematically to all the weeds in the garden.

Although contemporary systems theory was developed in the 1950s, there is no common definition of the word *system* as each writer on the subject tends to define the term in their own way. The word *system* is defined in ISO 9000 as a "set of interrelated or interacting elements", and there are no notes providing any more insight as to what a system might be. However, an operational definition of a system when used in sense (a) comes from a book published in association with The Open University (Carter, Martin, Mayblin, & Munday, 1983) and is shown in Box 8.2. We will use this definition as it has more practical use. Additional notes on the concept of a system are available on the companion website.

Box 8.2   A comprehensive definition of a system

System – A recognizable whole which consists of a set of interdependent parts. More specifically:

a)   A system is an assembly of components connected together in an organized way;
b)   The components are affected by being in the system, and the behaviour of the system is changed if they leave;
c)   This assembly of components does something;
d)   This assembly as a whole has been identified by someone who is interested in it (e.g. the agent, the client or the problem owner).

(Carter, Martin, Mayblin, & Munday, 1983)

Definition of a management system

The notion of quality systems emerged after WWII when the industrial practices were still largely based on scientific management as defined by Frederick Winslow Taylor. Taylor found that when the methods of work were left to the individual, it led to multiple ways of doing things and a wide variation in results. He believed that if a task was clearly defined and if those performing it can be trained and properly motivated, that productivity would be greatly improved (Taylor, 1911). This is doing things systematically not systemically. It was therefore inevitable that prescribing methods of work and inspection of work was central to prescriptions for managing quality that emerged in the late 1950s up to the mid-1990s.

One of the earliest definitions of a quality system comes from Armand Feigenbaum who defined it as "the network of administrative and technical procedures required to produce and deliver a product of specified quality standards" (Feigenbaum, 1961). Feigenbaum appears to be using the word *system* in the sense of being systematic rather than systemic as defined by Carter et al.

There are considerable differences in the way the terms *management system* and *quality management system* are defined as revealed by the ISO documents identified in Table 8.1.

First let us look at the similarities. All definitions refer to achieving objectives or their equivalent but in ways that are not consistent. Three of the five imply a management system is a way of managing something, and this reveals a distinction between the notion of a management system and a system of management as explained in Box 8.3.

Table 8.1 Different ISO definitions of a management system

ISO Document	Definition of a (quality) management system
1. ISO 9000:2015 clause 3.5.3	Set of interrelated or interacting elements of an organization to establish policies and objectives, and processes to achieve those objectives. (Note the comma was added by mistake to the ISO 9000 entry as it's omitted in the Annex SL entry.)
2. ISO Small Enterprise Handbook (2016):	A quality management system (QMS) is the way your organization directs and controls those activities which are related (either directly or indirectly) to achieving its intended results.
3. ISO Guide Reaping the Benefits of ISO 9001 (2015)	A quality management system is a way of defining how an organization can meet the requirements of its customers and other stakeholders affected by its work.
4. ISO Website (2016)	A management system describes the set of procedures an organization needs to follow in order to meet its objectives.
5. ISO 9000:2015 Clause 2.2.2	A QMS comprises activities by which the organization identifies its objectives and determines the processes and resources required to achieve desired results.

Box 8.3   Management system vs. system of management

Sometimes we reverse the terms *management* and *system* and refer to a system of management, but in doing so we switch the sense in which the word *system* is used. The term *management system* uses the word *system* in the same sense as in the compound noun *central heating system*, whereas the term *system of management* uses the word *system* in the same sense as a poker player has a system of playing poker.

The first definition is fundamentally different from the others because it takes the ISO 9000 definition of the term *system* that none of the others do. There is no explanation given in the notes in ISO 9000:2015 as to what the elements might be, but we will address this later. The ambiguity arises in the way the definition has been constructed. When we parse the sentence, we see that it consists of two statements: a management system is (a) a set of interrelated or interacting elements of an organization to establish policies and objectives and (b) processes to achieve those objectives. A set of interrelated or interacting elements is what ISO 9000 refers to as a system; therefore, the management system is a system + processes. If we now substitute the ISO 9000 definition of a process, we deduce that a management system is:

a) "a set of interrelated or interacting elements of an organization to establish policies and objectives" and;
b) "a set of interrelated or interacting activities that use inputs to deliver an intended result".

As an activity is an element of a system, it therefore looks as though a management system comprises a *management subsystem* and a *delivery subsystem*.

The second definition is fundamentally different because it's a way of directing and controlling activities. This implies it's a method because a method is *a way of doing something*. Inclusion of direction and control shows that this definition was derived from the definition of management in ISO 9000 which is *coordinated activities to direct and control an organization*. But for some reason it was not also derived from the definition of *system*. Had the authors used the substitution principle, they might have proposed: "A management system is a set of coordinated interrelated or interacting activities that direct and control an organization." This implies the QMS is a *management subsystem* but it doesn't include the delivery subsystem as in definition (1).

The third definition is different from the others by being a way of defining how an organization can do something rather than a way of doing something. The word *define* as used in ISO documents means: "state or describe exactly" (ISO Glossary, 2016) and as earlier *a way of doing something* is a method; therefore, this definition implies a management system is a method for producing a description, prescription or perhaps a specification.

The fourth definition is similar to definitions (2) and (3) by relating to a method but also different from the others by being a set of procedures that are used to achieve objectives. Feigenbaum's definition we referred to earlier also seems to be based on the same premise as definitions 2, 3 and 4 in Table 8.1.

The fifth definition from the same standard that includes the official definition (1) implies the QMS is a *management subsystem* because it is limited to identifying and determining things, thereby excluding the execution of those things.

These definitions are produced for different audiences, hence the difference in the style, but they should at least be consistent which sadly they are not. A table illustrating the evolution of definitions of a management system is available on the companion website.

Types of QMS

There are different types of systems and different ways of classifying them. Ackoff divided systems into two basic categories: abstract systems (systems of concepts) and concrete (systems of objects) (Ackoff, 1971) and then divided concrete systems into four basic types and classified them based on whether the parts and the whole were purposeful or purposive. These were deterministic systems, animated systems, social systems and ecological systems (Ackoff, 1998)

Box 8.4   Purposive and purposeful

The term *purposeful* refers to the capacity of a system to determine its own purpose (purpose with choice), whereas the term *purposive* refers to the capacity of a system to pursue a pre-set goal without the ability to change it (purpose without choice).

(Carter, Martin, Mayblin, & Munday, 1983)

Checkland suggested that the absolute minimum number of systems classes needed to describe the whole of reality is four: natural, designed physical, designed abstract and human activity systems (Checkland, 1981). Jackson defined a human activity system as "a model of a notional system containing the activities people need to undertake in order to pursue a particular purpose" (Jackson, 2003).

As a QMS is a human construct, it is neither an animated system nor an ecological system or a natural system, and as a QMS is not wholly composed of people, it's not a social system, which leaves the possibility that a QMS could be a designed abstract system, a designed physical system or a human activity system but an analysis of ISO 9000 and ISO 9001 reveal even more options (see Table 8.1).

When we encounter the term *quality management system* in ISO 9001, in almost all cases the requirement will make sense whichever definition from Table 8.1 we care to use. Only where the requirement refers to the performance of the QMS might we dismiss definitions (3) and (4). If the intended results of the QMS are products and services that meet customer and applicable legal requirements, we may dismiss all definitions except (1).

The QMS as a designed abstract system

An abstract system is one in which all the elements are concepts (Ackoff, 1971) so if users of ISO 9001 perceive a QMS to be a system of documentation (concepts), there is nothing in the introduction to the 2015 version that alludes to a change in the meaning of the term *management system*. In fact, statements that the organization shall implement a QMS and suggestions that organizations should adopt a QMS, together with explanations in guidance documents that a QMS is "a way of" directing and controlling an organization, all help to perpetuate the notion of a designed abstract system. The official definition (1) in Table 8.1 states the QMS to be set of interrelated or interacting elements to establish policies, etc. The QMS can't be an abstract system because abstract systems are passive and the phrase *to establish* implies the system is dynamic.

The QMS as a designed physical system

A designed physical system is a deterministic system, a system where neither the parts nor the whole are purposeful. Physical systems acquire their purpose from their designers. Systems in this class include automobiles, central heating systems, computer systems, transportation systems, etc. Evidence in ISO 9001 that a QMS could be a designed physical system is exemplified in clause 4.4.1 where it requires a QMS to be established and in Annex A.4 where it suggests "one of the key purposes of a QMS is to act as a preventive tool". There is a tendency for users of ISO 9001 to think of the QMS as the software applications used in processing information such as the IT systems used in sales, purchasing, inventory control and production, all of which help to perpetuate the notion of a QMS being a designed physical system. However, the official definition (1) in Table 8.1 clearly states the QMS to be a set of interrelated or interacting elements to establish policies, etc., and as policies are artefacts created by people, the system must include people so it can't be a designed physical system.

The QMS as a human activity system

There are as many different types of human activity systems as there are types of organization. For example, a fast food restaurant is a human activity system but it is intended to operate like a machine. Employees are frequently trained to interact with customers according to a detailed code of instructions and are monitored in their performance (Morgan, 1997). They don't have a choice and therefore function within a purposive system (purpose without choice). On the other hand, if we draw our system boundary around the whole organization including headquarters functions and its fast food restaurants, we extend the human activity system to encompass the brain of the organization and see that it's a purposeful system comprising purposive parts. As ISO 9001 implies the QMS produces products and services that deliver customer satisfaction, from Jackson's definition of a human activity system, we can deduce that the QMS could be "a model of a notional system containing the activities people need to undertake in order to produce products and services that deliver customer satisfaction". The requirements in ISO 9001 from which we can deduce that the QMS is a human activity system are those that refer to the performance of QMS, and there are 15 of these. We can therefore deduce that a QMS is not an abstract system or a physical system, but we need to explore what is meant by a *model*.

Mental models

It's a feature of our thinking that if we give something a name, it must exist. This is *reification*, that is, treating something which is not concrete, such as an idea, as a concrete thing (e.g. "the map is not the territory"). Many words we use for convenience refer to things that are not real, for example, *cold*. We draw the curtains at night to keep out the cold and the darkness, but neither *cold* nor *darkness* is a real object. Another often-quoted statement is that by George Box in Box 8.5.

Box 8.5   The utility of models

All models are wrong, but some are useful (Box, 1976), and the practical question is: How wrong do they have to be to not be useful?

(Box, 1986)

Both processes and systems are not tangible; they are mental constructs and are convenient ways of observing things, but they are not reality. When we look at something within a defined boundary and see a person working, we see activity. When we stand back and see a sequence of activities producing outputs, we are observing a process. When we look at the same things and see the buildings, the people, the relationships, the flow of information and the interactions, we are observing a system. We zoom in and out to select the object of interest. There is a limit to what we can capture so as we zoom in we lose sight of all the elements outside our range, and therefore from a single activity or process we cannot see what the system does. These different perspectives help us understand reality, and for practical purposes we can reveal most of what is of interest in three levels (see Figure 8.1).

An organization is a complex entity, and when we look at it we see buildings, workshops, offices, people, machines, activity, documents, etc. These are the tangible elements but there are many things we don't see, such as the spread of discontent, the effect of a dispute between colleagues or the consequences of an authoritarian style of management. We could

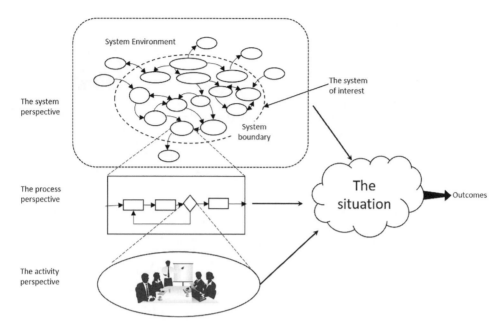

The system perspective

The process perspective

The activity perspective

System Environment

The system of interest

System boundary

The situation

Outcomes

Figure 8.1 Three levels of understanding

put these in our model if they exert such an influence that they need to be controlled. There-fore, when *establishing* a QMS we must be selective, we pick out the things that we believe are relevant to our system of interest and we can use ISO 9001 to confirm we have picked out the elements relevant to the management of quality. What we include in our model is strongly influenced by how we *see* things (i.e. our paradigm) which embodies core assump-tions that characterize and define our worldview. When two people share a paradigm, they will view reality in the same way and therefore understand each other's models of reality (Morgan, 1997). Models cannot be assessed as right or wrong but can only be judged accord-ing to their adequacy (Lisch, 2014).

Box 8.6   Paradigms and metaphors

We use the term *paradigm* here in the sense of a set of experiences, beliefs and values that affect the way an individual perceives reality and responds to that perception. They are alternative realities which are characterized by core assumptions.

Any paradigm may include different schools of thought or different ways of approaching or studying a paradigm which we call metaphors.

For further information, see Jackson, 2003.

We create these models through the process illustrated in Figure 8.2 which is based on Checkland's Soft Systems Methodology.

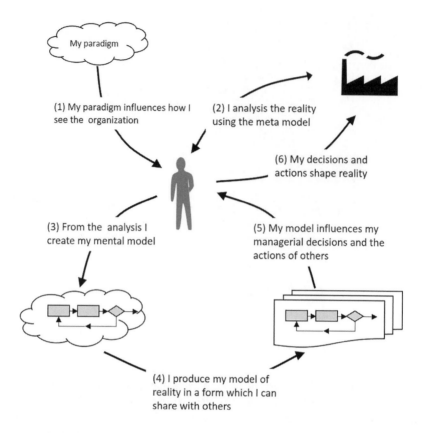

Figure 8.2 Creating and using a model of reality
(Reproduced with the permission of Winston Edwards)

If we consider a system to be a representation of reality from a particular perspective, a QMS would be *a systemic view of an organization from the perspective of how it creates and retains its customers.* Conversely, an environmental management system (EMS) would be *a systemic view of an organization from the perspective of how it protects the natural environment.* We would therefore find elements in a QMS that were also elements of an EMS because they influence the outcomes of both systems. Equally we would find elements of a QMS in an occupational health and safety management System (OHSMS); for example, production activities produce outputs for customers but also produce unintended outputs such as waste that affect the natural environment and affect the health and safety of the personnel involved.

The nature of systems

Systems possess different properties depending on their type, but all systems will have a boundary and a purpose or function, comprise elements, interconnections and structure and have emergence as explained next.

Boundary

In any study of a system there are things with have little or nothing to do with it and other things that are to be treated as part of the system in which we are interested. Meadows observes that systems rarely have real boundaries – there is no real boundary between the exhaust of an automobile and one's nose. There are only boundaries of word, thought, perception and social agreement – artificial, mental-model boundaries (Meadows, 2008). A boundary is therefore a convenience. A boundary can be defined as separating a system from its environment (Stacey, 2010). A system is a whole; therefore, when we draw the boundary we encircle a whole and not a bunch of unconnected parts, and if we move the boundary we create a different system. Where to draw the boundary around a system depends on the purpose of the discussion (Meadows, 2008).

What we include in a system are the elements that can be strongly influenced and controlled by the system because we need to understand how they work. Elements that influence the system but cannot be influenced or controlled by it should be placed outside the system (Carter, Martin, Mayblin, & Munday, 1983). Excluding things that have little influence on the system helps keep the system to a manageable size. A system boundary is shown in Figure 8.1, and further discussion on what should be included in the QMS is in Chapter 14.

System purpose or function

Systems do not necessarily have a purpose, that is, a reason for existence, but they do have a function, that is, what they do. The best way to deduce the system's purpose is to watch for a while to see how the system behaves (Meadows, 2008). A car braking system exists to enable a driver to stop the car, and the function of the respiratory system is to enable our bodies to take in oxygen. In both these cases the systems are purposive rather than purposeful, as neither is equipped with the ability to act independently. Both the car braking system and the respiratory system require a brain to activate them.

Box 8.7   Deducing the purpose of a system

If the government proclaims its interest in protecting the environment and then allocates little money or effort towards that goal, then protecting the environment is not in fact the government's purpose. Purposes are deduced from behaviours, not rhetoric or stated goals.

(Meadows, 2008)

We can deduce from the introduction to ISO 9001:2015 that the purpose of a QMS could be:

a)   to enable an organization to consistently provide products and services that meet customer and applicable legal requirements and enhance customer satisfaction, implying it's purposive and therefore a tool of management;

b)   to consistently provide products and services, etc., implying it's a purposeful goal-seeking system.

There is a fundamental difference between these two possibilities. The purposive system (a) has no means of setting or changing objectives and how they will be achieved. They are imposed by management. The purposeful system (b) has a means of setting and changing objectives and determining how they are achieved. It therefore includes the management subsystem. Parts of the management system act as tools but the whole system is not a tool of management. Also, parts of the system are implemented but the whole system is not implemented, again because the word implies it's a tool or a set of rules. Although the ISO 9000 definition of a management system lacks clarity in total, it is clear that the system it defines includes a goal-setting component and a goal-achieving component and therefore, for the purposes of this book we will assume that the management system is a purposeful goal-seeking system. We will come back to this concept later.

Elements

There is no explanation given in ISO 9000:2015 as to what the elements might be, but there is in Annex SL. Here it states that "the system elements include the organization's structure, roles and responsibilities, planning and operation", and although people are not mentioned, we must assume that people or at least their behaviours are an element of the system. In an earlier standard that wasn't withdrawn until 2000, a quality system was defined as "the organizational structure, responsibilities, procedures, processes and resources for implementing quality management" (ISO 8402, 1986), thus making it clear that people and other resources were part of a quality system.

Elements are what make a system what it is. Vary the element, and it changes the nature of the system, giving it different properties; therefore, the elements are the variables. Checkland refers to them as agents or actors, thus emphasizing the role of people in the system. These variables may be tangible or intangible. Meadows refers to school pride and academic prowess as two intangible elements of a university. She also identifies the tangible elements of a football team as including the players, coach, field and ball, and the elements of our digestive system as including teeth, enzymes, stomach and intestines (Meadows, 2008). This provides some insight into the nature of system elements.

The elements we include in our system are those which have the most influence on what it does A pragmatist would assume the tangible elements include people, products, services, tools, equipment, energy, facilities, etc., and the intangible elements include trust, fear, customer loyalty, reputation, core competences, etc., because they affect the outcomes, but neither the definitions nor the requirements are explicit.

Connectedness

As shown in the operational definition of the term *system* (Box 8.2) the parts or elements are connected in an organized way, and therefore connectedness is a key property of a system.

Many of the interconnections in systems operate through the flow of information. The example Meadows uses is of a football team where the elements are interconnected through the rules of the game, the strategy of the coach, the communication between the players and the laws of physics that govern the motion of the ball and the players (Meadows, 2008).

The way Sherwood explains connectedness in the following scenario illustrates very nicely the nature of a system:

When you drop a coin, the only entities involved are yourself, the coin and the ground. No one else, nothing else, is directly involved and the events take place in a very bounded context. But when you drop your price, the situation is very different. Many entities are involved and they are all connected together in one form or another. The event of dropping the price is not bound, but as ripple effects extending over space and time almost indefinitely. The ripple effect is a direct consequence of the connectedness between the various entities involved. If the connectedness were not present, the chain of cause and effect events would be bounded and stop quickly. It therefore becomes quite impossible to predict with any confidence what the outcome of a single action of dropping your price might be. It is far harder to predict the outcome of dropping your price than of dropping a small coin. It is all a question of connectedness.

(Sherwood, 2002)

We can see from Sherwood's example that changes in any variable within a system will influence other elements in the system and other systems of which it is a part.

Feedback

The term *feedback* refers to a situation in which two (or more) systems or system elements are connected such that each system or system element influences the other and their dynamics are thus strongly coupled. Feedback is fundamental to the way systems behave. It is the action of feedback that the system uses to control, to limit or to constrain its outputs or behaviour (e.g. the in-process controls in a production process). Sometimes feedback operates to exaggerate or to amplify behaviour; for example, a successful marketing strategy brings in more orders than the organization can cope with or secrecy within management builds distrust and leads to strike action by the workforce. An example of where feedback is used to control an output is shown in Figure 8.6. An example of where feedback operates to exaggerate or to amplify behaviour is shown in Figure 8.7.

Properties of management systems

For the management system to be a goal-seeking system, it needs to possess certain properties which come into existence through what Peter Senge refers to as balancing and reinforcing processes (Senge, 1990).

Resilience

A goal-seeking management system will need to possess resilience, which is the ability to survive, bounce back and persist within a variable environment, and this requires feedback mechanisms. This may be reflected in processes that enable objectives to be challenged and changed and processes for restoring stability after severe losses such as a major catastrophe. These feedback loops may trigger units that kick in when another unit fails (e.g. stand-by power supplies or rapid response teams).

Self-maintenance

A goal-seeking management system will need to possess an ability to correct, repair and maintain itself, and this requires feedback mechanisms. This may be reflected in processes for the maintenance of infrastructure, equipment, competence, documented information and system auditing and review. It will also be reflected in devolved power to local units to maintain their own infrastructure; for example, a system that switches off the heating centrally at the start of the summer may seem like a cost-saving measure, but it may cause distress in some offices where the sun doesn't shine and result in lower productivity.

Self-organization

A goal-seeking management system will need to possess the capacity to change its own structure in response to risks and opportunities, and this also requires feedback mechanisms. This may be reflected in processes for organization learning and development, knowledge creation and acquisition and the formation of new units and the dismantling of obsolete ones. These changes may be transformational, but changes may also be executed through improvements in culture, technology and management. It's likely to come up with whole new ways of doing things.

Hierarchy

One thing a goal-seeking management system will possess is hierarchy. The system will likely comprise subsystems having the same properties as the system. Subsystems may be arranged on a geographic, functional, project, product, service or team basis, each possessing self-maintenance, self-organization properties. As business expands new units will be formed possessing a similar management structure to the larger unit and attached to the centre so that information flows are maintained. But the greater the hierarchy, the greater the risk of malfunction as units get further away and out of sight of the centre.

Emergence

The property of emergence is the expected and the unexpected and unanticipated results that arise from interactions within a system. It's a property possessed by all systems. It's expressed by the axiom "the whole is greater than the sum of its parts", meaning that the essential properties of a system taken as a whole derive from the interactions of its parts, not their actions taken separately. When oxygen and hydrogen are combined in the correct proportions, water is produced, and with water comes wetness, a property not found in either oxygen or hydrogen, just as sound is not found in any of the components of a radio. It is implicit in ISO 9001:2015 that enhanced customer satisfaction is an emergent property of a QMS, as it is not found in any of the component parts, nor is it an output of any process but may be the outcome or consequence of possessing an output from the system.

Structure

Management systems are open systems that change their structures and behaviours in harmony with changes in their environments. The structure of a system is the way its elements are interconnected and held together. There are different types of elements and different

---

Box 8.8   QMS structure

The structure of a QMS is not the way its documentation is structured as in a hierarchy of policies, procedures, records, etc., but the way its elements are interconnected, the flow of information, the feedback loops and the cause-and-effect relationships.

---

ways in which they are connected, so it isn't a good idea to mix these up in any graphical representation as it makes them impossible to understand. The following models for a small restaurant represent the most common forms:

An external system map

An external system map (see Figure 8.3) shows the external factors that influence the organization, but the organization cannot easily influence in return. So you might find things like competitors or weather as an influence on the system.

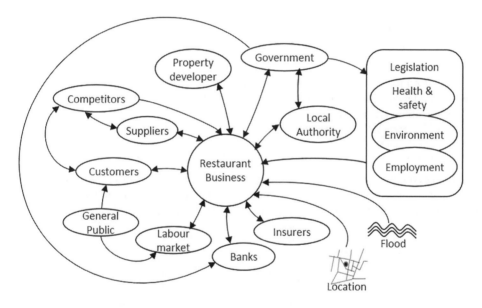

Figure 8.3 External system map

An internal system map

An internal system map (see Figure 8.4) simply shows the internal influences of different elements and their relationship. Occasionally a *hard* component will appear, such as a building but only when the building is having a direct influence on some other factor; for example, if there's a problem with storage space it will appear, but otherwise not. Soft components like food hygiene might appear because it's critical to success.

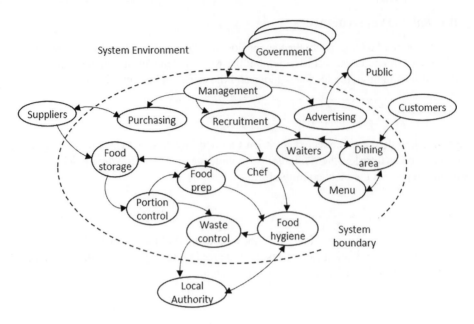

Figure 8.4 Internal system map

The purpose of these first two figures is to gain a richer understanding of the situation, so we can afford to be flexible. The next set of diagrams serves a different purpose. Although they aid understanding, their main function is to look for the most efficient way to manage the organization and its operations.

High-level process map

The high-level process map (see Figure 8.5) shows the network of interconnected processes as simple boxes with connecting arrows indicating the flow of information or materials. The processes included are those which achieve a common objective.

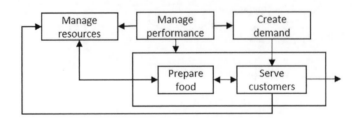

Figure 8.5 High-level process map

Activity-sequence diagram

The activity-sequence diagram (see Figure 8.6) shows the sequence of activities and the flow of information and material as it passes through a process.

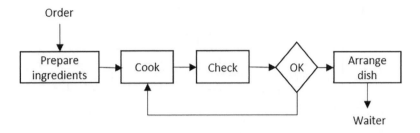

Figure 8.6 Activity-sequence diagram

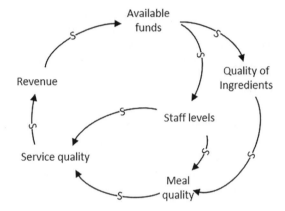

Figure 8.7 Causal loop diagram

Causal loop diagrams

Causal loop diagrams (see Figure 8.7) are used to assist in getting to the cause of an intractable problem or to help identify any possible unwanted *emergent properties*. The *S* on the line indicates the influence is in the same direction, and the *O* indicates it's in the opposite direction. The diagram describes the following scenario: If revenue is falling due to poor service quality, the budgets are cut, which affects the procurement process where cheaper ingredients are sourced. A reduced budget means no pay rises, so staff satisfaction decreases, staff leave and new staff are recruited on lower wages. Lower-quality ingredients together with less motivated staff affect the food preparation process resulting in lower meal quality. Servers provide customers with meals of lower quality which brings complaints, and service quality is reduced. Word gets about and reputation suffers with the result of fewer customers and lower revenue.

Organization chart

An organization chart (see Figure 8.8) shows the division of labour and the chain of command either as a hierarchy or a matrix and may show the relationship between locations, divisions, departments and what work they do. The owner of any outsourced processes should be shown.

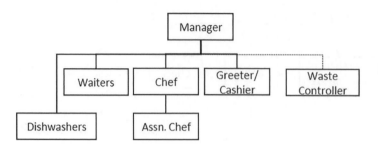

Figure 8.8 Organization chart

In addition there may be *infrastructure drawings* which show the relationship between the fixed structures including buildings, offices, plant, machinery, utilities, roads and pathways, etc., and *material conversion diagrams* which show the stages through which raw materials are converted into commodities such as chemicals, aggregates, paper, metals, etc.

Summary

Our task of explaining what a quality management system is has not been an easy one because there is little consistency in ISO documents as to whether the term is being used in the sense of how entities interact to produce results or in the sense of a set of methods for producing results. We were unable to remove doubt as to what a management system is by studying ISO documents, so we had to revert to first principles. We examined three types of systems and found evidence in the standards to support the view that a QMS could be a designed abstract system, a designed physical system or a human activity system, but dismissed the first two of these because they were inconsistent with the notion that the QMS is a set of interacting elements, one of which is people. We therefore concluded that it's more likely that a QMS is a human activity system which from Jackson's definition is a model rather than a physical system and settled on an operational definition that a *QMS is a systemic view of an organization from the perspective of how it creates and retains its customers*. To substantiate this, we looked at a restaurant from different perspectives showing the type of models that we can use as a basis for understanding how it functions, identifying problems and planning improvements.

Bibliography

Ackoff, R. L. (1971, July). Towards a System of Systems Concepts. *Management Science*, 17(11)..

Ackoff, R. L. (1998). A Systemic View of Transformational Leadership. *Systemic Practice and Action Research*, 11(1), 23–36.

Ackoff, R. M. (1999). *Ackoff's Best: His Classic Writings on Management: Chapter 1*. New York: John Wiley & Sons, Inc.

Box, G. E. (1976). Science and Statistics. *Journal of the American Statistical Association*. 71(356), 791–799.

Box, G. E. (1986). *Empirical Model-Building and Response Surfaces*. Hoboken, NJ: John Wiley & Sons.

Carter, R. C., Martin, J. N., Mayblin, B., & Munday, M. (1983). *Systems, Management and Change: A Graphic Guide*. London: Paul Chapman Publishing Ltd.

Checkland, P. B. (1981). *Systems Thinking, Systems Practice*. Chichester: John Wiley & Sons.

Feigenbaum, A. V. (1961). *Total Quality Control – Engineering and Management* – Second Edition. New York: McGraw-Hill.

ISO 8402. (1986). *Quality Vocabulary: International Terms*. Geneva: International Organization of Standardization.

ISO Glossary. (2016). Guidance on selected words used in the ISO 9000 family of standards. Retrieved from www.iso.org/iso/03_terminology_used_in_iso_9000_family.pdf

Jackson, M. C. (2003). *Systems Thinking: Creative Holism for Managers*. Chichester: John Wiley & Son.

Lisch, R. (2014). *Measuring Service Performance*. Abingdon: Routledge.

Meadows, D. M. (2008). *Thinking in Systems: A Primer*. White River Junction, VT: Chelsea Green Publishing.

Morgan, G. (1997). *Images of Organization*. London: SAGE Publications.

Senge, P. M. (1990). *The Fifth Discipline: The Art and Practice of the Learning Organization*. New York: Random House.

Sherwood, D. (2002). *Seeing the Forest for the Trees: A Manager's Guide to Applying Systems Thinking*. London: Nicholas Brealey Publishing.

Stacey, R. D. (2010). *Complexity and Organization Reality*. Abington Oxford, UK: Routledge.

Taylor, F. W. (1911). *The Principles of Scientific Management*. New York: W. W. Norton & Company.

9 Process and the process approach

Introduction

When we undertake work we do it for a reason and to achieve some objective. We acquire the things we need to achieve the objective; there is a series of actions we take; tools, equipment and information we use; energy and materials we consume; and decisions we make from the beginning to the end. These are the variables. The work progresses until completion. This progression is a process; it has a beginning and an end. It begins with an event or when we receive a command or reach a date and ends when we are satisfied with the resultant output. This output may be a tangible product or a service we provide to someone else. The output will possess certain features or characteristics, some of them needed, some of them not needed. The process we use determines these features or characteristics. Therefore, we can design or manipulate this process to produce any features or characteristics we so desire by altering the variables. If we start work with an objective and the determination to create an output possessing certain desired features or characteristics, are diligent in our planning and preparation and we do what we set out to do, we are more likely to produce outputs with those qualities than if we'd just thrown ourselves into the task without forethought. There's no guarantee we will achieve our objective because there may be variables outside our control that impede our best endeavours. However, if one manages processes effectively, many of the risks may be mitigated, leading to it being more likely that we'll be able to consistently and continually produce outputs of the desired quality.

In ISO 9001:2015 there are requirements to determine the sequence and interaction of processes, determine the resources needed for them, evaluate them, ensure these processes achieve their intended results and requirements for promoting the use of the process approach. It is therefore vital that we understand what a process is, what the process approach is and the context in which these terms are used in ISO 9001.

We touched on the process perspective in Chapter 8 as a way of understanding how the organization functions and thereby identify opportunities for improvement. In this chapter, we examine processes in greater depth, and in particular we examine:

- The common definition of a process and the way the word *process* is used
- The ISO definitions and how they have changed since 1987
- Perceptions of processes from different viewpoints
- The relationship between processes and procedures and processes and systems
- The process approach, how it developed, how it differs from a functional approach and how it's defined

- Several different process models
- Different types of processes

What is a process?

The common definition of process

The word *process* as a noun has been traced back to the 12th-century French word *procès* for legal contract and later for advance, progress, course or development of an action. Apart from specialist uses, the word is most commonly used today as a continuous and regular action or succession of actions occurring or performed in a definite manner, and having a particular result or outcome (OED, 2013).

Process evokes a sense of something changing over time or resisting change, whether it be in the mind (a thinking process) or the body (a digestive process) or in the office (a writing process) or in the workshop (an assembly process) or the natural environment (an isobaric process) or in politics (a peace process). Neave says that a process is anything that can be described by the use of a present participle (a verb ending in *-ing*), preferably with an object, for example, writing a report (Neave, 1990) but it's not an infallible method as shown by the examples earlier.

Defining an organizational process

As with other terms we have discussed so far, their meaning differs depending on the context in which they are used, and therefore we need to specify this context. There are different schools of thought on what constitutes a process in the organizational context but here we will confine ourselves to the ISO definitions in Table 9.1. Several additional definitions are presented on the companion website.

Table 9.1 Evolution of ISO definitions of process

Definition	Notes
A set of interrelated resources and activities which transform inputs into outputs ISO 8402:1994 clause 1.2	Resources may include personnel, finance, facilities, equipment, techniques and methods.
A set of interrelated or interacting activities which transforms inputs into outputs ISO 9000:2000 clause 3.4.1	Inputs to a process are generally outputs of other processes. A process where the conformity of the resulting product cannot be readily or economically verified is frequently referred to as a "special process".
A set of interrelated or interacting activities that use inputs to deliver an intended result ISO 9000:2015 clause 3.4.1	Whether the "intended result" of a process is called output, product or service depends on the context of the reference. Two or more interrelated and interacting processes in series can also be referred to as a process. Processes in an organization are generally planned and carried out under controlled conditions to add value.

A welcome change is the removal of the word *transforms* as it signifies a break with convention, but it also removes a 30-year anomaly. In ISO 9000:2000 clause 2.4 it stated that "Any activity, or set of activities, that uses resources to transform inputs to outputs can be considered as a process." An issue with definitions is that anything that does not match a definition cannot be the object being defined, and therefore this definition suggests that only if inputs were transformed into outputs was an object a process. This would imply that design is not a process because the inputs exist in their original form after the design is complete and can be used again – hence they are not transformed. Resources are used by a process rather than being inputs to a process, or as the 1994 definition suggests, part of a process. Many manufacturing work processes transform inputs, whereas most work processes in the service sector only use inputs and don't transform them. It was therefore about time the ISO 9000 definition changed and so it did in 2015 when the word *transformation* was dropped.

However, although the new definition removed one anomaly it introduced another: the notion that all processes use inputs to deliver an intended result. For an object to qualify as a process it does not need to deliver intended results or even have been designed to deliver intended results. It may in fact deliver unintended results but it's nonetheless a process. Perhaps the definition should be *A set of interrelated or interacting activities that use inputs to deliver a result.*

The concept of adding value and the party receiving the added value is perceived as important in the ISO 9000 definition, but whether a process adds value cannot qualify it as a process. A poorly managed process remains a process by virtue of the activities producing a result. Of course, the intent of a process should be to add value, and therefore the degree of value added can be a measure of the usefulness of a process.

It is easy to see how these definitions can be misinterpreted but it doesn't explain why, for many it results in flow charts they call processes. They may describe the process flow but they are not in themselves processes because they simply define steps in a sequence. A series of steps can represent a chain from input to output but it does not cause things to happen. Add the resources, the behaviours and the constraints and make the necessary connections, and we might have a model of a process that will cause things to happen. Therefore, any process description that does not connect the activities and resources with the objectives and results is not a process that delivers an intended result. Fraser identifies several pitfalls associated with the ISO definitions (see Box 9.1).

Perception of process

All work is accomplished by a process, but there are different scales of work from writing a letter (a micro-process) to building a cathedral (a macro-process), and the way we approach the work all depends on our perception of the process. If we allow ourselves to be persuaded that a single task is a process, we might well deduce that our organization has several thousand processes. If we go further and try to manage each of these nano-processes (they are smaller than micro-processes), we will lose sight of our objective very quickly. By seeing where the task fits in the activity, the activity fits within a process and the process fits within a system, we create a line of sight to the overall objective (see Figure 8.1). By managing the system, we manage the processes and in doing this we manage the activities and their consequences. However, system design is crucial. If the processes are not designed to function together to fulfil the organizational goals, they can't be made to do so by tinkering with the activities.

Box 9.1   Pitfalls to avoid with processes

Some surprising, yet widely held, beliefs are that:

- "processes" are a new concept
- "processes" have replaced "procedures"
- you can't manage both departments and processes in one organization
- to "transform something" means something other than to "change its form" or to "convert it" into something else
- all inputs to a process must be input at the start of a process
- all outputs from a process must be output at the end of a process
- any one activity cannot be part of more than one process
- if a definition is in an ISO standard then it must be "right".

There is also a lack of appreciation of the difference between the concepts of something being "put in" from outside a process, and "taken in" from within the process – and, in the same way, something being "put out" by a process and "coming out" from the process.

(Fraser, 2015)

Processes versus procedures

The procedural approach is about doing a task, conforming to the rules, doing what we are told to do, whereas the process approach is about understanding needs, finding the best way of fulfilling these needs, checking whether the needs are being satisfied and in the best way and checking whether our understanding of these needs remains valid. Some differences between processes and procedures are indicated in Table 9.2.

The ISO 9000:2015 definition of a procedures is a "specified way to carry out an activity or a process". Although machines, including computers, operate from instructions or procedures coded into the software, we assume the activities and processes referred to here are those carried out by people. A procedure need therefore only contain instructions to assist a person in performing a task. As such the procedure need not contain a description of the process, the resources and the cause-and-effect relationship that produce the process outputs, how risks are mitigated, etc. This level of detail may be reserved for process descriptions. An analogy would be the difference between a manual for operating a machine (the procedure) and the specifications, drawings, circuit diagrams and instructions for maintaining, stripping down, fault finding and repairing a machine (the process description).

Processes versus systems

The idea that processes and systems don't exist in the real world but are mental constructs and convenient ways of observing reality as explained in Chapter 8. This allows us to examine a phenomenon from different perspectives as explained in Box 9.2. This may be a difficult concept to understand but when we produce a flow diagram of a process we are creating a simple model of complexity. We assume the activities will always follow the same sequence without any deviations or interruptions, so in effect they are idealized versions of

Table 9.2 Process versus procedure

Procedures	Processes
Procedures are driven by completion of the task	Processes are driven by achievement of a desired outcome
Procedures are implemented	Processes are operated
Procedure steps are completed by different people in different departments with different objectives	Process stages are completed by different people with the same objectives; departments do not matter
Procedures are discontinuous	Processes flow to conclusion
Procedures focus on satisfying the rules	Processes focus on satisfying the customer
Procedures define the sequence of steps to execute a task	Processes generate results through use of resources
Procedures are used by people to carry out a task	People work through a process to achieve an objective
Procedures exist, they are static	Processes behave, they are dynamic
Procedures only cause people to take actions and decisions	Processes make things happen, regardless of people following procedures
Procedures prescribe actions to be taken	Processes function through the actions and decisions that are taken
Procedures identify the tasks to be carried out	People select the appropriate procedures to be followed

---

Box 9.2   Observing systems and processes

When we look at something within a defined boundary and see a person or a group of people working, we see activity. When we stand back and see a sequence of activities producing outputs, we are observing a process. When we look at the same things and see the way that a change in one element changes the behaviour of other elements, we are observing a system.

---

reality, but they help us determine how results are produced. The flow diagram is like a map and we know the dictum that *a map is not the territory* (attributed to Polish scholar Alfred Korzybski). We use models to represent reality but our models are only as good as what we can see, and what we see is influenced by our experiences, beliefs and values as explained in Box 8.6. Therefore, within a system processes are brought to life and run until the required outputs are produced. A process doesn't choose the goal, whereas a goal-seeking system such as a QMS chooses the goal and endures as long as the interconnections are maintained.

Process models

In the context of organizational analysis, a simple model of a process is shown in Figure 9.1. This appeared in ISO 9000:1994 but clearly assumes everything other than inputs and outputs are contained in the process. The process takes an input; work is carried out and outputs are produced, but the diagram does not in itself indicate whether these outputs are of added value or where the resources come from.

Now that ISO has changed the definition of a process, removing the word *transforms*, a more useful model might be that shown in Figure 9.2.

This model shows that the process is resourced to receive a demand and when a demand is placed upon the process, several pre-determined activities are carried out, pulling in the resources as and when they are needed. These activities will produce an output that satisfies the demand in a way that fulfils the constraints of the other stake-holders, assuming no deviations or interruptions. These activities have been deemed as those necessary to achieve a defined objective, and the results are reviewed and actions taken were appropriate to:

- improve the results by better control;
- improve the way the activities are carried out;
- improve alignment of the objectives and measures with current and future demands.

Figure 9.1 Simple process model

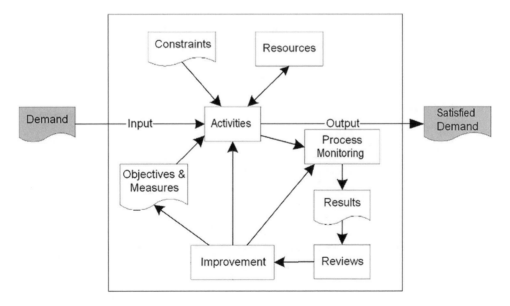

Figure 9.2 A managed process

If we go inside the box labelled "Activities" we may observe one of two types of controls. If there are planning, implementing and checking activities and feedback loops designed to control the outputs, as shown in Figure 9.3, this is closed loop control or adaptive control. Kaoru Ishikawa refers to this as vanguard control as it anticipates problems and prevents them before they actually occur. This is the intent of planning (Ishikawa, 1985). The output from the implement stage is checked and, if not acceptable, information is fed back to the implement stage to change the characteristics so when checked again the output is acceptable. When sampling is used to check outputs, it remains a closed loop control if the batch is held pending acceptable results from the checks. In such a process the person running the process is under self-control (i.e. the controller is inside the process).

If a lot of effort is put into planning in terms of error proofing and checked out in advance, subsequent checking of outputs may not be necessary and would in fact be wasteful. This is open loop control, or non-adaptive control, where any unacceptable variation is not picked up until something goes wrong downstream or is picked up during periodic monitoring as shown in Figure 9.4. In such a process the person running the process has no control and will continue producing unacceptable output until stopped by the controller outside the process. If the manager is not monitoring the process and only takes action after finding out from the downstream process that the output hasn't met the target, that is what Ishikawa refers to as "rear guard control" (Ishikawa, 1985).

It is important to recognize that in Figure 9.2 the activities associated with managing a process act *on* the process and do not act *in* the process; thus they are separate processes that are triggered by independent events. For example, the improvement cycles may be run on different timescales and in conjunction with particular initiatives. The frequency of the reviews will vary depending on the criticality of the process.

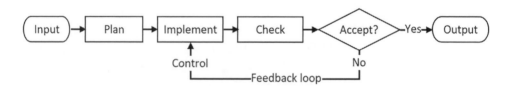

Figure 9.3 Closed loop control model

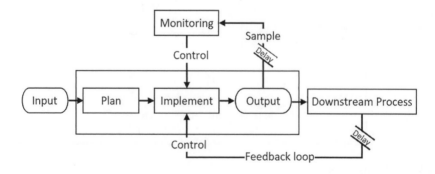

Figure 9.4 Open loop control model

Box 9.3   People and process

If a process is supposed to use inputs to deliver an intended output, it can't produce an output without the actions and decisions of people using other resources, which is why the people and other resources form part of the process but only while they are being used.

It is what people do that influence process dynamics. Their actions or inactions create flow, sequence, delay, breakdown, stability and many other process attributes.

Types of processes

We are focused on organization processes, and there are different types of processes, but the attributes that characterize them are not physical attributes.

Process characterization by purpose

All organizational processes are feedback processes, that is, processes where:

a)   a sensor measures output and feeds information to a comparator;
b)   a comparator transmits a signal to the action component; and
c)   an action component adjusts a parameter if necessary so that the output remains on target.

There are two types of feedback processes. Senge refers to these as reinforcing or amplifying processes and balancing or stabilizing processes (Senge, 1990). The reinforcing processes are engines of growth, and the balancing processes are engines of stability. If the target is to grow, increase or decrease the amount of something or widen the gap between two levels, reinforcing processes are being used. If the target is to maintain a certain level of performance, a certain speed, maintain cash flow, balancing processes are being used. The terminology varies in this regard. Juran refers to these as breakthrough and control processes where breakthrough is reaching new levels of performance and control is maintaining an existing level of performance (Juran, 1964).

There are many balancing processes in organizations; in fact most of the organization's processes are balancing processes as their aim is to maintain the status quo, keep revenue flowing, keep customers happy, keep to the production quotas, keep to the performance targets, etc. A few processes are reinforcing processes such as research and development processes, the process for expanding markets, building new factories, etc. All these place a burden on the balancing processes until they can no longer handle the capacity and something must change. Likewise, the reinforcing processes decrease orders and innovation and consequently the balancing processes have surplus capacity and again something must change.

In the balancing loop of Figure 9.5 when sales increase there is an upward pressure on price (S), but as price increases there is downward pressure on sales (O). Decreasing sales will reduce the price with the result that both sales and price will oscillate around some mean value (Edwards, 2013).

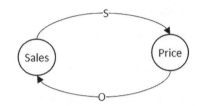

Figure 9.5 A balancing process

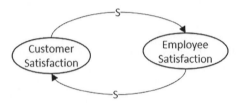

Figure 9.6 A reinforcing process

In his book *The Loyalty Effect* Frederick Reichfeld describes the strong correlation between customer satisfaction and employee satisfaction (Reichfeld, 2001). This is shown in Figure 9.10 as a reinforcing loop. Such a loop is known as a virtuous cycle of growth, but beware, for these loops can flip and become vicious cycles of decline (Edwards, 2013).

In these diagrams the arrows indicate the direction of causality. The causality is that *Employee satisfaction* directly drives the *Customer satisfaction*; whether this is increasing or decreasing depends on whether *Employee satisfaction* is increasing or decreasing. The symbol S implies that the elements connected by a specific cause-and-effect relationship move in the same direction. The symbol O implies that they move in opposite directions (Sherwood, 2002).

Process characterization by class

Processes are also characterized by class. As stated previously, all work is accomplished by a process, and all processes produce outputs; therefore, if we look at the organization as whole and ask, "What outputs will our stakeholders look for as evidence that their needs are being met?" we identify the organization's outputs. There must be processes producing these outputs, and we call these macro-processes. These processes are multi-functional in nature consisting of numerous micro-processes. Macro-processes deliver business outputs and are commonly referred to as *business processes*. For processes to be classed as business processes they need to be in a chain of processes having the same stakeholder at each end of the chain. The input is an input to the business, and the output is an output from the business.

Some people classify business processes into core processes and support processes but this distinction has limited benefit; in fact it may create in people's minds the perception that core processes are more important than support processes. All processes are

Table 9.3 Relationship of business process to work process

Scope	Business process	Work process
Relationship to organization hierarchy	Unrelated	Closely related
Ownership of process	No natural owner	Departmental head or supervisor
Level of attention	Executive level	Supervisory or operator level
Relationship to business goals	Directly related	Indirectly related and sometimes (incorrectly) unrelated
Responsibility	Multi-functional	Invariably single function (but not exclusively)
Customers	Generally external or other business processes	Other departments or personnel in same department
Suppliers	Generally external or other business processes	Other departments or personnel in same department
Measures	Quality, cost delivery	Errors, quantities, response time
Units of measure	Customer satisfaction, shareholder value, cycle time	% defective, % sales cancelled, % throughput

dependent upon each other to achieve the organization's goals but not all present the same level of risk.

If we ask of each of these business processes "What affects our ability to deliver the business process outputs?" we identify the critical activities which at this level are processes because they deliver outputs upon which delivery of the business output depends. These processes are the micro-processes, and they deliver departmental outputs and are task oriented. In this book these are referred to as work processes. A management system is not just a collection of work processes, but also the interaction of business processes. The relationship between these two types of processes is addressed Table 9.3. Sometimes work processes are referred to as sub-processes.

The American Quality and Productivity Center (AQPC) published a Process Classification framework in 1992 to encourage organizations to see their activities from a cross-industry process viewpoint instead of from a narrow functional viewpoint. The framework has five levels and 13 categories (see next). These appear to equate with business processes, and Version 7 published October 2015 identifies 1,622 process elements (APQC, 2015):

1   Develop Vision and Strategy
2   Develop and Manage Products and Services
3   Market and Sell Products and Services
4   Deliver Physical Products
5   Deliver Services
6   Manage Customer Service
7   Develop and Manage Human Capital

 8   Manage Information Technology (IT)
 9   Manage Financial Resources
10   Acquire, Construct, and Manage Assets
11   Manage Enterprise Risk, Compliance, Remediation, and Resiliency
12   Manage External Relationships
13   Develop and Manage Business Capabilities

This classification was conceived out of a need for organizations to make comparisons when benchmarking their processes. It was not intended as a basis for designing management systems. We can see from this list that several processes have similar outputs. For example, there are a group of processes with resources as the output. Also, some of these processes are not core processes but themes running through core processes. Similarly, with managing external relations, many processes will have external interfaces, so rather than one process there should be objectives for external relationships that are achieved by processes with external interfaces.

Perception of the organization

The word *process* either as a noun or a verb is commonly used in daily life without evoking the building of ships or automobiles, but add to it the words *control* or *management* and it's perceived to be synonymous with production – for example, the making of products in a factory or delivery of a service in a restaurant. There is a reason for this, and it's because we separate planning from doing. The brain work that goes into setting objectives and formulating strategies and plans for their achievement involves choices, making decisions, setting priorities; activities that require freedom to think so we wouldn't want to impose controls over it, or so we believe. We also probably see the organization as a machine for making money and therefore to be effective, a machine must be controllable by its operators, which is, of course, the role of management. Senge in the forward to Arie de Geus's book *The Living Company* writes that "seeing a company as a machine implies that it is created by someone outside. This is precisely the way in which most people see corporate systems and procedures – as something created by management and imposed on the organization." Seeing an organization as a machine implies that it's fixed and can change only if someone changes it, that its actions are reactions to decisions made by management, that it learns only as the sum of the learning of its individual employees (de Geus, 1999). As Morgan says "We talk about organizations as if they were machines, and as a consequence we tend to expect them to operate as machines: in a routinized, efficient, reliable, and predictable way" (Morgan, 1997).

   There are indeed processes within organizations that need to be routinized, efficient, reliable, and predictable, but there are also processes that need to be free from such constraints. Edwards referred to these two types of processes as cybernetic and exploratory processes (Edwards, 2013). He describes cybernetic processes as having pre-determined inputs, a sequence of activities and outputs or objectives to be met. Exploratory processes, he says, change their objectives and even the way they work and are often influenced by human prejudices, opinions and beliefs. Edwards compares cybernetic and exploratory processes in Table 9.4, which is reproduced with his permission.

Table 9.4 Cybernetic and exploratory processes

Feature	Cybernetic process	Exploratory process
Activities	1. Can be described as "machine like". 2. Can be subjected to cybernetic analysis and presentation. 3. Problems can be resolved by the use of mathematics/operational research and SPC. 4. Are objective. 5. Change will occur slowly. if at all.	1. Often influenced by human prejudices, opinions and beliefs. 2. Are not simple to describe. 3. Problems faced are ill defined. 4. Change can occur rapidly, though there will be periods of stability 5. Activities and sequences of activities can change during the process. 6. Are subjective.
Models	1. Commonly used activity- sequence diagrams. 2. Other sequential flow diagrams can also be used. 3. Diagrams show the flow of material or information only.	1. Causal loop diagrams 2. Influence diagrams 3. System maps 4. Diagrams show the set of interactions which take place.
Inputs	1. Known and unambiguous. 2. Incorrect input will cause failure (an undesirable result or no result). 3. The process cannot operate without them. 4. The process has no control over delivery. 5. The process must accept correct input.	1. Often ambiguous, can be affected by the source. 2. Inadequate input is a feature. 3. Are not always timely and may arrive in fragmented chunks. 4. Can differ each time the process is activated. 5. They may be accepted or rejected by the process. 6. The process may have to operate with incorrect inputs.
Transformation	1. Must follow a pre-ordained sequence of activities. 2. Is objective and can be analysed easily. 3. Nature of the transformation cannot be changed.	1. No set sequence. Activities will vary according to circumstances. 2. Is frequently not objective, influenced by personal beliefs, politics, etc. 3. Can change the way data are handled.
Objectives	1. Is purposive, cannot change its objectives/outputs. 2. Meets its objectives or fails. 3. Purpose without choice.	1. Purposeful, can change direction and objectives. 2. Time delays in feedback often mean it is not possible to know if objectives have been met. 3. Has freedom of choice.
Outputs	1. Pre-determined 2. Measurable, success or failure is recognizable 3. Cannot change its outputs 4. Fails if they are not achieved. 5. Clear measures of failure.	1. Not necessarily clear in advance 2. No clear measure of success 3. Can change direction mid-process and create new/different outputs 4. Limited measures of success/ failure.

(*Continued*)

Table 9.4 (Continued)

Feature	Cybernetic process	Exploratory process
Internal	1. Cannot repair itself 2. Is not autopoietic, cannot replicate itself. 3. Control is achieved through embedded feedback loops which will warn of failure or deviation. 4. Repetitive and repeatable.	1. Can self-repair 2. Is autopoietic, can "replicate" itself. 3. Control loops often subject to long delays. Warning of failure may come too late. 4. May change the sequence and direction of activities 5. Not easy to repeat.
Analysis	1. Is amenable to PDCA/Six Sigma/ mathematical type of problem-solving.	1. Is not amenable to PDCA-type approaches. 2. Requires a "soft-systems" methodology.
People	1. Ignores human nature. 2. People are extensions of the machine. 3. Individuals have limited control, must follow the "rules" of the process.	1. Almost exclusively about human interaction. 2. Experiences and thinks about the situation.

Fraser identified five types of processes which he says range from the most rigorously defined and controlled on the one hand to those subject to the greatest individual interpretation and choice (Fraser, 2015):

1　Mechanistic (as in a production line)
2　Responsive (to a generic event such as receiving a customer enquiry)
3　Developmental (where you choose to initiate action to create an outcome which is different /better than the last time you did it, such as business planning)
4　Special (as in "the peace process" in Northern Ireland)
5　Ongoing (as in implementing a policy or developing a *learning organization*)

Fraser does not distinguish between cybernetic and exploratory processes as does Edwards, but we can see that types 1 and 2 would be cybernetic and the others exploratory.

What is the process approach?

The process approach is a way of managing work which focuses on the objective of the work rather than on the objective of the function that performs it. An approach is a method of tackling an issue, a situation or achieving an objective, but TC 176 has declared the process approach to be a principle rather than a method.

To appreciate the differences between a process approach and a functional approach we need to explain the origins of each in the context in which they evolved.

Emergence of the functional approach

Around the same time that ISO 9001 was being revised for the first time in 1994 Hammer and Champy were promoting business process reengineering and in their book, they claim that "a set of principles laid down more than two centuries ago has shaped the structure, management and performance of American business throughout the nineteenth and twentieth centuries" and go on to say that "the time has come to retire those principles and to adopt a new set" (Hammer & Champy, 1993). They were referring to Adam Smith's seminal work *The Wealth of Nations* from the 18th century in which Smith describes how a pin-maker could make one pin in a day, and certainly could not make 20, but by dividing the making of pins into 18 distinct operations with each person making a part of the pin, 10 men might make 48,000 pins in a day (Smith, 1776). Pin making was a process, but dividing the work of making a pin into distinct operations and making each operation a speciality in itself created conditions for sub-optimization and fragmentation of work. These workers never complete a pin; they only work on part of the pin and optimization of their operation will not lead to optimizing the manufacture of the whole pin. Increase the scale of this and divide the work of the whole organization into distinct operations, and we have what has come to be known as the functional approach. What Smith writes about is the division of labour.

Functionalism, or organizing work by specialization or the skills people possess, continues to the present day but it's not where the work is done or who does it that matters. What matters is how work is managed.

Functional structures

Most organizations are structured into functions that are collections of specialists performing tasks. The functions are like silos into which work is passed and executed under the directive of a function manager before being passed into another silo. In the next silo the work waits its turn because the people in that silo have different priorities and were not lucky enough to receive the resources they requested. Each function competes for scarce resources and completes a part of what is needed to deliver a product or service to customers. When the division of labour theory was formulated by Adam Smith and later developed by Frederick Winslow Taylor at the turn of the 20th century, workers were not as educated as they are today. Technology was not as available and machines not as portable. Transportation of goods and information in the 18th and 19th centuries was totally different from today. As a means to transform a domestic economy to an industrial economy, the theory was right for the time. Mass production would not have been possible under the domestic systems used at that time.

Functions became major disciplines in an organization harnessing unique skills. A function is a collection of activities that make a common and unique contribution to the purpose of the business (Drucker, 1974). It is quite common to group work by its contribution to the business and to refer to these groupings as functions, and this results in there being a marketing function, a design function, a production function, etc. The marketing function in a business generates revenue, and the people contributing to marketing may possess many different skills such as planning, organizing, selling, negotiating, data analysis, etc. However, it should not be assumed that all those who contribute to a function reside in one department. The marketing department may contain many staff each with many skills, but often the design staff in a different department contribute to the product or service strategy which is one element of the marketing strategy. Likewise, the design function may have the major contribution from the design department but may also have contributors from research, test laboratory, trials and customer support. Therefore, the organization chart may in fact not define functions at all but a collection of departments that provide a mixture of contributions. In a simple structure the functions will be clear but in a complex organization, there could be many departments concerned with the marketing function, the design function, the production function, etc. Rummler and Brache observe that

> when we ask a manager to draw a picture of his or her business (be it an entire company, a business unit or a department) we typically get something that looks like a traditional organization chart with tiers of boxes and vertical reporting relationships. It doesn't show the customers, we can't see the products and services and we get no sense of the work flow.
>
> (Rummler & Branche, 1995)

When we organize work functionally, the hierarchy can be represented by the waterfall diagram (Figure 9.7).

In this diagram the top-level description of the way work is managed will probably be contained in a quality manual with supporting department manuals.

Balancing objectives

The combined expertise of all these departments is needed to fulfil a customer's requirement. It is rare to find one department or function that fulfils an organizational objective without the support of other departments or functions. However, the functional structure has proved to be very successful primarily because it develops core competences and hence attracts individuals who want to have a career in a particular discipline. This is the strength of the functional structure, but because work is always accomplished by a process it passes through a variety of functions before the desired results are achieved. No one person other than the CEO would carry responsibility for this chain of processes. If work is managed by function, one function will optimize its activities around its objectives at the expense of other functions. This will create gaps, overlaps, conflicts, bottlenecks and delays as managers pursue their own objectives. Functional optimization often contributes to the sub-optimization of the organization as a whole (Rummler & Branche, 1995).

One approach that aims to avoid these conflicts is what is referred to as "balancing objectives". On face value, this might appear to be a solution, but balancing implies that there is

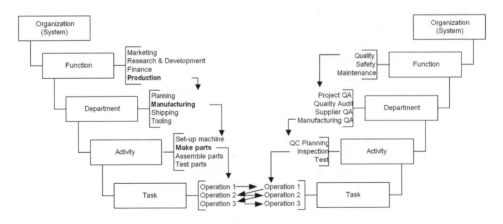

Figure 9.7 Functional decomposition of work

some give and take, a compromise or reduction in targets so that all objectives can be met. The result is often arrived at by negotiation, implying that quality is negotiable when in reality it is not. Customers require products and services that meet their requirements, not products and services that more or less meet their requirements. When objectives are derived from stakeholder needs, internal negotiation is not a viable approach. The only negotiation is with the customer, as will be explained in Chapter 11.

Emergence of the process approach

When the activities of the various functions that contribute to the achievement of a process objective are depicted in a flow chart with time and labour added, the inefficiencies of management by function will become clear. Instead of a job passing to each function for value to be added, it may be found that time and labour will be saved by keeping the job in the same location and moving the function to the job, but this is not a new idea.

As early as the 11th century the Venetian Arsenal operated similar to a production line with ships moving down a canal which were fitted by the various shops they passed. In this way, the workers could see the objective of their labour (Wikipedia (2), 2015). Henry Ford and his team borrowed concepts from watch makers, gun makers, bicycle maker, and meat packers; mixed them with their own ideas; and by late 1913 they had developed a moving assembly line for automobiles (Benson Ford Research Center, 2013).

Business outputs are generated by the combined efforts of all departments so processes tend to be cross-functional. Rarely does a single department produce a business output entirely without support from others, as shown in Figure 9.8. (For simplicity, not all outputs are shown.)

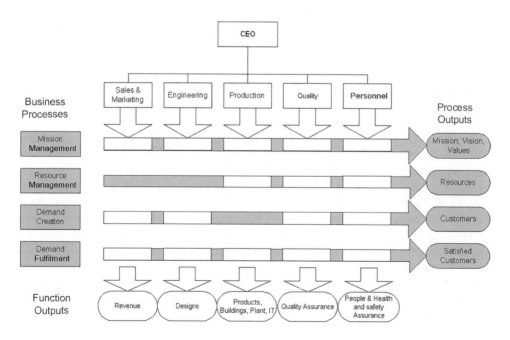

Figure 9.8 Managing by function or by process

Table 9.5 Function vs. process

Attribute	Functional approach	Process approach
Objective focused	Satisfying departmental ambitions	Satisfying stakeholder needs
Inputs	From other functions	From other processes
Outputs	To other functions	To other processes
Work	Task focused	Result focused
Teams	Departmental	Cross-functional
Resources	Territorial	Shared
Ownership	Departmental manager	Shared
Procedures	Departmental based	Task based
Performance review	Departmental	Process

In principle, the grey arrows indicate direction in which the process flows across departments. In reality it is probably not as simple as this because there will be transactions that flow back and forth between departments. The organization structure shows that functional outputs are indeed different from process outputs and obviously make an important contribution, but it is the output from business processes that is important to the business. When work is managed by process, the process activities are optimized around the process objectives, regardless of which function carries them out. Functional objectives become secondary to process objectives. This relationship is illustrated in Figure 9.8.

Some of the differences between a functional approach and a process approach are indicated in Table 9.5.

A common mistake when converting to a process approach is to simply group activities together and call them processes but retaining the function/department division. This perpetuates the practice of separating organization objectives into departmental objectives and then into process objectives. This is managing work as a process at a functional level, not at an organizational level. Another anomaly is that it assumes, for example, that the manufacturing department provides everything needed to perform the identified activities when in fact other departments are involved. A more effective approach ignores functional and departmental boundaries as represented Figure 9.9.

Superficially it may appear as though all we have done is to change some words but it is more profound than that. By positioning the business process at the top level, we are changing the way work is managed; instead of managing results by the contributions made by separate functions and departments, we manage the process which delivers the results regardless of which function or department does the work. This does not mean we disband the functions/departments; they still have a role in the organization of work. In fact, managing labour by process has been tried and it doesn't work except in process industries. As Thomas Davenport remarks in the foreword to Jeston and Nelis's book *Management by Process*, "not one major organization has adopted a horizontal organization structure; one composed entirely and only of processes" (Jeston & Nelis, 2008). Work can be organized in three ways: by stages in a process, by moving work to where the skill or tool is located or assembling a multi-skilled team and moving it to where the work is (Drucker, 1974). In

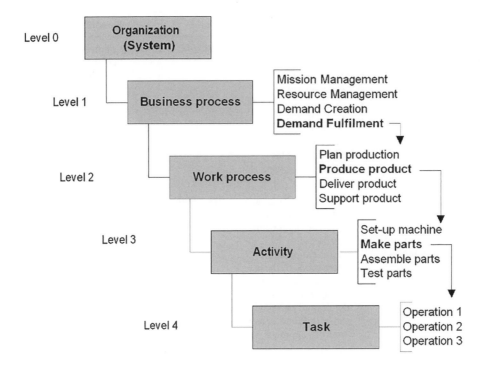

Figure 9.9 Process decomposition of work

these cases, we can still manage the work as a process or as a function. It comes down to what we declare as the objectives, how these were derived and how we intend to measure performance. If we ask three questions: *What are we trying to do, how will we make it happen and how will we know it's right?* we can either decide to make it happen through a process or through several functions/departments and measure performance accordingly. By making it happen through a process, we overcome the disadvantages of the functional approach.

Defining the process approach

The process approach was introduced into ISO 9000 in the 2000 version and revised in the 2015 version, as shown in Table 9.6 with a minor change in 2008.

The purpose of taking a process approach is clarified, but it misses the principal difference between a process approach and a functional approach which is brought out in the following explanation.

After formulating the strategy, it is not uncommon for the work required to implement the strategy to be divided among the various departments in an organization. These departments derive their objectives from the strategy, departments devise processes to achieve their objectives and report to management on their performance relative to these objectives.

Table 9.6 Definitions of the process approach

Definition	Explanation	Source	Remarks
A desired result is achieved more efficiently when activities and related resources are managed as a process.	The systematic identification and management of the processes employed within an organization and particularly the interactions between such processes is referred to as the "process approach".	(ISO 9000:2000)	For an approach to be a way of doing something, the definition does not inform us of a way of doing anything. It requires much more than this and a definition of a process to understand what the process approach is.
A desired result is achieved more efficiently when activities and related resources are managed as a process.	The application of a system of processes within an organization, together with the identification and interactions of these processes, and their management to produce the desired outcome, can be referred to as the "process approach".	(ISO 9001, 2008)	The definition remains the same but the explanation was changed to clarify the purpose of the process approach.
Consistent and predictable results are achieved more effectively and efficiently when activities are understood and managed as interrelated processes that function as a coherent system.	The QMS consists of interrelated processes. Understanding how results are produced by this system enables an organization to optimize the system and its performance.	(ISO 9000:2015)	This is one effect of using the process approach but it doesn't define it. Moreover, it appears to be describing a way of managing a system rather than a way of managing work as a process.

This is where the process approach is fundamentally different. Who does the work is of less importance in the process approach as long as the people doing it are competent. The objectives are derived directly from the strategy, thus ensuring their alignment and without considering which departments are involved in their achievement. The end-to-end processes are devised to achieve these objectives and only then is consideration given to who does what and finally performance is reported relative to the process objectives,not the departmental objectives. This approach forces departments to work together in teams to achieve the organization's objectives rather than in silos to achieve departmental objectives.

As a way of doing something, the process approach is better expressed as a series of actions, and this is to some extent defined in ISO 9000 clause 4.4.1 a) to h) and in the supporting guide (ISO N1289, 2015). Rather than repeat these requirements here, the following

10 actions summarize the approach with cross-reference to the relevant clauses of ISO 9001 in parentheses:

1 Deriving what it is you want to do from the needs and expectation of your stakeholders (i.e. the objectives you want to achieve or the outputs you want to deliver)

   (4.4.1a & 6.2)

2 Deriving measures of success from the constraints imposed by your stakeholders (i.e.
   the factors that will indicate whether the objectives have been achieved to the stakeholders' satisfaction or the outputs meet requirements)

   (referred to as criteria in 4.4.1c)

3 Defining the activities that are critical to achieving these objectives, the order in
   which they are carried out and the direction in which information flows of among
   them

   (referred to as methods in 4.4.1c)

4 Defining the resources, information and competences required to deliver the required
   outputs

   (4.4.1d)

5 Identifying the risks and putting in place measures that eliminate, reduce or control
   these risks

   (4.4.1f)

6 Determining how performance will be measured against the objectives and variation
   reduced

   (4.4.1c)

7 Executing the process as planned
8 Measuring what has been achieved and comparing it with the process objectives

   (4.4.1g)

9 Finding better ways of achieving the process objectives and improving process
   efficiency

   (4.4.1h)

10 Establishing whether the processes objectives remain relevant the needs of the stake-
    holders and if necessary changing them

   (9.3)

As will be evident from this, it was necessary to refer to other clauses of ISO 9001 than those in 4.4.1. A cross-reference is given in step 3 to clause 4.4.1c) assuming the methods referred to are the activities within a process. However, process activities are addressed in step 7 of the process approach in ISO Guide, The Process Approach in ISO 9001:2015 (ISO N1289, 2015).

Summary

Since the first edition of ISO 9001, three words have acted as tools to convey requirements. In the first edition, it was the word *procedure* that conveyed what organizations needed to document and implement to provide products and services that met requirements. The word *process* was less prominent but nonetheless present. However, it was in the 2000 version that *process* became prominent. It was recognized that outputs were more than the product of activities and the function that performed them, but a product of many elements combined in a cause-and-effect relationship. Managing a process is more important than implementing procedures or managing functions. In fact, who does what it is of less importance than the results to be achieved and the process which will deliver them. Viewing work as a process rather than a procedure or a function focuses everything associated with that work on the results to be achieved. Dividing results into the processes that will deliver them, rather than dividing work into specialisms, is the essence of the process approach, and it is the network of processes that holds the *system* together. A video on understanding the process approach to management is available on the companion website.

Bibliography

APQC. (2015, October 23). American Productivity & Quality Center. Retrieved from APQC Process Classification Framework (PCF): www.apqc.org/knowledge-base/documents/apqc-process-classification-framework-pcf-cross-industry-excel-version-70

ASQ. (2007). *Quality Glossary*. Retrieved from ASQ: http://asq.org/glossary/p.html

Benson Ford Research Center. (2013). The Innovator and Ford Motor Company. Retrieved from The Henry Ford: www.thehenryford.org/exhibits/hf/The_Innovator_and_Ford_Motor_Company.asp

Carter, R. C., Martin, J. N., Mayblin, B., & Munday, M. (1983). *Systems, Management and Change: A Graphic Guide*. London: Paul Chapman Publishing Ltd.

de Geus, A. (1999). *The Living Company Growth, Learning and Longevity in Business*. London: Nicholas Brealey Publishing.

Davenport, T. H. (1993). *Process Innovation: Reengineering Work through Information Technology*. Cambridge, MA: Harvard Business School Press.

Drucker, P. F. (1974). *Management, Tasks, Responsibilities, Practices*. Oxford: Butterworth-Heinemann.

Edwards, W. (2013, Winter). Thinking About Processes, Transformations and Control. *Journal of the United Kingdom System Society "Systemist"*, 34(3), 100.

Fraser, P. K. (2015, September, October and November). Business Process Principles. (P. Harding, Ed.) *SAQI e-Quality Edge* (193, 194 & 195). Brummeria, Pretoria: South African Quality Institute

Hammer, M., & Champy, J. (1993). *Reengineering the Corporation*. New York: Harper Business.

Ishikawa, K. (1985). *What Is Total Quality Control: The Japanese Way*. (D. J. Lu, Trans.). Englewood Cliffs, NJ: Prentice-Hall Inc.

ISO N1289. (2015). The Process Approach in ISO 9001:2015. Retrieved from International Organization for Standardization: www.iso.org/iso/iso9001_2015_process_approach.pdf

Jeston, J., & Nelis, J. (2008). *Management by Process: A Road Map to Sustainable Business Process Management*. Oxford: Butterworth-Heinemann.

Juran, J. M. (1964). *Managerial Breakthrough*. New York: McGraw-Hill.

Juran, J. M. (1992). *Juran on Quality by Design*. New York: The Free Press, Division of Macmillan Inc.

Morgan, G. (1997). *Images of Organization*. London: SAGE Publications.

Neave, H. R. (1990). *The Deming Dimension*. Knoxville: SPC Press.

OED. (2013). Retrieved from Oxford English Dictionary: oed.com

Reichfeld, F. (2001). *The Loyalty Effect*. Cambridge, MA: Harvard Business School Press.

Rummler, G. A., & Branche, A. P. (1995). *Improving Performance: How to Manage the White Space on the Organization Chart*. San Francisco: Jossey-Bass, Inc.

Senge, P. M. (1990). *The Fifth Discipline: The Art and Practice of the Learning Organization*. New York: Random House.

Sherwood, D. (2002). *Seeing the Forest for the Trees: A Manager's Guide to Applying Systems Thinking*. London: Nicholas Brealey Publishing.

Smith, A. (1776). *An Inquiry in the Nature and Causes of the Wealth of Nations*. Public domain.

Wikipedia (2). (2015, November 4). Venetian Arsenal. Retrieved from Wikipedia: https://en.wikipedia.org/wiki/Venetian_Arsenal#Mass_production

10 Risk and opportunity

> Uncertainty, in the presence of vivid hopes and fears, is painful, but must be endured if we wish to live without the support of comforting fairy tales.
>
> Bertrand Russell (1872–1970)

Introduction

ISO 9001:2015 includes the concept of risk in the form of definition, guidance and requirements. Previous editions included a clause on preventive action which aimed to prevent the occurrence of nonconformities, and to some extent this was risk mitigation but there were no requirements to determine probability of occurrence, assess the consequences and based on the anticipated impact, decide whether to avoid it, take it or manage it in some way. But how are these decisions made? On the basis of rule of thumb or soundly based risk analysis methods?

If we look at ISO 9001 through a *risk-tinted* lens we would see all requirements in ISO 9001 as risk treatments; therefore, risk and ISO 9001 is not a new combination. However, the decision as to whether to undertake these activities was not based on an analysis of the risk, a calculation of probability of occurrence and magnitude of the loss, but on the simple fact that they were requirements to which the organization had to conform to be granted certification. They were therefore not conceived as risk treatments.

In the 2015 version, there is not only a requirement to identify and address risk as a basis for planning but also a requirement to determine the effectiveness of actions taken to address risks, which implies some form of measurement. Although there is no requirement for formal methods for risk management or a documented risk management process, it is difficult to imagine how organizations will be able to demonstrate that risks have been identified and addressed and the effectiveness of the actions taken determined without using soundly based methods. However, these don't need to be quantitative methods. If we can't measure risk, we can't manage risk, but as Hubbard remarks "No matter how 'fuzzy' the measurement is, it's still a measurement if it tells you more than you knew before" (Hubbard, 2010).

The way in which the term *risk* is defined, used and explained in these standards creates some uncertainty as to what the term means in the context of ISO 9001 and has implications for users and so it's worth exposing the uncertainties in the terminology.

In this chapter, we explore the word *risk* through the following sections:

*   The common definition of risk
*   The new definition of risk

- What do we mean by uncertainty?
- The notion of positive and negative risk
- Types of risk
- Risks and opportunities
- Use of the term *risk* in ISO 9001:2015, including risk-based thinking
- The requirements for managing risk

The common definition of risk

If we look up the term *risk* in an English dictionary we will find that in the simplest terms, it is used to express the possibility of something bad happening; for example, "exposure to the possibility of loss, injury, or other adverse or unwelcome circumstance" (OED, 2013) and "the possibility that something bad or unpleasant (such as an injury or a loss) will happen" (MWD, 2015). There isn't one English dictionary in which the term is used to express the possibility of something good happening – that is until we look at standards in the field of risk management.

The new definition of risk

TC 176, the committee responsible for the development of ISO 9001, has been placed under an obligation from ISO to adopt a new common structure for management system standards, commonly referred to as Annex SL (ISO/IEC, 2015). This directive takes the definition of risk from the risk management vocabulary (ISO Guide 73, 2009) and modifies it to permit users of Annex SL to tailor the definition to the context of a particular management system standard.

In ISO Guide 73 and the ISO Risk Management Standard (ISO 31000, 2009), risk is defined as the "effect of uncertainty on objectives". In Annex SL and ISO 9000:2015 risk is defined as the "effect of uncertainty". These definitions pose the possibility that the uncertainty may be something good as well as something bad. In addition, ISO qualifies the use of the terms risk and opportunity as follows:

> Reference to 'Risks and Opportunities' is intended to broadly describe something that poses a threat having detrimental or negative effect, or alternatively, something that has the potential for a beneficial or positive effect. It is not intended to be the same as the technical, statistical, or scientific interpretation of the term risk.
>
> (ISO/TMB/JTCG N 360, 2013)

This statement is contradictory because it is explaining risks *and* opportunities and yet refers to opportunities as being *an alternative* to threats. It's also not grammatically correct, as risks and opportunities are two things, not one thing. More about this later.

What do we mean by uncertainty?

Uncertainty is simply something we are uncertain about, there is doubt, we are unsure. Now, not everything we are unsure about is important to us; for example, if we live in California we are unlikely to be concerned about whether we will be able to rent a car in Wales unless we are travelling to Wales and wish to rent a car. Also, if we did decide to rent a car in Wales and we learnt that we would have to drive on the left side of the road, something we hadn't done before, we may well have doubts over whether we could do that safely.

David Hillson makes the claim that "risk is uncertainty that matters" (Hillson, (1) 2016), and it is true that uncertainties that present neither risk nor opportunity to achievement of objectives are simply irrelevant uncertainties, but so what? What we are interested in are uncertainties that may result in loss and uncertainties that may result in gain. The two words used in Annex SL are risks (for uncertainties that may result in loss) and opportunities (for uncertainties that may result in gain). This seems to justify Hubbard's definition of risk as "a state of uncertainty where some of the possibilities involve a loss, catastrophe or another undesirable outcome" (Hubbard, 2009).

Types of uncertainty

Telling us to identify risks and opportunities is all well and good, but what kind of uncertainties would we be looking for? It's almost left to our imagination. If we can't imagine what could go wrong or what the future might bring, where do we start in this task? First, we remember that the risks and uncertainties being referred to are strategic and not tactical – so we are not concerned here with isolated incidents such as use of an obsolete document or delivery of the wrong order to a customer, but referring to risks and opportunities that can affect our ability to consistently satisfy our customers. When we talk about risks, more often than not we are referring to events, whether something will or won't happen, but there are other circumstances that we are exposed to which may affect our success, and to help us Hillson identifies four types of uncertainty which he labels with Greek words:

- Stochastic uncertainty – this is the uncertainty of events, that is, whether an event will or will not happen. We don't know when or even if an event that will affect what we are trying to achieve will occur, for example, when supply of a critical resource will cease, when a cyber-attack will cause significant operational disruption, when a major environmental disaster will affect our plants or supply chains.
- Aleatoric uncertainty – this is the uncertainty of variables, that is, whether results will be the same or different from those observed previously. We don't know which result of a range of possible results we will get, for example, how much something will cost, how long the job will take, how many people or how much material we'll need or will be able to acquire.
- Epistemic uncertainty – this is uncertainty of knowledge, for example, whether the knowledge we have is complete or incomplete and therefore ambiguous, whether we know what the customer wants or what we will learn from a survey or investigation, etc.
- Ontological uncertainty – this is uncertainty of the unknown, that is, whether everything that affects the results is inside or outside our frame of reference, things we haven't thought of, what are commonly referred to as blind spots or unknowns.

These are indeed useful in identifying more risks and opportunities, and it would certainly help if they were embodied into ISO 9001 although the Greek labels would probably be omitted.

The notion of positive and negative risk

Redefining the term *risk* does not end with the simple definition that risk is the "effect of uncertainty". ISO 9000:2015 appends several notes to the definition one of which is highly significant. An effect is defined as "a deviation from the expected – positive and/or negative". Although this might look like a definition of the word *effect*, it isn't (see later). If we focus on the word *deviation* rather than the word *effect*, there may be deviations that are positive and negative, for example, we expect the train to arrive on time but it varies around the expected time of arrival. If it's early we might regard this is as a positive deviation from the expected time of arrival and treat it as an opportunity to avoid the crowds at the other end. if it's late, we might regard this is as a negative deviation from the expected time of arrival and a risk of being delayed further by missing a connection at the other end. On the other hand, a delivery of goods earlier than expected might be a bad thing as there might be no space available to store them. A later-than-expected delivery of goods might also be a bad thing for the same reasons. Another example is variation in customer demand. We expect orders to arrive following product/service launch but they don't in sufficient number to sustain operations or perhaps too many orders arrive such that operations can't cope. These are both risks but there is a level above expectations where the pace of orders creates an opportunity to grow operational capacity.

Hillson and others in the LinkedIn Risk Management Group claim that we can trace the origin of the word *risk* back to Arabic, Greek, Chinese and Italian, to a word that was used to express good and bad, positive and negative. Languages evolve and meanings change, and from the *Oxford English Dictionary* the word *risk* has been used to express the possibility of loss or harm since the 13th century so it appears to have lost its use to express positive effects long ago, and there does not appear to be any justification for reintroducing this use in the 21st century. In fact, it may do more harm than good as it's highly unlikely that users of ISO 9001 will use the term *risk* to express the possibility of something good happening.

Box 10.1   Risk and risk taking

If I am in a situation where there is a possibility of increasing my wealth by taking advantage of a forecast rise in interest rates, this is an opportunity. The consequence is only positive if I take that risk and my wealth increases as a result. It is negative if I take the risk and my wealth declines as a result, that is, speculative risk taking. But risk is perceived as the possibility of loss. The idea that risk is perceived as the possibility of gain is counterintuitive.

In almost everything we do there are risks, and every day we make conscious and subconscious choices about them. We consciously weigh up the potential benefits and harms of exercising one choice of action over another. We also subconsciously do this based on our past familiarity with a situation. This can sometimes catch us out, which is why we develop certain habits like looking before we leap as a precaution or checking we have been given the right change by the cashier before putting it in our pocket. We have learnt ways of reducing risk to a level where we behave instinctively, and some of us navigate through life without befalling the risks that we face every day. Leonardo Buscaglia advocates that "Only the person who risks is truly free." (See Box 17.6 for the full quotation.)

Box 10.2   Decision analysis

Consciously weighing up the potential benefits and harms of exercising one choice of action over another.

   Decision analysis is a large body of theoretical and applied work that deals with making decisions under a state of uncertainty. It addresses decisions where trade-offs have to be made between uncertain costs, uncertain benefits, and other risks.

(Hubbard, 2009)

If having weighed up the potential benefits and harms, we may say to ourselves, nothing ventured nothing gained and choose to take the risk. If the risk that we thought would prevent us from achieving our objectives does not materialize we do indeed reap the benefits. Some people are now referring to this as *positive risk taking* (West Midlands, 2015) which is quite absurd because the risk remains a potentially negative effect. However, we should not be misled by the terminology because the motivation for this absurd term is in this example to discourage people, particularly the disabled, using risk as an excuse for not doing something of benefit to them.

   Someone who is pursuing an opportunity may be taking a risk and someone who by their actions may miss an opportunity is taking a risk. Thus, when we set out to identify risks, we should not only be looking at what could go wrong if we do X but what could go wrong if we don't do X. Rock climbing is hazardous but also presents opportunities for adventure, excitement and pleasure. People engaged in rock climbing are at risk of injury, perhaps even death, and may find it difficult to get insurance. Rock climbers don't insure against the possibility of pleasure, only against harm. Insurance companies are interested in opportunities to make money out of people taking risks but always define risk in the negative sense.

Types of risk

There are different types of risk and not all may be addressed by the QMS, but should the loss be incurred it may influence the performance of the QMS. In such cases, it's an external factor.

Strategic risk

Strategic risks result directly from operating within a specific industry at a specific time and include:

- Market risk – the risks present in the market and inherent to the industry or arising out of competition, for example, shifts in consumer preferences or emerging technologies that make the product line obsolete.

- Reputational risk – Loss of your company's reputation from product or service failures, lawsuits or negative publicity. According to Matt McGee (a search engine optimization consultant), "One negative blog post or review can spread online in a flash and change the direction of a company."
- IT risk – loss of business continuity due to certain inherent risks associated with the technologies.
- Environmental risk – Organizations that operate in or depend on suppliers from regions of the world prone to natural environmental disasters are exposed to risk of an unpredictable kind.
- Human capital risk – Organizations that depend on a particular source or type of labour may be exposed to risk of supply shortages or poaching from competitors.
- Health and safety risks – Organizations that operate in dangerous environments or provide services may expose members of the public to hazards.

Financial risk

Financial risks are associated with how the organization handles its financial assets, including:

- Debt and credit, interest rates and foreign exchange rates.
- The customer's ability to pay.
- The organization's ability to raise the necessary capital to fund improvements.

Operational risks

Operational risks are present in every enterprise and result from internal process failures such as:

- Product/service risk – You can't translate your concept into a working and compelling product/service.
- Technology risk – You can't build a good enough or, if necessary, breakthrough technology.
- Business development risk – You can't get deals with other companies that you depend on to build or distribute your product/service.
- Timing risk – You are too early or too late to the market or there are unforeseen external events, such as transportation breaks down, or a supplier fails to deliver a product or service when required.
- Margin risk – You build something people want but that you can't defend, and therefore competitors will squeeze your margins.
- Mistakes in execution – The formal plans and procedures are not implemented as intended.
- System failures – A common cause of failure reduces the ability of the system to consistently provide a conforming product or service.

Compliance risk

Risks associated with compliance are those subject to regulatory and statutory requirements, including legal infringements and rule breaches.

Risks and opportunities viewed from different perspectives

Box 10.3   Henry Ford on risks and opportunities

A hundred years ago Henry Ford established his company on four principles, the first of which was "an absence of fear of the future and of veneration for the past", which he expressed as follows: "One who fears the future, who fears failure, limits his activities. Failure is only the opportunity more intelligently to begin again. There is no disgrace in honest failure; there is disgrace in fearing to fail. What is past is useful only as it suggests ways and means for progress."

(Ford & Crowther, 1922)

Is the order significant?

In the phrase *risks and opportunities,* the words always appear in the same order throughout ISO 9001; it's never *opportunities and risks*. There is perhaps an issue as to whether the order of the words is significant. Putting *risks* first could imply that it's telling us to first look at what could go wrong and then look for reducing the likelihood and severity of these events so that the benefits outweigh the harms and the opportunity of taking a risk presents itself. In this way, opportunities arise out of risks.

But there is another way of looking at it. An organization first looks for opportunities such as a gap in the market, some new technology it can exploit, etc., and then looks for the risks that could affect its ability to realize these opportunities. This was the order preferred by TC 176 because it more accurately reflected a proactive approach to the management of quality. Had this issue been limited to discipline specific requirements, the order could have been changed. However, as the phrase was used in the Annex SL requirement, the order was probably correct for the other management system standards because they primarily address constraints such as health, safety, security, environment, etc.

A third way is to consider the risks as threats and so we have half of the analysis technique of strengths, weaknesses opportunities and threats, or SWOT. When we carry out this analysis, threats are not lost opportunities, and opportunities are not what is gained by taking a risk that does not occur; they are separate situations. This is the interpretation of JTCG Guide N360 mentioned previously. However, there is a difference between a risk and a threat. We tend to associate threats as anything that can exploit a vulnerability and therefore a threat exerts pressure, whereas a risk is present if we expose ourselves or our property to it. In the security industry, the terms *risk, threat, vulnerability* and *asset* have a particular relationship:

Risk = Asset + Threat + Vulnerability

This means that risk is a function of threats exploiting vulnerabilities to obtain, damage or destroy assets. Thus, threats (actual, conceptual or inherent) may exist, but if there are no vulnerabilities then there is little/no risk. Similarly, you can have a vulnerability, but if you have no threat, then you have little/no risk (TAG, 2016)

We are therefore left with these three interpretations, and as the order is not itself a requirement we can meet the intent in whichever order risks and opportunities are addressed.

Hillson says

> there are things in the future that could happen but might not happen but if they did happen they would be helpful. They would help us to save money, save time, increase value and benefits or enhance our reputation so we could look for these things and manage them proactively.
>
> (Hillson, (1) 2016)

But most of us would call these uncertainties opportunities, for example, we say, "there's a chance we'll win this new contract"; we don't say "there's a risk we will win this new contract" unless doing so is going to have undesirable consequences, in which case why did we bid for it? We say "this new technology will save us money and therefore we should not miss this opportunity." We don't say "this new technology will save us money and therefore we should not miss this risk."

A simple response would be:

- An uncertainty presents a risk if its occurrence may have a negative effect on an expected result and is therefore relevant.
- An uncertainty presents an opportunity if its occurrence may have a positive effect on an expected result and is therefore relevant.

That risk is now *an effect* is different from the way we normally use the word *risk* as in the sense of *exposure to a possibility*. But as an effect is the result of an action, we are now being told that it's more than a possibility and that it's a certainty. According to the OED we use the term *effect* to describe "an operative influence; a mode or degree of operation on an object" so ISO is also using the word *effect* differently from its normal use. If there is no action but the possibility of action, there is only the possibility of an effect, but ISO appears adamant that a risk is an effect and not the possibility of an effect. A possible explanation is that we can imagine an effect without experiencing it. We see a banana skin on the ground obstructing our path and we imagine the effect it would have if we proceeded to step on it. Therefore, ISO could be expecting us to refer to a situation as a risk where we are able to imagine that something good or bad could happen and may affect what we are trying to do. Wouldn't it be simpler to use the words *risks* and *opportunities* as we have always used them? Well, this is what ISO appears to have done in Annex SL.

The effect of different perspectives

One person's risk may be another person's opportunity, for example, the rock climber referred to previously takes out insurance from which both parties benefit; the rock climber receives protection and the insurance company receives revenue. The insurance company scans its environment and notices an increase in the number of people engaging in rock climbing and seizes the opportunity to sell policies to people taking such risks. This is not the positive side of risk. The risk remains negative. It simply the way a free market economy works.

Raise this to the organizational level and a risk from one perspective may be an opportunity from another perspective. A Chinese proverb clarifies the concept in Box 10.4.

Box 10.4   Bunkers and windmills

At times of great winds, some people build bunkers and others build windmills.

Use of the term *risk* in ISO 9001:2015

In every instance of the use of the term in the new standard the word *risk* is used in the negative sense and not once in the sense of a positive effect. In fact, other than in the guidance and definitions, the word *risk* is only used among the requirements in the form of the compound term *risks and opportunities* with one exception in clause 8.5.5 on post-delivery activities. So, it looks like ISO was taking no *risk* that the word *risk* could be misunderstood, but nonetheless retained the new definition so as to cause confusion and uncertainty. So much for ISO/IEC directives that require management system standards to be easily understood and unambiguous!

As mentioned in the introduction, the clause on preventive action has been removed and in its place a new clause added on: *Actions to address risk and opportunities*. If risk (effect of uncertainty) can indeed be positive, why would Annex SL refer to risks *and* opportunities? Could it be that not everyone on these committees thinks in the same way?

There are now two informative sections in ISO 9001 which address the subject of risk-based thinking, one in clause 0.3.3 and the second in Annex A4.

Risk-based thinking (RBT)

The inclusion of risk is a good thing because, for too long, the requirements have been treated by users as having to be met regardless of need. The only exceptions that were permitted were the requirements in section 7 and those clauses in which the word *appropriate* was used. Now, you are permitted to assess the risk, and if you can produce evidence to show that the actions taken to address risks and opportunities are proportionate to the potential impact on the conformity of products and services, it appears you don't need to meet a requirement that does not address a risk in the context of your organization.

A new guide to risk-based thinking has been released by TC 176 in which there is a novel interpretation of the word *opportunity*. It now appears that when faced with the risk of being injured crossing the road, the options you consider to reduce or eliminate the risk are opportunities. This isn't as crazy as it appears because it fits with the definition of an opportunity earlier, but these are not the only opportunities users of ISO 9001 should be identifying.

When we examine section 0.3.3 and Annex A.4 we won't find an explanation of the concept of RBT, but there is mention of it being implicit in previous editions of the standard. It is true that the requirements were intended to prevent past quality problems recurring but that is applying RBT in standards development not in managing quality. If it was implicit, why was it that auditors would require conformity with requirements in situations where nothing was at risk (e.g. document control)? There are 17 clauses where the word *appropriate* is used, for example, in clause 8.5.2 of the 2008 version where it states, "Corrective actions shall be appropriate to the effects of the nonconformities encountered", implying that organizations can weigh up the potential benefits and harms of exercising one choice of action over another.

Deming would ask for an operational definition as he believed you can't do business with concepts. An operational definition puts communicable meaning into a concept, and therefore as top management is required to promote RBT, they should produce an operational definition for it. One way of doing this is to seek answers to the following questions:

a)   What would we expect to see happening in an organization that was applying risk-based thinking?
b)   What would we expect to see happening in an organization that was not applying risk-based thinking?

This will create a range of answers that can be reduced to a few concise bullet points that express the essence of this way of thinking.

Requirements for managing risk

There are several requirements on the subject of risk scattered throughout ISO 9001:2015, some of which are duplicated under different headings:

- When planning for the QMS, the risks and opportunities that arise from an assessment of the context of the organization and the determination of stakeholder requirements are to be determined and addressed to give assurance that the QMS can achieve its intended results (6.1.1). This requirement is addressed in Chapter 21.
- When determining the processes needed for the QMS the organization shall address the risks and opportunities as determined in 6.1 (4.4.1f). This requirement is addressed in Chapter 21.
- Top management is to promote the use of risk-based thinking (5.1.1d). This requirement is addressed in Chapter 17.
- Top management is to ensure that risks and opportunities that can affect conformity of products and services and the ability to enhance customer satisfaction are determined and addressed (5.1.2b). This requirement is addressed in Chapter 17.
- The organization shall plan actions to address these risks and opportunities, including how to integrate and implement the actions into its QMS processes, and evaluate the effectiveness of these actions (6.1.2). This requirement is addressed in Chapter 21.
- Actions taken to address risks and opportunities are to be proportionate to the potential impact on the conformity of products and services (6.1.2). This requirement is addressed in Chapter 21.
- The results of analysing data and information arising from monitoring and measurement are to be used to evaluate the effectiveness of actions taken to address risks and opportunities (9.1.3). This requirement is addressed in Chapter 54.
- Management reviews are to be planned and carried out, taking into consideration the effectiveness of actions taken to address risks and opportunities (9.3.2e). This requirement is addressed in Chapter 56.
- When a nonconformity occurs, risks and opportunities determined during planning are to be updated if necessary (10.2.1). This requirement is addressed in Chapter 58.

Summary

In this chapter, we have attempted to expose some of the uncertainties about the meaning of the terms *risk* and *opportunity* in ISO 9001. We haven't addressed the requirements on risks and opportunities as these are dealt with elsewhere as indicated earlier. The good news is that you can ignore the definition of risk given in ISO 9000:2015 and assume the term is used in its negative sense and still understand the requirements because almost everywhere the term *risk* is used it is combined with the word *opportunity*.

Bibliography

Ford, H., & Crowther, S. (1922). *My Life and Work*. New York: Doubleday Page & Company.

Hillson, D. ((1) 2016, June 23). Managing Risk in Projects. Retrieved from www.youtube.com/watch?v=GO2rpxjbi_A

Hubbard, D. W. (2009). *The Failure of Risk Management: Why It's Broken and How to Fix It*. Hoboken, NJ: John Wiley & Son.

Hubbard, D. W. (2010). *How to Measure Anything: Finding the Values of Intangibles in Business*. Hoboken, NJ: John Wiley.

ISO 31000. (2009). *Risk Management – Principles and Guidelines*. Geneva: ISO.

ISO Guide 73. (2009). *Risk Management – Vocabulary*. Geneva: ISO.

ISO/IEC. (2015). *ISO/IEC Directives, Part 1—Consolidated ISO Supplement – Procedures Specific to ISO*. Geneva: ISO General Secretariat.

ISO/TMB/JTCG N 360. (2013, December 3). Concept Document to Support Annex SL. Retrieved from ISO Standards Development: http://isotc.iso.org/livelink/livelink?func=ll&objId=16347818&objAction=browse&viewType=1

MWD. (2015). Retrieved from Merriam-Webster Dictionary: www.merriam-webster.com/dictionary/system

OED. (2013). Retrieved from Oxford English Dictionary: oed.com

TAG. (2016, October). Retrieved from Threat Analysis Group: www.threatanalysis.com/2010/05/03/threat-vulnerability-risk-commonly-mixed-up-terms/

West Midlands. (2015, February 15). A Positive Approach to Risk & Personalisation. Retrieved from Think local act personal: www.thinklocalactpersonal.org.uk/Browse/safeguarding/?parent=8625&ch

11 Interested parties and stakeholders

Introduction

All organizations depend on the support of a wide range of other organizations and individuals to achieve their goals, the most obvious ones being customers, employees and suppliers, but there are others who play a very important role. These groups are referred to in ISO 9001 as interested parties (see Box 11.1).

Box 11.1   Interested parties

Person or organization that can affect, be affected by, or perceive itself to be affected by a decision or activity Examples include customers, owners, people in an organization, providers, bankers, regulators, unions, partners or society that can include competitors or opposing pressure groups.

(ISO 9000, 2015)

The problem with this definition and the explanatory notes is that it can include both benevolent and malevolent parties, for example, in the pharmaceutical industry some animal rights campaigners may be intent on closing down a research unit and use malevolent practices. ISO 9001 now obliges the requirements of interested parties relevant to the QMS to be determined and so it is important that we understand what and who they are, but the requirements appear to be pointing only to the benevolent parties whose relevant requirements are to be met. However, those parties whose interests are malevolent also need to be addressed, but rather than meeting their needs their potential impact needs to be anticipated and managed.

In the drafting of ISO 9000:2000 the term *stakeholder* was considered because it was a term used with the EFQM and the principles on which the model was based were being incorporated into ISO 9001 at the time. However, the traditional meaning of the term *stakeholder* as a person who holds money while the issue of ownership is being resolved still pertains in some countries so it was decided that the term *interested party* would be the preferred term, with the term *stakeholder* being an admitted term – meaning it has the same meaning as the term *interested party*. However, as we have shown earlier, the two terms are not used in the same way in business circles. To distinguish between benevolent and malevolent parties we will use the term *stakeholder* for interested parties with a benevolent interest.

In this chapter, we examine:

- Where the idea of a stakeholder came from
- The many definitions of stakeholders and how to make a rational choice
- The idea that all real stakeholders are beneficiaries
- These stakeholders in terms of their identity, importance and relationship with the organization

A historical perspective

As was stated in the introduction, no organization can accomplish its goals without the support of others. However, this belief has not always been so. In the 19th century, the only stakeholder of any importance was the owner who ran the business as a wealth-creating machine with himself as chief beneficiary. Workers had no influence, and customers bought what was available with little influence over the producer. Some suppliers were no more influential than the workers and society was pushed along with the advancing industrial revolution. The successful owners occasionally became philanthropic through guilt and a desire to go to heaven when they died. As a result, their wealth was distributed through various trusts and endowments.

Workers were the first to exert influence with the birth of trade unions but it took a century or more for worker's rights to be enshrined in law. Customers began to influence decisions of the organization with the increase in competition, but it was not until the Western industrialists awoke to competition from Japan in the 1970s that a customer revolution emerged. Then slowly in the 1980s and on through the millennium, the green movement began to exert influence resulting in laws and regulations protecting the natural environment. Although the Universal Declaration of Human Rights was proclaimed by the United Nations in 1948, it was not until 1976 that the International Covenant on Civil and Political Rights came into force but it remains controversial.

There is now a greater sensitivity to the impact of organizations' actions upon society and the planet. Workers are more aware of their rights and more confident of censuring their employer when they feel their rights have been abused. The wealth-creating organizations now distribute their wealth through their stakeholders rather than through philanthropy, although this remains a route for the biggest of corporations as evidenced by the philanthropy of Bill Gates and Warren Buffett.

What are stakeholders?

The concept of stakeholders is complicated by different meanings and uses dependent upon both context and association. In traditional usage, a stakeholder is a third party who temporarily holds money or property while the issue of ownership is being resolved between two other parties (e.g. a bet on a race, litigation on ownership of property).

In business usage definitions vary but these appear to have several traits. There is the notion of contributors to an organization's wealth-creation capacity being beneficiaries and risk takers, implying that those who put something into an organization, either resources or a commitment, are stakeholders but expect something in return.

It is difficult, if not impractical, in some cases for organizations to set out to satisfy the needs and expectations of all these interested parties. There needs to be some rationalization.

The following definition amalgamates the idea of contributors, beneficiaries, risk takers and voluntary and involuntary parties thus indicating that there is mutuality between stakeholders and organizations.

> The stakeholders in a corporation are the individuals and constituencies that contribute, either voluntarily or involuntarily, to its wealth-creating capacity and activities, and that are therefore its potential beneficiaries and/or risk bearers.
>
> (Post, Preston, & Sachs, 2002)

This accords with Peter Drucker's view that businesses are the wealth-creating organ of society and recognizes that there is more than one beneficiary. But ISO 9001 applies to all types of organization, and therefore we need to interpret *wealth* in other than financial terms and a more suitable term would be *value*. Drucker also advocated that the purpose of a business is to create customers, and it is the income generated by customers that cover the costs and pay taxes which create wealth for society. That there is more than one beneficiary shows that organizations cannot accomplish their mission without the support of other organizations and individuals who generally want something in return for their support.

By making a contribution, these beneficiaries have a stake in the performance of the organization. If the organization performs well, they get good value and if it performs poorly, they get a poor value at which point they can withdraw their stake.

Post, Preston and Sachs assume that organizations do not intentionally destroy wealth, increase risk or cause harm. However, had organizations not known of the risks, their actions might be deemed unintentional but some decisions are taken knowing that there are risks which are then ignored in the interests of expediency. Enron, WorldCom and Tyco are examples where serious fraud was detected and led to prosecutions in the United States. Volkswagen is a more recent example which has yet to come to court. There are many other much smaller organizations deceiving their stakeholders, each day some of which reach the local or national press or are investigated by the consumer association, trading standards and other independent agencies.

There is no doubt that customer needs are paramount as without income the organization is unable to benefit the other stakeholders. However, as observed by Post, Preston, and Sachs

> Organizational wealth can be created (or destroyed) through relationships with stakeholders of all kind – resource providers, customers and suppliers, social and political actors. Therefore, effective stakeholder management – that is, managing relationships with stakeholders for mutual benefit – is a critical requirement for corporate success.

Box 11.2   Food for thought

Just consider for a moment that if an interested party is anyone affected by the organization, would a burglar who gains access to an organization that makes security systems be one of its stakeholders?

Who are the interested parties?

As stated earlier there are two groups of interested parties: those that are benevolent and support the organization and those that are malevolent and may harm it. Some parties may be both depending on their prevailing interests, for example, a competitor may have a positive impact when it helps draw customers to an area but have a negative impact when it engages in a price war. We can place the stakeholders into five groups: customers, external providers, employees, investors and society – six if we separate regulators – but these are generally agents of society. The labels are not intended to be mutually exclusive. A person might perform the role of customer, supplier, shareholder, employee and member of society all at the same time. An example of this is a bartender. When serving behind the bar he can be an employee or a supplier if he contracts with an employment agency, but on his day off he might be a customer and since acquiring shares in the brewery he became a shareholder. He also lives in the local community and therefore benefits from the social impact the bar has on the community. If he doesn't pay for the drinks he takes, he could also be a criminal. He is a contributor; he affects the outcomes and is affected by those outcomes.

Although stakeholders have the freedom to exert their influence on the organization, before the advent of social media there was often little that one individual could do, whether or not that individual was an investor, customer, employee, supplier or simply a citizen. With the advent of social media an individual can make a complaint that would have otherwise been between two people becomes known worldwide. Reputations can be destroyed in an instant. However, this approach would be detrimental if the individual needed to maintain harmonious relationships. The individual may therefore choose not to exercise such power but lobby their parliamentary representatives, consumer groups, trade associations, etc., and collectively bring pressure to bear that will change the performance of the organization. Equally, an organization can influence stakeholders and potential stakeholders into supporting its products and services and when it becomes a major employer and/or exporter or of strategic significance, it may also influence government into changing legislation in its favour.

Each stakeholder brings something different to an organization in pursuit of their own interests, takes risks in doing so and receives certain benefits in return but is also free to withdraw support when the conditions are no longer favourable.

We address each of the groups of interested parties in more detail later indicating whether their interest is benevolent or malevolent. Both groups of interested parties are relevant to the QMS if their actions can favourably or adversely affect its outputs. If their interest is benevolent they can be expected to be treated as an opportunity and managed by understanding and satisfying their needs and expectations as addressed in Chapter 13. If their interest is malevolent, they can be expected to be treated as a threat and managed by understanding their motives and addressing the risks they present to the organization through countermeasures as addressed in Chapter 21.

Customers

The chief current sense of the word *customer* is a person who purchases goods or services from a supplier (OED, 2013). Turning this around, one might say that a customer is someone who receives goods or services from a supplier and therefore may not be the person who made the purchase. There must be intent on the part of the supplier to sell the goods and services and intent on the part of the receiver to buy the goods and services (a thief would not be classed as a customer as there is no intent to supply). This customer–supplier relationship is valid even if the goods or services are offered free of charge, but in general the receiver is either charged or offers something in return for the goods or services rendered. There is a transaction between the customer and supplier that has validity in law. Once the sale has been made, there is a contract between the parties that confers certain rights and obligations on both.

Customers provide money in return for the benefits that ownership of the product or service brings but may demand refunds if the product does not satisfy the need. They are also free to withdraw their patronage permanently if they are dissatisfied with the service. The UK Telecoms regulator Ofcom stated in the context of broadband services that "competition is only effective where customers can punish "bad" providers by taking their custom elsewhere, and reward "good" providers by staying where they are" (Ofcom, 2006).

Box 11.3   Customer

A person or organization that could or does receive a product or a service that is intended for or required by this person or organization.

(ISO 9000:2015)

The word *customer* can be considered as generic term for the person who buys goods or services from a supplier. In ISO 9000:2015 the term *customer* is defined differently (see Box 11.3). The examples appending that definition are explained as follows:

Consumer	A person who purchases goods and services for personal use. Alternatively, a shopper
Client	The customer of a professional service provider such as a law firm, accountant, consultancy practice or architect.
End user	The person using the products or services that have been purchased perhaps by someone else. This may be the person for whom the products and services were intended but may not be known at the time of purchase.
Retailer	An organization's customer which is offering products to consumers that have been supplied by the organization.
Receiver of product or service from an internal process	A product or service may be provided directly to a customer from an outsourced process which is classed as a process internal to the QMS.
Beneficiary	The customer of a charity.

It is relatively easy to distinguish customers and suppliers in normal trading situations but there are other situations where the transaction is less distinct. In a hospital, the patients are customers but so, too, are the relatives and friends of the patient who might seek information or visit the patient in hospital. In a school the parent is the customer but so, too, is the pupil unless one considers that the pupil is customer-supplied property!

In all cases the term *customer* is used in ISO 9001 without qualification, implying that a requirement could be referring to an internal or external customer. However, there is a note in clause 1.0 which states "the terms 'product' or 'service' only apply to products and services intended for, or required by, a customer" and the term *product* is defined as "output of an organization that can be produced without any transaction taking place between the organization and the customer". The organization being referred to here is the organization to which the QMS applies which means that an internal customer (co-worker) may receive outputs from internal suppliers but does not receive products or services from the organization and is therefore not deemed to be a customer in the context of ISO 9001.

External providers

An external provider is a provider that is not part of the organization that provides a product or a service. They provide products and services in return for payment on time, repeat orders and respect but may refuse to supply or cease supply if the terms and conditions of sale are not honoured or they believe they are being mistreated.

Several terms are used to designate an organization that provides something to another (e.g. supplier, retailer, contractor, dealer, merchant, vendor and service provider). In most of these cases we tend to assume that these are commercial transactions but that is not necessarily so. When TQM became popular in the 1980s, the notion that a worker was a supplier to another worker and a customer of another worker became fashionable even though there was no commercial transaction involved. It is therefore necessary to draw a distinction between internal and external provision as the nature of the risks would be different and consequently different requirements would need to be specified. However, the organization being referred to here is the organization to which the QMS applies which means that an internal supplier (co-worker) may provide outputs to internal customers but does not provide products or services of the organization and is therefore not deemed to be an external provider in the context of ISO 9001. See also Figure P3.2 in the introduction to Part 3.

In ISO 9001:2000 and 2008 outsourcing was treated separately from purchasing but it was realized that outsourcing is just another form of obtaining something from another organization. Also, as one of the aims of the revision to ISO 9001 was to make it more appealing to service organizations, it needed to adopt a more neutral language. The terms *supplier* and *purchasing* therefore needed to be dropped while acknowledging that, depending on the organizational context, they would continue in use by users of the standard.

Employees

Employees are people who work for an employer. In this sense, there is no distinction being made between manager and worker, but there is a distinction being made between people who are on payroll and people who are not. There are some legal issues with the terms *worker* and *employee* which are outside the scope of this book. Employees provide labour in return for pay and conditions, leadership and job security but are free to withdraw their labour if they have a legitimate grievance or may seek employment elsewhere if the prospects are more favourable.

Investors

Investors are those people and organizations that put money into an organization. They include shareholders, owners, partners, directors or banks and anyone having a financial stake in the business. Investors provide financial support in return for increasing value in their investment but may withdraw their support if the actual or projected financial return is no longer profitable.

Society

Society is an association or interaction with or between people and the culture, relationships, laws and economic circumstances distinguish one society from another. Society includes not only the people but the organizations set up on behalf of the people to govern, police and regulate the population and its interrelationships and therefore these bodies are interested parties.

Society provides the employees and the infrastructure in which the organization operates. Society provides a licence to operate in return for employment and other economic benefits to the community they bring such as support for local projects. Society can also censure an organization's activities through protest and pressure groups.

Regulators

We have stated that regulators are agents of society but they deserve a special category as there can be different regulators and not all may affect the QMS. National and local government issue statutes and regulations and use agencies to enforce compliance. The regulators that apply depend on the industry sector in which the organization operates such as those regulating advertising, charities, education, environment, finance, transport, health, law, social care, utilities, etc., and in addition, we can place certification bodies in this category. In general, regulators provide support to organizations in return for compliance but have the power to extract compliance or force closure.

Competitors

A common definition of a competitor is any person or organization which is a rival against another. The rival is commonly understood to be another organization in the same or similar industry which offers a similar product or service, but this is a very narrow definition and addresses only one type of competition. In reality, there are three types of competition:

- competition with similar products and services,
- competition with substitute products and services
- competition for consumer's purchases

Customers don't buy products and services; they buy the benefits they bring. For example, train companies compete with airlines for the same customers, postal services compete with e-mail, a football club competes with a pop concert for the money the customer has to spend on entertainment.

Competitors are not necessary harmful as their presence in a community may have a beneficial effect by bringing more customers into the community, whether they are physically located in one community as is the case in shopping centres or are picked up by the same search engines on the Internet. Competition drives improvement which is good for the stakeholders but it can also create stress and have a negative effect as weaker competitors try to hold their own in the market place.

Media

The media is the main means of mass communication (broadcasting, publishing and the Internet) regarded collectively (OED, 2013). The media may be helpful if the organization which employs the news gatherers is sympathetic to the organization's products and services and the way it serves the community. Favourable commentary in the media may have a beneficial effect but equally unfavourable commentary may adversely affect the organization's reputation whether the commentary is factual.

Politicians

Politicians are persons who are professionally involved in politics, especially as a holder of an elected office (OED, 2013). In a representative democracy, they are elected to represent their constituents which may be the local community, county/state or country. It arguably allows for efficient ruling by a sufficiently small number of people on behalf of the larger number.

Organizations need the support of those who represent them in local and national government so that they may open a channel of communication on any future legislation that may affect them. Organizations also need to express concerns to politicians on issues that are affecting the performance of the organization or will affect its prospects.

Interest groups

An interest group is a non-profit and usually voluntary organization whose members have a common cause for which they seek to influence public policy, without seeking political control. (Business Dictionary, 2016). They generally fall into four categories:

- Economic associations such as chambers of commerce and trade unions.
- Professional associations such as those for engineers, lawyers, doctors, quality professionals, etc.
- Public interest groups such as those campaigning for environmental protection, human rights, and consumer rights.
- Special interest groups such as non-governmental organizations (NGOs) and trade associations.

Organizations may have a mutual relationship with interest groups as a source of information, advice and training services. In general, the relationship is wholly benevolent but their allegiance to their cause may conflict with the commercial interests of the organization.

Pressure groups

A pressure group is a non-profit and usually voluntary organization whose members have a common cause for which they seek to influence political or corporate decision makers to achieve a declared objective. In general, pressure groups seek to change the status quo and may be political, social or religious. They may want to protect the environment, animals, children, the disabled, etc., and seek changes in the law to secure that protection and these changes may affect what the organization is trying to do. The needs and expectations of pressure groups are not relevant to the organization unless their support is needed and only relevant to the organization's ability to provide conforming products and services.

Criminals

Criminals are persons who have committed a crime, which is an action or omission which constitutes an offence and is punishable by law. Criminals can therefore be internal or external to an organization and their interest is solely malevolent. As illustrated earlier with the bartender, the roles interested parties play are not mutually exclusive so any one of them could become a criminal.

Summary

All organizations depend on the support of a wide range of other organizations and individuals to achieve their goals. Interested parties include those from which the organization seeks support but also includes those which have a malevolent interest. Organizations need to identify both types but treat each according to their impact. They should seek to understand the needs of those parties that are benevolent and seek to manage the risks posed by those parties that are malevolent. We identified, defined and analysed several different groups of interested parties and, in some cases, how the terminology had development. We noted that it's necessary to find out what the benevolent parties want in return for the contribution they can make and what risks are posed by those parties which intend to do the organization harm. The determination of interested parties and their requirements is addressed in Chapter 13.

Bibliography

Business Dictionary. (2016, January 12). Retrieved January 12, 2016, from www.businessdictionary. com

ISO 9000:2015. (2015). *Quality Management Systems – Fundamentals and Vocabulary*. Geneva: ISO.

OED. (2013). Retrieved from Oxford English Dictionary: oed.com

Ofcom. (2006, August 17). Broadband Migrations: Enabling Consumer Choice. Retrieved from Ofcom: www.ofcom.org.uk/consultations-and-statements/category-2/migration.

Post, J. E., Preston, L. E., & Sachs, S. (2002). *Redefining the Corporation: Stakeholder Management and Organizational Wealth*. Redwood, CA: Stanford University Press.

Key messages from Part 3

Chapter 6   Quality

1   The definitions in ISO 9000 form part of the requirements of ISO 9001, but terms used in ISO 9001 are not intended to substitute terms used within the standard user's organization.

2   Discussions over the meaning of words may be only semantics but we rely on the meanings we assign to words to influence our interpretations of events and so, ultimately our decisions and our actions.

3   The word *quality* as used in the ISO 9000 family of standards is as the standard or nature of something as measured against other things of a similar kind. It is not about perfection but a degree of excellence.

4   A product or service may not need to possess defects for it to be regarded as poor quality; it may not possess the features that customers expect for the price.

5   The decision as to whether something is of satisfactory quality rests with the receiver, not the producer.

6   The frequency with which the term *quality* is used in ISO 9001 has rapidly declined to avoid it being perceived as pertaining to the work of a quality department or implying perfection or excellence.

Chapter 7   Requirements

7   It's often believed that a requirement is a condition that is specified when, in fact, the way the term is used in ISO 9001 it is as "a need or expectation that is stated, customarily implied or obligatory".

8   A requirement does not need to have been documented for there to be a requirement that has not been met.

9   There are requirements that relate to the product or service to be provided and requirements that constrain how those products and services are to be provided. The latter cannot be treated independently of the former.

Chapter 8   Management system

10   There is little consistency in ISO documents as to whether the term *quality management system* is being used in the sense of how entities interact to produce results, or in the sense of a set of methods for producing results.

11 As organizations become more complex, we have become more aware of the dangers of a silo mentality in which a fix *here* simply shifts the problem to *there* and a fix now gives rise to a much bigger problem *later*.

12 Over the last 60 years or so new theories have emerged which help us understand complexity by looking at situations as wholes using the concept of a bounded system of linked components. This is the essence of the systems approach to management.

13 As a result of basing industrial practices on scientific management, the term *system* has been used predominantly as an orderly way of doing something, and quality management systems have been perceived as a set of documents that tell people what to do.

14 Some parts of organizations function like machines, but although use of such metaphors creates powerful insights, they also become distortions and a way of not seeing.

15 Seeing an organization as a machine implies that it's fixed and can change only if someone changes it, that its actions are reactions to decisions made by management.

16 Both processes and systems are not tangible; they are mental constructs or models and are convenient ways of observing things but they are not reality.

17 What we include in our model is strongly influenced by how we see things, that is, our paradigm which embodies core assumptions that characterize and define our worldview.

18 By including in our model those elements that serve to create and retain customers we establish a representation of reality from the perspective of how the organization manages product and service quality.

Chapter 9   Process and the process approach

19 In the first edition of ISO 9001 it was the word *procedure* that conveyed what organizations needed to document and implement to provide products and services that met requirements.

20 In the 2000 version, it was recognized that outputs were more than the product of activities and the function that performed them but a product of many elements combined in a cause-and-effect relationship.

21 Managing a process is more important than implementing procedures or managing functions. In fact, who does what is of less importance than the results to be achieved and the process by which they are produced.

22 Viewing work as a process rather than a procedure or a function focuses everything associated with that work on the results to be achieved.

23 Dividing results into the processes that will deliver them, rather than dividing work into specialities, is the essence of the process approach and it is the network of processes that holds the *system* together.

Chapter 10   Risk and opportunity

24 The way in which the term *risk* is used in ISO 9001 is in its negative sense because almost everywhere the term is used it is combined with the word *opportunity*.

25 Although there is no requirement for formal methods for risk management, the act of identifying, addressing and determining the effectiveness of the actions taken characterizes management.

26   The term *risk* is commonly used to express the possibility of something bad happening but the ISO definition of risk as the *effect of uncertainty* poses the possibility that the uncertainty may be something good as well as something bad.

27   Someone who is pursuing an opportunity may be taking a risk and someone who by their actions may miss an opportunity is taking a risk.

28   Risk-based thinking is simply the thinking one does by weighing up the potential benefits and harms of exercising one choice of action over another when faced with a possibility of something bad happening, including a missed opportunity

Chapter 11   Interested parties and stakeholders

29   All organizations depend on the support of a wide range of other organizations and individuals to achieve their goals.

30   Interested parties include those from which the organization seeks support but also includes those which have a malevolent interest.

31   Organizations need to identify both types but treat each according to their impact.

32   Organizations should seek to understand the needs of those parties that are benevolent and seek to manage the risks posed by those parties that are malevolent.

33   Although interested parties have the freedom to exert their influence on an organization, before the advent of social media there was often little that one individual could do, but now they can take a complaint worldwide and destroy an organization's reputations in an instant.

Part 4

Context of the organization

Introduction to Part 4

Imagine being invited to advise each one of the following organizations on how they could improve the quality of the products and services they provide:

- A retail bank
- A school for children
- An oil and gas production company
- An aircraft development and production company
- A city hospital
- A local agricultural society

Every one of these organizations is different:

- They operate in different sectors of the economy.
- The natural environment in which they operate is different.
- Their customers have different needs and expectations.
- Their employees possess different abilities and need different competences.
- The scale and nature of the physical resources they use are quite different.
- The scale of their operations is vastly different.
- The laws governing their operations are different.
- The uncertainties each face will be different.
- Their priorities will be different.

The circumstances that form the setting for an event, statement or idea and the terms in which it can be fully understood is referred to as *context*. The context of each of these organizations is different, and therefore only a foolish person would offer advice to any one of these organizations without understanding the context of the organization they are advising. But, hitherto, this was not a specified requirement of ISO 9001. Context was not a factor that featured prominently in the design of management systems. In fact, some organizations created their business model on the premise that a QMS could be cloned, and they offered identical manuals to their clients with the instruction that all that was needed was to change the logo in the page heading.

So why the change? Was it to curtail the ambitions of organizations offering cloned quality documentation? Such organizations will probably see their sales decline for other reasons such as there being no requirement now for a quality manual and documented procedures. In fact, the real reason is more likely to be that explained in Box 12.1.

There are number of reasons for the change:

- Documenting what you do is not what it's about anymore. First, documenting what you do froze the present, thereby creating resistance to change, and second it assumed an organization was a machine with a set of operating instructions, when in fact organizations are living entities responding to influences beyond its boundary.
- It's not about maintaining the status quo. Maintaining the status quo may be appropriate for some parts of an organization but other parts of the organization need to be innovative, embrace change and adapt to a new view of the world so that the organization may flourish.
- Today's problems are caused by yesterday's solutions. The future is uncertain; therefore, the solutions we created in the past were based on what we knew at that time and not on what we would know in the future.
- We live in a connected world. Actions have consequences, and the actions of others yesterday may affect our plans for today, and our actions today may affect the plans of others for tomorrow.

Probably the most significant reason is that the authors of ISO 9001 have realized that a QMS is not a bolt-on system or a system of documents that people use as shown in Figure P4.1, but a series of interacting elements that produce the organizations outputs as shown Figure P4.2. When driving a car, we don't drive towards our destination by looking behind to see where we have been and what mistakes we have made; we look ahead to see where we are going and what may impede or facilitate progress towards our destination. This is intuitive and so it should be in the pursuit of quality that we understand the context of our organization and adjust our course of action so we are able to avoid the obstacles and exploit those opportunities that are conducive to our success.

In Figure P4.2 the grey blobs represent the provisions that respond to the requirements of ISO 9001 and become elements of the QMS. This picture is merely symbolic as there would be far more of these than shown and even far more white blobs and interconnections in a real organization than shown.

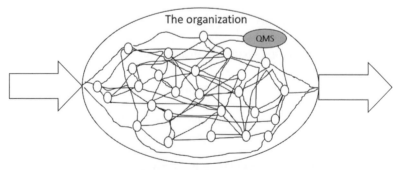

Figure P4.1 Bolt-on QMS

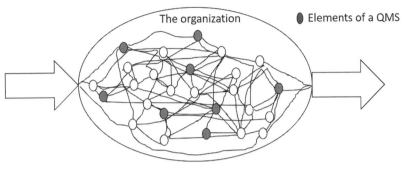

The external environment

Figure P4.2 Integrated QMS

As stated in the definition of a QMS in ISO 9000:2015 clause 3.5.4, the QMS is "part of a management system" but it is only a part which is formed from elements that have the common objective of enabling the organization to consistently provide products and services that meet customer and applicable statutory and regulatory requirements.

Although clause 4 of ISO 9001:2015 carries the title "Context of the organization" it contains four sub-clauses, each of which is related to the context of the organization but not all form part of the context. A case in point are the requirements to determine the scope of the QMS and establish, implement, maintain and improve the QMS, all of which are activities that should be carried out after understanding the context of the organization and not during it or before it. The requirements in clauses 4.1 and 4.2 in particular are more pertinent to the context of the organization than the others, but these should not be treated in isolation because one will need to understand the needs of interested parties before confirming that the stated purpose and direction is appropriate. It is also the case that some of the risks identified as a result of applying clause 6.1 will be derived from the same issues that affect the organization's ability to achieve the intended results of its QMS that are required to be determined in clause 4.1. There is also a statement from ISO that is buried in an obscure guide for standards writers using Annex SL (see Box P4.1). A video on quality and the context of the organization is available on the companion website.

Box P4.1   No implementation sequence is implied

There is no inherent assumption of sequence or order in which requirements are to be implemented by an organization. There is no inherent demand that all activities in a specific clause must be done before activities in another clause are started.

(ISO/TMB/JTCG N 359, 2013)

However, to remain consistent with the sequence of requirements, this part of the Handbook addresses the following clauses, each in a separate chapter but with cross-references to the others as appropriate:

- Clause 4.1 Understanding the organization and its context (see Chapter 11)
- Clause 4.2 Understanding the needs and expectations of interested parties (see Chapter 12)
- Clause 4.3 Determining the scope of the quality management system (see Chapter 13)
- Clause 4.4 Quality management system and its processes (see Chapter 14)

Bibliography

ISO/TMB/JTCG N 359. (2013, December 03). JTCG Frequently Asked Questions in Support of Annex SL. Retrieved from Annex SL Guidance documents: http://isotc.iso.org/livelink/livelink?func=ll&objId=16347818&objAction=browse&viewType=1.

12 Understanding the organization and its context

Introduction

Box 12.1   Not as new as one might think

Introduced as a new requirement in ISO 9001:2015 the introduction to the 2008 version did make the point that an organization's quality management system is influenced by its business environment and changes in that environment and was in effect suggesting that such influences were determined. I suggested in the sixth edition of this Handbook that auditors ask "What analysis has been conducted to determine the impact of changes in the business environment on your quality management system?" and those who did this were preparing their clients for the 2015 revision.

We should not set an objective and proceed to achieve it oblivious of what might happen that will affect what we are trying to do. The journey towards an objective is a journey we won't have taken before. For those organizations whose objective is to continue supplying the same products and services in the same markets, this statement may seem ridiculous. They may have taken this journey many times before without a problem, and therefore every time they receive a customer order they simply do what they have always done; they implement their proven policies and practices, and the customer is delighted with what they receive. However, every time they embark on that journey they are either consciously assuming that nothing has changed or will change during the course of their journey or they are simply oblivious to such changes. Eventually they might find their customers have found a supplier that offers something different that attracts them

Although the title of clause 4.1 is "Understanding the organization and its context" there is no actual requirement that contains these words, only a requirement "to determine external and internal issues". However, the result of undertaking an analysis to identify the factors that affect the ability of the organization to deliver its intended results and determining how significant they are should lead to an acute understanding of the context of the organization.

Many of the requirements are linked to clause 4.1 as shown in Figure 12.1.

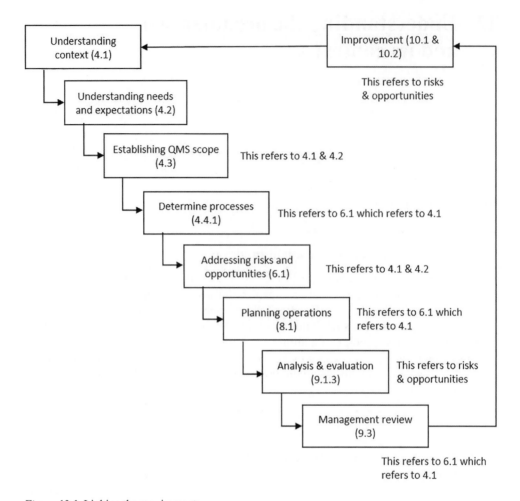

Figure 12.1 Linking the requirements

In this chapter, we examine the two requirements of clause 4.1, namely:

- Determining external and internal issues
- Monitoring and reviewing external and internal issues

Determining external and internal issues (4.1)

What does this mean?

The requirement contains several terms that need to be explained for its significance to be understood. These are external and internal issues, organization purpose, strategic direction and the intended results of the QMS.

External and internal issues

The word *issues* is often used to denote something that is a significant problem; therefore, the neutral word *factor"* would have been a better choice in this case, but Note 1 to this requirement explains that issues can include positive and negative factors or conditions for consideration, which is helpful. We will identify a range of issues later.

Organization purpose

The issues being identified are to be relevant to the organization's purpose which will prompt the question "what is our purpose? An organization's purpose is the reason for its existence but there will be differing views on this depending on how a person sees the organization; for example:

- The owner may believe the business exists to make money.
- The nurse may believe the hospital exists to heal the sick.
- The sales executive may believe the organization exists to put a computer on everyone's desk.

From the owner's viewpoint, making enough money to sustain the business is indeed important not only for the business but for society because it's the economic surplus from business that provides public services in society (see also Box 12.2). Drucker emphasizes that the function of a business is economic performance, which is probably what most people really mean when they say its purpose is to make money. Drucker also draws a distinction between business purpose (why it exists) and business function (what it does) and says that there is only one valid definition of business purpose and that is to create a customer (Drucker, 1974).

In this example, the sales executive is under the impression that his customers want desktop computers and therefore assumes the company is in the desktop computer business. But this way of thinking can be dangerous. When buggy whips were no longer needed some companies that made buggy whips went out of business because they couldn't see how they could put their skills to a new purpose. Seeing the purpose of a business beyond the products and services it currently produces is more likely to secure its survival.

Box 12.2   Drucker on business purpose

Business exists to supply goods and services to customers and economic surplus to society, rather than to supply jobs to workers and managers, or even dividends to shareholders. Jobs and dividends are necessary means but not ends. The hospital does not exist for the sake of doctors and nurses, but for the sake of the patients whose one and only desire is to leave the hospital cured and never come back. The school does not exist for the sake of teachers, but for the students.

For a management to forget this is mismanagement.

(Drucker, 1974)

It is not the products and services an organization produces that define its purpose but the value its customers derive from possessing them, and as this changes, so will the products and services the organization offers in response in attempting to deliver that value. It may have been desktop computers yesterday, but today it is pocket computers and tomorrow it may be wearable computers. But whether this business is in the computer business or the information business will depend on how its owners view it.

In organizations where economic performance is a constraint such as a hospital, church, university, charity or armed services the function is different. Their function is social performance rather than economic performance, but instead of there being customers that provide revenue, its taxation or private donations, fees, grants, subscriptions, etc., that fund the social service they provide. The nurse's perception of a hospital's purpose as healing the sick is therefore nearer the mark.

Box 12.3   Purpose, strategic direction and alternative terms

Mission = organization's purpose for existing as expressed by top management

(ISO 9000:2015)

Purpose = mission = reason for existing = what we do
Vision = aspiration of what an organization would like to become as expressed by top management

(ISO 9000:2015)

Strategic direction = vision = aspiration = where we are going
Values = The principles or moral standards held by a person or social group = what we stand for

---

Some organizations use the term *mission* to express their purpose, their reason for existence, but it does have a slightly different meaning. If we perceive business purpose as economic performance, that will never change; but if its purpose is to create a customer, what its customers value will change, so its mission will change in response to changes in the external environment. Therefore, statements of purpose or mission should be developed from the perspective of an organization's customers.

Strategic direction

A strategy is a broad plan of action; it's how the organization intends to accomplish its purpose or mission. However, a strategic direction is even broader. Of all the alternative paths that could be taken to fulfil our purpose or mission, the one that is chosen is the *strategic direction*. The term *vision* is used to express an organization's aspirations, what they want to become and therefore may also be used to express their strategic direction. However, ISO 9000:2015 clause 2.2.3 introduces some confusion by stating that "ways in which an organization's purpose can be expressed include its vision, mission, policies and objectives" but it doesn't mention strategic direction.

Box 12.4   When direction doesn't matter

"Would you tell me, please, which way I ought to go from here?" said Alice "That depends a good deal on where you want to get to," said the Cat. "I don't much care where –" said Alice. "Then it doesn't matter which way you go," said the Cat.

(Carroll, 1865)

A strong vision will connect goals to the company's underlying values and will make it more understandable about how to achieve each goal.

Intended results of the QMS

As explained Chapter 8, we revealed the inconsistencies in how several ISO documents explain what a QMS is, but by a process of deduction we concluded that a QMS is not a set of requirements or procedures but *a systemic view of an organization from the perspective of how it creates and retains customers.* The systemic view is a model created from viewing the organization as a goal-seeking system. As one of its goals is consistent provision of products and services that meet customer and applicable statutory and regulatory requirements, it's therefore likely that this is one of the intended results of a QMS. On studying ISO 9001 clause 0.1, other intended results may include:

- Demonstrated conformity to ISO 9001
- Enhanced customer satisfaction

In addition, there will be consequential results such as may be produced by the way the primary goals are achieved such as the satisfaction of stakeholders other than the customer.

Why is this necessary?

The primary reason why it is necessary to determine these issues is so that the organization has prioritized the appropriate actions and is thereby equipped to function as a system of appropriately interconnected elements to carry them out and achieve those results. This alignment is illustrated in Figure 12.2.

As stated in the introduction to this part of the Handbook every organization is different and it's the way that such issues affect a particular organization that makes its QMS unique. An organization's QMS cannot be cloned and expected to function effectively in another organization. Some routines will work effectively in other organizations because they each face the same situation but there are too many variables to expect all the strategies, policies, objectives and processes of one organization to work equally well in another organization unless it is cloned as it may be with a franchise operation such as McDonalds or KFC, but not all franchises are clones.

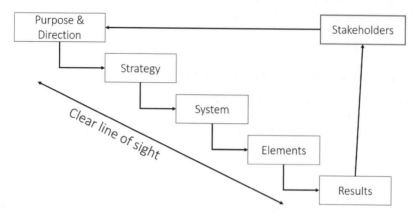

Figure 12.2 Alignment with purpose and direction

How is this addressed?

This new requirement recognizes that all organizations will be formed for a specific purpose, that they will be pursuing a particular strategy and be aware of some of the issues that may act as either drivers or barriers to fulfilling their goals. The business purpose, strategy and the identification of relevant issues are all outputs of a strategic planning process which may look something like that shown in Figure 12.3.

To develop an effective QMS it's necessary to take a wide view and understand the context of the organization system of which the QMS is a subsystem. By looking for issues that only affect the QMS, without having a broader perspective, there is a distinct possibility that significant influencing factors may be overlooked.

Clarifying purpose and direction

The first step is to clarify the purpose and direction of the organization and this may be revealed by seeking answers to some basic questions:

1   What do we do? This addresses the purpose for the organization's existence and what it seeks to accomplish (e.g. we run a national chain of fast food outlets).
2   How do we do it? This addresses the main method or activity through which the organization tries to fulfil its purpose (e.g. we prepare and serve safe and nutritious food to take away or for consumption on our premises that are located in strategic positions).
3   For whom do we do it? This addresses the target market for the organization's services (e.g. we serve those people seeking an inexpensive tasty meal who don't have the time to wait for it to be prepared).
4   What's critical to our success? This reveals the few things that must go well to ensure success (e.g. location, food safety, price and service quality).

The examples in (4) relate to a national chain of fast food outlets but if your organization is an airline, critical factors would include passenger safety, aircraft maintenance and aircrew competence.

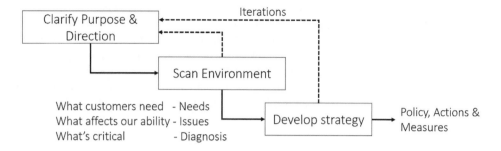

Figure 12.3 Strategic planning process outline

Clearly we are already reaching into ISO 9001 clause 4.2 to address this requirement because we cannot confirm the organization's purpose and direction without understanding the needs of customers we expect to buy our products and services which is why iteration loops are shown in Figure 12.3.

Scanning the environment

The second step is to scan the external and internal environments and this will involve:

- confirming who the stakeholders are;
- assessing changes in stakeholder needs relative to the organization's purpose and direction;
- confirming stakeholder success measures (i.e. what they will look for as evidence their needs are being met);
- identifying the external and internal factors or issues that may impede or facilitate fulfilment of purpose and direction.

Several techniques have emerged to identify these issues. The one most often used for identifying external factors is a PESTLE analysis and for internal factors, the other is a SWOT analysis (see http://pestleanalysis.com). However, neither PESTLE nor SWOT analysis provide a picture and neither show the systemic relationships between the various components, which is critical as we are modelling a system. PESTLE is a big picture tool but trying to show how these factors might interact within a 4 × 4 grid is well-nigh impossible, and it is those interactions that are likely to affect the organization the most. This calls for a different technique.

EXTERNAL SYSTEM MAP

In Chapter 8 on the subject of structure, there is an external system map (Figure 8.3) which was produced by seeking answers to the following questions:

a)   What or who in the external environment influences what we are trying to do?
b)   What's the relationship between the source of these influences?
c)   What's the nature of these influences?

In building the external system map we place the organization in the centre and through brainstorming in the management team identify the external interested parties with which the organization has a relationship, whether it's directly or indirectly. Within this list are some parties whose influence is more significant than others, and it is these that should feature in the map. Next, you work out how these parties are connected to your organization and to each other as this affects your relationship with them. Lastly include any significant influence there may be from the natural environment or infrastructure. Now we have a framework where we can use PESTLE analysis to determine the nature of the influences.

PESTLE ANALYSIS

PESTLE (Political, Economic, Social, Technological, Legal and Environmental) analysis measures the market relative to a particular organization or business proposition. It serves to identify what is going on in the external environment that could affect the future direction of the organization. The significance of the six factors may vary depending on the nature of the business. where, for convenience, the six factors have been merged into four. An example analysis using a fast food outlet chain in presented in Box 12.5.

Box 12.5   PESTLE on fast food outlets

Situation: Facing pressure to adapt to changing market conditions created by the obesity crisis and the rise of aggressive competitors.

Political: Government policies introduced to tackle the obesity crisis.

Economic: Declining customer base and less money to finance extensive research and developments efforts like our competitors do.

Social: Change in eating habits of younger people as they seek more nutritious and healthy food. Pressure groups succeeding in influencing planning authorities to reject applications for fast food outlets.

Technological: New contactless payment mechanisms require all outlets to be re-equipped with latest card reading devices.

Legal: New food hygiene and food waste regulations. Planning regulations prohibiting fast food outlets near schools. New living wage.

Environment: Several outlets located in areas exposed to flood risk.

INTERNAL SYSTEM MAP

In Chapter 8 on the subject of structure, there was an internal system map (Figure 8.3). This is a high-level description of the organization as a system which was produced by seeking answers to the following questions:

a)   What affects our ability to fulfil our purpose and pursue our strategic direction that we can influence?

b)   How are these components interrelated?

c)   How are these components related to the source of influence in the external environment?

The answers will be all the tangible and intangible components that constitute the organization as a system. Anything the organization is unable to influence or has little influence over should be in the external environment. Organization complexity is such that not everything can be included while at the same time retaining a coherent system map; therefore, only the significant connections are shown. If more detail is necessary to pinpoint the issues, create a larger scale map for that aspect. Now we have a framework where we can use SWOT analysis to determine the nature of the influences.

SWOT ANALYSIS

The SWOT (Strengths, Weaknesses, Opportunities and Threats) analysis looks at the organization itself. The PESTLE affects the SWOT but not vice versa. Without a clear understanding of an organization's strengths, weaknesses, opportunities and threats business plans may fail, goals will be missed and new product or service development programmes will fail to live up to their potential. The SWOT is akin to a capability assessment. The result enables management to act in a manner that does not leave the organization vulnerable. Strengths and weaknesses are internal to your organization, whereas opportunities and threats are external. The results are often very subjective and will vary depending on who does the analysis. SWOT should be used as a guide but use of current performance data and weighting factors can improve its validity.

The identification of strategic issues is the heart of the strategic planning process. Bryson covers strategic issues in depth and suggests that these issues fall into three main categories: (Bryson, 2004):

a)   Current issues that probably require immediate action.
b)   Issues that are likely to require action in the near future but can be handled as part of the organization's regular planning cycle.
c)   Issues that require no action at present but need to be continuously monitored.

The analysis can be performed at the enterprise, business and operational levels. At the enterprise level, the intended results of the QMS are one of the objectives to be achieved. At the business level, the intended results of the QMS will be the primary objective to be achieved, and at the operational level the intended results of the QMS concerned will be the intended results of the processes.

Box 12.6   SWOT of a chain of fast food outlets

Strengths

- Menus popular with young people;
- Some good locations;
- Strong brand image.

Weaknesses

- Inability to match sales volume of competitors;
- Premises in need of refurbishment;

- Heavy dependence on imported ingredients;
- Poor reputation for nutrition and healthy foods.

Opportunities

- New legislation on fighting obesity enables introduction of innovative food options;
- Increase in working from home creates demand for a home delivery service;
- Projected new shopping developments create opportunity to relocate vulnerable outlets.

Threats

- Some premises vulnerable to flooding;
- Today's parents are becoming fussy about what their children eat;
- Competitors have a head start in healthier food choices.

The standard does not require a PESTLE and SWOT Analysis nor any documented information, but it would be difficult to monitor and review these issues unless there was a record of what was found the last time the impact of internal and external issues was discussed.

It is important when using these tools to:

- identify the relevant factors that apply to your organization, the business unit or process;
- rate your organization, business or process relative to the factors;
- draw conclusions from this information relative to the declared purpose and direction, that is, is it critical or none critical to success (see the diagnosis later);
- validate these conclusions with others.

Identifying relevant issues

The analysis is done as part of strategic planning and therefore factors that affect the organization may not all affect the QMS. This should result in QMS development being perceived as part of strategic planning rather than as something done in isolation. If this analysis is performed at a key stage in the strategic planning process, it provides an opportunity to filter the results through the intended results of the QMS to reveal issues that are relevant to the QMS as shown in Figure 12.4.

As part of the strategic planning you might analyse competitor products or services and benchmark inside and outside the industry. There are many books and organizations you can turn to for advice on benchmarking. With benchmarking, you analyse your current position, find an organization that is performing measurably better and learn from them what they are doing that gives them the competitive edge. You then set objectives for change as a result of what you learn.

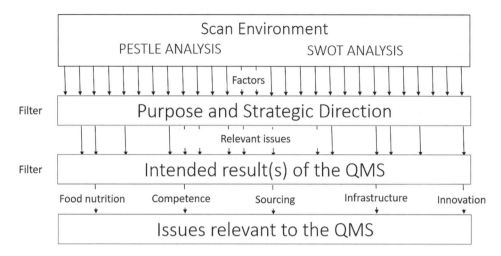

Figure 12.4 Identifying issues relevant to the QMS

It is imperative that the reasons why each of the issues represents a point of contention or significance are determined. This is the diagnosis and it should explain the nature of the challenge. A good diagnosis simplifies the often overwhelming complexity of reality by identifying certain aspects of the situation as critical. For example, it is pointless to identify food nutrition as being a weakness unless the specific processes have been identified and the root cause of their weakness found. It is also necessary for the factors such as management commitment and resource availability, that will turn food nutrition into a strength, to be also strengths and not weaknesses; otherwise, there will be tremendous difficulty in bringing about the change desired.

How is this demonstrated?

Demonstrating that relevant external and internal issues have been determined may be accomplished by:

a) presenting evidence of a process for determining:

    i how these issues are identified;
    ii how their criticality is determined;
    iii who is involved and what their responsibilities are;
    iv how the information is used to effect changes in the QMS.

b) presenting evidence that the organization's purpose and strategic direction have been communicated to those undertaking the analysis;
c) showing how the intended results of the QMS that have been determined align with the stated purpose and direction of the organization;
d) presenting evidence of the analysis that was undertaken to understand the context of the organization and what knowledge was gained from it;
e) selecting several identified issues and showing how it was decided which were critical and required action and which required no action;

f)   showing how the identified issues have been addressed in the policies and processes of the QMS (i.e. how the issues filter down to the point where they affect what people do).

Monitoring and reviewing external and internal issues (4.1)

What does this mean?

Monitoring and reviewing information about external and internal issues means observing changes in the information gathered when reviewing the system maps and revisiting the associated analysis and reviewing it to determine whether the changes have any significant effect on the result of the analysis. The monitoring is on-going – it's not a one-off event.

Why is this necessary?

The significance of external and internal factors continually changes because they are affected by changes in the global environment and therefore the system analysis needs to be reviewed periodically. How often depends on the volatility of the market sector in which the organization operates. In sectors where the technology or economic conditions change rapidly, the reviews will need to be performed more frequently. It will also be necessary to review these factors when there are changes in the organization that affect its strengths and weaknesses

How is this addressed?

There needs to be continual checking of the context of the organization, the issues and the risks and opportunities that they present.

In choosing the method of monitoring consideration needs to be given to the susceptibility of the data to change and the frequency of observation set accordingly. If there is a high number of issues, monitoring them all – even at different frequencies – may become a burden and not cost effective on more than an annual or once every-five-years basis. Consideration also needs to be given to responsiveness to change. Information provided by public bodies can lag behind the change by years and cannot be as responsive as someone on the scene of the action. It is important to retain information on issues so that comparisons can been made between periods to detect trends.

Knowledge about external factors can be gained through the media, news and current affairs and through web feeds or RSS (Rich Site Summary) for which content providers offer a subscription service. This enables subscribers to be notified of changes when they occur without having to visit the website. With critical factors, the data might be presented in chart form so that trends may be observed as is the case with the stock market, commodities and interest rates. Citing declining popularity, certain sites have reduced or removed support for RSS feeds. However, as of August 2015, Mozilla Firefox and Internet Explorer include RSS support by default, whereas Google Chrome and Microsoft Edge do not.

Knowledge about changes in internal factors is not as easy to monitor due to it being more apocryphal. Employees and managers in particular can acquire an inflated impression of the organization's strengths and weaknesses, often formed years ago and not recalibrated their impressions since they were first formed. What is needed is an unbiased means of determining strengths and weaknesses as indicated above.

How is this demonstrated?

Demonstrating that information about the relevant external and internal issues is being monitored and reviewed may be accomplished by:

a) presenting evidence of a process for determining:

    i how relevant external and internal issues are to be monitored;

    ii how criteria are established for alerting relevant managers of change;

    iii how often and by whom the information is to be reviewed.

b) selecting a representative sample from the initial analysis and showing that issues deemed critical are being monitored and that the results are being reviewed as planned;

c) showing how changes in the information gathered has changed the QMS.

Bibliography

Bryson, J. (2004). *Creating and Implementing Your Strategic Plan: A Workbook for Public and Non-Profit*. San Francisco, CA: Jossey Bass.

Carroll, L. (1865). *Alice's Adventures in Wonderland*. London: Macmillan & Co.

Drucker, P. F. (1974). *Management, Tasks, Responsibilities, Practices*. Oxford: Butterworth-Heinemann.

ISO 9000:2015. (2015). *Quality Management Systems – Fundamentals and Vocabulary*. Geneva: ISO.

13 Understanding the needs and expectations of interested parties

Introduction

Organizations need to attract, capture and retain the support of those organizations and individuals they depend upon for their success. All these are important but some are more important than others.

Organizations cannot survive without customers. Customers are one of the stakeholders, but unlike other stakeholders they bring in revenue which is the lifeblood of every business. Consequently, the needs and expectations of customers provide the basis for an organization's objectives, whereas the needs and expectations of the other stakeholders constrain the way in which those objectives are achieved. It follows therefore that the other stakeholders (investors, employees, suppliers and society) should not be regarded as customers as it would introduce conflict by doing so.

However, an organization ignores any one of these stakeholders at its peril which suggests that there must be a balancing act. There is a view that the needs of stakeholders must be balanced, as was discussed in Chapter 9, because it is virtually impossible to satisfy all of them, all the time. Managers feel they ought to balance competing objectives when in reality it is not a balancing act as that implies there is some give and take, win/lose, a compromise, a trade-off or reduction in targets so that all needs can be met. Organizations do not reduce customer satisfaction to increase safety, environmental protection or profit. The organization must satisfy its customers; otherwise, it would cease to exist, but it needs to do so in a way that satisfies all the other stakeholders as well – hence the cliché "customer first". If the organization cannot satisfy the other stakeholders by supplying X, it should negotiate with the customer and reach an agreement whereby the specification of X is modified to allow all stakeholders to be satisfied. If such an agreement cannot be reached the ethical organization will decline to supply X under the conditions specified.

Box 13.1   Revised requirement on stakeholders

ISO 9001:2008, required only customer and applicable statutory and regulatory requirements to be determined and met. This remains the case in the 2015 version, but in addition the requirements of those interested parties that are relevant to its QMS are to be determined and considered due to their potential impact on the organization's ability to consistently provide products and services that meet customer and applicable statutory and regulatory requirement.

In practice, the attraction of a sale often outweighs any negative impact upon other stakeholders in the short term with managers convinced that future sales will redress the balance. Regrettably, if unrestrained this approach ultimately leads to unrest and destabilization of the business processes as employees, suppliers and eventually customers withdraw their stake. In the worst-case scenario, it results in destabilizing the world economy as the credit crisis of 2008 demonstrated. The risks must be managed effectively for this approach to succeed.

In Chapter 11 we covered in some detail the subtle differences between interested parties and stakeholders, what they were and who they were defining the various sub-categories and the importance of each.

In this chapter, we examine the three requirements in clause 4.2 which relate to interested parties namely:

*   Determining the interested parties and their effect on the organization
*   Determining the requirements of interested parties
*   Monitoring and reviewing information about the interested parties

Although these are three separate requirements, to avoid repetition, the first two will be addressed together.

Determining the interested parties, their effects and their requirements (4.2a and b)

What does this mean?

There will be interested parties that are not relevant to the organization because they have nothing in common such as a customer seeking insurance services from an organization that is not in the insurance business. There may be interested parties with requirements that are relevant to the organization but not deemed relevant to the organization's products and services (e.g. employment legislation). However, this is a grey area because customer satisfaction is influenced by a diverse range of factors. Although employment legislation may not affect the product or service offered directly, if the company uses suppliers that employ child labour, it may result in their dissatisfaction with the company were they to learn about it. However, in this particular requirement reference is made only to products and services and not to enhancing customer satisfaction which may be an oversight as elsewhere in the standard (clauses 5.1.2, 6.2.1, 9.1.2 and 10.1) the organization is required to enhance customer satisfaction.

The organization seeks to determine and satisfy the needs and expectations of stakeholders, but there are other parties that intend to harm the organization, and therefore the organization does not seek to determine their requirements but seek to determine how it may counter their efforts to harm the organization. There is another group of interested parties who may help or harm the organization depending on whether their aims coincide with those of the organization. Therefore, the interested parties of concern are those that could positively or negatively affect the organization's ability to consistently provide products and services that meet customer and applicable statutory and regulatory requirements.

Box 13.2   Parties that help and those that harm

All interested parties influence the organization but in different ways. So as to avoid confusion we will refer to the benevolent parties (those that help the organization) as having requirements that need to be satisfied and malevolent parties (those that can harm the organization) as having intentions that have to be managed.

Why and when is this necessary?

The first part of this requirement contains the reason why it is necessary to determine the interested parties, and it is because of their effect or potential effect on the organization's ability to satisfy its customers. What is ambiguous is that it limits the interested parties to those that are relevant to the QMS as if there will be interested parties that can affect the organization's ability to consistently provide products and services, etc., but not affect the QMS.

Whereas in clause 8.2.2 there are requirements for determining requirements for the products and services to be offered to customers, the requirement in clause 4.2a goes further and aims to build a bigger picture and identify the interested parties that are enablers and disablers of the organization's ability to consistently provide products and services. Because they are enablers and disablers, the influence of interested parties that needs to be developed or discouraged should be determined and addressed well in advance of the stage where customer enquiries begin to come in because at that stage it will be too late to change their influence. It is therefore an activity performed during strategic planning.

It follows therefore that organizations must try to understand better the requirements and intentions of their interested parties then deal with them ahead of time rather than learn about them later.

How is this addressed?

If starting from the position of having a QMS certificated to ISO 9001:2008, it will be necessary to confirm the needs and expectations of the type of customer that the organization seeks to attract. In addition, it will now be necessary to ascertain the requirements and intentions of the other interested parties. To succeed, organizations need to build loyal, mutually beneficial relationships with all their stakeholders, but they also need to manage the interests of other parties to ensure their interest does not have a detrimental effect.

Identifying the relevant requirements of relevant interested parties

By posing a few key questions, an analysis can be carried out and presented in the form of a table as shown in Table 13.1. This table includes only a few of the interested parties to illustrate the technique.

Table 13.1 Analysis of interested parties to reveal relevant outcomes

Interested party	Nature of interest	Effect	Bias	Outcomes
Who are the parties on which we depend for our success or which threaten our success?	*For what reason do they have an interest in the organization?*	*What effect will their interest have on the organization's ability to satisfy its customers*	*Is this interest benevolent or malevolent*	*What are their requirements or intentions relative to the organization's purpose and strategic direction?*
Customers	To seek products and services that satisfy their needs	Provides opportunities to understand and satisfy customer needs or understand why they don't choose what your organization offers	Benevolent	Quality products and services On-time delivery Value for money Assurance A mutually beneficial relationship
External providers	To provide products and services that satisfy their needs	Provides the organization with the capability to create customers and satisfy their needs	Benevolent	On-time payment Certainty, integrity A mutually beneficial relationship
Employees	To utilize and develop competences	Provides the organization with the capability to create customers and satisfy their needs	Benevolent	Security, safety, integrity Good pay & conditions Job satisfaction A mutually beneficial relationship
Society incl. regulators	To create wealth without adversely affecting the environment	Attract resources and constrain activities	Benevolent	Employment opportunities A mutually beneficial relationship Protect the environment
Competitors	To discover our strengths and weaknesses	Reduce market share or take us over	Benevolent / Malevolent	Compete on price, delivery or product/ service features
Interest groups	To attract support for their campaigns	Alter legislation so as to be more favourable	Benevolent	Laws changed in organization's favour or against its interests
Criminals	To steal property or cause operations to cease	Disruption of operations leading to delayed delivery or suspension of services	Malevolent	Pursue cyber-attack, burglary, data theft or, vandalism, money laundering, etc.

CUSTOMERS

Determining customer needs and expectations to many is very different from determining customer requirements. The former implies that the organization should be proactive and seek to establish customer needs and expectations before commencing the design of products and services and offering them for sale. The latter implies that the organization should react to the receipt of an order by determining what the customer wants. However, from the definition in Box 7.1 we can see that the term *customer requirements* as used in ISO 9001 is not limited to a customer's specified requirements.

The organization's priority is to provide products and services that meet customer requirements, but they evolve and as a result an organization may no longer have the ability to meet them. Clearly in a contractual situation, any changes will be subject to mutual agreement, but when an organization sets out to create a demand it scans the environment in which it operates to establish customer buying behaviour, preferences and trends. It may miscalculate and begin to develop products and services that are not quite aligned with these preferences and trends or not select the most appropriate distribution channels for the products and services, and therefore the organization's ability to consistently provide conforming products and services will be compromised.

It may be useful to divide customers into different groups such as clients, wholesalers, distributors, retailers, consumers, end users and beneficiaries, as the influence each will have on the QMS will vary and it may warrant different techniques to determine their requirements (needs and expectations).

To discover what customers will require of the products and services they will purchase, you need to discover why people buy one product or service over another competing product or service. The answer requires an understanding of customer behaviour which is the process by which consumers and business to business buyers make purchasing decisions. This is a complex area of study which for consumers involves cultural, social and family influences and for business to business buyers involves economic, political, regulatory, technological and ethical influences. Boone and Kurtz's seminal work on marketing provides an in-depth coverage of the subject (Boone & Kurtz, 2013).

The marketing process is primarily concerned with finding out what customers want or what they could be enticed to want and attracting them to the organization so that their wants are satisfied. In this process, it is important to keep the organization's purpose and strategic direction in focus because all too easily, the organization may become entangled in pursuing opportunities that others may be far better equipped to satisfy. There are millions of opportunities out there. The key is to discover those that your organization can exploit better than any other and generate wealth in society.

To determine customer needs and expectations some key questions need to be answered:

* Who are our customers? (These are those persons or organizations that could or do receive your products and services.)
* Where are our customers? (This is the geographic and social areas from where their enquiries and orders originate.)
* What do they want from us? (This is what we think they want.)
* What do they want to achieve? (This is what we think they want to achieve.)
* What is value to the customer? (This is the benefit they tell us they want from our products and services.)

- What do our products and services do for our customers? (This is what we think they get).
- What do our customers get from our products and services? (This is what they tell us they get.)
- Which of the customer's wants are not adequately satisfied? (This is what they tell us we are not providing.)

The answers to these questions will enable marketing objectives to be established for:

- existing products and services in present markets;
- abandonment of obsolete products, services and markets;
- new products and services for existing and new markets;
- service standards and service performance;
- product or service standards and product or service performance.

The results of market research will be a mix of things. It will identify:

- enhancements to existing products and services;
- new potential customers for existing products and services;
- new potential markets;
- opportunities for which no product or service solution exists;
- opportunities for which no technology currently exist.

The organization needs to decide which of these to pursue and this requires a process that involves all the stakeholders. A process for developing the marketing strategy that only involves the marketers will not exploit the organization's full potential. The contributions from design, production, service delivery, legal and regulation experts are vital to formulating a robust set of customer requirements from which to develop new markets, new products and new services. The research may identify a need for improvement in specific products or services or a range of products or services, but the breakthroughs will come from studying customer behaviour. For example, research into telecommunications brought about the mobile phone, and technology has reduced it in size and weight so that the phone now fits into a shirt pocket. Further research on mobile phones has identified enhancements such as access to e-mail and the Internet, even TV through the mobile phone, but whether all these are essential improvements is debatable. To circumvent driving laws a breakthrough has arisen that eliminates manual interaction so that the communicator is worn like a hat, glove or a pair of spectacles, being voice activated and providing total hands-free operation. However, drivers still have accidents because conversations cause distractions. The solution is to take the driver away from the wheel and with robotics, this is now a possibility and will eventually make hands-free driving obsolete technology.

EXTERNAL PROVIDERS

Organizations depend on their external providers, and their ability to provide products and services that meet the organization's requirements clearly affect an organization's ability to satisfy their customers. The affect is intended to be always positive due to their mutual dependency, but relationships can be strained and problems in the supply chain can have a significant negative effect and compromise an organization's obligation to its customers. For this reason, it is essential that organizations understand the needs and expectations of their external providers which may include payment on time, information sharing, advanced notice of changes and help when requested.

Because of the shift from transactional-based marketing to relationship-based marketing, some organizations involve their external providers in their future planning to discover issues that may affect their effective implementation. This may require modification of those plans to synchronize the plans of the organization with those of external providers.

It may be useful to divide suppliers into manufacturers, service providers, consultants and contract labour as the influence and requirements each will have on the QMS will vary. The primary providers will be those supplying items that are used or embodied into the products and services offered to customers. However, there will be secondary providers whose performance can affect the core processes if it's not to the level expected such as cleaning services, equipment maintenance services, catering services (e.g. an outbreak of food poisoning in the staff canteen can disrupt customer deliveries).

EMPLOYEES

Organizations depend on their employees and the competences they bring to the organization can have both positive and negative effect on the ability of the organization to satisfy their customers. Their interpersonal skills can enhance the effectiveness of teams but when these skills are lacking they can erode the effectiveness of teams.

The very idea that employees should be satisfied at work is a comparatively recent notion, but clearly employee dissatisfaction leads to lower productivity. The measurement of employee satisfaction together with the achievement of the organization's objectives would therefore provide an indication of the quality of the work environment (i.e. whether the environment fulfils its purpose).

Many companies carry out employee surveys to establish their needs and expectations and whether they are being satisfied. It is a fact that unsatisfied employees may not perform at the optimum level and consequently product or service quality may deteriorate.

It may be useful to divide employees into managerial, professional, trade, clerical, part-time, cleaners, etc., as the influence each will have on the QMS will vary.

INVESTORS

Organizations depend on their investors to fund growth, development and improvement programmes. While the funding continues, their contribution is positive but on occasions funding may be withheld or withdrawn and thereby delay improvement initiatives. This will consequently have a negative effect on the ability of an organization to consistently satisfy its customers.

It may be useful to divide investors into shareholders, owners, partners, directors or banks as the influence each will have on the QMS will vary.

SOCIETY

Organizations depend on society as it is from society that the organization acquires consent to operate and acquires its employees and infrastructure services. In general, the influence of society on an organization is positive because it creates wealth but pressure groups and legislation may adversely affect what the organization is trying to do.

It may be useful to divide society into citizens (local, national and global), regulators, police, emergency services, health service as the influence each will have on the QMS will vary.

COMPETITORS

Understand your customers and why they buy from you, and you can keep them happy, continually satisfying their needs. Fail to satisfy these, and they will seek out an alternative – your competitors.

The strategies and tactics of your principle competitors need to be understood and evaluated to reveal their strengths and weaknesses relative to those of your own products and services, and this is the purpose of a competitive analysis. It will be more effective if data are gathered in a systematic manner rather than gathering opinions, conjecture, hearsay, etc. Sources of these data include:

*   Recorded data (e.g. annual reports, press releases, newspaper articles, analyst's reports and reports from government departments or regulators)
*   Observable data (e.g. pricing, advertising campaigns, promotions, tenders, and patent applications)
*   Opportunistic data (e.g. trade shows, seminars and conferences, recruiting ex-employees, meetings with suppliers, discussion with shared distributors or agents and social contacts with competitors)

There will be a lot of facts you know about your competitors but also much you don't know relative to quality such as:

*   Customer satisfaction and service levels
*   Customer retention levels
*   New product or service strategies
*   Productivity
*   Future investment strategy

OTHER INTERESTED PARTIES

With the other interested parties, you are primarily seeking to determine their intent because some of their interest may be malevolent and the risks these present will be addressed in Chapter 21.

Gathering the data

Decisions affecting the future direction of the organization and its products and services are made from information gleaned through market research. Should this information be grossly inaccurate, overoptimistic or pessimistic the result may well be the loss of many customers to the competition. It is therefore vital that objective data are used to make these decisions. The data can be primary data (data collected for the first time during a market research study) or secondary data (previously collected data). However, you need to be cautious with secondary data because they could be obsolete or have been collected on a different basis than needed for the present study.

The marketing information primarily identifies either problems or opportunities. Problems will relate to your existing products and services and should indicate why there has been a decline in sales or an increase in returns. To solve these problems a search for possible causes should be conducted. It may turn out that the products and services in themselves are fine but the means by which they are marketed, produced, distributed, provided or recycled may be creating problems for employees, the community or waste disposal services. Opportunities will relate to future products and services and should indicate unsatisfied wants.

There are three ways of collecting such data: by observation, survey and by experiment. Observation studies are conducted by actually viewing the overt actions of the respondent. In the industry, this can either be carried out in the field or in the factories where external providers can observe their customer using their materials or components.

Using surveys is the most widely used method for obtaining primary data. Asking questions that reveal their priorities, their preferences, their desires, their unsatisfied wants, their reaction to new ideas, etc., will provide the necessary information. Information on the profile of the ultimate customers with respect to location, occupation, lifestyle, spending power, leisure pursuits, etc., will enable the size of market to be established. Asking questions about their provider preferences and establishing what these providers supply that you don't is important. Knowing what the customer will pay more for is also necessary, because many may expect features that were once options, to be provided as standard.

A method used to test the potential of new products is the controlled experiment – using prototypes, alpha models, etc., distributed to a sample of known users. Over a limited period, these users try out the product and compile a report that is returned to the company for analysis.

A source of secondary data can be trade press reports and independent reviews. Reading the comments about other products or services can give you some insight into the needs and expectations of potential customers.

How is this demonstrated?

Demonstrating that the organization has determined the interested parties and their requirements or intentions that are relevant to the QMS may be accomplished by:

a)   presenting evidence of a process for determining:

   i   how the interested parties were to be identified;
   ii how the requirements and intentions were to be determined;
   iii how the effect on the organization's ability to consistently provide products and services that meet customer and applicable statutory and regulatory requirements were to be determined.

b)   presenting evidence that the interested parties and their requirements and intentions have been identified in accordance with described process;

c)   selecting a representative sample of the data and showing that they were generated in accordance with the designated process description.

Monitoring and reviewing information about the interested parties (4.2)

What does this mean?

Monitoring and review of information about these interested parties means checking periodically whether any of the data have changed and what this signifies. It does not mean that the organization's strategy should change.

Why is this necessary?

Determining the interested parties and their requirements is not a one-off event. Some of the information won't change very often but other information may change frequently due to fluctuations in the economy, technological advances, changes in government or corresponding to a company or national review cycle. No change may signify stability, but it might equally signify a party's interests have plateaued and may soon decline. This may be a good thing if the party's interest is malevolent but not if it's benevolent.

How is this addressed?

Within the analysis of the interested parties there will be some data that are stable and other data that are volatile, and it would aid monitoring if this distinction were to be identified so that the effort spent is proportional to criticality. Some things to monitor are:

* changes in customer preferences;
* the rate at which providers of essential resources are increasing or decreasing;
* the ease or difficulty in attracting investors and employees;
* the rate at which competition is increasing or decreasing and where it is coming from;
* the issues that attract media attention that are relevant to the organization;
* changes in the interests that get public attention and attract special interest or pressure groups;
* changes in the nature and concentration of crime.

The method used to obtain information about interested parties and their requirements is often a method that can be used repeatedly to monitor changes. Therefore, if a survey method was used to determine customer needs and expectations, repeating the survey with the same questions will reveal whether there have been changes. Change the questions and you'll never be sure you are observing a change or simply a need that existed previously.

How is this demonstrated?

Demonstrating that information about interested parties and their relevant requirements is being monitored and reviewed may be accomplished by:

a) presenting evidence of a process for defining the monitoring and review activities to be carried out in terms of what is monitored, when and by whom, who undertakes the review and deciding the actions to be taken on the results;

b)   selecting a representative sample of interested parties and presenting evidence that the prescribed monitoring and reviews are being carried out as planned.

Bibliography

Boone, L. E., & Kurtz, D. L. (2013). *Contemporary Marketing* – Sixteenth Edition. Boston: CENGAGE Learning Custom Publishing.

14 Scope of the quality management system

Introduction

The scope creates a context for the QMS. It defines its breadth and depth, what it applies to, what it includes and excludes and what it deals with so we can answer such questions as:

- Which of the organization's products and services are managed by this system?
- Which processes deal with the external factors that affect the ability of the organization to produce these products and services?
- Which processes ensure that the needs and expectations of the interested parties are met relative to these products and services?
- Which organizational units and locations are engaged in these processes?
- Which requirements of ISO 9001 are not applicable to our organization and why?

The definition of a QMS in ISO 9000 states it is "part of a management system", but there is no diagram or narrative about the management system showing the QMS as a part. This observation is also true for the management system as nowhere in ISO 9000 and ISO 9001 is there a diagram or narrative showing where the management system is a part of the organization. However, from a merely practical point of view, we need to define the scope of the QMS, if for no other reason than to separate it from other systems we care to name such as an environmental management system, a security system or a filing system!

Although it's important for conformity assessment purposes to know which requirements of ISO 9001 are deemed not applicable, the scope of the QMS has more to do with the elements that are strongly influenced and controlled by the system.

The scope of the QMS is not the same as the scope of certification or registration and may indeed be different depending on the needs of different markets or customers. This point is acknowledged in ISO TS 9002 and illustrated in Figure 14.1. This example has been created to illustrate different scopes. It is not an example from a real organization. The scope of certification is much less than the scope of ISO 9001 because only production, installation and servicing are services that the organization offers to its customers. These would appear on the certificate. It does not offer, for example, strategic planning and marketing services to its customers, but they do affect the services it does offer which is why they are included in the QMS and addressed in ISO 9001.

ISO 9001:2015 indirectly implies the QMS should include everything that *affects the ability of the organization to ensure the conformity of its products and services and the enhancement of customer satisfaction* and therefore bringing environmental, health, safety or financial issues that can affect product or service quality into the scope of the QMS. But once again we have a dilemma because there is a statement in clause 0.4 that the standard

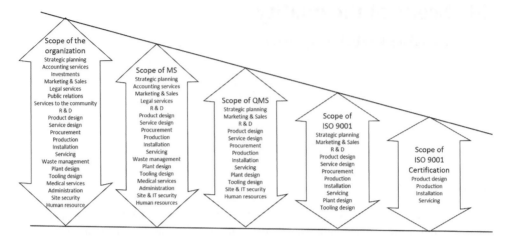

Figure 14.1 The scoping effect – how it varies depending on the focus

does not include requirements specific to other management systems. This could therefore be interpreted as ruling out of scope, environmental, health, safety or financial issues that can affect product or service quality, but if the organization can influence or control such issues placing them outside the scope of the QMS would be illogical. There are therefore things that affect the outputs and outcomes of the QMS which the QMS can control and things that it can't control but needs to mitigate and only the former are within the scope of the QMS.

In this chapter, we examine the four requirements of clause 4.3 which replace the single requirement in the 2008 version namely:

- Determining the boundaries of the QMS
- Determining the applicability of the QMS
- Applying the requirements of ISO 9001
- Documenting the scope of the QMS

Determining the boundaries of the QMS (4.3)

What does this mean?

Every system has a boundary (see Chapter 8). It's what separates the system from its environment. In fact, a system does not have boundaries in the plural because a system is a specific focus of interest. Here the standard is using the term *boundaries* to refer to the interfaces at the boundary where the organization interacts with its wider environment.

The system boundary is an imaginary line which envelopes what the system needs to control to deliver its outputs. The boundary may therefore envelop the whole organization, or be less than this if considering the QMS to be a subsystem, or it may extend beyond the organization if there are elements external to the organization it needs to control. The boundary is just like a line on a map, distinguishing one country from another, lines that would not be found on the ground if you were to go in search of them in reality.

This requirement is very different from that in the 2008 version (see Box 14.1).

Box 14.1   Revised requirement on scope

ISO 9001:2008 required to scope of the QMS including any exclusions to be established and maintained in a quality manual. The exclusions were related to the requirements of ISO 9001 rather than what the organization didn't do. However, other than the applicability of requirements, there were no criteria for determining what was included or excluded from the QMS. This omission has now been addressed in the 2015 version.

Why is this necessary?

It's important to know where the boundary of the QMS lies to minimize the relationships the participants in the system need to deal with.

How is this addressed?

Although the standard requires consideration to be given to the external and internal issues, the requirements of relevant interested parties and the products and services of the organization, the documented information is only required to addresses the types of products and services covered and justification for any requirement that are not applicable to the scope of the QMS. This implies that the scope statement *is not* required to:

a)   identify which processes form part of the QMS and which are part of other subsystems whose outputs influence the QMS;
b)   identify the interactions across the system boundary between external bodies and the QMS;
c)   identify interfaces with other parts of the organization's management system.

But to create an understanding of what the QMS is and what it controls, it would be helpful to include such aspects within the scope statement.

A way of determining the scope of the QMS and its boundary is to apply the method adapted from (Carter, Martin, Mayblin, & Munday, 1983) that is described here:

a)   Exclude components or relationships that have no functional effect on the system relevant to its descriptive purpose.
b)   Include items that can be strongly influenced or controlled by the system because you must understand how they work.
c)   Exclude items that influence the system but cannot in turn be easily influenced or controlled by the system. Put them in the environment as you only need to know their effects.
d)   Position the boundary either to enclose or to exclude complete clusters of relationships. Rather than cut across them. This minimizes the number of cross-boundary relationships and makes it easier to understand the effect of the environment on the system.

Influence and control are not the same (see Box 14.2 for the difference).

Box 14.2   Control and influence

To control means that an action is both necessary and sufficient to produce the intended outcome. To influence means that the action is not sufficient; it is only a co-producer (Gharajedaghi, 2011).

Necessity and sufficiency are implicational relationships between statements (see Wikipedia (3), 2015).

ISO 9001 also requires us to consider the issues referred to in clause 4.1, the requirements of relevant interested parties referred to in clause 4.2 and the products and services of the organization. So how might these affect the scope of the QMS? There follow four scenarios to illustrate how changes affect the scope:

1 Having determined the external issues, an organization in the confectionary business may find it can no longer guarantee a supply of cocoa as the world's supply is drying up so it decides to diversify. Thus, the scope of the system will change because new relationships will be formed across the boundary.

2 Having determined the internal issues, an organization heavily reliant on IT systems may find its major weakness is its vulnerability to cyber-attack and therefore needs to change its IT infrastructure to make its more resilient. Thus, the scope of the system will change because new relationships will be formed across the boundary.

3 Having determined the requirements of relevant interested parties, an organization that makes buggy whips finds there is no longer a market for them so they either need to redeploy their expertise or develop a new capability, and in both cases the scope of the system will change because of the new relationships formed across the boundary.

4 Having determined the products and services the organization needs to provide to satisfy its customer it finds it needs to develop new products and services that require capabilities beyond what the system can produce. Thus, the scope of the system will change because new relationships will be formed across the boundary.

In Chapter 8 we use an imaginary chain of fast food outlets as an instructive example when determining the structure of a system. A fast food outlet has the advantage of being a provider of both products and services, and it's one most readers will have experienced at one time or another so they will relate to it. Figure 8.4 is reproduced in Figure 14.2 for convenience. In this case the boundary is placed around the whole organization but also the outsource process of waste control. The interfaces with external bodies are shown by the lines passing through the boundary.

A map showing the external bodies with which the QMS interacts may also be useful as a means of identifying the external influences. An example is shown Figure 8.3. To be meaningful the nature of the interaction/influence needs to be identified, but this may be more easily achieved using a table rather than a diagram. How the entity functions to produce the

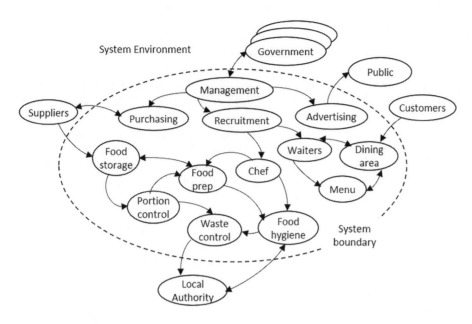

Figure 14.2 The internal system map

effect is of no concern. What is important is understanding the nature of the effect and being able to manage it.

How is this demonstrated?

Demonstrating that the organization has determined the boundaries of the QMS may be accomplished by:

a) presenting evidence of a process for determining the boundaries of the QMS and who has responsibility for determining them;

b) presenting the scope statement and showing that the system boundaries have been determined.

Determining the applicability of the QMS (4.3)

What does this mean?

Note 3 to the definition of a management system in ISO 9000:2015 clause 3.5.3 states that "the scope of a management system can include the whole of the organization, specific and identified functions of the organization, specific and identified sections of the organization or one or more functions across a group of organizations". The word *scope* here appears to be used in the sense of applicability, that is, what the QMS applies to, and this will depend on whether the QMS is intended to provide products and services to all the organization's customers or only the customers in certain locations or market sectors. For example, an organization may operate only from one site supplying a range of products and services to its customers and its QMS applies to those parts of the organization that can affect its ability to satisfy its customers. At the other end of the scale, a multinational operates in several countries, over many sites, with multiple divisions, multiple product and service ranges. There may be a QMS for headquarters, a QMS for each division or for each range of products or services because each range requires different capabilities. In general, different purposes or missions spawn different management systems because the purpose or mission of each division or business unit is different.

This might appear as though the scope of the QMS could be limited to a particular department such as a radiology unit in a hospital, and indeed it might if the customers concerned only interface with these functions and these functions have control of product and service conformity. To be in control of product and service conformity an organization needs to be able to determine the extent to which products and services conform to their specification as well as the applicable legal requirements.

If a customer is soliciting the services of a part of an organization and that part can treat other parts of the organization on which it depends as suppliers or external providers, the boundary of the QMS can be drawn around that part of the organization. If several organizations collaborate on a project and form a group comprising only certain functions within each of those organizations, the scope of the QMS would cross organizational boundaries.

The QMS may not cover all activities of the organization; therefore, those that are addressed by it or excluded from it need to be identified.

Why is this necessary?

Unless otherwise labelled, a QMS will normally apply to the whole organization because it is intended to deliver outputs that satisfy the organization's customers. However, in organizations with multiple sites, divisions, products, services, etc., its applicability needs to be defined to delineate between other systems in the same organization or perhaps more importantly designate what it doesn't apply to (e.g. in a hospital, the only unit having a formal QMS may be the radiology unit).

How is this addressed?

QMS applicability is generally designated in its name (e.g. dynamics division quality management system). If any further qualification is needed, it can be included in the scope statement. An organizational unit or process can only be outside the scope of the QMS if its performance can't be controlled by the QMS. A system map similar to that referred to in Figure 8.4 in conjunction with an organization chart (see Figure 8.8) can be used to show which organizational units or processes are internal or external to the QMS. Remember that what is in scope is that which can be strongly influenced or controlled by the QMS. The QMS will therefore exclude elements it depends on but cannot control such as suppliers and investors.

How is this demonstrated?

Demonstrating that the applicability of the QMS has been determined may be accomplished by presenting a scope statement that specifies the organization or part thereof to which the QMS applies. The applicability matrix referred to in the next section may be used to address this requirement and the requirement for declaring non-applicability of ISO 9001 requirements.

Applying the requirements of ISO 9001 (4.3)

What does this mean?

As is stated in clause 1 of ISO 9001 "all the requirements are generic and are intended to be applicable to any organization, regardless of its type or size, or the products and services it provides." It follows, therefore, that there would have to be a good reason for claiming that a particular requirement is not applicable.

Box 14.3   Revised requirement on exclusions

ISO 9001:2008 limited exclusions to the product realization requirements. The equivalent in ISO 9001:2015 is clause 8 on operation, but the applicability requirement now applies to the whole standard implying any requirement can be excluded or deemed not applicable providing it can be justified on the basis that it does not relate to anything the organization needs to do to satisfy its customers

---

One good reason would be that having weighed up the potential benefits and drawbacks of meeting a specific requirement you conclude there to be negligible benefit to customer satisfaction and conformity to legal requirements for the effort that would be expended.

ISO TS 9002 suggests that the applicability of the requirements of ISO 9001 determines the scope but a requirement would only be applicable if it relates to an element of the system, that is, the element determines which requirements are applicable not the other way around (e.g. if there is no work that is outsourced, the requirements concerning outsourcing don't apply, if the organization does not handle customer property, the requirements concerning customer property don't apply).

Why is this necessary?

If an organization were to declare conformity to ISO 9001 and knowingly not meet every requirement, it would be in breach of trust and possibly common law. When a potential customer comes to select a supplier and sees that it is certificated to ISO 9001, the expectation is that the organization meets all the requirements; therefore, so as not to mislead customers, any exceptions need to be declared. Declaring that a particular requirement of ISO 9001 is not applicable is also necessary for conformity assessment purposes so that the auditor does not waste time seeking evidence of conformity for requirements that are not applicable.

How is this addressed?

Most requirements of ISO 9001 are generic and applicable to all organizations regardless of size, complexity or the nature of its products and services. Some organization leaders might take the view that their organization is satisfying its customers and conforming to

ISO 9001:2008 without determining the context of the organization, determining risks and opportunities or determining organizational knowledge, so why should these new requirements apply to their organization?

It has never been the case that an organization needs to do no more than meet the requirements of ISO 9001 for it to produce products and services that satisfy its customers. Since 1987, ISO 9001 has included requirements which experience has shown to be the determining factors of product and service quality. Experience changes over decades and some factors reduce in importance while others increase. This is why there has been a decrease in requirements of a tactical nature like those for documentation and an increase in requirements of a more strategic nature like those for risks and opportunities, etc. It is more than likely that organization leaders have been addressing these issues but hitherto not formally or not as part of their QMS.

ISO 9001 won't be the only external standard or regulation which your organization must meet, and one way of mapping conformity is to produce an exposition or applicability table for each. These can take different forms. A simple applicability matrix would map the requirements to the processes in which they are fulfilled. This could be extended to include reference to the process description and the functions having responsibilities in that process as shown in Table 14.1. As some clauses of ISO 9001 contain multiple requirements, a correlation between requirements and processes at the level of detail in the example would be rather basic and would require a lot more digging to confirm conformity or non-applicability of a specific requirement.

The exposition carries more detail and is a response to the requirements with links to the documentary evidence that demonstrates either intent or results. An example of this is shown Table 55.1 where it is used in a requirements driven conformity audit. Expositions which provide a map through the management system for each management system standard (e.g. ISO 9001, ISO 14001, IATF 16949 may be a useful source of reference). These are not system descriptions so they can be relatively small documents.

Table 14.1 Extract from an ISO 9001:2015 applicability matrix

Clause and title	Applicable Processes	Ref document	Management	Marketing	Sales	Design	Operations	IT	Quality	HR	Maintenance
4 Context of the organization											
4.1 Understanding the organization and its context	Scanning the environment	PD001	•	•	•	•	•		•		
4.2 Understanding the needs and expectations of interested parties	Scanning the environment	PD001	•	•	•	•	•	•	•	•	•
4.3 Determining the scope of the quality management system QMS	System scoping	PD002	•	•	•	•	•	•	•	•	•
4.4 Quality management system and its processes QMS	Management system development	PD003	•	•	•	•	•	•	•	•	•

There are two requirements that apply when certain conditions prevail: clause 8.5.2 when traceability is a requirement and clause 7.1.5.2 when measurement traceability is a requirement. No exclusions need be declared for these if there are no requirements, but in circumstances where these requirements would normally be invoked in contracts or regulations, be prepared to demonstrate traceability is not required.

A few requirements won't apply if the work needed to satisfy customers will not involve the activity to which the requirement refers

Monitoring and measuring resources (7.1.5.1)

Clause 7.1.5.1 applies if you are monitoring or measuring products and services to verify their conformity. But if you are providing legal services, you are using your professional judgement to determine the advice you give. However, there will be aspects of the service that are monitored such as professional conduct.

Design and development of products and services (8.3)

Every organization provides outputs that can be in the form of tangible or intangible products (computers, materials, software or advice), services that may process product supplied from elsewhere or services that develop, distribute, evaluate or manipulate information such as in finance, education or government.

With every product, there is a service, or to put it another way, with some services there is a product. If the product is provided from elsewhere, a service still needs to be designed to process it. Although not every organization may provide a tangible product, every organization does provide a service. Some organizations manufacture products designed by their customers and therefore product design could be excluded. However, they still provide a service possibly consisting of sales, production and distribution and so these processes need to be designed. In franchised operations, the service is designed at corporate headquarters and deployed to the outlets but it is not common for an outlet alone to seek ISO 9001 certification – it would be more common for ISO 9001 certification to be a corporate policy and therefore the scope of certification would include service design. If you are not provided with the product or service characteristics necessary to plan product or service operations and must define those characteristics; this is product and service design and development.

It is therefore inconceivable that this clause could be excluded for anything other than for product design, meaning that it must be included for service design. However, there may be confusion about to which design activities the requirements apply. In deciding whether the design and development requirement applies or doesn't apply one simple test is the "purpose test" (see Box 14.4).

Box 14.4   Purpose determines applicability

If an organization's purpose is to design management systems and carparks, then clause 8.3 applies to the carparks and the management systems it designs. But if an organization's purpose is to provide banking services and its management need a car park or a management system, clause 8.3 does not apply to the design of the management system or the carpark but does apply to the design of the banking services.

Externally provided processes, products and services (8.4)

Every organization acquires products and services from external providers because no organization is totally self-sufficient except perhaps a monastery and it is doubtful that ISO 9001 would even enter the thoughts of a monk! Some acquisitions may be incorporated into products supplied to customers or simply passed onto customers without any further processing. Other acquisitions may contribute to the processes that supply product or deliver services to customers and there are perhaps some acquisitions that have little or no effect on the product or service supplied to customers but they may affect other stakeholders. There are requirements within the clause that may not apply such as those pertaining to externally provided processes where outsourcing is not carried out.

Validation of processes (8.5.1f)

This clause applies to processes where the resulting output cannot be verified by subsequent monitoring or measurement. If we apply the provisions of clause 4.4.1 to all processes, this clause is redundant but if there is a case for exclusion, justification needs to be given.

Customer property (8.5.3)

Not every organization receives customer property but such property does take a variety of forms. It is not only product supplied for use in a job or for incorporation into supplies, but also can be intellectual property, personal data or effects. Even in a retail outlet where the customer purchases goods, customer property is handed over in the sales transaction perhaps in the form of a credit card where obviously there is a need to treat the card with care and in confidence. In other situations, the customer supplies information in the form of requirements and receives a product or a service without other property belonging to the customer being supplied. Information about the customer obtainable from public sources is neither customer property nor is information given freely but there may be constraints on its use.

Preservation (8.5.4)

This clause applies to organizations handling tangible products and would include documentation shipped to customers. It applies in service organizations that handle products, serve food and transport products or people. It does not apply to organizations that deal in intangible product such as advice although if the advice is documented and the documents are transmitted by post or electronic means, preservation requirements would apply.

How is this demonstrated?

Demonstrating that all the requirements have been applied if applicable may be accomplished by presenting evidence of an applicability matrix which:

a)   identifies the processes and/or functions where conformity may be established;
b)   identifies which requirements are deemed not applicable together with their justification.

Documenting the scope of the QMS (4.3)

What does this mean?

This requirement requires no further explanation.

Why is this necessary?

The scope statement provides useful information that will enable people within the QMS to quickly establish if their activities will be affected by it and to ensure effective communication with interested parties because it is one area that can generate a lot of misunderstanding particularly when dealing with auditors, consultants and customers.

How is this addressed?

It is unlikely that you will be able to determine the boundaries of the QMS, its applicability and the requirements of ISO 9001 that apply and don't apply without documenting the information you gather. However, it's not a one-off activity and will need to be reviewed regularly to identify any changes in the internal and external environment.

How is this demonstrated?

Demonstrating that the scope of the QMS is available and maintained as documented information containing the relevant information may be accomplished by:
   Presenting a scope statement that:

a) identifies the products and services covered by the QMS;
b) provides justification for any requirement of ISO 9001 that are deemed not applicable to the scope of the QMS;
c) showing how the relevant external and internal issues have influenced the determination of the QMS scope;
d) showing how the relevant requirements of relevant interested parties have influenced the determination of the QMS scope;
e) presenting evidence that the QMS scope is reviewed when changes in the inputs have been detected.

Bibliography

Carter, R. C., Martin, J. N., Mayblin, B., & Munday, M. (1983). *Systems, Management and Change: A Graphic Guide*. London: Paul Chapman Publishing Ltd.

Gharajedaghi, J. (2011). *Systems Thinking: Managing Chaos and Complexity: A Platform for Designing Business Architectures*. Burlington, MA: Elsevier.

Wikipedia (3). (2015, January 23). Retrieved from Necessity and Sufficiency: https://en.wikipedia.org/wiki/Necessity_and_sufficiency

15 Quality management system

A map is not the territory it represents, but, if correct, it has a similar structure to the territory, which accounts for its usefulness.

Alfred Korzybski (1879–1950)

Introduction

The terms *establish*, *implement*, *maintain* and *improve* are used in the standard as though this is a sequence of activities. They imply that a QMS is first of all designed, then put into effect and once operational is maintained and subject to continual improvement. This is treating a QMS as a designed physical system. In reality, an organization that is consulting ISO 9001 is unlikely to be starting from scratch. It will already be well established and have been providing products and services to customers for some time, and therefore we need to interpret these terms from that perspective.

A system is a mental construct which is formed when we observe the interaction between elements within a boundary we have defined. If there is no interaction there is no system. Therefore, although a description of a QMS may exist in name which defines all the policies, procedures and processes and explains how customer requirements are met and their satisfaction enhanced, unless the organization is operational, no system exists – the documents define the QMS requirements, not the QMS. Similarly, if work is being undertaken in an organization and outputs are being produced and provided to customers, a system exists but it may not have been captured. Therefore, don't confuse the map with the territory or the system description with the system. The documents don't define everything that produces the results. People don't just follow instructions; they interpret situations and adapt their behaviour accordingly, hence the Korzybski quote at the beginning of this chapter. The formal documents are the map; the reality is a combination of the formal and the informal. Some might say that results are a combination of the system and the people but they assume the system is abstract when the system we are referring to includes the people.

When developing a management system, you should have in mind the cycle of sustained success as shown in Figure 15.1. This is what we think is happening, but the reality may be far from the case. In establishing, implementing, maintaining and improving a QMS, what you should be trying to do is to change the way the organization functions so that the cycle of sustained success in Figure 15.1 becomes a model of reality.

To establish consistency and avoid confusion (if that is at all possible with this subject) the following explanations are given:

Figure 15.1 Cycle of sustained success

- Establishing a QMS should be interpreted as creating a model of reality as explained in Chapter 8.
- Implementing a QMS should be interpreted as putting into effect the changes made to the model to improve organizational effectiveness. It is a mistake to refer to a system as being implemented when specific policies or procedures are being implemented. What such actions indicate are that interactions within the system are occurring.
- Maintaining a QMS should be interpreted as (a) maintaining the model as representative of reality and (b) maintaining the structure and processes that holds the system together.
- Continually improving the QMS should be interpreted as putting into effect changes made to the model on a recurring basis to enhance the performance of the organization.

In this chapter, we examine the four requirements of clause 4.4.1, namely:

- Establishing a quality management system including processes needed (4.4.1)
- Maintaining a quality management system (4.4.1)
- Implementing a quality management system (4.4.1)
- Continually improving a quality management system (4.4.1)

Establishing a quality management system including processes needed (4.4.1)

What does this mean?

In Chapter 8 we presented various contradictory ISO definitions explaining what a management system is. We dismissed them all as being unhelpful operational definitions and decided that a QMS is *a systemic view of the organization from the perspective of how it creates and retains customers.* We have now reached the point where we are required to establish a QMS so, with definition in mind how do we go about establishing a QMS?

a)   We look at the organization as if it were a system (i.e. a set of interacting elements that produces outputs greater than the sum of its parts).

b)   We put a boundary around those elements that through their interactions create and retain customers and put all other elements that are not influenced or controlled by this system outside it.

Box 15.1   Revised documentation requirement

The 2008 version required that the QMS be established, documented, implemented, maintained and its effectiveness continually improved. The QMS is not now required to be documented but is to include certain documentation. The QMS is also now to include the processes needed and their interactions (as well as a description of the processes when necessary), and these two changes imply a QMS is a dynamic system not simply a system of documents.

What we include in this model becomes a matter of identifying those elements that affect the organizations ability to create and retain its customers (see Chapter 8). Some of these elements will include management intentions in whatever form they are expressed, documented or undocumented, prescriptive or descriptive. These are referred to in ISO 9001 as *the quality management system requirements* (see also Chapter 17 in connection with cause 5.1.1c).

The standard requires that the established system includes the processes needed and their interactions. This phrase is interpreted as being the processes needed to determine and provide products and services intended for customers. The other point to bear in mind is that the QMS is not only composed of processes and interactions because it also contains both tangible and intangible elements. Although many will probably try to apply 4.4.1 to management processes, the phrase "processes needed for the QMS" is key to understanding. As the QMS cannot invent itself, it seems that the process by which it is invented is not part of the QMS and therefore not a process needed for (implementing!) the QMS and not subject to 4.4.1a) to h) (See also Chapter 16.)

Establishing a system in accordance with the requirements of ISO 9001 means that the characteristics of the system must meet the requirements of ISO 9001. However, the requirements of ISO 9001 are not expressed as system requirements of the form *the system shall,* but are expressed as organization requirements of the form *the organization shall.* It would therefore appear that the system model must be representative of the organization, so that whatever changes are made to the model are implemented in the organization (see Figure 8.2).

Why is this necessary?

ISO 9001 contains a series of requirements which, if met, will give the organization the capability of supplying products and services that satisfy the organization's customers. All organizations have a way of working, some methods may be documented and others not, and they may not have taken a systemic view of the organization to see how all the pieces come together to produce the results they observe. If an organization desires year after year success, it needs a formal mechanism to accomplish this – it won't happen by chance. This requires management to think of their organization as a set of interdependent elements which function together to produce the organization's outputs or in other words, as a system. These concepts are explained further Chapter 8.

How is this addressed?

We have established that to establish a QMS we need to create a systemic view of the organization from the perspective of how it creates and retains its customers. This is different from documenting what you do because it's not a set of procedures or instructions, although they may feature in it in the form of QMS requirements. What you are trying to capture is how work gets done, not how you'd like it to be done. All models are approximations so although it won't mirror reality exactly, you want it to be useful so make sure the models you create have a purpose and reflect a view of the organization that is shared by others.

Situation analysis

The first stage is to conduct a situation analysis, the objective of which is to get a clear understanding of how the organization creates and retains its customers, how it's performing and what the current drivers and barriers are. Although the model need not be any more detailed than is necessary to identify where a requirement of ISO 9001 would apply, it won't be apparent in some areas until you have drilled down into procedures for carrying out a particular activity. More detail will be required for operational processes than for management processes. Using techniques such as survey, audit or analysis of records, consideration needs to be given to revealing the following information as applicable:

RESULTS

What results are being achieved relative to:

- Customers (e.g. orders won and lost, satisfaction, loyalty, complaints).
- People under the organization's control (e.g. degree of satisfaction, loyalty, involvement, retention, motivation, grievances and accidents).
- Suppliers (e.g. degree of satisfaction, loyalty, performance on quality, cost and delivery)
- Society (e.g. degree of satisfaction within local community and types of complaints).
- Key issues from the SWOT analysis relative to the earlier points (see Chapter 12).

STRUCTURE

The structure of the QMS is the way its elements are interconnected, and this can be represented by work flows and influence diagrams. Examples are given by the system maps, process maps, activity sequence diagrams and organization charts in Chapter 8. Include key issues from the SWOT analysis relative to processes (Chapter 12).

CULTURE

Determine the values, beliefs, rituals and customs that characterize what it's like to work in the organization and key issues from the SWOT analysis relative to culture (see Chapter 12).

QMS description

All this information will represent the baseline from which subsequent improvements will be measured. If some of this information exists in the form of flow charts, procedures, charts, etc., it should be validated as being representative of the current situation. It's not helpful at this stage to use information that is speculative because the next step is to determine the gap between the state of the system now and the state it needs to be to conform to ISO 9001.

For those organizations upgrading their QMS to meet the requirements of ISO 9001:2015, their existing quality manuals may form the basis of a QMS description but as will be seen from the earlier points, a lot of additional information may be needed.

Gap analysis

Before implementing a QMS, it is necessary to determine the changes that may need to be made in the way the organization creates and retains its customers to become more effective and conform to the requirements of ISO 9001. To do this:

a)   review the QMS description against the requirements of ISO 9001 and identify the differences in both deeds and behaviours. This may be undertaken like a requirements-driven conformity audit (see Table 55.2).
b)   determine what needs to change, the benefits of making changes to the way quality is managed and revise the QMS description accordingly.
c)   define the objectives of the change and formulate a plan for achieving them that includes training and the reorientation of attitudes and beliefs, etc.
d)   get agreement to the plan from top management and those who will be affected by it.

How is this demonstrated?

Demonstrating that a QMS has been established may be accomplished by:

a)   presenting a model of the QMS from both a systems perspective and a process perspective showing the relationship between the elements that create and retain satisfied customers and their interface with external parties;
b)   presenting the results of the analysis of interested parties showing how the enabling processes have been determined;
c)   selecting a representative sample of customers and following a trail through the system from its outputs back to:

   i   the processes employed to produce the products and services from their inception;
   ii   the processes employed to attract customers;
   iii   the processes employed to supply resources to these processes;
   iv   the processes employed to manage these processes.

d)   confirming that the model of the QMS is fully representative of the way the organiza-
      tion creates and retains its customers.

Implementing a quality management system (4.4.1)

What does this mean?

As stated in the introduction to this chapter, implementing a QMS means putting into effect
any changes made to this model to bring about a change in reality (e.g. putting into effect any
new policies and practices). This means that QMS implementation is a process of changing
the status quo with all that entails.

Why is this necessary?

The model either reflects what is happening or what should happen and remains extant until
either reality changes or the model is changed. Changes made to the model won't have any
effect until they are implemented by the organization.

How is this addressed?

If what is happening is what should happen, no change is necessary but quite often some
assumptions may have been made when creating the model, for example, the model may
have been based on existing documented policies and practices without confirming they are
being implemented as intended. It is therefore important to confirm the model is representative of reality before making changes to it.

   You simply need to put into effect what is in the model which means implement the
plan, doing what you intended to do, keeping your promises, honouring your commitments,
changing the processes, etc. – simply said but extremely difficult for organizations to do
as can been seen from Box 15.2. Implementation is therefore change management, and an

---

Box 15.2   Why implementation fails

- Failure to obtain management commitment
- Failure to obtain employee commitment
- Failure to appoint the right leader
- Failure in communication
- Failure to prioritize action
- Failure to follow the plan and track progress
- Failure to take responsibility
- Failure to coordinate
- Failure to allow sufficient time
- Failure to deal with resistance to change
- Failure to educate and train staff affected by the change
- Failure to recognize unrealistic goals

effectively managed programme of introducing new or revised practices is a way of overcoming these failures.

How is this demonstrated?

Demonstrating that the QMS has been implemented may be accomplished by presenting evidence from an internal audit that the QMS is a valid representation of reality.

Maintaining a quality management system (4.4.1)

What does this mean?

As the QMS is a systemic view of an organization from the perspective of how it creates and retains its customers, maintaining a QMS could be interpreted as nothing more than maintaining a representation of reality, that is, if the processes are not delivering what they should, this should be reflected in the model; otherwise, the model is not useful except as a historical record. Although QMS documentation has not been regarded as a model, it has often been regarded as the QMS itself, for example, a request to update the QMS has resulted in updating documents rather than changing what people do.

However, what we observe, to which we assign the label QMS, is a bounded set of interdependent elements which create and retain customers. Therefore, maintaining this bounded set of interdependent elements must involve maintaining the structure and processes that holds the system together. This means maintaining the relationships, the interactions, the interconnections and fixing those that are broken.

Why is this necessary?

Without maintenance, any system will deteriorate. As the second law of thermodynamics (entropy) applies to all open systems, changes in the environment will cause system performance to decline as structure and processes fall apart unless specific action is taken to maintain them. A lack of attention to the elements and their relationships, interactions and interconnections will certainly result in a loss of capability and therefore poor quality performance, financial performance and lost customers.

How is this addressed?

In maintaining the QMS you need to keep:

- correcting special cause variation;
- physical resources operational;
- personnel competent;
- financial resources available for replenishment of consumables, replace worn out or obsolete equipment;

- the documented information up to date as changes in the organization, technology and resources occur;
- space available to accommodate input and output;
- buildings, land and office areas clean and tidy – remove the waste;
- benchmarking processes against best in the field.

In maintaining capability, you need to keep:

- replenishing human capital as staff retire, leave the business or are promoted;
- renewing technologies to retain market position and performance;
- surplus resources available for unforeseen circumstances;
- up to date with the latest industry practices;
- refreshing awareness of the vision, values and mission.

Another set of actions that can be used is the Japanese 5-S technique (Imai, 1986) are as follows:

1   Seiri (straighten up): Differentiate between the necessary and unnecessary and discard the unnecessary.
2   Seiton (put things in order): Make it easy to find things.
3   Seido (clean up): Keep the workplace clean.
4   Seiketsu (personal cleanliness): Make it a habit to be tidy.
5   Shitsuke (discipline): Follow the procedures.

How is this demonstrated?

Demonstrating that the QMS is being maintained may be accomplished by presenting evidence that confirms that the system performance remains stable.

Continually improving a quality management system (4.4.1)

What does this mean?

ISO 9000:2015 defines continual improvement as a "recurring activity to increase the ability to enhance performance requirements". Continually improving the QMS means putting into effect changes made to the model on a recurring basis to enhance the performance of the organization.

Why is this necessary?

See Chapter 59.

How is this addressed?

Further details are provided in Chapter 59.

How is this demonstrated?

Demonstrating that the QMS including the processes is being continually improved may be accomplished by presenting evidence of QMS performance (see 9.1.3c in Chapter 54) and explaining the observed changes in terms of the improvement initiatives that have been undertaken.

Bibliography

Imai, M. (1986). *KAIZEN: The Key to Japan's Competitive Success*. Singapore: McGraw-Hill.

16 Processes needed for the QMS

Introduction

Quality does not happen by chance; it must be designed into products and services, and any amount of inspection will not change their quality. It is therefore the processes which determine and provide products and services that holds the most potential for creating products or services of the utmost quality. Having a clear understanding of the factors that influence the capability of these processes is therefore crucial in managing product and service quality.

Clause 4.4.1 not only requires the processes needed for the QMS to be determine, but it also requires several actions to be undertaken relative to these processes. In the Committee Draft of ISO 9001:2015 this list of actions was under the heading of "Process Approach", implying that by taking these actions one was applying the process approach. The heading was later removed, but the list remains and in some respects, it can be interpreted as a summary of requirements that appear elsewhere within the standard in more detail. Indeed, this is the case for some of them, but not all, and therefore this list cannot just be ignored. Although there are eight statements (4.4.1a) to h), several contain more than one requirement, so in total there are 15. Of the 15 requirements, 7 are also addressed elsewhere in the standard as indicated here:

- Determining the resources needed (4.4.1d and 7.1) – see Chapters 24 to 28
- Assigning responsibilities and authority for processes (4.4.1e and 5.3) – see Chapter 20
- Addressing the risks and opportunities (4.4.1f and 6.1.1) – see Chapter 21
- Evaluating the processes and ensuring they achieve their intended results (4.4.1g and 9.1) – see Chapter 54
- Improving the processes and the QMS (4.4.1h and 10.1) – see Chapter 57
- Maintaining documented information (4.4.2a and 7.5.1) – see Chapter 32
- Retaining documented information (4.4.2b and 7.5.3.2d) – see Chapter 32

In this chapter, we examine the eight remaining requirements of clause 4.4.1 that are not addressed elsewhere, namely:

- Determining processes needed for the quality management system (4.4.1)
- Applying the processes of the QMS throughout the organization (4.4.1)
- Determining the inputs required (4.4.1a)
- Determining the outputs expected (4.4.1a)

- Determining process sequence(4.4.1b)
- Determining process interaction (4.4.1b)
- Determining the criteria and methods (4.4.1c)
- Determining methods needed to ensure effective operation (4.4.1c)

Determining processes needed for the quality management system (4.4.1)

What does this mean?

Which processes are needed for the QMS?

The standard does not require all processes employed in the organization to be determined. It requires only those processes needed for the QMS to be determined. So, what are these processes?

The activities from which the processes needed for the QMS are formed are already being carried out or are planned to be carried out. They comprise the organization's business processes, and it's simply a matter of determining which of these possess elements have a direct and indirect impact on product and service quality so they can be modified if necessary to bring about a system outcome of customer satisfaction. Business processes are the processes that are intended to produce outputs that achieve business goals.

These groups of activities form a chain that delivers value to customers and may include activities commonly referred to as strategic planning, market research, new product or service development, sales promotion, procurement, production, distribution, service delivery, sales and after sales but their labels are unimportant. The essential characteristic is that they contain elements and interconnections that positively influence the quality of the output and were they to be absent, the quality of the output would be unpredictable. Figure 16.1 shows what we would see if we were to imagine plucking out all these elements and interactions from the business processes and putting a boundary around them. The processes needed for the QMS are therefore not QMS processes but business processes containing elements and interconnections that positively or negatively influence the quality of the system outputs. The QMS does include processes, but these processes are the processes that enable the organization to function. There is no separate

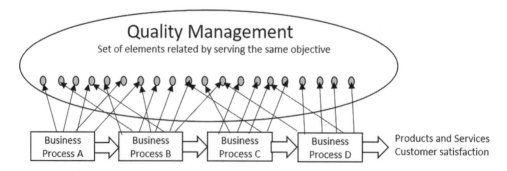

Figure 16.1 Elements of business processes that we imagine manage quality

set of processes that exist or that are planned that only serve product and service quality and nothing else.

Which types of processes?

As the standard is not specific to the types of processes, we must assume it is all types of processes and this would therefore include both business processes and work processes. In an unusual step ISO 9001 now refers to business processes requiring "QMS requirements to be integrated into the organizations business processes" blurring the distinction between processes needed for the QMS and business processes (see also 5.1.1c in Chapter 17).

Why is this necessary?

We cannot manage what we don't understand, and to manage outputs we therefore need to understand how outputs are produced. By definition, an output is the result of a process (Chapter 9); therefore, we can't manage outputs unless we know which processes are producing them.

How is this addressed?

What are we trying to do?

If our objective is to determine how people do their work, we could determine processes for cleaning floors, staking shelves, running a meeting, all of which may be important but we can't judge their importance from this level. We need to zoom out before we can see the contribution this work makes in the bigger picture.

If our objective is to determine how results of the QMS are achieved, we could determine a process for maintaining hygiene standards in the kitchen in which one activity is cleaning floors. We therefore need to know the relationship between hygiene standards in the kitchen and the intended results of the QMS, that is, what impact do they have on customer satisfaction, and this is where we need an appreciation of the context of the organization. To whom does the kitchen serve meals to? If it's customers, maintaining hygiene standards is a critical process as the risk of customer dissatisfaction is high, but if it's employees it's still important but not critical as the risk of customer dissatisfaction is low – perhaps unlikely if delivery of product or service is not affected.

There are different ways of determining the processes needed for the QMS but whichever method you choose, having determined what you think is a process needed for the QMS, validate your choice by getting an affirmative answer to the question: *What impact does the result of this process have on customer satisfaction?*

The stakeholder-driven method

An approach for aligning the mission, vision and values with the needs of the stakeholders and for defining appropriate performance indicators is a stakeholder analysis. This analysis addresses all stakeholders and by posing some key questions, the processes that deliver outputs that satisfy stakeholder needs are determined. By way of an example a fictional fast food business has been analysed and the results presented in Table 16.1. Although the

Table 16.1 Example of determining processes using stakeholder analysis

Stakeholders	Stakeholder needs (business outcomes)	Indicators of stakeholder satisfaction	Outputs that will deliver successful outcomes	Critical success factors	Enabling processes	Related business process
Who are the beneficiaries?	What are their needs relative to the mission?	What will they look for as evidence of satisfaction?	What outputs will deliver successful outcomes?	What factors affect our ability to deliver these outputs?	Which process deals with this factor?	What contribution does this enabling process make to the business?
	A financial return that meets target	Profitability	Positive cash flow	Maintaining demand	Maintain cash flow	Resource management
			Effective business strategy	Commitment to the planning process	Develop strategy	Mission management
			Growth in demand		Product promotion	Demand creation
			Growth in demand	Attracting customers	Product promotion	Demand creation
Investors	Growth	Sales	Keeping ahead of the competition	Market intelligence	Competitor analysis	Demand creation
				Service innovation	Service design	Mission management
		Market share	Consumer change in preference	Reputation	Performance review	Mission management
			Food and service quality and price	Value for money	Develop pricing strategy	Demand creation
				Customer satisfaction	Service design	Demand creation
					Service delivery	Demand fulfilment
Employees	Competitive pay and conditions	Staff grievances	Motivated workforce	Leadership	Create internal environment	Mission management
		Accident levels	High safety record	Ergonomics	Serving area design	Resource management
				Management style		
		Absenteeism	Low sickness record	Management style	Create internal environment	Mission management

Table 16.1 (Continued)

Stakeholders	Stakeholder needs (business outcomes)	Indicators of stakeholder satisfaction	Outputs that will deliver successful outcomes	Critical success factors	Enabling processes	Related business process
		Entry to exit time	Short queue length	Serving area design	Design food outlet Resource	management
				Staff competence	Staff development	
			Adequate dining area capacity	Dining area design	Design food outlet	Resource management
Customers	Fast, safe and nutritious food	After effects	Safe food		Store food Prepare food	Resource management
				Living by our values	Serve food	Demand fulfilment
			Food handled hygienically		Staff development	Resource management
			Cleanliness of equipment		Staff development	Resource management
		Food taste and texture	Food cooked at the right temperature	Competent staff	Staff development	Resource management
			Food stored in right environment			
	Safe, clean and hygienic environment	Hazard-free public areas	Efficient design of preparation, serving and dining areas	Competent design agency	Service design	Demand creation
			Clean and dry surfaces	Competent staff	Staff development	Resource management
		Appearance and practices in all areas	Competent staff	Training budget	Staff development	Resource management

(Continued)

Table 16.1 (Continued)

Stakeholders	Stakeholder needs (business outcomes)	Indicators of stakeholder satisfaction	Outputs that will deliver successful outcomes	Critical success factors	Enabling processes	Related business process
	Value for money	Competitiveness	Competitively priced menu	Market intelligence	Determine pricing strategy	Demand creation
			Attractive presentation of fresh food in consistent portions	Container design	Serving area design	Demand creation
				Delivery flow	Serve food	Demand fulfilment
				Portion control		
	Efficient counter service	Waiting time	Well-designed serving area	Competent design agency	Serving area design	Demand creation
			Competent staff	Training budget	Staff development	Resource management
		Staff courtesy	Selection criteria	Training budget	Staff development	Resource management
	Mutually beneficial relationship	Business continuity	Repeat purchases	Profit margins	Financial planning	Resource management
Suppliers	Prompt Payment	Payment on or before due date	On-time payment on invoices	Cash flow	Accounts	Resource management
Society	Compliance with statutory regulations and instruments	Compliance	Food handling meets hygiene regulations	Shared values	Prepare food	Demand fulfilment
					Serve food	
			Use of cleaning fluids meet COSHH regulations			
			Exit doors meet fire regulations			
			Sanitation meets building and public health regulations		Facility maintenance	Resource management

analysis is quite detailed, it is presented to demonstrate the technique and should not be assumed to represent any particular fast food outlet. This is only part of a full stakeholder analysis as you would also gather information on their judgment about your organization's performance.

Objective-driven method

The stakeholder-driven method provides a close alignment between stakeholder needs and process activities. It can be time consuming and difficult to do but worthwhile. If the quality management system is to be limited in scope to satisfying customers, you could simply take the business outputs that are derived from customer needs in the stakeholder analysis shown in Table 16.1.

Another method is to derive the processes from the strategic objectives, business objectives or project objectives but these will include objectives that serve more than customers. The processes identified in the system or organization model can be regarded as Level 0, implying there are further levels in a hierarchy as shown in Figure 16.2. It is important to remember that the purpose of any process is to achieve an objective, and therefore whether the objective is strategic such as a vision or mission, or is related to the completion of task such as serving a customer, there is still a process to achieve it.

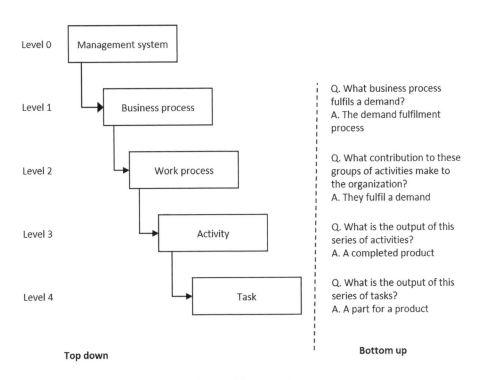

Figure 16.2 Process identification (top-down and bottom-up)

However, if the decomposition reaches a level where to go any further in the hierarchy you would be in danger of noting arm movements, you have gone a level too far! To identify the processes, sub-process or activities you need to know what objectives need to be achieved or what outputs are required, and for this you will need access to the business plans, project plans, etc. Objectives are simply outputs expressed differently. For example, if the output is growth in the number of enquires the process objective is to grow the number of enquiries. You can then ask several questions to determine the processes, sub-process or activities:

1   Ask what processes deliver these outputs or achieve these objectives;
2   Ask what activities produce these outputs or;
3   Ask what affects our ability to deliver these outputs which produces a list of things you need and the activities are how those things are acquired from others or created.

The second question may generate different answers from the third question, but asking both validates the answers to the second question. If you think you need to do X to achieve Y but when you pose question 3, there is no X in the list, you may have deduced that X is not critical to achieve Y and therefore is not a real process but part of another process. When you pose question 3 and find you need A (e.g. people with a specific skill), A is provided by another process and is therefore an input. However, if you find you need to complete G before starting H, G becomes a critical activity in the process (e.g. cleaning a surface before painting it).

Starting at Level 1 (see Figure 16.2) answering question 1 of the business outputs will identify the business processes in the call-out text, for example, an output of the business is a fulfilled demand; therefore, a demand fulfilment process is needed. Asking question 2 relative to the demand fulfilment process will identify the level 2 work processes, that is, the activities that produce the outputs such as plan production and produce product. Taking one of these work processes and repeating question 1, we identify level 3 processes such as set-up machine, make parts, etc. Answering question 2 of the make parts process, we identify the individual tasks at level 4.

Let us now suppose that when answering question 2 at level 3 you identified "inspect" as an activity but when answering question 3 you deduced that making conforming parts was the key factor, not inspection; thus inspection was an operation at level 4 and not a process at level 3 (Figure 16.2).

Deriving the business processes

Our working definition of a QMS is that it's *a systemic view of an organization from the perspective of how it creates and retains its customers*. We are therefore looking at the whole organization and selecting different groups of activities as follows:

a)   Activities that provide products and services that retain customers. This might be production but if the customer requirement is detailed in performance terms rather than in terms of a solution, it might also include product design. There are many other ways of satisfying a demand, and once again to avoid using labels that are also names of departments, a suitable name might be a demand fulfilment process.

b)   Activities that provide the resources needed by the other activities. The planning, acquisition, maintenance and disposal of resources would not be part of demand creation and fulfilment as resources are not an output of these processes. There is therefore a need for a process that manages the organization's resources, and so we might call this the resource management process.

c)   Activities that create the capability and the environment, provide direction and keep the organization on track to fulfil its vision consistent with its mission. We could call it a business management process, but we also might call the system the business management system so this could cause confusion. As the process plans the direction of the business and reviews performance against plan, we could call this process the mission management process.

These business processes and the purpose of each process explained as follows with Table 16.3 showing the stakeholders at each end of the process.

Mission management process	Determines the purpose and direction of the business, continually confirms that the business is proceeding in the right direction and makes course corrections to keep the business focused on its mission. The business processes are developed within mission management as the enabling mechanism by which the mission is accomplished.
Resource management process	Specifies, acquires and maintains the resources required by the business to fulfil the mission and disposes off any resources that are no longer required.
Demand creation process	Identifies new technologies, new markets, unsatisfied customer needs and exploits markets with products and services and a promotional strategy that influences decision makers and attracts potential customers to the organization. New product and service development would form part of this process if the business were market driven.
Demand fulfilment process	Converts customer requirements into products and services in a manner that satisfies all stakeholders. New product development would form part of this process if the business were order driven (i.e. the order contained performance requirements for which a new product or service had to be designed). These processes are interconnected as shown in Figure 16.3

If we take the AQPC processes and ask a simple question: *What contribution do these activity groups make to the business?* we can reduce the number of processes to four. If we now ask: *What name should we give to the process that makes this contribution?* we will identify four processes into which align with the 13 AQPC process categories as shown in Table 16.2.

Table 16.2 Process classification alignment

AQPC Category	Process classification framework AQPC classifications)	Contribution	Business process
1	Develop vision and strategy	These set the goals and enable us to achieve them	Mission Management
12	Manage knowledge, improvement and change		
10	Manage environmental health and safety		
11	Manage external relationships		
2	Design and develop products and services	These create a demand	Demand creation
3	Market and sell products and services		
4	Deliver products and services	These fulfil a demand	Demand Fulfilment
5	Manage customer service		
6	Develop and manage human capital	These provide capable resources	Resource management
7	Manage information technology		
8	Manage financial resources		
9	Acquire, construct and manage property		

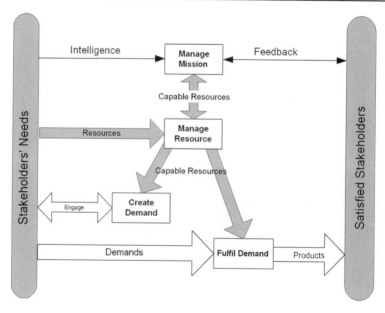

Figure 16.3 Business Process perspective of the organization

Activity driven approach (bottom-up)

Instead of coming at it from objectives, you can do it the other way around by identifying a sequence of activities and then:

1   Ask "what is the output or objective of this activity?" and thereby identify a stage output.
2   Ask "where does this output go to?" and thereby identify the next stage in the process.
3   Follow the trail until you reach the end of the chain of stages with several outputs.
4   Collect the answers from group to group, department to department and then.
5   Ask "what do these groups of outputs have in common?" and thereby identify a series of activity groups and then.
6   Ask "what contribution do these activity groups make to the business?" and thereby identify the business processes. For example, advertising creates a demand; therefore, it is part of the demand creation process.

Steps 5 and 6 are illustrated in Figure 16.4. Which groups of activities to include in a cluster depends on the way the questions are answered. There is no right or wrong answer, and each organization will be different depending on its context. It's an iterative process so at a later stage you might find that a group should be in a different cluster.

The bottom-up approach involves everyone but has some disadvantages. As the teams involved are focused on tasks and are grouping tasks according to what they perceive are the objectives and outputs, the result might not align with the organizational goals; these groups may not even consider the organization goals and how the objectives they have identified relate to these goals. It is like opening a box of components and stringing them together to discover what can be made from them. It is not very effective if one's objective is to satisfy the external customer – therefore, the top-down approach has a better chance of linking the tasks with the processes that will deliver customer satisfaction.

In every organization, there are sets of activities but each set or sequence is not necessarily a process. If the result of a sequence of activities adds no value, continue the sequence until value is added for the benefit of customers – then you have defined a business or work process.

There can be a tendency to drill down through too many layers such that at the lowest level you are charting movements of a person performing an activity or identifying pens and pencils in a list of required resources. For describing the processes needed for the QMS, it is rarely necessary to go beyond a task performed by a single individual. As a rough guide, you can cease the decomposition when the charts stop being multi-functional.

Table 16.3 Business process stakeholders

Business process	Input stakeholder (inputs)	Output stakeholder (outputs)
Mission management	Investors, owners (vision)	Investors, owners (vision achieved)
Demand creation	Customer (need)	Customer (demand)
Demand fulfilment	Customer (demand)	Customer (demand satisfied)
Resource management	Resource user (resource need)	Resource user (resource satisfies need)

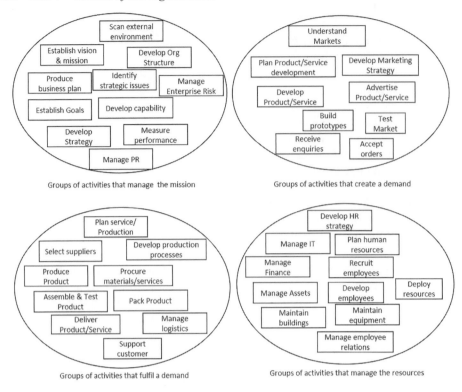

Figure 16.4 Clustering activity groups into business processes

Purpose and objective

From the definitions of a process it is clear that every process needs a purpose for it to add value. The purpose provides a reason for its existence, for example, the purpose of a recruitment process is to recruit people, and thus the purpose of a process is usually derived from the name by which it is known or vice versa. The purpose statement should be expressed in terms of what the process does and in doing so identify what if anything is to be converted or transformed, for example, the purpose of a sales process may be expressed either as "to sell products and services produced by the organization" or it could be expressed as "to convert prospects into orders for the organization's products and services".

The objective of a process is determined by the results to be achieved. Process analysis does not begin by analysing activities or operations. It begins by defining the results to be achieved. Those who commence process analysis by determining the results to be achieved will soon find themselves asking, "Why do we do this, and why do we do that?" There may be no answer other than "We have always done it that way."

The particular question to ask to reveal the objective of a process will vary depending on the answers you get to the question "What are you trying to do, what results are you trying to achieve, what is the end product or what is the output of this process?" This will vary for each instance of the process, for example, today you are trying to recruit graduates for work in new product or service development, but tomorrow you might be trying to recruit a new chef for the kitchen. The steps you take and the acceptance criteria will be influenced by the objective. You may want the chef to cook a meal to prove her competence but you would probably ask the graduate engineer for references and a resume or CV.

There is therefore a generic purpose for a process (why it exists or what it does) and a more specific result it is designed to achieve (what result is it intended to deliver or its objective).

Hierarchy

When using the top-down method, the key stages in the process are determined from the process objective. For each key stage charting every activity can make process maps appear very complex, but by layering the charts in a hierarchy, the complexity is reduced into more digestible proportions. The sequence of processes can also be demonstrated through documented information that define the direction of flow and the conditions for commencing and completing processes so that those engaged in a process are aware of the constraints which enable smooth transition between them

There is a hierarchy of processes from the business processes to individual operations such as "Measure dimension" as illustrated Figure 16.5. Depending on the level within the process hierarchy, an activity might be as grand as "Design product" or as small as "Verify drawing".

There are several activity levels. If we examine this hierarchy in a demand creation process the result might be as follows:

- A Level 1 activity might be "Develop new product". If we view this activity as a process, we can conceive a series of activities that together produce a new product design. These we will call Level 2 activities.
- A Level 2 activity might be "Plan new product development". If we view this activity as a process, we can conceive a further series of activities that together produce a new product development plan. These we will call Level 3 activities.
- A Level 3 activity might be "Verify new product development plan". If we view this activity as a process, we can conceive a further series of activities that together produce a record of new product development plan verification. These we will call Level 4 activities.

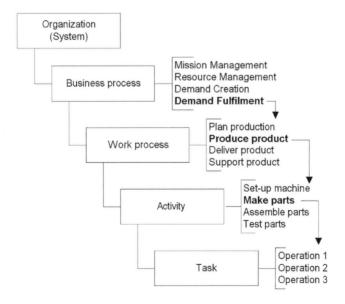

Figure 16.5 Process hierarchy

- A Level 4 activity might be "Select verification record blank". Now if we were to go any further in the hierarchy we would be in danger of noting arm movements. Therefore, in this example we have reached the limit of activities at Level 4.

If we now examine these series of activities and look for those having an output that serves a stakeholder's needs, we will find that there are only two. The demand creation process has "demand" as its output. This serves the customer, and the new product development process has "Product design" as its output, and this also serves the customer. The series of new product development planning activities has an output which is only used by its parent process so remains a series of activities. The activities of "Verify product design plan" and "Select verification record blank" only have any meaning within the context of a specific process so cannot be classed as processes.

Separating monitoring, measurement, analysis and evaluation processes

In the ISO Guide to the process approach there is a sequence of activities one of which is determining the processes of the organization and the guidance given is that "These processes include management, resources, operations, measurement, analysis and improvement" (ISO/ TC176/SC2/N1289, 2015). These are general groupings and should not be interpreted as being labels for specific processes or that, for instance, operations processes are separate from measurement processes. In fact, there are dangers in treating these activities as separate processes because they may be made to serve objectives that have not been derived from the parent process objective. Resource management, measurement, analysis and improvement processes are support processes and can't function separately from management and operations processes. These are the parent processes but sometimes the purpose of the support processes is forgotten and a situation as illustrated in Box 16.1 develops.

Box 16.1   Beware of reductionism

If the manager of an internal department starts to think of its analysis operations as a separate business unit, it might begin to promote its analysis capability beyond its original remit resulting in the resource allocation being depleted.

A managing director decides to outsource the internal quality audit function, and the new supplier becomes so focused on undercutting the competition that the information yielded by the audits is of little value other than to keep the ISO 9001 certification body content. The managing director, being unaware of the part played by the internal quality audit, tasks the financial director with a new job of determining whether the organization's policies and strategies are being implemented as planned, thus re-establishing the internal quality audit function under a new manager.

How is this demonstrated?

A process is formed when activity produces results; no activity, no process; therefore, a process that is yet to be activated can only be shown to have been determined by presenting a description of how it is intended to operate.

Demonstrating that the processes needed for the QMS have been determined may be accomplished by:

a) presenting evidence of an analysis of stakeholder requirements linking them to the businesses processes that produce outputs that satisfy them and showing which of these processes serve customer satisfaction or;

b) presenting evidence of an analysis linking objectives to the processes that achieve them and showing how the objectives serve customer satisfaction;

c) presenting evidence of an analysis linking activities being performed to the processes of which they are a part and showing how the objectives of these processes serve customer satisfaction;

d) presenting lists of processes can also demonstrate that the processes have been determined if the objectives and outputs of each of the listed processes are also defined and some means is provided which shows how they contribute to the intended results of the QMS.

Some organizations present process maps, process hierarchies, charts of process networks, but without additional information, these are not evidence that the processes needed have been determined. Who is to say all these processes are needed or if any are missing?

Applying the processes of the QMS throughout the organization (4.4.1)

What does this mean?

It is possible that the word *application* is an error and what is intended is "operation of processes" or "effective operation of processes" as in clauses 4.4c, 7.1.3 and 7.1.5.

Having determined that a process is needed to achieve an objective which serves the achievement of quality, the next step would be to ensure these processes are resourced and made operational. However, processes require activation, that is, an event that triggers them into operation and thus it is only when that trigger event occurs will these processes be operational. Until then they are models.

Why is this necessary?

Unless processes run or operate, the organization won't produce any outputs. However, the processes that are operating need to be those that have been determined as necessary rather than some other processes. Outputs are often the result of a network of processes, and therefore several processes may be operating in series and/or in parallel. However, there will be dormant processes that have yet to be activated or reactivated.

How is this addressed?

To ensure that a predetermined process is run when the event that would activate it occurs, the people that will respond to that event need to be aware that a process has been designed for such an event. This is the role of process maps and process libraries. They show relationships among the processes, their inputs and outputs and reference process descriptions where further information can be found.

Everyone should be aware of the processes in which they are engaged on a daily basis and be familiar with the actions, interactions and the supporting documented information. For processes in which they are engaged infrequently they should know where to access the information and resist the temptation to act on impulse. However, experience is a good teacher and this is an area where risk-based thinking is applied. Competent personnel know when they can take risks and when they shouldn't.

How is this demonstrated?

Demonstrating that the processes needed for the QMS are being applied throughout the organization may be accomplished by:

a)   presenting evidence that information on the processes needed for the QMS has been communicated to all personnel who need to be aware of them;

b)   selecting a representative sample of personnel and showing that they can:

    i   describe the process in which they are currently engaged;

    ii   locate this process on the process map;

    iii   locate the supporting documented information for this process;

    iv   show that the process they describe is as depicted on the process map and supporting documentation.

The operation of processes that run continually or continuously such as those in production and service delivery can easily be shown to be operating. There will be inactive processes that have yet to be reactivated such as a management review process where evidence of their operation can be demonstrated through records of previous cycles of the process, or for disaster recovery or business continuity processes where the only evidence of their operation is from past trials and exercises. Then there may be processes that have yet to be activated such as those designed to handle special events or new opportunities that cannot be simulated. The only way of demonstrating operation of these processes is through their similarity to other processes that have been operated successfully.

Determining the inputs required (4.4.1a)

What does this mean?

A process must be supplied with the items on which it is to act and the resources with which to carry out the actions – these are the process inputs but it has not always been the case as explained in Box 16.2.

Box 16.2   New requirement for process inputs

The 2008 version did not require the process inputs to be determined. It was probably taken as being self-evident, but since ISO 9000–1:1994 there has been a recognition that process inputs include information as well as items to be transformed and that resources were part of the process. However, in Figure 1 of ISO 9001:2015 resources are depicted as being inputs, which creates an anomaly because clause 4.4.1d) also requires resources to be determined implying they are not inputs.

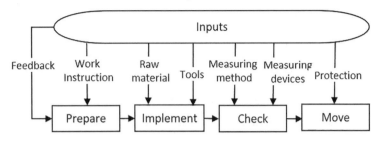

Figure 16.6 A process with multiple inputs

Multiple inputs

Within the process depicted in Figure 16.6 the work instruction is the trigger that activates the process, but the other inputs are taken in when necessary. People are input into every stage but are not shown. Other than demand-specific controls, constraints and resources all others are built into the process design. Feedback from another process is treated as an input and would be acted upon at the stage applicable.

There are therefore different types of process inputs:

- tangible items put into the process for conversion or transformation into tangible outputs;
- instructions that require work to be done or work to be re-done. these are not transformed or converted;
- resources that are taken into the process and used as and when needed.

Activators as inputs

Processes need to be activated to produce results. The activator or trigger can be an event based, time based or input based. With an event activated process operations commence when something occurs (e.g. a disaster recovery process). With a time-activated process operations commence when a date is reached (e.g. an annual review process). With an input activated process operations commence on receipt of a prescribed input (e.g. printed books are received into the binding process).

The concept of process activators enables us to see more clearly how processes operate and better understand the realities of process management.

Why is this necessary?

A process cannot function without inputs, and these need to be determined in advance so that provisions can be made either to make an appropriate response or to acquire the appropriate items at the appropriate stages so that the process runs as well as possible.

How is this addressed?

Process inputs are determined by answering the questions at each stage of the process:

- What do we need:

  - to commence each stage of the process?
  - to perform each activity and make each decision?
  - to close each stage of the process?

- Where will this come from?
- What criteria will we use to judge whether the inputs are fit for use?

How is this demonstrated?

Demonstrating that the inputs required to the processes needed for the QMS have been determined may be accomplished by presenting evidence such as a process map or process description that identifies the process inputs in term of:

a) what they are;
b) where they come from;
c) the acceptance criteria.

Determining the outputs expected (4.4.1a)

What does this mean?

Box 16.3   New requirement for process outputs

The 2008 version did not require the process inputs or outputs to be determined. It was probably taken as being self-evident but it was later recognized that not everything produced by a process is delivered to customers, and so a distinction had to be made between items produced for internal use and items produced for customers. The term *output* was used for items produced by a process and the term *product* re-designated for items produced by the organization. The term *output* was introduced into the ISO 9000 vocabulary in 2015 and defined as "result of a process".

The outputs of a process are considered to be the direct effects produced by a process. There is an assumption that all outputs from a process must be output at the end of a process and because ISO define a process as activities that use inputs to deliver an intended result, it is also assumed that the outputs will have value but both these assumptions are invalid. Anything that is absorbed by the environment beyond the process or that remains after the process is complete is an output, and therefore noise is an output. Also, information that is required to be captured and documented comprises outputs because they remain after the process is completed. Both noise and information can be produced and released at any stage from the process.

The outputs expected will be the results that the process is intended to produce. These will include the tangible and intangible results produced in pursuit of the process objectives. The tangible results could be conforming or a sales agreement. The intangible results could be ambiance in a restaurant or confidence in a transaction.

There will also be avoidable outputs and unavoidable outputs. The avoidable results will be unexpected should they arise but there will be some unavoidable outputs due to the inefficiencies in the process that are expected such as nonconforming items, noise and waste as illustrated in Figure 16.7.

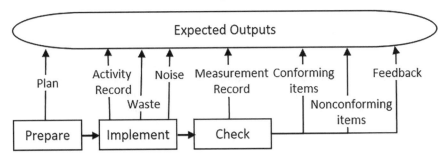

Figure 16.7 A process with multiple outputs

Note that the output is not simply product because product is not a result. Knowing that a process produced a product is unhelpful data. Knowing that a process produced 24 conforming products out of 25 that were planned is useful information.

As the term *product* is defined as "output of an organization that can be produced without any transaction taking place between the organization and the customer", there will be process outputs that become products when they exit the organization. This creates an anomaly because a process may produce intermediate outputs such as information in the form of a purchase order which becomes a product if it exits the organization. In practice it's not a problem but in theory, as a product, it is subject to the requirement in ISO 9001 relating to products such as those in clauses 8.2, 8.3 and 8.5. Hopefully common sense will prevail and no auditor would be expecting a purchase order to be subjected to such requirements.

Results

Results comprise outcomes and outputs see Box 16.4 and include:

- what is being produced in terms of the intended output?
- what is being produced in support of the intended output:

    - objective evidence of results achieved?
    - objective evidence of activities carried out?

- what unavoidable outputs are being produced due to process inefficiencies?

Box 16.4   Results

Results are the outcomes and outputs of a process and both have impacts on stakeholders.

Outputs are the direct effects produced by a process. These may have a direct effect upon a stakeholder.

Outcomes are the indirect effects of a process upon a stakeholder.

The outputs from business processes should be the same as the business outputs and these should arise out of an analysis of stakeholder needs and expectations (see Table 16.1).

The outputs expected may not be the outputs required. An example may clarify this. A current process output might be 50 units/week but this does not mean that 50 units/week was the objective. The objective might be to produce only 20 conforming units/week, so of the 50 produced, how many are conforming? If all are conforming, the process is producing surplus output. If fewer than 20 are conforming, the process is not capable. Therefore, the outputs you are currently producing are no indication of the outputs the process was designed to produce.

How is this demonstrated?

Demonstrating that the outputs expected from the processes needed for the QMS have been determined may be accomplished by presenting evidence such as a process map or process description that identifies the process outputs expected in term of:

a)   What they are?
b)   Where they go to?
c)   What information should be conveyed with them?

Determining process sequence (4.4.1b)

What does this mean?

Objectives are achieved through processes, each delivering an output that serves as an input to other processes along a chain that ultimately results in the objective being achieved. Sequence in this context therefore refers to the order in which the processes are executed to achieve a given output. The sequence maybe serial and/or parallel as shown in Figure 16.8.

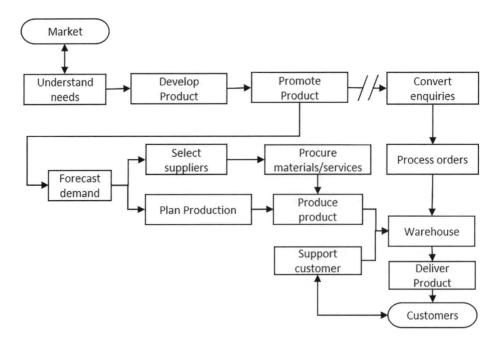

Figure 16.8 Sequence of processes

The requirement specifically refers to the sequence of processes rather than the sequence of activities within a process and in fact there is no requirement to determine the sequence of activities within a process. It is reasonable to assume that the sequence of activities is embodied within the requirement to determine processes, the argument being that if you don't know what activities are needed to produce the output and in what sequence they need to be performed, you can't have determined (found out) the process, where it starts and ends, and which activities are contained within its boundary.

Why is this necessary?

Those engaged in processes need to be aware of any constraints on the sequence of processes and the direction in which information and product flows through the process network. Such knowledge enables workers to plan their work, prevent and diagnose problems such as those arising from late delivery of inputs, bottlenecks, duplication of work and other inefficiencies that impair system effectiveness. Were people to be unaware of process sequence, abortive work might ensue and scarce resources wasted.

How is this addressed?

Sequential flow

Processes are often depicted as a flow chart representing a sequence of processes implying as one process is completed another starts but it is by no means always the case. If we examine Figure 16.8, we find that although the "Develop Product process" follows that of the "Understanding the market process", by presenting these activities as a flow it implies not only that one follows the other but also the latter does not commence until the former has been completed. This is clearly not the case. Understanding the market continues well after the product development is complete. Also, the "Promote product process" continues until the product is withdrawn from sale. In fact, several parts of this process may well be active at the same time but if we take one specific product and one specific customer, clearly when the customer makes enquiries, the promotion activities cease as they have attracted this customer to the organization.

Where the output depends upon work being executed in a defined sequence then it can be represented as a flow chart but when activities are activated by events or by time as opposed to inputs, there may be no flow between them as is the case between "Promote product" and "Convert enquiry". In the gap is not only a delay but the external environment where there are factors that affect the effectiveness of the promotion process.

Parallel flow

More than one process can run at the same time. A process may trigger more than one other process as shown by the "Forecast demand" process in Figure 16.8 creating a new sequential flow that returns to the main flow at some point.

Feedback loops

Within a process there will be loops feeding back information to a previous stage for action. Between processes there will also be feedback loops and between the organization and its customers there will be feedback loops. Only one feedback loop is shown in Figure 16.8 for clarity, and this is from the customer into the "Support customer" process which passes instructions to the warehouse to deliver a replacement product.

How is this demonstrated?

Demonstrating that the sequence of processes needed for the QMS has been determined may be accomplished by presenting: evidence such as process maps or process descriptions that identify the order in which the processes are activated

Determining process interaction (4.4.1b)

What does this mean?

Does interaction mean interconnection?

TC 176 have not defined what they mean by the phrase "process interaction" but in their document "Concept and use of the process approach for management systems" (ISO/TC176/SC2/N544R3, 2008), Figure 16.9 shows an example in which there are lines with arrows showing the direction of flow linking various shapes with no further explanation given. The arrows could be indicating a one-way or two-way flow of information or parts. An example of this type of interaction is shown in Figure 16.8. A better term for the linking of processes in sequence would be interconnection.

The verb *connect* means to join, fasten, or link together, whereas the verb *interconnect* means to connect each with the other (e.g. in Figure 16.9 there are processes A, B, C and D).

- process B is connected to processes A, C and D;
- process A is connected to process B but is not connected to processes C and D except through process B;
- processes A, B, C and D are therefore all interconnected.

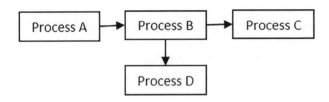

Figure 16.9 Interconnected processes

Does interaction mean reciprocal action?

The word *interaction* means a reciprocal or influencing action (OED, 2013) and for there to be a reciprocal action a change in A causes a change in B which has the effect of changing A. It may be a one-to-one relationship or a one-to-many relationship. The mechanism that causes this effect is known as feedback. As explained in Chapter 8, there are two types of feedback. One is where feedback is used to control an output as shown in Figure 8.6, and the other is where feedback operates to exaggerate or to amplify behaviour as shown in Figure 8.7.

Why is this necessary?

Assuming that the standard is referring to the sequence and interconnection of processes, it has become a convention to depict the sequence of processes with arrow-headed lines connecting the processes together with the arrows showing the direction of flow of information or material.

However, we also need to know what levers to pull to regulate the flow of outputs, their quantity and their quality. These levers are the variables, the outputs from processes and determining how they interact within the system enables us to understand how the organization works, what affects its performance and thereby manage it more effectively. Some variables have more of an impact on system performance than others. Understanding process interactions enables us to prioritize actions and avoid wasting effort on improving variables that add little value to the whole, putting more effort into improving those variables that add the most value.

When system performance is running at an optimum, the performance of its components may not necessarily be at an optimum but the interactions are being balanced. It is therefore incumbent upon managers to manage the interactions between processes to achieve the organization's goals and not unilaterally change their objectives, practices or performance thus destabilizing the organization.

How is this addressed?

Different ways of showing interactions

If we adopt the TC 176 interpretation of the term *process interaction*, what we need to do is produce a process map showing the direction of the inputs and outputs whether it's one way or two-way as illustrated in Figure 16.8. This will show the feedback loops that are used to control an output. However, these won't enable us to see the feedback loops that exaggerate or amplify behaviour. To reveal these we need to use causal loop diagrams. There are many excellent examples of causal loops diagrams in "Seeing the Forest for the Trees" (Sherwood, 2002).

Actually, it is not the processes that interact but the process outputs which are themselves variables; their quality and quantity can go up or down depending on what influences the processes. Interaction in this context therefore refers to the way process outputs affect other process outputs. Actions have consequences and within a system, changing one variable can have a positive or negative effect on other variables.

Which method to choose?

Process maps that show the sequence and interconnection of processes would indeed be useful providing the people engaged in the processes are involved in their creation. They are useful in explaining how work gets done, the sequence in which it should be done and where information flows through the organization. Producing causal loop diagrams as a means of studying the relationships among influences are useful in exploring what happens to the flow of outputs, their quantity and their quality when changes are made either to the process or to the demand on the process.

Managing interactions

There are two stages at which process interactions are determined, one is during process planning as explained earlier and the other is during the operation of processes. It is often at the interface between processes that problems arise in the timeliness or quality of inputs. Often personnel will compensate for process failure and hide the weaknesses in the interests of keeping the peace and maintaining work flow. Sometimes this is not possible and arguments and bottlenecks arise. Both the hidden and the visible interactions need to be determined and addressed and this is a role of a cross functional *process owner* (see 5.3b in Chapter 20).

How is this demonstrated?

Demonstrating that the interaction of processes needed for the QMS has been determined may be accomplished by:

a)   presenting evidence such as process maps or process descriptions showing the direction of the inputs and outputs and whether they are one way or two way or;

b)   presenting evidence such as causal loop diagrams which show the direction in which changes in the quantity or quality of the process outputs influence interdependent process outputs;

c)   presenting evidence that interactions between cross functional processes are actively being managed.

Clausal loop diagrams can show the consequences of altering a process variable which linear flow diagrams can't do but external auditors are likely to accept linear flow diagrams showing interconnections as evidence of process interaction.

Determining the criteria (4.4.1c)

What does this mean?

Determining criteria

The criteria for the effective operation and control of processes are the factors that affect its success. Determining the criteria means determining two things:

a)   the characteristics by which performance is judged. (These are the measures of process quality, for example, conformity of output, yield (ratio of conforming items to total produced), resource utilization, throughput, duration);

b)   the level of performance to be met. (These are the standards for starting, running and stopping the process, e.g. specification, requirement, budget, quota, plan.)

Box 16.5   Performance indicators

The quantifiable characteristics that indicate the extent to which an objective is being achieved.

A quality characteristic specifies what has to be produced and the performance indicator specifies how well it has to be produced (e.g. hygiene is a quality characteristic and a maximum bug count from a swab of a specified object would be the performance indicator).

Applying criteria

Applying the criteria means putting them into effect. Perhaps the reason why the words *and apply* were inserted in the 2015 version is to emphasize that determination is insufficient, but there is a requirement for the QMS and its processes to be implemented so the change should not have been necessary.

Effective operation, and control

A process that is operating effectively delivers the required outputs of the required quality, on time and economically while meeting the constraints that apply to the process.

Why is this necessary?

A process is not effective if it delivers the required quantity of outputs but they don't possess the required characteristics, are delivered late, waste resources or breach health and safety, environmental or other constraints. It is therefore necessary to determine the criteria for the acceptability of the process inputs and process outputs and the criteria for acceptable operating conditions sometimes referred to as standard operating conditions. Thus, it is necessary to ascertain the characteristics and conditions that must exist for the inputs, operations and outputs to be acceptable. There are starting conditions, running conditions and shutdown conditions for each process that need to be specified. If any one of these goes wrong, and whatever the sequence of activities, the desired result will not be achieved.

How is this addressed?

Measures

Measures are the characteristics used to judge performance. They are the characteristics that need to be controlled in order that an objective will be achieved. Juran refers to these as *the control subjects*.

There are two types of measures: stakeholder measures and process measures. Stakeholder measures respond to the question: *What measures will the stakeholders use to reveal whether their needs and expectations have been met?* Some call these key performance indicators. Process measures respond to the question: *What measures will reveal whether the process objectives have been met?* Profit is a stakeholder measure of performance (specifically the investors or stockholders) but would be of no use as a process measure because it is a lagging measure. Lagging measures indicate an aspect of performance long after the conditions that created it have changed. To control a process, we need leading measures. Leading measures indicate an aspect of performance while the conditions that created it still prevail (e.g. response time, conformity).

There are also output-driven measures and input-driven measures. Measures defined in verbs are more likely to be input driven. Those defined by nouns are more likely to be output driven. For example, in an office cleaning process we can either measure performance by whether the office has been cleaned when required or by whether the office is clean. The supervisor asks, *have you cleaned the office?* The answer might be yes because you dragged a brush around the floor an hour ago. This is an input-driven measure because it is focused on a task. But if the supervisor asks, *Is the office clean?* You need some criteria to judge cleanliness – this is an output-driven measure because it is focused on the purpose of the process. Governments often use input measures to claim that their policies are successful. For example, the success of a policy of investment in the health service is measured by how much money has been pumped in and not by how much service quality has improved.

The word *measure* does have different meanings. It can also refer to activities being undertaken to implement a policy or objective. For example, a government minister says *You will begin to see a distinct reduction in traffic congestion as a result of the measures we are taking.*

Process measures need to be derived from stakeholder measures, and a typical example of where they are not was the case in the UK National Health Service in 2005. Performance of hospitals was measured by waiting time for operations but the patient cares more about total unwell time. Even if the hospital operation waiting time was zero, it still might take two years to get through the system from when the symptoms first appear to when the problem is finally resolved. There are so many other waiting periods in the process that to only measure one of them (no matter how important) is totally misleading. Other delays started to be addressed once the waiting time for operations fell below the upper limit set by the government but in the interim period time was lost by not addressing other bottlenecks. Response is a performance measure in the UK Emergency Medical Service in 2005. The target was limited to a measure of response time. There were no targets for whether a life was saved by the crew's actions. There were also no targets for the number of instances where an ill-equipped ambulance got to the location on time and consequently a life was lost.

People naturally concentrate on what they are measured by. It is therefore vital that leaders measure the right things. Deming advocated in his 14 points that we should *Eliminate numerical goals and quotas for production*, as an obsession with numbers tends to drive managers into setting targets for things that the individual is powerless to control. A manager may count the number of designs that an engineer completes over a period. The number is a fact, but to decide about that person's performance based on this fact is foolish, the engineer has no control over the number of designs completed and even if she did, it tells us nothing about the quality of the designs.

Box 16.6   Wrong measures can have undesirable effects

Selecting the wrong measure can have undesirable effects. Somewhere there will be a measure that encourages people to take a shortcut, to deceive or cheat in order to get the job done or get a reward. With the wrong measures you can change good apples into bad apples. The person is either forced, coerced or encouraged to go down the wrong route by trying to achieve the measures upon which they are judged to reap praise or receive other rewards. However, the results you get might not be what they appear to be.

In the case of hospital waiting lists in the UK NHS, hospital administrators started to cheat in an attempt to meet the target. Patients were held in a queue waiting to get onto the waiting list, thus making it appear that the waiting lists were getting shorter. Another observation this time from the U.S. education system in the 1970s which remains as relevant in 2015 as it did back then:

> We measure the success of schools not by the kinds of human beings they promote but by whatever increases in reading scores they chalk up. We have allowed quantitative standards, so central to the adult economic system, to become the principal yardstick for our definition of our children's worth. Seems like the more we change the more we stay the same!
>
> (Kenniston, 1976)

There is interaction between measures, behaviours and standards as illustrated in Figure 16.10. This shows that measures produce behaviours that reflect the standards the group is actually following. If these standards are not the ones that should be followed, it is likely that the measures being used are incorrect.

This tells us that the quality of the output is not only dependent upon there being relevant standards in place and a process for achieving them, but that the measures must be in complete alignment with the standards, otherwise the wrong behaviours and hence the wrong results will be produced. All the measures should be derived from the stakeholder success measures or key performance indicators (KPIs); otherwise, they will influence people in the wrong direction.

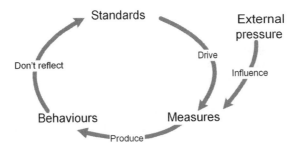

Figure 16.10 Standards, measures and behaviours

Targets

Measurements will produce data but not information. And not all information is knowledge. Managers need to know whether the result is good or bad. So, when someone asks *What is the response time?* and you tell them it's *10 minutes* they ask, *Is this good or bad?* You need a target value to convey a meaningful answer. The target obviously needs to be related to what is being measured, which is why the targets are set only after determining the measurement method. Setting targets without any idea of the capability of the process is futile. Setting targets without any idea what process will deliver them is incompetence, but it is not uncommon for targets to be set without any thought being given to the process that will achieve them. Staff might be reprimanded for results over which they have no control; staff might suffer frustration and stress trying to achieve an unachievable target.

Box 16.7   Unrealistic targets

Managers can only expect average results from average people and perhaps there are not enough extraordinary people to go around producing the extraordinary results they demand! This principle was expressed in another way by Sir Peter Spencer, Chief of Defence Procurement, UK MoD in the Bristol *Evening Post*, 30 July 2004 when he said:

> "A culture of unrealistic expectations leads to the setting of unachievable targets. We've got to change and stop being over-optimistic. We must set goals that we can be confident of achieving and then do just that."

A realistic method for setting targets is to monitor what the process currently achieves, observe the variation, then set a target that lies outside the upper and lower limits of variation – then you know the process will meet the target. There is clearly no point in setting a target well above current performance unless you are prepared to redesign the whole process. However, performance measurement should be iterative.

Defining process criteria

One essential element of the criteria for control is the requirements for the output, that is,

* the quality characteristics,
* the resources needed to produce those characteristics and
* the time it takes to produce them.

The other elements are deduced by asking the question: *What are the factors that affect our ability to deliver outputs of the quality required?* These are the constraints on the process for it to produce outputs of consistent quality.

Depending on the process concerned, these factors might be:

* the material type and condition, skill, depth of cut, feed and speed in a metal machining process;

- the physical, psychological and social environmental conditions for producing consistent outputs;
- the adequacy of the input requirement, designer competency, resource availability and data access in a design process;
- the dish description, portion size, quality and identity of ingredients, hygiene and staff competency in a food outlet;
- the audit objectives, method, timing, auditor competence, site access, data access and staff availability affect success in an auditing process.

Constraints may also arise out of a PESTLE and SWOT analysis carried out to determine the critical success factors (see Chapter 12 for further details). Values, principles and guidelines are also constraints that limit freedom for the benefit of the organization. After all it wouldn't do for everyone to have his or her own way! Some people call these things controls rather than constraints, but note Drucker's warning in Box 16.9. They include among the constraints the customer requirements that trigger the process, and these could just as well be inputs. Customer requirements, for the most part, are objectives, not constraints, but they may include constraints over how those objectives are to be achieved. For instance, they may impose traceability requirements that dictate labelling and record keeping.

Box 16.8   Types of results

- Results imply any results, good or bad.
- Specified results imply results that are communicated.
- Required results imply results that are demanded by stakeholders.
- Desired results imply results that are wanted by stakeholders.

How is this demonstrated?

Demonstrating that criteria needed to ensure the effective operation and control of processes have been determined and applied may be accomplished by:

a) presenting evidence of a process for defining the criteria for the process outputs and conditions under which they are produced;
b) presenting evidence that the criteria are being used by those operating the process;
c) presenting evidence that the application of these criteria produces outputs of the required quality.

Determining methods needed to ensure effective operation and control (4.4.1c)

What does this mean?

Determining the methods means determining the series of actions to define and deliver the results and not simply identifying a means to do something. Methods are ways of accomplishing a task or of doing something, and this can be expressed as a series of actions but can equally be a way of transmitting information, a way of preventing human error, a way of protecting the integrity of data, etc., and therefore not all methods are procedural in nature.

The methods that ensure effective operation are those regular and systematic actions that deliver the required results. In some cases, the results are dependent on the method used and in other cases, any method might achieve the desired results. Use of the word *method* in this context is interesting. It implies something different than had the standard simply used the word *procedure*. Procedures may cover both criteria and methods, but have often been limited to a description of activities to be carried out.

Why is this necessary?

Results won't happen by chance; a deliberate and systematic approach must be taken to the operation and control of processes to be confident that the required results will be produced each time the process is run.

How is this addressed?

Activities

Process activities are the actions and decisions that collectively deliver the process outputs. The activities are not determined by PDCA. Activities are determined by answering the question *What affects our ability to deliver these process outputs?* The sequence of these activities will be determined by simply asking *What do we/should we do next?*

At a high level the sequence might be that on receipt of a demand there are:

*   activities to assess the situation and if the demand is accepted, planning activities to establish how the deliverables will be produced and delivered;
*   doing activities that implement the plans;
*   checking activities to verify that the plans have been implemented as intended and to verify that outputs meet requirements;
*   activities resulting from the checking to correct mistakes or modify the plans.

This can be equated with a PDCA cycle but that would be a mistake because the PDCA cycle is a method of process improvement (see "The relevance and use of PDCA" in Chapter 5)

In principle, it should be possible to place all activities needed to achieve an objective into one of these categories. In reality, there may be processes where the best way of doing something does not follow exactly in this sequence.

Methods for operation and control of processes

Several different methods will be needed to ensure the effective operation and control of processes. Operation is not separate from control. For a process to run optimally, the controls should be embedded in the process and not separate from it. Although processes can operate without control and activities may not have to be under control to qualify as a process, it's essential that any process critical to product and service quality is under control; otherwise, the outputs would be unpredictable. Controls serve to bring about stability and a degree of predictability for reasons of economy as stated by Drucker in Box 16.9.

Box 16.9   Drucker on the control of work

Work is a process, and any process needs to be controlled. To make work productive, therefore, requires building the appropriate controls into the process of work.

Control is a tool of the workers and must never be their master. It must also never become an impediment to working.

It should always be remembered that control is a principle of economy and not of morality. The purpose of control is to make the process go smoothly, properly, and according to high standards.

The first question to ask of the control system is whether it maintains the process within a permissible range of deviation with the minimum effort. To spend a dollar to protect 99 cents is not control. It is waste. "What is the minimum of control that will maintain the process?" is the right question to ask.

(Drucker, 1974)

The following methods can be placed in three categories but are not exhaustive.

Methods for starting the process

There may be methods for:

- gathering information necessary to start to process, including the criteria that defines the operating conditions for the process, the standards to be met by both the activities, the facilities and the outputs (the key performance indicators);
- acquiring human capital;
- acquiring the physical resources such as tools, space, moving, making, monitoring and measuring equipment;
- setting up equipment and the facilities.

Methods for running the process

There may be methods for:

- doing the work;
- the safe operation of equipment and preservation of data and materials;
- informing the worker what the process is doing (monitoring);
- alerting the worker of problems;
- measuring, analysing and evaluating process outputs;
- monitoring parameters and variation in performance;
- recording and reporting results of measurement and yield;
- communicating with managers and co-workers of progress.

Methods for stopping the process

There may be methods for:

- dealing with conforming output and nonconforming output;
- disposal of waste, returning resources to designated areas;
- dealing with process interruptions, downtime, etc.

Many of these will be common for all processes of a particular type, and some may be unique to a particular process.

There are various methods of control:

- Supervisors control the performance of their work groups by being on the firing line to correct errors.
- Automatic machines control their output by in-built regulation.
- Manual machines control their output by people sensing performance and taking action on the spot to regulate performance.
- Managers control their performance by using information.

The method is defined by the words following the word *by* as in the earlier list.

You don't have to detail how a method is performed to have determined a method, for example, a method of preventing failure is by performing a Failure Mode and Effects Analysis (FMEA) but to apply the method consistently, a procedure or guide may well be needed see (AIAG, 2016). The method is therefore the way the process and constituent activities is carried out which together with the criteria contributes to the description of the process.

How is this demonstrated?

Demonstrating that methods needed to ensure the effective operation, and control of processes have been determined and applied may be accomplished by:

a)   presenting evidence of a process for defining:

    i   the methods for starting, running and shutting down processes;
    ii   the methods for monitoring and measuring process performance;
    iii   the methods for monitoring and measuring the quality of process outputs.

b)   presenting evidence that the methods identified are being applied;
c)   presenting evidence that the application of these methods produces outputs of the required quality.

Bibliography

AIAG. (2016, November). Potential Failure Mode and Effects Analysis Reference Manual. Retrieved from LEHIGH University: www.lehigh.edu/~intribos/Resources/SAE_FMEA.pdf

Drucker, P. F. (1974). *Management, Tasks, Responsibilities, Practices*. Oxford: Butterworth-Heinemann.

ISO/TC176/SC2/N1289. (2015). Retrieved from International Organization of Standardization: www.iso.org/files/live/sites/isoorg/files/archive/pdf/en/iso9001-2015-process-appr.pdf.

ISO/TC176/SC2/N544R3. (2008). ISO 9000—Quality Management. Retrieved from ISO: www.iso.org/iso/04_concept_and_use_of_the_process_approach_for_management_systems.pdf.

Kenniston, K. (1976, February 19). The 11 Year Olds of Today Are the Computer Terminals of Tomorrow. *New York Times*.

OED. (2013). Retrieved from Oxford English Dictionary: oed.com.

Sherwood, D. (2002). *Seeing the Forest for the Trees: A Manager's Guide to Applying Systems Thinking*. London: Nicholas Brealey Publishing.

Key messages from Part 4

Chapter 12   Understanding the organization and its context

1   Context was not a factor that was featured prominently in the design of management systems until ISO 9001:2015 because hitherto, organizations could be granted certification by having a bolt-on system or a system of documents that people use.

2   Every organization is affected differently by changes in the economy, markets, customer preferences, the natural environment, laws, scale of operations, uncertainties and their priorities.

3   Organizations must look ahead to see where they are going and what may impede or facilitate progress towards enhancing customer satisfaction.

4   No implementation sequence is implied by the order in which requirements are presented in ISO 9001.

5   Seeing the purpose of a business beyond the products and services it currently produces is more likely to secure its survival.

6   Of all the alternative paths that could be taken to fulfil an organization's purpose, the one that is chosen is its strategic direction.

7   A good diagnosis simplifies the often overwhelming complexity of reality by identifying certain aspects of the situation as critical.

8   Organizations that fail to periodically analyse the factors that affect its ability to consistently provide products and services that meet customer and applicable legal requirements may find their customers have found a supplier that offers a more attractive proposition.

9   Monitoring internal and external factors is on-going – it's not a one-off event.

Chapter 13   Understanding the needs and expectations of interested parties

10   The needs and expectations of customers provide the basis for an organization's objectives, whereas the needs and expectations of the other stakeholders constrain the way in which those objectives are achieved.

11   Relevant interested parties are those that could positively or negatively affect the organization's ability to consistently provide products and services that meet customer and applicable statutory and regulatory requirements.

12   Organizations must try to understand better the requirements and intentions of their interested parties then deal with them ahead of time rather than learn about them later.

13 To discover what customers will require of the products and services they will purchase, you need to discover why people buy one product or service over another
competing one.

14 Determining the interested parties and their requirements is not a one-off event.

Chapter 14   Scope of the quality management system

15 The scope of the QMS has more to do with the elements that are strongly influenced and
controlled by the system than with which elements are exposed to certification audit.

16 Anything the organization is unable to influence or has little influence over should be
in the external environment.

17 There are things that affect the outputs and outcomes of the QMS which the QMS
can control and things that it can't control but needs to mitigate and only the former
are within the scope of the QMS.

18 It's important to know where the boundary of the QMS lies to minimize the relationships the participants in the system need to deal with.

19 An organizational unit or process can only be outside the scope of the QMS if its
performance can't be controlled by the QMS.

20 The system element determines which requirements are applicable not the other way
around.

Chapter 15   Quality management system

21 Don't confuse the system description with the system because all the documents people
use to define and provide products and services that satisfy customers define the QMS
requirements not the QMS.

22 Establishing, implementing and maintaining a QMS is about creating a model of reality
and using it to bring about changes that improve organizational performance and
effectiveness.

23 Without maintenance, any system will deteriorate.

Chapter 16   Processes needed for the QMS

24 The processes needed for the QMS are not QMS processes but business processes
containing elements and interconnections that positively and negatively influence the
quality of the system outputs.

25 Process analysis does not begin by analysing activities or operations. It begins by
defining the results to be achieved.

26 There are dangers in treating monitoring and measurement activities as separate processes because they may be made to serve objectives that have not been derived from
the parent process objective.

27 Everyone should be aware of the processes in which they are engaged and be familiar
with the actions, interactions and the supporting documented information.

28 Process inputs are determined by determining what is needed to commence, perform
complete and close each stage of the process and where it will come from.

29 Process outputs will include the tangible and intangible results produced in pursuit of
the process objectives.

30 If you don't know what activities are needed to produce the output and in what sequence they need to be performed, you have not determined the process.

31 There are two types of interaction between processes: one is where feedback is used to control an output and the other is where feedback operates to exaggerate or amplify behaviour.

32 A process is not effective if it delivers the required quantity of outputs but of unsatisfactory quality, they are delivered late, waste resources or breach health and safety, environmental or other constraints.

33 The methods that ensure effective operation are those regular and systematic actions that deliver the required results.

Part 5

Leadership

Introduction to Part 5

Leadership without customer focus will drive organizations towards profit for its own sake. Leadership without involving people will leave behind those who do not share the same vision. If the workforce is unhappy, de-motivated and dissatisfied, it is the fault of the leaders. The vision, culture and motivation in an organization arise from leadership. It is the leaders in an organization who through their actions and decision create the vision and either create or destroy the culture and motivate or de-motivate the workforce thus making the organization's vision and culture key to the achievement of quality.

A search through clauses 5 and 9.3.1 will find what ISO 9001 requires of top management:

a) communicating the importance of effective quality management and of conforming to the QMS requirements;
b) engaging, directing and supporting persons to contribute to the effectiveness of the QMS;
c) ensuring that the:

   i   QMS achieves its intended results;
   ii   quality policy and quality objectives are established for the QMS and are compatible with the organization's context and strategic direction;
   iii resources needed for the QMS are available;
   iv   QMS requirements are integrated into the organization's business processes.

d) promoting improvement;
e) promoting the use of the process approach and risk-based thinking;
f) reviewing the QMS, at planned intervals, to ensure its continuing suitability, adequacy, effectiveness and alignment with the strategic direction of the organization;
g) supporting other relevant management roles to demonstrate their leadership as it applies to their areas of responsibility;
h) taking accountability for the effectiveness of the QMS.

It is interesting that in some of the bullet points top management is required to ensure something rather than do something. To ensure means to make certain and top management can't make certain that something will happen unless it is in their ability to control it or at least strongly influence it. However, it does mean that it can delegate to others the writing of the policy, the objectives and provision of resources but it can't escape responsibility for them and the way they are implemented and the results they achieve.

This part of the Handbook addresses the following clauses, each in a separate chapter but with cross-references to the others as appropriate

- Clause 5.1.1 Leadership and commitment – see Chapter 17
- Clause 5.1.2 Customer focus – see Chapter 18
- Clause 5.2 Policy – see Chapter 19
- Clause 5.3 Organizational roles, responsibilities and authorities – see Chapter 20

17 Leadership and commitment

Introduction

> Effective leadership doesn't depend on charisma. Dwight Eisenhower, George Marshall, and Harry Truman were singularly effective leaders, yet none possessed any more charisma than a dead mackerel.
>
> Peter F Drucker (1909–2005)

Defining leadership

In *Organizational Behaviour and Analysis* Rollinson draws on the work of many researchers who have studied leadership and cites a definition that is frequently used which is "the process whereby one individual influences other group members towards the attainment of defined group, or organizational goals" (Barron & Greenberg, 1990). Rollinson argues that this definition would cover situations in which coercion was used, which, he says, is not what most people associate with leadership and also that it neglects the reality that leadership is a two-way process in which there is reciprocal causality whereas the definition tends to imply it's a one way process so he proceeds to produce a definition of leadership that removes these issues which is

> "Leadership is the process in which leaders and followers interact in a way that enables the leader to influence the actions of the followers in a non-coercive way towards the achievement of certain aims and objectives".
>
> (Rollinson, 2008).

Defining top management

By using the term *top management*, ISO 9001 brings the actions and decisions of top management into the quality management system and makes them full partners in its success. Top management is a "person or group of people who directs and controls an organization at the highest level" (ISO 9000:2015). If the scope of the QMS is limited to a part of an organization, top management is the people who direct and control that part of the organization. All the requirements in this section of the standard commence with the words *Top management shall . . .* and although top management may delegate decisions to their subordinates, they carry responsibility for how their delegated authority is used.

In this chapter, we examine the 12 requirements of clause 5.1.1, namely:

* Leadership with respect to the QMS (5.1.1a)
* Commitment with respect to the QMS (5.1.1a)
* Taking accountability for the QMS (5.1.1a)

- Ensuring quality policy and quality objectives are established and are compatible (5.1.1b)
- Ensuring integration of QMS requirements into business processes (5.1.1c)
- Promoting use of the process approach and risk based thinking (5.1.1d)
- Ensuring resources are available (5.1.1e)
- Communicating the importance of effective quality management (5.1.1f)
- Ensuring the QMS achieves its intended results (5.1.1g)
- Encouraging contribution to the effectives of the QMS (5.1.1h)
- Promoting improvement (5.1.1i)
- Supporting other relevant management roles (5.1.1j)

Leadership with respect to the QMS (5.1.1a)

What does this mean?

Although the standard requires top management to undertake ten specific actions to demonstrate leadership with respect to the QMS, leadership is not so much about actions but the thinking that leads to action. As Nancy Kline expresses it, "Everything we do depends for its quality on the thinking we do first." (Kline, 1999). Demonstrating leadership with respect to the QMS means adopting a way of thinking about quality and how requirements for quality are met. It also involves exhibiting behaviours that influence the actions of its employees in a non-coercive way towards the achievement of customer satisfaction.

Why is this necessary?

Without leadership, the direction of an organization will be left to chance. Those managing it will be preoccupied with the status quo, keeping operations running, reacting to situations, which will be a challenge but leadership is required for organizations to change their performance and to consistently provide quality products and services in an ever-changing environment.

How is this addressed?

Any competent top management team will have identified its stakeholders, understood their needs and set the strategic direction in which the organization will travel over the next few years, the goals it will pursue and the values that will guide it on its journey. It should also be aware of how those goals will be achieved, the resources required etc. but it may not appreciate that achievement of its goals will depend on the quality of its products and services and how they are conceived and delivered.

   If your top management understands the importance of putting quality first when making decisions, they already possess the necessary mind-set to provide leadership with respect to the QMS. If they are of a mind to put profits first when making decisions, they have a learning opportunity that needs to be addressed. As can been seen from Figure 17.1, if integrity is a value and this is translated into the belief that you work with customers and prospects openly, honestly and sincerely, you would expect the company's customer service personnel to be open and honest when explaining a stock evaluation to customers. However, if there are people in the organization who by exerting pressure on others persuade them that loyalty is more important than integrity, the value is compromised by beliefs that are no longer a faithful translation of the value.

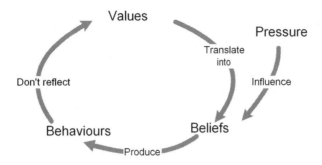

Figure 17.1 Interaction of values, beliefs and behaviours

Values are easily tested by examining actions and decisions and passing them through the set of values and establishing what they mean in practice. For example, if our value is *We treat others as we would like to be treated ourselves*, how come we pay part-timers below the minimum hourly rate? *If we value quality*, how come we allowed the installation team to commission a system that had not competed acceptance tests? If there's a misalignment between what you say is important and how people behave, it needs to be fixed immediately. Use examples and role models to get the message across rather than a series of rules, as invariably people learn better this way.

If there is one thing that top management will not be interested in, it is the detail of ISO 9001. So, it is not a case of expecting top management to understand the requirements and participate in the nitty-gritty. More importantly, it is a case of expecting top management to lead the drive for improvement in quality for the simple reason that if they lead, others will follow. If they show unwillingness to lead or appear indifferent to quality or how it is delivered (i.e. the QMS), no one else will care either.

The last thing you want is for top management to provide the funds and let you get on with it, issuing monthly reports that they scan but don't understand. If it gets to this stage, then you have failed to attract their attention. Permitting top management to take a back seat and act as observers with respect to quality and the QMS is no longer an option. They must participate in increasing the quality of the organization's outputs.

There follows a few key beliefs that top management need to share, which, were there to be any doubt, will impede the ability of the organization to create and retain satisfied customers:

- Profit is a requirement of a business and not an objective and is required to cover the risks of economic activity and thus to avoid loss.
- Economic activity arises from doing the right things right and this generates revenue.
- Revenue is directly proportional to the quality of the organizations products and services (i.e. the lower their quality the lower the revenue).
- Customers desire products and services that meet their needs and expectations in a way that poses no harm to society and they are the arbiter on what is good or poor quality.
- It's the application of quality management principles and practices that enables an organization to do the right things right and thereby produce products and services of a quality that creates and retains satisfied customers.
- Organizations which put quality first satisfy customers in a way that meets the needs of the other stakeholders.

- Quality is not the name of a department, a role or an activity but a result which if below what customers expect will drive them away.
- These results are produced by the interaction of people who are interconnected by shared values, policies, practices and resources applied through a network of processes which forms into a system.
- The system which manages product and service quality is the quality management system referred to by ISO 9001.
- That it is intended that users fit ISO 9001 to their organization and not fit the organization to ISO 9001. (provided by Paul Harding Managing, Director, South African Quality Institute)

The human factor

Many managers are well equipped to deal with technical issues but not as equipped to deal with human relationships, and yet to get anything done requires people. You need to develop a capability in handling people as well as technical capability to be able to achieve sustained success.

We can use the bicycle metaphor to illustrate these two capabilities (Hunsaker & Alessandra, 2009) as shown in Figure 17.2. Technical and people capability can be thought of as the two wheels of a bicycle.

The back wheel is the technical capability, providing the motive power, and the front wheel is the people capability taking the bicycle where the rider needs to go. This shows that you can have all the technical capability in the world but if the people won't cooperate or don't understand where to go, you won't get very far.

How is this demonstrated?

Demonstrating that top management is leading the organization with respect to the QMS may be accomplished by presenting evidence that they are:

a) establishing and communicating a clear vision of the organization's future;
b) establishing shared values and ethical role models at all levels of the organization;
c) being proactive and leading by example;
d) understanding and responding to changes in the external environment;
e) considering the needs of all stakeholders;
f) building trust and eliminating fear;

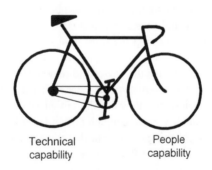

Technical   People
capability   capability

Figure 17.2 Capabilities of a manager

g)   providing people with the required resources and freedom to act with responsibility and accountability;
h)   promoting open and honest communication;
i)   educating, training and coaching people;
j)   setting challenging goals and targets that are aligned to the organization's mission and vision;
k)   communicating and implementing a strategy to achieve these goals and targets;
l)   using performance measures that encourage behaviour consistent with these goals and targets consistent with all the above.

Commitment with respect to the QMS (5.1.1a)

What does this mean?

Box 17.1   Commitment

Some men are born committed to action: they do not have a choice, they have been thrown on a path, at the end of that path, an act awaits them, their act.

Jean-Paul Sartre (1905–1980), French novelist and philosopher

---

A commitment is an obligation that a person (or a company) takes on to do something. There is a presumption that a management that is committed to the development of a quality management system will be committed to quality because it believes that the quality management system is the means by which quality will be achieved. This is by no means obvious because it depends on top management's perception of quality, the quality management system and its role in achieving the goals of the organization.

Many managers have given a commitment:

•   to quality without knowing what impact it would have on their business;
•   to meeting the requirements of ISO 9001 without really understanding what would need to change in the organization for conformity to be maintained;
•   without being aware that their own behaviour may need to change let alone the behaviour of their managers and staff.

Box 17.2   Revised requirement on commitment

In the 2008 version top management were required to provide commitment to the development and implementation of the QMS, which has in some quarters been interpreted as appointing someone to take charge of it, telling people it's important, allocating resources to it and chairing a management review once a year.

The new requirement places far more responsibility on top management for the ownership and performance of the QMS because it's not a system of documents but the means by which the organization creates and retains its customers, and that is management's responsibility.

The perceptions of quality may well be customer focused, but top management may perceive the QMS as only applying to the quality department and may therefore believe that a commitment to the development of a management system implies that they must commit to maintaining a quality department. This is not what is intended. A quality department is the result of the way work is structured to meet the organization's objectives. It has little to do with how quality is achieved (see Chapter 6). There are other solutions that would not result in the formation of a department dedicated to quality, and there is certainly no requirement in ISO 9001 for a department that is dedicated to quality.

Why is this necessary?

Staff expect their managers to honour their commitments, but it's easy to make promises to resolve immediate problems hoping that the problem may go away in due course. This is dishonest, and although managers may mean well, the problems will return to haunt them if they can't deliver on their promise.

How is this addressed?

Securing commitment is not easy. There is a road to commitment along which many will travel. Moving top management along that road is difficult; it may require transformational leadership because the motivation must come from within. You can't get commitment by holding the management to ransom or telling them what to do. The task is to move them along this road stage by stage, as indicated in Table 17.1, but the difficulty you will face is that you won't be sure which stage each of them is at until long after you have moved past that stage. Some people will appear to understand but don't, and others will tell you they have taken action but they haven't. There are several ways in which one might progress from Stage 0 to Stage 2 just as there are several ways in which people learn, of which there are four: feelers, thinkers, watchers and doers see Box 17.3.

The management must not knowingly ship defective products, give inferior service or in any other way knowingly dissatisfy its stakeholders. A manager who signs off concessions without customer agreement is not putting quality first whatever the reasons. It is not always easy for managers to honour all their commitments when the customer is screaming down the phone for supplies that have been ordered or employees are calling for promised pay raises. Priorities need to be set, as everything cannot be done at the same time but it will be evident from the performance data, providing the

Table 17.1 The road to commitment

Stage	Level	Meaning
0	Zero	I don't know anything about it.
1	Awareness	I know what it is
2	Understanding	I know what I have to do, why I should do it and what I need to do it.
3	Investment	I have the resources to do it and I know how to deploy them.
4	Intent	This is what I am going to do and how I am going to do it.
5	Action	I have completed the first few actions and it has been successful.
6	Commitment	I am now doing everything I said I would do.

key performance indicators are necessary and sufficient whether top management is committed to the QMS.

Box 17.3   Know the type of people you are dealing with

The feelers learn better by experiencing situations. They are unlikely to accept a hypothesis unless they have actual experience of a situation. They prefer examples, not theory.

The thinkers learn from logic, theory and abstract analysis and are more comfortable with abstract conceptualizations. They crave facts and often want to be left alone to work things out for themselves. They are less inclined to be influenced by examples and will often find the exception that makes the example inappropriate.

The doers learn better by experimentation. They want to get their hands dirty and get on with it. They don't need to have worked it out before rolling up their sleeves. They learn best in groups and don't relish instruction or theory.

The watchers learn by reflection and take a somewhat detached view. They come to their conclusions from careful observation and analysis. They don't need to get involved and can learn from audio-visual aids.

Adapted from (Hunsaker & Alessandra, 1986)

How is this demonstrated?

Each of the requirements in 5.1.1 identifies a way by which top management is to demonstrate its commitment and leadership with respect to the quality management system but they are the specifics and there some common behaviours that top management need to exhibit which may be revealed by presenting evidence that they:

- do what they say they will do and what they say they will do meets the needs and expectations of the stakeholders;
- don't tolerate substandard work;
- don't walk by problems;
- don't overlook mistakes;
- don't easily grant deviations from commitment;
- honour plans, procedures, policies and promises;
- listen to the customer and other stakeholders and this includes employees;
- progress, monitor and control the work they have authorized;
- divert people to resolve problems and encourage their staff to achieve performance standards.

Taking accountability for the QMS (5.1.1a)

What does this mean?

The phrase *accountability for the effectiveness* of the quality management system clearly implies that top management have to ensure that the QMS continually fulfils its purpose and hence need to understand what its purpose is (see Box 17.4).

Box 17.4   Change in the way QMS purpose is expressed

The 2008 version made frequent reference to achieving planned results which might have been interpreted as "achieve whatever the organization planned" – another way of expressing the axiom "say what you do and do what you say" that dominated use of the 1987 and 1994 versions. All this was in spite of the purpose of ISO 9001 being stated in clause 1 that it was for use where an organization needs to demonstrate its ability to consistently provide products that meet customer and applicable statutory and regulatory requirements and aims to enhance customer satisfaction.

   This purpose has not changed in the new version, but in place of "achieving planned arrangements" organizations are now required to "consistently provide products and services that meet customer and applicable statutory and regulatory requirements". It follows therefore that this is what the authors intend the purpose of the QMS to be, and although this is not explicitly stated among the requirements it is inferred in the introduction (0.1) and in lieu of any other statement in ISO 9001 this confirms the purpose of the QMS referred to in ISO 9001.

Accountability is "the quality of being accountable; liability to account for and answer for one's conduct, performance of duties" (OED, 2013). With respect to the QMS, the standard therefore requires top management to answer for the effectiveness of the QMS. So, if the QMS is fulfilling its purpose, top management should be able to explain what the organization is doing that has brought about and will sustain that success. If the QMS is not fulfilling its purpose, top management should be able to explain what the organization is doing that has brought about its failure and what if anything is being done to remedy that situation and prevent its recurrence (i.e. corrective action).

Why is this necessary?

Although top management may refer to the QMS by different names such as business management system or operating system, it is the system they have established for providing products and services that create and retain satisfied customers. It is supposed to make the right things happen and prevent the wrong things from happening and not simply be regarded as a set of documents that are created, maintained and audited by the quality department. Although the quality department or its equivalent (if one exists) may have taken the lead in developing the QMS, it is top management who own it because it's part of the organization they established. If the QMS isn't successful the organization won't be successful, revenue will fall and so will profits. They have a vested interest in the success of the QMS and should therefore willingly be able to account for its effectiveness.

   Were top management to be unable to account for the effectiveness of the QMS or whatever name they give it, it would demonstrate their ignorance of the system that creates and retains satisfied customers and put into question their ability to manage the organization.

How is this addressed?

For top management to be able to explain the performance of its QMS they need to have:

a) participated in the stakeholder analysis when the quality objectives were derived;
b) a good understanding as to how these objectives are achieved, what the critical parameters are, what risks have been addressed and what risks the organizations remains exposed to;
c) access to information on past and current performance (this information should be presented in a form that enables them to answer question 3 below).

How is this demonstrated?

Demonstrating that top management are taking accountability for the effectiveness of the quality management system may be accomplished by presenting evidence which responds to the following five questions:

1   What is it that we are trying to achieve with respect to quality? (These are the quality objectives and measures of success – the KPIs.)
2   What have we done to enable us to achieve these objectives? (These are the processes that have been designed to produce outputs that meet these objectives and which are being or will be executed as necessary.)
3   How are these processes performing relative to the objectives? (These are the results being achieved against the key performance indicators.)
4   What action was taken to prevent recurrence on the last occasion that performance in any one of these processes fell below the standard?
5   How do you know that these quality objectives are necessary and sufficient to address the needs and expectations of your stakeholders? (This should be evident from a recent stakeholder analysis which shows how the objectives have been derived from stakeholder needs and expectations.)

Ensuring quality policy and quality objectives are established and are compatible (5.1.1b)

What does this mean?

We will concentrate here on the role of top management in ensuring the quality policy and quality objectives are established and are compatible. Establishing policy and objectives that are compatible is addressed in Chapter 19. The context and strategic direction of the organization is explained in Chapter 12.

In the first version of ISO 9001, top management was required to define and document its policy for quality, including objectives for quality, and its commitment to quality. In the 2000 version, they were required to establish the quality policy and ensure quality objectives are established and in the 2015 version top management are required to ensure that the quality policy and quality objectives are established for the QMS. However, the 2000 wording remains extant as top management are required by clause 5.2.1 to establish the quality policy. But there has been a change from requiring quality policy and objectives for the organization to requiring quality policy and quality objectives for the QMS. As the term quality can apply to anything, this change limits the quality policy and quality objectives to serving the purpose of those parts of the organization contributing to the provision of products and services to customers see Box 17.4.

Why is this necessary?

It is not uncommon to find that quality policies and quality objectives have been defined by consultants or taken from books or a web page and put under the noses of top management to sign, which they do on the basis that they are needed to obtain an ISO 9001 certificate. Although it will take more than a change in the way requirements are phrased to bring about a change in thinking within top management, it is a step in the right direction to require them to demonstrate leadership with respect to the policy and objectives in conjunction with ensuring the QMS achieves its intended results. This basically says, *We the management have set these goals and you can count on us to support you 100% in achieving them.*

How is this addressed?

Quality policy

Organizations that have no policy regarding quality are allowing mediocre performance, mistakes, inefficiencies and low standards to prevail. It may also produce considerable variation in results as some managers pay attention to quality and others compromise quality by giving it a lower priority than other factors such as cost and delivery.

The first step is therefore for top management to recognize the importance of paying attention to quality, secondly for them to reflect on what policy they currently have regarding quality and then to determine what they want the quality policy to do. Afterwards they can assign someone to write it and these stages are addressed in Chapter 19.

Quality objectives

A quality objective is an objective that primarily benefits the customer by its achievement. It follows therefore that quality objectives are formulated when planning the QMS, that is, devising the system by which customers are created and retained (see also Chapter 8). To ensure quality objectives are established that are consistent with the organization's purpose and strategic direction top management must make sure that they come out of the strategic planning process as a result of identifying the issues relevant to the QMS, as shown in Figure 12.4, and this is addressed in Chapter 22.

How is this demonstrated?

Demonstrating that top management has shown leadership and commitment relative to establishing quality policy and quality objectives for the quality management system may be accomplished by:

a)   presenting evidence of a process for establishing the quality policy and quality objectives;
b)   showing where top management is involved in this process;
c)   showing that before approval of the quality policy and quality objectives they are reviewed by top management for compatibility with the context and strategic direction.

Ensuring integration of QMS requirements into business processes (5.1.1c)

What does this mean?

This requirement touches on three concepts that need to be explained.

Quality management system requirements

The term *quality management system requirements* is used in three ways. It's used in:

1   ISO 9001 to refer to the requirements of ISO 9001 itself;
2   ISO 9001 to refer to the requirements for an organization's QMS;
3   the organization to refer to the requirements of the QMS such as internal policies, procedures, standards.

There are four requirements of ISO 9001 that refer to quality management system requirements, namely:

•   5.1.1c) about their integration into business processes
•   5.1.1f) about conforming to them
•   7.3c) about not conforming to them
•   9.1.2 about auditing for conformity with them

Considering that a requirement is defined in ISO 9000:2015 as "a need or expectation that is stated, generally implied or obligatory", we are concerned not only with requirements stated in formal directives by management, customers and other stakeholders but general principles, norms and rules that are accepted for the industry sector in which the organization operates. These may include the instruments that guide what people under the organization's control do and how and when they do it. These may be contained in policies, process descriptions, procedures, and other directives and general principles, norms and rules that may not be stated in documented information.

These requirements will have been determined as a result of understanding the context of the organization and understanding the needs of interested parties, and therefore become the organization's own requirements for its QMS. These requirements may be stated in other management system standards such as IATF 16949 or ISO 13485 or in customer-specific requirements, but they may also arise from national legislation and legislation governing international trade.

If we assume that a QMS is a *quality management* system, (a system of managing quality) rather than a *quality* management system, (a management system for quality), the requirements are requirements for managing quality and therefore quality management requirements. This implies that if we satisfy these quality management requirements we will be deemed to be managing quality satisfactorily.

Box 17.5   New requirement for integration

Previous versions of ISO 9001 required a QMS to be established and implemented and this was often interpreted as requiring a new system that responded to the requirements of ISO 9001 thus separating this "system" from the business as shown Figure 3.1. The new requirement for integration of the QMS requirements into the business processes is intended to change this perception so that the system in focus is those parts of the organization that create and retain customers and direct and support them. The QMS is therefore not a separate system but a systemic view of the organization from the perspective of how it creates and retains customers.

Business processes

Business processes are the processes that are intended to produce outputs that achieve business goals and these are addressed in Chapter 16.

Integration

The requirement for integration of QMS requirements into business processes refers to requirements and not the system as a whole. It is therefore not suggesting that the QMS is to be integrated as a subsystem within the business processes. It is also not suggesting that the QMS is integrated with other systems such as an EMS or OHSMS.

Every requirement is implemented somewhere among the various business processes and not in isolation of them and this is illustrated by Figure 17.3. This shows each quality management requirement represented by a symbol being embedded into the part of the business process where it is to be implemented. A video on integrating a QMS is available on the companion website.

Why is this necessary?

Every requirement needs to get to the point of its implementation (i.e. conveyed to the people whose actions will demonstrate conformity with it). Implementation is not assured until one can see where a specific action or arrangement fulfils a specific requirement. This point of implementation is somewhere among the various business processes. It may be incorporated in a policy at a high level or buried in a procedure at a low level.

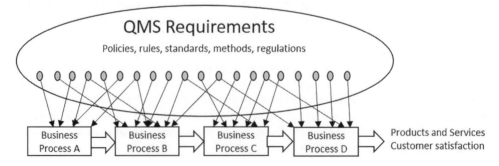

Figure 17.3 QMS requirements integrated into business processes

How is this addressed?

Top management need to understand the factors that impede the process of conveying requirements to the point of their implementation. One of the greatest barriers to integration has been treating the development of a QMS as a documentation project. It has led to the QMS being perceived as a set of documents with little relevance to the work that people do in marketing, design, production or sales. It exists to please the auditors.

Another barrier is the language. It's pointless placing policies within documents bearing the title "Quality Requirements" if that document does not pop up when it needs to be consulted. It's even worse if a user perceives from its title that it has nothing to do with what they are doing. For example, when referring to a document as quality requirements and including information on hygiene in food preparation, refer to it as hygiene requirements in food preparation, and it's more likely the requirement will get read by those preparing food in the kitchen. It's also pointless categorizing information by the department that created it when it's a different department that uses it.

Top management therefore needs to break down these barriers, and one solution is to encourage the people who work directly in particular processes to become involved in their development. It is they who implement the requirements but in a form that is relevant to what they do. The requirements must be written into training material and converted into habits or written into the information people use when doing their job.

How is this demonstrated?

Demonstrating that top management has shown leadership and commitment relative to ensuring the integration of the QMS requirements into the organization's business processes may be accomplished by:

a)   presenting evidence that the organization's requirements for managing quality have been documented;
b)   presenting evidence of the process by which the organization's requirements for managing quality are deployed;
c)   selecting a representative sample of business processes and showing that the actions and decisions taken within these processes are guided by the organization's requirements for managing quality and how they are to be achieved.

Promoting use of the process approach and risk-based thinking (5.1.1d)

What does this mean?

The process approach is explained in Chapter 9, and risk-based thinking is explained in Chapter 10, although risk-based thinking is now considered part of the process approach. These two concepts reflect a way of thinking that is different from (a) a functional approach to managing work and (b) an over-cautious approach to uncertainty.

Neither business process management nor risk management is new, and it's likely that thousands of organizations across the world have adopted such practices; however, they tend to be the larger enterprises. The process approach and risk-based thinking are less rigid than these formal practices but embody the same principles and so can be adopted by organizations of any size. Promoting the use of the process approach and risk-based thinking means that top management must have adopted the associated principles and concepts and adapted their style of management to reflect behaviours consistent with these principles and concepts that others will follow.

Why is this necessary?

Because work is always executed as a process, it may pass through a variety of functions before the desired results are achieved. If each function associated with the work serves a different objective this causes bottlenecks, conflicts and sub-optimization. Ultimately it results in strategies not being implemented and visions not being realized.

A rules-based approach to risk tends to promote an over-cautious approach when making decisions, a tendency to stick with the status quo, follow the procedures regardless of the consequences and we can end up a slave to certitude as Leonardo Buscaglia tells us in Box 17.6. There is a degree of uncertainly in everything we do. For an organization, doing nothing is not an option but neither is throwing caution to the wind. Risk-based thinking enables us to understand the risks involved in taking or not taking a certain course of action, weigh up the potential benefits and harms of exercising one choice of action over another and make an informed decision that is advantageous to the organization.

Box 17.6   Freedom or slavery

The person who risks nothing, does nothing, has nothing, is nothing, and becomes nothing. He may avoid suffering and sorrow, but he simply cannot learn, feel, change, grow or love. Chained by his certitude, he is a slave; he has forfeited his freedom. Only the person who risks is truly free.

Dr Felice Leonardo Buscaglia (1924–1998), former professor in the
Department of Special Education at the University of Southern California

---

How is this addressed?

Top management should not be promoting anything they don't understand because they need to reflect behaviours that others will follow. There is therefore a need for a promotion process which, in the context of managing quality, will include the following activities:

- reveal top management's understanding of:

  - processes and the way they believe they are currently managed;
  - risks and the way they believe they are currently identified and addressed.

- compare and contrast current understanding with the explanations of these concepts in Chapters 9 and 10;
- reach agreement on an interpretation of the process approach and risk-based thinking that is consistent with the context of their organization and also with the explanation in clause 0.3 of ISO 9001;
- consider what the people at upper, middle and lower levels would be doing if they were applying these principles in the organization and come up with some realistic examples. (see Chapter 5 for typical examples);

- compile a briefing paper that captures the key statements including the examples;
- use the process approach and risk-based thinking in planning a trial promotion;
- conduct the trial promotion with a sample of employees at each of the three levels, briefing them on the process approach and risk-based thinking, handing out the briefing notes and subsequently test understanding; if necessary, revise the briefing paper until acceptable to the sample group. this sample group may become a source of champions to promote these concepts;
- cascade the briefing paper through the layers of management requesting they run awareness events, test understanding and gather feedback;
- collect and analyse feedback and take timely appropriate action;
- continue to collect information as issues arise and undertake improvement in the promotion and awareness process.

This sequence of actions is a systematized approach which may not be appropriate in all situations. It is primarily a communication process (see Figure 31.1) that attempts to change ways of thinking, and the rate of adoption will differ depending on the different ways people learn (see Box 17.1) and the paradigms they hold (see Box 8.6).

How is this demonstrated?

Demonstrating that top management are showing leadership and commitment by promoting the use of the process approach and risk-based thinking may be accomplished by:

a) presenting evidence of a process for promoting the use of the process approach and risk-based thinking;
b) selecting a representative sample of managers at upper, middle and lower levels and seeking evidence of the extent to which they are aware of the process approach and as appropriate:

    i   how it affects the way they manage the work for which they are responsible;
    ii their interface with co-workers in the same department and in other departments

c) selecting a representative sample of employees at upper and middle levels and seeking evidence of the extent to which they are aware of risk-based thinking and how it affects the way they decide which course of action would be advantageous to the organization.

Ensuring resources are available (5.1.1e)

What does this mean?

Having determined requirements for the QMS, top management are under an obligation to ensure the resources needed for its establishment, implementation, maintenance and continual improvement are available. The word *ensure* means to make certain; therefore, top management should be in no doubt that the resources are available. However, this appears to contradict the requirement related to determining risks in clause 6.1.1. The availability of resources will be an uncertainty in many industries especially when there is a national shortage of certain skills and there is competition for natural resources.

As explained in Chapter 15, setting up a QMS includes provision of resources; otherwise, it cannot be claimed to exist except as a model. The specific resources to be considered are addressed in Part 7.

Why is this necessary?

Top management in particular need to ensure resources needed for the QMS are available because they direct the organization's resources to the endeavours which they believe will do the most to advance the organization towards its goals. Which endeavour takes priority may sometimes vary on a daily basis, often dependent on which of the stakeholders are exerting the most pressure.

Since the publication of ISO 9001 in 1987 there have been some unhelpful interpretations of what a QMS is, and this has influenced top management behaviour. Some of these interpretations are described in Chapter 3. If top management perceive the QMS to be a set of documents, it's likely they will think the resources needed for it are limited to providing the staff to maintain a set of documents and therefore not place it among their top priorities. If top management perceives the QMS to comprise all the elements that enable the organization to satisfy its customer, the resources needed will be those required for marketing, operations and support services – probably accounting for 80% of the resource budget.

How is this addressed?

Pitching the QMS as the engine that drives customers to the door and secures their loyalty is more likely to get top management to prioritize the availably of resources to the QMS but it will mean that every request for resources must be justified by the extent to which it does this. However, there will be many endeavours demanding resources that qualify on this basis, and therefore the process by which resources are procured and allocated needs to include provisions for presenting a sound business case. The business case should consider the impact on stakeholders from an assessment of the associated risks and opportunities and the process should give priority to those demands for resources that will bring benefit to customers – an approach which is consistent with the organization being customer focused (see Chapter 18).

It will, of course, be necessary to weigh the short-term and long-term benefits. Applying resources to fix an immediate problem may delay a long-term solution being funded which is a case for applying risk-based thinking (e.g. sacrificing a current customer's satisfaction for the opportunity of securing a long-term relationship with a larger group of customers). There may be situations where priority must be given to other stakeholders but these should not set a precedent.

In theory, top management could demonstrate that they are ensuring the availability of the resources needed for the QMS by including categories in the resource budgets that can be identified as serving the establishment, implementation, maintenance and continual improvement of the QMS but in practice it is not feasible. This would require the QMS to be distinguishable from other systems when in practice the QMS is merely a model and not a physical entity. One approach is to tag budget entries that serve the QMS but if 80% of the resource budget is allocated as suggested earlier, it might be simpler to tag entries that don't influence customer behaviour (see Box 17.7).

Box 17.7   Resources needed for the QMS

If it influences customer behaviour, supplier behaviour or employee behaviour, it counts.

How is this demonstrated?

Demonstrating that top management is showing leadership and commitment relative to the availability of resources needed for the QMS may be accomplished by:

a) presenting evidence of a process for procuring and allocating resources which includes provision for prioritizing allocation based on a sound business case;
b) presenting evidence that any resources needed for the QMS would be given priority;
c) selecting a representative sample of requests for resources needed for the QMS and showing that top management sanctioned the request where the business case was valid.

Communicating the importance of effective quality management (5.1.1f)

What does this mean?

Communicating the importance of effective quality management depends on what the sender and receiver of the message understand by the term *quality management*. Even when Ackoff used the term in an interview with Deming, it did not strike an immediate accord with Dr Deming, as seen in Box 17.8. ISO define the term *quality management* as shown in Box 17.9. Although this definition emphasizes that quality management must be led from the top and that its implementation involves all members of the organization, it doesn't have the simplicity of Ackoff's succinct statement. It's not that the ISO definition is wrong, but that Ackoff's definition is better for use by top management.

Box 17.8   Ackoff and Deming on quality management

DR. ACKOFF: How many universities that you are aware of have programs now in teaching quality management?

DR. DEMING: I don't know what quality management is. Quality is a product, not a method.

DR. DEMING: Put the question another way.

DR. ACKOFF: By quality management I mean how do you manage in such a way as to increase the quality of the output of an organization? I don't mean the management of quality; I mean the management of an organization to produce quality.

   Taken from the transcript of an interview with Dr. Ackoff and Dr. Deming in 1992 on "A theory of a system for educators and managers".

Box 17.9   ISO defines quality management

"Coordinated activities to direct and control an organization with regard to quality.

Quality management can include establishing quality policies and quality objectives, and processes to achieve these quality objectives through quality planning, quality assurance, quality control, and quality improvement."

(ISO 9000:2015)

The requirement is for communicating the importance of *effective* quality management not simply communicating the importance of quality management. Quality is a variable as we explained in Chapter 6, therefore even when an organization fails to satisfy some of its customers, it is managing quality but not doing it effectively. There is also no one way of increasing the quality of a person's, group's or organization's outputs. Not only is quality a variable but also, the ways of increasing the quality of outputs are also variable. What is appropriate in one situation may be totally inappropriate in another. It may also be the case that a person adopts a particular way of increasing the quality of their outputs which causes instability in the system.

The second part of the requirement is communicating the importance of conforming to the QMS requirements. These requirements were defined in Chapter 15.

Why is this necessary?

Telling people that quality is important is not enough on its own, as they also need to know how to work in such a way that will increase the quality of their outputs. This is expressed by the *term quality management*, and it's important that this message comes from top management.

How is this addressed?

In addition to this requirement there are three other requirements on the same subject, each expressed in different ways:

- the requirements in 5.1.2 for demonstrating leadership and commitment with respect to customer focus;
- the requirement in 5.2 for communicating the quality policy and;
- the requirements in 7.3 for persons doing work under the organization's control to be aware of the quality policy, quality objectives, their contribution to the effectiveness of the QMS and the implications of not conforming with the QMS requirements.

The task of educating the workforce to work in a way that will consistently increase the quality of their outputs rests with the individual managers but their task is eased a great deal if its importance has been emphasized by top management.

It is for top management to express the importance of effective quality management in their own words, and they may find it useful to emphasize the importance of developing a reputation for excellence in quality and what it means to the organization in terms of

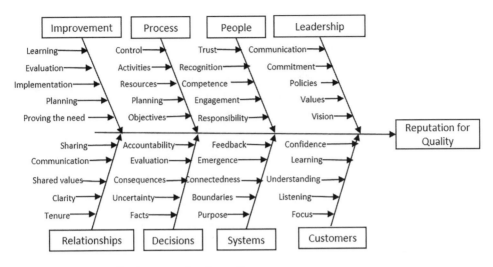

Figure 17.4 Factors affecting reputation for quality

attracting customers, creating employment, caring for the environment and contributing to society. From this standpoint, they could emphasize any number of factors on which effective quality management depends as illustrated in Figure 17.4.

How is this demonstrated?

Demonstrating that top management has communicated the importance of effective quality management and of conforming to the QMS requirements may be accomplished by:

a) presenting evidence that top management has informed personnel of the importance of quality management and of conforming to QMS requirements;
b) selecting a representative sample of personnel and establishing that they can identify the factors of their work they would focus on to increase the quality of their outputs and explain why these would have the most impact;
c) presenting the evidence top management have obtained that leads them to believe the workforce understands the importance of effective quality management and of conforming to QMS requirements;
d) selecting a representative sample of feedback from customers which includes positive and negative feedback and showing that it is consistent with these results.

Ensuring the QMS achieves its intended results (5.1.1g)

What does this mean?

For an ISO 9001 compliant QMS, the intended results of the QMS as explained in Chapter 12 are as a minimum:

a) consistent provision of products and services that meet customer and applicable statutory and regulatory requirements;

b)   demonstrated conformity to ISO 9001;
c)   enhanced customer satisfaction.

The implication in this requirement is that these results are to be achieved by the QMS rather than by something else. If the QMS is perceived as a set of rules for instance, these results could be achieved by a combination of playing by the rules or by breaking them if the rules disproportionately constrain what people do. If, as has been stated many times, the QMS is perceived as *a systemic view of the organization from the perspective of how it creates and retains customers*, the QMS includes everything that influences those results, whether people are playing by the rules or breaking them. Knowing what causes success or failure is a continuing issue in any organization. ISO 9001 is based on the premise that an organization can establish a system that will produce the previously noted results and can control the forces which influence this endeavour. However, if this was true, no ISO 9001 certificated organization would ever have unsatisfied customers. Top management cannot ensure (make certain) that the QMS will deliver these results because they don't control the universe but they can make every endeavour to enable the QMS to achieve its intended results.

Why is this necessary?

People cannot be held responsible for results over which they have no control, and this means having the ability to regulate performance in the event it is substandard. Clause 5.1.1a makes top management accountable for the effectiveness of the QMS, and with this brings the responsibility for results and hence the requirement for them to ensure the QMS achieves its intended results. Clause 5.1.1g is therefore a consequence of clause 5.1.1a.

How is this addressed?

For top management to regulate the performance of the QMS they need valid information on the performance of the QMS. This information is the same as would be subject to management review and is addressed in Chapter 56. However, top management should have access to such information on an on-going basis and not have to wait for a full review of the QMS to determine whether to intervene and issue directives for changes to be made.

There should be indicators that will alert top management to the current health of the QMS. Those indicators of performance relative to quality objectives are candidates for being the key performance indicators. However, there may be dozens of quality objectives, and so it's vital that the "key" performance indicators are exactly that. Regularly providing top management with a graphic representation of performance against each key objective is considered necessary for them to exercise control.

Although top management need this information and have a responsibility to intervene and cause performance to change, much of the information reaching top management is historical (i.e. the time when an intervention would correct performance has passed). The role of top management in this situation is to gain confidence that the controls in place on the front line are effective in maintaining performance. However, there may be situations where someone has taken their eye off the ball and it doesn't become apparent until top management become aware of the "big picture". With the use of modern IT aids, a real-time dashboard with information on KPIs can be accessible to top management whether they are in their office or travelling.

How is this demonstrated?

Demonstrating that top management has shown leadership and commitment relative to ensuring the QMS achieves its intended results may be accomplished by:

a) presenting evidence that the performance of the QMS against key performance indicators is being reported to top management;
b) presenting evidence that top management is reviewing this information and has intervened when performance is not as good as expected;
c) presenting evidence that the organization is consistently providing products and services that meet customer and applicable legal requirements.

Encouraging contribution to the effectiveness of the QMS (5.1.1h)

What does this mean?

Everyone whose work affects the quality of products and services provided to customers contributes to the effectiveness of the QMS in one way or another. Top management therefore needs to create conditions in which all these people are motivated to participate in improving the effectiveness of the QMS. The degree to which people are motivated to participate is referred to as employee engagement. Ghaleiw defines it as "the degree of an employee's positive or negative emotional attachment to their job, colleagues and organization that profoundly influences their willingness to learn and perform at work" (Ghaleiw, 2015). Employee engagement is a variable, as some people are not engaged or actively disengaged, and there are occasions when lack of engagement is solely attributable to the individual. The person may have a bad attitude or lack awareness of what impact his or her contribution makes to other workers, the organization and its customers and the other stakeholders. But the question to ask is this: Was the person hired in that state, or did something happen in the organization that prompted that attitude? If the person was hired that way, the organization should revisit its recruitment process (Rothwell, 2007). It is not a case, as implied in the requirement, that top management should be *directing* persons to contribute. Employees must be willing contributors as emphasized with point 8 of Deming's 14 points for management in which he advocated that they should "Drive out fear" (Deming W. E., 1982)

Why is this necessary?

Jon Choppin made the point that "any organization can become the best, but only with the full co-operation and participation of each and every individual contributor" (Choppin, 1997). A similar point was made by Steven Covey: "A cardinal principle of total quality escapes many managers: you cannot continually improve interdependent systems and processes until you progressively perfect interdependent, interpersonal relationships" (Covey, 1992).

   If we expand the system to include the organization and its customers, suppliers and members of the community with which it interacts, these individuals that Choppin refers to need to include everyone with whom members of the organization communicate. Creating anything but a harmonious relationship with its employees, customers, suppliers and the community can only be detrimental to the organization's success.

How is this addressed?

Research shows that workplaces in which employees have a high level of self-interest invested in the organization's success will be highly productive (Rothwell, 2007). Creating and retaining customers is a team effort, and top management does not have the monopoly on ideas for increasing the quality of organization's outputs. It is therefore necessary to engage with employees to encourage ideas for increasing the quality of a person's outputs from the individuals themselves or from the team in which they work.

Employers cannot employ a part of a person – they take the whole person or none at all. Every person has knowledge and experience beyond the job they have been assigned to perform. No one is limited in knowledge and experience to the current job they do.

What managers do and how they behave sets the tone for employee engagement. Closed-door management leads to distrust among the workforce. It is therefore not uncommon for those affected by decisions to be absent from the discussions with decision makers. Decisions that stand the test of time are more likely to be made when those affected by them have been involved. Managers should be seen to operate with integrity and this means involving the people.

Contributions from the workforce are unlikely to be forthcoming if employees don't feel valued. At one end of the spectrum are detached contributors, employees who see the value of work for its near-term economic benefits. At the other end are accomplished contributors, employees who see work as an opportunity to be a valuable part of a winning team (Rothwell, 2007), and it's the role of top management to manage a process that transforms detached contributors into accomplished contributors. This process is the employee engagement process the key stages of which are shown in Figure 17.5.

The cultural traits are the desired behaviours. They are what the leaders would expect to be happening in an organization in which the employees were engaged.

Box 17.10   Everyone is engaged

Everyone is engaged in something. They might just not be engaged in anything remotely of interest to the organization in which they are employed.

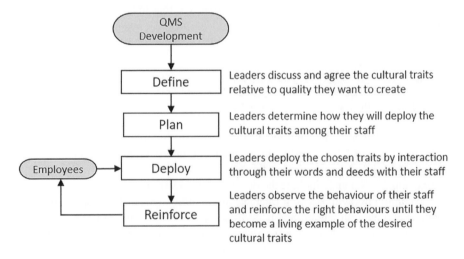

Figure 17.5 Employee engagement process relative to quality

Table 17.2 Factors affecting employee engagement

Factors that cause and enhance people engagement	*Factors that prevent/destroy people engagement*
• Helpful and efficient systems and structures • A workplace that is full of encouragement and support for the people who deliver the desired business results. • Positive and easy two-way communication. People are informed about issues that are directly related to their jobs. • People share their knowledge and focus on looking for and leveraging one another's strengths. • People are learning from their mistakes, because mistakes are seen as learning opportunities and quickly forgiven. • Positive energy and positive people. The spirit of teamwork is very strong. • People are willing to take on more tasks – they push boundaries. • People are eager to learn and their capability is being constantly assessed and improved through an effective training and development process. • Creativity and innovation are strongly encouraged. • People are treated fairly by their managers and they treat each other with dignity and respect. • Quality work is recognized and rewarded. • High loyalty and strong commitment to deliver a successful project. • Very cooperative, close relationships characterized by true caring and friendships.	• Aggressive line managers who demonstrate their lack of respect to their employees. • Unclear role and responsibility combined with unrealistic, unachievable objectives. • Lack of guidance and direction by the line managers. • Lack of openness and honesty which creates mutual distrust, hence constant worrying and suspicion becomes the norm. • Lack of knowledge and experience sharing. • Excessive time wasted defending positions and decisions. Hostile, energy-draining behaviours (yelling, blaming, accusing), that is, strong blaming culture. • Intense political atmosphere and disregard to cultural differences. • Ignoring people training and development. • Favoritism. • Lack of recognition. • Painful micro-management and bureaucracy (unhelpful management system). • Creativity and initiative are neither encouraged nor appreciated. • The individual's own negative behaviour, way of thinking, values and what work actually means to him or her contribute to his or her lack of engagement.

The key to employee engagement is for managers to focus on how things are done and how people work together, rather than the shorter but less effective approach of dictating the outcomes desired. An acute awareness of the factors that cause and prevent employee engagement is an essential prerequisite for managers when interacting with their workforce. A lot of research has been done to discover what engages and disengages employees in the workplace. One such study carried out in 2015 by Ghaleiw identified many of these factors which are reproduced with permission in Table 17.2.

How is this demonstrated?

Demonstrating that top management have shown leadership and commitment relative to engaging, directing and supporting persons to contribute to the effectiveness of the QMS may be accomplished by:

a) presenting evidence that top management has developed a process that stimulates employees to contribute to the effectiveness of the QMS;

b) presenting evidence that top management has agreed the cultural traits they want to create to produce accomplished contributors to the QMS;

c)   selecting a representative sample of employees and presenting evidence of their involve-
     ment with the employee engagement process;
d)   presenting evidence of employee surveys.

Promoting improvement of the QMS (5.1.1i)

What does this mean?

Promoting improvement of the QMS means that top management must adopt the improvement principle and the process approach (see Chapter 5) and adapt their style of management
to reflect behaviours consistent with these principles that others will follow. Improvement is
addressed in detail in Chapter 57, but in general improving the QMS means improving the
processes, products, services and the performance of the system.

Why is this necessary?

The reason why improvement in the QMS is necessary is that it is an open system, one
subject to external influences. There may be specific processes that are stable and delivering
outputs of the quality required and will remain stable provided none of the variable change.
But very little remains stable for long and so improvement is necessary.

It was Ishikawa's belief that standards quickly become obsolete (see Box 17.11) and
that organizations must continuously review their own quality standards, revise them and
improve them.

Box 17.11   Ishikawa on standards

There are no standards – whether they be national, international or company-wide –
that are perfect. Usually standards contain some inherent defects. Consumer requirements also change continuously, demanding higher quality year after year. Standards
that were adequate when they were first established quickly become obsolete.

(Ishikawa, 1985)

How is this addressed?

To demonstrate they are committed to improving the QMS, top management need to initiate improvements, not only when they are involved in the management review but on other
occasions also, for example, when market intelligence flags up an issue that exposes those
parts of the organization within the scope of the QMS to risk, or provides an opportunity to
pursue to the organization's advantage or when an issue from internal audit has been escalated to them.

The employee engagement process will promote improvement in the QMS by empower-
ing employees to contribute their ideas for improvement in products, services, processes.
However, it is important that everyone take heed of Ackoff's message in Box 20.4 to avoid
destabilizing a process and the system by encouraging improvement without consideration
of the consequence on the system.

How is this demonstrated?

Demonstrating that top management has shown leadership and commitment with respect to the QMS by promoting improvement may be accomplished by:

a)   presenting evidence that top management has initiated improvements in the QMS
b)   presenting evidence that:

    i   employees are empowered to contribute suggestions for improving products, services and processes;
    ii   processes are periodically reviewed and improved;
    iii policies, procedures, standards, codes, etc., are periodically reviewed and improved;
    iv actions from internal audits are undertaken without undue delay.

Supporting other relevant management roles (5.1.1j)

What does this mean?

All the requirements in clause 5.1.1 are addressed to top management and this final requirement points the finger at all the other managers and says "Now we are showing leadership and commitment with respect to the QMS, you need to follow our example in your area of responsibility." The only requirement in clause 5.1.1 for which the other managers don't have a remit is that for establishing the quality policy but they should be involved.

Why is this necessary?

To be a manager means sharing in the responsibility for the performance of the enterprise. Anyone who is not expected to take this responsibility is not a manager (Drucker, 1974). It is therefore only right that all managers exercise leadership relative to the QMS.

How is this addressed?

Top management may support their managers to exercise leadership relative to the QMS by:

*   mentoring and coaching;
*   developing the manager either through formal education and training or on-the-job training;
*   supporting self-development.

How is this demonstrated?

Demonstrating that top management has shown leadership and commitment with respect to the QMS by supporting other relevant management roles to demonstrate their leadership may be accomplished by:

a)   presenting evidence of a manager recruitment and development programme that includes aspects pertinent to increasing the quality of the organization's outputs;
b)   interviewing a selection of managers to establish:

    i   the extent of any education, training, coaching or mentoring provided;
    ii   that they understand the importance of all the aspects addressed in this chapter;
    iii that reasonable requests for education and/or training have been granted.

Bibliography

Barron, R. A., & Greenberg, J. (1990). *Behaviour in Organizations*. Needhan Heights, MA, USA: Allyn and Bacon.

Choppin, J. (1997). *Total Quality through People*. Leighton Buzzard: Rushmere Wyne, England.

Covey, S. R. (1992). *Principle-Centred Leadership*. London: Simon & Shuster UK Ltd.

Deming, W. E. (1982). *Out of the Crisis*. Cambridge, MA: The MIT Press.

Drucker, P. F. (1974). *Management, Tasks, Responsibilities, Practices*. Oxford: Butterworth-Heinemann.

Ghaleiw, M. A. (2015). Delivering Outstanding Performance through Highly Engaging Quality Culture. 2015 ASQ MENA Regional Quality Conference. Dubai: ASQ. Retrieved from www.asqmea.org/download/Highly-Engaging-Quality-Culture-Paper-Rev-3.pdf

Hunsaker, P. L., & Alessandra, A. J. (2009). *The New Art of Managing People*. New York: Simon & Schuster, Inc.

Ishikawa, K. (1985). *What Is Total Quality Control: The Japanese Way*. (D. J. Lu, Trans.). Englewood Cliffs, NJ: Prentice-Hall Inc.

ISO 9000:2015. (2015). *Quality Management Systems – Fundamentals and Vocabulary*. Geneva: ISO.

Kline, N. (1999). *Time to Think: Listening to Ignite the Human Mind*. London: Octopus Publishing Group.

OED. (2013). Retrieved from Oxford English Dictionary: oed.com

Rollinson, D. (2008). *Organisational Behaviour and Analysis*. Harlow, Essex: Pearson Education Limited.

Rothwell, W. J. (2007). Beyond Rules of Engagement. Retrieved from Dale Carnegie Training: www.dalecarnegie.co.uk/assets/1/7/Beyond_Employment_Engagement.pdf

18 Customer focus

Introduction

Some organizations are driven by the needs of the shareholders to maximize shareholder value. In this situation, top management communicates the organization's goals and the lower levels respond by adopting practices that enhance financial performance. Within such a culture, each level responds to the needs of the level above and its performance is measured in financial terms. This shareholder centric relationship is illustrated in Figure 18.1 where the customer-facing staff are at the base of the pyramid and the customers beneath it. Proponents of financial performance measures argue that they are necessary because of the primary objective of companies to maximize shareholder value.

The premise on which maximization of shareholder value is based is that if organizations pursue this goal, both shareholders and society will benefit. This is a flawed premise because a given variable can only be optimized, subject to other constraints. It cannot maximize both customer value and shareholder value. It must pick one main objective and treat the others as constraints – a principle of linear programming. Organizations must choose between making shareholder value the primary goal, subject to meeting a basic customer value hurdle, and making customer value the main goal, subject to creating a minimum shareholder value (Martin, 2010).

In organizations that are driven by the needs of the customer their goal is to maximize customer value. They put the customer first and in this situation, top management views itself as serving the needs of middle management who serve the needs of the customer-facing and support staff who in turn serve the needs of the customer (Miller, 2016). This customer-centric relationship turns the pyramid referred to earlier on its head as illustrated in Figure 18.2.

The customer-focused organization continually aligns its people, its processes, its products and services, in fact all aspects of the organization towards satisfying customer needs and expectations. It's a belief that a strong focus on customers, paying attention to their needs, their expectations, their preferences and their opinions will lead to customer loyalty and outstanding success. If people were to ask themselves before making a decision – how will this decision keep our customers happy and our bottom line healthy? – the organization would begin to move its focus firmly in the direction of its customers. It's an AND, not an OR, relationship. Successful organizations don't keep their customers happy OR keep their bottom line healthy – it's a cause-and-effect relationship. As explained earlier they can only optimize one variable so by keeping their customers happy they keep the bottom line healthy as long as the right things are done right.

The primary instrument for aligning the organization towards satisfying customer needs and expectations is the quality policy, and that is addressed in Chapter 19 along with the axioms of customer first and quality first which is the objective that a customer-focused culture achieves.

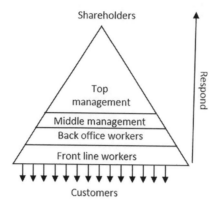

Figure 18.1 Shareholder- focused organization

Figure 18.2 Customer-focused organization

In this chapter, we examine the three ways by which top management are required to demonstrate leadership and commitment with respect to customer focus in clause 5.1.2 namely:

*   Demonstrating leadership regarding customer and applicable legal requirements
*   Demonstrating leadership regarding risks and opportunities
*   Maintaining focus on enhancing customer satisfaction

Demonstrating leadership and commitment (5.1.2)

What does this mean?

Customer-focused leadership is outwardly focused. It focuses on discovering and satisfying the needs and expectations of those who could or do receive a product or a service that is intended for or required by them. Demonstrating customer-focused leadership means that when the lower ranks *see* that their leaders (from their direct supervisors to those in the executive suite) are committed to keeping the customer in their sights, they are more likely to strive for the same focus; they'll follow their lead. The emphasis here is on *seeing*, and this means not only seeing their leaders are customer focused by the way they express themselves in information they release either orally or written, but also seeing how their leaders behave on a day-to-day basis both when operations are running smoothly and when problems are encountered. Leaders who put the customer's interest first are customer focused; they;

- listen to their customers and seek to understand their needs and expectations before offering their organization's services;
- look for opportunities to help their customers be more successful;
- look for risks that may impede their customers' success and address these risks in ways that are to their customers' advantage;
- look for ways of enhancing customer satisfaction;
- assure their customers that any offerings will meet all applicable statutory and regulatory requirements;
- don't look for ways of deceiving their customers;
- don't look for ways of profiting from a customer's lack of expertise in a particular technology.

Why is this necessary?

Where individuals in an organization come face to face with customers a lack of awareness, attention or respect may lead to such customers taking their business elsewhere. Where individuals are more remote from customers, an understanding of customers in the supply chain will heighten an awareness of the impact of their actions and decision on customers.

Organizations exist to create and retain satisfied customers – those that do not do so fail to survive. Creating and retaining satisfied customers should be the principle concern of top management for it is from satisfied customers that a revenue stream is sustained. Organizations that help their customers resolve their problems, avoid risks, exploit opportunities and comply with the law are more likely to engender loyalty from their customers.

Not-for-profit organizations have customers even though they may not refer to them as customers. Although the people the organization serves may not purchase anything, they are affected by what the organization does and if the organization fails to fulfil their needs, it ceases to exist. Governments are a prime example. If they fail to satisfy the voters, they fail to be re-elected. It is therefore essential for the survival of an organization that it determines and meets customer requirements and ensures they are understood.

How is this addressed?

When communicating its vision and strategy for the organization, top management will need to emphasize the importance of listening to customers, understanding their needs and mobilizing the organization around satisfying those needs.

Ensuring customer and applicable legal requirements are determined

For top management to ensure customer and applicable legal requirements are determined, leaders will have in place processes to research the market to reveal customer, preferences, applicable legal requirements and opportunities for developing new products and services. They will also have processes in place to determine what requirements customers are seeking to satisfy when they make enquires and place orders. It does not mean top management have themselves to determine customer and applicable legal requirements – it's an activity they can delegate but they remain responsible for the results, therefore they need to be confident that the processes they have commissioned are effective.

Ensuring customer and applicable legal requirements are understood

For top management to ensure customer and applicable legal requirements are understood leaders will put in place:

a)   processes that will deploy these requirements to their point of implementation and ensure understanding by those responsible for meeting them;
b)   review processes that confirm understanding with the customer before acceptance of an order;
c)   further reviews throughout the fulfilment processes as a continual check on understanding and will;
d)   include customer awareness skills in any training provided to enable everyone to know who the customers are and be attentive to their needs and expectations.

Ensuring customer and applicable legal requirements are consistently met

For top management to ensure customer and applicable legal requirements are consistently met they will have in place processes for reporting on performance against customer and applicable legal requirements at regular board meetings.

Ensuring relevant risks and opportunities are determined and addressed

For top management to ensure that relevant risks and opportunities are determined and addressed they will have in place processes for identifying and addressing risks and opportunities at all stages of process, product and service development. They will also put in place formal reviews throughout the fulfilment processes as a continual check on the effectiveness of processes for identifying and addressing risks and opportunities.

Ensuring that the focus on enhancing customer satisfaction is maintained

For top management to ensure that the focus on enhancing customer satisfaction is maintained they should regularly ask their direct reports:

•   What lessons have you learnt from a customer recently?
•   When was the last time you witnessed a customer using our products or services?
•   What have you learnt recently of your staff's attitude towards customers?
•   What risks have been identified recently that that could potentially impact customer satisfaction?
•   What opportunities have been identified recently to make our customer more successful?
•   What new statutory and regulatory requirements have been identified to protect our customers when using our products and services?

- What have you learnt from recent analysis of customer satisfaction levels?
- When was the last time a customer complaint was escalated to you to resolve?
- What initiatives are being deployed to identify and share customer insights and surface and resolve customer issues?
- What actions have you taken recently to directly or indirectly boost engagement with customers?

For every response to each of these questions top management should seek assurances as to the follow-up actions taken to reach a customer focused outcome. The link between employees, customers and profitability is explained in Box 18.1. Top management want their direct reports to succeed who want their office staff to succeed who want their front line to succeed who want their customers to succeed. The Cunard Line focuses all personnel under their control on their customers through use of the cornerstone statement in Box 18.2.

Box 18.1   Connecting the success factors

Employee satisfaction drives loyalty, which in turn drives productivity, because replacing experienced workers is costly. Productivity drives value, value drives customer satisfaction, customer satisfaction drives customer loyalty and, ultimately, customer loyalty drives profitability and growth.

(Field, 2008)

Box 18.2   Cunard Line's cornerstone statement

At one point in every day one of our guests will come into contact with one of us, the Cunard employee, and at that moment in time we will be Cunard Line. Our entire reputation as a company will be in our hands and we will make a lasting impression. The impression will either be good or it will be bad and we will have spoken to our guests more loudly than all of our community involvement, advertising and public relations put together. (Captured from literature available to Cunard passengers)

How is this demonstrated?

Demonstrating that top management exhibit customer focused leadership and commitment may be accomplished by:

a) presenting evidence that top management have expressed a commitment to putting customers' first;
b) presenting evidence that top management has put processes in place:

  i   that determine current and future customer and applicable statutory and regulatory requirements and deploy this information to the point at which it is implemented;
  ii that confirm understanding of a customer's declared requirements, both with the customer and with those responsible for their implementation;

     iii   that measure the extent to which customer and applicable statutory and regulatory requirements are being met at each stage of demand fulfilment;

     iv   that identify and address risks and opportunities and assess the effectiveness of measures taken.

c)   presenting evidence:

     i   that managers and their direct reports have regular contact with customers;

     ii   that top management reward customer-focused behaviour;

     iii   that performance relative to customers is visible for all employees to see how they are doing;

     iv   that managers and their direct reports are aware of the statutory and regulatory requirements that apply and can show how well they are being met;

     v   that managers and their direct reports have not sanctioned deviations from statutory and regulatory requirements;

     vi   that top management is holding their direct reports accountable for performance relative to customers rather than for adherence to procedures;

     vii   that managers and their direct reports are measuring parameters that have been derived from customer and applicable statutory and regulatory requirements;

     viii   that personnel are acting in way that is consistent with being customer focused.

Bibliography

Field, A. (2008, February 29). Customer Focused Leadership. Retrieved from Harvard Business Review: https://hbr.org/2008/02/leadership-that-focuses-on-the-1.html.

Martin, R. L. (2010, January-February). The Age of Customer Capitalism. Retrieved from Harvard Business Review: https://hbr.org/2010/01/the-age-of-customer-capitalism.

Miller, R. (2016, April 11). What Is Customer-Focused Leadership. Retrieved from The Training Bank: www.thetrainingbank.com/what-is-customer-focused-leadership/

19 Policy

For if the trumpet give an uncertain sound, who shall prepare himself to the battle?
The First Epistle of Paul the Apostle to the Corinthians Verse 14:8

Introduction

This is one of the most important requirements in ISO 9001 because it is intended to encourage the desired behaviour relative to quality.

Policies are enacted where there are choices to be made and often set one organization apart from another. All organizations will therefore have policies. They will exist at all levels in an organization but they might not be referred to as policies. Any guide to action or decision whether or not it is documented, can be classed as a policy. At the corporate level policies generally act as constraints on how its objectives are to be accomplished or may be instruments for translating the organizations shared values into artefacts. These may include policies on a wide range of issues examples of which are listed in Chapter 32. In any particular situation, a policy will influence which factors are to be given priority, for example, in clinching a deal with a customer, should the salesperson give priority to selling the organization's services over understanding customer needs, or promise a deliver date the customer wants regardless of what the organization can achieve? As was explained in Box 6.2, the decision as to whether something is of satisfactory quality rests with the receiver not the producer and therefore a quality policy is applied to producers in order that they may act in a way that satisfies the recipients of their work.

There is some duplication in these requirements:

- The quality policy is required to be maintained in both clause 5.2.1 and 5.2.2a.
- The quality policy is required to be available in both clause 5.2.2a and 5.2.2c.
- The quality policy is required to be implemented in 5.2.1 but also required to be applied in 5.2.2b.
- Communication conveys meaning; otherwise, it's just sending messages, and yet clause 5.2.2b requires both communication and understanding of the quality policy.

In this chapter, we examine the eight requirements of clause 5.2 in five separate sections, namely:

- Establishing a quality policy that is appropriate (5.2.1a), c) and d))
- Providing a framework for quality objectives (5.2.1b)

- Maintaining the quality policy (5.2.2a)
- Communicating and applying the quality policy (5.2.2b)
- Availability of the quality policy (5.2.2a) and c))

Establishing a quality policy that is appropriate (5.2.1)

Box 19.1   Quality policy

Quality policy is defined as the "intentions and direction of an organization as formally expressed by its top management related to quality" (i.e. the degree to which a set of inherent characteristics of an object (i.e. anything perceivable or conceivable) fulfils requirements (i.e. needs or expectations).

(ISO 9000:2015)

What does this mean?

Quality policy

ISO 9000:2015 defines a quality policy as a "policy related to quality" which only reveals its hidden meaning after substituting the ISO 9000 definitions for the words *policy*, *quality*, *requirement* and *object* (see Box 19.1). This implies that a quality policy is not limited to top management's intentions and directions with respect to the organization's products and services and can equally apply to anything that is needed or expected. Whereas other policies will focus on specific aspects of management such as pricing, investment, procurement or safety, the quality policy focuses on how well activities are performed, whether it's producing an output, running a meeting or dealing with a dispute, there are needs and expectations to be met. The quality policy establishes the organization's position on how those needs and expectations are to be met. However, the quality policy required by ISO 9001 is limited to that needed for the QMS, which the standard confines to the products and services provided to customers. Therefore, at a minimum the quality policy must address customer needs and expectations, but this depends on many other needs and expectations also being met such as the outputs of processes within the scope of the QMS.

Organization purpose, context and strategic direction are explained in Chapter 12. Establishing a quality policy that is appropriate to the purpose of the organization means that it must be consistent with what the organization has been set up to do and where it is going.

Establishing a quality policy

For a policy to become established, it must reflect the beliefs of the organization and underpin every conscious thought and action. A quality policy that is posted in the entrance hall is published, but not established. However, the requirement to *establish a quality policy* means something different. It means *formulating* a quality policy because clause 5.2.2 addresses the communication and application of the quality policy etc.

Commitment to continual improvement

There are three requirements for improvement in clause 10.1, one of which addresses improvement of the performance and effectiveness of the QMS and a requirement for continuous improvement of the QMS in clause 10.3. As explained in Chapter 59 improvements in the QMS are not improvements in products and services unless they arise as a consequence. Therefore, the quality policy is not required to include a commitment to improve products and services.

Commitment to satisfy applicable requirements

Applicable requirements are the requirements of customers and other interested parties that apply to the provision of products and services the organization offers to its customers. In the context of the quality policy they are not intended to be requirements other than the above but as the definition of the term *quality* widens it to the quality of anything, the policy may be extended to apply to the quality of anything.

Why is this necessary?

It is necessary for top management to impress on their workforce that the organization has entered into certain obligations that commit everyone in the enterprise. Such commitments need to be communicated through policy statements to ensure that when taking actions and making decisions, staff give top priority to meeting the requirements of the relevant interested parties.

As was explained in Chapter 8, emergence is the unexpected and unanticipated results that arise from interactions within a system. It is therefore necessary to create conditions in which people can be relied upon to exhibit desired behaviours so that the results they produce will be, in most cases, expected and anticipated.

When effort is not expended on keeping things in order, keeping them up to date, organizations will slowly go into decline and therefore continual improvement is necessary to not only maintain the status quo but also equip the organization to face new challenges.

How is this addressed?

Box 19.2   Guiding principles at Toyota

1   Honour the language and spirit of the law of every nation and undertake open and fair business activities to be a good corporate citizen of the world.
2   Respect the culture and customs of every nation and contribute to economic and social development through corporate activities in their respective communities.
3   Dedicate our business to providing clean and safe products and to enhancing the quality of life everywhere through all of our activities.
4   Create and develop advanced technologies and provide outstanding products and services that fulfil the needs of customers worldwide.
5   Foster a corporate culture that enhances both individual creativity and the value of teamwork, while honouring mutual trust and respect between labour and management.
6   Pursue growth through harmony with the global community via innovative management.
7   Work with business partners in research and manufacture to achieve stable, long-term growth and mutual benefits, while keeping ourselves open to new partnerships.
(Toyota-Global, 2015)

Reflecting on the current policy with regard to quality

Before writing a quality policy, the top management team should reflect on the shared values, principles or philosophy that has guided the organization to where it is today. If not already documented, they should attempt to write down their guiding values, principles or philosophy and reach a consensus on what they are. Toyota, for example, has seven guiding principles (see Box 19.2). Principles 3 and 4 could be regarded as the quality policy, but Toyota also has a corporate social responsibility policy that is an interpretation of the guiding principles at Toyota that takes into consideration Toyota's relations with stakeholders. Among the statements is "Based on our philosophy of 'Customer First', we develop and provide innovative, safe and outstanding high quality products and services that meet a wide variety of customers' demands to enrich the lives of people around the world (Guiding Principles: 3 and 4)."

Among the existing values and principles might be found statements referring to quality, customers, standards, regulations, improvement or innovation that avoid the need to create a document carrying the label *Quality Policy*. Even so, the wording will need to be reviewed to confirm whether it provides sufficient guidance to encourage the desired behaviour relative to quality.

What do we want the policy to do?

In clause 5.2.1b) the standard requires the policy to provide a framework for establishing quality objectives, and this is addressed later, but it's not the only thing the policy needs to do.

As stated previously, policies are enacted where there are choices to be made. It's doubtful that any organization would not want to satisfy its customers but individuals are faced with having to balance what they see as competing objectives both corporate and personal (see also balancing objectives in Chapter 9). What they may think may be in the organization's short-term interests may not be in the organization's long-term interests but short-term solutions often have long-term effects. Deming referred to this as *a lack of constancy of purpose* (Deming, 1994). For an organization to achieve its vision it needs constancy of purpose, and the quality policy is one such artefact to create constancy of purpose relative to quality. For example, do you want:

- customers to receive defective products even if they will accept them?
- employees to put quality first and consequently potentially stop the production line?
- an environment where employees hide their mistakes or freely own up to them?
- an environment where employees walk by problems they find or alert the area supervisor?
- employees to meet targets beyond the capability of the process by cheating or do you want them to get the targets changed to align with the process's assessed capability?
- downstream processes to compensate for the poor quality provided by upstream processes?
- employees to tolerate poor performance of materials, equipment, facilities, processes, people or initiate improvement?

If your top management is not serious about quality, whether it's good or bad, the people under its control and your customers will soon get the message, whatever the quality policy may state.

It is not uncommon to find quality policy statements similar to the following from the Internet:

> We will consistently provide products and services that meet or exceed the requirements and expectations of our customers. We will actively pursue ever improving quality through programs that enable each employee to do their job right the first time and every time.

Taken in isolation we can't assess whether this statement is compatible with the context and strategic direction of the organization. However, it ticks the boxes for conforming with clause 5.2.1 b), c) and d) as it can be used as a framework for setting quality objectives, it commits the organization to satisfy customer requirements and expectations and commits it to continual improvement. However, how will it affect what employees do?

Employees may believe that unless they are responsible for delivering products or services to customers, it's the quality department's job to check that customer requirements are being met. Employees may also believe that they are to continue doing what they always do until required to engage in an improvement program. It may therefore not do what the organization wants it to do because it appears to be addressed to management.

Top management may therefore need to address these and many other issues before being satisfied that the quality policy is compatible with the context and strategic direction of the organization.

Expressing the quality policy

Policies are not expressed as vague statements or emphatic statements using the words *may*, *should* or *shall*, but clear intentions by use of the words *we will* – thus expressing a commitment – or by the words *we are*, *we do*, *we don't* and *we have* expressing shared beliefs. Very short statements tend to become slogans which people may not understand how they affect what they do. Their virtue is that they rarely become outdated. Long statements confuse people because they contain too much for them to remember. Their virtue is that they not only define what the company stands for but how it will keep its promises.

Policies are more easily understood when expressed in terms that are understood by the employees. Terms such as *interested parties*, *quality management system* and even *quality policy* are ISO speak and not common among organizations that have not encountered ISO 9001 and may not be readily understood. Spell it out if necessary, use language that is understood in your organization – in fact it is highly desirable where relevant to state exactly what you mean rather than use the specific words from the standard.

Customer first or quality first

To make a quality policy readily understood many organizations turn to a single axiom that will focus the mind. Two such axioms are *customer first* and *quality first*.

As stated earlier, the Toyota philosophy is based on *customer first*. It started when Shotaro Kamiya became company president in 1935. Toyota lives by Kamiya's philosophy of "Customer first, dealer second, manufacturer third" (Toyota, 2015). What this meant for the company was that to offer true customer satisfaction, they needed to be flexible and respond to customers' changing needs, thus subordinating the needs of manufacturing and dealerships to those of customers. Toyota does not depend for its survival on it making cars, but on satisfying its customers as illustrated in another example in Box 19.3.

Box 19.3   The misguided poultry farmer

Imagine you are a poultry farmer and you are in business to make money. You want to sell more eggs so as to make more money. If you perceive the process output to be money and you focus on that, you won't get more eggs out of the hens just by shouting at them, and you would quickly run out of eggs and your hens would die. You'd make money alright but not for long, and even if hens are cheap and in plentiful supply the business would be unsustainable.

If you perceive the output as being class 1 eggs rather than money, and you focus on feeding the hens (the process) using some of the money you make to improve the process to consistently produce class 1 eggs, you will find that the profits rise, sufficient to purchase and feed more hens and thus grow the business. Thus focusing on the quality of the process and not the quality of the profit is a more sustainable strategy.

We have known for a long time that the reputation of an organization stands on its approach to quality. Reputations are hard to win and easy to lose. Whatever top management proclaim, it is what individuals do that builds a reputation and therefore:

A. If the organization has a reputation for providing fault-free products or superior services that possess the features customers want, it is likely to become a world leader.
B. If the organization has a reputation for providing products or services that possess most of the features customers want but which might occasionally be faulty or below expectations, it is likely to become an average player in the market; being lower or higher in the ratings depending on the level of complaints it receives.
C. If the organization has a reputation for providing faulty products or inferior services that don't possess the features customers want but appear as if they do, it is likely to become notorious for being a rogue trader.

It's likely that top management will want their organization to emulate Type A but their customers may look upon them as Type B or even C.

Box 19.4   Advocates of quality first

- 1890 Newport News Shipbuilding, builder of the Liberty Ships in WWII, their motto was "We will build good ships here; at a profit if we can, at a loss if we must, but always good ships." (Fox 1986)
- 1910 Henry Ford said that, "We have had just one main purpose during these years, and that is to give the people transportation of the most dependable quality at the lowest possible cost." (Benson Ford Research Center, 2017)
- 1919 Robert Bosch declared that "it has always been an unbearable thought to me that someone should inspect one of my products and find it inferior. I have therefore always tried to ensure that only such work goes out as its superior in all respects." (Bosch 1885)
- 1981 Kaoru Ishikawa wrote in his book on total quality control that "the pursuit of short-term profit loses competitiveness and ultimately long-term profit whereas putting quality first increases profits in the long run." (Ishikawa, 1986)
- 1986 Masaki Imai famous for his book on Kaizen wrote that "if you take care of the quality the profits will take care of themselves." (Imai, 1986)
- 1990 Ford Q101 Standard – "Quality comes first – To achieve customer satisfaction, the quality of our products and services must be our number one priority." (FMC,1990)
- 2008 The entire Canon Group is working to uphold the president's and COO's policy of making a total commitment to "quality first" (Canon, 2008).
- 2010 The BBC in its Strategic Review is putting quality first as a way of showing that in spite of the cuts it must make, programme quality will not suffer. (BBC 2011)
- 2010 Toyota is putting quality first in the new Auris production line in the UK. They say that "quality has never before been such a consideration for car buyers." (Toyota, 2010)
- 2012 Chief Designer at Apple Computers Sir Jonathan Ive said that "our goal is not to make money but to make great products." (Daily Telegraph 2012)

Not everyone under the control of an organization will know what the organization's customers want. The standards they are working to likely have been derived from customer and applicable legal requirements and are, in effect, substitute requirements. For these people *quality first* will be a better axiom than *customer first. Quality first* or *making quality a priority* has been a management axiom for over 100 years, as indicated in Box 19.4. However, the terms can convey very different meanings. A video on putting quality first is available on the companion website.

Using the quality management principles

In the ISO 9000:2015 definition of quality policy it is suggested that the eight quality management principles be used as a basis for establishment of the quality policy, and an example of such policies is presented in Box 19.5. If we take just any one of these policy statements they will look good in the lobby – visitors will be impressed – but the bottom line is whether actual performance meets the expectations set. If no one thinks through the process for ensuring the links between the policy, the objectives and the outputs, the policy won't be met with any consistency.

Box 19.5   A quality policy based on quality management principles

On customers
We will listen to our customers, understand their needs and expectations and endeavour to satisfy them in ways that meet the expectations of our other stakeholders.

On leadership
We will establish and communicate our vision for the organization and through our leadership exemplify core values to guide the behaviour of all to achieve our vision.

On people
We will involve our people in the organization's development, utilize their knowledge and experience, recognize their contribution and provide an environment in which they are motivated to realize their full potential.

On processes
We will manage work as cross-functional processes so as to ensure the results they produce achieve pre-defined objectives.

On systems
We will manage the organization as a system of interdependent elements that are organized in such a way that they produce results that satisfy all our stakeholders.

On continual improvement
We will provide an environment in which every person is motivated to continually improve the efficiency and effectiveness of our products, processes and our management system.

On decisions
We will base our decisions on the logical and intuitive analysis of data collected where possible from accurate measurements of product, process and organizational characteristics.

On relationships
We will develop alliances with those parties on which we depend for our success and work with them to jointly improve performance. We will also endeavour to manage those interested parties whose motives are malevolent without

On profits
We will satisfy our stakeholders in a manner that will yield a surplus that we will use to develop our capabilities and our employees, reward our investors and contribute to improvement in our society.

On the environment, health and safety
We will operate in a manner that safeguards the environment and the health and safety of those who could be affected by our operations.

Expressing a commitment to satisfy applicable requirements (5.2.1c)

The quality policy can state explicitly that the organization is committed to meeting applicable requirements, but it would prompt people to ask what was meant by *applicable requirements*. In the example given earlier is the statement "We will consistently provide products and services that meet or exceed the requirements and expectations of our customers." This limits the requirements to those of customers and (probably unintentionally) ignores legal requirements applicable to the organization's products and services. It also ignores the requirements of other interested parties that are relevant to the organization's products and services (e.g. the workers don't want to incur injury or contract disease by handling the product). The principle of *quality first* resolves this to some extent as it's about fulfilling expectations related to the object.

Expressing a commitment to continual improvement (5.2.1d)

The quality policy can state explicitly that the organization is committed to continual improvement of its QMS but doing that requires there to be an explanation somewhere of what a QMS is. To avoid use of the term *QMS* the policy could be expressed as *continually improving the system for creating and retaining customers*.

Whenever top management are faced with a choice of pursuing an opportunity to enhance the performance of the QMS, they weigh up the advantages and disadvantages of accepting the status quo and find ways of achieving their objectives that are consistent with the policy, and they also resist attempts to change the policy. This may involve delaying some improvement until the conditions are more favourable.

Making the quality policy appropriate

Many statements in quality policies are generic in that they are appropriate to the purpose, context and strategic direction of any organization – there is nothing about them that makes them specific. Were every organization to echo the same quality policy it would become unoriginal or hackneyed, a cliché or platitude that is supposed to be meaningful but has become a motherhood statement. Everyone says it, but it will never happen like, world peace.

Consider the first statement in Box 19.4. It tells us what the company does and why it puts quality first, and although those particular words are not used, it's plain for all to see that is what it means.

The words chosen may create ambiguity or conflict between purpose, context and strategic direction. A quality policy that is appropriate for a bank is unlikely to be appropriate for a fast food outlet. For one thing, the purpose of a bank is to borrow and lend money, and the purpose of a fast food outlet is to provide a cheap meal quickly so they will place emphasis on different things. The context is also totally different and although both will seek to build a brand that customers can trust, their strategic direction will be different. A policy crafted for predictable situations such as mass production would not be appropriate for an innovative high-tech company that must take risks to make breakthroughs. By necessity it doesn't get it right first time – in fact it thrives on learning from mistakes.

The quality policy is not a mission statement but it may be embedded within a mission statement and by doing so the issue of it being appropriate can be resolved. If it is believed necessary to expand on the quality policy through explanation or examples of typical behaviours a separate quality policy statement may be more effective. Whatever solution is chosen, have your mission statement to hand and ask: *In what way might this/these statement(s) be inappropriate to our purpose or our context or not support our declared strategic direction?*

You may need more than one quality policy if the organization offers fundamentally different products and services to its customers. A high-tech company may offer standard products from its production line but also offer artificial intelligence solutions from its research unit or a calibration service from its metrology laboratory. It would probably want the quality policies to be different to create the conditions appropriate to the purpose of the unit.

Box 19.6   Quality first principle

When quality is made the first priority in all actions and decisions the long-term needs and expectations of all stakeholders are assured.

How is this demonstrated?

Demonstrating that a quality policy has been established that conforms to the requirements of clause 5.2.1 may be accomplished by:

a)   presenting the quality policy statement together with statements of the organization's mission and vision and showing that the quality policy;

    i   is consistent with the organization's purpose as stated in the mission statement;
    ii   is consistent with the organization's context as stated in the mission statement;
    iii   supports the organization's strategic direction as stated in the vision statement.

b)   showing where the policy includes a commitment to continual improvement;
c)   showing where the policy includes a commitment to meeting applicable requirements.

Providing a framework for quality objectives (5.2.1b)

What does this mean?

There are two different interpretations of this requirement. One interpretation is that it implies that each statement within the quality policy should have an associated quality objective and each lower-level quality objective should be traceable to higher-level quality objectives that have a clear relationship with a statement within the quality policy. Through this relationship, the objectives deploy the quality policy. In this respect the quality policy would appear to be the mission statement of the organization.

Box 19.7   Policies and objectives

Policies differ from objectives in that policies remain effective until changed, whereas objectives remain effective until achieved.

The other interpretation is that the quality policy represents a set of guiding principles, values or constraints and, therefore when setting as well as reviewing quality objectives, these principles should be employed to ensure that the objectives are appropriate to or consistent with the purpose of the organization. This means that the policies are used to measure or frame (put boundaries around) the objectives. It does not mean that the words used in the quality policy should somehow be translated into objectives.

The relationship between policy and objectives is important because ISO 9001 implies one is derived from the other but in reality, the relationship is quite different as illustrated in Figure 19.1.

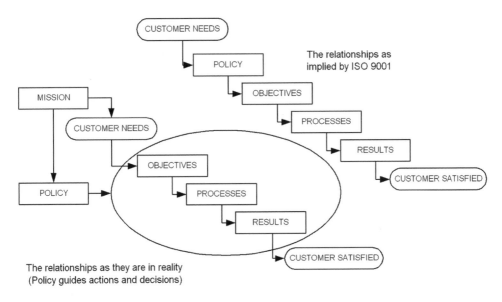

Figure 19.1 Policy – objectives – process relationship

Why is this necessary?

There needs to be a link between policy and objectives otherwise the processes designed to achieve the objectives would be unlikely to be consistent with the policy. A consequence of displaying a quality policy and impressing on everyone how important it is, is that it becomes disconnected from real work. It does not get used unless it is also linked to the work that people do and such a link is made by using the policy to frame objectives or measure their achievement. For example, If an objective is to reduce toxic emissions to atmosphere by 10% and the quality policy was to put customers first, the objective would be valid if stakeholder analysis showed that customers wanted a reduction in emissions but didn't quantify it. However, if a legitimate pressure group wanted a reduction in toxic emissions of greater than 10%, this could legitimately be interpreted as a customer expectation because it would not be unreasonable to assume that customers would share the same concerns as the pressure group. Thus, the objective would need to change to a reduction in toxic emissions of greater than 10%.

How is this addressed?

Without being linked to the business processes, the quality policy remains a dream. There must be a means to make these policies a reality and it is by using the policies as measures of success for objectives that this is accomplished. By deriving objectives from customer needs and expectations and the policies from the constraints governing how these objectives will be achieved you will produce a series of objectives and measures for the enabling processes.

Purely as an exercise and not as a real policy, we could fabricate a policy that does provide a framework for setting objectives but it is not recommended.

How is this demonstrated?

Demonstrating that a quality policy provides a framework for setting quality objectives may be accomplished by:

a)   presenting the quality policy statement together with statements of quality objectives;
b)   showing that the results to be achieved are consistent with the quality policy;
c)   showing how the measures of success for each objective align with the quality policy.

Communication and application of the quality policy (5.2.1b)

What does this mean?

For a policy to be communicated it must be brought to the attention of personnel. Personnel must be made aware of how the policy relates to what they do so that they understand what it means before action is taken. Without action, there is no demonstration that communication has been effective. If you are already doing it, publishing the policy merely confirms that this is your policy. If the people do not exhibit the right behaviours, there will be a need to change the culture to make the policy a reality.

For a policy to be applied, it suggests that an action or decision is planned and then *before* being taken, the policy is applied to produce a result consistent with the policy. If the policy was applied after the fact a result would be produced that was inconsistent and then an attempt made to correct the result which in many cases would be impossible, for example, after you have served a disgusting meal to a customer, you now take it away and serve a quality meal, following which your customer is bound to question your commitment to quality!

Why is this necessary?

A policy in a nice frame positioned in the lobby of an organization may impress the visitors but unless it is understood and applied when people carry out actions or make decisions, it will have no effect on the performance of the organization.

How is this addressed?

Communicating the quality policy

Top management need to establish an employee engagement process that creates a socially cohesive corporate culture in which people are fully engaged. Communication is a sub-process within this process and is a process in which information and its meaning is conveyed from a sender to a receiver, the output of which is understanding and acceptance. For a quality policy to become established it must be communicated and managers need to become the role model so that by their actions and decisions they exemplify the policy. Belief in the policy is unlikely if the quality policy is merely perceived as something written only to satisfy ISO 9001.

Although it is important that management shows commitment towards quality, policy statements can be one of two things – worthless or obvious. They are worthless if they do not reflect what the organization already believes in and is currently implementing. They are obvious if they do reflect the current beliefs and practices of the organization. It is therefore foolish to declare in your policy what you would like the organization to become.

This is perhaps the most difficult requirement to achieve. Any amount of documentation, presentations by management, and staff briefings will not necessarily ensure that the policy is understood. Communication of policy is about gaining understanding. It is not just the sending of messages from one source to another as is explained in Chapter 31. So how do you ensure that the policy is understood?

One method to ensure understanding is for top management to consider the following actions:

- Start by explaining where the policy fits relative to the purpose and strategic direction of the organization.
- Debate the policy together and thrash out all the issues. Don't announce anything until there is a uniform understanding among the members of the management team. Get the managers to face the questions: *are we all agreed about why we need this policy? Do we intend to adhere to this policy?* And remove any doubt before going ahead.
- Ensure the policy is presented in a user-friendly way.
- Produce several examples showing the desired behaviours in terms of what would be happening in the organization if this policy was being met (see Chapter 5 in the section on the quality management principles for some examples).
- Announce to the workforce that you now have a quality policy and how it affects everyone from the top down, why it's necessary and where it fits within the context of the organization.
- Publish the policy so that everyone becomes aware of it.
- Display the policy in key places to attract peoples' attention.
- Arrange and implement training or instruction for those affected.
- Use the policy as measures of success in the process descriptions (see Chapter 32).
- Test understanding at every opportunity (e.g. at meetings, when issuing instructions or procedures, when delays occur, when failures arise and when costs escalate).
- Audit the decisions taken that affect quality and go back to those who made them if they do not comply with the stated policy.
- Take action every time there is misunderstanding. Don't let it go unattended and don't admonish those who may have misunderstood the policy. It may not be their fault!
- Every time there is a change in policy, go through the same process. Never announce a change and walk away from it as the change may never be implemented!
- Give time for the understanding to be absorbed. Use case studies and current problems to get the message across.

Applying the quality policy

Once the quality policy has been understood by those whose work it affects, its application is embedded in every action and decision they take. Applying the quality policy is thinking and behaving in line with the policy. Nancy Kline expresses it very well with "Everything we do depends for its quality on the thinking we do first" (Kline, 1999).

There will be many difficult decisions where the short-term interests of the organization may need to be subordinated to the needs of customers or shareholders. Internal pressures may tempt people to cut corners, break the rules and protect their own interests. Committing the organization to meet applicable requirements and continual improvement may be easy decisions to take – but difficult to honour. Whenever management are faced with a choice of action or decision where the expedient course would be inconsistent with the policy they find ways of achieving their objectives in ways that are consistent with the policy or get the policy changed.

How is this demonstrated?

Demonstrating the extent to which the quality policy has been communicated, understood and applied within the organization may be accomplished by:

1   presenting evidence of an employee engagement process that addresses awareness of the quality policy;

2   selecting a sample of people under the control of the organization including managers, professional staff, operators, contract labour and newly recruited staff and asking the following questions:

   a)   How does the quality policy relate to the purpose and strategic direction of the organization?
   b)   How does the quality policy affect what you do?
   c)   What happens if you can't accomplish all the tasks in the allotted time?
   d)   What would you do if you discovered nonconformity immediately prior to delivery?
   e)   What would you do if a problem kept recurring despite attempts at corrective action?
   f)   What would you do if instructed to ignore a requirement that was applicable to the work for which you were responsible?
   g)   What would you do if you were put under pressure to deliver a product or provide a service to a customer knowing that all the requirements had not been or could not be fulfilled?
   h)   How would you treat a customer who continually complains about your products and services?
   i)   What action would you take if someone asked you to undertake a task for which you were not trained?
   j)   What action would you take if you noticed that a product for which you were not responsible was in danger of being damaged or contaminated?
   k)   What would you do if you ran out of the approved material and can't get a delivery in time to honour your commitment to a customer but you do have an untested alternative?
   l)   What would you do if you were notified about a batch of defective product at relatively low value and decided to scrap the lot and later you were notified of another batch and then another batch – the value now runs into thousands?
   m)   What would you do if you find a subordinate manager is rewarding production teams for record runs related to volume/waste/delay and during the run and was not considering quality (including safety)?

Maintenance of quality policy (5.2.2a)

What does this mean?

Maintaining the quality policy as documented information means that the quality policy is to be classified as "information required to be controlled and maintained by an organization" (ISO 9000:2015). Controlling information and ensuring it's available is addressed in Chapter 32.

Why is this necessary?

The policy is required to be appropriate for the organization's purpose (mission) and while the mission may not change, the environment in which the organization operates does change. These changes may affect the quality policy.

To be sure that the agreed quality policy is conveyed in an uncorrupted form and available for reference by those it affects, it's necessary to documented and control it. However, the policy itself is the overall intentions of management with regard to quality; therefore, documenting the policy is as much for the benefit of top management as others as it serves as a reminder as to what has been agreed and to ensure consistency in communication. Were the policy to remain undocumented, over time it may be conveyed differently as each person remembers it differently or puts their own spin on it.

How is this addressed?

The policy will need to be reviewed considering changes in the economic, social and technological environment for its suitability to enable the organization to fulfil its purpose.

The review may conclude that no change is needed to the actual words, but the way they are being conveyed might need to change. If the environment or the organization has changed, the policy might be acceptable but needs to be interpreted differently and conveyed to different people using different examples than were used previously.

Changes in policy have wide impact and therefore should not be taken lightly. They should be reviewed by top management during strategic planning reviews, business reviews or process reviews. We are not talking about tinkering with the wording but a real change in direction. Changes in technology might mean that the workforce ceases to be predominantly on site as it becomes more effective to promote home or remote working. This change will impact the policy regarding leadership and people. Changes in the economic climate might mean that the workforce ceases to consist primarily of employees as it becomes more effective to outsource work to subcontractors and consultants. This change will affect the policy regarding leadership, employees and external providers.

How is this demonstrated?

Demonstrating that the quality policy is available and being maintained as documented information may be accomplished by:

a) presenting documented information purporting to be the quality policy that as a minimum addresses the aspects identified in clause 5.2.1;
b) presenting evidence that this policy has been agreed by top management and under control as per clause 7.5.3;
c) interviewing persons doing work under the organization s control and confirming they have access to the quality policy.

Availability of quality policy (5.2.2c)

What does this mean?

This means making the quality policy available to customers, investors, suppliers, employees, regulators, etc.

Why is this necessary?

Certain parties will be interested in knowing what the organization's policy is relative to the quality of its products and services. They will want to be assured that it takes quality seriously, either because they intend to obtain such products and services or to invest in the organization.

How is this addressed?

The quality policy can be included in company brochures, annual reports, or on the organization's website along with other information about themselves. Some organizations have a web page that includes information about its approach to quality, health and safety; corporate social responsibility; and relevant legislation. It's treated as part of public relations.

How is this demonstrated?

Demonstrating that the quality policy is available to relevant interested parties may be accomplished by:

a)   identifying the methods used for making the policy available;
b)   presenting the documents or accessing the web site where the policy may be viewed;
c)   presenting evidence which shows that relevant interested parties have been informed where they can access the quality policy.

Bibliography

BBC. (2011, December 2) Delivering Quality First. Retrieved from http://www.bbc.co.uk/aboutthebbc/insidethebbc/howwework/reports/deliveringqualityfirst.html

Benson Ford Research Center. (2017, April). Henry Ford Quotations: Retrieved from https://www.thehenryford.org/collections-and-research/digital-resources/popular-topics/henry-ford-quotes/

Bosch, R (1885) Retrieved (2017) from http://www.the-financedirector.com/contractors/business-process-outsourcing/bosch-communication-center/bosch-communication-center3.html

Canon, (2008, July) Canon Sustainability Report 2008. Retrieved from http://www.canon.com/csr/report/pdf/sustainability2008e.pdf

Daily Telegraph. (2012, July 30) Apple design chief: 'Our goal isn't to make money'. Retrieved from Daily Telegraph: www.telegraph.co.uk/technology/apple/9438662/Apple-design-chief-Our-goal-isnt-to-make-money.html

Deming, E. W. (1994). *The New Economics for Industry, Government and Education* – Second Edition. Cambridge, MA: The MIT Press.

FMC, (1990) Worldwide Quality Q-101, Dearborn MI: The Ford Motor Company

Fox, W. A. (1986) Always Good Ships – Histories of Newport News Shipbuilding: Virginia Beach: Donning Company Publishers

Imai, M. (1986) KAIZEN: The Key to Japanese Competitive Success. New York: McGraw-Hill.

Ishikawa, K. (1985). What Is Total Quality Control: The Japanese Way. (D. J. Lu, Trans.). Englewood Cliffs, NJ: Prentice-Hall Inc.

ISO 9000:2015. (2015). *Quality Management Systems – Fundamentals and Vocabulary*. Geneva: ISO.

Kline, N. (1999). *Time to Think: Listening to Ignite the Human Mind*. London: Octopus Publishing Group.

Toyota. (2015, February 26). No More Than One Cup's Worth of Water Can Fit into a Single Cup: To Hold More Water, You Need More Cups. Retrieved from Toyota-Global: www.toyota-global.com/company/toyota_traditions/philosophy/jan_feb_2007.html

Toyota-Global. (2015, February 24). Guiding Principles at Toyota. Retrieved from Toyota-Global: www.toyota-global.com/company/vision_philosophy/guiding_principles.html

Toyota. (2010, May 12). Toyota Auris hybrid production: quality first and foremost Retrieved from Toyota Media Site: http://media.toyota.co.uk/2010/05/toyota-auris-hybrid-production-quality-first-and-foremost/

20 Organizational roles, responsibilities and authorities

Introduction

Having a well-defined quality policy and well-defined quality objectives will change nothing. The organization needs to empower its people to implement the policy and achieve the objectives through effectively managed processes. Only then will it be able to accomplish its mission, realize its vision and achieve its goals. This starts by assigning responsibility and delegating authority for the work to be done and from this determining the competences of those concerned. This will be a cascading process so that, as the processes are developed and the activities identified, the responsibilities and authority for carrying them out can be assigned and communicated.

In this chapter, we examine the six requirements in Clause 5.3, namely:

- Assigning and communicating responsibilities and authorities (5.3)
- Responsibility and authority for ensuring conformity to ISO 9001 (5.3a)
- Responsibility and authority for ensuring process performance (5.3b)
- Responsibility and authority for reporting QMS performance (5.3c)
- Responsibility and authority for promoting customer focus (5.3d)
- Responsibility and authority for ensuring QMS integrity (5.3e)

A historical perspective

Requirements for authority and responsibility to be defined relative to quality have been in ISO 9001 since its first publication in 1987 and were a common feature of its predecessors such as the American military standard Mil-Q-9858 and the British standard, BS 5750. The term *management representative* appeared in NATO quality assurance publications on which BS 5750 and other national standards were based.

A requirement for the appointment of a management representative to ensure the requirements of the standard were implemented and maintained was introduced into ISO 9001:1987. In the 1994 version, they added a responsibility for reporting on the performance of the quality system as a basis for its improvement. Then in the 2000 version the term *management responsibility* was only used in the clause heading, thus removing the requirement for a specific appointment and the responsibilities were changed. It was required that a member of management had responsibility for ensuring that processes needed for the QMS were established, implemented and maintained; for reporting on the performance of the QMS and any need for improvement; and for ensuring the promotion of awareness of customer requirements throughout the organization. In the 2008 revision, it was emphasized that the person appointed had to be a member of the organization's management.

There is now no mention of a management representative either in a clause heading or a requirement (see Chapter 2 for the rational). The responsibilities are to be assigned within the organization thus prohibiting the outsourcing of these responsibilities but permitting any number of people to discharge them.

The requirement for reporting to top management on the performance of the QMS and its improvement remains. The requirement from the 1987 and 1994 versions for ensuring the QMS conforms to the requirements of the standard has been reintroduced and the requirement in the 2008 version for maintaining the integrity of the QMS when changes are made has been moved from planning into clause 5.3. The responsibility for promoting awareness of customer requirements has changed to that of ensuring the promotion of customer focus which is a much broader concept. The responsibility for ensuring that processes needed for the QMS were established, implemented and maintained has now changed to ensuring that the processes are delivering their intended outputs. This responsibility can now be placed on process owners. So in essence the responsibilities remain and the only significant change is that they are no longer to be assigned to a management representative or a member of the organization's management. They can be assigned to any competent person within the organization. There is no requirement for independence or freedom to resolve matters pertaining to quality as there was in the Defence Standards that preceded ISO 9001.

If a person carries the title of management representative, the 2015 revision presents an opportunity to reconsider whether it's appropriate as not all the responsibilities defined in clause 5.3 may be assigned to one individual. However, for small organizations this may still be a convenient way of identifying the focal point for quality matters.

Assigning and communicating responsibilities and authorities (5.3)

What does this mean?

Box 20.1   Revised requirement for responsibility and authority

In the 2008 version responsibility and authority was required to be defined and communicated. In the 2015 version it is now required to be assigned and communicated. In order to communicate responsibility and authority one would need to have defined it so in some ways the word *define* was superfluous. Communication is about ensuring the implications of carrying a particular responsibility are understood by the person to whom is has been assigned and therefore the first step is to decide who is to carry the responsibility and then communicate that assignment to them.

Responsibility

Responsibility is, in simple terms, an area in which one is entitled to act on one's own accord. It is the obligation of staff to their managers to perform the duties of their jobs. It is thus the obligation of a person to achieve the desired results or conditions for which they are accountable to their manager. If you cause something to happen, you must be responsible for the result just as you would if you caused an accident. So to determine a person's responsibility ask: *What can they cause to happen?*

Authority

Authority is, in simple terms, the right to take actions and make decisions. In the management context, it constitutes a form of influence and a right to take action, to direct and co-ordinate the actions of others and to use discretion in the position occupied by an individual, rather than in the individuals themselves. The delegation of authority permits decisions to be made more rapidly by those who are in more direct contact with the problem. So to determine a person's authority ask: *On what matters are you authorized to exercise your discretion?*

The requirement applies to all persons under the organization's control responsible for carrying out activities in the processes included within the scope of the QMS. This will include the top management, heads of department, professional and operational staff whose work directly or indirectly affects the quality of the organization's outputs to its customers.

Roles

A role has been defined by Rollinson as "a set of expectations and obligations to act in a specific way in certain contexts" (Rollinson, 2008). The term *role* in this context can be ascribed to an individual or a group of individuals. In general, the term is one of several used to describe what people do at any given time. For example, a quality manager in an engineering company who chairs a meeting on quality improvement has a profession, an occupation, a job, a position and a role.

- The profession is engineering for which there are numerous occupations.
- The occupation is quality management for which there are several positions.
- The position is quality manager which comprises several jobs.
- The job is change management which embraces several roles.
- The role is leader at the present time which embraces several expectations and obligations to act in a specific way.

To illustrate how roles change, the quality manager collects data to reveal facts. The role changes from leader to analyst. Later the quality manager takes the facts and searches for causes. The role changes from analyst to investigator. After finding the causes, the quality manager encourages others to produce solutions – the role is now facilitator. If the quality manager produced the solutions, the role undertaken would be designer or innovator. Once the problems are solved, the quality manager goes to see the customer to show how performance has improved – the role changes again, this time to ambassador.

For some people, all these roles are rolled up into a position because they perform them continually. For others, they are transitory. Even being a quality manager, project manager or auditor may be a role if it's transitory, and this is the direction in which this requirement in ISO 9001 has travelled. Rather than specify responsibility and authority for a particular position, by requiring certain responsibility and authority to be assigned, it permits anyone with the necessary competence to perform a particular role. Top management may have therefore assigned responsibility and delegated authority for 5.3a) to 5.3e) to one person or to several persons.

Assigning responsibilities and authorities

Assigning responsibility means transferring tasks, activities or jobs to other people to lighten the burden on management without losing accountability for results. Authority is delegated rather than assigned because, unlike responsibility which passes upwards in a hierarchy, authority passes downwards. Therefore, when a manager delegates authority to others to makes decisions that managers loses the right to make those decisions.

Communicating responsibility and authority

Communication of responsibility and authority means that those concerned are not only informed of the activities they and others are expected to carry out and the corresponding results they are expected to achieve, but also understand their obligations so that there is no doubt on either side about what they and others will be held accountable for.

Why is this necessary?

There are several reasons for why is it necessary to communicate this information:

- to convey consistency and avoid conflict;
- to show which functions make which contributions and thus serve to motivate staff;
- to establish channels of communication so that work proceeds smoothly without unplanned interruption;
- to indicate from whom staff will receive their instructions, to whom they are accountable and to whom they should go to seek information to resolve difficulties.

In the absence of the delegation of authority and assignment of responsibilities, individuals may assume duties that may duplicate those duties assumed by others while jobs that are necessary but unattractive will be left undone. Where managers hang on to their authority and don't delegate, it encourages decisions to be made only by them which can result in an increased management workload but also engender a feeling of mistrust by the workforce.

How is this addressed?

When a process is activated something or someone must undertake the first step. Unless responsibilities for the actions have been assigned and authority for decisions delegated nothing will happen. Even if the first step is undertaken by a machine, someone will have responsibility for ensuring that it can operate when called upon to do so. In some respects, it matters not who does what provided they are competent and indeed, in an emergency, managers may undertake tasks normally performed by others if they are competent to do so. However, work and labour are divided in organizations to make work productive and the worker achieving. Activities are often grouped by speciality or discipline rather than by process. For example, the buyers may be situated in the purchasing department, production planners maybe situated in the production department and supplier QA assessors may be situated in the quality department but they work together in the procurement process to ensure delivery of components of the right quality, cost and delivery Just-In-Time to the production line.

Roles

The roles will be identified primarily through the organization structure and it is not untypical for there to be role descriptions for director, manager, team leader, etc., so that anyone performing these roles has the same responsibilities and authority. This may be varied only by the level in the hierarchy and the speciality, for example, the management component of a production manager's role description is the same for all managers but supplemented by a unique description for the production speciality. Process descriptions allocate actions and decisions to roles.

Assigning responsibility and authority

Although a person may occupy a particular position, they may be eligible to perform any number of roles if they possess the necessary competence.

A role can be divided into two components: actions and decisions. Responsibilities and authority should therefore be described in terms of the actions assigned to an individual to perform and discretion delegated to an individual, that is, the decisions they are permitted to take together with the freedom they are permitted to exercise. Each role should therefore have core responsibilities that provide a degree of predictability and innovative responsibilities that in turn provide the individual with scope for development.

In assigning responsibilities and authority there are some simple rules that should be followed:

*   Through the process of delegation, authority is passed downwards within the organization and divided among subordinate personnel whereas responsibility passes upwards.
*   A manager may assign responsibilities to a subordinate and delegate authority; however, he or she remains responsible for the subordinate's use of that authority.
*   When managers assign responsibility for something, they remain responsible for it. When managers delegate authority, they lose the right to make the decisions they have delegated but remain responsible and accountable for the way such authority is used. Therefore, it is necessary for managers to ensure the competence of those to whom they entrust their authority. Accountability is one's control over the authority one has delegated to one's staff.
*   It is also considered unreasonable to hold a person responsible for events caused by factors over which they are powerless to control see Box 20.2.

Box 20.2   Conditions for self-control

Before a person can be in a state of control they must be provided with three things:

*   Knowledge of what they are supposed to do, that is, the requirements of the job, the objectives they are required to achieve;
*   Knowledge of what they are doing, provided either from their own senses or from an instrument or another person authorized to provide such data;
*   Means of regulating what they are doing in the event of failing to meet the prescribed objectives. These means must always include the authority to regulate and the ability to regulate both by varying the person's own conduct and varying the process under the person's authority. It is in this area that freedom of action and decision should be provided.

(Adapted from Juran, 1974)

The person given responsibility for achieving certain results must have the right (i.e. the authority) to decide how those results will be achieved, otherwise, the responsibility for the results rests with those who stipulate the course of action.

Individuals can rightfully exercise only that authority which is delegated to them and that authority should be equal to that persons' responsibility (not more or less than it) and to their competence. If people have authority for action without responsibility, it enables them to walk by problems without doing anything about them. Authority is not power itself. It is quite possible to have one without the other! A person can exert influence without the right to exert it. It is also irresponsible to assign responsibility and delegate authority to a person who has demonstrated a lack of necessary competences. However, where a person is appointed to a role for which they are not yet competent it is customary for this person to be under close supervision, to have a coach, mentor or role model who is steering their development into the role and to whom they can turn to for advice at any time.

Communicating responsibility and authority

There are several ways in which responsibilities and authority can be communicated:

- By word of mouth
- In an organization structure diagram or organigram
- In function descriptions
- In job descriptions
- In terms of reference
- In procedures
- In process descriptions and flow charts

The standard does not stipulate which method should be used. In very small companies a lack of such documents defining responsibility and authority may not prove detrimental to quality provided people understand their responsibilities and are competent to execute them. However, if you are going to rely on training, you need to consider how training can be carried out in a consistent manner without written material.

In organizations that undertake projects rather than operate continuous processes or production lines, there is a need to define and document project related responsibilities and authority. These appointments are often temporary, being only assigned for the duration of the project. Staff are assigned from the line departments to fulfil a role for a limited period. To meet the requirement for defined responsibility, authority and interrelationships for project organizations you will need project organization charts and project job descriptions for each role. Because project structures are temporary, processes need to be in place that control the interfaces between the line functions and project teams.

Some organizations have assigned responsibility for each element of the standard to a person, but such managers are not thinking clearly. There are 51 clauses and many are interrelated. Few can be taken in isolation therefore such a practice is questionable. When auditors ask: *Who is responsible for purchasing?* ask them to specify the particular activity they are interested in. Remember you have a system in which authority is delegated to those competent to do the job.

We have only addressed the message carriers and not the process of transmission, reception and decoding but this is addressed in Chapter 31.

How is this demonstrated?

Demonstrating that responsibilities and authority have been assigned may be accomplished by:

a) presenting evidence of intent through policies declaring general and specific responsibilities and authority;
b) presenting evidence of artefacts such as job descriptions, flow charts, etc., that are compatible with the policies;
c) selecting a representative sample of projects, processes, objectives, activities or other element of work and presenting evidence that responsibility and authority for its execution and authority has been assigned.

Demonstrating that responsibilities and authority have been communicated may be accomplished by:

a) interviewing a representative sample of people and establishing:

    i what work they do and whether it is within the scope of the QMS;
    ii what responsibilities and authority they believe have been assigned to them;
    iii how they were made aware of these responsibilities and authority;
    iv whether the source of their knowledge is legitimate and confirms what they say;
    v whether they have been assigned sufficient responsibility and authority for self-control (see earlier);
    vi whether they understand the responsibility and authority allocated to their co-workers.

b) analysing a representative sample of decisions where quality could be compromised and establishing that the decisions were consistent with the stated responsibilities and authority thereby providing evidence that the assigned responsibilities and authority were understood, for example,

    i resource allocation;
    ii release of design, product or service into use or to the customer;
    iii disposition of nonconformities;
    iv handling problems with external providers;
    v handling of customer complaints.

Responsibility and authority for ensuring conformity to ISO 9001 (5.3a)

What does this mean?

As stated in clause 1 of ISO 9001, the standard specifies requirements for a quality management system when an organization needs to demonstrate its ability to consistently provide products and services that meet customer and applicable statutory and regulatory requirements, and aims to enhance customer satisfaction. It follows therefore that someone in the organization needs to have responsibility for determining whether or not the QMS conforms to the requirements of ISO 9001 and also have the authority to require changes to the QMS when necessary to bring it into conformity with the requirements of ISO 9001.

Clause 9.2.1 requires that internal audits are conducted to provide information on whether the QMS conforms to the requirements of ISO 9001. Clause 5.3a) requires top management to give the person or group conducting these audits the responsibility and authority it needs to do its job effectively.

This role is similar to the role of the organization's accountant, lawyer, employment law adviser and occupational health and safety adviser etc. One of the responsibilities these roles have in common is that they identify relevant requirements, determine whether the organization conforms and if not they work with other personnel in bringing policies and practices into conformity with the requirements. It will not be their only job, and they need not be permanent employees.

The authority being assigned is to ensure conformity (meaning making certain) implying that the person appointed will be able to demand conformity if necessary, so it is more than an advisory role where the person's advice can be ignored in the interests of expediency. However, whether something conforms or not is not an exact science and can often be challenged which is why the person appointed needs to understand the intent of the requirements and possess appropriate interpersonal skills.

Why is this necessary?

The organization operates in an environment in which the requirements with which it needs to conform may be constantly changing. To meet this challenge, the organization needs access to specialists who understand these requirements and can advise on the actions that are needed to bring about conformity. It is necessary to provide these specialists with the freedom they need to execute their responsibilities without fear or favour.

How is this addressed?

As with finance, employment, health and safety law, etc., it is not necessary for all employees or even managers to understand these laws although an appreciation of them leads to more informed decisions. It is likewise unnecessary for all employees, including managers and executives, to understand the requirements of ISO 9001 and be able to determine conformity. In fact, if it was decided to make conformity to ISO 9001 a responsibility of all managers, it would result in multiple interpretations and consequently a dysfunctional system.

However, the requirement does not confine the responsibility to one person and so depending on the size of organization and the locations from which it operates, the role may be split by location.

Top management may have chosen to delegate authority for conformity to ISO 9001 to a compliance manager to look after all sites and the compliance manager may delegate local authority to a site compliance manager. Alternatively, top management may have chosen not to appoint a compliance manager and have delegated authority for conformity to ISO 9001 to the local site compliance managers. Both arrangements satisfy the requirement. However, local site compliance managers may be appointed by the local site manager and if left without a coordinating function and focal point, discontinuities may arise but these may be reduced through networking and social media without the need for a head office position. There may be of course other considerations that justify having one central authority for conformity to ISO 9001 such as dealings with the chosen certification body and for interfacing with customers on quality matters.

There is no requirement for those assigned specific responsibility and authority to carry a title that includes the words quality, compliance or ISO 9001 in the role or job title. What is important is that the words in the role title convey the holder's responsibility without creating ambiguity.

How is this demonstrated?

Demonstrating that responsibility and authority for ensuring that the QMS conforms to the requirements of ISO 9001 has been assigned may be accomplished by:

a)   presenting a policy or job description that confers on a particular person or persons this responsibility;

b)   interviewing the designated person(s) and confirming their awareness of their responsibility and authority for ensuring conformity of the QMS to ISO 9001;

c)   selecting a representative sample of internal and external QMS audit reports and showing that whenever a nonconformity with a requirement of ISO 9001 had been detected and the due date for correction action not met, the intervention of the person accountable for conformity to ISO 9001 has been effective.

Responsibility and authority for ensuring process performance (5.3b)

What does this mean?

Although ISO 9001 refers to *QMS processes* and to *processes needed for the QMS*, as explained in Chapter 8, the QMS is a systemic view of the organization from the perspective of how it creates and retains customers and therefore the processes of interest are those that serve this purpose. They are the organization's business processes as depicted in the process classification framework referred to in Chapter 9.

Box 20.3   Revised requirement on the responsibility for processes

In the 2008 version responsibility to ensure processes needed for the QMS were established, implemented and maintained was required to be assigned to a management representative. This could have been interpreted as requiring the management representative to establish, implement and maintain the processes needed for the QMS or for that person to have authority over those who undertook such work.

In the 2015 version, the requirement for a management representative has gone and with it the requirement to assign anyone with responsibility to ensure processes needed for the QMS were established, implemented and maintained. The requirement has been reduced to assigning responsibility and authority for ensuring that the processes are delivering their intended outputs. If we assign responsibility for results to a particular person, it is a basic principle of management that we must provide them with the authority to determine how those results are to be achieved and to acquire the resources necessary to achieve them. It is not necessary that this person establish and implement a process but this person does need to be able to maintain and improve the process so that it continues to deliver the intended results.

Responsibility and authority for ensuring that the processes are delivering their intended outputs implies it is intended that someone can be held accountable for the performance of a process or a group of processes, regardless of which functions are engaged in their execution. It also implies that this person be delegated the authority to require changes should the process outputs not be as planned. This denotes self-control as explained in Box 20.2.

Another implication is that the process has a performance that not only can be measured but that it's important to measure.

Why is this necessary?

As explained in Chapter 9, processes involve everyone who contributes to results regardless of their specialism. They are often cross-functional but not necessarily so as it depends on the division of labour in the organization. This dimension creates a problem for organizations that are structured on a functional basis because "as work passes between functions it crosses no-man's land, the white space on the organization chart", as Geary Rummler puts it (Rummler & Branche, 1995). As each function strives to meet its objectives, it optimizes its own performance which contributes to the sub-optimization of the organization as a whole. It is therefore necessary to manage work as a process across functional boundaries and optimize the process rather than the work of each function, but this requires we assign responsibility and authority for the process separate from the functions involved in it.

How is this addressed?

When it comes to assigning responsibility and authority for the performance of processes, we need to be selective. As a process uses inputs to deliver an intended result, there are a range of results, from how satisfied our customers are to whether an agenda for a meeting went out on time. There might be a temptation to produce metrics for every output but this would result in micro-management at the expense of macro-management. Every process output contributes to the output of other processes which eventually produces an output of value to the customer. It is at this level that there needs to be accountability for process performance and this lies with a person who is answerable for performance (takes the credit if it works, and must sort it out if it doesn't), but does not need to be involved in the day-to-day activities. This is the role of the process owner, but views differ on the terminology. The term *process owner* could imply that those engaged in the process report to them or that the process owner makes all the decisions associated with the process. Jeston and Nelis suggest the term *process steward* is better because it implies the role of a custodian who has to work in collaboration with other process stakeholders to achieve business outcomes (Jeston & Nelis, 2008).

Whether we refer to processes as business processes, management processes, support processes, work processes or any other type of processes for the purpose of applying this requirement, we need to ask, *in what way does this process output create value for our customer?* If the process we are considering only produces an agenda for an internal meeting, the agenda alone does not create value for the customer and is therefore not a candidate for assigning responsibility. However, if the meeting is part of another process that produces a quotation that is required by the customer, the process for producing quotations qualifies as a process that creates value for the customer. All processes should create value but the value they create maybe for the benefit of stakeholders other than the customer. Some processes may not create value at all for any stakeholders and are candidates for elimination but some may produce outputs that are only necessary for satisfying a regulation unrelated to a customer.

If we use the process classification framework as a guide, we can see where responsibility for a process can be assigned, for example:

> Level 1—Category (e.g. Manage customer service). This process will definitely be intended to add value for customers as that's its primary purpose. Responsibility for this category of processes may have been assigned to the head of customer services. This process qualifies for assignment of a process owner.
>
> Level 2—Process Group (e.g. Service products after sales). This process may encompass the management of warranty claims, servicing and repairing products under warranty, and replacement or recovery and therefore will definitely be intended to add value for customers. Responsibility for this group of processes may have been assigned to a customer service group leader. This process qualifies for assignment of a process owner.
>
> Level 3—Process (e.g. Process warranty claims). This process will definitely be intended to add value for customers. Responsibility for this process may have been assigned to a section leader. This process qualifies for assignment of a process owner.
>
> Level 4—Activity (e.g. Investigate warranty issues). As this is part of the level 3 process it does not add value on its own so doesn't qualify for assignment of a process owner.
>
> Level 5—Tasks (e.g. Define issue). As this is part of the level 3 process it does not add value on its own so doesn't qualify for assignment of a process owner.

Having assigned a process owner, the responsibilities and authority need to be determined. Typical responsibilities might be:

- collection and reporting process performance to those managing the work;
- finding areas of improvement and running projects to close the gaps;
- enforcing process standards.

Without process owners for cross-functional processes, the interfaces tend to be ignored and therefore process owners of cross-functional processes will need, in addition, responsibility and authority for:

- leading a cross-functional process management team that sets the process objectives, plans the process, monitors performance and undertakes process improvement;
- facilitating the resolution of interface problems among the functions that contribute to a process.

The role of process owner is one that anyone with the necessary competences can perform. If the process is unifunctional the process owner may be the functional manager. But if the process is cross-functional the process owner should be someone who:

- has the most to gain if the process succeeds and the most to lose if it fails (may have rewards linked to performance);
- manages the largest number of people engaged in the process;
- understands the impact the process has on the organization's performance and the impact the external influences have on the process.

Chapter 13 of Rummler and Branches' book on improving performance provides useful guidance on managing processes and the various roles involved, as does Jeston and Nelis's book. These and other authors identify different roles or different labels for the same role (e.g. process owner, process steward, process executive, process analyst, process designer, process architect, process sponsor etc.).

How is this demonstrated?

Demonstrating that responsibility and authority for ensuring that the processes are delivering their intended outputs has been assigned may be accomplished by:

a) presenting evidence of process maps that denote where accountability for each process lies;
b) presenting role description for process owners or whatever term is used to denote accountability for process performance;
c) selecting a representative sample of process performance reports and presenting evidence that whenever there's a gap between expected and achieved performance the intervention of the person accountable for process performance has been effective.

Responsibility and authority for reporting QMS performance (5.3c)

What does this mean?

When integrated into the organization's business processes, QMS requirements no longer refer to a distinct entity. Function managers and process owners should report on the performance of their functions and processes. However, to report on the performance of the QMS, a person must collect and analyse factual data across all company operations to determine whether the quality objectives are being achieved and if not, to identify opportunities for improvement. This person is the one to whom top management assign responsibility and authority for reporting on QMS performance.

Why is this necessary?

Each manager cannot measure the performance of the organization relative to quality from information generated by their function alone. Individually they carry responsibility for the utilization of resources within their own area. The performance of the organization relative to quality can only be measured by someone who has the ability and authority to collect and analyse the data across all company operations. All managers may contribute data, but this needs to be consolidated to assess performance against corporate objectives just as a finance director consolidates financial data.

How is this addressed?

To report on QMS performance and identify opportunities for improvement in the QMS the person needs the right to:

• determine the effectiveness of the QMS;
• report on the performance of the organization relative to the quality of its products and services;

- identify opportunities for improvement in the QMS;
- cause beneficial changes in quality performance.

This person may be anyone with the available capacity and necessary competence. If top management has established a position in the hierarchy with responsibility for quality matters, just as they may have done for finance, health, safety and security, it would make sense to include responsibility for reporting on the performance of the organization relative to quality. However, responsibility for identifying opportunities for improvement in the QMS should be included in everyone's job description with a secondary responsibility to notifying the person with responsibility for quality matters.

By installing data collection and transmission nodes in each process, relevant data can be routed to the person responsible for analysis, interpretation, synthesis and assessment. It can then be transformed into a language suitable for management action and presented at the management review. However, this requirement imposes no reporting period; therefore, performance should also be reported when considered necessary or on request of top management.

How is this demonstrated?

Demonstrating that responsibility and authority for reporting on the performance of the QMS has been assigned may be accomplished by:

a) presenting a policy or job description that confers on a particular person this responsibility;
b) interviewing the designated person and confirming their awareness of their responsibility and authority for reporting on the performance of the QMS.

Responsibility and authority for promoting customer focus (5.3d)

What does this mean?

In a customer-focused organization, all managers and team leaders would have responsibility for ensuring the promotion of customer focus. Customer focus as a concept is explained in Chapter 5 under the quality management principles and in Chapter 18.

Why is this necessary?

Unless staff are aware of customer needs and expectations and how important the customer is to the organization it is likely that the organization will be inundated with customer complaints. Customer satisfaction is the aim of the QMS and hence it is important that all staff at all levels do not lose sight of this. Clearly all managers are responsible for promoting behaviours that show awareness of customer needs and expectations, but this does not mean it will happen as internal pressures can cause distractions. Constant reminders are necessary when making decisions in which customer satisfaction may be directly or indirectly affected. Staff in a customer-facing role are in the firing line and were they to exhibit inappropriate behaviours, this can immediately result in lost orders, customer complaints, etc. Heightened awareness of customer requirements and the role people play in achieving them can inject a sense of pride in what they do and lead to better performance.

How is this addressed?

Top management may assign responsibility and authority for ensuring the promotion of customer focus verbally or through policies and job descriptions.

How is this demonstrated?

Demonstrating that top management has assigned responsibility and authority for ensuring the promotion of customer focus may be accomplished by:

a) presenting evidence of policies and job descriptions where this responsibility and authority is stated or;
b) interviewing a representative sample of people and establishing that they are aware of their responsibility to promote customer focus throughout the organization;
c) selecting a representative sample of management review records and presenting evidence that whenever an issue where customer satisfaction has been compromised had been discussed and the due date for correction action not met, the intervention of the person accountable for ensuring promotion of customer focus has been effective.

Responsibility and authority for ensuring QMS integrity (5.3e)

What does this mean?

Box 20.4   Ackoff on improvement

Doing the wrong thing right is not nearly as good as doing the right things wrong
    Until managers take into account the systemic nature of their organizations most of their efforts to improve their performance are doomed to failure.
    If we have a system of improvement that is directed at improving the parts taken separately you can be absolutely sure that the performance of the whole will not be improved. We don't improve the quality of a part unless by doing so we improve the quality of system of which it forms a part.

(Ackoff, 1994)

The QMS is required to be subject to continual improvement which means that some elements of it will be changing periodically. Responsibility for QMS integrity therefore means that those planning and implementing such changes have a responsibility to ensure the integrity of the system is not adversely affected when changing elements of it. The QMS may have lots of flaws to start with, but whatever its effectiveness, changes should always improve its effectiveness and not make it worse.

Why is this necessary?

The QMS consists of many elements and changing any one of these in isolation may inadvertently affect its performance even though the change may have been intended to improve its performance. The outcome from many reengineering and TQM initiatives of the 1980s and 90s was patchy for the reasons Ackoff gives in Box 20.4.

How is this addressed?

This requirement does not mean that responsibility and authority for ensuring system integrity can or should be assigned to a single individual. The QMS is a dynamic system. We are not dealing with a set of documents and ensuring their integrity when something changes. There will be documents that capture or model the system and responsibility and authority for maintenance of that model may be assigned to a specific individual. We may designate this person as the system architect who will need an understanding of systems theory. However, responsibility and authority for changing any element of the QMS and the way it is interconnected to other elements should be assigned to the person who plans and implements the change but any changes should be planned and implemented with the approval of the system architect.

How is this demonstrated?

Demonstrating that responsibility and authority has been assigned for ensuring the integrity of the QMS when changes are planned and implemented may be accomplished by:

a) presenting typical role or job descriptions that define the responsibilities and authority for planning and implementing changes that affect the performance of the QMS;

b) showing where in these descriptions the specific responsibility and authority is defined for preserving system integrity when planning and making changes;

c) interviewing personnel to confirm awareness and familiarity with their responsibilities and authority for planning and implementing changes;

d) selecting a representative sample of changes to the QMS and presenting evidence that whenever a change had been planned and the due date for its implementation not met, the intervention of the person accountable for ensuring the integrity of the QMS has been effective.

Bibliography

Ackoff, R. L. (1994). *Learning and Legacy of Dr. W. Edwards Deming*. Retrieved from www.youtube. com/watch?v=OqEeIG8aPPk

Jeston, J., & Nelis, J. (2008). *Management by Process: A Road Map to Sustainable Business Process Management*. Oxford: Butterworth-Heinemann.

Juran, J. M. (1974). *Quality Control Handbook* – Third Edition. New York: McGraw-Hill.

Rollinson, D. (2008). *Organisational Behaviour and Analysis*. Harlow, Essex: Pearson Education Limited.

Rummler, G. A., & Branche, A. P. (1995). *Improving Performance: How to Manage the White Space on the Organization Chart*. San Francisco: Jossey-Bass Inc.

Key messages from Part 5

Chapter 17   Leadership and commitment

1   Leadership is the process in which leaders and followers interact in a way that enables the leader to influence the actions of the followers in a non-coercive way towards the achievement of certain aims and objectives.
2   Everything we do depends for its quality on the thinking we do first.
3   If the top management are of a mind to put profits first when making decisions, they have a learning opportunity that needs to be addressed.
4   Top management's perception of quality is vital to whether it is committed to the development of a quality management system.
5   For top management to account for the effectiveness of the QMS they need to be able to explain its performance.
6   Organizations that have no policy regarding quality are allowing mediocre performance, mistakes, inefficiencies and low standards to prevail.
7   Ensuring integration of QMS requirements into business requires top management to overcome the barriers that prevent requirements reaching the people who will meet them in a form they will understand.
8   Top management must have adapted their style of management to reflect behaviours consistent with the process approach and risk-based thinking so that others will follow.
9   Any organization can become the best, but only with the full co-operation and participation of each and every individual contributor.
10   Improvement in the QMS is necessary becasue it is an open system and therefore subject to external influences.
11   To be a manager means sharing in the responsibility for the performance of the enterprise, which includes taking accountability for the QMS.

Chapter 18   Customer focus

12   Organizations cannot maximize both customer value and shareholder value. They must pick one main objective and treat the others as constraints.
13   In organizations that are driven by the needs of the customer, their goal is to maximize customer value in a way that creates a minimum shareholder value and leads to customer loyalty and outstanding success.
14   Leaders who put the customer's interest first are customer focussed.

15 Top management want its direct reports to succeed, and they want their office staff to succeed, and they want their front-line staff to succeed who want their customers to succeed.

Chapter 19 Policy

16 The primary instrument for aligning the organization towards satisfying customer needs and expectations is the quality policy.
17 Policies are enacted where there are choices to be made; therefore the quality policy exists to guide personnel in taking actions that are consistent with the organization's commitment to customers and its strategic direction.
18 Whatever top management proclaim, it is what individuals do that builds a reputation for quality.
19 Communication of policy is about gaining understanding and not just the sending of messages from one source to another.
20 Front-line personnel may be empowered by making the customer the first priority, whereas personnel remote from customers may be empowered by making quality the first priority.
21 The quality policy is applied *before* action is taken to produce a result consistent with the policy.

Chapter 20 Organizational roles, responsibilities and authorities

22 Although a person may occupy a particular position, they may be eligible to perform any number of roles if they possess the necessary competences.
23 Those assigned responsibility for ensuring the requirements of ISO 9001 are met determine whether the organization conforms, and if not, they work with other personnel to bring policies and practices into conformity with the requirements.
24 If it was decided to make conformity to ISO 9001 a responsibility of all managers, it would result in multiple interpretations and consequently a dysfunctional system.
25 As each function strives to meet its objectives, it optimizes its own performance, which contributes to the sub-optimization of the organization as a whole.
26 It is at the level where a process produces an output of value to the customer that there needs to be accountability for process performance, and this lies with a person who is answerable for its performance.
27 The performance of the organization relative to quality can only be measured by someone who has the ability and authority to collect and analyse the data across all company operations just as a finance director consolidates financial data.
28 Heightened awareness of customer requirements and the role people play in achieving them can inject a sense of pride in what they do and lead to better performance.
29 Responsibility and authority for changing any element of the QMS and the way it is interconnected with other elements should be assigned to the person who plans and implements the change.

Part 6

Planning

Introduction to Part 6

Clause 6 addresses planning, but not all planning. The context is QMS planning rather than operational planning, which is addressed in clause 8.1. However, as the QMS is formed from processes, clause 6 also applies to the planning of the high-level processes rather than the processes for providing specific products and services addressed in clause 8.1. Remember that the order in which requirements are stated is not significant and therefore when we place the requirements in the order in which they are likely to be carried out we get a cycle as shown in Figure P6.1.

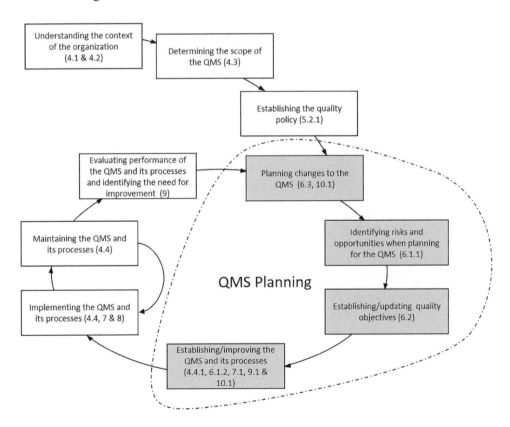

Figure P6.1 Planning of the QMS

The planning stages are represented by the grey boxes. Note the clause numbers particularly those associated with establishing the QMS:

- Clause 4.4.1 applies because this requires the QMS and its processes to be established (i.e. planned).
- Clause 6.1.2 applies because this requires actions to address risks and opportunities to be integrated into the QMS, which is done during planning.
- Clause 7.1 applies because this requires resources for the QMS to be determined which is done during planning.
- Clause 9.1 applies because this requires monitoring and measurement activities and methods to be determined which is done during planning.
- Clause 10.1 applies because this requires actions to improve the QMS to be determined and implemented which is done during planning.

The external and internal environments are continually monitored and as a result it may be necessary to change the QMS or the scope and the quality policy on which it is based. This need for change feeds into QMS planning as shown in Figure P6-1.

21 Actions to address risks and opportunities

> I have approximate answers and possible beliefs about different things, but I'm not absolutely sure about anything.
>
> Richard Feynman (1918–1988 Nobel Prize in physics 1965)

Introduction

When we take action to address risks and opportunities, we are not reacting to circumstances that have already happened but trying to deal with circumstances that have yet to happen so that we are adequately prepared for the favourable or unfavourable consequences.

The way an organization creates and retains customers is very much dependent on its ability to determine and address external and internal factors that impede or facilitate its performance. In general organizations prefer certainty to uncertainty, particularly when making investments, but having assessed the risks and addressed them organizations can move forward with confidence. And if the results turn out differently, this enables them to learn from their experience and continually improve.

In Chapter 12 we showed how issues relevant to the QMS can be identified by filtering the results of the PESTLE and SWOT Analysis through the purpose and mission and then the intended results of the QMS. The requirements in clause 6.1 focus on the results of these analyses and require the organization to determine which of the identified issues present a change, a risk or an opportunity.

Many of the issues uncovered by the analysis may be facts and are therefore not risks or opportunities per se because there is no uncertainty about them – they have already happened. These are changes that just have to be assessed and addressed. However, their effect on the organization may well be uncertain, and therefore the issues may be a cause or source of risk or opportunity. This creates three pathways into planning the QMS as shown in Figure 21.1.

Risk assessment is the overall process of risk identification, risk analysis and risk evaluation and is one of the processes carried out when planning for the QMS. Whether one is using methods of risk assessment where a risk is identified, analysed and evaluated in a few seconds or a few weeks, the steps are the same, the difference is in the rigour of the assessment and the magnitude of the decisions it informs. Therefore, the bigger the decisions (as judged by its consequences if it's wrong), the greater the effort that is put into the assessment of risk.

A link between opportunities and risks is shown because there may be risks associated with not pursuing an opportunity and risks involved in pursuing an opportunity. When we

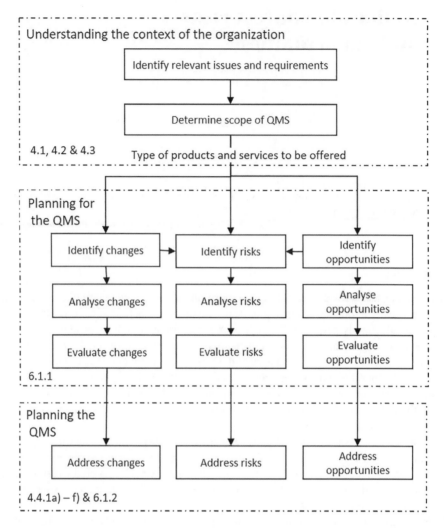

Figure 21.1 Outline risk assessment activity sequence

look for risks we are looking for potentially undesirable outcomes. When we look for opportunities we look for desirable outcomes and these incur taking risks but we should do both. Pursuing a strategy of only looking for undesirable outcomes is a pessimistic approach to quality management, whereas looking for both risks and opportunities is a balanced approach to quality management.

We cannot assume that the order in which the requirements are presented in the standard is the order in which they are intended to be, or indeed, can be implemented (see Box P4.1).

In this chapter, we examine the four requirements of clause 6.1, namely:

- Risks and opportunities that need to be addressed (6.1.1)
- Planning actions to address risks and opportunities (6.1.2a)

- Integrating actions into QMS processes (6.1.2b (1))
- Evaluating the effectiveness of actions that address risks and opportunities (6.1.2b (2))

Risks and opportunities that need to be addressed (6.1.1)

What does this mean?

Clause 4.4.1f) requires risks and opportunities determined in accordance with the requirements of 6.1 to be *addressed* but it's only clause 6.1.1 that requires risks and opportunities to be *determined*, implying that the requirements of 6.1.1 are to be met before evaluating the processes and implementing any changes (as per 4.4.1g). Clause 6.1.1 also requires we consider the issues referred to in clause 4.1 and the requirements referred to in clause 4.2 and determine the risks and opportunities that need to be addressed when planning for the QMS. However, clause 5.1.2b) requires top management to ensure the risks and opportunities that can affect conformity of products and services and the ability to enhance customer satisfaction are determined and addressed. Such risks may have their source outside the QMS or be inherent in the design of the QMS and the products and services it produces and therefore may have their source within the QMS. This leads to the conclusion that there are extrinsic and intrinsic risks, that is, risks the sources of which are external to the QMS (extrinsic risks) and risks the sources of which are internal to the QMS (intrinsic risks). This relationship is explained further in Box 21.1. A video on quality and risk is available on the companion website.

The conclusion we reach is that clauses 4.1, 4.2, 4.3 and 6.1.1 are intended to be implemented before we establish the QMS (4.4), and then we address the risks and opportunities (6.1.2) and this is why planning for the QMS is separate from planning the QMS in Figure 21.1. Basically, it's treating the QMS as a black box. This interpretation is confirmed by JTGC Guide N360 where it states:

> The intent of the clause on Actions to address risks and opportunities is to specify the requirements for the planning needed as a prerequisite to establishing the MS. It specifies what needs to be considered and what needs to be addressed. The planning is performed at a strategic level, versus the tactical planning done for Operational planning and control.

Box 21.1   Intrinsic and extrinsic risk in ISO 9001

The risks referred to in clause 6.1.1 are extrinsic (their source is from outside the QMS), whereas the risks referred to in clause 5.1.2b) are intrinsic (their source is from within the QMS). The problem is that clause 5.1.2b) only refers to top management *ensuring* (intrinsic) risks are determined and addressed. There is no equivalent clause to clause 6.1.1 requiring the intrinsic risks and opportunities *to be* determined and addressed. In the absence of a requirement in clauses 8.1, 8.3 and 8.5 to determine and address intrinsic risks, users will therefore have to apply clause 5.1.2b) and 5.1.1d) on risk-based thinking to derive such a requirement.

Why is this necessary?

The system that was effective in enabling the organization to successfully create and retain customers in the past may not be so successful in the future, and therefore it is necessary to assess the impact of the anticipated changes in the external and internal environment and identify those which will have a negative effect and those that will have a positive effect. It is also necessary to assess the impact of changes in stakeholder requirements and identify those new requirements which are outside the scope of the current QMS and those features and characteristics stakeholder no longer require.

How is this addressed?

Process overview

In order to address this requirement, we need to examine the relevant issues and requirements that emerge from understanding the context of the organization and stakeholder requirements and identify the risks and opportunities that they present. If we step back and take a view of the process of which identifying risks and opportunities is a part, we will get a better idea of how issues and interests are converted into actions and this is shown in Figure 21.2.

   In the sequence of steps below the diagram, it will be observed that they do not follow the order in which the requirements are stated and apply requirements from clauses 4, 6 and 7.

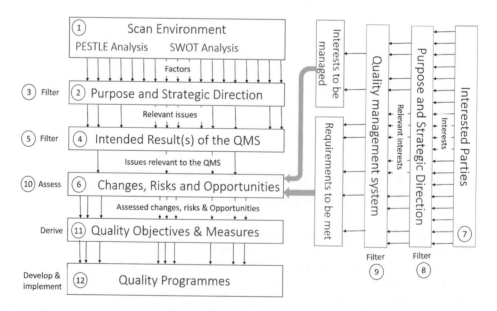

Figure 21.2 Converting issues and interests into actions

1 Scan the internal and external environment to gather information against factors that could affect the future direction of the organization. (no requirement)
2 Confirm the organization's purpose and strategic direction. (no requirement)
3 Filter the results through the organization's purpose and strategic direction to identify relevant issues that may impede or facilitate the organization's ability to fulfil its purpose or proceed in its chosen strategic direction. (clause 4.1)
4 Confirm the intended results of the QMS. (no requirement)
5 Filter the identified issues through the intended results of the QMS to identify issues relevant to the QMS. (clause 4.1)
6 Determine which of these issues present a change (clause 6.3), a risk or an opportunity (clause 6.1.1) to the ability of the QMS to fulfil its purpose.
7 Determine the interested parties and their interests. (clause 4.2)
8 Filter the interests of these interested parties through the organization's purpose and strategic direction to identify the relevant interests. (clause 4.2)
9 Filter these interests through the QMS scope to separate those interested parties having relevant requirements to be met, for example, customers from those with interests that need to be managed (e.g. competitors, criminals and pressure groups). (clause 6.1.1)
10 Analyse and evaluate the changes (clause 6.3), risks and opportunities including the interests of interested parties that need to be managed. (clause 6.1.1)
11 Select the results of the evaluation and the requirements to be met and derive quality objectives and success measures which if achieved will address these changes, risks, opportunities and requirements satisfactorily (clause 6.2.1)
12 Develop quality programmes for achieving these quality objectives. (clauses 6.2.2, 6.3, 7.1, 7.2, 7.3, 7.4 and 7.5)

Risk assessment is the overall process of risk identification, risk analysis and risk evaluation so we will examine the elements of this process as it applies to "planning for the QMS".

Identifying risks and opportunities

There are seven key questions we need to answer to identify and address the risks and opportunities that will either impede or facilitate the effectiveness of the QMS:

1 What are we trying to do? – the objective, the goal (Step 4 in Figure 21.2).
2 What might affect what we are trying to do? – the uncertainties that might help or hinder (Steps 4 and 5 in Figure 21.2).
3 Which of these are most important? – the risk assessment.
4 What can we do about it? – risk treatment.
5 Have we taken the action we planned to take? – implementation.
6 Did the action work – risk monitoring.
7 What's changed since the last time we took action? – risk review.

Whether we are planning to undertake a minor or a major task, asking and answering these questions will help us undertake those tasks more successfully. In this section we are interested in questions 2 and 3.

Many of the risks and opportunities will be external to the organization arising from uncertainties in political, economic, social and technological environment, but there will also be risks and opportunities arising from within, from the people, processes, technologies, environment and the culture, all of which could either impede or facilitate success. Opportunities tend to arise when some period, situation or circumstance either comes to an end or is about to begin.

DISTINGUISHING RISKS AND OPPORTUNITIES FROM CAUSES AND EFFECTS

Having looked at hundreds of risk registers throughout the world, the risk management consultant David Hillson found that over half of them identify causes or effects as risks, implying that the producers are trying to manage things that are not risks at all (Hillson, (2) 2016). It is important to distinguish between issues that are facts and issues that are risks, but often, we sometimes make statements that we present as describing risks when they are describing issues, problems, facts, causes or effects.

Causes are existing conditions such as the sun is shining today; risks are what might happen such as we might not sell many umbrellas if the sun is shining, and the effects are what could follow (e.g. the takings could be down today).

If we look at this from the perspective of what we are trying to do, which is this case might be to provide customers with protection from the natural environment, and ask what affects our ability to do this, it's tempting to say that the risk is the uncertainty about the weather; but we can't manage the weather. It's a fact that the weather changes so variation in weather conditions is the cause of uncertainty about future umbrella sales. In selling umbrellas, we are taking a risk that sales may fluctuate due to weather conditions. The environment in which organizations operate provides the source of their risks but the actual risks rather depend on what they do in that environment.

We therefore express:

- the existing condition in terms of what is, what has/has not happened, what does/does not occur. These are facts;
- the uncertain event in terms of what may, might or possibly happen;
- the effect in terms of what could or would follow.

We can therefore construct a statement template that separates cause, risk and effect and this would be as follows:

As a result of <an existing condition>, <an uncertain event> might occur which would lead to <an effect on the objective>. Expressed in simple terms we ask, what do we know, what uncertainty does that present us with and why does it matter? Some examples follow:

- There is a shortage of labour with the skills we need (cause) so we might not be able to take on any more work (risk) and will therefore lose business to our competitors (effect). In this case the risk we manage is *work load*.
- Revenue is down 20% on last year (cause) but we have identified a source of lower cost materials of the same quality (opportunity) and as a result we can reduce our prices (effect). In this case the opportunity we manage is *material costs*.

- There is an increase in cybercrime (cause) and our IT infrastructure may be vulnerable to attack (risk) which may impede our ability to process customer orders (effect). In this case the risk we manage is the *vulnerability of our IT system to cybercrime.*

One way of identifying risks and opportunities is to examine the issues and establish which issues are causes and which are risks. Many issues may arise because a change has already occurred and the organization needs to embrace the change (e.g. a change in legislation). What may be uncertain is the organization's ability to respond to the change in a timely and effective manner. Others may present a risk or an opportunity depending on whether or not the organization chooses to address them.

Uncertainty presents both risk and opportunity, with the potential to erode or enhance value. Therefore, in seeking to identify risks and opportunities we need to be looking for the sources of uncertainty.

Many organizations maintain a risk register which contains details of all categories of risk. Often risk categories include strategic, operational, finance and compliance risks and risks affecting the intended results of the QMS could be included in any category. To trace risks through to the provisions made to control them it may be necessary to assign codes in the existing risk register or compile a separate risk register and include the following information:

a) The intended results of the QMS (or a process if being undertaken at that level) – this would be stated in the header as a point of reference for all entries (i.e. all entries either potentially impede or facilitate achievement of these results)
b) A unique identifier for each issue
c) The relevant issues as deduced from the PESTLE and SWOT analysis
d) Whether the issue poses a risk or an opportunity
e) The category of risk or opportunity for reporting purposes (e.g. strategic, operational, finance, compliance)
f) Description of the uncertainty that might create, enhance, prevent, degrade, accelerate or delay achieving the intended results of the QMS or one of its processes. These should include not just the events that may or may not happen but also variables, ambiguities and blind spots (see "Types of uncertainty" in Chapter 10).

Identification should include risks whether or not they are under the control of the organization. Risk identification methods can include:

- Informal inductive reasoning techniques such as posing questions such as:
  - how will the mitigation of (risk) give assurance that the QMS can achieve its intended results?
  - how will pursing this (opportunity) give assurance that the QMS can achieve its intended results?
  - how will the mitigation of (risk) prevent, or reduce the potential for QMS failure?
  - how will mitigation of (risk) improve our ability to satisfy our customers?
  - how will pursuing (opportunity) enhance our ability to satisfy our customers?

- Evidence-based methods, such as checklists and reviews of historical data.
- A team of experts following a structured set of prompts or questions.

The people identifying risks and opportunities need to possess sufficient knowledge about the organization, the QMS and the issues to make credible judgements. Access to historical data will be beneficial in determining whether the issue has arisen previously and how it was assessed, addressed and what the outcome was.

Box 21.2   Drucker on risk

Risk: The likelihood of success versus the likelihood of failure for any undertaking. Assessing a risk is the process of trying to quantify or judge which likelihood is the larger and by how much.

The end result of successful strategic planning must be capacity to take a greater risk, for this is the only way to improve entrepreneurial performance. To extend this capacity, however, we must understand the risks we take. We must be able to choose rationally among risk-taking courses of action rather than plunge into uncertainty on the basis of hunch, hearsay, or experience, no matter how carefully quantified.

(Drucker, 1974)

COMMON CAUSE FAILURE

A catastrophic situation can arise when random and systematic events cause multiple devices, systems, or layers to fail simultaneously. This is referred to as common cause failure (CCF). The individual failure of a component may not destabilize a system and the probability of several components failing independently of each other at the same time due to an external event is extremely unlikely. Common cause failure is an engineering term and normally applies to engineered systems but can be applied to organizations, for example,

- If the organization outsources production to a country in a region of the world that is prone to earthquakes, in a single event, its source of product could be cut off.
- If a plant is isolated, feeding off power from a single source it is vulnerable to interruption due the weather conditions, mechanical failure or an illegal action.
- If an organization depends on its connection to the Internet to transact business a failure of the ISP may disrupt business continuity.

A catastrophic situation can also arise when there is simultaneous failure of two or more identical objects in the same mode. This is referred to as common mode failure and is a subset of common cause failure. In an organizational context, an example of common mode failure is where there are identical computer systems on the same network and a computer virus attacks all of them causing simultaneous failure.

Analysing risks and opportunities

Hubbard advocates that a weak risk management approach is effectively the biggest risk in an organization, and he also wisely states that if risks are not properly analysed they cannot be properly managed (Hubbard, 2010). Although ISO 9001 doesn't require risk management, it does require the effectiveness of actions to address risks to be determined, and this requires some form of measurement.

Risk is analysed by determining the causes and sources of risk, their consequences and the likelihood those consequences can occur taking into account the effectiveness of any existing controls. Hubbard identifies several methods in use for risk analysis.

1   Expert intuition – Pure gut feel unencumbered by structured rating systems of any kind.
2   Expert audit – External experts develop comprehensive check lists and may or may not use formal scoring or stratification methods.
3   Simple stratification methods – These use green-yellow-red or high-medium-low rating scales to assess likelihood and consequence in a two-dimensional matrix (see later).
4   Weighted scores – There are also more elaborate scoring methods with dozens of "risk indicators," each on some scale, which are then multiplied by some "weight" so they can be added up to a "weighted risk score".
5   A calculus of preferences – Methods such as multi-attribute utility theory (MAUT), multi-criteria decision-making (MCDM), and analytic hierarchy process (AHP) are more structured than the weighted scores but ultimately still rely on the judgments of experts.
6   Probabilistic models – These determine the odds of various losses or gains and their magnitudes are computed mathematically. It is the basis for modelling risk in the insurance, financial and engineering industries. They could use subjective inputs but also historical data and the results of empirical measurements. Examples of this method include fault tree analysis, failure mode and effective analysis and the Monte Carlo method.

Hubbard argues that each of these is flawed in some important way, and most of them are no better than astrology. It rather depends on what they are to be used for. The qualitative methods (1–4) are useful as a means to prioritize action but not for making big decisions as they may make many decisions far worse than they would have been using merely unaided judgments. For such decisions it's necessary to use quantitative methods.

CALIBRATING THE ESTIMATORS

With physical measurements, we calibrate the measuring instruments to be confident that the results are valid. When it comes to measuring risk, we need an equivalent form of calibration otherwise it's just guesswork that anyone can do. Why should we accept an estimate of probability from one person over another? One way of determining if a person is good at quantifying uncertainty is to assess their predictions to discover how many fall within 90% of the true result. Unfortunately, extensive studies have shown that very few people are naturally calibrated estimators. Leading researchers in this area have been Daniel Kahneman, winner of the 2002 Nobel Prize in Economics, and his colleague Amos Tversky. The research shows that everyone is biased either towards overconfidence or underconfidence. Research done by Fischhoff, Phillips, and Lichtenstein, and published in 1980 under the title *Calibration of Probabilities*, showed that bookies were rather better at assessing the odds of events than others and concluded that assessing uncertainty is a general skill that can be taught with measurable improvement. So, for example when a calibrated project manager says they are 80% confident that funding for a major project will be forthcoming, there really is an 80% chance the funding will be forthcoming. Hubbard provides further information on calibrating estimators in his book How to measure anything (Hubbard, 2010)

It is recommended that the reader consult other works for an in-depth appreciation of risk analysis techniques. However, as it's such a commonly used method outside the insurance sector further explanation of a simple stratification method is provided.

THE STRATIFICATION METHOD

The method defines consequence and likelihood using an "ordinal scale" – a scale that indicates a relative order of what is being assessed; not actual units of measure. Consequence and likelihood can be graded as very high, high, medium, low and very low.

The level of risk or opportunity is the magnitude of a risk or opportunity expressed in terms of the combination of consequences and their likelihood. In the matrix of Figure 21.3 the cells indicated by H imply a high level of risk or opportunity and the cells indicated by L imply a low level of risk or opportunity. By use of different colours to represent each rating (e.g. green, orange and red) the matrix can be turned into a "heat map". This is not scientific, it's not a measurement of risk or opportunity but it does provide a means of prioritizing action.

- Where the level of risk is high, it's telling us to stop and consider the harm that will be done if we don't address a risk that is very likely to happen.
- Where the level of opportunity is high it's telling us to stop and consider the benefit that will be lost if we don't address an opportunity that is very likely to happen.
- Where the level of risk is low, we don't need to waste time trying to reduce or eliminate risks that are unlikely to happen.
- Where the level of opportunity is low, we don't need to waste time pursuing opportunities that are unlikely to materialize.
- Where the level of risk or opportunity is medium we attend to these only after addressing those risks and opportunities that are high.

The scales are meaningless unless we define what we mean by high, medium and low and these definitions are specific to the objectives we are trying to achieve. The following two examples of low risk are comparable because the population size is similar and the skill about equal but clearly there are more variables not under the control of the surgeon in case A than of the pilot is case B.

A. The risk of serious complications developing as a result of cataract surgery is very low which, based on current statistics, means there is a 2% or less likelihood of complications (According to the American Society of Cataract and Refractive Surgery (ASCRS), 3 million Americans undergo cataract surgery each year, with an overall success rate of 98% or higher.

(Knobbe, 2016)

B. The risk of an aircraft accident on landing or take off at London Heathrow is also very low which, based on current statistics, means that there is a 0.0000337% likelihood of an accident (In the last 20 years there have only been three accidents in 8.9 million movements.)

(CAA, 2016)

If the likelihood of an aircraft accident on landing at London Heathrow was equal to the likelihood of serious complications developing as a result of cataract surgery, there would have been over 9482 accidents in 2015, that's 26 accidents a day! It's doubtful if Heathrow would remain open for business if there were any more than one accident a month let alone 26 a day.

An alternative to scoring likelihood and consequence by categories such as high, medium and low, is to use a 1–5 or 1–8 point-based scoring method with associated descriptions. However, this does not make it a quantitative method. Adding or multiplying scores that have been made based on opinion does not quantify the scores.

RATING LIKELIHOOD

At the time of writing this book no relevant examples of strategic quality risk assessment could be found; therefore, examples have been taken from the health care sector.

Likelihood can be estimated quantitative by using probability data or qualitatively using subjective levels. In an industry that is mature lots of data exists on accidents and incidents and so there is less guesswork involved. An example of likelihood levels used in the health sector is shown in Table 21.1.

RATING CONSEQUENCE

Table 21.2 is an example from the health sector showing how the degree of judgement about consequences has been refined by providing some objective criteria.

Table 21.1 Likelihood levels used in the health sector

Level	Frequency
1	Less than once every 10 years
2	Once a year to once every 10 years
3	Once a month to once a year
4	Once a week or once a month
5	Once a day to once a week
6	More than once a day

Table 21.2 Consequences rating used in the health sector

Rating	Description	Definition
5	Catastrophic event	Death or serious physical or psychological injury or the risk thereof. Serious injury specifically includes loss of limb or function. Must meet two of the three criteria: 1. Results in unanticipated death or major permanent loss of function 2. Associated with a significant deviation from the usual process 3. It has the potential for undermining the public confidence
4	Major event	Injury or permanent loss of bodily function (sensory, motor, physiologic, or intellectual), disfigurement, surgical intervention required, increased length of stay, increased level of care
3	Moderate event	An event, occurrence, or situation involving the clinical care of a patient in a medical facility which could have injured the patient but did not cause an unanticipated injury or require the delivery of additional healthcare services
2	Minor event	Failure is not noticeable to the patient and would not affect delivery of care. Failure can be overcome with modifications to the process; failure may cause minor injury.
1	Near miss	A process variation that does not affect the outcome but for which a recurrence carries a significant chance of a serious outcome. No injury, no increased length of stay or level of care.

RISK MATRIX

A risk matrix is a visual representation of the analysis which shows likelihood against consequences. By placing likelihood on a vertical axis and consequence on a horizontal axis we can create a risk matrix as shown in Figure 21.3 that helps visualize the priorities. One of the advantages of this type of presentation is that it can be used to show how an action taken to address the risk has changed the estimate of its probability and/or impact (illustrated by the black dots in Figure 21.4).

If the consequence of uncertainty is negative it becomes a risk matrix and if the consequence is positive, it becomes an opportunity matrix. By positioning the risk matrix and opportunity matrix side by side, we create a zone in the middle of the diagram which represents the very worst risks and the very best opportunities and thereby the priorities for action as shown in Figure 21.5.

An example of a risk matrix in the healthcare sector with associated criteria is shown in Figure 21.6. The probability of occurrence might be deduced intuitively but more likely to be derived from research into national health statistics, thus giving it far more credibility.

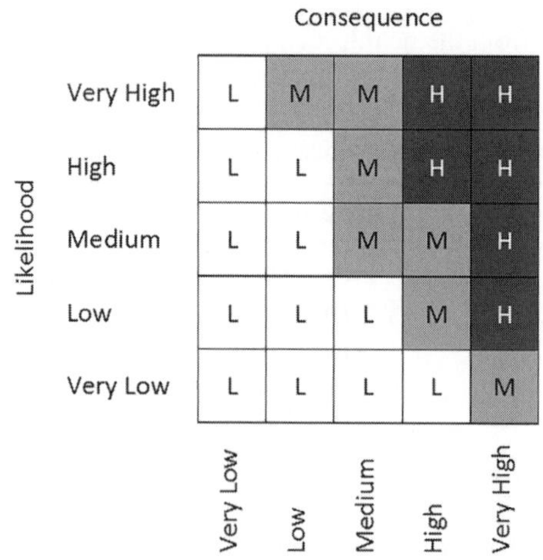

Figure 21.3 Risk matrix

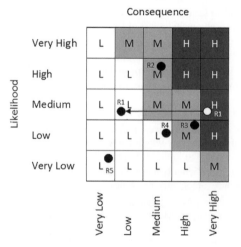

Figure 21.4 Risk matrix after risk treatment

Negative Consequence   Positive Consequence

Likelihood	Very Low	Low	Medium	High	Very High	Very High	High	Medium	Low	Very Low
Very High	L	M	M	H	H	H	H	M	M	L
High	L	L	M	H	H	H	H	M	L	L
Medium	L	L	M	M	H	H	M	M	L	L
Low	L	L	L	M	H	H	M	L	L	L
Very Low	L	L	L	L	M	M	L	L	L	L

Figure 21.5 Matrix of risk and opportunity assessment

Likelihood			A	B	C	D
More than once a day	6		High	High	High	High
Once a day to once a week	5		Medium	High	High	High
Once a week or once a month	4		Medium	Medium	High	High
Once a month to once a year	3		Low	Medium	Medium	High
Once a year to once every 10 years	2		Low	Low	Low	Medium
Less than once every 10 years	1		Low	Low	Low	Low
			A	B	C	D
			Low	Moderate	Severe	Fatality
			(Minimal harm)	(Short-term harm)	(Permanent or long term harm)	(Death of one or more people)
			Consequence			

Figure 21.6 National Patient Safety Agency (UK) Risk Assessment Scoring Matrix

Evaluating risks and opportunities

Risk evaluation is the process of comparing the results of risk analysis with risk criteria to determine whether the risk and/or its magnitude is acceptable or tolerable. The organization should define criteria to be used to evaluate the significance of risk. Risk criteria are the terms of reference against which the significance of a risk is evaluated (ISO 31000, 2009)

According to ISO 31010 on risk assessment techniques, defining risk criteria involves deciding:

- the nature and types of consequences to be included and how they will be measured;
- the way in which probabilities are to be expressed;

- how a level of risk will be determined;
- the criteria by which it will be decided when a risk needs treatment;
- the criteria for deciding when a risk is acceptable and/or tolerable;
- whether and how combinations of risks will be considered.

RISK APPETITE

The amount and type of risk that an organization is prepared to pursue, retain or take to meet their strategic objectives are referred to as *risk appetite* and it will vary from sector to sector, organization to organization. An organization with an aggressive appetite for risk might set aggressive goals, whereas an organization that is risk averse, with a low appetite for risk, might set conservative goals. It follows therefore that an organization should establish its risk appetite before setting its goals, and this will inevitably shape its strategy. There needs to be a consensus across all functions and at all levels on the organization's risk appetite otherwise decision-making will continually run into problems. Everyone needs to agree on how much risk is acceptable. For each major group of objective as the risk appetite will be different for different groups of objectives.

The management of a high-tech organization, for example, may assign a high-risk appetite to product innovation, a more moderate-risk appetite to issues of personnel development, a low-risk appetite to information system security and a very-low-risk appetite to its reputation for product and service quality. Putting this into practice means that the organization:

a) exhibits a higher-risk appetite when approving new product and service development;
b) expresses a moderate-risk appetite for authorizing staff promotions, education and training;
c) expresses a very-low-risk appetite for risks that would significantly reduce the quality of its products and services or the health and safety of its staff.

RISK TOLERANCE

The acceptable level of variation relative to achievement of a specific objective is referred to as risk tolerance. Risk tolerance is a practical concept for dealing with tactical issues where not all inputs to a process are the same and necessitate equal treatment and is best measured in the same units as those used to measure the related objective, for example, a health services organization has a low-risk appetite related to patient safety but a higher appetite related to response to all patient needs, and although it strives to treat all emergency room patients within two hours and critically ill patients within 15 minutes, it accepts that in rare situations (5% of the time) patients in need of non-life-threatening attention may not receive that attention for up to four hours (Rittenberg & Martens, 2012).

Another example from Rittenberg and Martens's paper is one that helps balance profit objectives and quality objectives. Managers of an aerospace company want to improve a product's profitability but know the company has a low-risk appetite for not meeting customer expectations. They know they cannot reduce product costs if such changes decrease performance (e.g. they might use new technology, but cannot use inferior components). Top management may have set specific profit requirement by product line but have also communicated a low-risk appetite for product failure for loss of customers because of product quality or delivery issues. The articulation of risk tolerances helps guide the company's operational development in this case.

EVALUATION RESULTS

The result of the evaluation should be a series of decisions, based on the risk apetite and risk tolerance as to whether the risk, as defined in terms of its likelihood and consequances, is acceptable or tolerable, or needs to be treated. This information should be added to the risk register.

Box 21.3   The leaky valve

Arthur Rudolph, one of 118 German rocket engineers who developed the giant Saturn 5 rocket, is reported to have said. "You want a valve that doesn't leak and you try everything possible to develop one but the real world provides you with a leaky valve. You have to determine how much leaking you can tolerate."

(New York Times 1996)

Box 21.4   Fail to prepare, prepare to fail

In a conversation between Eric "Winkle" Brown, (1919–2016) Navy test pilot in WWII, and Kirsty Young on the BBC's Desert Island Discs in 2014 he said "There was always that aura of risk; you came to value life in a different sort of way and I always made a point of preparing myself very well indeed before a flight." In response to Kirsty Young's suggestion that "Kick the tyres, light the fires and last one off's a sissy" was not his motto, he replied, "That sort of attitude cost an awful lot of lives." (BBC, 2014)

How is this demonstrated?

Demonstrating that the risks and opportunities that need to be addressed have been determined may be accomplished by:

a)   presenting evidence of a process for identifying risks and opportunities that arise from an analysis of the internal and external issues that are relevant to the QMS;
b)   presenting evidence that the identified risks and opportunities have been categorized, analysed and evaluated for their effect on the QMS and what their likelihood of occurrence and consequence might be;
c)   selecting a representative sample from each category and presenting evidence to support the ratings given (e.g. the source of the data and the competence of the estimators).

Planning actions to address risks and opportunities (6.1.2a)

What does this mean?

Planning actions to address risks and opportunities means figuring out what to do about the risks and opportunities that have been identified and quantified.

Planning also includes determining how to incorporate the actions deemed necessary or beneficial into the QMS, either through objective setting (6.2), operational control (8.1) or other specific elements of the QMS, for example, resource provisions (7.1) and competence (7.2).

Why is this necessary?

The purpose of planning is to anticipate potential scenarios and consequences, and as such is preventive in addressing undesired effects before they occur. Similarly, it looks for favourable conditions or circumstances that can offer a potential advantage or beneficial outcome and includes planning for those worthy of pursuit.

How is this addressed?

There are several actions that can be taken once risks and opportunities have been identified, analysed and evaluated.

Whether a risk is to be avoided, eliminated, reduced, taken, shared or accepted depends on an organization's risk appetite.

Box 21.5   Cost-benefit relative to risk

Effort spent on of risk management should always be less than the value at risk.

Action to address risks

There are several possibilities as to the action to be taken to address risks and these are described next. Once a plan of action has been agreed it should be added either directly or by reference to the risk register along with the name or title of the person who will manage it. Additional columns can be added to accommodate progress reporting and changes to the probability and consequence of the risk. However, before considering the actions to address risk look at Box 21.6 and read the article.

Box 21.6   The six mistakes executives make in risk management

1   We think we can manage risk by predicting extreme events.
2   We are convinced that studying the past will help us manage risk.
3   We don't listen to advice about what we shouldn't do.
4   We assume that risk can be measured by standard deviation.
5   We don't appreciate that what's mathematically equivalent isn't psychologically so
6   We are taught that efficiency and maximizing shareholder value don't tolerate redundancy.

(Taleb, Goldstein, & Spitznagel, 2009)

AVOIDING RISKS

Avoiding risk is deciding not to carry on with the proposed activities due to the risk being unacceptable or finding an alternative that is more acceptable.

In day-to-day operations, everyone should manage their exposure to risk so that where possible risks are avoided commensurate with achieving the stated objectives. The type of risks encountered on a daily basis probably won't be included in the list of risks identified as a result of assessing the internal and external issues when planning for the QMS. They are more likely to be identified when planning operations.

It might be thought natural to avoid risks at all costs but that is not always practical. The risks have been identified from analysing the context of the organization and therefore to avoid a risk may require changing the context i.e. changing the type of business in which the organization is engaged. In other cases, the risk may arise as a result of particular technologies the organization has adopted and therefore if the risk is of sufficient magnitude it may present an opportunity to change these technologies. The risk may arise due to certain operations being located in regions susceptible to danger either from natural or manmade causes. Moving the location may avoid the risks.

RETAINING RISK BY INFORMED DECISION OR ACCEPTING RISKS

Accepting a risk is a decision to accept the risk or the residual risk after mitigation. Accepting a risk means no action can be taken to eliminate, share or transfer the risk or reduce it further and reliance is placed on contingency plans if the risk materializes. This decision is likely when information provided shows that a risk is not adequate enough to warrant the added cost it will take to avoid it (e.g. risk of an earthquake may be accepted because it is too expensive to do anything more about it).

TAKING RISKS

Where the potential benefits of taking a course of action outweigh the losses that may be incurred, a risk may be taken to pursue an opportunity. Taking a risk is different to accepting a risk in that there may be no choice but to accept certain risks if a particular objective is to be achieved, whereas when taking a risk, you are deliberately playing the odds (i.e. choosing one option over another) to seize an opportunity. Some industries thrive on taking risks whereas others are more conservative. Where the losses may be high, contingency plans may be deployed to mitigate them.

ELIMINATING THE RISK SOURCE

Where the source of the risk is external to the organization there may be little one can do to eliminate its source. If the risk is posed by pending legislation that happens to be quite onerous for the organization, putting pressure (with or without assistance from pressure groups) on the legislators may delay or even cause abandonment of the legislation. This course of action is more likely with public bodies and multinational corporations.

Where the source of risk is internal there will be more opportunities to eliminate it, for example, outdated and inefficient operating practices that can be replaced, but again, your options may be limited by the context of the organization (e.g. cannot solve a skill shortage with robotics because the technology is not mature enough).

RISK REDUCTION OR MITIGATION

Risk reduction is an action that reduces the likelihood and/or consequence of the risk. The likelihood might be changed simply by obtaining more reliable data or increasing the skills of those who are assessing probability of occurrence (see earlier on calibrating the estimators). Other solutions are to make the QMS less vulnerable to the risk by design. Whether it's an uncertain event, uncertain variation or uncertain knowledge, design the QMS so that it's less likely to be affected.

Consequences can be changed by building redundancy into the system so that the organization is not wholly dependent on one person or group of people, one technology or one piece of infrastructure. Eliminate the potential for common cause failures.

SHARING OR TRANSFERRING THE RISK

Risk sharing occurs when two parties identify a risk and agree to share the loss upon the occurrence of the loss due to the risk. Risks can be shared by forming partnerships, joint ventures or outsourcing processes to organizations that are in a better position to ensure the risk is mitigated. However, with outsourcing the risk rests primarily with the service provider but the organization retains some of the risk because they are managing the service provider. There may be little inherent risk in the process but considerable risk in managing it from a distance.

Actions to address opportunities

The adoption of new practices, launching new products, opening new markets, addressing new clients, building partnerships, using new technology and other desirable and viable possibilities to address the organization's or its customers' needs may all incur risk. However, there is a difference between:

a)   deciding a course of action to achieve an objective and looking for things that could go wrong, reducing the risk and sticking with the course of action at reduced risk and;
b)   having an objective and looking for innovative ways of achieving it that had hitherto not been thought of and might not happen but if they did they would assist achieve the objective.

The first of these is addressing risks, and the second is addressing opportunities. There's a risk the opportunity will not happen, and therefore actions could be taken that will make it more likely that the opportunity will happen. It therefore has the reverse effect to actions to address risk. The actions needed to make things happen should be planned and these plans implemented. As they are not routines, they should be regarded as projects and therefore each of the possibilities identified in the first paragraph earlier (Note 2 of clause 6.1.2) will be designated as a project and subject to project management practices. Taking another look at Note 2 we find that the types of opportunities being referred to are strategic in nature. They are in a different category to tactical opportunities such as when deciding where to store material waiting use when planning the construction of a building.

How is this demonstrated?

Demonstrating that actions to address the identified risks and opportunities are planned may be accomplished by:

a) presenting evidence of the risks that have been identified and plans for dealing with them;
b) presenting evidence of the opportunities that have been identified and plans for making them happen.

Integrating actions into QMS processes (6.1.2b (1))

What does this mean?

Having identified the risks and opportunities, their likelihood and consequences and prioritized the action to be taken, this requirement is prompting us to think about how, when where and by whom these actions will be taken by the QMS processes rather than by a separate initiative that is taken independently of those processes.

Why is this necessary?

It is not uncommon for weaknesses, opportunities and threats to be dealt with through managerial exhortation, a plea by management for everyone to do better, seize every opportunity or simply just work harder. However well meaning the exhortations may be, a properly orchestrated plan will be necessary to change what people do and the way they work so that action to address the risks and opportunities is taken in conjunction with the organization's business processes. These processes may need to change or be supplemented by new processes to mitigate the risks and facilitate the opportunities on a continual basis and it is this that is referred to in clause 4.4.1f).

How is this addressed?

There is some duplication in the standard regarding planning. Integrating actions to address risks and opportunities into the QMS is done when establishing the QMS (4.4.1) as explained in Chapters 15 and 16 but also when planning how to meet quality objectives (6.2) as explained in Chapter 22 and planning of changes (6.3) in Chapter 23.

In planning how to integrate the actions into the QMS processes, we start by identifying the processes where such actions would take place, then analysing the process to locate the stage where the conditions need to change to reduce the risk or exploit the opportunity. In some cases, a branch of the process network may need to be redesigned, in others all it may need is a change to a checklist. In this way, on executing a process, the actions intended to address risks or opportunities will be implemented. We also need to update the process risk register so that there is a record of provisions built into the process to mitigate risk or exploit the opportunity.

How is this demonstrated?

Demonstrating that plans have been made to integrate and implement actions to address risk into the organization's processes may be accomplished by:

a) presenting evidence that the plans for addressing risk and opportunity include detail on the processes affected and the action to be taken on each process;
b) selecting a representative sample of the processes identified in the risk reduction plans and presenting evidence that the risks these processes have been designed to mitigate include those identified in the risk register.

c)   selecting a representative sample of the processes identified in relevant planning docu-
      ments and presenting evidence that the opportunities these processes have been designed
      to exploit include those identified in the Risk Register as opportunities.

Planning how to evaluate the effectiveness of actions (6.1.2b (2))

What does this mean?

In addition to planning the actions to mitigate a risk or facilitate an opportunity the standard
requires thought be given to how the effectiveness of those actions will be evaluated i.e. to
what extent did the actions taken mitigate the identified risk or facilitate the identified opportunity? It appears that it is asking us to plan to evaluate not only the actions taken but the methods
we used to analyse the risks and opportunities. We are therefore not in breach of this requirement if we failed to identify a risk or opportunity that did materialize – that is an issue that may
be raised with the way the organization addressed clause 6.1.1. However, if we underestimated
the likelihood of a risk or its consequences and as a result the action taken turned out to be woe-
fully inadequate, the action cannot be deemed to have been effective. Therefore, in evaluating
the actions taken we can't divorce the action from the method used to determine it.

Why is this necessary?

Whatever managers do they carry responsibility for their actions and decisions and its axiomatic that successful managers employ methods that work, whether they be informal or
formal methods, they owe their success to the methods they use. It follows therefore that
the methods managers use to identify and address risks and opportunities must also work.
With many decisions they get immediate feedback from the control systems they have put
in place if something doesn't work. However, when dealing with uncertainties, managers
tend to assume that their methods of managing risks must be working if the things they were
not expecting don't happen and the things they were expecting do happen, but that is their
perception not the result of measurement.

   If your organization is investing its scarce resources in reducing risks and increasing
opportunities the management need to know whether it's money was well spent.

How is this addressed?

Even under the best circumstances, where the effectiveness of the action itself was tracked
closely and measured objectively, adequate evidence may not be available for some time.
Also, if the methods used do not actually measure the risks in a mathematically and scientifically sound manner, management doesn't even have the basis for determining whether a
method works (Hubbard, 2009). It simply gives the appearance that it works, which may be
acceptable for low level risks. For high risks, the measure of effectiveness should therefore
be based on whether and by how much risk was actually reduced or opportunity increased
and whether the risk was acceptable for a given investment.

   When planning how to evaluate the effectiveness of the actions, we need to answer some
basic questions:

a)   If the action taken was to avoid a risk, over what timescale will we monitor the situ-
      ation to detect if the risk has been avoided and?

b) If the action taken was to take a risk to pursue an opportunity, how will we know if the anticipated adverse effects materialized and if they did what impact they had?

c) If the action taken was to eliminate the source of risk, how will we know if we've eliminated it?

d) If the action taken was to reduce the likelihood of a risk, how will we know if we have done this?

e) If the action taken was to increase the likelihood of an opportunity, how will we know if we have done this?

f) If the action taken was to change the magnitude of the consequences how will we know if we have done this?

g) If the action taken was to share the risk, how will we know if that was a wise decision?

h) If the action taken was to accept the risk, when will we know if that was a wise decision?

Hubbard cites four methods of determining effectiveness in risk management:

1   Statistical inferences based on large samples
2   Direct evidence of cause and effect
3   Component testing of risk management
4   A "check of completeness"

Statistical inferences based on large samples

If we are trying to reduce the risk of a rare event it's doubtful that we could use actual results from one organization as to the effectiveness of a particular risk treatment as they would be statistically insignificant. We'd have to look further afield and engage other organizations in the experiment so that the sample is big enough.

Direct evidence of cause and effect

There are situations where the risk treatments detect the risk before any harm is done such as the fire walls and virus protection in an IT system. There are also those treatments that reduce the magnitude of the consequences such as disaster recovery processes. We therefore know those risks would have done harm had it not been for the effectiveness of the risk treatments. There are also disastrous situations which we know could have been prevented if only we'd applied a known risk treatment. However, there are situations where one risk was averted only for another to arise as a consequence, for example, storing data in the cloud means you don't lose your data if your laptop is stolen but if your network connection goes down you can't access the cloud.

Component testing of risk management

If we can't use other methods because we lack data, we can test the validity of the methods we have used to analyse and evaluate the risks or look at research in the same sector and different sectors to validate the methods we are using (i.e. benchmarking).

A check of completeness

This is simply comparing the items evaluated against a list of known risks for a company. It helps us determine whether our approach to risk is or is not too narrowly focused.

Evaluating the effectiveness of actions to address opportunities

The plans made to address opportunities are relatively straight forward. The objective is to increase your chances of exploiting an opportunity and therefore we examine the evidence and if the objectives was achieved the actions were effective.

Constructing a plan

A plan for evaluating the effectiveness of the actions taken to address risks and opportunities simply needs to identify the risks and opportunities in the risk register and state how effectiveness will be evaluated, who will do it, when it will be done, how the results will be reported and how progress will be reviewed.

How is this demonstrated?

Demonstrating that that you have a plan to evaluate the effectiveness of the actions taken to address risk and opportunity may be accomplished simply by presenting evidence of a plan:

Bibliography

BBC. (2014, November 14) Desert Island Discs – Captain Eric 'Winkle' Brown: Retrieved from http://www.bbc.co.uk/programmes/b04nvgq1

CAA. (2016, June 19). Airport Data 2015. Retrieved from Civil Aviation Authority: www.caa.co.uk/Data-and-analysis/UK-aviation-market/Airports/Datasets/UK-Airport-data/Airport-data-2015/

Drucker, P. F. (1974). *Management, Tasks, Responsibilities, Practices*. Oxford: Butterworth-Heinemann.

Hillson, D. ((2) 2016, June 20). Managing Risk in Practice – Workshop. Retrieved from www.youtube.com/watch?v=fVIqy51oyS4

Hubbard, D. W. (2009). *The Failure of Risk Management: Why It's Broken and How to Fix It*. Hoboken, NJ: John Wiley & Son.

Hubbard, D. W. (2010). *How to Measure Anything: Finding the Values of Intangibles in Business*. Hoboken, N.J.: John Wiley.

ISO 31000. (2009). *Risk Management: Principles and Guidelines*. Geneva: ISO.

Knobbe, C. A. (2016, June 18). Cataract Surgery Complications. Retrieved from All About Vision: www.allaboutvision.com/conditions/cataract-complications.htm

New York Times. (1996) Arthur Rudolph, 89, Developer Of Rocket in First Apollo Flight: Retrieved from http://www.nytimes.com/1996/01/03/us/arthur-rudolph-89-developer-of-rocket-in-first-apollo-flight.html

Rittenberg, L., & Martens, F. (2012). Understanding and Communicating Risk Appetite. Retrieved from Committee of Sponsoring Organizations of the Treadway Commission: www.coso.org/Documents/ERM-Understanding-and-Communicating-Risk-Appetite.pdf.

Taleb, N. N., Goldstein, D. G., & Spitznagel, M. W. (2009, October). The Six Mistakes Executives make in Risk Management. Retrieved from Harvard Business Review: https://hbr.org/2009/10/the-six-mistakes-executives-make-in-risk-management.

22 Quality objectives and planning to achieve them

> Asking for a strategy that is guaranteed to work is like asking a scientist for a hypothesis that is guaranteed to be true – it is a dumb request.
>
> Richard Rumelt

Introduction

Several terms are used to define the ends or outcomes an organization seeks to achieve. These terms include *mission, vision, goals, aims, objectives, targets* and *milestones*, and all tend to create confusion. The confusion is compounded when qualifying words are added to these terms such as *quality objectives, environmental targets, project goals*, etc. In a generic sense they could all be classed as objectives but what generally sets them apart are timescales, levels and perspectives.

Timescales are addressed by expressing objectives as being strategic (long term), or tactical (short term). Levels are addressed by expressing objectives as being corporate (company-wide), business process, work process, product or service. Perspectives are addressed by labelling the objectives as quality objectives as opposed to financial or marketing objectives.

Quality objectives arise from multiple sources within ISO 9001 as illustrated in Figure 22.1. None of these are mutually exclusive because they are derived from the clause structure of the standard which does not align with the way organizations function. For example, in undertaking a SWOT analysis as described in Chapter 12, we may identify operational issues and improvement opportunities. In determining the causes of nonconformity, we may reveal systemic problems that require a coordinated programme of corrective action for which there will be quality objectives. A video on planning for quality is available on the companion website.

In this chapter, we examine the 10 requirements of clause 6.2.1, namely:

- Establishing quality objectives (6.2.1)
- Ensuring consistency with quality policy (6.2.1a)
- Ensuring quality objectives are measurable (6.2.1b)
- Taking account of applicable requirements (6.2.1c)
- Ensuring relevance of quality objectives (6.2.1d)
- Monitoring quality objectives (6.2.1e)
- Communicating quality objectives (6.2.1f)
- Updating quality objectives (6.2.1g)
- Maintaining information on quality objectives (6.2.1)
- Planning to achieve quality objectives (6.2.2)

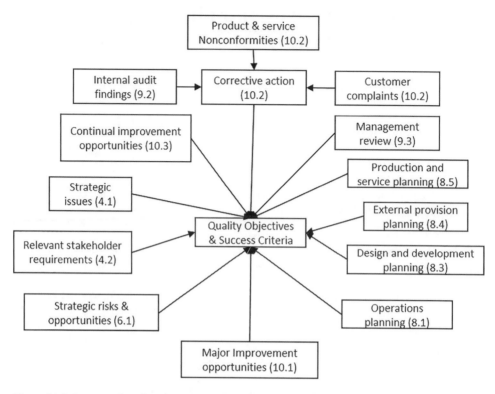

Figure 22.1 Sources of quality objectives within ISO 9001

The sequence in which the requirements are addressed in this chapter is not indicative of a sequence of action, for example, the monitoring, communicating and updating of quality objectives comes after planning and ensuring consistency, measurability, taking account of requirements and ensuring relevance is all part of establishing quality objectives and not separate to it.

Establishing quality objectives (6.2.1)

What does this mean?

Box 22.1   What are objectives?

- Objectives are the "result to be achieved". (ISO 9000:2015, 2015)
- Objectives are not the path on which we travel but the compass bearing by which we navigate.
- Objectives can be strategic, tactical, or operational.
- Objectives can be long term, medium term or short term.
- Objectives can apply at different levels (such as corporate, departmental, project, product, process and personal).
- Objectives can relate to different organizational perspectives (such as finance, health, safety, environment, security, social responsibility and quality).

Quality objectives

ISO 9000:2015 defines a quality objective as an "objective related to quality", but when we substitute the terms with the definitions it becomes *a result to be achieved related to the degree to which a set of inherent characteristics of an object fulfils requirements*. This might be technically correct, but it's not a definition that people will find easy to work with.

Discussions with users of ISO 9001 in 2016 revealed several different interpretations as to what a quality objective is depending on how the case is argued.

a)   That all objectives are quality objectives.

b)   That all quality objectives are business objectives but all business objectives are not quality objectives.

c)   That any objective that increases the degree to which the inherent characteristics of an object fulfils the requirements is a quality objective.

d)   If we measure performance in terms of quality, cost and delivery, the target for the degree of conformity to requirements becomes the quality objective.

e)   If the primary beneficiary of meeting an objective is the customer, the objective is a quality objective.

f)   Everything a business does must directly or indirectly affect the condition of its outcomes and therefore all business objectives are quality objectives.

The first argument we could use is that when we look at the ISO definition of quality management and the associated terms (ISO 9000:2015 clauses 3.3.4 to 3.3.8), it rather suggests that it's all about setting and achieving objectives and that the only thing that distinguishes quality objective from other objectives is that quality objectives are focused on fulfilling needs and expectations and other objectives are not. An objective that is not focused on fulfilling someone's needs or expectations is hard to imagine; therefore, we could reach the conclusion that all objectives are quality objectives. If we accept this argument the QMS becomes the enterprise management system and will address all stakeholders and their needs and expectations but it's outside the scope of ISO 9001.

The second argument we could use is that the pursuit of quality is a strategic priority but it's not the only strategic priority. An organization may want to pursue improvement in profitability, growth in market share, improvement in operational efficiency, development of its people and increase in its social capital. The pursuit of quality has traditionally been associated with operational efficiency through measures that maintain or improve product and service conformity. As a consequence, the perceived value in maintaining or improving product and service quality is balanced against other priorities and so we reach the conclusion that all quality objectives are business objectives but all business objectives are not quality objectives. If we accept this argument the QMS will address customers but only in so far as responding to their current needs and expectations. Although this was sufficient to meet the requirements of the 2000 and 2008 versions, it is not sufficient for the 2015 version because of the requirements for determining the future needs and expectations of potential customers (i.e. market research).

The third argument we could use is that the words *an object* now appears in the ISO 9000 definition of "quality" and the words *customer* and *product and service* are nowhere to be seen in the definition. On this basis, "reducing emissions to atmosphere" would be a quality objective because it increases the degree to which the inherent characteristics of the organization fulfil the requirements of environmental legislation. If we accept this argument the scope of the QMS will expand beyond satisfying requirements for products and services and take it outside the scope of ISO 9001 altogether.

The fourth argument we could use is that we expect an output to meet requirements, that it has been produced economically and in compliance with applicable regulations, and that it is delivered when required. We can therefore measure performance in terms of quality, cost and delivery (QCD). This means that whatever we wish to achieve we can judge performance using the measures of QCD. We set a target value for the degree to which the inherent characteristics of an object fulfils requirements and this target becomes the quality objective. On this basis, we would add "quality objectives" in the form of success measures to each organization objective. If we accept this argument the QMS would become an Enterprise QMS, viewing the whole organization from a quality perspective.

The fifth argument we could use is that we need a way of qualifying objectives related to quality in the context of ISO 9001 and one such way is to determine who the primary beneficiary is. If we ask who is the primary beneficiary of reducing emissions to atmosphere we are likely to conclude it is society rather than investors, employees or customers, although they may be secondary beneficiaries. If were to ask who is the primary beneficiary of bringing a more resilient lithium battery to market, we are likely to conclude it would be customers. As ISO 9001 is primarily about satisfying customers, it follows therefore, that in the context of ISO 9001, "a quality objective" is an objective that will primarily benefit an organization's customers" when its achieved. If we accept this argument, it means that the QMS becomes a systemic view of an organization from the perspective of how it creates and retains its customers and this is certainly consistent with the intent of ISO 9001.

There is no right or wrong answer to: *what is a quality objective?* However, of the five arguments the one that aligns more closely with the intent of ISO 9001 is the fifth. Therefore, in the context of ISO 9001 a quality objective is *an objective that primarily benefits the customer by its achievement.*

Relevant functions

Quality objectives are required to be established at relevant functions and these will be those organizational units (e.g. divisions, groups, departments, sections, etc.) that are engaged in the processes that are instrumental in providing products and services for customers from their conception to their obsolescence.

Relevant levels

As "levels" are identified in clause 6.2.1 as a separate category to function and process, we could assume the levels being referred to are levels within the functional and process hierarchy or we could take a different view that it's referring to levels beneath the mission and vision of the organization, for example:

* Level 0 Mission and Vision
* Level 1 Strategic objectives
* Level 2 Cross-functional process objectives or initiatives (project objectives)
* Level 3 Individual department or process objectives
* Level 4 Team objectives
* Level 5 Personal objectives

Relevant processes

The relevant processes are those that have been determined through the analysis of interested parties (see Chapter 16) that collectively create and retain customers. These will include the business processes and the sub-processes of which they are comprised.

Why is this necessary?

The requirement for defining objectives is one of the most important requirements. Without objectives, there can be no means of measuring how well the organization is performing. If you don't know where you are going, any destination will do! Objectives are therefore necessary as a basis for measuring performance, to give people something to aim for, to maintain the status quo to prevent decline and to advance beyond the status quo for the enterprise to grow.

All work serves an objective and it is the objective that stimulates action. The reason for top management ensuring that quality objectives are established is to ensure that everyone is stimulated to produce work of an acceptable quality thereby enabling the organization to produce outputs that satisfy its customers. Organizations that strive to satisfy their customers build a reputation for quality and this creates more customers and brings in more revenue. Setting quality objectives at relevant levels aims to improve operational efficiency and productivity by reducing nonconformity. It reduces the cost of quality, thereby generating a greater surplus to invest in further improvement.

How is this addressed?

Process for establishing objectives

Achievable objectives do not necessarily arise from a single thought. There is a process for establishing objectives. At the strategic level, the subjects that are the focus for setting objectives are the outputs that will produce successful outcomes for the organization's stakeholders. Customer needs, regulations, competition and other external influences shape these objectives and cause them to change frequently. The measures arise from an analysis of current performance, the competition and the constraints of customers and other stakeholders and there will emerge the need for either improvement or control. This process is outlined in Chapter 21 where objectives are established at step 11 in Figure 21.2.

The steps in the objective setting process should include:

- Identifying the need (see below for different methods)
- Nominating candidates for strategic or tactical quality objectives together with the measures of success
- Proving the need to the appropriate level of management in terms of:

  - whether the climate for change is favourable
  - the urgency of the improvement or controls
  - the size of the losses or potential losses, gains or potential gains
  - the priorities

- Identifying or setting up the forum where the question of change or control is discussed

- Conducting a feasibility study to establish whether the objective can be achieved with the resources that can be applied and whether the measures are realistic
- Selecting achievable objectives for control and/or improvement
- Deploying and communicating the objectives

The standard does not require that objectives be achieved but it does require that their achievement be planned, resourced, communicated and monitored. It is therefore prudent to avoid publishing objectives for meeting an unproven need which has not been rigorously reviewed and assessed for feasibility. It is wasteful to plan for meeting objectives that are unachievable as it diverts resources away from more legitimate uses.

Identifying the need

Quality objectives are required to be compatible with the context and strategic direction of the organization, and this can be achieved in different ways.

They can result from deploying the strategy as shown in Figure 22.2, in which case they should be compatible with it. This is a top down approach and referred to as strategy deployment.

They can be derived from an analysis of current performance, risks or opportunities then subject to an assessment for compatibility with the context and strategic direction of the organization. This is a bottom-up approach.

Objectives for control and improvement

All managerial activity is concerned either with maintaining performance or with making change. Change can retard or advance performance. That which advances performance is beneficial. In this regard, there are two classes of quality objectives, those serving the control of quality (maintaining performance by mitigating risks) and those serving the improvement of quality (making beneficial change by exploiting opportunities).

The objectives for quality control should relate to the standards you wish to maintain or to prevent from deteriorating due to the issues identified when scanning the internal and external environment. To maintain your performance and your position in the market you should continually seek improvement to avoid or mitigate the risks. Remaining static at whatever level is not an option if your organization is to survive. Although you will be striving for improvement it is important to avoid slipping backwards with every step forward. The effort needed to prevent regression may indeed require innovative solutions. While to the people working on such problems, it may appear that the purpose is to change the status quo, the result of their effort will be to maintain their present position not raise

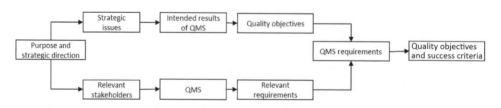

Figure 22.2 Deriving quality objectives from organization context

it to higher levels of performance. Control and improvement can therefore be perceived as one and the same thing depending on the standards being aimed for and the difficulties in meeting them.

Strategic objectives

From the first step of establishing internal and external issues (see Chapter 10) several issues are likely to emerge that are more tactical than strategic. It is important to capture the tactical issues as people will think these have the most impact on their work and dealing with these often engages people in the strategic planning process. Also, addressing the tactical issues often removes barriers to confronting the strategic issues.

Strategic objectives need to be deployed through the processes; otherwise, they will remain pipedreams, and therefore the strategic plan should show the linkage between processes and objectives in a way as shown in Table 16.1 for business outputs. The statements of objectives may be embodied within business plans, product development plans, improvement plans and process descriptions.

Constraints as objectives

Many constraints are expressed as objectives such as reducing waste and absenteeism but as explained in Chapter 7 under demands and constraints. The true objective is to create and satisfy customers under conditions that constrain waste and absenteeism. Also, treating profit as a constraint rather than an objective causes the designers and producers to look for ways of reducing costs while satisfying customers.

Process objectives

At the process level the objectives are concerned with process performance – addressing process capability, efficiency and effectiveness, use of resources, and controllability. As a result, objectives for control may focus on reducing errors and reducing waste, increasing controllability but may require innovative solution to achieve such objectives. Objectives for improvement might include increasing throughput, turnaround times, response times, resource utilization, environmental impact, process capability and use of new technologies, etc.

Product or service objectives

At the product or service level, objectives are concerned with product or service performance addressing customer needs and competition. Again, these can be objectives for control or improvement. Objectives for control might include removing nonconformities in existing products (improving control), whereas objectives for improvement might include the development of new products with features that more effectively satisfy customer needs (improving performance), use of new technologies, and innovations. A product or service that meets its specification is only of good quality if it satisfies customer needs and requirements. Eliminating all errors is not enough to survive – you need the right products and services to put on the market. Objectives for satisfying the identified needs and expectations of customers with new product features and new service features will be quality objectives.

Departmental objectives

At the departmental level objectives are concerned with organizational performance – addressing the capability, efficiency and effectiveness of the organization, its responsiveness to change, the environment in which people work, etc. Control objectives might be to maintain expenditure within the budget, to keep staff levels below a certain level, to maintain moral, motivation or simply to maintain control of the department's operations. Objectives for improvement might be to improve efficiency by doing more with less resources, improving internal communication, interdepartmental relationships, information systems, etc.

Personal objectives

At the personal level, objectives will be concerned with worker performance addressing the skills, knowledge, ability, competency, motivation and development of people. Objectives for control might include maintaining time keeping, work output and objectivity. Objectives for improvement might include improvement in work quality, housekeeping, interpersonal relationships, decision-making, computer skills, etc.

How is this demonstrated?

Demonstrating that the quality objective has been established at relevant functions, levels and processes may be accomplished by:

a)   presenting evidence declaring the quality objectives the organization is committed to achieve;

b)   mapping these objectives onto the organization's functional and process structures;

c)   showing how the objectives are arranged hierarchically.

Ensuring consistency with quality policy (6.2.1a)

What does this mean?

Quality objectives that are consistent with the quality policy are those which serve the same intent and are not obviously in conflict. The relationship between policy and objective is addressed in Chapter 15.

There is a tendency to interpret this requirement as though it is referring only to consistency between documented statements of policies and objectives, but the intent is surely that the objectives being pursued are consistent with the policy that is understood because it is this combination that is influencing results rather than the written statements.

Why is this necessary?

Policies are a guide to action, they express principles and beliefs and are intended to unify the organization purpose. Actions from the top to the bottom in an organization therefore need to be consistent with these principles and beliefs. Objectives drive actions therefore unless the objectives are consistent with policies it is likely that work being undertaken in pursuit of objectives will not be aligned with the policies.

How is this addressed?

When objectives conflict with policy there may be disastrous consequences which was evident in the case of the Enron scandal revealed in October 2001. Enron had as one of its values Integrity which it expressed as "We work with customers and prospects openly, honestly and sincerely. When we say we will do something we do it. When we say we cannot or will not do something then we won't do it." But its actions of hiding debts and falsely reporting revenue were clearly inconsistent with this policy and therefore the executives who masterminded such actions had objectives that were inconsistent with their policy (Wikipedia (4), 2015).

Say you have a policy that addresses customer focus. Your objectives might include marketing objectives that were customer focused thereby linking the policy with the objectives. You may have a human capital objective for improving employee motivation. However, in this instance the process designed to achieve this objective would need to demonstrate adherence to a policy for the involvement of people. Here the process and not the objective links with the policy.

If the policy is "We will listen to our customers, understand their needs and expectations and endeavour to satisfy those needs and expectations in a way acceptable to our other stakeholders", an objective which penalizes suppliers for poor performance would be inconsistent with this policy.

For quality objectives to be consistent with the quality policy access to the documented statements is necessary, but as the Enron example shows, it is not written statements of intent alone that will confirm the reality; evidence of actions taken is also necessary to reveal whether the objectives being pursued are actually consistent with the policy that is understood. So, although an examination of the documented statements will reveal any obvious inconsistencies, it will be the examination of practices that reveal the significant inconsistencies which, if ignored, may have a detrimental impact on long-term performance.

How is this demonstrated?

Demonstrating that quality objectives are consistent with the quality policy may be accomplished by:

a) presenting the documented quality policy;
b) selecting a representative sample of documented quality objectives;
c) comparing the objectives with the policy and showing they are not in conflict;
d) selecting a representative sample of people to interview;
e) capturing what these people say it is they are trying to achieve and how they are going about achieving it;
f) showing that what they are trying to achieve is consistent with the quality policy.

Ensuring quality objectives are measurable (6.2.1b)

What does this mean?

Measurable quality objectives are objectives that are expressed in a way that lend themselves to a practical means of measurement.

Why is this necessary?

Having set an objective management need a means of deciding whether it has been achieved. Without such means, there is uncertainty, achievement becomes a matter of opinion and is variable; therefore, a means of measurement removes that uncertainty. Measurement provides consistency and predictability and produce facts on which decisions can be made. It is the purpose of measurement to inform decision makers.

How is this addressed?

Hubbard tells us that "Anything can be measured. If a thing can be observed in any way at all, it lends itself to some type of measurement method." He goes on "No matter how 'fuzzy' the measurement is, it's still a measurement if it tells you more than you knew before" (Hubbard, 2010). There is therefore no need to be daunted by making objectives measurable.

The published interpretation of the 2008 version RFI 035, states that objectives having Yes/No criteria are deemed measurable. An example would be: *Develop a new product to meet the requirements of the "YYYYY" market by March 2018.*

There should be a tangible result from meeting an objective and a defined period should be specified when appropriate. The objective should therefore be expressed in the following form: what is to be achieved and what will success look like (i.e. what will be happening or will have happened as a result of achieving this objective?). The success measures indicate when the objective has or has not been achieved – namely the passing of a date, a level of performance, or the absence of a problem, a situation or a condition that currently exists or the presence of a condition that did not previously exist.

All of the organization's objectives should in some way serve to fulfil requirements of customers and other stakeholders. Objectives at lower levels should therefore be derived from those at higher levels and not merely produced to satisfy the whim of an individual.

If you have an objective to be world class, what measures will you use that indicate when you are world class? You may have an objective for improved delivery performance. What measures will you use that indicate delivery performance has improved? You may choose to use percent delivered on time. You will also need to set a target relative to current performance. Let us say that currently you achieve 74% on-time delivery so you propose a target of 85%. However, targets are not simply figures better than you currently achieve. The target must be feasible, and therefore it is necessary to take the steps in the process described previously for setting objectives.

A technique has evolved to test the robustness of objectives and is identified by the letters SMART meaning that objectives should be Specific, Measurable, Achievable, Realistic and Timely. Although the SMART technique for objective setting is used widely, there is some variance in the words used. The S of SMART has been used to denote Small, meaning not too big to be unachievable – one small step at a time. The A of SMART has been used to denote Attainable, Accountable and Action oriented, and the R of SMART has been used to denote Resource-consuming action and Relevant.

In the last 40 years or so there has emerged an approach to management that focuses on objectives. Management by objectives or management by results has dominated boardrooms and management reports. In theory management by objectives or results is a sensible way to manage an organization but in practice this has led to internal competition, sub-optimization and punitive measures being exacted on staff that fail to perform. Deming advocated in the 11th of his 14 points "Eliminate management by objectives" for the simple reason that management often derives the goals from invalid data.

They observe that a goal was achieved once and therefore assume it can be achieved every time. If they understood the process they would realize that the highs and lows are a characteristic of natural variation. They observe what the competition achieves and raise the target for the organization without any analysis of capability or any plan for its achievement. Management sets goals and targets for results that are beyond the capability of staff to control. Targets for the number of invoices processed, the number of orders won, the hours taken to fix a problem. Such targets not only ignore the natural variation in the system but are also set without any knowledge about the processes that deliver the results. If a process is unstable, no amount of goal setting will change its performance. If you have a stable process, there is no point in setting a goal beyond the capability of the process – you will get what the process delivers.

How is this demonstrated?

Demonstrating that your quality objectives are measurable may be accomplished by:

a) presenting the objectives and showing how their achievement will be measured;
b) selecting a representative sample of objectives and presenting the results of measurement.

Taking account of applicable requirements (6.2.1c)

What does this mean?

Applicable requirements are requirements that govern the conformity of products and services and these will be customer, regulatory and the organization's own requirements for its products and services. Requirements that feature either in the objective or the criteria for success are objectives that take into account applicable requirements.

Why is this necessary?

The requirement is the source of the objective, the reason for establishing the objective and therefore one is derived from the other. An objective that does not take into account requirements is an objective without a purpose. Although it may appear absurd for an objective to serve no purpose, if we formulate objectives by thinking of things we'd like to do regardless of their achievement adding value, we will have objectives without a useful purpose. This requirement therefore serves to ensure there is alignment between objectives and requirements.

How is this addressed?

When defining quality objectives, we are creating an instrument for achieving either internal or external requirements and therefore the objective maybe a translation of the requirement, be derived from the requirement or its success criteria maybe derived from the requirement.

How is this demonstrated?

Demonstrating that quality objectives take into account applicable requirements may be accomplished by:

a) presenting the objectives and identifying the requirements from which they or their success measures have been derived;

b)   showing how they relate to the requirements;
c)   identifying the source of these requirements.

Ensuring relevance of quality objectives (6.2.1d)

What does this mean?

Quality objectives that are relevant to conformity of products and services and the enhancement of customer satisfaction are objectives that serve to increase the quality of the organization's outputs that it provides to its customers both in the quality of conformity and the quality of design.

Why is this necessary?

There is a possibility that objectives that serve purposes other than customer satisfaction may be deemed to be quality objectives as was indicated in the section on establishing quality objectives (e.g. some people may regard all business objectives as quality objectives). Clearly what is intended by this requirement is that for objectives to qualify as quality objectives they must serve customer satisfaction.

How is this addressed?

If we use our working definition of a quality objective, which was that it's an objective that primarily benefits the customer by its achievement, our quality objectives will be relevant to conformity of products and services and the enhancement of customer satisfaction. Also, if we derive our quality objectives as shown in Figure 21.2, or from customer-specific requirements they will conform with this requirement.

How is this demonstrated?

Demonstrating that the quality objectives are relevant to conformity of products and services and the enhancement of customer satisfaction may be accomplished by asking them: *How relevant is this objective to conformity of products and services and the enhancement of customer satisfaction?* If the answer we get is "it isn't" or "not much", what we have doesn't qualify as a quality objective.

Monitoring quality objectives (6.2.1e)

What does this mean?

Although the requirement in clause 6.2.1 is for quality objectives to be monitored, this is not strictly correct syntax. A quality objective is a statement that expresses the results to be achieved relative to quality and the intent won't be to monitor a statement but to monitor the achievement of quality objectives.

Monitoring (the achievement of) quality objectives means checking periodically and systematically that work in pursuit of the objectives is being undertaken as planned.

However, this requirement is misplaced because the other requirements in clause 6.2.1 apply to the setting of objectives. Monitoring of objectives comes after they have been set and plans for achieving them have not only been prepared but are being carried out.

Why is this necessary?

Communicating objectives and plans for their achievement is obviously necessary, but people are likely to have many demands made upon them. As a result, they will inevitably have to prioritize their work, and unless periodic checks are made of progress, one cannot be certain that work is proceeding as planned.

How is this addressed?

When planning to achieve objectives, it is wise to build into the plan provision for monitoring. This may be informal if the team or the level of work is small or if otherwise formal. The project leader will normally schedule periodic progress reviews to determine if there are any issues with:

- the project objectives and measures of success;
- the plan including the division of work and the schedule of work;
- the acquisition of resources;
- completion of tasks;
- team dynamics;
- external providers;
- the quality of information;
- the quality of work.

When choosing the method of monitoring quality objectives, you need to consider for whom the information is being produced and what form of presentation is suitable. Some people like tables of figures, others like graphical presentations. Objectives differ but often there are intermediate stages of achievement and these can be plotted on a graph or histogram so that progress towards the objective can be conveyed. If the objective is to reduce a quantity and there is a target to meet, daily, weekly or monthly quantities can be displayed showing progress towards the target. If the objective is to achieve something by a certain date, the percentage achievement can be plotted.

In an A&E department of a hospital the objective might be for casualties to be seen within 4 hours and so they erect a digital display in the waiting area which indicates the average waiting time. In a telecommunications data centre where service availability is a critical to quality characteristic, a digital display indicates the current service availability as a percentage and a monitor sits on the CEOs desk so he or she is always kept informed. In a call centre, the number of calls waiting, the call response time and the number completed may all be displayed on digital displays. In these three examples, there are computer programmes capturing data in real time from processes that are running continuously and recording key parameters.

How is this demonstrated?

Demonstrating that quality objectives are being monitored may be accomplished by:

a) presenting evidence declaring the quality objectives the organization is committed to achieve;
b) selecting a representative sample of objectives and retrieving the plans;

c) presenting the provisions made for monitoring achievement of the objectives;

d) presenting evidence that the planned monitoring provisions are in place and that progress is being actively monitored.

Communicating quality objectives (6.2.1f)

What does this mean?

Objectives are not established until they are understood and therefore communication of objectives must be part of this process. Communication is incomplete unless the receiver understands the message but a simple yes or no is not an adequate means of measuring understanding. Measuring employee understanding of appropriate quality objectives is a subjective process.

Communication is addressed in general in Chapter 31, and from Figure 31.1 it will be seen that in the context of quality objectives, communication is not simply about publishing them or telling people what they are; that is data transmission. When quality objectives have been communicated, the people who will be involved in their achievement, will understand what they are and what their role is in their achievement.

Why is this necessary?

It is a fallacy that saying something will be done will result in it being done. Many speeches have been made and plans been produced that resulted in inaction. The doers must be just as motivated as the planners for them to even want to take action, which is why it is necessary to communicate quality objectives.

How is this addressed?

Getting commitment to achieve objectives starts with gathering together those who will be involved in their achievement.

Through the data analysis carried out to meet the requirements of Clause 9.1.3 you will have produced metrics that indicate whether your quality objectives are being achieved. If they are being achieved, you could either assume your employees understand the quality objectives or you could conclude that it doesn't matter. However, results alone are insufficient evidence. The results may have been achieved by pure chance and in six months' time your performance may have declined significantly. The only way to test understanding is to check the decisions people make. This can be done with a questionnaire but is more effective if one checks decisions made in the work place. Is their judgement in line with the objectives or does their behaviour have to be adjusted repeatedly?

One can audit the decisions people make and ascertain whether they were consistent with the objectives and the associated measures. A simple example is where you have an objective of producing conforming product in a way that decreases dependence on inspection (the measure). By examining corrective actions taken to prevent recurrence of nonconformities you can detect whether a person decided to increase the level of inspection to catch the nonconformities or considered alternatives. Any person found increasing the amount of inspection has clearly not understood the objective.

How is this demonstrated?

Demonstrating that objectives have been communicated may be accomplished by:

a) presenting evidence declaring the quality objectives the organization is committed to achieve;
b) selecting a representative sample of objectives and retrieving the plans for achieving the selected objectives;
c) selecting from the plans personnel to interview;
d) showing that the personnel engaged on the project:

   i know what they trying to achieve;
   ii can explain how what they trying to achieve relates to the quality objective stated in the plan.

Updating quality objectives (6.2.1g)

What does this mean?

Requirements from which objectives have been derived may change thereby signalling a need to update the objectives. Work carried out in pursuit of objectives may also change direction due to prevailing circumstances and this may require the original objectives to be modified if those circumstances remain permanent. It is for this reason that the words "as appropriate" are included in the requirement.

Why is this necessary?

Objectives are established and communicated before work to achieve them commences. Once work is underway, it will proceed in the pre-defined direction unless prevailing circumstances necessitate a change. Those involved will be aware of the changes and why they are necessary but anyone joining the team and studying the plans will only be aware of the original objective. It is necessary to maintain constancy of purpose (Deming's first point; see Glossary) and to therefore update objectives.

How is this addressed?

It's likely that quality objectives will be stated in several documents, for example, strategic plans, quality manuals, quality programmes, quality plans, product or service development plans, personal development plans, websites, etc., and therefore if the requirements change for whatever reason, the relevant objectives need to change and the corresponding documented information also needs to change.

How is this demonstrated?

Demonstrating that quality objectives have been updated as appropriate may be accomplished by:

a) presenting evidence declaring the quality objectives the organization is committed to achieve;

b) selecting a representative sample of objectives and retrieving the available documented information;
c) retrieving the plans for achieving the selected objectives and selecting personnel to interview;
d) interviewing personnel engaged on the project to reveal that the objectives they claim to be pursuing are those stated in the plan.

Maintaining information on quality objectives (6.2.1)

What does this mean?

Maintaining documented information on the quality objectives means assembling the information produced by the objective setting process and keeping it current.

Why is this necessary?

The reason why information on quality objectives needs to be retained is that:

* Before work commences it enables effective communication to those assigned responsibility for achievement them.
* During execution of work it enables confirmation that what staff are striving to achieve is consistent with what they should be striving to achieve.
* After completion of work it provides a point of reference for measuring performance.
* When the objectives are being reviewed it's important to know how and why they were originally set to ensure any decision for change is soundly based.

How is this addressed?

The process for setting objectives will generate a lot of information and not all of it will be worth retaining. As a minimum the following information should be documented:

* The quality objectives that top management have agreed to resource.
* The criteria by which success will be measured.
* The source of the objectives (i.e. customer requirement, strategic issue, operational issue, risk or opportunity).
* The reasons why achievement of the objectives is necessary (i.e. What happens if they are not achieved?).
* Identity of the plans for their achievement.
* Identity of records showing the results achieved.
* Date when progress was last reviewed.
* Date when the objective and success measures were last reviewed or changed.

How is this demonstrated?

Demonstrating that the organization is maintaining documented information on its quality objectives may be accomplished by:

a) presenting evidence declaring the quality objectives the organization is committed to achieve;

b) selecting a representative sample of objectives and retrieving the available documented information;

c) showing that information retrieved is up to date and is sufficient for tracking achievement and currency of the objectives.

Planning to achieve quality objectives (6.2.2)

What does this mean?

Planning is performed to achieve objectives and for no other purpose and therefore the requirement clearly indicates that the purpose of the management system is to enable the organization to meet its quality objectives. This is reinforced by the definition of quality planning in ISO 9000:2015 which states that it is "part of quality management focused on setting objectives and specifying necessary operational processes and related resources to fulfil the quality objectives".

Why is this necessary?

For objectives to be achieved the processes for achievement need to be planned. This means that the management system should be result-oriented with the objectives employed to drive performance.

How is this addressed?

Objectives are not wish lists. The starting point is the mission statement and although ISO 9001 suggests that the quality objectives should be based on the quality policy, is it more likely to be the strategic issues arising from the PESTLE and SWOT analysis that drives the setting of objectives.

For each objective, there should be a plan that defines how that objective is to be achieved and in particular:

a) The scope and applicability of the work (i.e. what the plan covers and what it applies to).

b) The strategy (i.e. of all the ways of achieving the objective, this is the approach that has been chosen as the most appropriate).

c) The processes that will deliver the results required and how they are interconnected.

d) The risks that need to be managed and the measures taken to mitigate them.

e) The activities to be undertaken in each process to deliver the results.

f) The resources required to execute the activities.

g) The roles and responsibilities (i.e. who's on the team and what role they play, who leads the team, who does what, who makes what decisions).

h) The timeline and bar chart indicating when activities are expected to start and finish.

i) The dependencies (i.e. what must be complete or provided before and activity can start or finish).

j) The information requirements (i.e. what information is to be produced, to what standards and for what purpose).

k) The provisions for monitoring and measurement performance.

l) The reporting and communication requirements.

m) The reviews which are to track progress and evaluate results.

n)   The provisions to be put in place to maintain the level of performance that is to be achieved by executing the plan.

It is very important that ownership of the plan for achieving an objective is vested in those who will have responsibility for achieving it and therefore these people should participate in the planning process.

How is this demonstrated?

Demonstrating that the organization has undertaken adequate planning for meeting quality objectives may be accomplished by:

a)   presenting evidence declaring the quality objectives the organization is committed to achieve;

b)   selecting a representative sample of objectives and presenting the plans made to achieve them;

c)   showing that the plans define the work to be carried out, those responsible, the resources, timescales and provisions for evaluating the results.

If conformity with this requirement has been demonstrated, it might mean that the evidence presented is also deemed to not only satisfy conformity with 4.4.1d because objectives are achieved through processes but also satisfy those needed to design and evaluate the processes as being fit for purpose.

Bibliography

Wikipedia (4). (2015, February 15). Retrieved from: http://en.wikipedia.org/wiki/Enron_scandal#Causes_of_downfall

Hubbard, D. W. (2010). *How to Measure Anything: Finding the Values of Intangibles in Business.* Hoboken, NJ: John Wiley.

ISO 9000:2015. (2015). *Quality Management Systems: Fundamentals and Vocabulary.* Geneva: ISO.

23 Planning of changes

Introduction

In Managerial Breakthrough, Juran says that: "All managerial activity is directed at either breakthrough or control. Managers are busy doing both things and nothing else" (Juran, 1964). In this context, Juran defined control and breakthrough as the organized sequence of activities by which organizations prevent change or achieve change, respectively.

Change is a constant. It exists in everything and is caused by physical, social or economic forces. Its effects can be desirable, tolerable or undesirable. Desirable change is change that brings positive benefits to the organization and results in improvement. Tolerable change is change that is inevitable and yields no benefit or may have undesirable effects when improperly controlled. The challenge is to cause desirable change and to eliminate, reduce or control undesirable change so that it becomes tolerable change. It is desirable change that is the subject of this chapter.

As will be seen in Figure P6.1, the need for changes to the QMS will arise from an analysis of internal and external issues and changes in policy or from an analysis of performance.

The standard does not require any formal change management process but if a change is to be made that affects the QMS it would seem logical to treat it with at least the same level of rigour as was used in establishing the QMS to begin with. Remember, we are changing the system not changing individual documents or products and services; that may be a consequence of the change.

In this chapter, we examine the requirement of clause 6.3 for carrying out changes to the QMS in a planned manner and address the five constraints namely:

- Considering the purpose of changes to the QMS (6.3a)
- Considering potential consequences of changes to the QMS (6.3a)
- Considering the integrity of the QMS (6.3b)
- Considering the availability of resources (6.3c)
- Considering the allocation responsibilities and authorities (6.3d)

Carrying out changes to the QMS in a planned manner (6.3)

What does this mean?

The requirement is expressed in terms of when a need for change has been determined, implying that the need is an output from other processes. It may be either as a result of an evaluation of the effectiveness of the QMS, the result of management review or a reported opportunity for improvement (OFI), etc., but this does not mean it's a foregone conclusion that the change should go ahead. When planning to carry out a change difficulties may be revealed that either warrant reconsideration of the need for or the nature of the change, delay in its implementation or abandonment of the change altogether.

Carrying out changes in a planned manner means having a clear idea of:

a) what is to be changed;
b) why it has to be changed;
c) what results are to be achieved;
d) how success will be measured;
e) what effect the changes will have;
f) how the effects of the change will be managed;
g) what work needs to be undertaken to bring about the desired result;
h) when the work needs to start and be complete;
i) who will do the work;
j) what resources will be needed to undertake this work;
k) how the integrity of the system will be maintained during the change.

Why is this necessary?

It would be unwise to carry out a change without carefully considering the implications, whether such a change is feasible and planning in detail how, when and by whom it is to be carried out.

As was stated in Chapter 8, everything is connected to everything else therefore changing one thing in an organization will invariably influence other things with which it is related. It is therefore considered irresponsible to change anything without having a plan which addresses those aspects on which its successful execution depends.

How is this addressed?

Identifying a change to the QMS

The word *change* itself can be used to denote different actions such as to alter, amend, adjust, correct, modify, refine or revise. If we look at the way the word *change* is used in ISO 9001 we find that there are four types of changes that are considered:

a) Changes in the organization purpose, its strategic direction, the internal or external issues or stakeholder requirements that expose the QMS to a different set of circumstances than were previously planned for. These types of changes are referred to in:

   (9.3.2b)   Changes in external and internal issues that are relevant to the QMS.

b) Changes in the performance of the QMS that cannot be corrected by using the established processes. These types of changes are referred to in:

   (10.2.1f)   When a nonconformity occurs, including any arising from complaints, the organization shall make changes to the QMS, if necessary;
   (9.3.3a)   The outputs of the management review shall include opportunities for improvement;

(9.3.3b)   The outputs of the management review shall include decisions and actions related to any need for changes to the QMS;

(10.1c)   Improving the performance and effectiveness of the QMS.

c) Changes in specific contracts, projects, products or services that the QMS is designed to handle. These types of changes are referred to in:

(8.1)   The organization shall control planned changes to the processes needed to meet the requirements for the provision of products and services;

(8.2.1b)   Handling enquiries, contracts or orders, including changes;

(8.2.4)   Changes to requirements for products and services;

(8.3.6)   Design and development changes;

(8.5.6)   Control of changes for production and service provision;

(10.1a)   Improving products and services to meet requirements as well as to address future needs and expectations.

d) Changes to practices to bring them into line with management's intentions. These types of changes are referred to in:

(4.4.1g)   Implement any changes needed to ensure that these processes achieve their intended results;

(9.3.3a)   Opportunities for improvement;

(10.1b)   Correcting, preventing or reducing undesired effects.

There follow some changes that have little or no impact on performance:

- We alter the document numbers so that they align with the clause number of ISO 9001:2015.
- We adjust settings on a machine to reduce wear.
- We correct an instruction so that it is more readily understood.
- We modify the layout of a storage area so its contents are more easily accessed.
- We refine the justification for outsourcing services to make it easier for managers to make more consistent decisions.

The very nature of a system is that it consists of interacting parts that form a whole; therefore, changes in any part that affects the way it interacts with other parts will influence the performance of the whole. We therefore need to ask, will the alteration, amendment, adjustment, correction, modification, refinement or revision we are going to make (or are making) change the performance or behaviour of the part in a way that changes the QMS as a whole so it either produces different outputs or the same outputs in a different way? If not, it's not a change to the QMS.

An example illustrates this concept:

Our research indicates that customers now seek assurance of the origin and identity of ingredients. We therefore need to change the way the integrity of ingredients is maintained from their origin to the point of sale to our customers.

This is a system change because it changes the vetting processes from negative to positive, that is, instead of a system that accepts ingredients based on the labels attached to them, the system is changed such that ingredients won't be accepted without proof of origin and identity.

The purpose of the change

The purpose of a change is the reason why it is deemed necessary or is a response to the question "why are we doing this or why do we have to do this?" In the earlier example the purpose of the change is "to provide assurance to customers of the origin and identity of the ingredients we use".

The results to be achieved

The result to be achieved is the objective, it's what we want to happen. In the earlier example the objective is: *To redesign the system such that only ingredients that are traceable to their origin and identity are purchased, used and provided to customers.*

The measures of success

The measures of success are what we'd look for as evidence that the objective has been achieved. In the earlier example the success measure might be as follows:

a) Ingredients without supporting objective evidence of origin and identity would not be purchased or enter the manufacturing or delivery processes.
b) Processes in place for determining the identity of ingredients.
c) Provision in place to prevent cross contamination.
d) Provisions in place that preserved traceability of data at each process stage.
e) Heightened awareness of employees to traceability.
f) Cost of change is less than the value gained.
g) Change to be introduced and its effectiveness proven by January 2017.

The consequence of the change

The term *consequence* is used instead of the term *outcome* because of difficulty in its translation. Consequences like outcomes are the indirect effect of a system or process on its surroundings. As Fraser remarks, outcomes come out of a process. They are not put out of a process (see Box 9.1). There is no one action that produces an outcome. Outcomes emerge from the interaction of elements and these may be intended or unintended, for example, the intended consequences of a service delivery process might be customer satisfaction but an unintended consequence might be employee stress. It follows therefore that when considering the consequence of a change we not only need to think about the desirable consequences but also the unintended consequences.

In the example the potential consequences might be:

a) Increase in sales as customers begin to recognize brand value.
b) Termination of contracts with suppliers that cannot provide the required proof of origin and identity.
c) Resistance to new work practices.
d) Lost revenue from stock deemed unsaleable due to inability to certify origin and identity.
e) Fines imposed by regulators when nonconforming products are detected.
f) Increase in manufacturing costs.

To bring about a change in performance, it may involve changing the organization structure, the technology, the plant, machinery, the processes, the policies, the procedures, the competency levels of staff and perhaps the culture. It may be a change that is made all at once, over time and may affect only one process or all processes. A change that eliminates bottlenecks or reduces staffing levels in one area may change the timing of inputs into other processes and thus change system performance. It is therefore necessary to consider multiple consequences, both short-term and long-term consequences, as some effects are not felt for a considerable time and not in the places that one might expect. A technique for studying these influences is the causal loop diagram (see Figure 8.7).

It will therefore be necessary to carry out a feasibility study before detail planning is undertaken to determine what the effects of the change, as it is envisaged, will be and whether the organization has the capability to execute it.

Managing the consequences

Managing the consequences is about maximizing the positives and minimizing the negatives. The positives won't materialize until the negatives are under control. In our example there won't be increased revenue until the organization has acquired a reputation for brand integrity, and that won't happen if it continues to use suppliers that are unreliable, or if staff are slow in adapting to the new practices, and certainly not until any suspect stock is out of the supply chain.

System changes will have both a technical component and a social component because the system contains people. Any change will affect what people do and the way they do it and so are a form of threat to habits, beliefs and status. The workplace will comprise different types of people. Some will see change as an opportunity and embrace it, others will be cautious of change and want to be persuaded that change is indeed necessary and there will be some who see change as a threat and will be resistant to it. It's essential for the person assigned to manage the change that they possess the appropriate competences to handle resistance to change. Two useful sources of information on managing change are *Managerial Breakthrough* by Dr J. M. Juran who devotes a chapter to resistance to change and Robbins and Finley's book *Why Change Doesn't Work*.

Box 23.1   Seven unchangeable rules of change

- People do what is in their best interest, thinking as rationally as circumstances allow them to think.
- People are not inherently anti-change. Most will in fact, embrace initiatives provided the change had positive meaning for them.
- People thrive under creative challenge, but wilt under negative stress.
- People are different. No single "elegant solution" will address the entire breadth of these differences.
- People believe what they see. Actions do speak louder than words, and a history of previous deception intensifies present suspicion.
- The way to make effective long-term change is to first visualize what you want to accomplish and then inhabit this vision until it comes true.
- Change is an act of the imagination. Until the imagination is engaged no important change can occur.

(Robbins & Finley, 1998)

The schedule of work

A plan of action will be needed to undertake the change and this will vary depending on the magnitude of the change. Although the need for change will have been determined by management, there will still be other decisions to take regarding the strategy and mechanism of change.

The first step will be to retrieve and study the system model and associated information including policies, process descriptions, flow charts and causal loop diagrams which, hopefully, represents the *as is* situation and produce one of more strategies which, if implemented, will achieve the change objectives. Each of the strategies should be reviewed by management against the success criteria.

The next step is to choose the most appropriate strategy and then produce a detail description of what has to change and in what order.

The success of any change depends on how it is managed. Managed poorly, it creates more problems than it solves. Managed well, it accomplishes the vision of those who sponsored it.

Some useful questions to answer are:

- Who and what is affected by this change?
- How will our stakeholders react to it?
- How much of this change can we achieve ourselves?
- What parts of the change do we need help with?
- How will each of our stakeholder's benefit from the change?

Some dos and don'ts are:

- Do secure the active participation of those who will be affected by the change, both in the planning and in the execution.
- Do listen to people's reactions and act sensitively to their concerns.
- Do treat people with dignity.
- Do put yourself in the other person's position and see the change from his or her perspective.
- Do allow time for people to come on board.
- Don't sell change to people as a way of getting agreement.
- Do plan to measure performance before, during and after the change.
- Do plan to start small so that the original plan can be modified as experience is gained.
- Don't plan to remove the old process infrastructure entirely until the new processes have been proven effective.
- Do provide sufficient time for the mental changes to take place.
- Don't chose a time to commence change when the organization is already undergoing change unless there are distinct technical and social advantages of riding on the back of an agreed change.
- Do build on the success of previous changes when appropriate by utilizing proven strategies.
- Do plan and execute the change concurrently with associated changes to documentation.

The responsibilities

There are two groups of people: those who will manage the change and those who will be affected by the change whose responsibility may change. Responsibility for managing the change needs to be assigned before planning for change, and this person should, if necessary, form a team to plan and organize execution of the change. In specifying what needs to change there may be changes in roles and responsibilities and these need to be handled with care. At one level, the change may not affect roles or responsibilities and simply change working practices but at another level, the change may result in roles being made obsolete and people laid off as no other suitable work is available.

The resources

If changes were to be planned and their implementation attempted without giving due consideration to the availability of resources needed, it is highly likely that the change will not achieve its objectives. The change may have been planned and the resources determined but until their availability has been established no date for implementation of the change can be announced with any confidence.

When considering the availability of resources, it may be possible to accommodate the change within the present work load. However, other proposed changes may require either additional resources of the same type or different resources.

There is likely to be competition for scarce resources therefore each proposed change will be funded on its merits. This may create role conflict between those whose primary function is to bring in revenue and those who work on what are perceived to be support activities. Making changes to maintain the status quo may be perceived as a support function, but making breakthrough changes is top management work and should receive the appropriate priority.

Preserving system integrity

If changes in the QMS are permitted to take place without consideration of their impact on other elements of system, there is likely to be deterioration in performance. In the past it may have been common for changes to be made and some months later the organization charts and procedures to be updated. Sometimes an updated document is the first and only news a person affected by a change has of it. Such situations are indicative of a lack of attention of system integrity.

To meet this requirement, change management processes need to be designed and put in place. The integrity of the management system will be maintained only if these processes are made part of the system so that in planning the changes, due consideration is given to the impact of the change on the organization, its resources, processes, practices and products and any documentation resulting from or associated with these processes.

How is this demonstrated?

Demonstrating that changes to the QMS have been carried out in a planned manner may be accomplished by:

a) presenting the results of strategic planning and performance reviews where issues requiring a change to the QMS were identified and the need for action agreed;
b) selecting a representative sample of proposed changes and presenting plans for their determination and execution;
c) showing where consideration has been given to the consequences of change;

d)   showing where consideration has been given to the availability of resources;

e)   showing where consideration has been given to roles and responsibilities and authority;

f)   showing where consideration has been given to system integrity;

g)   showing how the change has been addressed in the policies and processes of the QMS (i.e. where the change affects what people do and how they do it);

h)   showing the provisions made to evaluate the change.

Bibliography

Juran, J. M. (1964). *Managerial Breakthrough*. New York: McGraw-Hill.

Robbins, H., & Finley, M. (1998). *Why Change Doesn't Work*. London: Orion Business Books.

Key messages from Part 6

Chapter 21 Actions to address risks and opportunities

1 When we take action to address risks and opportunities, we are not reacting to circumstances that have already happened but trying to deal with circumstances that have yet to happen so that we are adequately prepared for the favourable or unfavourable consequences.

2 Pursuing a strategy of only looking for undesirable outcomes is a pessimistic approach to quality management, whereas looking for both risks and opportunities is a balanced approach to quality management.

3 There are risks the sources of which are external to the QMS (extrinsic risks), and these are addressed in clause 6.1 before a QMS is established. Then there are risks the sources of which are internal to the QMS (intrinsic risks), and these are addressed in clause 5.1.2b) during and after a QMS is established.

4 If risks are not properly analysed they cannot be properly managed.

5 When it comes to measuring risk, we need a form of calibration; otherwise, it's just guesswork that anyone can do.

6 Converting a qualitative method of risk analysis into a points-based scoring method does not make it a quantitative method, neither does adding or multiplying scores that have been made on the basis of opinion.

7 Whether a risk is to be avoided, eliminated, reduced, taken, shared or accepted depends on an organization's risk appetite, and this should be established before its goals are set as this will inevitably shape its strategy.

8 Taking a risk to pursue an opportunity is different to accepting a risk in that there may be no choice but to accept certain risks if a particular objective is to be achieved, whereas when taking a risk, you are deliberately playing the odds to seize an opportunity.

9 There's a risk an opportunity will not happen and therefore actions could be taken that will make it more likely to happen.

10 A properly orchestrated plan that changes the way people work will be far more successful than managerial exhortation to do better, seize every opportunity or simply work harder.

11 If the methods used to evaluate the effectiveness of actions to mitigate risk do not actually measure the risks in a mathematically and scientifically sound manner, management doesn't even have the basis for determining whether a method works.

Chapter 22   Quality objectives and planning to achieve them

12   A quality objective is an objective that primarily benefits the customer by its achievement; therefore, for objectives to qualify as quality objectives they must serve customer satisfaction.

13   There are two classes of quality objectives: those serving the control of quality (maintaining performance by mitigating risks) and those serving the improvement of quality (making beneficial change by exploiting opportunities).

14   Objectives drive actions; therefore, unless the objectives are consistent with policies it is likely that work being undertaken in pursuit of objectives will not be aligned with the policies.

15   It is not written statements of intent alone that will confirm the reality; evidence of actions taken is also necessary to reveal whether the objectives being pursued are actually consistent with the policy that is understood.

16   Without a means of deciding whether an objective has been achieved, its achievement becomes a matter of opinion.

17   If a thing can be observed in any way at all, it lends itself to some type of measurement method and it's a measurement if it tells you more than you knew before.

18   If we formulate objectives by thinking of things we'd like to do regardless of their achievement adding value, we will have objectives without a useful purpose.

19   When choosing the method of monitoring quality objectives, you need to consider for whom the information is being produced and what form of presentation is suitable.

20   The doers must be just as motivated as the planners for them to even want to take action, which is why it is necessary to communicate quality objectives.

21   Although ISO 9001 suggests that the quality objectives should be based on the quality policy, it is more likely to be the strategic issues arising from the PESTLE and SWOT analysis that drive the setting of objectives.

Chapter 23   Planning of changes

22   The need for change may arise from of an evaluation of the effectiveness of the QMS, the result of management review or a reported opportunity for improvement.

23   Everything is connected to everything else; therefore, changing one thing in an organization will invariably influence other things with which it is related.

24   In ISO 9001 the word *change* refers to changes to purpose and objectives, to QMS performance, to specific projects, contract, products and services and to practices.

25   All changes should have a reason, a purpose, a measure of success and will have consequences some of which may be undesirable.

26   Managing the consequences is about maximizing the positives and minimizing the negatives. The positives won't materialize until the negatives are under control.

27   Any change will affect what people do and the way they do it and may threaten their established habits, beliefs and status.

28   The success of any change depends on how it is managed. Managed poorly, it creates more problems than it solves. Managed well, it accomplishes the vision of those who sponsored it.

Part 7

Support

Introduction to Part 7

This part of the Handbook addresses section 7 of ISO 9001 covering more topics than was covered by the section on resource management in the 2008 version as shown in Figure P7.1

Ackoff remarked that "resources are anything physical or mental that can be used to obtain something else one needs or desires and therefore information, knowledge and understanding are resources as much as is money" (Ackoff, 1999). The resources therefore needed go beyond those that are specifically addressed by the requirements of clause 7.1.1.

Resource management is a key business process in all organizations. In practice, resource management is a collection of related processes that are mostly departmentally oriented but information, knowledge and understanding is ubiquitous; it's not confined to any one department or function.

- Financial resources might be controlled by the Finance Department.
- Purchased materials, equipment and supplies might be controlled by the Purchasing Department.
- Measuring equipment maintenance might be controlled by the Calibration Department.
- Plant maintenance might be controlled by the Maintenance Department.
- Personnel might be recruited, developed and dismissed by the Personnel Department.
- Building maintenance might be controlled by the Facilities Management Department.
- Knowledge management is usually controlled by the custodian.
- Documented information is usually controlled by the custodian.
- Understanding is developed through effective communication, education and training by good teachers wherever they happen to be.

These departments might control the resources in as much that they plan, acquire, maintain and dispose of them, but they do not manage them totally because they are not the sole users of the resource. They might therefore only perform a few of the tasks necessary to manage resources.

Whatever the resource, first it has to be planned, then acquired, deployed, maintained and eventually disposed of. The interconnections are shown in Figure P7.2.

Disposal is not a term we use when knowledge or people are no longer needed. People have their employment or contract terminated, and organizational knowledge is lost when staff leave or the technology makes it obsolete. The detail of each process will differ depending on the type of resource being managed.

The standard does not address financial resources specifically (see ISO 9004) but clearly they are required to implement and maintain the management system and hence run the

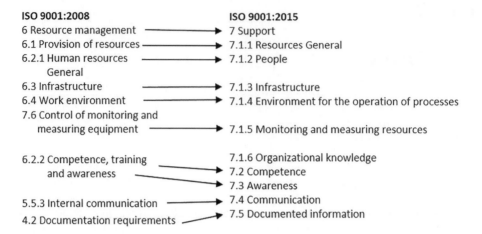

Figure P7.1 Clause alignment between 2008 and 2015 versions in Section 7

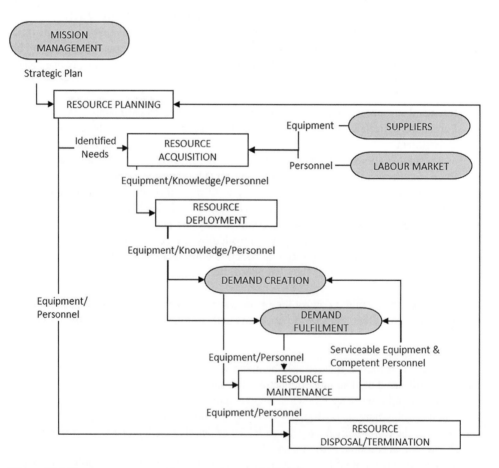

Figure P7.2 Resource management process stages and interfaces

organization but they are an input to the QMS from the financial management system. Hence, finance has an influence on the QMS but is not controlled by it.

Purchasing is not addressed under resource management but under operations because it is primarily concerned with purchasing items directly associated with deliverable product or service. Competence and awareness (clauses 7.2 and 7.3) are allied to clause 7.2 on people resources, and although information is a resource it's considered to be separate in the standard.

Requirements to determine or consider resources

There are ten separate requirements to determine the resources needed. If each of these referred to the resources needed for specific processes, it would not be difficult to understand what was required but there are general requirements which if met could imply all the specific requirements have been met depending on how the general requirements are interpreted. Apart from the six requirements in section 7 there are four others:

* 4.4.1d) the resources needed for processes needed for the quality management system
* 6.2.2b) resources that will be required to achieve the quality objectives
* 8.1c) the resources needed to achieve conformity to the product and service requirements
* 8.3.2e) internal and external resource needs for the design and development of products and services

Some inconsistencies among the support requirements

The omission of a requirement to maintain human resources could be an oversight by TC 176 but it could also be a device to enable competence and awareness to be addressed separately in clauses 7.2 and 7.3. In most cases the requirements address the quantity of the resource with the words *necessary* and *needed*, and they address the quality of the resource with the words *for the operation of its processes and to achieve conformity of products and services*.

The resource requirements for infrastructure (7.1.3), process environment (7.1.4) and organizational knowledge (7.1.6) require resources to be determined "for the operation of processes and to achieve conformity of products and services", whereas the resource requirement for people in 7.1.2 requires resources for "the operation and control of processes". It's assumed that resources for the operation of processes and to achieve conformity of products and services, are the same as resources for the operation and control of processes and that all these requirements were intended to be consistent.

Another inconsistency is that clause 7.1.2 requires resources necessary for the effective implementation of the quality management system but the other clauses in 7.1 don't, for example, infrastructure (7.1.3), process environment (7.1.4) and organizational knowledge (7.1.6) are not specifically required for the effective implementation of the quality management system so its inclusion under 7.1.2 creates and anomaly which may be another oversight.

If the QMS is the system which consistently provides products and services that meet customer and applicable legal requirements and enhances customer satisfaction, we are referring to the internal and external resources including but not limited to; the people, infrastructure, process environment, information and knowledge needed:

* to establish, implement, maintain and continually improve the QMS;
* for the processes of the QMS;

- for monitoring and measurement;
- to achieve its quality objectives;
- for the operation and control of its processes;
- for the design and development of products and services;
- to achieve conformity to the product and service requirements.

None of these are mutually exclusive of the others.

This part of the Handbook comprises nine chapters (Chapters 24–32) that address clauses 7.1 to 7.5 of ISO 9001:2015.

- Clause 7.1 Resources – see Chapters 24–28
- Clause 7.2 Competence – see Chapter 29
- Clause 7.3 Awareness – see Chapter 30
- Clause 7.4 Communication – see Chapter 31
- Clause 7.5 Documented information – see Chapter 32

Bibliography

Ackoff, R. M. (1999). *Ackoff's Best: His Classic Writings on Management: Chapter 1*. New York: John Wiley & Sons, Inc.

24 People

Introduction

It is often said by managers that people are their greatest asset, but in reality, it is not the people exactly but the contribution they can make and the number of them that qualifies them as assets. If we adopt Ackoff's definition of resources as "anything physical or mental that can be used to obtain something else one needs or desires" and categorize people as a resource, it does tend to infer that people are what we use to get what we want. Certainly, some people use other people for such ends, but that is not how they should be perceived in an organization. Unlike physical resources, people are adaptable, they are flexible and can learn new skills, reach new levels of performance when the right competences are developed and when working in the right environment. But like physical resources, the human is affected by the environment and even when equipped with the right competences may not function as well as expected if the conditions are not appropriate. As people are vital to enable organizations to achieve sustained success perhaps they should be perceived as capital and not resources (see Box 24.1).

Box 24.1   From human resources to people to human capital

The 2008 version categorized personnel performing work as human resources in clause 6.2 of that standard. In the 2015 version the category has changed to *people* and *personnel* changed to *people* but remains as a category of resource.

When the workforce is treated like a form of resource, the organization finds the best way to exploit it and use it for its purposes. These resources are used and depleted, and when they are no longer useful, the remuneration stops and their employment terminated.

When a workforce is looked upon as a form of capital, the incentive is for the organization to get the best return from its investment. Unlike a human resource, a human capital grows over time with productive use and its competences, knowledge and creativity grow and multiply in value (adapted from an article by Jas Chong 2012).

Perhaps at the next revision people won't be classed as a resource but as capital.

Building and retaining a workforce of the capacity and capability to fulfil the organization's objectives is a difficult task, one which carries many risks but also many opportunities. Many factors influence the availability of human capital. If your organization is in an emerging sector, it will attract young graduates seeking exciting employment opportunities.

But if your organization is in a declining sector, it may have great difficulty attracting young graduates and be reliant on an aging workforce. Fortunately, not all customers want products and services at the forefront of technology and so the older industries sometimes have a vital role to play in our societies.

The economic climate affects the availability of capital and the sustainability of remuneration levels, and therefore a project that was going to employ many more people may not now go ahead or the competition makes it no longer economical to undertake certain work in-house. The work either has to be therefore outsourced into a lower wage economy or the customers will go elsewhere.

In this chapter, we examine the two requirements of clause 7.1.2 namely:

• Determining the people needed
• Providing the people needed

Note: Evidence presented to demonstrate conformity with 7.1.1 might mean that it is also deemed to satisfy conformity with 7.1.2, but not vice versa because there are types of resources other than people. Also, evidence presented to demonstrate conformity with 7.1.2 might mean that it is also deemed to satisfy conformity with 7.1.1, 4.4.1d, 6.2.2b, 7.1.5.1, 8.1c and 8.3.2e relative to persons necessary.

Determining the people needed (7.1.1 and 7.1.2)

What does this mean?

Determining the people needed is an act of planning: anticipating the number of people of various abilities that are needed to achieve the organizational objectives as planned. The numbers and abilities needed in this case are those needed to execute the work that is within the scope of the QMS. There may therefore be three different groups of people:

a)   Those needed to create, maintain and improve the model of the QMS and assess any gap between the model and the requirements of ISO 9001. These people may be specialists in quality management but may also include staff from departments that operate the processes represented by the model (we'll refer to these as the system improvement team).

b)   Those needed to determine and carry out the planned work represented in the model. These people will be the managers and staff from the departments that operate the processes included in the model and will vary depending on demand. This is the estimated demand resulting from marketing, sales and planned changes in strategy and objectives. (we'll refer to these as the process operation teams).

c)   Those needed to carry out unplanned work which results from failure to do work right first time. Some organizations set up special units to handle customer returns and call centres to handle problems that customers have in completing forms for the provision of services. This work has been planned but is a result of poor management, a belief that failure is a certainty. Failure demand is a common characteristic in public services and is addressed in The Whitehall Effect (Seddon, 2014) (we'll refer to these as the process contingency teams).

Making provision for contingencies is sensible because of uncertainty about the future, but making provision to repeat known failures would appear careless.

Why is this necessary?

By anticipating the number of people needed to carry out work well in advance and initiating the acquisition process in good time, it is more likely that personnel will be available to carry out the work required when needed and thereby avoid delays and customer's dissatisfaction.

How is this addressed?

The standard does not require personnel budgets but a common way of determining the people needed is to produce personnel budgets. However, the personnel budget should not be just the head count but the people needed for achieving specific objectives such as those for the organization individual department, projects, processes or contracts. Without the focus that an objective provides, a budget has no purpose. Objectives may be for improvement or control and therefore there needs to be budgets for improvement projects and budgets for maintaining the status quo which may involve maintaining capacity and throughput.

To be useful in the management of quality, personnel budgets need to indicate:

- The objective to be achieved
- The numbers of persons required
- The competence required which can be expressed as a job title and grade or skill level
- When they will be needed and over what period will they be needed
- Whether the persons needed will be permanent staff, part time staff, agency staff or outsourced
- Any assumptions about the numbers, competences, availability and source of persons

When preparing budgets, allowance should be made for the *certainties* such as staff reaching retirement, holidays and reorganizations, as well as new work that has been committed but yet to started. Allowance also must be made for the *uncertainties* such as absenteeism and project overruns and the availability of the right calibre of personnel when needed. At the time of preparing the budget, the personnel planned to undertake the work maybe available, or a recent turnaround in the fortunes of a competitor may provide an opportunity to recruit the right calibre of personnel. In the event of those personnel not being available when approval to start work has been given contingency plans need to be prepared for the development of the necessary competences.

The requirement is for the persons necessary for the effective implementation of the QMS, and for the operation and control of the organization's processes to be determined but it would be unusual to find within a departmental budget for instance, a line item for implementing a QMS against which the persons required are specified. However, where cost collection codes are used it should be possible to determine the proportion of human capital allocated to certain processes or activities and aggregate these for cross-functional processes, including the proportion to be engaged in correcting errors. If process mapping software is being used, it may contain features for adding task completion time and for computing the total human capital of various abilities needed to run a particular process. The danger with these tools is that it forces you into a lot of guesswork at a detail level that you would not otherwise do. You therefore need to do a simple cost–benefit analysis before expending the effort involved. If you have used such tools, it would be prudent to compare the estimate with the actual expenditure once the process is running as expected and readjust the estimate so that when it is used in bids for new work, at least it reflects reality and not guesswork.

When developing a process, the competences required to carry out either the entire process or part of a process should be determined (see Chapter 29). As the process matures, the number people of specified abilities required for a given process output should be recorded and retained as reference data so that every time a new job/project has to be estimated, the estimates will be based on objective evidence. For exploratory processes that vary each time, a manager will often produce an estimate based on his or her experience from the perceived size of the project and allow a surplus for contingency.

How is this demonstrated?

Demonstrating that persons necessary for the effective implementation of the QMS and for the operation and control of the relevant processes have been determined may be accomplished by:

a)   presenting evidence of a process for determining the people required;
b)   selecting a representative sample of quality objectives and presenting the personnel budgets for achieving them;
c)   presenting evidence that in preparing the budgets consideration has been given to the capabilities of, and constraints on, existing internal resources and what needs to be obtained from external providers.

Providing the people needed (7.1.1 and 7.1.2)

What does this mean?

Providing the people needed means acquiring and deploying the human capital that has been identified as being needed in the budgets.

Why is this necessary?

Without providing the people, the work required won't get done, but having a budget for the people doesn't mean they will be provided, as that will depend on management's priorities.

How is this addressed?

The personnel acquisition process (see Figure 24.1), should deliver the personnel in the right quantity and competence when they are needed. The deployment process should prepare the personnel for taking up their duties (see Figure 24.2). Therefore, if there is an identified need for people, they have been provided to the process that needs them only when they are in position to assume their duties (i.e. deemed competent or ready to take up a position under supervision). The HC development process (see Figure 29.1) should maintain competences, so that there is no shortage of supply of people. If only it was this simple but it's not because of the uncertainties.

Risks

Providing people in the quantity and calibre required when they are needed is a process with a lot of risks. When budgets are prepared the decision to undertake the work may not

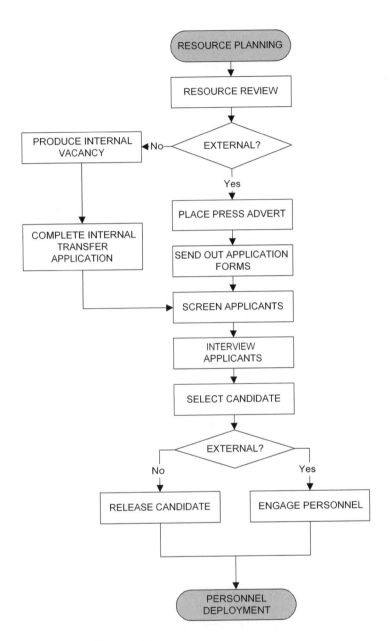

Figure 24.1 Personnel acquisition activity sequence

have been made, the contract may not have been awarded. There is often a time lag between submitting the budget and getting approval to recruit people and many things may change between these two events. Priorities change and so a planned quality improvement may be delayed. There may also be changes in the labour market. An opportunity that was available when the budgets were prepared may no longer be available. These issues and many more will need to be addressed and as each will be unique, there is no pre-defined process

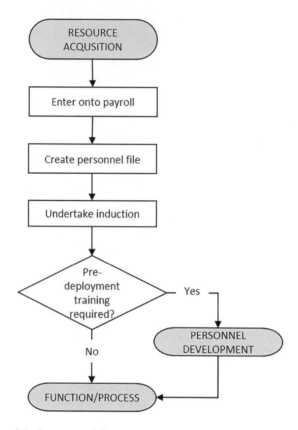

Figure 24.2 Personnel deployment activity sequence

to follow. One thing managers can do is to keep abreast of changes in the labour market by studying national surveys. For example, the UK's HR professional body CIPD produces a quarterly Labour Market Outlook that provides a set of forward-looking labour market indicators, highlighting employers' recruitment, redundancy and pay intentions.

Quality of recruitment

When recruiting people, it's important that they will not only fit in the organization but also be a good fit to the job they are required to do; otherwise, the organization takes a risk that the person may be totally ineffective or cause major problems by not possessing the appropriate characteristics.

- Person–organization fit refers to the degree to which a person's personality, values, goals, and other characteristics match those of the organization.
- Person–job fit is the degree to which a person's knowledge, skills, abilities, and other characteristics match the job demands.

A proactive and creative person may be a good fit for an organization where risk taking is valued but may be a poor fit for an organization where predictable behaviour is required.

Because we all differ in our individual abilities, some types of people are better at some types of jobs than others, regardless of their technical capability. This requires managers to appreciate the different behavioural styles that people exhibit. Place an amiable person in a job that requires a person to take decisions quickly and confidently and you will be disappointed by their performance.

Box 24.2   Behavioural styles

Hunsaker and Alessandra adapted David W. Merrill's original research of 1964 and postulated four behavioural styles: Amiable, Analytical, Expressive and Driving.

The Amiable person tends to be slow in taking action and decisions, likes close relationships but works well with others and therefore has good counselling skills.
The Analytical person tends to be cautious in taking actions and decisions, likes structure, prefers working alone but good problem-solving skills.
The Expressive person tends to be spontaneous in taking actions and decisions, likes involvement, works well with others and has good persuasive skills.
The Driving person tends to take firm actions and decisions, likes control, works quickly by himself or herself and has good administrative skills.

(Hunsaker & Alessandra, 1986)

Unless the recruitment process recognizes the importance of matching people with the culture, mavericks may well enter the organization and either cause havoc in the work environment or be totally ineffective due to a lack of cultural awareness

The objective of the personnel acquisition process will be to provide sufficient numbers of personnel with the desired characteristics the organization needs to fulfil its goals

The objective of the personnel deployment process will be to provide personnel with the required competences to where and when they are needed.

Personnel records

There is no requirement in clause 7.1.2 for records of the provision of persons, but there is a requirement in clause 4.4.2 for the organization to retain documented information to have confidence that the processes are being carried out as planned.

How is this demonstrated?

Demonstrating that persons necessary for the effective implementation of the QMS and for the operation and control of the relevant processes have been provided may be accomplished by:

a) presenting evidence that the processes for acquiring the personnel required and deployment them to where they are needed have been designed;
b) selecting a representative sample of quality objectives and presenting evidence that the HC acquisition and deployment processes were implemented as planned;

c)   where problems were encountered in providing the personnel necessary, presenting evidence of the actions taken to resolve them.

Bibliography

Chong, J. (2012, December). Human Capital vs. Human Resources: A Word and a World of Difference. Retrieved from Dignity of Work: https://thedignityofwork.com/2012/12/04/human-capital-vs-human-resources-a-word-and-a-world-of-difference/#comments

Hunsaker, P. L., & Alessandra, A. J. (1986). *The Art of Managing People*. New York: Simon & Schuster, Inc.

Seddon, J. (2014). *The Whitehall Effect*. Axminster: Triarchy Press.

25 Infrastructure

Introduction

Infrastructure is defined in ISO 9000:2015 as the "system of facilities, equipment and services needed for the operation of an organization". Whether it was intentional or an oversight is not known, but unlike clause 7.1.2 requiring the organization to "determine and provide the persons necessary for the effective implementation of its quality management system and for the operation and control of its processes", this clause does not require infrastructure necessary for the effective implementation of its quality management system to be determined. However, it does qualify the processes it applies to as those to achieve conformity of products and services. There's an inference here that the infrastructure referred to is only for *operations* and not for *management* activities, but it is not clear. We will therefore assume that it applies to infrastructure that can affect the organization's ability to create and retain customers and therefore includes that required for management and operations.

There is no difference, except in what they are used for, between the infrastructure necessary for the operation and control of processes associated with products or those associated with services. Telecommunications present the same issues whether they are used in production or service delivery.

These items are elements of the QMS because they are connected through a structure that enables the organization to function. Remove any item in this infrastructure and it disables the organization in some way. Infrastructure also includes the buildings and utilities such as electricity, gas, water and telecommunications. Within the buildings it would include the office accommodation, furniture, fixtures and fittings, computer software and hardware, networks, data storage, dining areas, medical facilities, laboratories, plant, and machinery. Outside the buildings it would include the access roads, signage, transport and their maintenance facilities. In the offices and workshops, it would include the IT systems for planning and resourcing operations and the computer systems for processing, storing and displaying information. In fact, everything an organization needs to operate other than the money, people and consumables. In many organizations infrastructure is classified under the heading of capital expenditure because it is not order driven, that is, it does not change on receipt of an order but might change before a big contract is signed that was bid based on changes in the infrastructure such as new IT systems, buildings for new assembly lines, etc.

In this chapter, we examine the three requirements in Clause 7.1.3, namely:

- Determining the components and systems that comprise the infrastructure
- Providing the components and systems that comprise the infrastructure and
- Maintaining the components and systems specifically plant and facilities and planned, preventive and corrective maintenance

In this case the processes for determining, providing and maintaining the infrastructure are work processes within the resource management process. The process objective would be to provide an infrastructure that enabled the organization to achieve its objectives. The demand comes from the mission management process and the output goes through every process back to the mission management process where it is assessed.

Note: Evidence presented to demonstrate conformity with 7.1.1 might mean that it is also deemed to satisfy conformity with 7.1.3 but not vice versa because there are other types of resources than infrastructure. Also, evidence presented to demonstrate conformity with 7.1.3 might mean that it is also deemed to satisfy conformity with 7.1.1, 4.4.1d, 6.2.2b, 7.1.5.1, 8.1c and 8.3.2e relative to the infrastructure necessary.

Determining the infrastructure needed (7.1.1 and 7.1.3)

What does this mean?

The emphasis in this requirement is on the infrastructure needed for the operation of the organization's processes and to achieve conformity of products and services. The words *for the operation of the organization's processes* are superfluous because it's the same infrastructure that achieves conformity of products and services as is used for the operation of the organization's processes.

The requirements of clause 7.1.1 also apply, and therefore the infrastructure being referred to is also that needed to establish, implement, maintain and continually improve the QMS.

As conforming products and services is the organization's output, it follows that most of the infrastructure exists for this purpose. However, there will be areas, buildings, facilities, etc., that may not be dedicated to this purpose but to meeting requirements of stakeholders other than the customer of the organization's products and services (e.g. staff welfare, pensions, investment management, public relations). The requirement is not implying that these other facilities do not need to be identified, provided and maintained, but that such provision is not essential to demonstrate conformity to ISO 9001. As with determining resources previously, ask: *Why would we want to exclude particular infrastructure from the quality management system? What business benefit is derived from doing so?*

Why is this necessary?

The design, development and supply of products and services do not exist in a vacuum. There is always an infrastructure within which these processes are carried out and on which these processes depend for their results. Without an appropriate infrastructure, the desired results will not be achieved. A malfunction in the infrastructure can directly affect results.

How is this addressed?

In determining the infrastructure needed to develop and operate the processes some are product, project, contract or order specific, others are needed for maintenance and growth of the organization. These are likely to be classified as capital assets. The management of the infrastructure is a combination of asset management (knowing what assets you have, where they are, how they are depreciating and what value they could realize) and of facilities management (identifying, acquiring, installing and maintaining the facilities). Organizations in general maintain registers for fixed and liquid assets. A fixed asset register would identify assets and property that cannot easily be converted into cash and this register would identify items of infrastructure but only in financial terms. Facilities management may maintain other registers of tangible assets identifying their status and assigning a unique reference code for traceability purposes.

As the infrastructure is a critical factor in the organizations capability to meet customer requirements, and ability to continually meet customer requirements, its management is vital to the organization's success. Within the resource management process there are therefore several work processes related to the management of the infrastructure. It would be impractical to put in place one process because the processes will differ depending on the services required. Based on the generic model for resource management (see Figure P7.2), several planning processes will be needed for identifying and planning the acquisition, deployment, maintenance and disposal of the various assets. In describing these processes, you need to cover the aspects addressed in Chapter 32 on process descriptions, and in doing so, identify the impact of failure on the organization's ability to achieve conforming product.

There will be infrastructure that is in the general pool of capital assets that is used by all processes such as the energy supplies, communication equipment, transport and buildings. The budgets for these won't be specific to the QMS and it's unlikely other than for dedicated plant, to calculate the proportion of overhead that is allocated to any particular process or to achieve any particular objective. However, infrastructure needed in addition to achieve product, project or contract objectives should be identified in separate infrastructure budgets.

To be useful in the management of quality, these infrastructure budgets need to indicate:

- The objective to be achieved;
- The rationale for acquiring new items of infrastructure;
- The nature of the additional items of infrastructure required to achieve the objectives;
- When the new items of infrastructure will be needed and over what period it will be needed;
- Whether the items of infrastructure are to be capital or rented assets;
- Any assumptions about the quantities, availability and source of the additional items of infrastructure.

When designing specific processes for the design, development, production and delivery of products and services the infrastructure required for the effective operation of the processes needs to be determined. Account needs to be taken of the constraints the existing capability may impose on the performance required of the process, for example, the buildings may not have the capacity to store the process output or in the case of services may not have the space to safely handle the anticipated demand. Additional assets may need to be built or rented from external providers. Although the assets required for the process to function as intended should be specified in the process description, any changes required to the current infrastructure should be the subject of a separate analysis as it will vary depending on current demand. It may therefore be necessary as part of the management review to review this analysis to determine if it remains valid for the foreseeable future.

In selecting such equipment, you should determine whether it can produce, maintain or handle conforming product in a consistent manner. You also need to ensure that the equipment can achieve the specified dimensions within the stated tolerances. Process capability studies can reveal deficiencies with equipment that are not immediately apparent from inspection of the first off.

How is this demonstrated?

Demonstrating that the infrastructure has been determined for the operation of the operational processes may be accomplished by:

a) presenting process descriptions that identify the items of infrastructure on which the operation of the processes depends;
b) presenting the results of an analysis which shows the extent to which the current infrastructure satisfies the needs of the processes and what additional infrastructure is required.

Providing the infrastructure needed (7.1.1 and 7.1.3)

What does this mean?

Providing infrastructure simply means acquiring and deploying the infrastructure that has been determined as being necessary for the operation of its processes to achieve conformity of products and services.

Why is this necessary?

Unless the required infrastructure is provided the desired results will not be achieved. However, it is often the case that resources are not available to fund infrastructure that has been planned and as a result contingency plans have to be brought into effect or else difficulties will be encountered. A failure to provide adequate infrastructure can adversely affect the organization's performance. In many cases the plans may address the worst case not taking account of human ingenuity to get by with less than is needed or, "rob Peter to pay Paul", when their backs are against the wall.

Customers may have bought tickets for a flight to a new destination, a rail journey through a new tunnel, a football match at a new stadium relying on the airport, the tunnel or the stadium being open for business on time. A new micro-processor, television or automobile may be dependent on a new production plant and orders may be taken based on projected completion dates. With major capital works, plans are made years in advance with predictions of completion dates based on current knowledge.

The example in Box 25.1 illustrates the importance of infrastructure on performance and the link with customer satisfaction.

Box 25.1   The importance of infrastructure to customer satisfaction

A new air traffic control centre at Swanwick in the UK was planned to replace West Drayton ATC outside London because when the plans were made in the late 1980s it was predicted that capacity would be exceeded by the mid-1990s. The operational handover date was set as December 1996. After many delays, by the summer of 2000 it was still not operational and the problem that it was designed to solve was getting worse. Air traffic congestion continued to delay flights and dissatisfy customers until it was finally brought into service two years later in January 2002. It now controls 200,000 square miles of airspace among the busiest and most complex in the world.

How is this addressed?

Providing the infrastructure is associated with the acquisition and deployment of resources. It follows therefore that processes addressing the acquisition and deployment of buildings, utilities, computers, plant, transport etc. need to be put in place. Many will use the purchasing process but some require special versions of this process because provision will include installation and commissioning and all the attendant architectural and civil engineering services. Where the new facility is required to provide additional capability so that new processes, products or services can be developed, the time to market becomes dependent on the infrastructure being in place for production or service delivery to commence. Careful planning is often required because orders for new products may well be taken based on projected completion dates and any delays can adversely affect achievement of these goals and result in dissatisfied customers.

Documentation may be available from the supplier of the equipment that adequately demonstrates its capability; otherwise, you may need to carry out qualification and capability tests to your own satisfaction. In the process industries, the plant is specially designed and so needs to be commissioned and qualified by the user. Your procedures need to provide for such activities and for records of the tests to be maintained.

How is this demonstrated?

Demonstrating that the infrastructure has been provided for the operation of the operational processes may be accomplished by:

a)   presenting the results of checks performed when processes were deemed operational that testify that the specified infrastructure has been provided;
b)   presenting evidence that the processes are operating in accordance with the agreed process descriptions without work-around plans instituted because planned infrastructure was unavailable.

Maintaining the infrastructure needed (7.1.1 and 7.1.3)

What does this mean?

The identification and provision of the infrastructure needs little explanation but in maintaining it the implications go beyond the maintenance of what exists. Maintenance is more to do with maintaining the capability the infrastructure provides. Plant and facilities can be relatively easily maintained providing funds are made available, but maintaining their capability means continually providing a capability even when the existing plant and facilities are no longer serviceable. Such situations can arise due to man-made and natural disasters. Maintaining the infrastructure means maintaining output when there is a power cut, a fire, a computer virus, a flood or a gas explosion. Maintaining the infrastructure therefore means not only retaining something in a serviceable condition but also making provision for disaster recovery and therefore maintaining business continuity.

The contingency actions required by customers are addressed by clause 8.2.1e and are dealt with in Chapter 34, but there is an implication in the scope statement in clause 1.0 that to *consistently* provide products and services, the contingency actions need to also apply to the infrastructure.

Why is this necessary?

Unless the infrastructure is maintained or contingency actions taken when necessary, the desired results will not be achieved. A failure to maintain adequate infrastructure and contingency plans can adversely affect the organization's performance.

How is this addressed?

Maintenance of plant and facilities

There are two aspects to maintenance as addressed previously. Maintaining the buildings, utilities and facilities in operational condition is the domain of planned preventive and corrective maintenance. Maintaining the capability is the domain of contingency plans, disaster recovery plans and business continuity provisions. In some industries, there is no obligation to continue operations as a result of force majeure including natural disasters, war, riots, air crash, labour stoppage, illness, or disruption in utility supply by service providers etc. However, in other industries, provisions have to be made to continue operations albeit at a lower level of performance despite force majeure.

Although such events cannot be prevented, their effects can be reduced and in some cases eliminated. Contingency plans should therefore cover those events that can be anticipated where the means to minimize the effects are within the organization's control. What may be a force majeure situation for your suppliers does not need to be the same for your organization.

Start by doing a risk assessment and identify those things on which continuity of business depends, namely power, water, labour, materials, components, services, etc. Determine what could cause a termination of supply and estimate the probability of occurrence. For those with a relatively high probability (1 in 100) find ways to reduce the probability. For those with lower probability (1 in 10,000 chance) determine the action needed to minimize the effect (see also Chapter 21).

If you are located near a river and it floods in the winter, can you claim it to be an event outside your control when you chose to site your plant so close to the river? (OK the land was cheap – you got a special deal with the local authority – but was it wise?) You may have chosen to outsource manufacture to a supplier in a poorer country and now depend on them for your supplies. They may ship the product but because it is seized by pirates it doesn't reach its destination – you may therefore need an alternative source of supply! A few years ago, we would have thought this highly unlikely, but after 300 years and equipped with modern technology pirates have returned to the seas once again.

When plant is taken out of service either for maintenance or for repair, it should not be reintroduced into service without being subject to formal acceptance tests which are designed to verify that it meets your declared standard operating conditions. Your procedures need to provide for such activities and for records of the tests to be maintained

Maintenance of equipment

In a manufacturing environment, the process plant, machinery and any other equipment on which process capability depends need to be maintained and for this you will need:

- A list of the equipment on which process capability depends.
- Defined maintenance requirements specifying maintenance tasks and their frequency.

- A maintenance programme which schedules each of the maintenance tasks on a calendar.
- Procedures defining how specific maintenance tasks are to be conducted.
- Procedures governing the decommissioning of plant prior to planned maintenance.
- Procedures governing the commissioning of plant following planned maintenance.
- Procedures dealing with the actions required in the event of equipment malfunction.
- Maintenance logs that record both the preventive and corrective maintenance work carried out.

In a service environment if there is any equipment on which the capability of your service depends, this equipment should be maintained. Maintenance may often be subcontracted to specialists but nevertheless needs to be under your control. If you can maintain process capability by bringing in spare equipment or using other available equipment, your maintenance procedures can be simple. You merely need to ensure you have an operational spare at all times. Where this is not possible you can still rely on the call-out service if you can be assured that the anticipated down time will not reduce your capability below that which you have been contracted to maintain.

The requirement does not mean that you need to validate all your word-processing software or any other special aids you use. Maintenance means retaining in an operational condition and you can do this by following some simple rules.

There are several types of maintenance; planned maintenance, preventive maintenance, corrective maintenance and predictive maintenance (see Appendix B for definitions).

An effective maintenance system should be one that achieves its objectives in minimizing down time (i.e. the period in which the equipment is not in a condition to perform its function). To determine the frequency of checks you need to predict when failure may occur. Will failure occur at some future time, after a certain number of operating hours, when being operated under certain conditions or some other time? An example of predictive maintenance is vibration analysis. Sensors can be installed to monitor vibration and thus give a signal when normal vibration levels have been exceeded. This can signal tool wear and wear in other parts of the machine in advance of the stage where nonconforming product will be generated.

The manuals provided by the equipment manufacturers should indicate the recommended preventive maintenance tasks and the frequency with which they should be performed covering aspects such as cleaning, adjustments, lubrication, replacement of filters and seals, inspections for wear, corrosion, leakage, damage, etc.

Drawings should be provided for jigs, fixtures, templates and other hardware devices and they should be verified as conforming with these drawings prior to use. They should also be proven to control the dimensions required by checking the first one to be produced from such devices. Once these devices have been proven, they need checking periodically to detect signs of wear or deterioration. The frequency of such checks should be dependent on usage and the environment in which they are used.

Another source of data is from your own operations. Monitoring tool wear, corrective maintenance, analysing cutting fluids and incident reports from operators you can obtain a better picture of a machine's performance and predict more accurately the frequency of checks, adjustments and replacements. For this to be effective you need a reporting mechanism that causes operators to alert maintenance staff to situations where suspect malfunctions are observed. In performing such monitoring, you cannot wait until the end of the production run to verify whether the tools are still producing conforming product. If you do you will have no data to show when the tool started producing nonconforming product and will need to inspect the whole batch.

An effective maintenance system depends on it being adequately resourced. Maintenance resources include people with appropriate skills, replacement parts and materials with the funds to purchase these materials and access to support from original equipment manufacturers (OEMs) when needed. If the OEM no longer supports the equipment, you may need to cannibalize old machines or manufacture the parts yourself. This can be a problem because you may not have a new part from which to take measurements. At some point, you need to decide whether it is more economical to maintain the old equipment than to buy new. Your inventory control system needs to account for equipment spares and to adjust spares holding based on usage.

For the system to be effective there also has to be control of documentation, maintenance operations, equipment and spare parts. Manuals for the equipment should be brought under document control. Tools and equipment used to maintain the operational equipment should be brought under calibration and verification control. Spare parts should be brought under identity control and the locations for the items brought under storage control. The maintenance operations should be controlled to the extent that maintenance staff should know what to do, know what they are doing and be able to change their performance if the objectives and requirements are not being met. Although the focus should be on preventive maintenance, one must not forget corrective maintenance. The maintenance crew should be able to respond to equipment failures promptly and restore equipment to full operational condition in minimum time. The function needs resourcing to meet both the preventive and corrective demands because it is downtime that will have the most impact on production schedules.

An outline facility maintenance process flow is illustrated in Figure 25.1. The objective of this process is to maintain facilities in a condition such that are continually fit for their intended purpose.

Infrastructure records

There is no requirement in clause 7.1.3 for records of the maintenance of infrastructure but there is a requirement in clause 4.4.2 for the organization to retain documented information to have confidence that the processes are being carried out as planned.

Any records you maintain need to be useful, and it would be wise to maintain records particularly of plant, facility and equipment maintenance so that in the event of a problem, a pending prosecution or simply a customer complaint you can carry out a proper investigation. You would want to either demonstrate you acted in a reasonable and responsible manner or find the root cause of a problem – you can't demonstrate you are compliant with clauses 8.7 and 10.2 on corrective action unless you have records to analyse, but don't go overboard; think about what you might need and conduct a process FMEA, and this might reveal exactly what records you need to keep.

How is this demonstrated?

Demonstrating that the infrastructure has been maintained for the operation of the operational processes may be accomplished by:

a)   presenting evidence of planned, preventive, corrective and predictive maintenance;
b)   presenting evidence of contingency plans and evidence of any validation tests conducted;
c)   presenting evidence of real incidents that made the infrastructure inoperable and the effectiveness of contingency plans;
d)   presenting evidence of product and service nonconformity and showing that none were caused by items of infrastructure that had not been maintained.

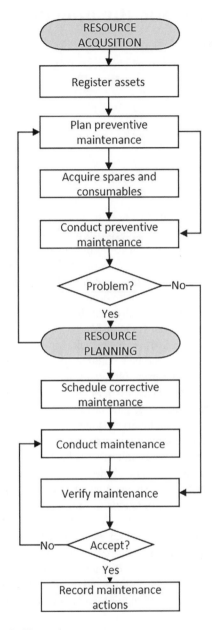

Figure 25.1 Key stages of a facility maintenance process

26 Environment for the operation of processes

Introduction

Prior to the 2000 version of ISO 9001 this requirement was concerned with providing a suitable working environment for process control. In the 2000 version, a separate clause was included requiring the work environment to be managed to ensure conformity to product requirements. The term *work environment* was defined in ISO 9000 as a set of conditions under which work is performed and a note was added advising that these conditions can include physical, social, psychological and environmental factors. However, as this was only advisory, the conditions were limited to physical factors, an assumption confirmed when a note was added in the 2008 version of ISO 9001.

The note which appears in the 2015 version makes it quite clear that a combination of human and physical factors can indeed be a suitable environment in which people work. In addition, the engagement of people principle would indicate that it is intended that several factors should be addressed. These factors would include those that could potentially affect the competence, empowerment and engagement of people and their ability to enhance the organization's capability to create and deliver value.

However, there is reluctance in some quarters to introduce the behavioural aspects of management into the management systems audit process because of the difficulties in obtaining objective evidence in short periods of time allowed for third party audits. Longer periods of observation are often necessary to come to any conclusions about the culture and are therefore omitted from the audit. Despite the limitations of third-party audit, this should not be cause for management to ignore the social and psychological factors, which have been known to influence productivity since the behavioural school of management emerged in the 1920s with Elton Mayo's experiments in industrial research. These experiments concluded that performance of employees is influenced by their surroundings and by the people that they are working with as much as by their own innate abilities and the physical conditions of the workplace.

In this chapter, we examine the three requirements of clause 7.1.4, namely:

- Determining the environment needed
- Providing the environment needed
- Maintaining the environment needed

Note: Evidence presented to demonstrate conformity with 7.1.1 might mean that it is also deemed to satisfy conformity with 7.1.4 but not vice versa because there are other types of resources than the process environment. Also, evidence presented to demonstrate conformity with 7.1.4 might mean that it is also deemed to satisfy conformity with 7.1.1, 4.4.1d, 6.2.2b, 7.1.5.1, 8.1c and 8.3.2e relative to the process environment necessary.

Determining the environment needed (7.1.1 and 7.1.4)

What does this mean?

The operating environment

The environment must be appropriate for the product being produced, for the service being provided and for the equipment being used. The environment must also be appropriate for the people producing the product, providing the service and using the equipment which means that in some cases there are two different environments. There is the environment under which plant, machinery and equipment function and product is produced and service delivered, and there is the environment under which the people engaged in those processes work and these are not necessarily the same. Some processes operate under conditions that are hostile to humans and are only accessible to humans when non-operational or through a protective interface, for example, the product of a nuclear power station is electricity and the environment for its production is hostile, but the environment where the plant engineers work when running the process is expected to be benign.

The processes

The requirement applies to all processes, and as the outputs of all processes within the scope of the QMS affect the outputs of the QMS, the environment in which these outputs are produced is critical to their quality. These processes will include management processes and operations processes, including the associated measurement processes.

Human and physical factors

The standard refers to human and physical factors and then divides these human factors into social and psychological factors implying that physical factors that affect humans are not human factors. There are physical factors that affect products and the non-human element of services (e.g. equipment and facilities) and physical factors that affect human performance (e.g. light level, noise, temperature, chemicals and dust) but there are also social and psychological factors that affect human performance (e.g. fear, isolation, stress and bullying). A search on the term *human factors* produces a variety of definitions (see Box 26.1). In some standards, human factors and ergonomics are used as synonyms, which is not helpful if one's perception of ergonomics is that it's about the interrelationship between humans, the tools and equipment they use in the workplace. This sets human factors apart from those factors of the work environment that are not related to the tools and equipment people use, but arise as a result of the organizational culture and climate in which they work. In the absence of a universal definition of human factors we will simply refer to the separate categories of physical, social and psychological factors.

Box 26.1   Human factors – definitions

Characteristics of a person having an impact on an object under consideration (ISO 9000:2015).

Scientific discipline concerned with the understanding of interactions among human and other elements of a system, and the profession that applies theory, principles, data, and methods to design in order to optimize human well-being and overall system performance (ISO 27500:2016).

Physical or cognitive characteristics, or social behaviour, of a person (ISO 10018:2012)

Environmental, organizational and job factors, and human and individual characteristics, which influence behaviour at work in a way which can affect health and safety.

The study of the interrelationship between humans, the tools and equipment they use in the workplace, and the environment in which they work (Kohn, Corrigan, & Donaldson, 1999).

The study of all the factors that make it easier to do the work in the right way (World Health Organzation).

Although this grouping serves to identify related factors, it is by no means comprehensive or exclusive. Each has an influence on the other to some extent. Relating the identification of such factors to the achievement of conformity of product tends to imply that there are factors of the work environment that do not affect conformity of product. Whether people produce products directly or indirectly, their behaviour affects their actions and decisions and consequently the results of what they do. It is therefore difficult to exclude any factor on the basis that it does not influence the capability of the process to produce conforming outputs in some way or other as it's a question of risk and how that risk is managed.

Why is this necessary?

The work environment is critical to all process outputs, to the product and service and to worker motivation and performance and extends beyond the visible and audible factors commonly observed in a workplace. Noise levels do not need to cause harm for them to be a factor that adversely affects worker performance. Libraries are places of silence simply to provide the best environment in which people can concentrate on reading. No physical harm arises if the silence is broken!

In a report published by the UK Office of National Statistics in 2017, it was estimated that 137.3 million days were lost due to sickness absences in the UK in 2016, down from 178 million days in 1993, a reduction from 7.2 to 4.3 days per worker. Minor illnesses were the most common reason given for sickness absence (34.0 million). Slightly fewer days, (30.8 million) were lost to musculoskeletal conditions, and 15.8 million days were lost due to stress, anxiety and depression. Workers in process plants, machine operations were estimated to be 80% more likely to be off work due to sickness than those in professional occupations. (ONS, 2017). These people were employed to do a job that in their absence is either not done at all or added to another person's work load which may have a detrimental effect on the ability of co-workers to cope and consequently the ability of processes to sustain prior performance levels. There is also an unrecoverable cost to the organization, money that could have been put to better use.

Box 26.2   Absence management survey

Ben Willmott, head of Public Policy at the CIPD, the professional body for HR and people development, commented: This is the fifth year in a row in which 30% or more of employers have reported an increase in employees coming into work when they are ill (known as presenteeism). It's a real concern that the problem of presenteeism is persisting, as we might have expected it to drop during the economic recovery as people tend to feel more secure in their jobs. The problem may well be a hangover from the recession but we need to address the issue of presenteeism head-on. The message to businesses is clear: if you want your workforce to work well, you have to take steps to keep them well and this means putting employee health above operational demands.

(CIPD, 2015)

It is the duty of management to control the physical factors first within the levels required by law, second within the levels necessary to prevent deterioration of the product and third as necessary for people to perform their jobs as efficiently and effectively as possible. It is also the task of management to create conditions in which personnel are motivated to achieve the results for which they are responsible and therefore remove or contain any de-motivating elements such as friction and conflict in the workplace.

How is this addressed?

For a solution, we can use a similar approach to that taken towards the natural environment. Environmental management is the control over activities, products and processes that cause or could cause environmental impacts. The approach taken is based on the management of cause and effect where the activities, products and processes are the causes or aspects, and the resulting effects or potential effects on the environment are the impacts. In the operating environment, the effort should be focused on eliminating negative impact and creating positive or beneficial impacts that also lead to an improvement in performance based on risk. The aspect and impacts register is a risk assessment for the environment. It can also be used to link to legislation and procedures and so facilitate review when there is a change in any one of these factors.

Taking a risk-based approach

We can never be sure of what will unfold as each day commences because the work place is full of uncertainty by virtue of the many variables within it. However, some variables may have little effect so it's important to first identify those processes where the processing environment may be critical to the quality of the process outputs. We can therefore ask three questions:

a)   What ambient conditions are maintained in the workplace regardless of the type of work being performed? (The ambient conditions are those established to comply with the relevant legal requirements for the workplace such as temperature, humidity, cleanliness, noise, lighting, drainage, ventilation and emissions.)

b)   To what extent is the quality of the process output dependent on the control of environmental factors beyond those regulated by ambient conditions?

c)   How likely is it that the environment needed will be controlled by the workforce without having to introduce additional controls?

An initial risk assessment should be undertaken to prioritize action by identifying those processes where further study of the environment is needed. It can be revisited and specifically focused studies undertaken following receipt of adverse reports from process reviews. FMEA can be used to assess risks to products see (AIAG, 2016) and health and safety risk assessment can be used to assess risks to people see (Bateman, 2006).

Where the materials used to create process outputs are sensitive to variation in the environment, consideration needs to be given to undertaking a detailed risk assessment, identifying the risks and determining the likelihood of conditions detrimental to the product arising and stipulating the environment to be created.

Where the workers are sensitive to variation in the environment, consideration needs to be given to undertaking a detailed risk assessment, identifying the risks and determining the likelihood of conditions detrimental to worker productivity arising and stipulating the environment to be created. In addition to health and safety issues there may be factors that cause workers to make mistakes (e.g. long time between breaks, passively observing tests, cold draughts and electromagnetic radiation).

Identifying the physical factors affecting the product in the work environment

To identify the physical factors of the work environment affecting the product requires a study of the product's characteristics to reveal whether they are sensitive to temperature, light, humidity, mishandling, vibration and contamination etc. and whether the environment in which the product or components of it will be produced contains any of these risk factors. Where material degrades on exposure to light or air, the production processes should be designed to provide the protection required when the material exits the process. In addition to visible light, other types of radiation across the whole spectrum may impact the product.

The factors of the operating environment that affect product conformity should be identified through a design FMEA. Basically, you are asking the question: *In what way could variation in a specific factor increase the likelihood of nonconformity?* If variation in any of the factors has to be restrained the environment needs to be controlled.

Identifying the physical factors affecting measurement in the work environment

Clause 7.1.5 on monitoring and measuring resources does not address the environment and therefore clause 7.1.4 also applies to measurement environment. To maintain the integrity of measurement, physical measurements need to be undertaken in an environment which is controlled or known to the extent necessary to ensure valid measurement results. The controlled environment consists of a workspace in which the temperature, the rate of change of temperature, humidity, pressure, dust, cleanliness, electromagnetic interference, access and other factors are controlled. The environments will differ depending on the degree of accuracy required. The environment for calibrating reference devices will be to a standard that may be higher than that for undertaking measurements in production.

Identifying the physical factors affecting people in the work environment

To identify the physical factors requires a study of the anatomical, anthropometric, physiological and biomechanical characteristics as they relate to the work people are required to carry out. These may include space, temperature, noise, position, light, humidity, hazards, hygiene, vibration, pollution, accessibility, physical stress and airflow. These factors may be perceived as primarily affecting safety, but workers who are in physical discomfort doing their job because of its physical demands are at risk of producing nonconforming product or delivering a poor service.

The study of the relationship between a person and his or her job is referred to as *ergonomics* (ISO 26800, 2011) and it deals principally with the relationship between a person, the job, the equipment used and the environment. A job exists within an organization and there are factors that arise from performing the job such as physical movement, the user interface, the physical environment and the psychological factors of the job itself (the psychological factors of the organization are dealt with separately).

Once again you are asking the question in what way could variation in a specific factor increase the likelihood of nonconformity. If variation in any of the factors has to be restrained the environment needs to be controlled. These factors can influence individual behaviour by causing fatigue, distraction, accidents and a series of health problems. There are laws governing many of the physical factors such as noise, air pollution, position, space and safety. There are also laws related to the employment of disabled people that affect the physical environment in terms of access and ergonomics. There are also laws that prohibit vulnerable people being employed in conditions injurious to their health.

The layout of the workplace, the distances involved the areas of reach, seating, frequency and type of movement all affect the performance of the worker. These factors require study to establish the optimum conditions that minimize fatigue, meet the safety standards while increasing productivity.

Where people are an integral part of a mechanized process the man–machine interface is of vital importance and must be carefully considered in process design. The information on display panels should be clear and relevant to the task. The positioning of instruments, and input, output and monitoring devices should allow the operator to easily access information without abnormal movement. The emergency controls should be within easy reach and the operating instructions accessible at the workstation. Legislation and national standards cover many of these aspects and can be downloaded from the HSE website for free.

- Research legislation and associated guidance literature to identify those factors that could exist in the work environment due to the operation of certain processes, use of certain products or equipment. We do X therefore from historical and scientific evidence there will be Y impact. (VDUs, RSI, airborne particles, machinery, etc.).
- Determine the standard for each factor that needs to be maintained to provide the appropriate environment.
- Establish whether the standards previously determined can be achieved by workspace design, by worker control or by management control or whether protection from the environmental impact is needed (protection of ears, eyes, lungs, limbs, torso or skin).
- Establish what could fail that would breach the agreed standard using FMEA or hazards analysis and identify the cause and the effect on worker performance.
- Determine the provisions necessary to eliminate, reduce or control the impact such as finding a solution that reduced or eliminated the need to do the work.

Identifying the psychological factors

To identify the psychological factors requires a study of the cognitive and climatic factors affecting the worker. This means identifying the mental factors that help or prevent people from being in the right frame of mind to perform well.

Cogitative factors are concerned with mental processes, such as perception, memory, reasoning, and motor response. Workers who become unable to cope with their job because of its mental demands or the stress brought on by having to make difficult decisions are at risk of producing nonconforming product or delivering a poor service. It is therefore necessary to study the mental demands of the job and identify the cognitive skills needed to complete a task to the standard required. The only way we can be effective is to know our people and regularly communicate with them. Mark Horstman provides useful guidance in this field in *The Effective Manager* (Horstman 2016).

Box 26.3   Human factors – what to consider

When assessing the role of people in carrying out a task, consider the following:

- Treat operators as if they are normal humans who make mistakes and are generally unable to intervene heroically in emergencies.
- Accept that an operator won't always be present to detect a problem or be able to take appropriate action immediately.
- Accept that people won't always follow procedures.
- How the training provided for operators relates to accident prevention or control.
- Recognize that training will not tackle all slips/lapses effectively.
- Recognize that operators are prone to unintentional failures, may take short cuts and may have differing motivation to management expectations.
- Recognize the human component and discuss human performance in risk assessments.
- Apply techniques appropriately by targeting resources where they will be most effective.
- Highlight the significant issues in your risk assessments.
- Provide soundly based probabilities of human failure in quantitative risk assessment that are traceable to legitimate data sources rather than just stating they are low risks e.g. if there is a chance of electrocution then it will always be a high risk due to death being the more obvious outcome.

Adapted from (HSG48, 1999)

Climatic factors are concerned with a set of conditions to which people react. Unlike culture which is more permanent, climate is temporary and is thought of as a phase the organization passes through. In this context, therefore, the work environment will be affected by a change in the organizational climate. Several external forces cause changes in the climate such as economic factors, political factors and market factors. These can result in feelings of optimism or pessimism, security or insecurity, complacency or anxiety. Employees naturally take the lead from the leader and can easily misread the signals. They can also be led by a manager who does not share the same ethical values and when under threat of dismissal; an otherwise law-abiding citizen can be forced into falsifying evidence.

Once again you are asking the question in what way could variation in a specific factor increase the likelihood of nonconformity. If variation in any of the factors has to be restrained the environment needs to be controlled. The problem with psychological factors is that there is no easy fix (e.g. you can't install a thermostat to regulate the heat in discussions!). There are two solutions one lies in building effective teams and other lies in effective job design.

Managers need to understand and analyse human behaviour and determine the conditions that need to be created in which employees are motivated to achieve the organization's objectives. One would think that people whose job it is to get results through other people would understand human behaviour, but invariably they don't. Not only do they expect extraordinary results from ordinary people, they continually make decisions without any thought as to their effect on the people they expect to implement them and those who will be affected by them. Deming advocated that "a manager of people needs to understand that all people are different and the performance of anyone is largely governed by the system in which they work." (Deming, 1994) A technical approach to management places all the emphasis on the goal and getting the job done, regardless of the human cost. A behavioural approach to management places the emphasis on the interaction between the people so that they are motivated to achieve the goal of their own volition.

There have been many studies on worker motivation, for example, Maslow's hierarchy of needs, Hertzberg's two-factor theory and McClelland's theory of learned needs, and Rollinson compares each of them (Rollinson, 2008). Probably the most popular theory is Frederick Herzberg's two-factor theory which he and his co-workers developed in 1959, and although it has many critics it remains highly popular with managers. Hertzberg theorized that people are motivated by things that make them feel good about their work, which he referred to as motivating factors, and that people have an aversion to things that make them feel bad, which he referred to a hygiene factors because they stop dissatisfaction. However, he didn't believe their removal created satisfaction so motivation factors and hygiene factors were not opposites – they simply had different roles in the work environment. Hygiene factors include such things as pay, conditions, job security, quality of supervision, interpersonal relations and company policies and procedures. Motivation factors include sense of achievement, recognition, responsibility, the nature of work, personal growth and advancement.

Box 26.4   The motivation process objective

A measure of employee satisfaction might be staff turnover and management style may be considered a critical success factor. The output the employee is looking for as evidence that management have adopted an appropriate style is a motivated workforce. Motivation is a result but there is no process that produces motivation. It is an outcome not an output. Instead of expressing the objective of the process as to motivate the workforce, it becomes 'To maintain conditions that sustain worker motivation' and the process thus becomes one of developing, maintaining and refreshing these conditions.

Identifying the social factors of the work environment

To identify the social factors requires a study of the organizational policies, practices and the culture in the workplace. Although the policies may be companywide, workplaces may have different cultures due to the ethnic diversity among co-workers (organization culture is expressed by the values, beliefs, rituals and customs that characterize what it's like to work in an organization). These factors may include social norms, values and behaviours, religion, sexuality, social class, special needs due to disabilities, age, gender, experience, etc.

The way people behave is largely a product of their upbringing, their life experiences and the core assumptions that characterize and define their worldview. The more diverse the work environment, the more behaviours will vary. As people from difference cultural backgrounds will hold different values, these differences will be reveal in the way they react to situations, their situational awareness, the attitude they take to the offer of help or the imposition of rules. The thinking patterns and approach to problem solving will also vary and their readiness to work beyond finishing time to solve an urgent problem will vary as some workers will put family before company. There will also be communication issues and apart from difficulties in understanding what people are saying due to different dialects and accents, there will be differences in interpretation and differences in non-verbal signals. Unless common forms of communication are agreed, a casual remark spoken in jest, as a compliment or a rebuke may turn a calm situation into a quarrel, a fight or a disaster.

It is therefore necessary to study the social interactions in the workplace such as the worker–boss, worker–subordinate, worker–colleague and worker–peer relationship and means of communication such as e-mail, mobile devices and social media to identify influences that may affect worker motivation and harmony in the workplace. As observed by Covey, "you cannot continually improve interdependent systems and processes until you progressively perfect interdependent, interpersonal relationships" (Covey, 1992, p. 267) and you won't have created conditions in which your staff are motivated to do a good job and avoid human error if they are not engaged which is brought about by adopting the people engagement principle see also clause 5.1.1h in Chapter 17.

The square shown in Figure 26.1 provided in "Why Teams Don't Work" (Robbins & Finley, 1998) shows behavioural differences. It puts the behavioural styles identified by Hunsaker and Alessander in Box 24.2 in another form. One of these squares is where others see you as occupying. From left to right it measures assertiveness, from asking to telling, and from top to bottom it measures responsiveness, whether we react in a controlled (top) or emotional (bottom) fashion. This is quite a rough guide because people are far more complex in general.

Knowing which style your co-workers fit enables you to adjust the way you communicate with them, make them feel comfortable and reduce their tension level. Were you to ignore behavioural styles the risk of the relationship breaking down increases with a consequential effect on product and service quality. In fact, it is in the service sector where success depends more on relationships than technical competences that relationship building is key to success.

Once again you are asking the question in what way could variation in a specific factor increase the likelihood of nonconformity. If variation in any of the factors has to be restrained the environment needs to be controlled. The influence of social factors on the

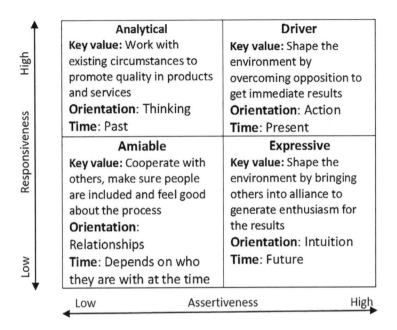

Figure 26.1 Zones of behaviour

work environment is managed by building effective teams, and therefore in any particular process, consideration needs to be given to the type and balance of people you want on the team.

How is this demonstrated?

Demonstrating that the environment necessary for the operation of its processes has been determined may be accomplished by:

a) presenting the results of process risk assessments that support decisions taken on the processing environment needed;

b) presenting evidence of studies that have been performed to identify the physical factors of the processing environment relating to the product that would increase the risk of nonconformity;

c) presenting evidence of studies that have been performed to identify the physical, social and physiological factors of the processing environment relating to the people involved that would increase the risk of nonconformity;

d) presenting evidence of process descriptions which define the processing environment to be maintained;

e) presenting evidence that regular communication take place with the workforce that allow them to bring up their issues so that they know they are being listened to and that you understand.

Providing and maintaining the environment needed (7.1.1 and 7.1.4)

What does this mean?

Providing and maintaining the environment necessary for the operation of the processes simply means creating and maintaining the environment that has been determined as being necessary for the operation of its processes to achieve conformity of products and services. This may not only involve maintaining the status quo but keeping pace with internal and external changes that affect the processing environment.

Why is this necessary?

If the environment determined as necessary for the operation of the processes is not provided and maintained, the risks that were identified may materialize and nonconforming product may be produced and customer satisfaction reduced. It may not be a foregone conclusion unless physical damage to the product or disruption of the service will certainly follow if the stipulated environment is not provided.

How is this addressed?

To provide and maintain the necessary environment managers need to create, monitor and adjust conditions so that the physical and human factors of the work environment act positively towards achievement of the planned results. Some of the factors affecting the work environment are constraints rather than objectives, that is, they exist only because we have an objective to achieve. Noise in the workplace occurs because we need to run machines to produce product that satisfies the customer. If we didn't need to run the machines, there would be no noise. However, some of the constraints are of our own making. If the style of management created an environment that was more conducive to good industrial relations, the work force would be more productive.

Providing the environment needed to protect products

To achieve high performance from electronic components particle and chemical contamination must be minimized during fabrication and assembly. Cleanrooms are often built in which product is manufactured, assembled and tested. To produce food and drugs to the regulatory standards, high levels of cleanliness and hygiene need to be maintained during production and food preparation. For these and many other reasons, the work environment needs to be controlled. If these conditions apply you should:

a)   document the standards that are to be maintained;
b)   prohibit unauthorized personnel from entering the areas;
c)   provide training for staff who are to work in such areas;
d)   provide alarm systems to warn of malfunctions in the environment;
e)   provide procedures for maintaining the equipment to these standards;
f)   maintain records of the conditions as a means of demonstrating that the standards are being achieved.

Providing the environment needed to sustain worker productivity

To provide an environment in which psychological and social factors affecting the worker are taken into account managers need to build effective teams, invest in effective job design and underpin these by a culture in which quality of product and of service is the first priority. Communication is one way that management can improve productivity even if they cannot address the de-motivating elements directly.

One of the problems with motivation theories as that some of them assume all people are alike. Although most people have a hierarchy of needs, they are not all alike and the order of priority is different in different cultures. What are hygiene factors in one culture may be motivating factors in another. Not everyone wants challenging work. Some people only go to work to earn money to finance their leisure activities and are content with their level of development. The important thing here is that they care enough about the product of their labour that they are motivate to do the right things right and that there's nothing in their environment that puts the quality of their work at risk.

- **Temperature** can have a noticeable impact on worker performance, although we each differ in our tolerance to high and low temperatures, differences that can lead to conflict. Although it may not have been necessary to control temperature in certain work areas, the change in climate may make provision of supplementary cooling equipment necessary so that workers remain comfortable and able to maintain their standard of performance.
- **Noise** within the legal limits can be a distraction and therefore lead to mistakes being made. Soundproofing may be required between offices or workshops and the noise generator. Scheduling building maintenance or improvement work when staff are on holiday may lessen the impact of noise on the working environment.
- **Lighting** enables people to see what they are doing and if insufficient can lead to mistakes being made. The light intensity and colour required varies depending on the degree of accuracy and precision that is required but also on the eyesight of employees. Enabling those with impaired vision to access brighter light will reduce their mistakes. Even those with normal eyesight can suffer eye strain and headaches and become irritable which may lead to conflict in the workplace.
- **Air quality** within legal limits can cause problems for asthmatics and those with allergies and any pollutants that cause people undertaking precision work to cough or sneeze may have a detrimental effect on what they are doing. Regulated air-quality monitoring schemes may be applicable.
- **Space** provided to workers should not only meet anatomical standards but also satisfy social needs. People also need a degree of privacy to get on with their job without intrusion from their neighbouring co-workers. Placing workers in too close proximity may induce distraction by tempting them to join discussions not relevant to the job they are doing.
- **Working patterns** may be standardized and lead to frustrations coming to or leaving the workplace all at the same time. Productivity may be increased by staggering the working day, employing flex time or telecommuting. Although these have social benefits and cost benefits, they may remove some of the risks of human error as they can eliminate distractions.
- **Fear** in the workplace is detrimental to quality as people will hide their mistakes. Managers need to create conditions in which their employees are at ease and feel comfortable seeking guidance if in doubt and not afraid of reprisals for inadvertent mistakes or inadvertently deviating from policies.

For more on job design, motivation building effective teams and on building or changing organization culture see the bibliography.

How is this demonstrated?

Demonstrating that the environment necessary for the operation of its processes has been provided may be accomplished by presenting evidence from the analysis of process data that:

a)   show conformity with applicable environmental legislation;
b)   show the extent to which environment factors are having a detrimental effect on product and service conformity.

Bibliography

AIAG. (2016, November). *Potential Failure Mode and Effects Analysis Reference Manual*. Retrieved from LEHIGH University: www.lehigh.edu/~intribos/Resources/SAE_FMEA.pdf
Bateman, M. (2006). *Tolley's Practical Risk Assessment Handbook*. Abingdon, Oxon: Routledge.
CIPD. (2015, October 12). Press Office. Retrieved from Chartered Institute for Personnel Development: https://www.cipd.co.uk/about/media/press/121015-absence-survey
Covey, S. R. (1992). *Principle-Centred Leadership*. London: Simon & Shuster UK Ltd.
Deming, E. W. (1994). *The New Economics for Industry, Government and Education* – Second Edition. Cambridge, MA: The MIT Press.
Horstman, M. (2016). *The Effective Manager*. Hoboken, NJ: John Wiley & Sons.
HSG48. (1999). *Reducing Error and Influencing Behaviour*. London: HSE.
ISO 10018. (2012). *Quality Management: Guidelines on People Involvement and Competence*. Geneva: International Organization of Standardization.
ISO 26800. (2011). *Ergonomics – General Approach, Principles and Concepts*. Geneva: International Organization for Standardization.
ISO 9000:2015. (2015). *Quality Management Systems – Fundamentals and Vocabulary*. Geneva: ISO.
Kohn, L. T., Corrigan, J. M., & Donaldson, M. S. (1999). To Err Is Human: Building a Safer Health System. Institute of Medicine. Retrieved from www.nap.edu/download/9728
ONS. (2017, March 9). *Sickness Absence in the Labour Market*. Retrieved from Office of National Statistics: https://www.ons.gov.uk/employmentandlabourmarket/peopleinwork/labourproductivity/articles/sicknessabsenceinthelabourmarket/2016#main-points
Robbins, H., & Finley, M. (1998). *Why Change Doesn't Work*. London: Orion Business Books.
Rollinson, D. (2008). *Organisational Behaviour and Analysis*. Harlow, Essex: Pearson Education Limited.
World Health Organization. (2009). *What Is Human Factors and Why Is It Important to Patient Safety?* Retrieved from www.who.int/patientsafety/education/curriculum/who_mc_topic-2.pdf

27 Monitoring and measuring resources

Introduction

There have been requirements for measuring resources in ISO 9001 since the first edition in 1987 in clause 4.11 when it was referred to as "inspection, measuring and test equipment". The same title was retained in the 1994 edition. In the 2000 edition, the title was changed to "Control of monitoring and measuring devices and placed under Product Realization in clause 7.6. In the 2008 revision, it was changed to "Control of monitoring and measuring equipment", and now in the 2015 edition it has been changed once again, and this time it's placed under Resources with the title Monitoring and measuring resources as clause 7.5.1. However, if we look closely we'll see that 7.5.1 is now split into two parts. The first (7.1.5.1) addresses monitoring and measurement resources, and the second (7.1.5.2) addresses traceability but refers to measuring equipment.

The change from equipment or devices to resources is not significant but it emphasizes the fact that to carry out measurement we need not only suitable devices but in addition competent people, suitable methods and a suitable environment which collectively constitute the resources. The competence of the people undertaking measurement is addressed in response to clause 7.2 in Chapter 29. The methods of measurement are addressed in response to clause 9.1.1b) in Chapter 52 and the environment is addressed in response to clause 7.1.4 in Chapter 26. It is therefore the measuring devices that are addressed in this chapter.

With the 2015 edition, the clause has been removed from 'Product Realization/Operations' and located under resources so there is no implication it only applies to tangible characteristics. This is true of the first part (7.1.5.1) but the second part (7.1.5.2) is certainly only applicable to tangible characteristics.

In this chapter, we examine the 10 requirements of clause 7.1.5, namely:

* Determining resources for ensuring valid and reliable results (7.1.1 and 7.1.5.1)
* Providing resources for ensuring valid and reliable monitoring and measuring results (7.1.1 & 7.1.5.1)
* Ensuring suitability of selected monitoring and measuring resources (7.1.5.1a)
* Maintaining integrity of monitoring and measuring resources (7.1.5.1b)
* Retaining evidence that monitoring and measurement resources are fit for purpose (7.1.5.1)
* Calibration and verification of measuring equipment (7.1.5.2a)
* Recording the basis for calibration (7.1.5.2a)
* Indicating calibration status (7.1.5.2b)
* Safeguarding monitoring and measuring instruments (7.1.5.2c)
* Determining and addressing the impact of defective instruments (7.1.5.2)

Note: Evidence presented to demonstrate conformity with 4.4.1d. 6.2.2b and 7.1.1 might mean that it is also deemed to satisfy conformity with 7.1.5.1 but not vice versa because there are other types of resources than monitoring and measuring resources. Also, evidence presented to demonstrate conformity with 7.1.5.1 might mean that it is also deemed to satisfy conformity with 7.1.1, 4.4.1d, 6.2.2b, 7.1.5.1, 8.1c and 8.3.2e relative to the monitoring and measuring resources necessary.

Determining and providing resources for ensuring valid and reliable results (7.1.1 and 7.1.5.1)

What does this mean?

There is an important distinction to be made when interpreting this requirement which is implied in the phrase "when monitoring or measuring is used to verify the conformity of products and services". There are monitoring requirements in the standard that are intended to inform decisions other than decisions about conformity of products and service and the requirements of 7.1.5.1 do not apply to those monitoring and measurement activities (e.g. monitoring information about external and internal issues in clause 4.1).

The integrity of products and services depends on the quality of the resources used to create and assess their characteristics. This clause implies there are ways to verify the conformity of products and services other than monitoring or measuring and this may be rooted in the perception that only physical quantities are measurable and that intangibles are not. The quality of a product or service is judged by the extent to which it fulfils our needs and expectations. Some of these needs and expectations can be expressed objectively and can therefore be quantified, for example, when you buy a kilogram of sugar, you expect it to weigh a standard kilogram within the limits specified by law. Weight is an objective quality of the product and is thus a physical quantity. However, when you transact business with your bank you expect the cashier to be courteous but you have no weights and measures to judge the level of courtesy, it's a subjective quality of the service and is thus a non-physical quantity, but it's a quantity none the less. In his book *How to Measure Anything*, Douglas Hubbard writes, "if you can observe a thing at all, you can observe more or less of it" and "If we can observe it in some amount, it must be measurable" (Hubbard, 2010). ISO 9000:2015 defines measurement as "process to determine a value" and in a note to the definition "the value determined is generally the value of a quantity". The term *measurement* is therefore not restricted to physical quantities. It also does not mean certainty, an exact quantity – if it did, few things would be measurable. Many quality characteristics are not measured in SI units but in percentages, and these percentages are often averages based on samples with a confidence level.

Box 27.1   Numbers and quantities

Numbers are the product of counting. Quantities are the product of measurement.
Gregory Bateson 1904–1980, U.S. scientist, philosopher

Monitoring and measuring resources comprises the monitoring or measuring instruments, software, measurement standard, reference material or auxiliary apparatus, the people who monitor and measure product and service characteristics, the methods used and the environment in which measurements are undertaken.

The measuring and monitoring resources should encompass the sensor, the transmitter and the receiver because the purpose of measurement is to take decisions and without receipt of the information no decisions can be taken. Also, you need to be aware that the transmitter and receiver may degrade the accuracy and precision of the measurement.

Why is this necessary?

It is necessary to determine the measuring and monitoring resources needed to ensure valid and reliable results because:

a)   if the device you use to create or measure characteristics is inaccurate, unstable, damaged or in any way defective you won't know if the product or service possesses the required characteristics;

b)   if the people undertaking the measurements are not competent, you won't know if the factors that can affect measurement accuracy and precision were considered

c)   if the method the people are using is not soundly based, you won't know if you can rely on the results

d)   if the environment was not suitable you won't know if you can rely on the results.

You know nothing about an object until you can measure it, but you must measure it accurately and precisely. The devices you use therefore need to be fit for their purpose and be controlled to ensure continued fitness for purpose.

How is this addressed?

Planning

When determining measuring and monitoring devices, you need to identify the characteristic to be measured, the unit of measure, the target value and then choose appropriate measuring or monitoring device. As the type of process outputs, product or service may vary considerably, the range of measuring devices also varies widely. It is relatively easy to identify the measuring and monitoring devices for hardware product and processed material but less easy for services, software and information.

In many cases, suitable devices may be available to do the measuring, but you can't be sure unless a needs analysis has been carried out. A monitoring and measurement needs analysis scans the products to be produced, services to be provided and the processes that will deliver them to identify monitoring and measurement requirements. This is done during product and service design and is an output of the detail design stage (see Table 37.1). The type of information that needs to be assembled is:

a)   Product/service/process identity

b)   Characteristics to be measured

c)   Specification including tolerance

d)   Monitoring or measurement method

e)   Monitoring or measurement devices currently available

A checklist should be created which seeks to establish whether there are any characteristics identified in the assembled information that cannot be measured to the accuracy and precision required with the devices that are currently available. Where no current devices are available a plan for their development or acquisition will be needed.

If a new unit of measure is proposed, a new sensor may be required. The monitoring and measuring devices should be selected during process design and reviewed when there is change to a process. It may be necessary to prepare a monitoring and measurement plan with an associated budget that identifies the resources required and a schedule showing when they will be required. For a project involving new products and services this may be a substantial item of work such as the development of a new measurement capability involving new facilities, equipment and personnel or may be a routine as the purchase of additional devices with no additional staff required.

Devices for measuring physical quantities

Devices for measuring physical quantities are those which measure mass, length (metre), time (second), electric current (ampere), thermodynamic temperature (kelvin), luminous intensity (candela) and the amount of a substance (mole). There are also several units derived from the base units.

There are two categories of equipment that determine the selection of physical equipment: general-purpose and specialized equipment. It should not be necessary to specify all the general-purpose equipment needed to perform basic measurements that should be known to competent personnel. However, you will need to tell them which equipment to use if the measurement requires unusual equipment or special environmental conditions. In such cases the equipment to be used should be specified in the verification procedures. To demonstrate that you selected the appropriate equipment at some later date, you should consider recording the actual equipment used in the record of results. With mechanical equipment, this may not be necessary because wear will be normally detected by periodic calibration well in advance of a problem with the operation of the equipment.

With electronic equipment subject to drift with time or handling, a record of the equipment used will enable you to identify suspect results in the event of the equipment being found to be outside the limits at the next calibration. Away of reducing the effect is to select items of equipment that are several orders of magnitude more accurate than is needed.

Devices for measuring non-physical quantities

There are many measurements that cannot be made using physical equipment. The quality of a service is often wholly dependent on the competence of the person providing it such as with consultancy services, legal services, medical services and the characteristics to be measured are subjective such as empathy, courtesy, trust, responsiveness, appearance. Hubbard remarks "No matter how 'fuzzy' the measurement is, it's still a measurement if it tells you more than you knew before."

The most common non-physical quantity in the field of quality management is customer satisfaction, and the device most commonly used is the customer survey either used directly by an interview with customers or by mail. Many service organizations develop metrics for monitoring service quality relative to the type of service they provide. Some examples are provided in Table 27.1.

Table 27.1 Service quality measures

Service provided	Measures
Laboratory	Turnaround time
	Conformity with requirements
	Calibration accuracy
	Time to respond to complaints
Telephone	Line availability
	Callout response time
	Time to reply to complaints
Distribution	Time to respond to complaints
	Supply delivery time
	Received condition
Water	Time to reply to complaints
	Supply connection time
	Water quality
Data analysis	Report accuracy
	Conformity with requirements
	Time to respond to complaints
Education	Class size
	Percentage of pupils achieving pass grades

How is this demonstrated?

Demonstrating that the resources needed to ensure valid and reliable results have been deter-mined and provided may be accomplished by:

a) presenting evidence of an analysis performed to determine monitoring and measure-ment requirements;
b) presenting plans for the development or acquisition of additional monitoring and measurement devices;
c) selecting a representative sample of devices planned for development or acquisition and presenting evidence that they have been provided as planned.

Ensuring suitability of selected monitoring and measuring resources (7.1.5.1a)

What does this mean?

Monitoring and measurement activities vary depending on the nature of what is to be checked and what accuracy and precision is required. Measuring hygiene levels in a food process-ing plant require different devices, competences, methods and environment than monitoring the flow of traffic through temporary road works. It also means that account is taken of any uncertainty in the measurement system.

Why is this necessary?

When monitoring or measurement is required to be undertaken the devices, the people using them, the methods used and the environment need to be capable of the accuracy and

precision that is required otherwise the results will not be valid or reliable and therefore decisions made using these results will be flawed.

How is this addressed?

The arrangements for ensuring that only resources suitable for the specific type of monitoring and measurement activities are employed will vary depending on the risks. Several risk
factors need to be considered:

a) The inherent capability of the device – this might vary from *not known* to *proven
    capability*.
b) The competency of the person undertaking the measurement – this might vary from
    no known competence to *assessed competence*.
c) The choice of devices available – this might vary from *none* to *an abundance*.

Devices for measuring physical quantities

Where new monitoring or measurement techniques are developed the measurement process
should be treated as a new process and subject to evaluation to verify the device is capable
of producing results with the accuracy and precision required. In some cases, it is left to the
discretion of the competent person responsible for undertaking the measurement. In other
cases, detail planning will be needed to make sure each work station is equipped with the
appropriate tools and measuring devices needed and controls put in place that will ensure
that only the specified resources are used.

  Where it is impractical or too risky to leave the choice of monitoring and measurement
resources to the person undertaking the activity, details may need to be specified in the
product or service specifications, operations plan, process descriptions or procedures. The
process description may provide guidance on the conditions to be maintained and control
plans or procedures may indicate the type of devices to be used.

MEASUREMENT UNCERTAINTY

There is uncertainty in all measurement processes. There are uncertainties attributable to
the measuring equipment being used, the person carrying out the measurements and the
environment in which the measurements are carried out. When you make a measurement
with a calibrated instrument you need to know the specified limits of permissible error (how
close to the true value the measurement is). If you are operating under stable environmental conditions, you can assume that any calibrated equipment will not exceed the limit of
permissible error. Stable conditions exist when all variation is under statistical control. This
means that all variation is due to common causes only and none due to special causes. In
other cases, you will need to estimate the amount of error and take this into account when
making your measurements. Test specifications and drawings, etc., should specify characteristics in true values, that is, values that do not take into account any inherent errors. Your
test and inspection procedures, however, should specify the characteristics to be measured
taking into account all the errors and uncertainties that are attributable to the equipment, the
personnel and the environment when the measurement system is in statistical control. This
can be achieved by tightening the tolerances to be confident that the actual dimensions are
within the specified limits.

VARIATION IN MEASUREMENT PROCESSES

Measurement processes used in production must be in statistical control so that all variation is due to common cause and not special cause variation. It is often assumed that the measurements taken with a calibrated equipment are accurate and indeed they are if we take account of the variation that is present in every measuring system and bring the system under statistical control. Variation in measurement processes arises due to bias, repeatability, reproducibility, stability and linearity (see the Glossary in Appendix B).

It is only possible to supply parts with identical characteristics if the measurement processes as well as the production processes are under statistical control. In an environment in which daily production quantities are in the range of 1,000 to 10,000 units, inaccuracies in the measurement processes that go undetected can have a disastrous impact on customer satisfaction and consequently profits.

ACCURACY AND PRECISION

Accuracy and precision are often perceived as synonyms but they are quite different concepts. Accuracy is the difference between the average of a series of measurements and the true value. Precision is the amount of variation around the average. So, you can have a measuring device that gives a large variation around the true value with repeated measurements but whose average is the true value (see Figure 27.1).

Alternatively, you could have a device which gives small variation with repeated measurements around a value which is wide of the true value. The aim is to obtain both accuracy and precision. The difference in accuracy and precision can cause expensive errors. You should not assume that the result you have obtained is both accurate and precise unless the device has been calibrated immediately prior to use and the results of its accuracy and precision provided.

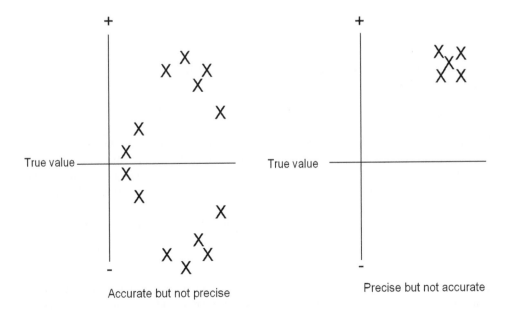

Figure 27.1 Accuracy and precision

Devices for measuring non-physical quantities

Where the quantity being measured is non-physical the device being used to measure it needs to be fit for its purpose. The device may be a technique or method including the manipulation of data collected from observation or analysis.

Service characteristics may include such quantities as courtesy, flexibility, reliability, accuracy, and some of these can be measured simply by using set criteria and observation. The service designer forms a focus group either from members of the organization or potential users of the service and they establish, for example, what a courteous service looks like and what a discourteous service looks like or what constitutes a flexible service and an inflexible service. A series of questions or a checklist is derived from the results of the focus group and a device for measuring a service quality characteristic is born.

Where appropriate, trials need to be carried out to test the reliability of the device and these may be done by testing the device on a sample having known characteristics. Planting the sample with a known quantity and discovering whether the device detects the known quantity.

How is this demonstrated?

Demonstrating that the resources provided are suitable for the specific type of monitoring and measurement activities being undertaken may be accomplished by:

a)   presenting evidence of the policies governing the determination of monitoring and measurement resources;

b)   where the choice of resources is at the discretion of the person undertaking the measurement, presenting evidence of their competence;

c)   where the choice of resources is constrained by prescribed requirements present examples and select a representative sample of monitoring and measurement activities from the examples presented showing that the resources being used are consistent with those requirements.

Maintaining integrity of monitoring and measuring resources (7.1.5.1b)

What does this mean?

Maintaining monitoring and measuring resources to ensure their continuing fitness for purpose means not only that the devices are to be maintained but also maintaining the level of competence in the people undertaking the measurements, the methods used and the environment in which the measurements and monitoring are undertaken. As stated in the introduction to this chapter the people, methods and environment are addressed elsewhere in the book, so we will only address the maintenance of the devices used in this chapter.

Why is this necessary?

If the measurements of product and service are to have any meaning, they must be performed in a manner that provides results of integrity – results others inside and outside the organization can respect and rely on as being accurate and precise. If the integrity of measurement is challenged and the organization cannot demonstrate the validity of the measurements, the quality of the product remains suspect.

How is this addressed?

Physical monitoring or measurement device may become unfit for its purpose when it loses its integrity, that is, it no longer functions, it's damaged, degraded, its power source dies or its verification status can no longer be verified. Non-physical devises such as surveys may become unfit for their purpose when the nature of what they are monitoring changes to an extent that invalidates the method of measurement.

Equipment maintenance schedules

In addition to equipment which requires calibration, maintenance registers should be kept for other equipment used for monitoring and measurement to track preventive and corrective maintenance. There should be a maintenance schedule for each item of equipment. Such schedules may include periodic checks for damage, deterioration, wear and also cleaning, and replenishment of consumables where necessary.

Monitoring equipment

Equipment that is used to monitor process parameters may be exposed to severe conditions if located in the natural environment or close to plant and machinery that gets dirty, dusty and greasy. The gauges, sensors, transmission lines and connectors need periodic checks for signs of damage or deterioration to preserve their integrity.

Comparative references

Comparative references are devices that are used to verify that an item has the same properties as the reference. They may take the form of colour charts or materials such as chemicals which are used in spectrographic analysers or those used in tests for the presence of certain compounds in a mixture or they could be materials with certain finishes, textures, etc. Certificates should be produced and retained for such reference material so that their validity is known to those who will use them. Materials that degrade over time should be dated and given a use-by date. Care should be taken to avoid cross-contamination and any degradation due to sunlight (as can happen with colour charts). Such devices should be protected and stored in conditions that preserve their integrity. A specification for each reference material should be prepared so that its properties can be verified.

Other comparative references are those which have form or function where the criteria are either pass or fail, that is, there is no room for error or where the magnitude of the errors does not need to be taken into account during usage. Verification of such devices includes checks for damage, loss of components, function, etc.

Software

Although software does not degrade or wear out, it can be corrupted such that it no longer does the job it was intended to. Any bugs in software have always been there or were introduced when it was last modified. Software therefore needs to be checked prior to use and after any modifications have been carried out, so you cannot predetermine the interval of such checks. In many cases software malfunction will be apparent by the absence of any result at all, but in some cases, a spurious result may be generated that appears to the observer as correct. Re-confirmation is necessary therefore after a period where the equipment may have been used in situations where intended or unintended changes to the configuration could have been made.

Indicators

Some equipment may be used solely as an indicator such as a thermometer, a clock, a tachometer or steel tapes, Accuracy may not be an issue but they need to be kept in good condition so they remain fit for purpose. Steel rules, tapes and other indicators of length should be checked periodically for wear and damage and although accuracy of greater than 1 mm is not normally expected, the loss of material from the end of a rule may result in inaccuracies that affect product quality.

Maintaining devices that measure non-physical quantities

Devices which measure non-physical quantities such as surveys can remain a valid means of monitoring while the questions or criteria remain relevant, for example, if a customer survey tool is used to monitor customer satisfaction and it's based on the current product and service offerings in the current market, it won't remain fit for purpose if those parameters on which it is based change. Therefore, any such devices need to be regularly reviewed to ensure they remain relevant.

How is this demonstrated?

Demonstrating that resources provided are maintained to ensure their continuing fitness for purpose may be accomplished by:

a)   presenting evidence of preventive maintenance plans;
b)   selecting a representative sample of physical monitoring and measuring devices and confirming:

   i   that the selected devices have designated maintenance schedules;
   ii   that there is evidence that preventive maintenance is being carried out on these selected devices.

c)   searching through the nonconformity database and finding no evidence that the cause of a nonconformity was confirmed to be defective monitoring and measurement equipment;
d)   presenting process, system and service KPIs and plans for their monitoring and measurement;
e)   selecting a representative sample of devices used for measuring non-physical quantities and confirming that they have been maintained in line with changes in what they are measuring.

Retaining evidence that monitoring and measurement resources are fit for purpose (7.1.5.1)

What does this mean?

Appropriate documented information is the record of the results of calibration and verification indicating the accuracy or integrity of the device prior to any adjustment and records after adjustment. These records apply not only to equipment designed and produced by the organization but also those owned by the organization and those owned by employees and customers when being used for product acceptance.

Why is this necessary?

It is important to record the results of calibration and verification to determine whether the device was inside the prescribed limits when last used. It also permits trends to be monitored and the degree of drift to be predicted. Calibration records are also required to notify the customer if suspect product or material has been shipped.

How is this addressed?

Physical devices

Calibration and verification records are records of activities that have taken place. Records should be maintained not only for proprietary devices but also for devices you have produced and devices owned by customers and employees.

These records should include where appropriate:

- The precise identity of the device being calibrated or verified (type, name, serial number, configuration if it provides for various optional features).
- The modification status if relevant (applies to specially designed test equipment and gauges).
- The name and location of the owner or custodian.
- The date on which calibration or verification was performed.
- Reference to the calibration or verification procedure, its number and revision status;
- The condition of the device on receipt.
- The results of the calibration or verification in terms of readings before adjustment and readings after adjustment for each designated parameter (e.g. any out-of-specification readings).
- An impact assessment of any out of specification conditions.
- The date fixed for the next calibration or verification.
- The permissible limits of error.
- The serial numbers of the standards used to calibrate the device.
- The environmental conditions prevailing at the time of calibration.
- A statement of measurement uncertainty (accuracy and precision).
- Details of any adjustments, servicing, repairs and modifications carried out.
- The name of the person performing the calibration or verification.
- Details of any limitation on its use.
- Notification to the customer if suspect product has been shipped.

Clearly not all this information would be presented on one record but the records should be indexed so that all this information is traceable both forwards and backwards. For example, the record containing the results of an assessment of out of specification conditions should carry a reference to the related calibration record and vice versa.

Devices for measuring non-physical quantities

Evidence from focus groups, trials and experiments that serves to validate the integrity of the measuring device should be retained.

How is this demonstrated?

Demonstrating that evidence of fitness for purpose of the monitoring and measurement resources has been retained may be accomplished by:

a)   presenting a register of physical monitoring and measurement devices;
b)   selecting a representative sample of these devices and retrieving the records which testify their integrity or calibration status;
c)   presenting a register of non-physical monitoring and measurement devices;
d)   selecting a representative sample of these devices and presenting evidence of validation.

Calibration and verification of measuring equipment (7.1.5.2a)

What does this mean?

Measurement traceability means that there is an unbroken chain of measurements from the measuring instrument being used back to international or national measurement standards (Figure 27.2). It is a requirement when stated in a contract, regulation or the organization's own policies. Measuring equipment are devices used for measuring a physical quantity such as mass, length and time and their derivatives.

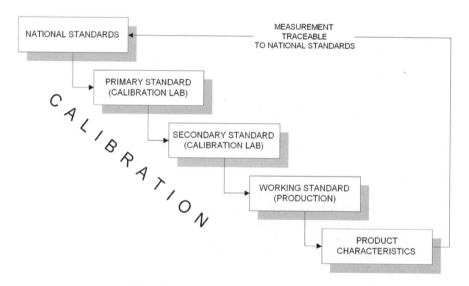

Figure 27.2 Traceability of standards

Box 27.2   Revised requirement on traceability

Criteria is now included for assessing when calibration of measuring instruments is applicable, implying that it is not applicable to all measurements.

In a measurement system, the physical signal is compared with a reference signal of known quantity. The reference signal is derived from measures of known quantity by a process called calibration. The known quantities are based on standards that in most cases are agreed internationally.

Calibration applies to all measuring instruments used for providing evidence of conformity at any stage throughout operations and not just at the product release stage.

There are two systems used for maintaining the accuracy and integrity of measuring equipment – a calibration system and a verification system. Calibration is a process of comparing a physical signal with a reference signal of known quantity whereas verification is establishing the correctness of a quantity. Certain variables might be calibrated such as length or capacitance but attributes might be verified such as form and function. (The presence of a substance is verified not calibrated). Therefore, depending on the equipment being examined, calibration, verification or both may be necessary. The terms *calibration* and *verification* are not mutually exclusive.

Why and when is this necessary?

Calibration is necessary to reduce the uncertainty of repeatable and accurate measurements to an acceptable level. Measurement traceability may be a statutory or regulatory requirement, a stakeholder expectation or where it is perceived to be a commercial advantage for customers to be able to reproduce the same results themselves on verifying the characteristics of products provided.

How is this addressed?

Calibration

Calibration is concerned with determining the values of the errors of a measuring instrument and often involves its adjustment or scale graduation to the required accuracy (see "Accuracy and precision"). You should not assume that just because equipment was once accurate it would remain so forever. Some equipment if well treated and retained in a controlled environment will retain their accuracy for very long periods. Others if poorly treated and subjected to environmental extremes will lose their accuracy very quickly. Ideally, measuring equipment should be calibrated before and after use (a) to prevent inaccurate equipment being used and (b) to confirm that no changes have occurred during use. However, this is often not practical and so intervals of calibration are established which are set at such periods as will detect any adverse deterioration. These intervals should be varied with the nature of the equipment, the conditions of use and the seriousness of the consequences should it produce incorrect results.

Variations can arise in measurements taken in different locations due to the measuring equipment not being calibrated to the same standards as other equipment. With the introduction of the SI system of units, this variation could be eliminated provided the quantity used to calibrate the measuring equipment was traceable to national or international standards.

> **Note: Use – not function – determines the need for calibration.**

Further guidance on the calibration process can be found in ISO 10012.

How is this demonstrated?

Demonstrating that measuring equipment is calibrated or verified against traceable measurement standards when measurement traceability is a requirement may be accomplished by:

a) presenting evidence that measurement traceability is a stakeholder requirement;
b) presenting a description of the process by which measurement equipment is calibrated or verified and in particular:

   i   the criteria for determining which equipment requires calibration or verification;
   ii the criteria for the determination of calibration intervals;
   iii how measurement traceability is assured.

c) selecting a representative sample of equipment being used for measurement purposes and providing evidence that its calibration is controlled by this process;
d) retrieving records for the equipment sampled which provide evidence to trace the validity of measurement results to national or international standards of measurement.

Recording the basis for calibration (7.1.5.2a)

What does this mean?

For physical and chemical measurements that are based on the fundamental units of measure (metre, kilogram, second, ampere etc.) there are national or international standards but for other measures no national or international standard may exist against which to calibrate a measuring instrument. Each industry has developed a series of measures by which the quality of its products and services are measured and has accordingly developed standards that represent agreed definitions of the measures.

Why is this necessary?

Without a sound basis for comparison, the effort of measurement is wasted.

How is this addressed?

A sound basis for calibration can be devised by bringing together a group of experts within the organization or its associated trade association who have established by investigation, experimentation and debate what constitutes the standard and that a device or number of devices has been produced which can be used to compare the product or result with the standard using visual, quantitative or other means. The device may be a physical instrument but could be information such as a set of agreed criteria and the method of measurement.

Where you devise original solutions to the measurement of characteristics, the theory and development of the method should be documented and retained as evidence of the validity of the measurement method. Any new measurement methods should be proven by rigorous experiment to detect the measurement uncertainty and cumulative effect of the errors in each measurement process. The samples used for proving the method should also be retained to provide a means of repeating the measurements should it prove necessary.

How is this demonstrated?

Demonstrating the basis used for calibration or verification may be accomplished by:

a) identifying those standards of measurement being used for which there are no national or international standards;

b) presenting evidence of the process by which these standards were validated;

c) selecting a representative sample of these standards and providing evidence that they have been validated in accordance with this process.

Indicating calibration status (7.1.5.2b)

What does this mean?

Calibration status is the position of a measuring device relative to the period between calibrations. If the date when calibration is due is in the future, the device can be considered calibrated – if the current date is beyond the date when calibration is due, the device is not necessarily inaccurate but remains suspect until verified. However, devices can also be suspect if dropped or damaged even when the date of calibration is due is in the future. The requirement only applies to physical devices subject to wear, drift or variation with use or time.

Why is this necessary?

While a robust calibration system should ensure no invalid measurement instruments are in use, system failures are a possibility. As the consequences of failure are greater than the effort involved in checking the validity of devices before use, it is prudent to provide a means for checking calibration status.

How is this addressed?

All measurement instruments subject to calibration need to display an identification label that either directly or through traceable records, indicates the authority responsible for calibrating the device and the date when the next calibration is due. Measuring equipment should indicate its calibration status to any potential user. Measuring instruments too small for calibration status labels showing the due date may be given other types of approved identification. It is not mandatory that users identify the due date solely from the instrument itself, but they must be able to determine that the instrument has been calibrated. Serial numbers alone do not do this unless placed within a specially designed label that indicates that the item has been calibrated or you can fix special labels that show a circular calendar marked to show the due date. If you do use serial numbers or special labels, then they need to be traceable to calibration records that indicate the calibration due date.

Devices used only for indication purposes or for diagnostic purposes should also display an identity that clearly distinguishes them as not being subject to calibration. If devices are taken out of use for prolonged periods, it may be more practical to cease calibration and provide a means of avoiding their inadvertent use with labels indicating that the calibration is not being maintained. You may wish to use devices that do not fulfil their specification either because part of the device is unserviceable or because you were unable to perform a full calibration. In such cases, you should provide clear indication to the user of the limitation of such devices.

How is this demonstrated?

Demonstrating that the status of measuring equipment is identified may be accomplished by:

a) presenting evidence of the process by which the status of calibrated or verified measuring equipment is identified;

b) selecting a representative sample of measuring equipment and providing evidence that their calibration or verification status has been clearly identified in accordance with the process requirements.

Safeguarding monitoring and measuring instruments (7.1.5.2c)

What does this mean?

Each measuring and monitoring device has a range within which accuracy and precision remain stable – use the device outside this range and the readings are suspect. Once a device has been calibrated or verified, safeguards need to be in place to prevent unauthorized or inadvertent adjustment, avoidable damage or deterioration.

Why is this necessary?

Measuring instruments are very sensitive to vibration, contamination, shock and tampering and thus it is necessary to protect them to preserve their integrity during use, maintenance and storage.

How is this addressed?

Safeguarding against tampering

To provide adequate safeguards against any deliberate or inadvertent adjustment to measuring devices, there should be evidence that seals have been applied to the adjustable parts or where appropriate to the fixings securing the container. The seals should be designed so that tampering will destroy them. Such safeguards may not be necessary for all devices. Certain devices are designed to be adjusted by the user without needing external reference standards, for example, zero adjustments on micrometers. If the container can be sealed, then you don't need to protect all the adjustable parts inside.
   Your procedures will need to specify:

•   those verification areas that have restricted access and how you control access;
•   the methods used for applying integrity seals to equipment;

- the authority permitted to apply and break the seals;
- the action to be taken if the seals are found to be broken either during use or during calibration.

Safeguarding against damage

To provide adequate safeguards in place against damage and deterioration, measuring devices should always be stored in the special containers provided by the manufacturer when not in use. Handling instructions should be readily available for instruments that may be fragile or prone to inadvertent damage by careless handling. Instruments prone to surface deterioration during use and exposure to the atmosphere should be protected and moisture absorbent or resistant materials used.

When transporting measuring instruments adequate protection needs to be provided. Should you employ itinerant service engineers, ensure that the instruments they carry are adequately protected as well as calibrated.

How is this demonstrated?

Demonstrating that adequate safeguards are in place to prevent invalidation of calibration status and subsequent measurement results may be accomplished by:

a) presenting evidence of the process by which the integrity of calibration status is safeguarded;
b) selecting a representative sample of measuring equipment and providing evidence that:

   i   the devices used to prevent inadvertent adjustment or tampering are intact;
   ii warning notices are displayed when applicable and the designated protection devices are being used;
   iii the equipment is undamaged;
   iv it is stored in the protective containers when not in use.

Determining and addressing the impact of defective instruments (7.1.5.2)

What does this mean?

This is perhaps the most difficult of requirements to meet for some organizations. It is not always possible or practical to be able to trace product to the particular instruments used to determine its acceptability. The requirements apply not only to your working standards but also to your calibration standards. When you send calibration standards away for calibration and they are subsequently found to be inaccurate, you may need a method of tracing the devices they were used to calibrate. A calibration standard that is found inaccurate within limits for a specified measurement may not be inaccurate for the range of measurement for which it is being used. Action would be needed only if the inaccuracies rendered the results obtained from previous use to be inaccurate.

Why is this necessary?

If a measurement has been taken with an instrument that is subsequently found inaccurate, the validity of the measurement is suspected and therefore an assessment is needed to establish the consequences. In most cases the instrument used is accurate to an order of magnitude greater than that required, therefore, if found outside tolerance, it may not mean that the product measured is nonconforming. However, if measurements are taken at the extreme of instrument accuracy, the product may well be nonconforming if the device is found to be inaccurate.

How is this addressed?

Risk mitigation

To reduce the impact of defective measuring instruments, you can select devices that are several orders of magnitude more accurate than your needs so that when the devices drift outside the tolerances, they are still well within the accuracy you require. There remains a risk that the device may be wildly inaccurate due to damage or malfunction. In such cases, you need to adopt the discipline of recalibrating devices that have been dropped or are otherwise suspect before further use.

Contingency planning

There may still be a degree of uncertainty as to whether measuring equipment is fit for use and therefore adequate provisions need to be made for dealing with the consequences of defective measuring instruments. Accordingly, you would need the means of:

- tracing the products that have been accepted using the defective instrument;
- determining the significance of the errors on intended use;
- locating the products affected ensuring appropriate action is taken;
- reducing the impact of defective measuring instruments.

You need to carefully determine your policy in this area paying attention to what you are claiming to achieve. You will need a procedure for informing the custodians of unserviceable measuring devices and one for enabling the custodians to track down the products verified using the unserviceable device and assess the magnitude of the problem. You will need a means of ranking problems in order of severity so that you can resolve the minor problems at the working level and ensure that significant problems are brought to the attention of the management for resolution. It would be irresponsible for a junior technician to recall six months' production from customers and distributors based on a report from the calibration laboratory without an assessment of significance being carried out.

Traceability

Showing that the type and serial number of the measuring instruments used to conduct measurements is recorded is a first step, but you will also need to record the actual measurements made. Some results may be made in the form of ticks or pass or fail and not by recording actual readings. In these cases, you will have a problem in determining whether the amount by which the equipment is out of specification would be sufficient to reject the product. In extreme circumstances, if the product is no longer in the factory or in service delivery, this situation could result in product recall from your customer or distributor, or in the case of services a local or national alert (e.g. if the quality of food or medicines are suspect).

Correction

You need to assess what would have happened if you had used serviceable equipment to carry out the measurements. Would the product have been reworked, repaired, scrapped or the requirement merely waived. If you suspect previously shipped product to be noncon-forming and now you have discovered that the measurements on which their acceptance was based were inaccurate, you certainly need to notify your customer if known. In your report to your customer, state the precise amount by which the product is outside specification so that the customer can decide whether to return the product – remember the product specification is but an interpretation of what constitutes fitness for use. Out of 'spec' doesn't mean unsafe, unusable, unsaleable, etc., but national or international regulations may make it illegal and so take precedence.

How is this demonstrated?

Demonstrating that appropriate action has been taken if the validity of previous measure-ment results has been adversely affected may be accomplished by:

a)   presenting evidence of the process that is initiated when measuring equipment is found to be unfit for its intended purpose;
b)   retrieving incident logs, selecting a representative sample of incidents and presenting evidence that the prescribed process was followed.

Bibliography

Hubbard, D. W. (2010). *How to Measure Anything: Finding the Values of Intangibles in Business*. Hoboken, NJ: John Wiley.

28 Organizational knowledge

Introduction

In Chapter 5 of Management Challenges in the 21st Century on the subject of Knowledge - Worker Productivity, Peter Drucker writes: "The most important contribution of management in the 20th century was its 50-fold increase in the productivity of the manual worker in manufacturing. The most important contribution management need to make in the 21st century is similarly to increase the contribution of knowledge work and the knowledge worker" (Drucker, 1999). Organizational knowledge is therefore of prime importance to sustain them in the years ahead. It is a resource, like data, information and understanding, but unlike other resources:

- Using knowledge does not consume it.
- Transferring knowledge does not result in losing it.
- Knowledge is abundant, but the ability to use it is scarce.
- Much of an organization's valuable knowledge walks out of the door at the end of the day which makes knowledge rather paradoxical in nature.

Knowledge management emerged as a scientific discipline in the early 1990s and it has been defined in many ways, one of the simplest being *the process of capturing, developing, sharing, and effectively using organizational knowledge*. Although there is no explicit requirement for knowledge management in ISO 9001, by this definition of it, the requirements of the standard are consistent with it. However, as Michael Polanyi wrote in *The Tacit Dimension*, in 1967 we should start from the fact that *we can know more than we can tell* and therefore what is uncertain about the requirements in clause 7.1.6 is whether the authors believed all knowledge can be captured explicitly or whether they believed that the knowledge they require to be determined, maintained, acquired and accessed can be both embodied in media and embedded in people. Sharing knowledge becomes as much about sharing people as sharing media – it's simply that the knowledge carrier is different and must be treated differently.

In this chapter, we examine the four requirements of clause 7.1.6, namely:

- Determining organizational knowledge
- Maintaining organizational knowledge
- Availability of organizational knowledge
- Acquisition of and access to additional knowledge

Note: Evidence presented to demonstrate conformity with 4.4.1d. 6.2.2b and 7.1.1 might mean that it is also deemed to satisfy conformity with 7.1.5.1 but not vice versa because there are types of resources other than the organizational knowledge.

Determining organizational knowledge (7.1.6)

What does this mean?

Knowledge

Knowledge is a step in the continuum from data to wisdom as shown in Box 28.1. From the ISO dataset, knowledge is cognizance which is based on reasoning (ISO/TS 19150–1:2012).

Box 28.1   From data to wisdom via knowledge

Data are symbols that represent the properties of objective and events.

Information consists of processed data with the processing being directed at increasing its usefulness. It answers questions with such words as *who, what, where, when* and *how many.*

Knowledge results from analysing the relationship between various pieces of information and is conveyed by answers to "how-to" questions. It allows us to describe things. Knowledge proceeds from parts to the whole.

Understanding results from synthesizing information and is conveyed by answers to "why" questions. It allows us to explain things. Understanding proceeds from the whole to the parts.

Wisdom is the ability to increase effectiveness. It deals with values and involves the exercise of judgement.

One can know something without understanding it but one cannot understand something without knowing it.

(Ackoff, 1999)

An organization may possess tons of information but without knowledge of how to use it, its worthless, it simply remains unused or used clumsily. Therefore, determining the knowledge necessary for the operation of processes means determining the pieces of information that are needed to develop, resource, operate and manage the processes, knowing where and from who to get the information, knowing when it's needed, knowing whether it's the right information and of the right quality and knowing what to do with it but it also means recognizing what we don't know.

The theory of knowledge (*epistemology*) is concerned with the question of what it is to know something. The philosopher, Ray Billington suggests that "to the non-philosophical minded this may be perceived to be spending precious time on the pursuit of the obvious". He writes,

> Surely I either know something or I don't. If I enter a general knowledge quiz, I can either answer the questions or I cannot. (I may have forgotten the answer but that is just a matter of recall). The significant point about questions asked in a quiz is that their answers can be found in an encyclopaedia or dictionary. But there remain many issues about which people claim to have knowledge which cannot be verified in this way.
>
> (Billington, 1988)

Billington asks us to consider the following statements:

a)   I know that California is on the west coast of the United States.
b)   I know that water at sea level boils at 100 degrees Centigrade.
c)   I know my wife loves me.
d)   I know my dead uncle still counsels me whenever I have problems.

Clearly there are two categories of statements here, and that the status of the verb "to know" differs in each. Billington asks "how and on what basis does one test the accuracy of one's assertions?" The first two statements can be verified as true but the latter two are statements based on individual perception.

   Bertrand Russell in his book *The Problems of Philosophy* writes "Is there any knowledge in the world which is so certain that no reasonable man could doubt it? In our daily lives, we assume as certain many things that on closer scrutiny are contradictory." He describes how a table differs in the way it appears to an observer as distinct from its reality, the difference, between what things seem to be and what they are (Russell, 1998). We believe others see the same things we see. Obviously if several people are looking at an object, no two people will see the object from exactly the same point of view as some will look at it from a different angle. Its colour will appear different to each person; it will appear different to the touch and to the ears when each taps it. The surface will appear different when looked at by the naked eye as opposed to the view seen through a microscope. This is the knowledge of *things*. But there is another type of knowledge, the knowledge of *truths*. An assertion, a statement, a proposition ort indeed a set of requirements will be interpreted differently depending on how a person looks at it, making it difficult to reach consensus on a universal truth.

   When it comes to knowledge needed for the operation of the organization's processes, there will be facts but there will also be PHOG (*Perceptions* such as the influence of a particular policy, *Hearsay* such as the boss says it's true so it must be true, *Opinion* such as the judgement of an expert witness and *Guesswork* such as the likelihood of something happening or not happening).[1]

   Separating the facts from the PHOG is a matter of belief, truth and justification.

Organizational knowledge

The term *organizational knowledge* is not defined in 9000 but in Note 1 to the requirement it is "knowledge specific to the organization and is gained by experience. It is information that is used and shared to achieve the organization's objectives."

   If we interpret these statements as conditional, rather than individually exclusive the definition contains contradictions.

a)   If the knowledge has to be specific to the organization, it rules out any knowledge that is common to other organizations, technologies, industries, alliances or cultures. However, among the notes it includes external sources which are not necessarily specific to the organization.
b)   If information is used to achieve the organization's objectives, it can't be the only information that is used because not all information needed will be specific to the organization.

   The intent must surely be that the organization possesses the knowledge necessary for the operation of its processes whatever its source or form, and therefore the definition is flawed. It might be more sensible to define organizational knowledge as Alan Frost does: "all the knowledge resources within an organization that can be realistically tapped by that organization. It can therefore reside in individuals and groups, or exist at the organizational level" (Frost, 2016).

Why is this necessary?

Much of an organization's valuable knowledge walks out the door at the end of the day and therefore, despite the volume of information that may have been accumulated since the organization was founded, is only that which helps the people left behind to continue the work that is useful. Building this knowledge bank is therefore critical to an organization's success which is why it's necessary to determine the knowledge necessary for the operation of its processes.

How is this addressed?

First, it's necessary to appreciate that there are different types of knowledge.

Explicit knowledge

Explicit knowledge is the knowledge captured or codified in a tangible form – in what ISO 9000 refers to as *documented information*. This is knowledge that is easy to identify, store, retrieve, transmit and change. However, most of what is in *documented information* is not knowledge but remains as information (see Box 28.1).

Tacit knowledge

Tacit knowledge is "knowledge that inhabits the mind of the individual" (Polanyi, 2009). It's know-how, know-what, know when it's the right time, know-who and know where to – knowledge gleaned from experience that is difficult to articulate. In fact, the more difficult knowledge is to articulate, the more valuable it is. Because of this, tacit knowledge is often context dependent and personal in nature. It is hard to communicate and deeply rooted in action, commitment, and involvement (Nonaka, 1994). At the highest level, tacit knowledge is also *know-why*; it's the knowledge that creates understanding.

A comparison between tacit and explicit knowledge taken from *Knowledge Management in Theory and Practice* is given in Table 28.1.

Table 28.1 Comparing tacit and explicit knowledge (Dalkir, 2011)

Properties of tacit knowledge	Properties of explicit knowledge
Ability to adapt, to deal with new and exceptional situations	Ability to disseminate, to reproduce, to access and to reapply throughout the organization
Expertise, know-how, know-why, and care-why	Ability to teach, to train
Ability to collaborate, to share a vision, to transmit a culture	Ability to organize, to systematize, to translate a vision into a mission statement into operational guidelines
Coaching and mentoring to transfer experiential knowledge on a one-to-one, face-to-face basis	Transfer knowledge via products, services, and documented processes

Implicit knowledge

Implicit knowledge is the knowledge that is not made explicit but could be. In some texts this is deemed to be tacit knowledge, but this view implies that all tacit knowledge can be made explicit which leaves no room for the knowledge that can only be shared by experience.

Embedded knowledge

Embedded knowledge is locked in the organization's processes, products, services, culture, routines, artefacts, or structures. Some of the knowledge may have been implicit knowledge that left with the individuals when they departed the organization. It may be characterized by the expression *we have always done it this way*. Over time, an organization learns what works and what doesn't work and this gets embedded in the way it operates.

Individual knowledge

The knowledge possessed by an individual will be a combination of the earlier points. Some of it will have been gleaned from reading (explicit knowledge), some of it by learning on the job (tacit and implicit knowledge), some of it by socializing in the workplace (embedded knowledge).

Determining knowledge needed for the operation of processes

There is no requirement for a knowledge determination process or that such a process be documented but there are requirements for determining processes and for planning the QMS and operations and organizational knowledge is the result of these activities.

The knowledge will be partially contained in the documented information produced to support the operation of the processes, that is, the instructions, procedures, standards, etc., that are to be used by those engaged in the process. It will also be contained in the minds of the people undertaking the actions and making decisions required by the instructions and procedures, etc., so that by stipulating people of a certain competence, you determine the knowledge you want them to bring to the process (e.g. you require a receptionist for a hotel booking process). You provide the explicit knowledge through the procedures that must be followed, but you also provide knowledge through training and coaching and rely on the tacit knowledge the person possesses to determine how to handle the awkward customer and to make the right tactical decisions such as what technique to employ, when to escalate a problem, what to disclose or not to disclose to the customer.

When planning to achieve an objective, its intuitive to ask: *What do I need to know?* To get the answer it is likely that you will first access your memory and retrieve all you think is relevant. The paradigm you hold will drive you in a certain direction in search of the knowledge you want. You'll then gather together the explicit knowledge you are aware of and if necessary form a group of people each possessing knowledge you need to work out how to achieve the objective and what you need to know. These people may not share your paradigm and so bring in different and possibly opposing knowledge. In bringing these people together you have determined the type of knowledge you need. They share their knowledge; you consult the explicit knowledge and in so doing you can prepare a plan of action which conveys knowledge to those who will implement the plan. They are assigned to achieve the objective and retrieve the plan which determines the explicit knowledge they need. The rest is up to them using their tacit knowledge to repeat the process.

When planning a process, product or service you'll encounter situations where the knowledge you need is not contained in the organization and either you search for explicit knowledge or you search for someone with the appropriate tacit knowledge (i.e. the *know-how*).

How is this demonstrated?

Demonstrating that the knowledge necessary for the operation of the organization's processes has been determined may be accomplished by:

a) presenting the documented information supporting the processes needed for the QMS which conveys the explicit knowledge which those engaged in the process need to have in their possession;

b) presenting the documented information that defines the competences required of those engaged in the processes who can convey the tacit knowledge.

Maintaining organizational knowledge (7.1.6)

What does this mean?

Maintaining organizational knowledge means keeping the organizational knowledge in the organization, keeping it up to date, relevant and useful to accommodate staff turnover and organizational development. It also means allowing some existing knowledge to become obsolete as the needs for it declines. As knowledge is explicit, implicit, tacit and embedded, it means not only keeping the documented knowledge current but also the tacit knowledge that inhabits the minds of the people up to date relevant and useful.

Why is this necessary?

Within any organization there will be change and it is therefore necessary to maintain the level of knowledge in the organization when people join and people leave, when systems become obsolete and when new systems are brought on stream so that performance is not adversely affected.

How is this addressed?

Imagine that there is a pool of all the knowledge the organization needs which is being drained of knowledge that is no longer required and supplemented by additional knowledge that is required. Maintaining organizational knowledge is therefore accomplished through several different processes.

Maintaining knowledge through document control

The explicit knowledge will be captured in documented information and therefore whenever new knowledge is acquired that can be captured explicitly. The change controls should be activated to keep the documents up to date.

When products, services and processes are no longer being maintained the documented information associated with them is often archived, thus removing explicit knowledge from the knowledge pool. This obviously does not remove knowledge in people's heads. A different process is needed for doing that.

Maintaining knowledge through process review

Periodically, each process should be reviewed not only to establish that it is performing as required, but also that its objectives, methods and resources remain relevant. One of the resources is the knowledge required. It's important the people know of the current and past performance of the process; therefore, maintenance of these records is necessary so that those concerned have an accurate picture. It is also necessary to review the knowledge required to operate and manage the process by asking, do we need to know X any more or have the lessons learnt from recent problems been conveyed to all those engaged in this process? As a result of the review consideration may be given to archiving certain documents and updating others.

Maintaining knowledge through project review

On completion of a project, time should be set aside to review the whole project with those engaged on it, assessing each stage and capturing what worked well and what could be done better. It should not be a process for apportioning blame for things that went wrong but a process to learn from successes as well as failures. Concentrating on how things go right tends to encourage rightness. Measures of rightness always encourage better behaviour as there is no longer fear of a blame culture as happens when we focus on failure.

Maintaining knowledge during staff changes

If employees possess specialist knowledge crucial to the organization's operations, provisions need to be made to either keep those individuals in the employment of the organization or to enable them to transfer their knowledge before they leave. This presents a risk that needs to be managed. We never know when someone might leave so we must assume they will at some point and take precautions.

When staff leave the organization, they will take with them everything they have learnt. Trying to capture this knowledge after they have indicated their intention to leave is too late as it's unlikely they will be cooperative therefore you should attempt to capture implicit knowledge while staff are fully engaged and exhibit no signs of wanting to leave.

Staff movement can cause particular problems with legacy systems (the systems that are outdated or of a previous version). Customers are likely to be using systems that are no longer in current production but nonetheless their contracts place obligations on the producer to maintain a certain capability to provide customer support. There may be distinct advantages of retaining knowledge of sound architectures not only to support customers that are still using them, but also to improve more modern architectures.

If staff simply move position in the same organization, there may be opportunities to capture the knowledge they will no longer be using in their new position. When new staff join, they bring knowledge into the organization and some of it may be contradictory but you won't know until a situation arises which prompts their knowledge to be revealed. So, there's not much you can about it. It's important that the induction process has provision for making new staff aware of the explicit knowledge they need and for embedding tacit knowledge. The former is a relatively simple matter of pointing the person in the right direction to find the information they need, but the latter will require a very different process. A common approach is on-the-job training but formal training may be required followed by competence assessment.

How is this demonstrated?

Demonstrating that organizational knowledge is maintained to the extent necessary may be accomplished by:

a)   presenting evidence of a document control process that provides for documented information to be promptly updated following a change in a process;

b)   presenting evidence of a process for reviewing the continued relevance of process knowledge and provision for its update;

c)   presenting evidence of a process for capturing trends in process performance;

d)   presenting evidence of a recruitment process and showing how process knowledge is maintained when staff leave and when new staff are engaged.

Availability of organizational knowledge (7.1.6)

What does this mean?

Making organizational knowledge available to the extent necessary means enabling those with a need to know to access the information. Therefore, if a person needs to know certain information to discharge their responsibilities effectively, that information should be made available to them.

Why is this necessary?

A person cannot be held accountable for results without being in possession of all the essential information they need to produce those results.

How is this addressed?

Making organizational knowledge available is accomplished in two ways:

a)   by capturing and distributing knowledge in the form of documented information (explicit knowledge) to those having a need for it and enabling them to access it;

b)   by placing those with the knowledge among those who haven't and putting them under an obligation to part with that knowledge when it's prudent to do so.

Organizations can either leave knowledge in the heads of those who discover it or convert as much as possible into explicit knowledge so others can access it in the form of documented information (e.g. in guides, methods, codes of practice, training materials, procedures, case studies). The development of this information passes through a PDCA cycle which after planning, it's used and then the results of its use are studied to see what lessons can be learnt. Knowledge is improved and the documented information updated and disseminated to those who need to know. These studies may take the form of formal reviews to learn about what went wrong and how it can be improvement plus what went right and what needs to be reinforced.

Knowledge is power and some people guard their knowledge as if they were protecting the crown jewels, perceiving themselves as indispensable. They are under an illusion that were they to part with their knowledge, their influence in the organization would decline, so they hang on to it. What they seem to be unaware of is that they acquired the knowledge while in the employment of the organization and therefore they don't have exclusive rights to it. One solution is to make the person a mentor so that they retain their recognition and status as an expert but are rewarded based on the extent to which their knowledge has been transferred and continuity maintained.

Those individuals who have acquired specialist knowledge that can't be captured explicitly need to be identified and the information added to their individual competence records.

With documented knowledge, particularly when stored in digital form, it can be relatively easy to search for the information you are looking for. However, with tacit knowledge, by its very nature it's not documented or stored in digital form so access to it is more difficult. You must create conditions in which after an incident, the perpetrator cannot legitimately claim *I didn't know who to ask*. In every organization, there is the hidden network, the grape vine, in which word gets about as to *who knows what*. Each person creates a personal contact list which is often better than any formal mechanism. However, adding specialist knowledge to competence records and making these records searchable by managers makes the knowledge accessible.

How is this demonstrated?

Demonstrating that organizational knowledge has been made available to the extent necessary may be accomplished by:

a)   presenting evidence of a process for disseminating explicit knowledge of:

   i   how to undertake activities and make decisions;
   ii   lessons learnt relative to a person's job.

b)   selecting a representative sample of people engaged in a process and showing that everyone can access the explicit knowledge that has been made available;
c)   presenting evidence of a process for disseminating tacit knowledge to those who need it;
d)   selecting a representative sample of people engaged in a process and showing that everyone has access to the people whose knowledge they need that has been made available (e.g. asking the name of the person they would contact in the event of a problem they couldn't solve).

Acquisition of and access to additional knowledge (7.1.6)

What does this mean?

As knowledge is a resource organizations use to fulfil their goals, it follows that as needs and trends change the current knowledge may be found insufficient and additional knowledge may be required to continue to fulfil organizational goals.

Why is this necessary?

An organization's capability is built on its acquisition and application of knowledge. Change can arise at any level in the organization and threaten its capability. If the organization is resistant to change, it won't seek to extend its knowledge and as a result will eventually lag behind its competitors and slowly cease to exist.

How is this addressed?

In preparing any plan to achieve an objective for control or improvement, an assessment of current knowledge needs to be made to identify additional knowledge that is needed and provisions made for its acquisition or access.

This knowledge may be accessed from outside the organization using such sources as libraries, trade associations, universities or other legitimate sources or may be acquired through hiring people who possess the required knowledge or by creating it from internally funded research. Additional planning may therefore be necessary to work out how the search for new knowledge will be carried out.

When necessary depending on the significance of the new knowledge required, the search for knowledge may be a simple action in the record of a review meeting or it may warrant the designation of a special project with a budget and resources, for example,

- The search for knowledge about ISO 9001 may be satisfied with a phone call.
- The search for knowledge about a specific market may be satisfied with literature research.
- The search of knowledge about success and failure may be satisfied with a diagnostic team.
- The search for knowledge about customer preferences may be satisfied with a focus group.
- The search for knowledge about a material lighter than paper but as strong as steel would probably be a multi-million-dollar project.

There are countless ways of searching for knowledge, and at some point you will need to decide whether it's more economical to hire the knowledge, buy the book and teach yourself or if you can't find it in people or books, undertake the research and discover the knowledge by scientific or other means.

How is this demonstrated?

Demonstrating how the organization had acquired or accessed the necessary additional knowledge may be accomplished by:

a) presenting evidence of a process for acquiring or accessing the necessary additional knowledge;
b) selecting a representative sample of changes in the internal or external environment and presenting the results of the analyses which established the need for additional knowledge;
c) presenting plans for acquiring or accessing the planned additional knowledge.

Note

1 Contributed by the late Tony Brown.

Bibliography

Ackoff, R. M. (1999). *Ackoff's Best: His Classic Writings on Management: Chapter 1*. New York: John Wiley & Sons, Inc.

Billington, R. (1988). *Living Philosophy: An Introduction to Modern Thought*. New York: Routledge.

Dalkir, K. (2011). *Knowledge Management in Theory and Practice* – Second Edition. Cambridge, MA: The MIT Press.

Drucker, P. F. (1999). *Management Challenges for the 21st Century*. Oxford: Butterworth Heinemann.

Frost, A. (2016, July 28). Introducing Organizational Knowledge. Retrieved from KMT (Knowledge Management Tools): www.knowledge-management-tools.net/introducing-organizational-knowledge.html

Nonaka, I. (1994). A Dynamic Theory of Organizational Knowledge Creation. *Organizational Science*, 5(1), 14–37.

Polanyi, M. (2009). *The Tacit Dimension*. Chicago: University of Chicago Press.

Russell, B. (1998). *The Problems of Philosophy*. Oxford: Oxford University Press.

29 Competence

Introduction

When seeking resources to satisfy the organization's needs managers tend to look for those that can be put into use without the need for additional investment. One wouldn't expect to buy a new piece of equipment and have to rebuild and modify it before it was fit for use. But people are often acquired based on their potential. They can be developed and continually redeveloped to suit the organization's needs although, as with most things, there are limits. People possess natural traits, some of which are not possible to change. Until the 2000 version of ISO 9001, the standard did not address competence. Although personnel were required to be qualified based on appropriate education, training, the organization was required only to identify training needs and provide training. Competence is a much broader concept, and is associated with what people can do rather than what courses they have taken.

Box 29.1   Drucker on competence

To take on tasks for which one lacks competence is irresponsible behaviour. It is also cruel. It raises expectations which will then be disappointed.

(Drucker, 1974)

In this chapter, we examine the five requirements of clause 7.2, namely:

- Determining competence (7.2a)
- Assessing competence (7.2b)
- Developing competence (7.2c)
- Evaluating effectiveness of actions taken (7.2c)
- Retaining evidence of competence (7.2d)

Determining competence (7.2a)

What does this mean?

Applicability of requirement

The requirement makes competence a condition for those personnel whose work affects the performance and effectiveness of the QMS. Competence therefore applies to any work that either directly or indirectly affects the organizations ability to consistently provide products and services that meet customer and applicable statutory and regulatory requirements and enhance customer satisfaction (see Box 29.1). Certainly, anyone who encounters customers (potential or otherwise) is in the frame, and those in the back office who interface with them or are involved in defining and providing the organizations products and services. It therefore includes employees (directors, managers, professional and non-professional staff) as well as personnel engaged under contract and those undertaking outsourced processes (see Chapter 42).

Box 29.2 Revised requirement on competence

The competence requirement in the 2015 version is no different in principle than in the 2008 version but their applicability has been extended. In the 2008 version the competence requirement applied to personnel performing work affecting conformity to product requirements which with a narrow interpretation limited its applicability to those people producing product or delivering services. Now this has changed to person(s) doing work under its control that affects the performance and effectiveness of the quality management system which, considering the purpose of the QMS, will apply to almost everyone in the organization as well as those engaged by the organization whether remunerated or not.

The definition of competence has also changed. A person is no longer deemed competent solely on the basis of appropriate education, training and skills, but on what they can achieve.

Although this may appear an onerous requirement, users are expected to take a risk-based approach as is the case with all requirements in ISO 9001. This means determining the potential impact on the conformity of products and services and enhancing customer satisfaction of not applying competence requirements in certain cases.

Defining competence

Competence is the "ability to apply knowledge and skills to achieve intended results" (ISO 9000:2015). Competence is therefore not a probability of success in the execution of one's job; it is a real and demonstrated capability. A person may claim to have certain ability or attributes but proof of competence is only demonstrated if the desired results are achieved.

Although the opposite of competence is incompetence, this is perceived as a pejorative term, so we tend to use the phrase *not yet competent* to describe those who have not reached the required standard of competence. If for some reason a competent person became incapacitated, he or he would no longer be deemed competent to perform the job they were performing prior to the incapacity.

Determining competence

Determining the competence necessary means determining the results or outcomes required of a job, position, task or role, the performance criteria or standards to be achieved, the evidence required and the method of obtaining it.

Difference between competence and qualification

There is a note to the definition of competence in ISO 9000:2015 which states that "demonstrated competence is sometimes referred to as qualification". The term *qualification* is used to denote the completion of a course or training programme which confers the status of a recognized practitioner of a profession or activity (OED, 2013); therefore, if a person has the appropriate education, training and skills to perform a job, the person can be considered qualified. In some regulations, particularly those concerned with health and safety, there are situations that require a qualified person and situations that require a competent person and the two are not considered equivalent. The competent person might not possess the qualifications of the qualified person and the qualified person may not possess the attributes to be competent. As illustrated in Box 29.3, a competent person would be one who was aware of the consequences of their actions.

Box 29.3   Patient competency in a hospital

A patient demonstrates mental competency if they can show appropriate understanding, insight and judgement regarding the decision they make and the immediate consequences that may result.

Education is concerned with the acquisition of knowledge; training is concerned with the acquisition of skills to perform a task and putting the two together in real situations rather than in the classroom produces experience. If a person demonstrates the ability to achieve the desired results, the person can be considered competent. Qualified personnel may not be able to deliver the desired results under the prevailing conditions. Although they may have the knowledge and skill and can present their certificates, they may exhibit inappropriate behaviours (the expert who knows everything but whose interpersonal skills cause friction with staff to such an extent that it adversely affects productivity!). A competent person is therefore one that demonstrates appropriate behaviours. A competent person may not be able to demonstrate academic achievements or formal training and therefore cannot be a substitute in situations that require a qualified person but has somehow acquired the knowledge, the skills and the behaviours to get the job done although can't explain how he or she does it.

Intended results or outputs and outcomes

In any organization there are positions that people occupy, jobs they carry out and roles they perform in these positions. For each of these, certain outputs and outcomes are required from the occupants. The starting point is therefore to define the outcomes required from a job and then define what makes those performing it successful and agree these with the role or jobholder. Having set these standards for both outputs and outcomes, they become the basis for competence assessment and competence development. The education and training provided should be consistent with enabling the individual concerned to achieve the agreed standards.

The intended results include the expected outputs and outcomes. The outputs are *put out* of the process, they are what the process is intended to produce. These can be controlled directly by the process. The outcomes *come out* of the process, they arise as a consequence of the output and the way it is produced such as behaviours, influence and stamina. The

outcomes are also dependent on the conditions or context in which the role or job is performed. These cannot be controlled directly and often only emerge long after the output has
been released. Two examples might help to clarify this point:

• A manager of a large enterprise may produce the same outcomes as the manager of
a small enterprise but under entirely different conditions. It follows therefore that a
competent manager in one context may not be competent in another.
• A person who occupies the position of production manager in a glass factory performs
the role of a manager but also performs several different jobs concerned with the production of glass and therefore may need to be competent to negotiate with suppliers,
use computers, produce process specifications, blow glass, drive trucks, test chemicals,
administer first aid, etc., depending on the scale of the operations being managed.

Each job comprises several tasks that are required to deliver a particular result. Having a
forklift truck driving certificate is not a measure of competence. All this does is to prove
that the person can drive a forklift truck. What the organization may need is someone who
can move four tonnes of glass from point A to point B safely and efficiently using a forklift
truck. The management role may occupy 100% of a person's time and therefore the number of competencies needed is less than those who perform many different types of jobs in
addition to that of management. For this reason, there is no standard set of competencies for
any particular position because each will vary but there are national schemes for assessing
competence relative to specific occupations.

Influence of the system

Competence is the ability to meet standards that apply in the particular job not in a classroom or examination but an ability to perform in the real working environment with all the
associated variations, pressures, relationships and conflicts. However, this is where we run
into difficulties because we can't hold a person accountable for results that are beyond their
ability to control and we can't ignore the influence of the *system* in which a person functions.
The constraints imposed by the prevailing culture, management style, policies and access to
resources can affect the ability of individuals to achieve intended results. Some people are
better at achieving results under difficult circumstances than others. Some will get results
regardless of the long-term consequences and it is these who may get rewarded and others
may perform best when the negative influences in the work environment are minimal. If a
person depends on the cooperation of certain individuals to deliver the required outputs,
and those individuals withdraw their cooperation because of a petty squabble over a parking space in the company parking lot, that person becomes no longer competent in the eyes
of the manager who is waiting for that output. This is but one example, and there will be
countless examples where a person becomes unable to deliver the required outputs due to
influences of the system in which they work i.e. common cause variation (see Chapter 52).

Why is this necessary?

Creating the line of sight from goals to results

The jobs that people perform must be related to the organization's objectives and as these
objectives are achieved through processes, these jobs must contribute to achievement of the

process objectives. In the decomposition from the system level where the business processes are identified through to work processes and sub-processes you will arrive at a level where the results are produced by a single person or group of people doing a particular job. It is these results on which the organization depends for its success. Mistakes at this level can affect performance at every level all the way up to the business goals.

Producing role clarity

Establishing competencies for all roles in the business helps to focus the minds of the managers and employees on what are the important criteria for the performance of a particular role. This significantly helps with recruitment and development and eliminates doubt as to what contribution a person makes to the organization.

Competence and the professions

Competence is particularly important in the professions because the outputs result from an intellectual process rather than an industrial process. We put our trust in professionals and expect them to be competent but methods of setting standards of competence and their evaluation have only been developed over the last 15 years. It was believed that education, training and experience were enough, but incidents of malpractice particularly in the medical profession have caused the various health authorities to look again at clinical competence. The financial crisis of 2008 has triggered a similar review and attempts are still being made eight years on to change the culture in the financial sector.

The assumptions we make about a person's ability

When assigning responsibility to people, we often expect that they will determine what is needed to produce a good result and perform the job right first time. We are also often disappointed. Sometimes it is our fault because we did not adequately explain what we wanted or more likely, we failed to select a person that was competent to do the job. We naturally assumed that because the person had a college degree, had been trained in the job and had spent the last two years in the post, that they would be competent. But we would be mistaken, primarily because we had not determined the necessary competence for the job and assessed whether the person had reached that level of competence. In theory, we should select only personnel who are competent to do a job but in practice we select the personnel we have available and compensate for their weaknesses either by close supervision or by providing the means to detect and correct their failures. A principal benefit of teams is that they compensate for individual weaknesses. A competent team may include team members who are not yet competent individually but with the support of others within the team they can be highly effective.

How is this addressed?

Different approaches to competence determination

The competence-based movement developed in the 1960s from work undertaken by David McClelland an American Psychologist, out of a demand from businesses for greater accountability and more effective means of measuring and managing performance. This led to research into what makes people effective and what constitutes a competent worker. Two distinct approaches emerged which made a clear distinction between competences and competencies. Competence and competences were concerned with effect and output and therefore described what people need to do to perform a job well, whereas competency and competencies were concerned with effort and input and described the behaviour that lies behind competent performance they therefore described what people bring to the job. (CIPD, 2016)

Two approaches to competence-based systems have emerged. The British model focuses on standards of occupational performance and the American model focused on competency development. In the UK, the standards reflect the outcomes of workplace performance. In the United States, the standards reflect the personal attributes of individuals who have been recognized as excellent performers (Fletcher, 2000) but what individuals achieved in the past is not necessarily an indication of what they achieve in the future; they age, they forget, their eyesight deteriorates and they may not be as agile both physically and mentally as they once were.

When competence needs to be determined

Requirements for new competencies arise in several ways as a result of the following:

- New job specifications.
- New process specifications, maintenance specifications, operating instructions etc.
- Development plans for introducing new technologies.
- Project plans for introducing new equipment, services, operations, etc.
- Marketing plans for launching into new markets, new countries, new products and services.
- Contracts where the customer will only permit trained personnel to operate customer-owned equipment.
- Corporate plans covering new legislation, sales, marketing, quality management etc.
- An analysis of nonconformities, customer complaints and other problems.
- Developing design skills, problem-solving skills, risk management skills or statistical skills.
- Business improvement through gradual change, step change or transformation.

An approach to determining competence

Determining the competence necessary for performing a job is a matter of determining the results required of a job, the principal tasks the individual is expected to perform and the performance criteria or standards to be achieved. It is important that the individuals whose performance is to be assessed are involved in the setting of these standards for the following reasons:

- To determine the results required, ask: What must be achieved by this process?
- To determine the principal tasks individuals are expected to perform to achieve these results, ask" What must be done for this to be achieved? These are the units of competence.
- To determine the performance criteria including the expected behaviours the individual is to exhibit, ask: How well must this be achieved?

The jobs that people perform must be related to the organization's objectives and as these objectives are achieved through processes, these jobs must contribute to achievement of the process objectives.

Several units of competence will be necessary to achieve a given outcome. For example, a front-line operator's primary output is conforming product. The operator needs to possess several competences for conforming product to be produced consistently (see Box 29.4).

Box 29.4   Units of competence – examples

Several units of competence will be necessary to achieve a given outcome therefore an operator might need the ability to:

- understand and interpret technical specifications,
- set up equipment and create safe working conditions,
- decide when the work is ready to start,
- operate the equipment so as to produce the required output,
- determine when the equipment is malfunctioning and stop the process,
- determine when to measure characteristics,
- select the correct measuring instruments,
- undertake accurate measurements,
- apply variation theory to the identification of problems,
- apply problem-solving methods to maintain control of the process,
- communicate problems and when to seek their resolution,
- decide when and when not to adjust the process,
- determine that the output meets the specification,
- document the information required,
- determine when other people should know the work has been completed,
- apply quality, health, safety, security and environmental policies.

Simply possessing the ability to operate a machine is therefore not a mark of competence.

How is this demonstrated?

Demonstrating that the necessary competence of people doing work under its control that affects the performance and effectiveness of the quality management system has been determined may be accomplished by:

a) presenting evidence of a process for identifying the outputs required and the outcomes expected (these indicate what must be achieved or the key results for which a person is responsible);

b)   presenting information that defines the principal tasks the person is expected to perform to achieve these results (these are the units of competence);

c)   presenting information that defined how well these tasks must be carried out, including the expected behaviours the individual is required to exhibit (these are the performance criteria);

d)   presenting evidence that the individuals whose competency is to be assessed were involved in setting these standards.

Assessing competence (7.2b)

What does this mean?

The requirement is specific as to how competence is to be judged and creates an anomaly. The definition of competence refers to the ability to apply knowledge and skills to achieve intended results and the requirement refers to appropriate education, training, or experience and makes no mention of results which is probably an oversight. Appropriate education, training, or experience relates to how a person acquired the knowledge and skills to achieve intended results, and the inclusion of *or* implies the requirement can be satisfied by people who are competent based on experience; therefore, they can achieve the intended results based on their experience alone.

Although ISO 9001 does not require vocational qualifications, the implication of this requirement for competence is that managers will need to select personnel based on their ability to deliver the results required. Selecting a person simply because of the qualification they hold or that they are a member of the same club would not be appropriate unless, of course, they can also deliver the required results!

The current method of selection may be based on past performance, but without performance standards in place and a sound basis for measurement, this method is not capable of delivering competent people to the workplace.

Why is this necessary?

Traditionally, personnel have been selected based on certificated evidence of qualifications, training and experience rather than achievement of results, although in selection interviews a person's track record will normally be examined. Here are some examples showing the inadequacy of this method which the selection process should address:

*   A person may have received training but not have had the opportunity to apply the knowledge and skills required for the job.
*   A person may have practiced the acquired skills but not reached a level of proficiency to work unsupervised.
*   A person may possess the knowledge and skills required for a job but may be temporarily or permanently incapacitated (a professional footballer with a broken leg is not competent to play the game until his leg has healed and he is back on top form).
*   A person may have qualified as a chemist 30 years ago but not applied the knowledge since.

- Airline pilots who spend years flying one type of aircraft will require some period in the flight simulator before flying another type because they are not deemed competent until they have demonstrated competency.
- A person may have been competent in maintaining particular air traffic control equipment but has not had occasion to apply the skills in the last 12 months. In this industry, the engineers need to demonstrate competence before being assigned to maintain a particular piece of equipment because flight safety is at risk.

The above examples illustrate why qualifications, training certificates and years of experience are not necessarily adequate proof of competence because, above all competence is judged by the ability to achieve results. Also, two people with the same qualifications for a job can be performing to very different standards. In addition, it should be noted that a person may be competent to make decisions in one sphere of life but not another.

How is this addressed?

This may appear an onerous requirement; however, users are intended to take a risk-based approach as is the case with all requirements in ISO 9001. The competence of those whose work has the most effect on the performance, and effectiveness of the QMS should be assessed but the competence of others whose work has immeasurable effect would not.

In addition to performance criteria several methods may be used to collect evidence of competence. The evidence should be against the unit of competence not against each performance criterion because it is competence to deliver the specified outcome that is required not an ability to produce discrete items of evidence.

Establishing soundly base methods of competence assessment

It is important that the competence assessment methods are soundly based and to do this we need to determine whether the proposed method is:

a) based on the use of explicit statements of performance;
b) focused on the assessment of outputs or outcomes of performance (i.e. the results);
c) independent of any specified learning programme;
d) based on a requirement in which evidence of performance is collected from observation and questioning of actual performance as the main assessment method;
e) one which provides individualized assessment;
f) one which contains clear guidance to auditors regarding the quality of evidence to be collected;
g) one which contains clear guidelines and procedures for assuring the quality of the results.

Key questions to ask in assessing competence

Table 29.1 shows the key questions to be asked, the terms used and one example. It should be noted that there may be several performance criteria and a range of methods used to collect

Table 29.1 Competence-based-assessment

Key question	What this is called?	Example
What must be achieved?	Outcome	Conforming product
What must be done for this to be achieved?	Unit of competence	The ability to apply variation theory to the identification of problems (this is one of several)
How well must this be done?	Performance criteria	Distinguishes special cause problems from common cause problems (this is one of several)
How should assessment be conducted?	Assessment method	Observation of performance
What evidence should be collected?	Evidence requirement	Run charts indicating upper and lower control limits with action taken only on special causes (this is one of several)

the evidence. It should also be noted that the evidence should be against the unit of competence not against each performance criteria because it is competence to deliver the specified outcome that is required not an ability to produce discrete items of evidence.

Box 29.5   Key features of competence-based assessment

- Focus on outcomes,
- Individualized assessment,
- No percentage rating,
- No comparison with other individual's results,
- All standards must be met,
- Ongoing process,
- Only "competent" or "not yet competent" judgements made.

(Fletcher, 2000)

Competence assessment method

To operate a competence-based approach to staff selection, development and assessment, it is necessary to set standards, collect evidence of competence and match evidence to standards.
   Several questions arise when considering the collection and assessment of evidence.

- What do we want to assess?
- Why do we want to assess it?
- Who will perform the assessment?
- How will we ensure the integrity of the assessment?
- What evidence should be collected?
- Where will the evidence come from?
- How much evidence will be needed?
- When should the assessment commence?

- Where should the assessment take place?
- How will we conduct the assessment?
- How will we record and report the findings?

How is this demonstrated?

Demonstrating that persons doing work under its control are competent based on appropriate education, training, or experience may be accomplished by:

a) presenting evidence that managers have access to relevant competency records and can select people for specific assignments based on their competence;

b) selecting a process and identifying the person(s) responsible for producing the outputs;

c) presenting evidence that competence requirements have been determined for the person(s) producing those outputs;

d) presenting evidence that a competence assessment process is employed to validate competence;

e) presenting evidence that the method of assessment requires:

   i   set criteria for the required performance;
   ii the collection of evidence of competence;
   iii matching evidence to standards;
   iv a development plan for areas in which a "not yet competent" decision has been made.

f) presenting evidence that this process:

   i.   will reveal any physical and mental changes that may impair a person's competence;
   ii. will find out what people care about because it is this that conditions their behaviour.

g) presenting evidence that the competence of selected individuals has been assessed and found satisfactory by the designated authority (i.e. the person the organization has delegated the authority to conduct competence assessment);

h) presenting evidence that periodic reviews of competence records are undertaken to identify personnel development needs.

Developing competence (7.2c)

What does this mean?

Box 29.6   Revised requirement on competence development

In the 2008 version there were no notes qualifying the types of actions referred to, and so the actions could have been limited to development of the individuals who lack competence. Now the added notes open up a range of different actions beyond training and development of individuals and relate it to the actions of the organizations.

Having identified the competence needs and assessed competence, this requirement addresses the competence gap. If clause 7.2c is read in isolation the competence referred to could be that of people or that of the organization as a whole as the heading to 7.2 is simply Competence. However, it's doubtful that this was the intent so we can assume that the actions to be taken are those taken to acquire person(s) with the necessary competence.

Why is this necessary?

When it is found that staff are not yet competent in some aspect of their work it is indicative that they are not performing as expected and consequently the organization's performance is likely to be adversely affected, if not immediately it will be in the long term. Often other staff will compensate for any weaknesses, sometimes to the extent that the weaknesses are obscured but this is unproductive as it draws resources away from productive work. Action is essential to avoid a deterioration in standards and adverse impact on customers.

Box 29.7   Levels of competence

The beginner is an unconscious incompetent – he doesn't know what he needs to know to do the job.

The learner is a conscious incompetent – he knows what he needs to know but has not yet acquired the knowledge or the skill.

The professional is a conscious competent – he has acquired the skills and knowledge and can do the job well but may have to be consciously aware of what he is doing from time to time.

The master is an unconscious competent – he can do the job well without thinking about it.

How is this addressed?

Range of actions

When it is found that people lack the necessary competences to deliver the results required by the process in which they are engaged, the organization has a choice whether to:

- Develop the education, skills and proficiency of the personnel concerned through internal or external courses.
- Provide mentoring where a more senior person acts as a point of contact to give guidance and support.
- Provide coaching on the job where a more experienced person transfers knowledge and skill.
- Job rotation where a person is temporarily moved into a complementary job to gain experience.
- Provide special assignments where a person is given a project that provides new experiences.

- Provide action learning where a group of individuals work on their own but share advice with others and assist in solving each other's problems.
- Arrange on the job learning where the individual explores new theories and matches these with organizational experience.
- Re-assign currently employed persons to a different job.
- Hire competent personnel.
- Subcontract the work to an organization that employs people with the required competences.
- Redesign the process so it can be executed by current staff applying their existing competences.

Whatever decision is taken to acquire the necessary competence it needs to be documented and a plan for implementing the decision prepared, agreed with those it affects, implemented and records of actions taken retained. Subsequently this plan is used in evaluating the effectiveness of the actions taken.

How is this demonstrated?

Demonstrating that actions have been taken to acquire the necessary competence may be accomplished by presenting the plan of action and records of the actions taken.

Evaluating effectiveness of actions taken (7.2c)

What does this mean?

Whatever action has been taken to acquire the necessary competence the effectiveness of the action in delivering the required competence is required to be evaluated.

Why is this necessary?

Where the action taken was for the individual to take a course of education training, the mere delivery of education or training is not proof that it has been effective. Many people attend school only to leave without gaining an education. Some may pass the examinations but are not educated because they are often unable to apply their knowledge in a practical way except to prescribed examples. The same applies with training. A person may attend a training course and pass the course examination but may not have acquired the necessary proficiency – hence the necessity to evaluate the effectiveness of the actions taken.

How is this addressed?

Acquired competence is assessed from observed performance and behaviours in the workplace, not from an examination of education and training programmes far removed from the workplace. There are three parts to the evaluation:

- An evaluation of the performance activity before development.
- An evaluation of the performance immediately on completion of development activity.
- An evaluation of the development activity within weeks of its completion.

Development activity evaluation (the initial stage)

Activity evaluation by the students themselves can only indicate how much they felt motivated by the event. It is not effective in evaluating what has been learnt. This is more likely to be revealed by examination at the end of the event or periodically throughout the development period. However, the type of examination is important in measuring the effectiveness of the personnel development, for example, a written examination for a practical course may test the theories behind the skills but not the personal mastery of the skills themselves. A person may fail an examination by not having read the question correctly, so examination by itself cannot be a valid measure of training effectiveness. You need to examine the course yourself before sending your staff on it. If you want information to be conveyed to your staff, a lecture with accompanying slide show may suffice. Slide shows are good for creating awareness but not for skill training. Skills cannot be acquired by any other means than by doing.

Development activity effectiveness – short term (the intermediate stage)

We often think of training as a course away from work. We go on training courses. But the most effective training is performed on the job. Training should be primarily about learning new skills not acquiring knowledge – that is education.

It would be doubtful whether the necessary competences would be acquired if the external training was a series of "talk and chalk" sessions where the tutor lectures the students, runs through hundreds of slides and asks a few questions!

On returning to work or normal duties after a course it is important that the skills and knowledge learnt are put to good effect as soon as possible. A lapse of weeks or months before the skills are used will certainly reduce the effectiveness. Little or no knowledge of skill may have been retained. Training is not about doing something once and once only. It is about doing something several times at frequent intervals. One never forgets how to ride a bicycle or drive a car regardless of the time-lapse between each attempt, because the skill was embedded by frequency of opportunities to put the skill into practice in the early stages.

Therefore, to ensure effectiveness of training you ideally need to provide opportunities to put into practice the newly acquired skills as soon as possible. The person's supervisor should then examine the students' performance through sampling work pieces, reading documents he or she produces, observing the person doing the job and reviewing the decision they make. If you have experts in the particular skills then in addition to appraisals by the supervisor, the expert should also be involved in appraising the person's performance. Pay particular attention to the person's understanding of customer requirements. Get this wrong and you could end up in trouble with your customer!

Development activity effectiveness – long term (the final stage)

After several months of doing a job and applying the new skills, a person will acquire techniques and habits. The techniques shown may not only demonstrate the skills learnt but also those being developed through self-training. The habits may indicate that some essential aspects of the training have not been understood and that some reorientation is necessary. It is also likely that the person may have regressed to the old way of doing things and this may be due to matters outside their control. The environment in which people work and the attitudes of the people they work with can have both a motivating and de-motivating effect

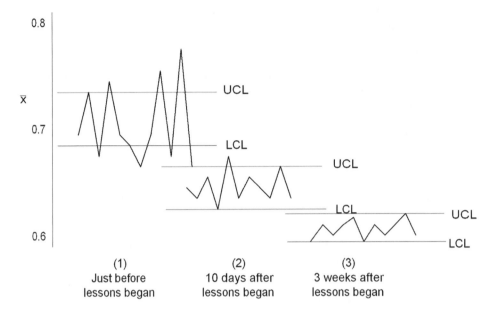

Figure 29.1 Average daily scores for a manual worker learning a new method

on an individual. Again, the supervisor should observe the person's performance and engage the expert to calibrate their judgement. Pay particular attention to customer requirements and whether the trainee really understands them. If there are significant signs of regression you will need to examine the cause and take corrective action. This can be illustrated as a run chart an example of which is shown in Figure 29.1.

How is this demonstrated?

Demonstrating that the effectiveness of actions taken to acquire the necessary competence has been evaluated may be accomplished by:

a) presenting evidence that the competence of those individuals who have been subjected to competence development has been assessed and found satisfactory by the designated authority (i.e. the person or organization the organization has delegated the authority to conduct competence assessment);

b) presenting evidence that in cases where the results of competence assessment were unsatisfactory, a programme has commenced to enable the person to acquire the necessary competence or the person has been reassigned duties;

c) presenting evidence that the actions taken to re-assign work, hire personnel, subcontract the work or redesign the process were evaluated and deemed effective;

d) presenting evidence of on the job training showing that:

    i   it was structured and programmed;

    ii it took place under close supervision;

iii   it was conducted in an environment in which the individuals were free to learn and not frightened of doing something wrong;

iv   it was subject to periodic monitoring.

e)   presenting evidence of classroom training that:

i   it was directly relevant to the deficient units of competence that enabled the participants to learn by doing or to learn by self-discovery and insight;

ii   any practical aids used during training:

- represent the equipment or facilities that in use either on the production line or in service locations;
- adequately simulate the range of operations of the production equipment or the activities of the service;
- are designated as training aids and only used for that purpose;
- simulate or inject fault conditions to teach diagnostic skills and judgement in making decisions;
- are recorded and maintained indicating their serviceability and their design standards, including records of repairs and modifications.

Retaining evidence of competence (7.2d)

What does this mean?

Box 29.8   Revised requirement removes inconsistency

The 2008 version was somewhat inconsistent in this requirement by requiring personnel to be competent, and the necessary competence to be determined but only requiring records of education, skills, training and experience and not records of competence. Now that inconsistency has been removed by requiring evidence of competence.

Evidence of competence extends beyond lists of training courses, academic qualifications and periods of experience because these only record actions taken and not whether they were planned or whether they achieved the desired result. Evidence of competence is evidence of results achieved and, most importantly, evidence showing how competence was assessed.

This evidence will therefore provide evidence of:

- the extent to which a person's abilities fulfil certain competence requirements;
- activities performed and criteria used to determine an individual's competence.

Why is this necessary?

Evidence of competence is necessary for two fundamental reasons:

- To prove to an interested party that a particular person has been deemed competent to perform a particular job, role or task.
- To assist in any investigation in which the competence of an individual or the method of or authority for a competence assessment is challenged.

How is this addressed?

Typical types of evidence include:

- The job specification or process description may identify the competences needed.
- The personnel development plan (PDP) identifies the education, training and behavioural development required to bridge the gap in terms of courses of study, training and development together with dates.
- Re-verification provides evidence of education, training and behavioural development undertaken together with dates completed.
- Competence assessment records, including:

    - Name of the person concerned
    - Results for which the person is responsible
    - Units of competence required
    - Performance criteria
    - Assessment method
    - Evidence generated
    - Conclusions
    - Date of assessment
    - Person responsible for the assessment

- The certification of competence provides evidence that the actions were effective. However, certification of competence is not required unless it is necessary for regulatory purposes. With this method, you will also need to maintain separately in the personnel records, historical records of education, training and experience to provide a database of capability that can be tapped when searching for potential candidates for new positions.
- Personnel files provide evidence of the courses taken, the training authority the dates, their duration and the examination results (if taken).
- Certificates of training provide evidence of training, but these are not necessarily evidence of competence.
- Proficiency badges carried on the person provide evidence of competency for certificated personnel performing special processes in regulated industries when records are held at some distance away from an individual's workplace.

How is this demonstrated?

Demonstrating that evidence of competence has been retained may be accomplished by:

a) retrieving the competency records of selected personnel for the work they are doing;
b) showing that the selected competency records indicate:

    i   the level of competence that has been attained against the corresponding competence requirements for the work;
    ii the methods used to assess each unit of competence;
    iii the authority conducting and certifying the last assessment and the date of that assessment;
    iv the date scheduled for when a review of the person's competence will be conducted.

Bibliography

CIPD. (2016, August). Competence and Competency Frameworks. Retrieved from CIPD: www.cipd.co.uk/hr-resources/factsheets/competence-competency-frameworks.aspx

Drucker, P. F. (1974). *Management, Tasks, Responsibilities, Practices*. Oxford: Butterworth-Heinemann.

Fletcher, S. (2000). *Competence-Based Assessment Techniques*. London: Kogan Page.

ISO 9000:2015. (2015). *Quality Management Systems – Fundamentals and Vocabulary*. Geneva: ISO.

OED. (2013). Retrieved from Oxford English Dictionary: oed.com

30 Awareness

Introduction

The requirements of clause 7.3 address some of the issues that an employee engagement process should be designed to resolve. The process should ensure employees are:

a) aware of what the organization is trying to achieve;
b) aware of how their contribution enables the organization to achieve its objectives and;
c) aware of the implications of not conforming to requirements, so that they are more likely to realize their full potential.

More details of employee engagement are given in Chapter 17 in the section that addresses clause 5.1.1h. and in the quality management principles explained in Chapter 5.

In this chapter, we examine the four requirements of Clause 7.3, namely:

*   Awareness of quality policy (7.3a)
*   Awareness of quality objectives (7.3b)
*   Awareness of contribution (7.3c)
*   Awareness of implications (7.3d)

As the requirement containing the phrase "persons doing work under the organization's control" is rather long to keep repeating, in this chapter we will refer to these persons as *personnel* in italics.

Awareness of quality policy (7.3a)

What does this mean?

The quality policy is addressed in Chapter 19, and were it to be communicated and made available as required by clause 5.2, persons within the organization will be aware of the quality policy. This leaves other persons that are not within the organization but perhaps doing work for the organization either on its premises or elsewhere.

Why is this necessary?

It is necessary for all those who can affect achievement of the organizations goals and whom it can strongly influence are not only aware of its quality policy but also understand it so that they are motivated to exhibit the desired behaviour towards quality.

How is this addressed?

It is recommended that the approach described in response to clause 5.2.1b) in Chapter 19 is adopted for all *personnel*, ensuring that they have a good understanding of the purpose and strategic direction of the organization from the outset.

How is this demonstrated?

Demonstrating that *personnel* are aware of the quality policy may be accomplished in the same way as described in by in response to clause 5.2.1b) in Chapter 19.

Awareness of quality objectives (7.3b)

What does this mean?

As explained in Chapter 22 in the context of ISO 9001, a quality objective is an objective that primarily benefits the customer by its achievement and therefore persons doing work under the organization's control need to be aware of the quality objectives that pertain to what they are doing.

When undertaking work, a person may be told that it needs to be completed by a certain date and certain resources may be made available for their use in undertaking the work. Although the time taken and the utilization of resources are important, they are constraints or measures of success and not its objective. The work will be required to be undertaken to deliver an intended result which will satisfy a customer. Most of these intended results will be expressed in terms of the criteria that the output needs to meet.

Why is this necessary?

Every activity *personnel* are required to perform should serve the organization's objectives either directly or indirectly. All activities affect the organization in some way, and the quality of results depends on how these activities are perceived by the *personnel* performing them and the degree of control they have over producing them. In the absence of clear direction, *personnel* use intuition, instinct, knowledge and experience to select the activities they perform and how they should behave. Awareness of objectives means that individuals are more able to select the right activities to perform in a given context.

How is this addressed?

For each quality objective, there should be a plan that defines the processes involved in its achievement. Assess these processes and determine where critical decisions are made, who is assigned to make them and then make the *personnel* involved aware of these objectives and what the organization is looking for as evidence that they have been achieved i.e. the measures of success. One way is for managers to advise the *personnel* through examples, samples or case studies of the type of actions and behaviours that are considered appropriate.

How is this demonstrated?

Demonstrating that *personnel* are aware of the relevant quality objectives may be accomplished by:

a) presenting evidence of an employee engagement process that addresses awareness of quality objectives;

b) selecting a representative sample of *personnel* and establishing that they can:

    i   articulate what it is they are trying to achieve;

    ii   identify the process in which they are engaged;

    iii point to the results they are achieving;

    iv explain how these results relate to the quality objectives that have been established for the process in which they are engaged;

    v   explain how the quality of their output is being measured.

Awareness of contribution (7.3c)

What does this mean?

Some activities make a significant contribution to the achievement of objectives and others make less of a contribution but all contribute. Awareness of this contribution means that individuals can apportion their effort accordingly. Awareness of the importance the contribution an individual makes means that individuals can approach the activity with the appropriate behaviour.

Why is this necessary?

Other than those planning the work, *personnel* carrying out those plans often don't know why they are required to do things, why they don't do other things, why they should behave in a certain way and why they should or should not put a lot of effort into a task. Some people may work very hard but on activities that are not important, not relevant or not valued by the organization. Other people may not work productively and not realize the impact their behaviour has on the organization's performance. *Working smart* is much better and more highly valued, and therefore awareness of the relevance and importance of activities and their contribution to the organization's objectives is essential for enabling an organization to function effectively.

    Awareness of contribution also puts a value on the activity to the organization and therefore awareness of the contribution that other people make puts a job in perspective and can overcome grievances and discontent. Personnel can sometimes get carried away with their sense of self-importance that may be based on a false premise. When managers make their *personnel* aware of the context in which activities should be performed, it helps redress the balance and explain why some jobs are paid more than others, or are more highly valued than others.

How is this addressed?

Explaining the relationship between what people do and the effect that has on customers can have a remarkable impact on how *personnel* approach the work they perform. Awareness creates pride and a correct sense of importance. It serves to focus everyone on the organization's objectives.

There are perhaps thousands of activities that contribute to the development and supply of products and services, some of which create features that are visible to the customer or are perceived by the customer as important. These features may be associated with the appearance, odour, sound, taste, function or feel of a product where the activity that creates such features is focused on a small component within the product the customer purchases. They may also be associated with the actions, appearance or behaviour of service delivery personnel where the impact is immediate because the personnel come face to face with the customer.

There are several ways of creating this awareness:

• Identifying the critical to quality (CTQ) characteristics in the documented information that is provided for the *personnel* or the team of which they are a member.
• Photographs or diagrams of the final product or service showing where these CTQ characteristics are located in context with the product or service.
• Videos explaining the importance of CTQ characteristics and showing situations in which customers are using a product or service.
• Market research information which shows what customers value in the types of products and services they buy, and why they buy one product or service over another competing product or service.

However close to or remote from the customer and seemingly significant or insignificant, the result of an activity has the potential to impact customer satisfaction. The probability and degree of impact need to be assessed, but to assess the impact of a result it must be observable (see Chapter 21 for further explanation).

How is this demonstrated?

Demonstrating that *personnel* are aware of their contribution to the effectiveness of the QMS and the benefits of improved quality performance may be accomplished by:

a) presenting evidence that of an employee engagement process that addresses awareness of contribution;
b) selecting a representative sample of *personnel* and establishing that they can:

    i   articulate what it is they are trying to achieve;
    ii explain the consequences of what they achieve;
    iii explain the effect that has on the process in terms of output quality, resource utilization and time to produce the output;
    iv explain how that makes the QMS more or less effective;
    v they can explain the benefits to be gained from improvement in their performance.

Awareness of implications (7.3d)

What does this mean?

The term *quality management system requirement* is explained in Chapter 17 in connection with cause 5.1.1c), and it will be apparent that there will be many such requirements that could apply to the work personnel do. The implications of not conforming with such requirements refer to the consequences, effects or repercussions that could arise. There may be cases where the *personnel* have no choice but to conform due to the error proofing controls, for example, where the software prevents information being processed unless complete and accurate. But there will be many cases where error proofing is not possible or indeed practical and where conformity depends on the diligence and judgement of the *personnel*.

Box 30.1   New awareness requirement on implication of nonconformity

The 2008 version required personnel to be aware of the relevance and importance of their activities with an implication that this was with reference to achieving quality objectives. However, in the 2015 version this has been contained in a requirement about awareness of contribution to the effectiveness of the QMS and a new requirement introduced for an awareness of the implications of not conforming with the QMS requirements.

Why is this necessary?

Personnel may have no idea of where their role fits in the big picture and consequently are insensitive to its impact. Such insensitivity can be dangerous as it can put lives, property, the environment and the success of the organization at risk. Oversensitivity can also be dangerous as it may result in risk averse behaviour, causing *personnel* to do nothing unless they are certain of the outcome. The QMS requirements should be constructed in a way that allows for individual creativity and risk taking but also applies a degree of control that detects and regulates work within boundaries appropriate to the organization's context. Making *personnel* aware of the implications of not conforming with QMS requirements is part of the employee engagement process and is necessary to instil in them the confidence they need to realize their full potential.

How is this addressed?

In many organizations this sensitivity is low. The manager's task is to heighten sensitivity so that everyone is in no doubt what effect nonconformity with QMS requirements has on the customer and on organizational performance.

Where necessary process descriptions should invoke all the applicable QMS requirements at the appropriate stages in the process, thereby alerting the person performing an activity to the process specific requirements that apply. A process risk assessment should examine the immediate and long term consequences of a specific activity not being carried out as specified and the probability of it being detected at the next or subsequent step in the process. There may be so many QMS requirements that it is impractical to apply risk assessment on every one in every possible scenario. In such cases one applies risk-based thinking (see Chapter 10) and focuses on those requirements which if not met would have the greatest impact on customer satisfaction. However, it is sometimes not enough to explain the consequences of failure; you may need to enable *personnel* to see for themselves the effect by using samples, simulations, prototypes or case studies.

Making *personnel* aware of the quality issues and how important these issues are to the business and to themselves and the customer may not motivate certain individuals. Perhaps as indicated in the introduction to this chapter, you might have hired the wrong person and the only thing you can so with such people is either not engage them in work that affects the quality of the organization's products and services or keep their work under close supervision until they become fully engaged in what the organization is trying to do. The intention should be to build an understanding of the collective advantages of adopting a certain style of behaviour.

Measuring a person's understanding of QMS requirements is a subjective process but measuring the outputs *personnel* produce is not. Competence assessment would therefore be an effective means of measuring the effectiveness of the engagement process. In this way, you don't have to measure it twice. Competence assessment serves to indicate whether *personnel* can do the job and to take appropriate action on detecting a nonconformity, and this therefore also serves to demonstrate the engagement process has or has not been effective.

Through the data analysis carried out to meet the requirements of Clause 9.1.3 and the internal audits of clause 9.2.1, metrics will be produced that indicate whether the QMS requirements are being met. If they are being met, you could either assume *personnel* understand the consequences of nonconformity and have taken action to prevent their occurrence or have taken remedial action, or you could conclude that it doesn't matter. However, the standard requires a measurement. Results alone are insufficient evidence. You need to know how the results were produced. The results may have been achieved by pure chance and in six months' time your performance may have declined significantly. Nonconformity with certain requirements may not have any observable effect until no one conforms with them. A few isolated nonconformities may not have an observable effect. The only way to test understanding is to check the decisions people make. This can be done with a questionnaire but is more effective if one checks decisions made in the work place. Is their judgement in line with your objectives or do you have to repeatedly adjust their behaviour? Take a walk around the plant or service outlet and observe what people do, how they behave, what they are wearing, where they are walking and what they are not managing. You might conclude that:

- A person not wearing eye protection obviously does not understand the safety objectives.
- A person throwing rather than placing good product into a bin obviously does not understand the product handling policy.
- A person handling food without wearing sterilized gloves does not understand the hygiene policy.
- A person operating a machine equipped with unauthorized fittings obviously does not understand either the safety objectives, the control plan or the process instructions.
- An untidy yard with evidence of coolant running down the public drains indicates a lack of understanding of the environmental policy.

There are many ways by which awareness of the of the implications of not conforming with the QMS requirements may be observed.

How is this demonstrated?

Demonstrating that personnel under the organization's control are aware of the implications of not conforming with the QMS requirements may be accomplished by:

a) presenting evidence of an employee engagement process that addresses awareness of the implications of not conforming with the QMS requirements;

b) selecting a representative sample of *personnel* and establishing that they can:

    i   identify the QMS requirements apply to their work;

    ii explain the immediate consequences not conforming with these requirements;

    iii point to a subsequent process stage where the impact of their not conforming to a specific QMS requirement will be felt;

    iv explain how they would be alerted to the impact of any nonconformity;

    v explain the long-term consequences of their not conforming with these requirements;

    vi also explain what the effect would be if everyone didn't conform to such requirements.

31 Communication

The most important thing in communication is hearing what isn't said.

Peter F. Drucker (1909–2005)

Introduction

The standard no longer requires a specific communication process by name as this was deleted at the FDIS stage. The reason for this could be that naming processes in the standard led some users to only define the named processes and no others. However, anywhere there is activity being undertaken to achieve an intended result there is a process. There is therefore a communication process whether or not one is required by ISO 9001. Although communication processes are important they are not generally business processes but processes which every process rely upon to be effective.

Box 31.1   Revised requirement on communication

The 2008 version required top management to ensure that appropriate communication processes are established within the organization and this appeared under the heading of "Internal Communication". In the 2015 version this has now been extended to include both internal and external communications relevant to the QMS and also changed so that communication is everyone's responsibility.

The requirements are expressed in a way reminiscent of Kipling's poem (see Box 31.2) except for why and where to communicate. Determining *why to communicate* should be a precursor to determining *what to communicate* and determining where to communicate is partially addressed by the requirement for determining *with whom to communicate*, but it is also important where the communication is transacted and this may be addressed by the requirement for determining *how to communicate*.

Box 31.2   Kipling's six honest serving men

In his poem, "The Elephant Child" Rudyard Kipling refers to the words What, Why, When, How, Where and Who as six honest serving men that taught him all he knew. They initiate open questions which will explore a subject to a degree most unlike closed questions which may elicit nothing more than a yes/no answer.

The communication process

As soon as you receive or transmit a message you have potentially established a process for communicating with people, but if the process merely transmits and receives information, it is not a communication process.

Rollinson defines communication as "a process in which information and its meaning is conveyed from a sender to receiver(s)" (Rollinson, 2008). This definition implies that whatever means are used to convey the information the sender and the receiver must have consensus about its meaning; otherwise, communication has not taken place and the sender has not got his or her message across. The effectiveness of a communication process is therefore assessed by the extent to which the sender and the receiver of a message agree about its meaning.

Because communication pervades every process it's not a process like other processes that produce tangible outputs. The output is *meaning* and this gets passed through other processes where it may become distorted. Stories from WWI tell of a brigadier giving the order *Send reinforcements we are going to advance* and by the time it had passed through several intermediaries and reached the trenches it was received as *Send three and four pence we are going to a dance*. In the 21st century we can ensure the same words as transmitted are received but there remain many barriers as highlighted in Box 31.3.

Box 31.3   Barriers to effective communication

Message formulation – people differ considerably in their capabilities to express their thoughts.

Perception – a sender's perception of the receiver can obscure the real meaning.

Encoding – people differ in the meanings they attribute to the same word, abbreviation or image

Physical noise – extraneous signal drowning or masking the message.

Psychological noise – interference with transmission or meaning by virtue of the setting, the way the message is conveyed or the timing of its delivery.

Inappropriate channel – using written material to convey honesty and commitment when verbal would be more convincing.

Inappropriate media – using verbal for long complex messages, written for controversial messages or unsecured media for commercially sensitive messages.

Decoding – recipient's pre-conceived ideas can distort intended meaning.

Information overload – receiver is unable to cope with the flood of messages to which he or she must attend.

Adapted from Rollinson, 2008 Chapter 15

There will be several processes between seeking to understand the needs of potential customers to resolving issues that existing customers have long after they took delivery of the product or used the service. The form of communication in each of these processes will differ in some respect. Every contact with a customer is likely to proceed through a different process. The specific inputs, activities, resources, constraints and outputs will differ but each communication process will have some common features (Rollinson, 2008).

- Sender – all messages originate in the brain of the sender, for whom they have a meaning.
- Encoding – the sender encodes the message from thoughts into words (spoken or written) or pictures using a set of common symbols (e.g. language, dictionaries, glossaries).
- Channel and medium – messages are transmitted in one or more ways (e.g. verbal or written) through different media (e.g. face to face, telephone, e-mail, video, letters, written reports, formal specifications, spreadsheets).
- Decoding – receiver decodes the symbols into thoughts to be able to understand the message.
- Receiver – takes in the message using his or her sensory processes and processes the information to attribute it with meaning.
- Feedback – the sender receives a signal that his or her message has been understood.
- Technology – the tools used by the sender and receiver to convey the message.
- Barriers – anything that interferes with or masks the message.
- Organization structure – the levels through which the message must pass to reach the intended recipient.
- Organizational culture – the acceptable patterns of behaviour, norms and values that influence the way communication takes place.

There are however, overt and covert communication processes. The overt communication processes are those described earlier, but there are covert communication processes where signals or lack of signals are inadvertently sent because of involuntary behaviours (e.g. an inordinate delay in response might be interpreted as *they don't care*). Procrastination over issues might be interpreted as *a lack of commitment*. Consumers are not fools; they can sense when organizations are more loyal to other stakeholders than the people who buy their products and services.

The effectiveness of communication, whether it is internal communication or external communication, is vital to organization success and therefore it is necessary to pay attention to the process by which it is carried out. What should be observed is illustrated in Figure 31.1.

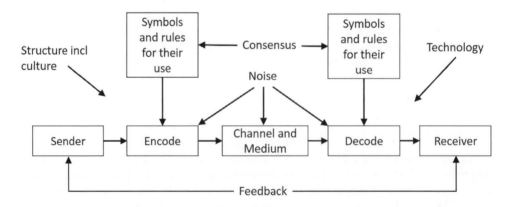

Figure 31.1 Generic communication process (adapted from Rollinson 2008)

In this context, the nature of the message is unimportant. What is of prime importance is *meaning* and whether the sender and receiver both attribute the same meaning to the messages being transmitted from both sides. This outcome depends on there being a high degree of consensus about the meaning of the terms used and the rules for their use.

Interactive communication skills

Organizations are formed by bringing people together to create something that none of them could do on their own. It's therefore axiomatic that these people need to be able to communicate with one another to get things done and fulfil the purpose of the organization. If you are not a good communicator, learn good communication skills. There are several skills that enable people to become better communicators, and the eight skills described next are a summary of those described in the *Art of Managing People* (Hunsaker & Alessandra, 2009)

1   Questioning skills – asking the right questions elicits crucial information, secures people's participation, stimulates thinking, clears obstacles to understanding, enables people to be guiding in the right direction and builds trust.
2   Listening skills – listening is often taken for granted, but it is 50% of communication and therefore barriers to effective listening need to be removed such as the listener's attitude either towards the speaker or the message, their subject knowledge, the setting, control of emotions, concentration and opportunities to engage with the speaker (see Box 31.4).
3   Projecting the right image – the first impressions a person projects, their depth and breadth of knowledge, flexibility, enthusiasm and sincerity all hasten or retard the development of trust and rapport.
4   Communicating through voice tones – the resonance, rhythm, speed, pitch, volume, inflection and clarity of the voice all influence the meaning that is communicated.
5   Using body language effectively – a person's body movements, facial expressions, gestures, eye contact and posture reveal much more about their attitudes and emotional state than their words.
6   Creating the right space – a person's territory, environment, possessions and proximity to others influences the way people interact and act as drivers and barriers to communication.
7   Use of time – a person's use of time can communicate their attitude, what they value, their priorities, etc., and being such an expressive language can facilitate communication with others.
8   Using feedback – communication depends on reaching a consensus, and to do this the parties involved need to give and get definitions of the words and phrases used, avoid making assumptions, ask questions, speak the same language, read the signals, provide feedback on behaviour not the person and recognize when to withhold feedback.

Box 31.4   Active listening

Active listening is a form of listening in which the listener makes it abundantly clear to the speaker that the message being conveyed is being heard, understood and reflected upon.

In this chapter, we provide an overview of the communication process before examining the five requirements of clause 7.4, namely:

- Determining what to communicate (7.4a)
- Determining when to communicate (7.4b)
- Determining with whom to communicate (7.4c)
- Determining who communicates (7.4e)
- Determining how to communicate (7.4d)

Determining what to communicate (7.4a)

What does this mean?

Determining what to communicate means, giving thought to what it is you want to say and what meaning you want to convey.

Internal communication refers to the flow of information within the network of processes in the QMS. Depending on the organization, the communication maybe within a single site or across multiple sites and possibly countries. External communication refers to the flow of information between the QMS and other parts of the organization and between external parties. This communication may be within a single country or cross-national.

There will be lots of communication in an organization not all of which is relevant to the QMS. If the communication is intended to be about anything that is within the scope of the QMS (see Chapter 14), it's relevant to the QMS.

Why is this necessary?

There is no person in an organization that is not affected by communication, and serious problems can arise if communication is not effective. Therefore, it is vital that thought is given to what to communicate before any message is transmitted. Sometimes quality problems attract media attention, as was the case with the Deepwater Horizon oil spill in the Gulf of Mexico in 2010. Tony Hayward, then CEO of BP, made the fatal mistake of saying he wanted his life back, which showed a blatant lack of respect for those who had lost their lives in the explosion. This is a result of making the wrong mistakes, but if you are going to make mistakes, make the right ones as illustrated by Ackoff's statement in Box 31.5.

Box 31.5   Making the right mistakes

If you are going to make mistakes, make the right ones.

Doing the wrong thing right is worse than doing the right thing wrong because a mistake in doing the right thing wrong can be corrected sometimes with little effort, whereas a mistake in doing the wrong thing right requires one to start again and waste all the effort already spent.

Inspired by Russ Ackoff

How is this addressed?

What to communicate is fundamentally determined by the objective to be achieved, and every process should have an objective. However, in the digital age its seems organizations are awash with information, and some of it might not have been sanctioned by the organization. Employees are inundated with e-mail and it may carry formal and informal communication. Here we will deal with the formal information. Informal information is addressed under "What should be documented" in Chapter 32.

External communication

What the organization communicates externally is vital to its survival as it needs to attract customers, investors, employees, etc., but also to do so in a way that is not detrimental to other interested parties, for example, a company in the energy business may announce to its customers its plans for exploiting fracking technology which will bring down the price of oil and gas for its customers. The local community gets to learn about this and launches a protest outside the company gates, blocking deliveries. The bad publicity depresses share price and deters investors, and the company struggles to make a success of its new venture so its customers go elsewhere.

Deciding what to communicate externally is likely to arise in many externally facing processes such as public relations, sales, marketing, service delivery and customer support. Whether it's the CEO addressing the shareholders or a server in a restaurant it's important to first understand the purpose of what you want to say (i.e. why you want to communicate at this time) and then what message you want to communicate.

Some typical whys and whats with external communication that illustrate relevance to the QMS are:

*   To inform the recipients about something in which they may have an interest, for example, a new service, upgrades to existing services, the way the organization manages quality, etc.
*   To alert the recipients about something they need to know so that they can, if necessary take action, for example, hazards, product recall, amended specifications, user instructions, disposal instructions, etc.
*   To advise the recipients about matters that may affect their use of the organization's products and services, for example, recently discovered side effects of medicines, amended returns procedure, etc.
*   To announce changes that may affect the recipient's relationship with the organization, for example, business results, contact details, terms and conditions, business closure, new owners, etc.
*   To explain how you intend to deal with a problem and resolve it to the recipient's satisfaction, for example, customer complaint, reported nonconformity, etc.

Internal communication

Deciding what to communicate internally is likely to arise in all processes at the interface between co-workers, departments and divisions. There will be a multitude of things to communicate internally. From the communication that triggers an action, to the communication that enables a worker to interact in the process, to the communications that pass among the processes in the network and the communication that terminates the process, people are continually deciding what to communicate. Some typical whys and whats with internal communication that illustrate relevance to the QMS are:

- To inform workers as to the results the organization and its processes are expected to achieve and enable them to determine how their performance will be measured (e.g. objectives, specifications, targets, KPIs).
- To inform workers as to what they are required to do and enable them to participate in the planning (e.g. plans, process descriptions, job descriptions, work instructions).
- To advise workers as to how they are expected to do the work and enable them to improve such methods (e.g. policies, procedures, guides, codes of practice).
- To enable workers to discover how the process in which are involved is performing (e.g. metrics, reports, instrumentation).
- To inform workers as to the action they are expected to take if work does not proceed as planned and to enable workers to inform managers as to the actions taken (e.g. policies, plans, procedures process descriptions, codes of practice, reports).
- To inform workers as to the effect of any action they have taken to correct or improve performance.

How is this demonstrated?

Demonstrating that the organization has determined what to communicate both internally and externally that is relevant to the QMS may be accomplished by:

a)   presenting evidence of policies that address internal and external communication;
b)   selecting a representative sample of processes and the associated process descriptions and showing that at the internal interface between processes and the interface with external parties, consideration is given to the nature of the information that is to be communicated;
c)   showing that at these stages in the process, the information being communicated is relevant to the purpose of the QMS.

Determining when to communicate (7.4b)

What does this mean?

When a reason to communicate has been established and it's been decided what to communicate, the next step is to decide when to communicate, and this will depend on a range of factors that will affect its priority.

Why is this necessary?

The timing of communication is very important to its effectiveness. If it's the right message, to start with, transmit it too early, and the significance of it won't be understood. Transmit the message too late, and the moment has passed for it to make any difference.

How is this addressed?

To determine when something should be communicated, we need an appreciation of the consequences to set its priority. The timing of a communication can be critical to its success. For instance,

- Launching a quality improvement initiative just as redundancies are being announced will brand it with the kiss of death.
- Holding a management review on a Friday afternoon usually results in an early finish before the important business has been transacted.

There will be information that is to be communicated:

- on a recurring basis (e.g. monthly reports, agendas for regular meetings);
- immediately because it's urgent (e.g. responses to customers, hazard alerts, instructions to start or stop a process, to deliver a package);
- a certain number of days prior to an event (e.g. criteria for a test, schedule for an audit);
- a certain number of days after an event (e.g. the results of a test);
- at a pre-determined time (e.g. a public announcement);
- as and when necessary (e.g. policies, procedures).

An executive who demands to be kept informed of progress will soon stop reading the reports and if the process continues without change, the reports will just pile up in his or her office. This is not an uncommon phenomenon. A manager may demand reports following a crisis but fail to halt further submissions when the problem has been resolved. The opposite is also not uncommon where a local problem is not communicated outside the area and action is subsequently taken which solicits the response: *Why didn't you tell us you had a problem?*

How is this demonstrated?

Demonstrating that the organization has determined when to communicate both internally and externally information relevant to the QMS may be accomplished by:

a) presenting evidence of policies that address internal and external communication;
b) selecting a representative sample of processes and the associated process descriptions and showing that at the internal interface between processes and the interface with external parties, consideration is given to the timing of the information that is to be communicated;
c) showing how the barriers to effective communication been identified and addressed.

Determining with whom to communicate (7.4c)

What does this mean?

Communication involves a sender and a receiver and this requirement is about determining who the receiver should be. The receiver may be a person, a select group of co-workers or managers, an organization, an audience at a meeting or conference, a group of stakeholders or indeed the general public.

Why is this necessary?

Without identifying a receiver, there is no communication at all but the greater the number of recipients the greater the potential for misunderstanding as was the case illustrated in Box 31.6.

Box 31.6   Making the wrong mistakes

Gerald Ratner, former CEO of the major British jewellery company Ratners Group, who in 1991 addressed an Institute of Directors conference, jokingly denigrated his company's products "We also do cut-glass sherry decanters complete with six glasses on a silver-plated tray that your butler can serve you drinks on, all for £4.95. He went on to say his stores' earrings were "cheaper than an M&S prawn sandwich but probably wouldn't last as long" (Daily Telegraph, 2007).

The value of the Ratner group plummeted by around £500 million. Ratner was not addressing the customers who bought his cheap jewellery but fellow businessmen. His mistake was making a judgement about the quality of the merchandise his company sold. He had forgotten why his customers bought the jewellery and that it's the customer not the producer that judges quality, see also Box 6.2.

How is this addressed?

The job descriptions should provide the incumbent with the necessary authority to achieve the results for which they are responsible, and this will include the authority to decide with whom they should communicate. However, in addition to job descriptions an individual will be constrained by the culture, for example, some organizations have a tradition where there's no barrier on whom you can speak with, whereas with other organizations, it is considered a career-limiting offence to communicate with your manager's manager.

Selecting the right recipient is important as it influences how the message should be transmitted. Although you may think your message is only reaching its intended audience, you need to assess the risk of it reaching further and what you say coming back to bite you as it did with Gerald Ratner.

Therefore, actions have consequences and so job descriptions should restrain authority so that a person's authority does not exceed their responsibility (e.g. a server in a restaurant receives a complaint from a customer). In responding to that complaint, the server's role changes from receiver to sender. The server's authority only goes so far as to permit them to engage with the customer who complained. It does not permit the server to broadcast the nature of the complaint to everyone else in the restaurant, or to the media.

Everyone will have a sphere of influence that is partly constrained by the role they perform but also by the respect they attract and this enables their messages to reach much further. This can have a detrimental effect if, in an unguarded moment they say something inappropriate.

Where there are established protocols, these should he adhered to. In the absence of any protocols, who you communicate with depends on several factors:

- The specificity of information, that is, only one person needs it, or will understand it so why send it to anyone else?

- The sensitivity of the information which influences their need to know, that is, if the information is not needed for their job why send it to them?
- Including anyone who might be interested does not improve the chance that the message will be understood or even acted upon.
- The people who are more likely to act on the message are those to whom it is addressed and the fewer of these there are the better.
- Those who are copied may like to know but won't do anything.
- Don't copy the manager of the addressee without the consent of both as otherwise it's telling the addressee you don't have authority to send the message.
- Only send blind copies to those who you know expect to be informed without the addressee knowing.

Clearly, not everything can be or should be communicated to all levels because some information will be sensitive, confidential or simply not relevant to everyone. Managers therefore need to exercise a *need to know* policy that provides information necessary for people to do their job as well as creating an environment in which people are motivated. Other than national and commercial security, too much secrecy is often counterproductive and creates an atmosphere of distrust and suspicion that affects worker performance. If you don't want somebody to have information say so, don't give them a reason that is untrue just to get them off your back for it will return to bite you!

An issue that arises with external communication is cross-national communication where the language, customs and culture differ from that of the originating organization. This can affect what is to be communicated to whom, for example, in France managers see information as power rather than something to be shared with their staff, whereas in Sweden, which is more egalitarian, there is a greater propensity to share information. However, confidentiality overrides culture and will determine what information may be communicated to whom.

How is this demonstrated?

Demonstrating that the organization has determined to whom information relevant to the QMS is to be communicated may be accomplished by:

a) presenting evidence of policies that address internal and external communication;
b) selecting a representative sample of processes and the associated process descriptions and showing that at the internal interface between processes and the interface with external parties, consideration is given to selecting the recipients of the information that is to be communicated;
c) showing how the barriers to effective communication been identified and addressed.

Determining who communicates (7.4e)

What does this mean?

Communication involves a sender and a receiver and this requirement is about determining who the sender should be.

Why is this necessary?

Without identifying a sender, there is no communication, but often the identity of the sender gives credibility to the communication. We all present a risk when put in a position we are ill equipped to deal with. Choose the wrong sender, and the message may fail to have any impact or may have disastrous consequences.

How is this addressed?

The job descriptions should provide the incumbent with the necessary authority to achieve the results for which they are responsible. This should include the authority to decide who should communicate on which subjects and to whom certain information should be communicated.

In the absence of any protocols, who communicates what will depend on several factors:

- The communication skills of the individual
- The knowledge and experience in the subject
- The ability to handle negative feedback and turn it to an advantage
- The extent of their delegated authority
- The degree of respect they command from others
- The demands of the situation

In the example earlier of a sever in a restaurant, the point was made that the server's authority would not extend to permitting a customer complaint to be broadcast to everyone else in the restaurant or the media. This means that others will have that authority. However, there are some situations where a person in a lowly role has authority to communicate with everyone and that is in an emergency but as already stated, actions have consequences and false alarms carry a great responsibility. The message may not be about a fire or flood but about the quality of something on which a process relies and careless talk can cost lives but also lead to human error (e.g. informing a co-worker that checks were complete when they weren't).

How is this demonstrated?

Demonstrating that the organization has determined who is to communicate information relevant to the QMS may be accomplished by:

a) presenting evidence of policies that address internal and external communication;
b) selecting a representative sample of processes and the associated process descriptions and showing that at the internal interface between processes and the interface with external parties, consideration is given to selecting the sender of the information that is to be communicated;
c) showing how the barriers to effective communication been identified and addressed.

Determining how to communicate (7.4d)

What does this mean?

Communication involves a sender and a receiver and this requirement is about determining how the sender should communicate with the receiver.

Why is this necessary?

There are many ways of communicating a message and even if it's the right message, sent by the right sender to the right receiver, its effectiveness will depend on how its delivered as some methods are far more effective than others.

How is this addressed?

The job description won't stipulate how information is to be communicated, that is often the role of policies but even policies will defer to the sender to assess the situation and select an appropriate means of delivery.

In determining how information is to be transmitted consideration needs to be given to the audience and their location along with the urgency, sensitivity, impact and permanency of the message.

- Audience influences the language, style and approach to be used (Who are they?).
- Location influences the method (Where are they?).
- Urgency influences the method and timing when the information should be transmitted (When it is needed?).
- Sensitivity influences the distribution of the information (Who needs to know?).
- Impact influences the method of transmission and the competency of the sender (How should they be told and who should tell them?).
- Permanency influences the medium used (Is it for the moment or the long term?)

Oral communication

The way someone says something can have a great effect on what meaning is being communicated. Information transmitted orally is fast and permits immediate feedback, permitting the recipient to signal whether the message has been understood and to seek clarification if necessary. The interaction between sender and receiver continues until a consensus is reached.

Oral communication is best for inspiring people, building relationships, explaining what a person is to do, creating aspirations, building teams, explaining the consequence of an employee's performance, giving feedback to management, dealing with contentious issues etc. Oral communication is best for anything that does not need to be written and if it were written it would not be as effective.

Oral communication provides information that is lost when speech is written rather than spoken. The oral parts don't always communicate the same meaning as the written parts. It provides colour and sentiment and in a face-to-face situation has the advantage of adding nuances, emphasis, anecdotes and other techniques to put a message across in different ways and align a wider range of listeners.

With oral communication, not only has there to be a skilful speaker but also a skilful listener. Hunsaker and Alessandra provide 19 rules of listening which are summarized in Box 31.7.

Box 31.7   The rules of listening

1   Remember that it is impossible to listen and talk at the same time.
2   Listen for the speaker's main ideas.
3   Be sensitive to your emotional deaf spots.
4   Fight off distractions.
5   Try not to get angry.
6   Do not trust to memory certain data that may be important – take notes.
7   Let your employees tell their own story first.
8   Empathize with your employees.
9   Withhold judgement.
10 React to the message not the messenger.
11 Try to appreciate the emotion behind the words more than their literal meaning.
12 Use feedback.
13 Listen selectively.
14 Relax.
15 Try not to be critical of someone else's point of view even if it's different from your own.
16 Listen attentively, making intermittent eye contact.
17 Try to create a positive listening environment.
18 Ask open questions.
19 Be motivated to listen.

Adapted from (Hunsaker & Alessandra, *The Art of Managing People*, 1986)

Oral communication is prone to physical noise which masks the message and psychological noise by virtue of the setting, the way the message is conveyed or the timing of its delivery. Long speeches run the risk of losing the concentration of the listener, much of what is said will be forgotten and there is a high likelihood that what is spoken will be misinterpreted. But if the speech is recorded on audio or video media these weaknesses can be overcome as it can be played over and over. Handing out notes of the speech afterwards affords the listener the opportunity to go over what was said.

Written communication

Information can be transmitted in writing more reliably than information delivered orally but it is not spontaneous. If an immediate response is required oral communication is vital.

However, written information is more permanent and it enables the recipient to go over it at their convenience, revisit passages that were not clear the first time and refresh their memory later. It also enables the recipient to pick apart the information, note questions they'd like answered and affords them with the facility to play back the words in the order they were transmitted.

From the sender's point of view, it enables the sender to be more precise about what is said, to overcome the obstacle with oral communication which relies on the receiver remembering what was said. In this way, written communication is more appropriate for lengthy messages, technical information and for information that is to be used as a source of reference.

However, written communication is less effective at transmitting feelings, trust, integrity. It's too rigid to show emotion and supress the other party's tension.

When feedback is also given in writing the interaction between sender and receiver may continue for days, weeks or months before a consensus is reached. Sometimes a consensus is not reached. Sometimes the parties become frustrated by the difficulty in making themselves understood. All that generally happens is that tensions get so high that one of the parties stop giving feedback. In such cases and depending on the importance of the information, a face to face meeting usually resolves the issues. The best approach is not to let the exchange reach this stage before meeting face to face, either in person or by video link.

Language

When communicating with top management its best to use the language of money, whereas when communicating with those lower down the hierarchy, the language changes from technical to non-technical and about things not money.

How is this demonstrated?

Demonstrating that the organization has determined who is to communicate information relevant to the QMS may be accomplished by:

a)   presenting evidence of policies that address internal and external communication;
b)   selecting a representative sample of processes and the associated process descriptions and showing that at the internal interface between processes and the interface with external parties, consideration is given to method of communicating information;
c)   showing how the barriers to effective communication been identified and addressed.

Bibliography

Hunsaker, P. L., & Alessandra, A. J. (1986). *The Art of Managing People*. New York: Simon & Shuster, Inc.

Hunsaker, P. L., & Alessandra, A. J. (2009). *The New Art of Managing People*. New York: Simon & Schuster, Inc.

Rollinson, D. (2008). *Organisational Behaviour and Analysis*. Harlow, Essex: Pearson Education Limited.

32 Documented information

Introduction

In a supporting ISO document is the following statement "It is stressed that ISO 9001 requires (and always has required) a "documented quality management system", and not a "system of documents" (ISO/TC 176/SC2/N1276, 2015). This may have been true until the 2015 revision, but whereas in previous versions the organization was required to establish, document, implement and maintain a quality management system, the word *document* was removed from the 2015 version.

Many ISO 9001 certified organizations have missed the opportunity all along to exercise their prerogative in the first place to deliberately consider what information the organization needs – some have only documented what ISO 9001 required them to document and now as the requirement for a quality manual has been removed, many may be seen throwing their quality manual away, which would be rather foolish to say the least! ISO 9001 is not intended to dictate what organizations do. What it does is lay down criteria against which an organization may demonstrate it can consistently provide products and services of the quality required. The key word here is *consistently*, and unless the organization is a machine, it's unlikely people will do the right things right unless they have reliable communication processes. Since the Sumerians invented writing in about 3100 BCE the written word has been one of the most important tools of communication humans have used to transmit complex messages. The message may not always be correct, but unlike verbal communication, the written word can pass among millions of people and remain intact provided it remains in its original form.

One of the problems with previous versions of ISO 9001 has been the considerable emphasis on documentation as explained in Chapter 3. Having been required to produce documented procedures, users did so in their millions, often without thinking of the message they were conveying and why that message needed to be conveyed in the first place.

The 2015 version is very much different as no specific documented procedures are required. The documentation to be produced is that which the organization deems necessary for the effectiveness of its quality management system. This gives organizations the flexibility to decide the documentation they need and no longer produce documentation only because it was a requirement of the standard. There are, however, 33 requirements for documented information in the 2015 version as opposed to the 30 requirements for documents or records in the 2008 version. So why on the one hand say the 2015 version gives the organization flexibility to decide on the documentation needed, and on the other require 33 specific documents? The answer lies in the ISO Guidance on the requirements for documented information (ISO/TC 176/SC2/N1276, 2015) which states:

> To claim conformity with ISO 9001:2015, the organization has to be able to provide objective evidence of the effectiveness of its processes and its quality management system [This is implied by clause 9.1.1]. Clause 3.8.3 of ISO 9000:2015 defines *objective*

evidence as "data supporting the existence or verity of something" and notes that "objective evidence may be obtained through observation, measurement, test, or other means. Objective evidence does not necessarily depend on the existence of documented information, except where specifically mentioned in ISO 9001:2015."

In this chapter, we examine the 12 requirements of clause 7.5.1, namely:

- Documented information required by ISO 9001 (7.5.1a)
- Determining documented information necessary for the QMS (7.5.1b)
- Identifying and describing documented information (7.5.2a)
- Determining appropriate format and media (7.5.2b)
- Review and approval of documented information (7.5.2c)
- Availability and suitability of documented information (7.5.3.1a & 7.5.3.2a)
- Controlling distribution, access, retrieval and use (7.5.3.2a)
- Protection of documented information (7.5.3.1b)
- Storage and preservation of documented information (7.5.3.2b)
- Controlling changes to documented information (7.5.3.2c)
- Retention and disposal of documented information (7.5.3.2d)
- Controlling documented information of external origin (7.5.3.2)

Note: As it has been 15 years since the dawn of the digital age, guidance on controlling paper documents has been drastically reduced from that of previous editions of this Handbook.

Documented information required by ISO 9001 (7.5.1a)

What does this mean?

The standard refers to specific documented information among the requirements so at a minimum this information is required to be either available, maintained or retained. The selected dictionary meanings of these three words as defined in (ISO Glossary, 2016) are as follows:

- Available = able to be used or obtained.
- Retained = keep possession of, not abolish, discard or alter.
- Maintain = cause or enable (a condition or state of affairs) to continue.

It follows therefore that it is intended that documented information that is required to be retained shall not be altered and that which is to be maintained shall be altered as necessary to enable its continued use.

A list of this documented information required by the standard is provided in Table 32.1. As will be seen from this table, nearly all the documentation requirements from the 2008 version have been carried over into the 2015 with a few exceptions:

- The QMS is no longer required to be documented.
- A quality manual is no longer required.
- The six documented procedures are no longer required.
- Records of competence are now required instead of records of education, training, skills and experience.
- Records of the validity of the previous measuring results are no longer required.
- Records of the results of preventive actions taken are no longer required.

Table 32.1 Documented information required by ISO 9001

#	Clause	Purpose of documented information	Action	Required by ISO 9001:2008
1.	4.3	Documented information describing the scope of the QMS	Maintain	4.2.2a)
2.	4.4.2	Documented information supporting the operation of processes	Maintain	4.2.1
3.	4.4.2	Documented information providing confidence that the processes are being carried out as planned	Retain	4.2.4
4.	5.2.2	Documented information defining the quality policy	Maintain	4.2.1a)
5.	6.2.1	Documented information defining the quality objectives	Maintain	4.2.1a)
6.	7.1.5.1	Documented information describing evidence of fitness for purpose of monitoring and measurement resources	Retain	7.6
7.	7.1.5.2	Documented information describing the basis used for calibration or verification	Retain	7.6
8.	7.2	Documented information describing evidence of competence	Retain	None
9.	8.1e1)	Documented information providing confidence that the processes have been carried out as planned	Maintain and Retain	7.1d)
10.	8.1e2)	Documented information to demonstrate conformity of products and services to requirements	Maintain and Retain	7.1d)
11.	8.2.3.2a)	Documented information describing the results of the review of requirements related to products and services	Retain	7.2.2
12.	8.2.3.2b)	Documented information describing any new requirements for products and services	Maintain	None
13.	8.3.3	Documented information on design and development inputs	Retain	7.3.2
14.	8.3.4a	Documented information on the results to be achieved by the design and development process	Retain	None
15.	8.3.4b/e	Documented information recording the results of design reviews and records of actions taken	Retain	7.3.4
16.	8.3.4c/e	Documented information on design and development verification activities, the results and records of actions taken	Retain	7.3.5
17.	8.3.4d/e	Documented information recording the results of validation activities and records of actions taken	Retain	7.3.6
18.	8.3.5	Documented information on design and development outputs	Retain	None
19.	8.3.6	Documented information on design and development changes	Retain	7.3.7
20.	8.4.1	Documented information describing the results of the evaluations, monitoring of the performance and re-evaluations of the external providers	Retain	7.4.1
21.	8.5.1a)1	Documented information that defines the characteristics of the products and services shall be available	Maintain	7.5.1a)
22.	8.5.1a)2	Documented information that defines the activities to be performed and the results to be achieved	Maintain	7.5.1b)

#	Clause	Purpose of documented information	Action	Required by ISO 9001:2008
23.	8.5.2	Documented information necessary to maintain traceability of outputs	Retain	7.5.3
24.	8.5.3	Documented information describing what has occurred to customer property	Retain	7.5.4
25.	8.5.6	Documented information describing the results of the review of unplanned changes essential for production or service provision	Retain	None
26.	8.6a)	Documented information providing evidence of conformity with the acceptance criteria	Retain	8.2.4
27.	8.6b)	Documented information shall provide traceability to the person(s) authorizing release of products and services to the customer	Retain	8.2.4
28.	8.7.2	Documented information describing nonconforming outputs the actions taken, the concessions granted and the identifying the authority for acceptance	Retain	8.3
29.	9.1.1	Documented information as evidence of the results of performance and the effectiveness of the QMS	Retain	None
30.	9.2.2f	Documented information as evidence of the implementation of the audit programme and the audit results	Retain	8.2.2
31.	9.3.3	Documented information as evidence of the results of management reviews	Retain	5.6.1
32.	10.2.2	Documented information as evidence of the nature of the nonconformities and any subsequent actions taken	Retain	8.3
33.	10.2.2	Documented information as evidence of the results of any corrective action	Retain	8.5.2e)

Although these listed documents are no longer required by ISO 9001, this does not mean that they are not necessary for the effectiveness of the organization's QMS. Such documentation is not necessarily the same as the objective evidence needed to demonstrate the QMS is effective, for example, in a large organization, information may need to be documented so as to be reliably transmitted to those who need it but in a small organization it may be sufficient to transmit the information verbally.

Why is this necessary?

Why should the QMS include the documented information required by ISO 9001? Why does ISO 9001 need to require any specific documents at all, why not leave it to the organization to decide what it needs? Had the standard not contained this specific requirement, the requirement in 7.5.1b would have been sufficient because it is likely that an organization practicing risk-based thinking would generate a list of documents that include all of those in Table 32.1 and many more. However, ISO 9001 is used by organizations that need to demonstrate their ability (see ISO 9001 clause 1) and therefore a balance had to be reached between leaving it to chance and what customers should expect to be available. There has certainly been a dramatic shift in emphasis from the first version in 1987 which required 30 specific procedures and 13 records to be documented.

Box 32.1   Risk-based-thinking in documenting information

Risk-based thinking permits users to get away from producing documents and ticking boxes for the sake of it. But it comes with a warning. If you abandon methods that have proved successful but costly in the past, you will need to present evidence that doing so won't result in undesirable consequences.

How is this addressed?

Including documented information in the QMS is accomplished when a process and its interface with the external environment is brought within the scope of the QMS.

How is this demonstrated?

Demonstrating that the prescribed documented information is included in the QMS may be accomplished by presenting the process descriptions and showing where this information is generated or captured, maintained or retained as applicable.

Determining documented information necessary for the QMS (7.5.1b)

What does this mean?

A QMS functions through the interaction of its elements and this interaction is caused by the transmission of information between them. Every process is triggered by information, and each activity within it is dependent on receiving the correct information to produce the correct outputs whether the information is conveyed by instruments, verbally, non-verbally or in a document.

> Documented information is "information required to be controlled and maintained by an organization and the medium on which it is contained".
>
> (ISO 9000:2015)

The requirement therefore focuses only on that information which needs to be documented and controlled for the QMS to be effective. It excludes information that does not need to be retained or maintained or made available for others to use during their work such as that captured temporarily to aid the memory. It will be information that to be conveyed accurately, reliably, repeatedly and consistently from one source to another, it needs to be captured on some medium (e.g. paper, magnetic, electronic or optical computer disc, photograph or master sample, or combination thereof) and its development, use, maintenance, storage and disposal controlled.

Rather than stipulate the documents that need to be controlled, other than what is required elsewhere in the standard, ISO 9001 provides for the organization to determine what information needs to be controlled for the QMS to be effective. This would include several different types of documents. Some will be product, service and process specific, and others will be common to all processes.

Why is this necessary?

To document everything you do would be impractical and of little value. The standard explains that the extent of quality management system documentation can differ due to the size of the organization and its type of activities, processes, products and services, the complexity of the processes and their interactions, and the competency of personnel.

How is this addressed?

Reasons for not documenting information

There are several reasons for not documenting information:

*   If the course of action or sequence of steps cannot be predicted, a procedure or plan cannot be written for unforeseen events.
*   If there is no effect on performance by allowing freedom of action or decision, there is no mandate to prescribe the methods to be employed.
*   If it cannot be foreseen that any person might need to take action or make a decision using particular information from a process, there is no mandate to require the information to be documented. (However you need to look beyond your own organization for such reasons if demonstrating due diligence in a product liability suit requires access to evidence.)
*   If the action or decision is intuitive or spontaneous, no manner of documentation will ensure a better performance.
*   If the action or decision needs to be habitual, documentation will be beneficial only in enabling the individual to reach a level of competence.

Apart from those aspects where there is a legal requirement for documentation, the rest is entirely at the discretion of management but not all managers will see things the same way. Some will want their staff to follow rules and others will want their staff to use their initiative.

Factors affecting the amount of information documented

The documents in regular use need only detail what would not be covered by education and training. If you need something to be done in a particular way because it is important to the outcome, or to ensure it's done in the most economical way, the method will need to be documented so that others may learn the method.

The documents should not be so short as to be worthless as a means of instruction. They need to provide an adequate degree of direction so that the results of using them are predictable. Staff should be trained for routine activities, making documented procedures unnecessary. However, when dealing with activities that are not routine, if you neglect to adequately define what needs to be done and how to do it, don't be surprised that staff don't know what to do when called upon or constantly make mistakes.

Controlled documents

There are three types of controlled documents:

*   Policies and practices, including process descriptions, guides, operating procedures and internal standards.

- Documents derived from these policies and practices, such as drawings, specifications, plans, work instructions, technical procedures, records and reports.
- External documents referenced in either of the first two.

There will always be exceptions, but in general most documents used in a management system can be classified in this way. Although all these types of documents are controlled, they are not subject to the same level of control.

- **Derived documents** are those that are derived by executing processes, for example, audit reports result from using the audit process, drawings result from using the design process, procurement specifications result from using the procurement process. There are, however, two types of derived documents: prescriptive and descriptive documents.
- **Prescriptive documents** are those that prescribe requirements, instructions, guidance, etc., and may be subject to change. They have issue status and approval status and are implemented in doing work.
- **Descriptive documents** result from doing work and are not implemented. They may have issue and approval status. Specifications, plans, purchase orders, drawings are all prescriptive whereas audit reports, test reports, inspection records are all descriptive.

Policies and practices

POLICIES

The only policy ISO 9001 requires is a quality policy but many other policies will constrain the structure and effectiveness of the QMS. Policies are guides to both thinking and action. As a result, they don't tell a person how to do something; they merely channel the decision making along a particular line by delimiting the span of consideration (Hodgetts, 1979). Any statement made by management at any level that is designed to constrain the actions and decisions or choices of those it affects is a policy. Policies don't have to be written down, but those that matter usually are. All policies set boundary conditions but many see policies as requirements to be met – they are requirements but only in so far as constraining an action or decision. By guiding the thinking and action of managers and others policies become essential in ensuring the effectiveness of the QMS; in fact, one could say that policies can make or break a QMS.

Different types of policies may affect the business processes:

- Government policy, which when translated into statutes applies to any commercial enterprise.
- Corporate policy, which applies to the business as a whole and expresses its intentions on particular strategic issues such as the environment, quality, financial matters, marketing, competitors, etc.
- Investment policy – how the organization will secure the future.
- Expansion policy – the way in which the organization will grow, both nationally and internationally.
- Personnel policy – how the organization will treat its employees and the labour unions.
- Safety policy – the organization's intentions with respect to hazards in the workplace and to users of its products or services.
- Social policy – how the organization will interface with society.

- Operational policy, which applies to the operations of the business, such as design, procurement, manufacture, servicing and quality assurance. This may cover, for example,

    - Pricing policy – how the pricing of products is to be determined
    - Procurement policy – how the organization will obtain the components and services needed
    - Product policy – what range of products the business is to produce
    - Inventory policy – how the organization will maintain economic order quantities to meet its production schedules
    - Production policy – how the organization will determine what it makes or buys and how the production resources are to be organized
    - Servicing policy – how the organization will service the products its customers have purchased

- Department policy, which applies solely to one department, such as the particular rules a department manager may impose to allocate work, review output, monitor progress, etc.
- Industry policy, which applies to a particular industry, such as the codes of practice set by trade associations for a certain trade.

Policies can take many forms; the purpose and mission of the organization when expressed by the management become a policy, so do the principles or values guiding people's behaviour – what is or is not permitted by employees, whether or not they are managers. In organizations that have a strong value-based culture, policies are often undocumented. Rules are appropriate to a command and control culture. In all cases you need to ask, *what would be the effect on our performance as an organization if this were not documented?* If the answer is nothing or a response such as "well, somebody might do xyz", *what effect will that have? Can we prevent them doing it through appropriate training?* And if we do all those things *how likely will it happen?* If you cannot predict with a degree of certainty that something will happen that should be prevented, leave people free to choose their own path unless it's a legal requirement. Even in a command-and-control culture one does not need to write everything down, as policies are needed only for important matters where the question of right or wrong does not depend on circumstances at the time, or when the relevance of circumstances only rarely comes into the picture.

A common practice is to paraphrase the requirements of ISO 9001 as operational policy statements and include them in a Quality Manual (see also Chapter 2). Although this approach does provide direct correlation with ISO 9001, it renders the exercise futile because users can read the same things by referring to ISO 9001. Operational policies should respond to the needs of the organization, not the standard. In fact, this is deprecated in clause 0.1 where it states that "it is not the intent of this International Standard to imply the need for alignment of documentation to the clause structure of this International Standard."

PROCESS DESCRIPTIONS

Process descriptions are required indirectly by ISO 9001 in clause 4.4 and are necessary in ensuring the effectiveness of the QMS because they contain or reference everything that needs to be known about a process. See Companion web site for a guide to their content.

PROCEDURES

Procedures are required indirectly by ISO 9001 in clause 4.4.1c) and are necessary in ensuring the effectiveness of the QMS because they lay out the steps to be taken and methods to be used in setting up, operating and shutting down a process. They are often mistaken for processes but in general only prescribe those actions to be taken by people.

A procedure is a sequence of steps to execute a routine task. It informs a user how a task should be performed. Procedures prescribe how one should proceed in certain circumstances to produce a desired output. Procedures are documented when it is important for the routines to be undertaken in the same way each time.

Procedures can only work where judgment is no longer required or necessary. Once you need to make a judgement, you cannot prescribe what you might or might not do with the information in front of you. A form of judgement-based procedure is a decision tree that flows down a chain of questions to which either a yes or a no will route you down a different branch. The chart does not answer your questions but is a guide to decision-making.

There remains confusion between processes and procedures, and this is addressed in Chapter 9.

The relationship between processes and the defining documentations such as procedures, instructions, standards and guides is illustrated in Figure 32.1.

STANDARDS

Standards are required indirectly by ISO 9001 in clauses 4.4.1c), 8.1, 8.3.3, 8.3.5, 8.4.1, 8.5.1 and 9.2.2 and are essential in ensuring the effectiveness of the QMS because they define the criteria required to judge the acceptability of the measured process capability and product quality.

Standards define the acceptance criteria for judging the quality of an activity, a document, a product or a service. There are national standards, international standards, standards for a particular industry and company standards. Standards may be in diagrammatic form or narrative form or a mixture of the two. Standards need to be referenced in process descriptions or operating procedures. These standards are in fact your quality standards. Product and service standards describe features and characteristics that your products and services

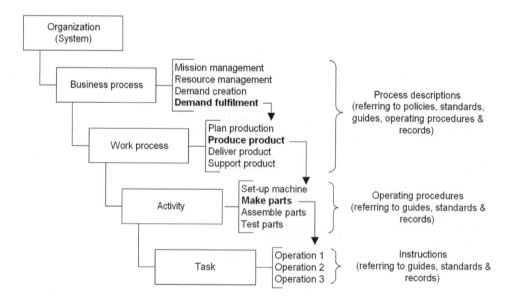

Figure 32.1 The relationship between processes and the defining documents

must possess. Some may be type specific, others may apply to a range of products or types of products and services some may apply to all products and services whatever their type.

GUIDES OR CODES OF PRACTICE

Guides or codes of practice are required by ISO 9001 in clause 8.3.3 and implied in 7.1.6. They are necessary in ensuring the effectiveness of the QMS because they provide information of use during the preparation, operation, shut down and troubleshooting of processes.

Guides are aids to decision-making and to the conduct of activities. They are useful as a means of documenting your experience and knowledge and should contain examples, illustrations, hints and tips to help staff perform their work as well as possible. Without such guides, organizations become vulnerable after laying off staff in a recession. Unless you capture best practice, staff take their experience with them and the organization loses its capability see also Chapter 28.

Derived documents

There are several types of derived documents.

SPECIFICATIONS

Specification are indirectly required by ISO 9001 in clauses 8.2, 8.3, 8.4 and 8.5 These are crucial in ensuring the effectiveness of the QMS because they govern the characteristics of the inputs and outputs and are thus used in process and product design and monitoring and measurement.

PLANS

Plans are required by ISO 9001 in clauses 6.2, 6.3, 8.1, 8.3.2, 9.2 and 9.3. Plans are crucial in ensuring the effectiveness of the QMS because they lay out the work that is to be carried out to meet requirements. Unfortunately, the word *plan* can be used to describe any intent, will or future action so that specifications, procedures and process descriptions could be called plans when they are part of what you intend to do. A plan is a statement of the provisions that have been made to achieve a certain objective. It describes the work to be done and how the work will be done and managed.

REPORTS

Reports are indirectly requirement by ISO 9001 in clauses 5.3, 8.5.3, 9.1.3 and 9.2.2. Reports are useful in ensuring the effectiveness of the QMS because they contain information about the process or the product being processed or the service being provided, the performance of the QMS and feedback from stakeholders. They may be used to guide decision-making both in the design and the operation of processes.

RECORDS

Records are "documents stating results achieved or providing evidence of activities performed" (ISO 9000:2015). Records are required by ISO 9001 in all clauses where documented information is to be retained but this is not indicative that no others are necessary.

Records are essential in ensuring the effectiveness of the QMS because they capture factual performance from which decisions on process performance can be made. The reason for establishing records is to provide information necessary for managing processes, meeting objectives and demonstrating compliance with requirements – both customer requirements and legal requirements. It is from records that reports are compiled.

Records are produced during an event or immediately afterwards. Records do not arise from contemplation. They contain facts, the raw data as obtained from observation or measurement and produced manually or automatically.

There is no requirement to produce records solely to satisfy an auditor. The records concerned are those for the effective operation of the organization's QMS and they are records that need to be controlled (i.e. prevented from loss and alteration). If a record has no useful purpose within the management system, there is no requirement that it be established or maintained.

INSTRUCTIONS

Instructions are indirectly required by ISO 9001 in clause 8.5.1. Instructions are crucial in ensuring the effectiveness of the QMS because they cause processes to be initiated and define variables that are specific to the date and time, location, product or customer concerned. Work instructions define the work required in terms of who is to perform it, when it is to commence and when it is to be completed. They also include what standard the work must meet and any other instructions that constrain the quality, quantity, delivery and cost of the work required. Work instructions are the product of implementing an operating procedure or a document standard (see further explanation below).

In simple terms, instructions command work to be done, procedures define the sequence of steps to execute the work to be done. Instructions may or may not refer to procedures that define how an activity is performed. In many cases an instruction might be a single command such as: *Pack these goods* and not be documented except perhaps on a Post-it Note you find on your desk on return from lunch. Such instructions are transient and unnecessary to retain. However, the manager may issue an instruction for certain goods to be packed in a particular way on a specified date and the package to be marked with the contents and the address to which it is to be delivered; such detail that it needs to be written down. So that the task is carried out properly, the methods of packing may be specified in a procedure. The procedure would not contain specific details of a particular package – this is the purpose of the instruction. The procedure is dormant until the instruction to use it is initiated or until personnel are motivated to refer to it. If the instruction is destroyed after being implemented, there is nothing to refer to if an error is subsequently detected.

INTERNAL REFERENCES

These documents are useful in ensuring the effectiveness of the QMS because they will contain data and information relevant to the equipment, people, facilities or other factors on which an activity or the setup or operation of a process depends. Reference documents differ from other types of documents in that they simply contain data or information that is useful in carrying out a task. They are not used to prescribe requirements or describe results

External reference documents

These documents are those not produced by the organization but used by the organization as a source of data and information, consequently the categories of external reference documents are identified by their source. There are several types, including national and international standards, public data, customer data, supplier data and industry data.

Competence and documentation

Competence may depend on the availability of documentation. For example, a designer will refer to data sheets to assist in selecting components not because of a lack of competence but because of a person's limited memory and a desire for accuracy. The designer can remember where to look for the relevant data sheets, but not the details. If the document containing the relevant data cannot be found, the designer is unable to do the job and therefore cannot demonstrate competence.

When personnel are new to a job, they need education and training. Documentation is needed to assist in this process for two reasons: first to make the process repeatable and predictable and second to provide a memory bank that is more reliable than the human memory. As people learn the job they begin to rely less and less on documentation to the extent that in some cases, no supporting documentation may be used at all to produce the required output. This does not mean that once the people are competent you can throw away your documentation. It may not be used daily, but you will inevitably have new staff to train and improvements to make to your existing processes. You will then need the documentation as a source of information to do both.

How is this demonstrated?

Demonstrating that documented information necessary for the effectiveness of the QMS has been determined may be accomplished by presenting process descriptions that list the documented information that is to be used in and generated by the process.

Identifying and describing documented information (7.5.2a)

What does this mean?

A document is appropriately identified if it carries some indication that will quickly distinguish it from documents of the same type or purpose and enables its timely retrieval from storage media.

Why is this necessary?

Confusion with document identity could result in a document being misplaced, destroyed or otherwise being unobtainable. It can also result in incorrect documents being located, changed or used. The consequences may be trivial if the error is easily corrected but may be quite serious if it goes undetected, particularly if the document contains information that people are reliant on for its accuracy and legitimacy. Any document that requires a reader to browse through it looking for clues as to what it is, is clearly not appropriately identified.

How is this addressed?

Where documented information is frequently referenced in other documents or in conversation it would be appropriate if the reference contained meaningful elements that enables timely retrieval, but inappropriate if it was just a random number with no meaning and difficult to locate. However, it may be sensible to assign meaningless numbers automatically in electronic storage system to prevent inadvertent overwriting of current versions. In such cases, there, is an identification by which a document is known and a technical identification which is used electronically.

All documents should carry a date and where documents are subject to change it is also prudent for the identification to include the version and its date of release. In addition, the issuing authority should be denoted so that users are aware of the documents origin. A document description is the title it carries on the front and/or in the header or footer and may include a subtitle for further identification.

How is this demonstrated?

Demonstrating the appropriate identification and description of documented information may be accomplished by:

a)   presenting documentation standards showing the identification conventions that are to be used and the provisions made to enable users to distinguish between similar documents;

b)   selecting a sample of documents and presenting evidence that the conventions are being followed.

Determining appropriate format and media (7.5.2b)

What does this mean?

Box 32.2   New format and media requirement

The only reference to document format and medium in the 2008 version was a note in clause 4.2.1. It has now been recognized that the format and medium of documented information does have a bearing on its usability and therefore becomes a quality characteristic. It is also an important element in the communication process because if the chosen format and medium is inappropriate the information cannot be decoded by the recipient.

At one time, we only had two ways of transmitting information, oral and written, but then came photography and audio recording on shellac, vinyl, magnetic tape, video tape, floppy disc and later audio and video recording on optical disc and semiconductor memory. Apparently, 2002 marked the beginning of the digital age for information storage, when more information was being stored on digital than on analogue storage devices (Wikipedia (5), 2015).

Why is this necessary?

When we create and transmit information, we often assume it can be decoded by the intended recipient but we may find to our cost that the assumption was unwise. The recipient may not speak the same language, may not possess the version of software to open the file or may not even have a computer. Every organization differs in the devices it uses to create, transmit and access documented information. Small suppliers run their operations on a tight budget so often can't afford to upgrade their software every time there is a new version. There is a risk that recipients won't be able to access the information thereby causing disruption in processes and jeopardizing delivery from suppliers or to customers.

How is this addressed?

Standardizing tools and platforms

To minimize the risk of recipients within the organization not being able to access information provided to them, organizations often standardize on the tools and platforms used to create and transmit documented information. As indicated earlier, this can present a financial burden that causes compromises to be made, for example, the sales and marketing department has upgraded to the latest operating system and association applications because they interface with customers but the facility maintenance department is still using old computers running an operating system and applications that are incompatible with those of sales and marketing. However, they still need access to policies and procedures and if they can't receive them electronically, they will revert to using uncontrolled paper copies.

Any standardization can usually only apply within the organization. Customers and other stakeholders would not take kindly to having to purchase new devices to access the information provided by the organization. For this reason, organizations tend to adopt a common standard for documents transmitted externally unless by prior arrangement. The Portable Document Format (PDF) was invented by Dr John E. Warnock in 1991 for this purpose, and he has guided and shaped not only Adobe technology, which he co-founded with Charles Geschke, but the industry of information technology as a whole.

Designing processes to mitigate risk

In any process that produces information there needs to be a stage where consideration is given to the format in which it is prepared, the format and medium on which it is transmitted and the devices available to the recipient. This means that producers of information need to consider the devices the intended recipients of the information have available and what information needs to be provided prior to opening files that will alert recipients to take the actions needed to enable them to access the file. It is commonplace on websites to provide various options for downloading information (e.g. PDF, WORD, HTML) and different size files thereby accommodating different download speeds. See also Chapter 18 on customer focus.

Updating

Information may have been documented on paper, magnetic tape, electronic or optical computer disc, photographs or a combination thereof and when it comes to updating it consideration needs to be given to its format and medium. Whether the opportunity should be taken to transfer the information from one medium to another and whether to do this for all or only for the information that requires updating, is at the discretion of the organization's management.

Archiving

Archives are collections of documents or records which have been selected for permanent preservation because of their value as evidence or as a source for historical or other research. Consideration should be given to the format and medium of documented information that is destined for archiving because after a few years it may only be accessible on obsolete devices.

Language issues

The language in which documented information is created becomes an issue when the information is needed in other languages and translation services need to be used. Depending on the nature of the information and the users, it may be necessary to require reverse translation by another agency to confirm the accuracy of the first translation. Maintaining synchronization between different language versions can consume a lot of resource, and so translations are not often commissioned until the information has been approved.

Equality issues

Consideration needs to be given to the abilities of the recipient and the regulations that apply such as legislation for the disabled as this may affect font size and colour contrast in paper documents and on computer displays.

Environmental issues

Account needs to be taken of the different locations and environments where the information will be used as information transmitted on paper may be accessible in ambient conditions but in an environment where moisture, dirt, grease is commonplace it won't be durable. Information transmitted electronically may be accessible on a handheld device by a worker inside a building but not be visible when the worker needs it outside the building in bright sunlight.

Security issues

Information transmitted electronically may have embedded security features that prevent copying and printing, etc. These may get in the way of legitimate use by prevent a user printing a copy to take to a meeting.

How is this demonstrated?

Demonstrating that the appropriate format and media is used when creating and updating documented information may be accomplished by:

a) presenting evidence of policies and/or codes of practice that mitigate risks associated with multi-format and multimedia;
b) presenting evidence showing that in process design, consideration is given to format and media issues before documented information is released and updated;
c) selecting a representative sample of documented information of different formats and media and presenting evidence that the declared policies and/or codes of practice have been followed.

Review and approval of documented information (7.5.2c)

What does this mean?

A review is a "determination of the suitability, adequacy or effectiveness of an object to achieve established objectives" (ISO 9000:2015), and translating this into plain English using the definitions in ISO JTCG Guide N360 (see Box 56.1) we get:

a) The suitability of documented information is judged by how well it fits its intended use.
b) The adequacy of documented information is judged by how well it covers the subject matter.
c) The effectiveness of documented information is judged by how well it fulfils its objective.

Appropriate review and approval means that the relevant interested parties have been given the opportunity to determine the suitability and adequacy of the document; that the designated approval authorities have considered the comments from the reviewers and checked there are no outstanding issues and have agreed the documented information may be released.

Why is this necessary?

By subjecting documented information to an approval process prior to its use you can ensure that the documents in use are fit for their purpose and make a positive contribution to the organization. Such a practice will also ensure that no unapproved documents are in circulation, thereby preventing errors from the use of invalid documents

How is this addressed?

When creating or updating a document, it goes through several stages, and a review followed by its approval will be carried out at some point before the document or an updated version of it is released.

Review process

Judging what an appropriate review is depends on the type of document, its purpose and on who the stakeholders are. A review by the author may be appropriate for a document with a narrow interest but a panel of experts including customers might be appropriate for a document with wide interest such as one specifying product or service characteristics.

One of the difficulties in soliciting comments to documents is that you will gather comment on what you have written but not on what you have omitted. A useful method is to ensure that the procedures requiring the document specify the acceptance criteria so that the reviewers and approvers can check the document against an agreed standard. Users should be the prime participants in the review process so that the resultant documents reflect their needs and are fit for the intended purpose.

When judging the suitability of documented information, consideration needs to be given to the contribution it makes to the organization. If its provisions save resources, are a good fit with other policies and practices and advance the organization towards its vision it would be judged as making a positive contribution.

Provision will need to be made for assessing the comments, reconciling differences of opinion and reaching a consensus on the final draft before the document is released for approval.

Approval process

In some cases, it may not be necessary for anyone other than one person to approve a document. In others, it may be necessary for several people to approve a document each approving it for a particular aspect such as technical authority, financial authority, project authority. It all depends on whether the subject matter comes within the scope of one person's authority or several.

The process descriptions or procedures should identify who the approval authorities are, by their role or function, preferably not their job title and certainly not by their name because both can change. The procedure need only state that the document be approved (e.g. by the chief designer) prior to issue. Another method is to assign each document to a custodian. The custodian is a person who takes responsibility for its contents and to whom all change requests need to be submitted. A separate list of document custodians can be maintained, and the procedure need only state that the custodian approves the document.

The standard doesn't require that documents visibly display approval but it is necessary to be able to demonstrate through the process controls that only approved documents can be released into the user domain. With electronic systems, indication of approval is accomplished by electronic signature captured by the software as a function of the security provisions. These can be set up to permit only certain personnel to enter data in the approval field of the document. The software is often not as flexible as paper-based systems and therefore provisions need to be made for dealing with situations where the designated approval authority is unavailable. If you let competency determine authority rather than position, other personnel will be able to approve documents because their electronic signature will provide traceability.

With most electronic file formats, you can access the document properties from the Toolbar. Document properties can tell you when the document was created, modified, accessed and printed. But it can also tell you who the author was and who approved it provided the author entered this information before publication.

How is this demonstrated?

Demonstrating that documented information is subject to appropriate review and approval for suitability and adequacy may be accomplished by:

a) presenting evidence of a process for the review and approval of documented information that is created and updated;

b) showing the provisions made for document review and approval, selection of reviewers and approval authorities, the review criteria, collection and assessment of comments and for document approval when comments have been satisfactorily resolved;

c) selecting a representative sample of documents in use and presenting evidence of their review and approval and how reviewers reached a consensus on the document's suitability and adequacy.

d)   selecting a representative sample of documents in use and providing evidence that they were approved in accordance with declared process prior to release for use.

Availability and suitability of documented information (7.5.3.1a and 7.5.3.2a)

What does this mean?

Availability for use means the users have access to the documents they need at the location where the work is to be performed and when they need access to it. It does not mean that users should possess copies of the documents they need, in fact this is undesirable because the copies may become outdated and not withdrawn from use. Suitable for use means that the version of a document that is available is the version that is to be used for a task and that it's complete and readable using the devices at that location. It may not be the latest version because there may be a reason to use a different version of a document such as when building or repairing different versions of the same product.

Controlling documented information in this context means putting in place provisions that ensure that only valid documents are available for use where and when they are needed and that users are unable to inadvertently gain access to any invalid versions of those documents.

Why is this necessary?

Information essential for the performance of work needs to be available to those performing it otherwise they may resort to other means of obtaining what they need that may result in errors, inefficiencies and hazards.

How is this addressed?

The document availability requirement applies to internal and external documents alike even though there is a separate requirement for external documents to be controlled. Customer documents such as contracts, drawings, specifications and standards need to be available to those who need them to execute their responsibilities. Often these documents are only held in paper form and therefore distribution lists will be needed to control their location. If documents in the public domain are required, they only need be available when required for use and need not be available from the moment they are specified in a specification or procedure. If you provide a lending service to users of copyrighted documents, you would need a register indicating to whom they were loaned so that you can retrieve them when needed by others.

How is this demonstrated?

Demonstrating that documented information is available and suitable for use, where and when it is needed may be accomplished by:

a)   selecting a representative sample of process descriptions;
b)   showing that personnel engaged in these processes can access the specified documented information when and where they need it and that it's suitable for the activities they are required to carry out.

Protection of documented information (7.5.3.1b)

What does this mean?

Box 32.3   New requirement for protection

The 2008 version only required protection of records, which applied to the protection of records from dirt and moisture in the workplace environment. The 2015 version now recognizes the external threats to information technology and extends the protection requirement accordingly.

There are two requirements concerning the protection of documented information. One refers to protection from loss of confidentiality, improper use, or loss of integrity and the other refers to protection against the unintended alteration of records. The second is self-explanatory, but the first does require some explanation as to what it means.

Document information is vulnerable to unauthorized denial of use, modification and release contrary to the desire of the organization controlling the information. An unauthorized person may be able to read and take advantage of information, an intruder can prevent an authorized user from referring to or modifying information and an unauthorized person may be able to make changes in stored information even without reading that information. These intrusions are characterized as computer viruses, malware, phishing, cybercrime, spoofing, hacking, PC hijack, spyware, etc.

Why is this necessary?

The complexity of the modern age means that organizations rely heavily on documented information to function and with that face a range of risks that may affect their ability to function efficiently and effectively. All information held or processed by an organization is subject to threats of attack, error, nature (e.g. flood or fire) etc., and is subject to vulnerability inherent in its use (ISO/IEC 27000, 2016).

There are legal requirements on data protection with which organizations must comply but documented information and the platform on which it is stored is not inherently secure unless made so. The alteration of records by the organization's personnel may be due to human error. However, there are also competitors that would gain an advantage were they to have access to proprietary information and there are other interested parties that seek to acquire information for personal gain or to cause harm and so protection of documented information is necessary to prevent such inadvertent and unauthorized actions.

How is this addressed?

Protection against unauthorized denial of use, modification and release

Most of the protection mechanisms are associated with IT security systems which are outside the scope of this Handbook. For solutions to information protection issues, it is recommended that ISO/IEC 27000:2016 is consulted for general guidance and that ISO/IEC 27002 in particular is consulted for guidance on dealing with particular threats and platforms. The topics ISO/IEC 27002:2013 addresses include:

- Mobile devices and teleworking
- Media handling
- Access control
- Cryptographic controls
- Protection from malware
- Information backup
- Network security
- Information security in supplier relationships
- Compliance issues

ISO/IEC 27002 is designed not only for organizations to use as a reference for selecting controls within the process of implementing an information security management system based on ISO/IEC 27000 but also as a guidance document for organizations implementing commonly accepted information security controls (ISO/IEC 27002, 2013)

From a quality management perspective, it is important that an appropriate information security strategy is adopted that will provide an adequate degree of protection. This strategy should be derived from soundly based principles and practices and provide for the information security provisions to be periodically assessed and the results addressed in the management review see Chapter 56.

Protecting against inadvertent alteration

A way of avoiding inadvertent change in digital media is to publish documents in a portable document format (PDF), CD-R or store them in a database that locks any documents that are retrieved as this provides built-in security measures. Users access the documents through a web browser or directly from a server with controlled access. The users can't change documents but may be permitted to print them and naturally, printed versions would be uncontrolled.

If unprotected digital media are available for users, inadvertent change can be a real problem. A document that has been approved might easily be changed simply because the *current date* has been used in the approval date field. This would result in a change in the *approval date* every time a user accesses the document. Proprietary software claiming to meet the requirements of ISO 9001 for document control may not contain all the features you need. Whatever the controls, they need rigorous testing to ensure that the documents are secure from unauthorized change.

How is this demonstrated?

The objective of a secure system is to prevent all unauthorized use of information. It is hard to prove that this objective has been achieved, for one must demonstrate that every possible threat has been anticipated and a corresponding effective solution put in place. Thus, an expansive view of the problem is most appropriate to help ensure that no gaps appear in the strategy. In contrast, a narrow concentration on protection mechanisms, especially those logically impossible to defeat, may lead to false confidence in the system as a whole (Saltzer & Schroeder, 1975). ISO/IEC 27004 provides guidance on information security measurement.

Demonstrating that documented information is adequately protected may be accomplished from an analysis of the data logs generated by the installed security software together with reports from users.

Controlling distribution, access, retrieval and use (7.5.3.2a)

What does this mean?

Distribution, access, retrieval and use are actions taken to bring documented information to those who require it and therefore in addressing each of these actions we would be deciding the what and why and when and how and where and who of distribution, access, retrieval and use.

Why is this necessary?

To control documented information, it is necessary to control their development, approval, release, security, change, distribution, storage, access, retrieval, use, maintenance, obsolescence and disposal. There are many actions one needs to take with documented information. If any of these actions are not controlled, the information may not be available when required or its quality may be compromised.

How is this addressed?

This is a case for document control procedures but because the distribution, access, retrieval and use of each document may differ the decisions will probably be document specific. What a general document control procedure can do is provide the policies for guiding thinking and action. Distribution, access, retrieval and use are closely related overlapping subjects but to focus on specific aspects of each they are addressed separately.

Releasing documents

The term release in the context of documents means that the digital file is uploaded onto the server to be accessed by authorized users. You will, of course, wish to release draft documents for comment but obviously, they cannot be reviewed and approved beforehand. Your draft documents need to be denoted as such and held on a different server or in a different directory to approved versions and provisions made to prohibit draft documents being released into the user domain.

Distribution of documented information

For documented information to be available to those who need it, the intended user needs access rights to the server where it is located and access to the appropriate devices and software to read the file.

Access to documented information

There will be several reasons why a person needs access to documented information; most of them will be legitimate but it's also necessary to make provision to prevent illegitimate access and unauthorized change.

Where hard copies are taken from digital versions watermarks can be imprinted which alert users that the copy in their possession may not be the latest version and this will prompt a user to check the database.

Access to digital media can be controlled through the server on which they are stored and there are a range of options such as access to view, to print, to comment, to change or to save on another device. These can be restricted so it's necessary to set access permissions that are appropriate to the role of the individual user to prevent undesirable effects, for example, access to archive copies, or *insurance copies*, may be limited to an administrator and access to records will be limited to view and print so that deliberate or inadvertent change is denied. A balance should be attained between security of the information and their accessibility. You may need to consider those who work outside normal working hours and those rare occasions when the troubleshooters are working late, perhaps away from base with their only contact via a computer link.

A question often asked by auditors is: *How do you know you have the correct version of that document?* The question should not arise with an electronically controlled documentation system that prohibits access to invalid versions.

With digital media, access to obsolete documents can be barred to all except the chosen few. All it needs is for operational versions to be held in an operational directory and archived versions to be transferred into an archive directory automatically when a new version is released. On being transferred the revision status should be changed automatically in a way that indicates that later versions have been released.

Retrieval of documented information

You may grant access to documented information but that's of no use if the user cannot retrieve it on demand. For digital media, the most common approach is to use search engines that display results from which a user can select the information wanted. A unique identity, recognizable codes and a date make retrieval easier so that users will know they are not only looking at the right document but it's also the version of it they are looking for. It's important not to put obstacles in the way of people who simply want to do their job. This can happen for the best of reasons such as security and confidentiality, but if a process requires the information, a way should be found to enable its retrieval without undue delay.

In providing record retrieval you need to consider two aspects. There's a need to enable authorized retrieval of records and prohibit unauthorized retrieval. If records are held in a secure directory, you need to nominate certain persons with access rights and ensure that these people can be contacted in an emergency. Your procedures should define how you provide and prohibit access to the records. With digital media, password protection will accomplish this objective provided you control the enabling and disabling of passwords in the records database. For this reason, it is advisable to install a personnel termination or movement process that ensures that passwords are disabled on departure of staff from their current post. The last thing you'd want is to install a user-friendly retrieval mechanism that enabled disgruntled employees to remove the corporate memory!

Another issue is the application software needed to display digital information. You need to be vigilant and assess whether any information is vulnerable to obsolescence in the application software and take appropriate action (e.g. there may be names of departments, locations and other data that may change following a reorganization).

Use of documented information

In pre-planned processes the documented information to be used should be referenced in the process descriptions or in derived documents so that those documents not specified are not essential. However, there may be a need to impose limitations or restrictions as to its use for commercial or security purposes. In exploratory work, such as studies, investigations and surveys, one cannot predict what documents might be needed, and it is these cases that one should proceed with caution.

It is a commonly held belief that objects should be used only for their intended purpose and it's your own fault if there are undesirable consequences but if we had adhered to that policy many innovations would not have come about. In fact, we cannot predict to any degree of accuracy what documentation information will be used other than that which has been prescribed. There is, of course, its intended use, but then there will be other uses and some of these may cause problems. A standard such as ISO 9001 is intended to be used to assess an organization's QMS but we often find that it is used to design the organization's QMS and as a result the QMS turns out to be less effective than it might have been because the standard provides few explanations and no methods. Some documents may be used in investigations, others in litigation and care should be taken to avoid taking statements out of context as the terminology used will be appropriate to the intended context and possibly no other. Language changes over decades so when using documents, research papers from many decades ago, again one should proceed with caution because the terms used may have had a meaning different to the current day. Another misuse is using records as evidence of conformity when if one digs beneath the surface, one finds their quality to be suspect (e.g. the criteria does not match their intended use and even though the results show conformity), there's a lack of objective information such as the prevailing conditions at the time the observations were made.

How is this demonstrated?

Demonstrating that distribution, access, retrieval and use of documented information have been addressed may be accomplished by:

a)   presenting evidence of document control policies that address document distribution, access, retrieval and use to the degree appropriate to the type and class of documents;

b)   selecting a representative sample of key activities covering the scope of the QMS and showing that the personnel:

    i   can access the information they need to do their job (i.e. get into the library);

    ii   can retrieve the specific information they need (i.e. find what they're looking for);

    iii know what authority they have to access, print, change or store offline the information they need;

    iv   are aware of any limitations and restrictions pertaining to use of the information.

Storage and preservation of documented information (7.5.3.2b)

What does this mean?

As documented information can be contained on a range of media, the requirement applies not only to the storage and preservation of the media containing the information, but also to the selection of appropriate media on which the information is to be stored.

The storage and preservation of documented information are actions taken to maintain the integrity of information awaiting use and therefore in addressing each of these actions we would be deciding the what and why and when and how and where and who of document storage and preservation. Legibility is one aspect of preservation and refers to the ease with which the information in a document can be read or viewed by the human eye.

Why is this necessary?

The conditions under which documented information is stored are crucial to its integrity. Incorrect temperature and relative humidity can cause significant damage to books and documents in hard copy. Warm, damp conditions provide more energy and so increase the speed of decay. Sources of magnetic radiation can cause loss of data on digital media. Also, if an inappropriate media is selected, the information may not be retrievable on all devices from which it needs to be accessed. The means of transmission and use of documents may also cause degradation such that they fail to convey the information originally intended.

How is this addressed?

Storage conditions

In the digital age, most information can be stored on digital media of one form or another, and users should be made aware of the media manufacture's storage conditions. In both cases, you need to control the process and the conditions of storage. Advice on preservation of books and documents in hard copy, including the environmental conditions that should be maintained, is available from (British Library Preservation Advisory Centre, 2013).

Staff in possession of hard copy and digital media should be alert to changes in storage conditions and take appropriate action when the conditions may be compromised. Staff responsible for the areas in which hard copy and digital media are stored should check periodically the conditions are being maintained.

Storage of records

With paper archives, you will need to maintain records of what is where and if the server is under the control of another group inside or outside the organization, you will need adequate controls to prevent loss of identity and inadvertent destruction. A booking in/out system should be used for completed records when they are in storage to prevent unauthorized removal.

You will need a means of ensuring that you have all the records that have been produced and that none are missing or if they are, you know the reason. One solution is for the file name to contain a sequential number so that not only are they listed in numerical order in the directory, you can also easily detect if any are missing and easily retrieve them. Records should be secured from inadvertent deletion by protection methods that prevent their deletion.

Preservation of legibility

Generally, any document that is printed or photocopied should be checked for legibility before distribution. Legibility is not often a problem with electronically controlled documents. However, there are cases where diagrams cannot be magnified on screen so it would be prudent to verify the capability of the technology before releasing documents. Photographs and other scanned images may not transfer as well as the original and lose detail so care must be taken in selecting appropriate equipment for this task. Not every user will have 20:20 vision! Many organizations now scan records to store in Portable Document Format (PDF) and make them available over a network, but there are variations in the quality of scanned images which can render the records illegible. Checks on image quality should therefore be performed before destroying the original. Documents used in a workshop environment may require protection from oil and grease. Unlike prescriptive documents, records may contain handwritten elements and therefore it is important that the handwriting is legible. Signatures are not always legible, so it is prudent to have a policy of printing the name under the signature. Documents subject to frequent photocopying can degrade and result in illegible areas.

In service areas where staff interface with the customer, till receipts need to be legible for the customer to return goods and exercise their rights to a refund.

Preventing loss

The action taken depends on whether the media is paper or digital and how the information is lost. Paper media can be lost by damage or lost by theft, the former being less serious than the latter. To avoid losing originals they shouldn't be used for work and should be stored securely so that a copy damaged in the work environment is not critical.

To guard against loss of originals it is prudent to produce additional copies of critical information as an insurance against inadvertent loss. These insurance copies should be stored in a remote location under the control of the same authority that controls the original documented information.

Records, especially those used in workshop environments, can become soiled and therefore provisions should be made to protect them against attack by lubricants, dust, oil and other materials which may render them unusable. Plastic wallets can provide adequate protection whilst records remain in use. You will also need to consider loss by fire, theft, and unauthorized removal.

Back-up and offsite data storage

Insurance copies of computer disks should also be kept in case of problems with the hard disk or the file server. Data back-up at defined periods should be conducted and the backed-up data stored securely at a different location than the original data. It's important to have back-up hard drives either disconnected from computers when not being used or password protected so that ransomware cannot install unauthorized encryption.

Outsourced data storage

Another option is to outsource data storage. *Cloud storage* is a model of data storage in which the digital data are stored in logical pools, the physical storage spans multiple servers and often locations, and the physical environment is typically owned and managed by a hosting company. However, outsourcing data storage increases the attack surface area. When data have been distributed, they are stored at more locations, increasing the risk of unauthorized physical access to the data. The number of people with access to the data who could be compromised (i.e. bribed, or coerced) increases dramatically (Wikipedia (6), 2015). But cloud storage does have many advantages so it's a question of managing the risks (see Chapter 21). ISO/IEC 27017 also deals with cloud storage.

How is this demonstrated?

Demonstrating that the storage and preservation of documented information has been addressed may be accomplished by:

a) presenting evidence of the procedures, work instructions or training manuals that specify the storage and presentation policies for documented information;
b) selecting a representative sample of documents in hard copy and digital media within key processes, and verifying:

    i   that the prescribed storage conditions are being maintained;
    ii that the information, particularly any graphics they contain, is legible throughout.

Controlling changes to documented information (7.5.3.2c)

What does this mean?

There are five clauses that address change of one form or another and how they are related is shown graphically in Figure 32.2. Clause 7.5.3.2c only relates to changes to the information carrier (the media) the others relate to changes to the information which is why the requirement refers to version control and not configuration control.

    Changes to documented information are under control when:

*   authorized changes are implemented;
*   unauthorized changes are prevented from being implemented;
*   revised documents are subject to re-approval as verification of continues fitness for use;
*   users can distinguish between controlled and uncontrolled documents;
*   users can determine the version they are looking at;
*   users can establish what has been changed in a document following its revision.

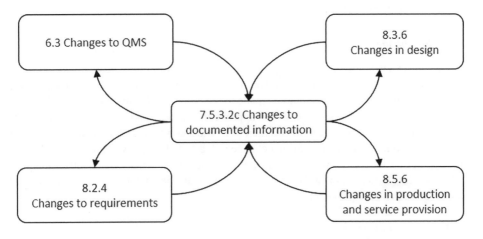

Figure 32.2 Relationship between clauses addressing change

Why is this necessary?

Were there to be no control of changes to documented information, it would quickly lose its integrity and the quality of work that depends on its integrity would deteriorate. It is necessary to denote the revision status of documents so that first and when necessary, planners can indicate the version that is to be used and second, that users can clearly establish which version they are using or which version they require to avoid inadvertent use of incorrect versions.

How is this addressed?

What is a change?

In controlling changes, it is necessary to define what constitutes a change to documented information. If you allow any markings on hard copies of documents, you should specify those that must be supported by change notes and those that do not. Markings that add comment or correct typographical errors are not changes but annotations. Alterations that modify instructions are changes and need prior approval. The approval may be in the form of a change note that details the changes that have been approved.

Depending on the security features of the PDF files it may be possible to annotate documents and save them locally. These are not changes, but the software may also allow text and object changes which are more serious and should be disabled in an approved document.

The change process

The document change process may consist of several key stages such as:

- Request for change
- Permission to change
- Revision of document
- Recording the change
- Review of the change
- Approval of the change
- Issue of change instructions
- Issue of revised document

Request for change

Anyone can review a document but approved documents should only be changed, revised or amended under controlled conditions. When it is decided that documented information needs to change, a request for change should be made to the issuing authority. Even when the person proposing the change is the same as would approve the change, other parties may be affected and should therefore be permitted to comment. The most common method is to employ document change requests. By using a formal change request, it allows anyone to request a change to the appropriate authorities in a way that captures essential information in a retrievable form.

Most document control software provides the necessary fields but the fields you may need in a change request are provision to enter:

- the document title, issue and date;
- the originator of the change request (who is proposing the change, his or her location or department);

- the reason for change (why the change is necessary);
- what needs to be changed (which paragraph, section, etc., is affected);
- the changes in content required where known (this could be supplied by marked-up copy or on comment sheets).

By maintaining a register of such requests, you can keep track of who has proposed what, when and what progress is being made on its approval. You may, of course, use a memo or phone call to request a change, but this form of request becomes more difficult to track and prove you have control. You will need to inform staff where and how to submit their requests.

Permission to change

On receipt of change requests, you need to provide for their review by the change authority. The change request may be explicit in what should be changed or simply report a problem that a change to documented information would resolve. Someone needs to be nominated to draft the new material and present it for review but before that, the approval authorities need to determine whether they wish the document to be changed at all. There is merit in reviewing requests for change before processing to avoid abortive effort. You may also receive several requests for change that conflict and before processing you will need to decide which change should proceed. While a proposed change may be valid, the effort involved may warrant postponement of the change until several proposals have been received – it rather depends on the urgency (see later).

You need to be careful about how you control changes to data. Data that have not been issued to anyone do not require approval if changed. Only the data that have been issued to someone other than its producer need be brought under change control. If you are using data provided by someone else, in principle you can't change it without that person's permission. However, there will be many circumstances where formal change control of data is unnecessary and many where it is vital as with scientific experiments, research, product testing, etc. One way to avoid seeking approval to change data is to give the changed data a new identity, thereby creating new data from old data. It is perfectly legitimate for internal data (but not copyrighted data) because you have not changed the original data provided that others can still access them. If you use a common database for any activities, you will need to control changes to the input data.

Making the change

The technology available for producing and controlling documents has changed dramatically over the last 50 years. There are five levels of technology in use:

- Documents produced, stored and distributed on paper (handwritten or typed).
- Documents produced and stored electronically but distributed on paper.
- Documents produced and stored electronically but distributed on computer disc.
- Documents produced, stored, distributed locally and controlled electronically (intranet).
- Documents produced, stored, distributed worldwide and controlled electronically (Internet).

Each technology requires its own controls such that the controls applied to one type of technology may be totally inappropriate for another technology. Although we live in an age of information technology, all five types operate concurrently. The pen and paper are not obsolete and have their place alongside more sophisticated technologies. Maintenance personnel require documentation that may only be available in paper form although many might be equipped with portable devices with a mobile data link to a central database. Document controls therefore need to be appropriate to the technology used to produce, store, distribute and control the documented information.

Changing hard copy media

Many organizations still use paper – and for good reasons. Paper does not crash without warning! Paper can be read more easily at a meeting, on a train or in sunlight. Comments can be added more easily to paper even if they are electronically produced. However, as the use of hard copy is declining rapidly, it is suggested that hard copy is not subject to change control. It's either valid or invalid, and if invalid it should be destroyed.

Changing digital media

With digital media, a macro can be run on the database to update all references to a particular aspect thus updating automatically all affected documents. Where this mechanism gets complicated is in cases where there are different forms of data capture and storage. For example, the computer-aided design (CAD) data will probably not be generated using the same software tools as the management procedures. Advertising literature may be generated using drawing packages or DTP software and not word processing software. Flow charts may not be generated using word processing software. The technology is not yet available to search and replace information held in different forms on multiple platforms.

Revision conventions

When a document is revised, its status changes to signify that it is no longer identical to the original version. Every change to a document should revise the revision index. Version 1 may denote the original version and Version 1.1 a minor change and Version 2 a major change. Software documents often use a different convention to other documents such as release 1.1, or version 2.3. A convention often used with draft documents is letter revision status whereby the first draft is Draft A, second draft is Draft B and so on. When the document is approved, the status changes to Version 1. During revision of an approved document, drafts may be denoted as Version 1A, 1B, etc., and when approved the status changes to Issue 2. Whatever the convention adopted, it is safer to be consistent to prevent mistakes and ambiguities.

Revision letters or numbers indicate maturity but not age. Dates can also be used as an indication of revision status but dates do not indicate whether the document is new or old and how many versions there have been. In some cases, this is not important, but in others there are advantages in providing both date and revision status.

Identifying changes

There are several benefits in identifying changes:

- Approval authorities can identify what has changed and so speed up the approval process.

- Users can identify what has changed and so speed up the implementation process.
- Auditors can identify what has changed and so focus on the new provisions more easily.
- Change initiators can identify what has changed and so verify whether their proposed changes were implemented as intended.

There are also several ways in which you can identify changes to documents by:

- sidelining, underlining, emboldening or similar technique;
- a change record within the document (front or back) denoting the nature of change;
- a separate change note that details what has changed and why;
- appending the change details to the initiating change request.

With digital media, your problems can be eased by the versatility of the software. Using a database, you can provide users with all kinds of information regarding the nature of the change, but be careful. The more you provide, the greater the chance of error and the harder and costlier it is to maintain. Staff should be told the reason for change if it enables them to understand its relative importance and you should employ some means of ensuring that, where changes to documents require a change in practice, adequate instruction is provided. A system that promulgates change without concern for the consequences is out of control. The changes are not complete until everyone whose work is affected by them both understands them and are equipped to implement them when necessary. Although not addressed under document control, the requirement for the integrity of the management system to be maintained during change in clause 6.3 implies that changes to documented information should be reviewed before approval to ensure that the compatibility between information is maintained. When evaluating the change, you should assess the impact of the requested change on other areas and initiate the corresponding changes in the other information.

Re-approval

Depending on the nature of the change, it may be necessary to provide the approval authorities with factual information on which a decision can be made. The change request and the change record should provide this information. The change request provides the reason for change, and the change note provides details of what has changed.

The change should be processed in the same way as the original document and be submitted to the appropriate authorities for approval. The approval does not have to be by the same people or functions that approved the original. The criteria are not whether the people or functions are the same, but whether the approvers are authorized. With digital media, archived versions provide a record of approvals provided they are protected from revision.

Notification of changes

With digital media, changes can be made to documents in a database without anyone knowing and therefore it is necessary to provide an alert so that users are informed when a change has been made that may affect them. If the information is of a type that users invariably access rather than rely on memory, change instructions may be unnecessary.

Version control of digital media

Regarding digital media, arranging them so that only the current versions are accessible is one solution. In such cases and for certain type of documents, document numbers, versions and dates may be of no concern to the user. If you have configured the security provisions so that only current documents can be accessed, providing release status, approval status, dates, etc., adds no value for the user, but is necessary for those maintaining the database. It may be necessary to provide access to previous versions of documents. Personnel in a product-support function may need to use documentation for various models of a product as they devise repair schemes and perform maintenance. Often documentation for products no longer in production carries a different identity but common components may still be utilized in current models.

How is this demonstrated?

Demonstrating that changes to documented information are controlled may be accomplished by:

a) presenting process descriptions showing how required changes to documented information are processed and unauthorized changed prevented;
b) selecting a representative sample of requests to change documented information and establishing that:

   i the requests have been processed according to the approved procedures;
   ii where a change is approved, the documents concerned have been changed accordingly and not otherwise;
   iii where a change has been rejected the documents concerned have not been changed.

c) selecting a representative sample of documents in use that have been changed and establishing that:

   i the version in use is visible;
   ii evidence can be retrieved authorizing the change in version;
   iii evidence can be retrieved authorizing any visible changes made to the information (hard copy only);
   iv evidence can be retrieved indicating the nature, reason and extent of the change.

Retention and disposal of documented information (7.5.3.2d)

Box 32.4   Records

A record is a statement of an event that has passed. In the context of a QMS, any documented information can constitute a record if it contains evidence of results achieved or activities performed or is evidence supporting those results and activities such as policies, specifications and procedures which define what *was* required to be achieved and how the work *was* required to be performed and conformity measured. In the 2015 version records are referred to as documented information retained as evidence of conformity.

What does this mean?

Documented information has a life cycle. It is created, retained in storage, retrieved, used and maintained. The documented information that is to be retained differs from that which is to be maintained as shown in Table 32.1. When documents that are required to be maintained cease being maintained they become records that apply to work performed up to the date they were superseded. To avoid confusion, we will refer to any documented information that is no longer maintained as records. Some records will contain evidence of results achieved and activities performed and others will provide the supporting evidence for those records (see Box 32.4).

Records are retained if they are useful or required by law. When their usefulness has lapsed, a decision is made as to whether to retain them in an archive or to destroy them. This is the meaning of the word *disposition* in this context (it has nothing to do with the personality of the information!). Customers may specify retention times for certain records as might regulations applicable to the industry, process or region in which the organization operates.

Why is this necessary?

Documented information that contains evidence of activities performed or results achieved is often needed in the management of processes, the diagnosis of problems, decision-making, compilation of reports, demonstration of conformity and to support a defence against prosecution in a court of law. It is therefore important that the provisions are made for their retention and disposition. It may be a criminal offence to dispose of certain records before the limit specified in law. Being unable to retrieve this information may put the situation for which it is needed at risk and therefore retaining such information is a means of risk mitigation.

Documented information that is no longer maintained may become obsolete if it ceases to be used but nonetheless may be retained for use in research, investigations in defending a product liability suit.

How is this addressed?

Retention of records

It is important that records are not destroyed before their useful life is over. There are several factors to consider when determining the retention time for records.

- *The duration of the process* – process control records will need to be retained whilst the process is running and for the life of its outputs to facilitate diagnosis of quality issues in the event of problems downstream, for example, the consequence of an incorrect setting of a process parameter may not be revealed until the product containing the output is in use.
- *The duration of the contract* – some records are only of value whilst the contract is in force.
- *The life of the product* – access to the records will probably not be needed for some considerable time, possibly long after the contract has closed. In some cases, the organization is required to keep records for up to 20 years and for product liability purposes, in the worst-case situation (taking account of appeals) you could be asked to produce records up to 17 years after you made the product.

- *The period between management system assessments* – auditors may wish to see evidence that corrective actions from the last assessment were taken.

You will also need to take account of the external provider's records and ensure that adequate retention times are invoked in the contract.

The document in which the retention time is specified can present a problem. If you specify it in a general procedure you are likely to want to prescribe a single figure, say five years for all records. However, this may cause storage problems – it may be more appropriate therefore to specify the retention times in the procedures that describe the records. In this way you can be selective.

You will also need a means of determining when the retention time has expired so that if necessary you can dispose of the records. The retention time doesn't mean that you must dispose of them when the time expires – only that you must retain the records for at least that stated period. Not only will the records need to be dated but also the files that contain the records need to be dated and if stored in an archive, the shelves, drawers or folders also dated. It is for this reason that all documents should carry a date of origin, and this requirement needs to be specified in the procedures that describe the records. If you can rely on the selection process a simple method is to store the records in bins or digital media that carry the date of disposal.

Should the customer specify a retention period greater than what you prescribe in your procedures, special provisions will need to be made and this is a potential area of risk. Customers may choose not to specify a time and require you to seek approval before destruction. Any contract that requires you to do something different creates a problem in conveying the requirements to those who are to implement them. The simple solution is to persuade your customer to accept your policy. You may not want to change your procedures for one contract. If you can't change the contract, the only alternative is to issue special instructions.

Disposition of records

Disposition in this context means the disposal of records once their useful life has ended. The requirement should not be confused with that on the retention of records. Retention times are one thing and disposal procedures are quite another.

Records are the property of the organization and not personal property so their destruction should be controlled. Controls should ensure that records are not destroyed without prior authorization and, depending on the medium on which data are recorded and the security classification of the data, you may also need to specify the method of disposal. The management would not be pleased to read details in the national press of the organization's performance, collected from a waste disposal site by a zealous newspaper reporter – a problem often reported as encountered by government departments and civic authorities!

How is this demonstrated?

Demonstrating that the organization has addressed the retention and disposal of documented information may be accomplished by:

a)   presenting evidence of the provisions made for the retention and disposal of documented information;

b)   selecting a representative sample of current contracts/projects, selecting records identified in the relevant process descriptions and verifying they have been retained accordingly;

c) selecting a representative sample of contracts/projects that were completed less than n-years ago (where n is the record retention period) then selecting records identified in the relevant process descriptions and verifying they have been retained accordingly;

d) selecting a representative sample of contracts/projects that were completed more than *n* years ago (where *n* is the record retention period), then selecting records identified in the relevant process descriptions and verifying disposal decisions have been made and carried out accordingly.

Controlling documented information of external origin (7.5.3.2)

What does this mean?

An external document is one produced externally to the organization's QMS. There are two types of external documents, those in the public domain and those produced by specific customers and/or suppliers. The requirement does not apply to all external documents that may come into the organization's possession. It only applies to those documents needed for planning the QMS and those it receives from external sources that are used in the processes identified as being within the scope of the QMS. Controls over external documents will be limited to their distribution, security, change, storage, access, retrieval, use, and disposal as all other actions are the responsibility of the issuing authority.

Why is this necessary?

External documents used by the organization are as much part of the QMS as any other document and hence require control so that invalid information is not used and valid information is made available to those who need it for effective operation of the processes.

How is this addressed?

The control that can be exercised over external documents is somewhat limited. You cannot for instance control the revision, approval or identification of external documents but you can ensure such documents are appropriately identified before being made available. You can control the use of external documents by specifying access rights and which versions are to be used and you can remove invalid or obsolete external documents from use or identify them in a way that users recognize them as being invalid or obsolete. If amendment instructions are sent by the issuing agency you can control when the amendments are made and ensure they are made.

Those external documents that are necessary for the planning and operation of the quality management system need to be identified within the relevant documented information and this is not limited to policies and procedures but includes any product-related documents such as drawings and specifications. External documents are likely to carry their own identification that is unique to the issuing authority. If they do not carry reference numbers, the issuing authority is normally indicated which serves to distinguish them from internal documents. Where no identification is present other than a title, the document may be invalid. This sometimes happens with external data and forms. If the source cannot be confirmed and the information is essential, it would be sensible to incorporate the information into an appropriate internal document.

Distribution control

To control the distribution of external documents you need to designate the custodian in the appropriate process descriptions or procedures and introduce a mechanism for being notified of any change in ownership. If the external documents are classified, prior approval should be granted before ownership changes. This is particularly important with military contracts because all such documents must be accounted for. Unlike the internal documents, many external documents may only be available in paper form so that registers will be needed to keep track of them. If digital versions are provided, you will need to make them read-only and put in place safeguards against inadvertent deletion from the server.

Version control

In some cases, the versions of public and customer-specific documents are stated in the contract, and therefore it is important to ensure that you possess the correct version before you commence work. Where the customer specifies the issue status of public domain documents that apply you need a means of preventing their withdrawal from use if they are revised during the term of the contract.

Where the version of public domain documents is not specified in a contract, you may either have a free choice as to the version you use or, as more likely, you may need to use the latest version in force. Where this is the case you will need a means of being informed when such documents are revised to ensure you can obtain the latest version. The ISO 9000 series, for instance is reviewed every five years so could well be revised at five-year intervals. With national and international legislation, the situation is rather different because this can change at any time. You need some means of alerting yourself to changes that affect work being undertaken and there are several methods from which to choose:

* Subscribing to the issuing agency of a standards updating service;
* Subscribing to a general publication which provides news of changes in standards and legislation.
* Subscribing to a trade association which provides bulletins to its members on changes in the law and relevant standards.
* Subscribing to the publications of the appropriate standards body or agency.
* Subscribing to a society or professional institution that updates its members with news of changes in laws and standards.
* Joining a business club which keeps its members informed of such matters.
* As a registered company, you will receive all kinds of complementary information from government agencies advising you of changes in legislation.
* As an ISO 9001–registered company you will receive bulletins from your certification body on matters affecting registration.

The method you choose will depend on the number and diversity of external documents you need to maintain and the frequency of usage.

How is this demonstrated?

Demonstrating that documented information of external origin is identified and controlled may be accomplished by:

a)   presenting evidence of a process for identifying and controlling documented information of external origin;

b) presenting evidence that provision has been made to control the distribution, security, change, storage, access, retrieval, use, and disposal of these documents;

c) selecting a representative sample of current contracts/projects, recognizing documents of external origin, locating them and establishing whether they are identified as such and that their distribution, security, change, storage, access, retrieval, use, and disposal is being controlled in accordance with the governing procedures.

Bibliography

British Library Preservation Advisory Centre. (2013). Managing the Library Archive Environment. Retrieved from www.bl.uk/aboutus/stratpolprog/collectioncare/publications/booklets/managing_ library_archive_environment.pdf

Hodgetts, R. M. (1979). *Management, Theory, Process and Practice*. Philadelphia: W. B. Saunders Company.

ISO 9000:2015. (2015). *Quality Management Systems – Fundamentals and Vocabulary*. Geneva: ISO.

ISO/IEC 27000. (2016). *Information Technology: Security Techniques*. Geneva: International Organization for Standardization.

ISO/IEC 27002. (2013). *Information Technology: Security Techniques*. Code of practice for information security controls. Geneva: International Organization for Standardization.

ISO/TC176/SC2/N1276. (2015). Guidance on the Requirements for Documented Information. Retrieved from ISO/TC 176 Home Page: http://isotc.iso.org/livelink/livelink/fetch/2000/2122/-8835176/-8835848/8835872/8835883/Documented_Information.docx.

ISO Glossary (2016) Guidance on selected words used in the ISO 9000 family of standards Retrieved from www.iso.org/iso/03_terminology_used_in_iso_9000_family.pdf

Saltzer, J. H., & Schroeder, M. D. (1975). *The Protection of Information in Computer Systems*. Cambridge, MA: The MIT Press.

Wikipedia (6). (2015, March 31). Cloud Storage. Retrieved from Wikipedia: https://en.wikipedia.org/wiki/Cloud_storage

Wikipedia (5). (2015, March 31). Data Storage Device. Retrieved from Wikipedia Free Encylopedia: https://en.wikipedia.org/wiki/Data_storage_device

Key messages from Part 7

Chapter 24   People

1 Unlike physical resources, people are adaptable and flexible; they can learn new skills and reach new levels of performance when the right competences are developed and when working in the right environment.
2 As people are vital to enable organizations to achieve sustained success they should be perceived as capital and not resources, which implies that people are what we use to get what we want.
3 The people needed will include those needed to model the QMS, those needed to determine and carry out the planned work represented in the model and those needed to carry out work which results from failure to do work right first time.
4 When recruiting people, it's important that they will not only fit in the organization but also be a good fit to the job they are required to do.
5 Because we all differ in our individual abilities, some types of people are better at some types of jobs than others, regardless of their technical capability.

Chapter 25   Infrastructure

6 The management of the infrastructure is a combination of asset management and facilities management.
7 Account needs to be taken of the constraints the existing infrastructure may impose on processes for the design, development, production and delivery of products and services.
8 Where a new facility is required to provide additional capability the time to market becomes dependent on the infrastructure being in place for production or service delivery to commence.
9 Maintenance of infrastructure goes beyond maintenance of what exists and is more to do with maintaining the capability the infrastructure provides.
10 Maintaining the capability is the domain of contingency plans, disaster recovery plans and business continuity provisions.

Chapter 26   Environment for the operation of processes

11 The performance of employees is influenced by their surroundings and by the people that they are working with as much as by their own innate abilities and the physical conditions of the workplace.
12 The environment under which the infrastructure functions and process outputs are produced is different from the environment under which the people engaged in those processes work.

13 Managing the processing environment is about identifying and addressing factors that will impede or improve process output quality and worker productivity.

14 Workers who become unable to cope with their job because of its mental demands or the stress brought on by having to make difficult decisions are at risk of producing nonconforming product or delivering a poor service.

15 A manager of people needs to understand that all people are different and the performance of anyone is largely governed by the system in which they work.

Chapter 27   Monitoring and measuring resources

16 The integrity of products and services depends on the quality of the resources used to create and assess their characteristics.

17 The measuring and monitoring resources should encompass the sensor, the transmitter and the receiver because the purpose of measurement is to take decisions and without receipt of the information no decisions can be taken.

18 You know nothing about an object until you can measure it, but you must measure it accurately and precisely for the decisions to be soundly based.

19 It is only possible to supply parts with identical characteristics if the measurement processes as well as the production processes are under statistical control.

20 Calibration is necessary where the uncertainty of repeatable and accurate measurements needs to be reduced to an acceptable level.

Chapter 28   Organizational knowledge

21 Organizational knowledge is all the knowledge resources within an organization that can be realistically tapped by that organization.

22 Knowledge can be both embodied in media and embedded in people; therefore, sharing knowledge becomes as much about sharing people as sharing media.

23 Much of an organization's valuable knowledge walks out of the door at the end of the day.

24 There is the knowledge of things which will be different depending on the environment under which they are examined and the knowledge of truths which will be interpreted differently depending on how a person looks at it.

25 Separating the facts from perceptions, hearsay, opinion and guesswork is a matter of belief, truth and justification.

26 There is knowledge that is in a tangible form (explicit), that inhabits the mind of the individual (tacit), that is not made explicit but could be (implicit), that is locked in the organization's artefacts or structures (embedded).

27 Organizations have a knowledge pool which is filled and drained when people join or leave and when media is created, archived and destroyed.

28 An organization's capability is built on its acquisition and application of knowledge and therefore management need to be vigilant to threats which exploit its vulnerability.

Chapter 29   Competence

29 People are often acquired based on their potential and developed and continually redeveloped to suit the organization's needs.

30 Competence is not a probability of success in the execution of one's job but a real and demonstrated capability to achieve intended results.

31 Although a qualified person may have the requisite knowledge and skill and can present their certificates, they may exhibit inappropriate behaviours and be therefore not
competent for certain roles.

32 A person may no longer be competent when they are unable to deliver the required
outputs due to influences of the system in which they work.

33 Competence is particularly important in the professions because the outputs result from
an intellectual process rather than an industrial process.

Chapter 30   Awareness

34 Awareness of policy and objectives means that individuals are more able to select the
right activities to perform in a given context.

35 Awareness creates pride and a correct sense of importance. It serves to focus everyone
on the organization's objectives.

36 People who are unaware of the consequences of their actions are not competent.
People who are aware of the consequences of their actions and ignore them are
reckless.

Chapter 31   Communication

37 The output of a communication process is meaning and this gets passed through other
processes where it may become distorted.

38 With every overt communication, there is covert communication where signals or lack
of signals are inadvertently sent because of involuntary behaviours.

39 Organizations are formed by bringing people together to create something that none
of them could do on their own; therefore, it's vital that these people are able to communicate with one another to get things done.

40 In determining how information is to be transmitted, consideration needs to be given
to the audience and their location along with the urgency, sensitivity, impact and
permanency of the message.

Chapter 32   Documented information

41 Many ISO 9001–certificated organizations have missed the opportunity all along to
exercise their prerogative in the first place to deliberately consider what information
the organization needs.

42 To claim conformity with ISO 9001:2015, the organization must be able to provide
objective evidence of the effectiveness of its processes and its quality management
system which is why the standard still stipulates requirements for certain documented
information.

43 Organizations will need to generate far more information that is required by ISO
9001 to provide conforming products and services but only that which needs to
be documented and controlled for the QMS to be effective needs to be
documented.

44 The complexity of the modern age means that organizations rely heavily on documented
information to function which exposes them to threats that may exploit their
vulnerabilities.

45 From a quality management perspective, it is important that an appropriate information security strategy is adopted that will provide an adequate degree of protection.

46 There are many actions one needs to take with documented information and if any of these actions are not controlled, the quality of processes, products and services may be compromised.

47 The conditions under which documented information is stored are crucial to its integrity.

48 Were there to be no control of changes to documented information it would quickly lose its integrity and the quality of work that depends on its integrity would deteriorate.

49 External documents used by the organization are as much part of the QMS as any other document.

50 The organization cannot control the revision, approval or identification of external documents but it can ensure such documents are appropriately identified before being made available and it can control their use.

Part 8

Operation

Introduction to Part 8

Operations management is an area of management concerned with designing, and controlling the process of designing and producing products and designing and providing services. It basically covers the demand fulfilment process referred to previously that has interfaces with demand creation, resource management and mission management processes. It is also the order to cash process implying that the inputs are orders and the output is cash; therefore, it would include the invoicing and banking activities if we were addressing the whole management system. However, the requirements of ISO 9001 clause 8 also include requirements for the control of externally provided products, services and processes, a process that could fit as comfortably under resource management because its use is not limited to the acquisition of components, but is a process that is used for acquiring all physical resources as well as services and personnel such as contract labour.

Differences between production and service delivery processes

---

Box P8.1   Product quality and service quality

A poor-quality product remains a poor-quality product for the rest of *its* life, whereas a poor-quality service may remain a poor-quality service for the rest of *your* life.

Workers on a production line may never make face-to-face contact with the organization's customers. The interaction there is between workers and customers is generally at the professional and executive levels in a production organization. However, in a service organization, workers often make face-to-face contact with the organization's customers on a more intimate, and regular basis. The operational processes of production often only interact with the customer at the beginning and end, whereas with service processes the customer is involved throughout the process and this creates many more opportunities to instantly delight and distress customers than with production processes. With production processes, the impact of delays, problems with the paperwork and packaging and other grumbles are forgotten if the product delights the customer, but with services, as they deliver more intangible benefits, wait times, queuing, courtesy, empathy and trust have immediate significance. If a product possesses the wrong features or executes those features using unreliable materials, techniques or processes, it won't sustain a market but if a service delivers the wrong advice, a wrong decision, the wrong product it may not sustain a business. You can dispose of the wrong product if all you planned to do was use it and maybe recover your financial loss, but it's more difficult, sometimes impossible, to recover from the wrong advice, a wrong decision or the wrong product if it was ingested as it may stay with you for the rest of your life.

Determination of customer requirements

There are four requirements for determining customer requirements in ISO 9001:

1   Clause 4.2 where it requires the requirements of interested parties (customers are one of the interested parties) that are relevant to the quality management system to be determined due to their effect or potential effect on the organization's ability to consistently provide products and services that meet customer requirements.
2   Clause 5.1.2a where it requires top management to ensure that customer requirements are determined as part of demonstrating leadership and commitment with respect to customer focus.
3   Clause 8.1a where it requires the requirements for the products and services to be determined as one of the actions needed to plan, implement and control the processes needed to meet the requirements for the provision of products and services.
4   Clause 8.2.2a where it requires the organization to ensure that the requirements for the products and services are defined when determining the requirements for the products and services to be offered to customers.

Clause 4.2 provides the reason for doing it; clause 5.1.2a directs that it's done; clause 8.1a) positions the activity as an input to planning, implementation and control processes; and clause 8.2.2a) requires that it's done when determining the requirements for the products and services to be offered to customers.

Externally provided products and services

In clause 8.4 of ISO 9001:2015 there are 21 requirements, some general, some specific, that are not presented in any logical sequence but are placed under three headings that neither reflect a PDCA sequence nor a process sequence nor address externally provided processes, products and services separately. Clause 8.4.1 consists of three unnumbered paragraphs which makes cross-references awkward. All this has made the task of addressing individual requirements in a logical sequence while avoiding repetition rather difficult. Clause 8.4 is therefore addressed through three chapters.

This part of the Handbook comprises 19 chapters (Chapters 33–51) that address clauses 8.1 to 8.7 of ISO 9001:2015.

- Clause 8.1 on operations planning is addressed by Chapter 33
- Clause 8.2 on requirements for products and services is addressed by Chapters 34, 35 and 36
- Clause 8.3 on design and development is addressed by Chapters 37, 38, 39, 40 and 41
- Clause 8.4 on external providers is addressed by Chapters 42, 43 and 44
- Clause 8.5 on production and service provision is addressed by Chapters 45, 46, 47, 48 and 49
- Clause 8.6 on release of products and services is addressed by Chapter 50
- Clause 8.7 on control of nonconforming outputs is addressed by Chapter 51

33 Operational planning and control

Introduction

In Part 4 we addressed the requirements for establishing a quality management system, and meeting these requirements should result in a system of processes that will enable the organization to provide products and services that satisfy customers and other requirements. In many cases, the work processes will have been designed and will form part of the management system. However, the nature and complexity of specific projects, contracts or orders may require these work processes to be tailored or enhanced to suit particular needs. The demand fulfilment process is likely to enable the organization to deliver conforming product and services regardless of the specific features and characteristics of the product or service. It will generate product specific documents therefore the business process won't change. These product or contract-specific documents are what are addressed by these requirements.

There is some duplication in clause 8.1 because the determination of product and service requirements is addressed in 8.1a) and 8.2.2 and therefore we will address both these requirements in Chapter 35.

In this chapter, we examine the 10 requirements of clause 8.1, namely:

- Planning operational processes (8.1)
- Determining product and services requirements (8.1a)
- Establishing process, product and service criteria (8.1b)
- Determining, maintaining and retaining resources for operational activities (8.1c)
- Controlling processes (8.1c)
- Determining, maintaining and retaining documented information (8.1e)
- Operations planning output (8.1)
- Control of planned changes (8.1)
- Review of unintended changes (8.1)
- Control of outsourced processes (8.1)

Planning operational processes (8.1)

What does this mean?

Operations planning is the overall planning activity which determines the work and resources needed to undertake a project, execute a contact or order or provide a service. The processes needed to meet the requirements for the provision of products and services are the processes needed to specify, develop, produce and supply the products or services required.

Why is this necessary?

The business processes which provide the organization's products and services should have been developed so that planning to meet specific orders does not commence from scratch. These processes provide a framework that aids the planners in deciding on the specific processes, actions and resources required for specific projects, contracts or orders. The process descriptions may not contain details of specific products, dates, equipment, personnel or product characteristics. These may need to be determined individually for each product or service – hence the need to plan and develop processes needed to meet the requirements for the provision of (specific) products and services.

How is this addressed?

Typical processes

Several operational processes may be needed including where applicable:

- **A sales process** which determines which of the organization's products and services are required by its customers.
- **A project, contract or order planning process** which plans the provision of the identified products and services.
- **A product design and development process** which designs and validates the hardware and software products needed.
- **A service design process** which designs the services needed.
- **A facility development process** which designs and builds the infrastructure and environment needed to support the production and service delivery processes.
- **A process development process** which designs and develops the processes needed to produce the product and service features that customers require.
- **A service transition process** which develops service delivery capabilities up to full operational status.
- **A procurement process** which acquires the materials, components and services needed to accomplish the design and/or generate or deliver the product or operate the service.
- **A preproduction process** which develops production capability up to full operational status.
- **A production process** which produces the product.
- **A distribution process** which supplies product from stock.
- **A service delivery process** which provides the service.
- **An installation process** which installs the product on customer premises.
- **A maintenance process** which maintains equipment the organization has installed on the organization's or the customer's premises.
- **A customer support process** which provides post-delivery support and warranty services.

The names and scope of these processes will vary dependent on the industry (e.g. in the IT industry, service development may be referred to as service transition). The sales process may be as simple as selling products from stock which is a service or as complex as bidding for a major national infrastructure project which is also a service of selling its capability.

There are too many variations in operations processes to provide much more than an overview. Where the organization designs products and services in response to customer requirements, design would be part of the demand fulfilment process and the relationship between the processes might be as shown in Figure 33.1. Some organizations install the products they design and produce (e.g. an air traffic control system or double glazing), which is why products from production are shown entering the service delivery process. Depending on the nature of the order, orders for existing services and products may not enter operations planning and go direct to service delivery, production or distribution.

For proprietary products, operations planning is driven by the demand creation process which searches for new opportunities that will result in the development of new products and services and thus its output will lead into the demand fulfilment process with proven products and service to sell.

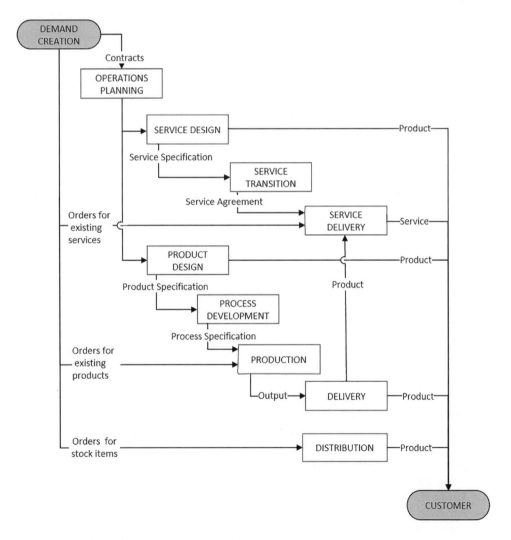

Figure 33.1 Relationship between operational processes

Planning the processes

In planning operations processes, several factors are involved – tasks, timing, responsibilities, risks, opportunities, resources, constraints, dependencies and sequence. The flow charts for each process that were developed in establishing the QMS identify the tasks. The planner's job is to establish whether these tasks, their sequence and the process characteristics in terms of throughput, resources, capacity and capability require any modification to meet the requirements of a particular project, contract or order.

Tools often used in operations planning are Gantt charts and PERT charts. The Gantt chart depicts the tasks and responsibilities on a timescale showing when the tasks are to commence and when they are to be completed. PERT charts display the same information but show the relationship between the tasks. These tools are useful in analysing a programme of work, determining resources and determining whether the work can be completed by the required end date using the allocated resources.

When planning processes, there are several elements that need to be defined. Many of these are addressed in clause 4.4.1, but there are few in the following list that aren't:

- The processes through which the work flows.
- The interconnections among the processes that form the network.
- The objectives and measures of success for the network of processes.
- The event that triggers the network into action, the inputs that are required and what outputs are produced.
- The objectives and measures of success for each process in the network.
- The capacity/throughput required of each process.
- The inputs and outputs for each stage in each process.
- The activities to be performed at each stage and the criteria the outputs are required to meet.
- The process owners/stewards for each process and the chain of processes.
- The allocation of responsibilities in each process.
- Provision for tracking progress through the various processes so that you know what stage the work has reached at any one time.
- The special tools, equipment and other physical resources required to produce the process output (general-purpose tools and equipment need not be specified because your staff should be trained to select the right tool for the job).
- The methods to be used to produce the output.
- The information required for each activity to be performed as planned
- The labour resources required to generate the outputs.
- The environment to be maintained during production of the output if anything other than ambient conditions are needed.
- The process specifications and workmanship standards to be achieved.
- The stages at which verification is to be performed and the methods to be used.
- The handling, packaging, marking requirements to be met.
- The monitoring and measurement to be undertaken.
- The methods of ensuring the integrity of measurement.
- The methods to be used to operate any equipment which generates the outputs.
- Provisions for indicating there are problems and pre-planned solutions for dealing with them so that the impact of delays, bottlenecks, etc., are minimized.
- Provisions for indicating that process outputs are available for use, transfer or delivery.
- The precautions to be observed to protect health, safety and the environment.

Risks and opportunities arising from internal and external issues

The process design requirements should address those risks and opportunities that were identified from an analysis of the internal and external issues that affect the QMS (see Chapter 21), for example,

- If new legislation was detected as a result of scanning the external environment, this legislation would be cited in the process design requirement where it applies.
- If a scarcity of certain materials was detected, constraints would be imposed in the product or process requirements.
- If an opportunity to modernize existing products, services or processes has been identified as a result of new technology, a product, service or process improvement project would be initiated.
- If contactless payment mechanism were detected as being a growing trend in over-the-counter sales, this would be a requirement in the sales process design requirement.
- If the inefficiencies in division of work between front-office and back-office functions was creating a failure demand at the customer interface and therefore becoming a significant risk to customer satisfaction, this would form the basis of a system improvement project as if affects several processes.

Box 33.1   Addressing an opportunity

A major construction project on a city site had very little land for storing materials and so needed many costly lorry deliveries. There was space on an adjacent site where another developer was working. If a deal could be made, it would be possible to use that space to store materials. This possibility was recorded as an opportunity and evaluated.

Risks arising from potential process failures

Not all risks arise from outside the process; many arise from inherent design weaknesses. In order that a process delivers conforming output every time, it is necessary to build protection into it. After designing a process, a way of detecting design weaknesses that has been used successfully in many industries is the Failure Mode and Effects Analysis (FMEA). The basic approach is to answer the questions:

HOW COULD THIS PROCESS FAIL TO ACHIEVE ITS OBJECTIVES? This results in the identification of failure modes. Next ask for each failure mode:
WHAT EFFECT WOULD THIS FAILURE HAVE ON THE PERFORMANCE OF THIS PROCESS? This results in identifying the effects of failure. Then ask:
WHAT COULD CAUSE THIS FAILURE? This identifies the causes of failure of which they may be several so you need to get to the root cause by using the five Whys technique. Now ask:
HOW LIKELY IS IT THAT THIS WILL OCCUR REGARDLESS OF ANY CONTROLS BEING IN PLACE? This results in a list of priorities.

Now examine the process and establish what provisions are currently in place to prevent, control or reduce the risk of failure. Some failure modes might be removed by process redesign.

In other cases, review or inspection measures might be needed and finally, strengthening of the routines, process instructions, training or other provisions may reduce the probability of failure into the realms of the unlikely.

This methodology can be used to test processes for compliance with requirements of national or international standards. Even if no provision has been made to remove the risk, it could be that the risk is so unlikely that no action is needed.

When performing risk assessment, the failure modes should be realistic. They should be based on experience of what has happened – possibly not in your organization but somewhere else. A potential failure is not the one that might never happen otherwise you will never get out of bed in a morning, but it is one that either has happened previously or the laws of science suggest it will happen when certain conditions are met.

The changes that you make as a result of the risk assessment should reduce the probability of process failure within manageable limits. It is recommended that the provisions made to mitigate risk are described in the process description so that in the event they prove ineffective, you can locate and modify them. Examples will be provided on the companion website.

How is this demonstrated?

Demonstrating that the processes needed to meet the requirements for the provision of products and services have been planned may be accomplished by:

a) presenting evidence of plans for the design and production of products and the design and provision of services which identify the network of processes that will deliver the required outputs;
b) presenting evidence of new or changed product, service and process design requirements showing that they take account of risks and opportunities identified as a result of reviewing the context of the organization.

Determining product and services requirements (8.1a)

This requirement is addressed in Chapter 35.

Establishing process, product and service criteria (8.1b)

What does this mean?

The requirement for criteria for the processes is duplicated in clause 4.4.1c, and this aspect of this requirement is therefore addressed in Chapter 16 under the heading "Determining the criteria (4.4.1c)". In this section, we will therefore address only product and service acceptance criteria.

The way the requirements are expressed in clause 8.1 implies there is a difference between requirements for products and services and acceptance criteria for products and services. It all depends on how the customer requirements are expressed. In some cases they may be expressed in definitive and quantifiable terms either directly or by reference to standards. In other cases, they may be expressed in unquantified performance terms (e.g. the service provided shall be reliable, the product shall be safe). It is therefore necessary to establish how reliable is *reliable*, how safe is *safe*. Specifications often contain subjective statements and require further clarification in order that an acceptable standard can be attained. In some cases, the requirements can be verified directly such as when a measurable dimension is stated. In other cases, the measurements to be made must be derived, such as in the food industry where there is a requirement for food to be safe for human consumption. Standards are established for levels of contamination, microbes, etc., which, if exceeded, are indicative food is not safe for human consumption.

Why is this necessary?

To verify that the products or services meet the specified requirements there needs to be unambiguous standards for making acceptance decisions. These standards need to be expressed in terms that are not open to interpretation so that any qualified person using them would reach the same decision when verifying the same characteristics in the same environment using the same equipment.

How is this addressed?

A common method of determining acceptance criteria is to analyse each requirement and establish measures that will indicate that the requirement has been achieved. These are what Deming referred to as operational definitions (see introduction to Part 3). In some cases, national or international standards exist for use in demonstrating acceptable performance. The secret is to read the statement and then ask yourself, *Can I verify we have achieved this?* and if you can ask: *How will I do this?* If not, select a standard that is attainable, unambiguous and acceptable to both customer and supplier that if achieved will be deemed as satisfying the intent of the requirement.

The results of some processes cannot be directly measured using gauges, tools, test and measuring equipment and so an alternative means of determining what conforming product or service is must be found. The term given to such means is *workmanship criteria*, criteria that will enable producers, inspectors or supervisors to gain a common understanding of what is acceptable and unacceptable. Situations where this may apply in manufacturing are soldering, welding, brazing, riveting, deburring, etc. It may also include criteria for finishes, photographs, printing, blemishes and many others. Samples indicating the acceptable range of colour, grain and texture may be needed and if not provided by your customer, those that you provide will need customer approval.

In the service sector standards, will be needed to define the acceptable or unacceptable level of service for each of the service quality characteristics. SERVQUAL a method developed in the 1980s to measure quality in the service sector is a useful starting point (Zeithami, Parasuraman, & Barry, 1990). The authors identified ten determinants of service quality and subsequently reduced them to the following five characteristics:

1 Reliability: the ability to perform the promised service dependably and accurately.
2 Assurance: the knowledge and courtesy of employees and their ability to convey trust and confidence.
3 Tangibles: the appearance of physical facilities, equipment, personnel and communication materials.
4 Empathy: the provision of caring, individualized attention to customers.
5 Responsiveness: the willingness to help customers and to provide prompt service.

Acceptance criteria for these characteristics would define what an acceptable level of service would look like, and this will vary depending on the type of service offered. Analysis of stakeholder needs will identify which characteristics are more important than others and risk assessments will help identify those characteristics which are likely to vary.

The criteria need to be defined by documented standards or by samples and models that clearly and precisely define the distinguishing features that represent both conforming and nonconforming product and service. To provide adequate understanding, it may be necessary to show various examples of workmanship or behaviour from acceptable to unacceptable so that the producer doesn't strive for perfection. These standards, like any others, need to be controlled. Samples of behaviour can be illustrated in a narrative or video. They should be certified as authentic samples and measures taken to preserve their integrity. Ideally they should be under the control of someone other than the person responsible for using them so that there is no opportunity for them to be altered without authorization. The samples represent your company's standards; they do not belong to any individual, and if used by more than one person you need to ensure consistent interpretation by training the users.

How is this demonstrated?

Demonstrating that criteria for the acceptance of products and services have been established may be accomplished by:

a) presenting evidence that the essential product and service characteristic have been defined;

b) presenting evidence that for each defined characteristic, criteria for determining the acceptability of exhibited characterizes have been defined;

c) presenting evidence that these criteria are appropriate controlled to prevent unauthorized modification.

Determining resources for operational activities (8.1c)

What does this mean?

Resources are anything physical or mental that can be used to obtain something else one needs or desires. This requires detailed planning and logistics management and may require many lists and sub-plans so that the resources are available when required. Research, inventory management and information technology are elements of such planning.

Why is this necessary?

All businesses are constrained by their resources. No organization has an unlimited capability. It is therefore necessary when planning new or modified products and services to determine what resources will be required to design, develop, produce and supply the product or service. Even when the requirement is for existing products or services, the quantity or delivery required or the level of demand might strain existing resources to an extent where failure to deliver becomes inevitable.

How is this addressed?

Successful implementation of this requirement depends on managers having current details of the capability of the processes at their disposal. At the higher levels of management, a decision will be made as to whether the organization has the inherent capability to meet the specific requirements. At the lower levels, resource planning focuses on the detail, identifying specific equipment, knowledge, people, materials, capacity and most important, the time required. A common approach is to use a project-planning tool such as Microsoft Project that facilitates the development of Gantt and PERT charts and the ability to predict resources levels in terms of manpower and programme time. Other planning tools will be needed to predict process throughput and capability. The type of resources to be determined may include:

- People
- Infrastructure

  - Special equipment tools, test software and test or measuring equipment
  - Equipment to capture, record and transmit information internally or between the organization and its customers
  - New technologies such as computer aided design and manufacturing (CAD/CAM).
  - Fixtures, jigs and other tools
  - New instrumentation either for monitoring processes or for measuring quality characteristics
  - New measurement capabilities
  - New research and development facilities
  - New handling equipment, plant and facilities, clean rooms, laboratories

- Knowledge and skills

  - New knowledge necessary to undertake new contracts and projects
  - New skills required to operate the processes, provide the service, design new equipment and perform new roles

- Process environment

  - Cleanrooms
  - New style of management

How is this demonstrated?

Demonstrating that the resources needed to achieve conformity to the product and service requirements have been determined may be accomplished by:

a) presenting evidence of a capability assessment undertaken to establish the gap between what resources the organization needs to provide the products and services to be offered to customers and the resources it currently possesses;

b) presenting evidence of resource budgets detailing the additional resources to be acquired and showing how they will be delivered in time for when they are needed – this is often presented in PERT charts that show the resource acquisition timings overlaid on the project bar chart.

Evidence presented to demonstrate conformity with 4.4.1d, 7.1.1 and 6.2.2b might mean that it is also deemed to satisfy conformity with 8.1c but not vice versa because there are other types of resources than those needed achieve conformity to the product and service requirements.

Controlling processes in accordance with the criteria (8.1c)

What does this mean?

Process criteria establishes the standard to be met by the process and implementing control in accordance with the criteria means that when process performance is compared against those standards it will be found to conform. Were a decline in performance to be detected, action would be taken to restore performance to level required to meet the standard.

Why is this necessary?

Processes that are not controlled produce outputs of variable quality because of the natural variation in resources and the process environment. The performance and behaviour of people is probably the most significant variable but variations in the quality of information provided into the process, the materials, equipment and the facilities used by the process all contribute to variation in output quality.

How is this addressed?

Implementing control of processes should be as simple as carrying out the activities that have been defined in the process description. However, when planning a process, we are setting out what we want to happen, anticipating what might go wrong and building in provisions for avoiding those risks. Everything might not go to plan, and therefore when planning processes that are intended to be repeatable, it's prudent to undertake studies to establish just how capable a process is at producing outputs of consistent quality every time it is run (see Chapter 16 under 4.4.1g).

How is this demonstrated?

Demonstrating that process controls are being implementing in accordance with the criteria may be accomplished by presenting the same evidence of process performance as is presented in response to clause 4.4.1g.

Determining, maintaining and retaining documented information (8.1e)

What does this mean?

Many activities will be carried out at various levels of product and service development, production and delivery. These activities will generate data, and these data needs to be collected in a form that can be used to demonstrate that processes have been carried out as planned and that products and services fulfil requirements. This does not mean that every activity or every result needs to be recorded but when planning processes, what data need to be recorded, how they are recorded and when they are to be recorded should be determined as part of the planning activity.

Why is this necessary?

Documented information collected under controlled conditions provides confidence to people not able to observe activities and results for themselves. When investigating failures and plotting performance trends, records are also needed for reference purposes. On particular contracts, only those procedures that are relevant will be applied and therefore the records to be produced will vary from contract to contract. Special conditions in the contract may make it necessary for additional records to be produced.

How is this addressed?

Process records

The records required to provide confidence that processes have been carried out as planned should be identified during process development. With some processes, it may be necessary to record each step in the process but with others the resultant output may indicate that the process can only have been carried out in a certain way.

Continued operation of the process should generate further records that confirm that the process is functioning properly (i.e. meeting the requirements for which it was designed). Process records should indicate the process objectives and exhibit performance data showing the extent to which these objectives are being achieved. These may be in the form of bar charts, graphs, pie charts etc. From the process hierarchy in Figure 32.1, you will observe that there would be process records for business processes, work processes and for activities but not tasks.

Product and service records

By assessing the product or service requirements and identifying the stage in the process where these requirements will be verified, the type of records needed to capture the results should be determined. In some cases, common records used for a variety of products may suffice. In other cases, product-specific records may be needed that prescribe the characteristics to be recorded and the corresponding acceptance criteria to be used to indicate pass or fail conditions.

Record maintenance and retention

For record maintenance and retention see Chapter 32.

How is this demonstrated?

Demonstrating that the evidence necessary to have confidence that the processes have been carried out as planned and products and services conform to requirements may be accomplished by:

a)   presenting process descriptions that specify the documented information that is to be maintained and retained;

b)   indicating in these process descriptions the data that are to be recorded:

   i   to have confidence that the processes have been carried out as planned;
   ii   to demonstrate products and services conformity to the specified requirements.

Operations planning output (8.1b)

What does this mean?

For planning outputs to be suitable for the organization's operations they need to match the input requirements of the processes they feed. The planning process involves making choices, deciding what to do, when to do it, where to do it, how to do it, who will do it and what resources will be needed to do it and therefore will produce different outputs each having a different purpose and it will vary depending on the nature of the product, project, contract or service and its complexity.

Why is this necessary?

The standard does not impose a particular format for the output of the planning activity or insist that such information carriers are given specific labels. Each product and service is different, but unless the planning output is in a form that can be used by the execution processes the plans will remain unimplemented. For those engaged in the executing processes, there will be an expectation that all necessary provisions have been made, for example, that the resources needed have been acquired and that any necessary training has been completed.

How is this addressed?

Some of the decisions may involve a trade-off between two or more possible solutions such as deciding whether to do the work in-house or subcontract it. If the decision is to subcontract part of the work the planning process needs to produce an output that feeds into the procurement process and the planning process won't be complete until a contract has been agreed with a suitable contractor. For simple products, the planning output may be a single document such as a project plan. For complex products, the planning output may be in the form of a project plan and several supplementary plans each being in the form of a manual with several sections.

   Discrete plans are needed when the work to be carried out requires detailed planning beyond that already planned for by the QMS. The planning outputs are dependent on the work that is required but may include:

* Project plans
* Product or service development plans
* Production or service delivery plans
* Configuration management plans
* Risk management plans
* Product assurance plans
* Quality plans
* Procurement plans
* Reliability and maintainability programme plans
* Control plans or verification plans

- Installation plans
- Commissioning plans
- Performance evaluation plans
- Contingency plans

It is not necessary to produce a separate quality plan if the processes of the QMS that are to be utilized are identified in the project plan. Sometimes the project is so complex that a separate quality or quality assurance plans may be needed simply to separate the subject matter into digestible chunks. A useful rule to adopt is to avoid giving documents a title that reflects the name of a department wherever possible.

How is this demonstrated?

Demonstrating that the outputs from planning are suitable for the organization's operations is primarily about communication but there are two aspects to consider:

a) whether an execution process exists for each of the plans and in which case:

   i   Which process will execute the activities defined in the plan?
   ii What checks have been carried out to verify compatibility between the planning outputs and the process inputs of their receiving process?

b) whether there is a planning output for each of the processes we expect to execute on this project or contract and if not:

   i   From where will the process receive its inputs?
   ii What checks have been carried out to establish that there are no project specific conditions that apply to this process?

Control of planned changes (8.1)

What does this mean?

There are several requirements in the standard referring to the control of changes in operations:

- Handling changes to contracts in clause 8.2.1
- Changes to requirements for products and services in clause 8.2.4
- Control changes made during, or after, the design and development of products and services, in clause 8.3.6
- Control changes for production or service provision in clause 8.5.6

It is therefore assumed that the changes being referred to here are intended to be changes to operational plans as a result of changes to contracts, products or services.

Why is this necessary?

Plans are made for developing and providing products and services so that objectives are achieved. If we fail to plan, we are planning to fail and therefore any situation that could affect the integrity of the plans we have made needs to be brought under control.

A lot of the effort put into operations planning goes into determining resources, and therefore if the internal or external situation changes the resources may be insufficient or in the wrong place and thus jeopardize success.

How is this addressed?

Any planned change to a project, contract or order should be passed through the same process as produced the initial operational plans. The planning process is therefore restarted, and on large projects there may be a function dedicated to handling such changes.

The change may arise at any stage in the product or service life cycle and should be reviewed for impact on cost, schedule and quality. This is often handled by making provision on the change proposal form for these elements to be considered and if any are affected, further assessments are to be provided. Following agreement to change proposals, work commences on changing the plans and these changes approved before the revised plans are implemented.

Perhaps the most important issue with planned changes is communication, making sure those potentially impacted by the change are aware of it and given an opportunity to submit impact statements and be involved in discussions about changing the plans. It's therefore important to track who has been contacted and what responses have been received. Proceeding with a planning change without positive confirmation from those affected may be risky so careful consideration needs to be given to the potential consequences.

How is this demonstrated?

Demonstrating that the organization is controlling planned changes to operational plans may be accomplished by:

a)   presenting evidence of a pre-defined process for handling changes to operations planning:
b)   showing how this process maintains the integrity of the operational plans;
c)   presenting a record of the planned changes that are being or have been made;
d)   selecting a representative sample of planned changes and showing that they are being undertaken in accordance with the pre-defined process and specifically that:

  i   those involved in the initial planning (or their successors) were consulted and their concerns addressed;
  ii   changes to the documented information are traceable to the planned change that was authorized.

Review of unintended changes (8.1)

What does this mean?

It is not clear whether the requirement is referring to unintended changes to operations planning or to operations. It's doubtful that plans would be unintentionally changed unless by cyber-attack and it's more likely that events take a different course than were planned, for example, events don't start when scheduled or are not completed on schedule and that these have consequences. There may be several reasons for this, some of which are described in Box 33.2.

Why is this necessary?

It's not uncommon for situations to arise that were unexpected and result in there being an unintended change to the plan. As a plan is generally constructed based on dependencies, any delay or alteration in one process will inevitably have a knock-on effect on dependent processes. If chaos is to be avoided, action needs to be taken to mitigate any adverse effects.

Box 33.2   Unintended consequences

The finance management process uses cash flow as the measure of performance and as a consequence delays paying suppliers on time. This has a knock-on effect on production because in retaliation, the suppliers withhold further deliveries until outstanding invoices have been paid.

Another example is where the packer gets a deduction from his wages if there is a customer complaint because the packer is the last person to check the product before delivery. As a consequence, the packer won't let anything through the gate until everything has been checked, including the most trivial of issues.

A component left to its own devices will attempt to kill off other components in a competitive environment whether inside or outside an organization.

How is this addressed?

With planned changes, we have a luxury of considering the potential consequences before the change is made. With unintended changes, they have already occurred and we have to deal with the consequences immediately. The actions taken will depend on what has happened and might vary from a minor adjustment to a complete rethinking of the way the project will proceed.

The planning process needs to be running continuously so that both planned and unplanned changes, including tasks completed early, are put into the schedule and rescheduling undertaken to reveal the impact.

Local managers need discretion to exercise their judgement on making minor adjustments to schedules, logistics, staff assignments and routines to overcome bottlenecks and disputes that upset the smooth running of a process. An informal or formal mechanism for alerting process owners of issues that have arisen needs to be in place. When the unintended changes are more serious such as the loss of a key supplier, a realization that the planned design solution won't work or the project just runs out of money, there needs to be a major project review where all the issues are brought out and serious attention given to the best way to proceed.

How is this demonstrated?

Demonstrating that the consequences of unintended changes are reviewed and appropriate action taken may be accomplished by:

a) presenting evidence of the original project schedule and the current schedule;
b) selecting a representative sample of changes and presenting evidence of reviews that have been conducted and the actions undertaken to mitigate adverse effects;

c)   presenting the results of an analysis undertaken to verify the actions were effective in mitigating the consequences of unintended changes.

Control of outsourced processes (8.1)

In clause 8.1 is a requirement to ensure that outsourced processes are controlled which is duplicated in clauses 8.4.1 and 8.4.2a); this is addressed in Chapter 42.

Bibliography

Zeithami, V. A., Parasuraman, A., & Barry, L. L. (1990). *Delivering Quality Service: Balancing Customer Perceptions and Expectations*. New York: The Free Press.

34 Customer communication

Introduction

The way communication with customers takes place will influence their choice of where to place orders or contracts and whether to sustain or terminate a relationship.

In general, customers communicate with an organization at six stages in their relationship:

1 When expressing their preferences in marketing surveys
2 When making enquiries, placing an order or negotiating a contract
3 When seeking assurance of progress, involvement in design and development or requesting change
4 When taking delivery of product or using the service
5 When making a formal complaint
6 When expressing their views in satisfaction surveys

ISO 9001:2015 prescribes requirements at each of these interfaces, and these are addressed in clauses 8.2.1 and 8.3.2.

In ISO 9001:2008 customer communication was designated a customer-related process along with determination and review of product requirements but customer satisfaction, another customer-related process, was addressed as part of monitoring and measurement. In ISO 9001:2015 the category of customer-related processes has gone which resolved the anomaly with customer satisfaction being omitted, but customer communication is now considered under the general heading of "Requirements related to products and services" as though all communication with customer is about requirements. The quality of communication with customers is vital as its effectiveness is directly proportional to organizational success. Without some form of communication with customers an organization will be invisible to them and fail to attract their interest, so this should not be subordinated to any particular element of a QMS but be a primary element of it.

Box 34.1 Change in definition of who a customer is

ISO 9000:2015 defines a customer as a "person or organization that could or does receive a product or a service that is intended for or required by this person or organization." Previously the person or organization had to receive a product to qualify as a customer, and therefore a potential customer now counts as a customer.

From the reference made to customers one might assume that the requirements apply only to communication with current customers, but the anomaly is resolved by referring to the definition of a customer (see Box 34.1).

Enquiries, orders and contracts begin with potential customers and can continue after they become customers and throughout the relationship until your obligation to them ceases (i.e. when they dispose of the product or cease using the service). The requirements therefore apply before, during and after provision of product or service to potential customers, current customers and former customers still in possession the products. After expiry of warranty a supplier still has an obligation to its customers to alert them to anything that may be a health or safety hazard.

Although the ISO 9000:2015 definition of a customer appends a note that states "a customer can be internal or external to the organization" and advocates that *the next process is the internal customer of the preceding process*, the term *customer* in clause 8.2.1 is obviously external to the organization as it would be absurd to treat the interaction between co-workers as customer communication.

In this chapter, we examine the eight requirements of clause 8.2.1, namely:

- Providing information relating to products and services (8.2.1a)
- Handling customer enquiries (8.2.1b)
- Handling customer contracts, including changes (8.2.1b)
- Handling orders, including changes (8.2.1b)
- Obtaining customer feedback (8.2.1c)
- Communicating with customers in relation to customer complaints (8.2.1c)
- Communicating with customers in relation to customer property (8.2.1d)
- Establishing requirements for contingency action (8.2.1e)

Providing information relating to products and services (8.2.1a)

What does this mean?

Considering the explanation of communication earlier, in this requirement we are not dealing with the technical content of product and service information but dealing with the quality of information being communicated, whether passed verbally or in writing. If the information provided doesn't communicate the intended meaning, it's of poor quality. Depending on the type of relationship the organization has with its customers, there may be many processes that interface with customers from which such information is transmitted, and this requirement refers only to the customer communication element of these processes.

Why is this necessary?

Customers are only aware of product and service information that is accessible to them, but whether they receive it or retrieve it (as is the case if the information is posted on the Internet), it must accurately represent the products and services offered; otherwise, it is open to misrepresentation and liable to prosecution in certain countries. It is also necessary to ensure that the information does not create expectations which the organization is unable to satisfy. This is especially pertinent to the dialogue between sales staff and potential customers, who being eager to clinch a sale, inflate the virtues of the product or service being offered, obscure the real costs of ownership or use and make promises on delivery that the organization is unable to honour.

How is this addressed?

As stated earlier this requirement is concerned with the quality of information available to customers and has two dimensions. There is the misleading of customers into believing a product or service provides benefits that it cannot deliver (accuracy), and there is the relationship between information available to customers and information as it would need to be to properly represent the product (compatibility).

Accuracy depends on getting the balance right between imaginative marketing and reality. Organizations naturally desire to present their products and services in the best light – emphasizing the strong points and playing down or omitting the weak points. Providing the omissions are not misleading to the customer this is legitimate. The advertising process should ensure information accurately represents the product or service and does not infringe advertising regulations and sale of goods and services laws (see Chapter 35).

Compatibility depends on maintaining the information once it has been released. Product and service information takes many forms, and keeping it all compatible is not an easy task. An information control process is therefore needed to ensure that information compatibility is maintained when changes are made.

As the requirement applies before, during and after sales the range of information relating to products and services that is transmitted will extend beyond advertising material to include contract/project deliverables as well as price and availability information, when delivering the service and notifying customers of service disruptions and changes. It can also include information transmitted both verbally and in writing or in graphics in meetings, by telephone or in electronic communication.

How is this demonstrated?

Demonstrating that effective processes have been established for communicating with customers about products and services may be accomplished by:

a) presenting evidence that provisions have been made for conveying product and service information to customers and how its quality is assured through the channels and media that are used;

b) inviting auditors to sit in on meetings with customers to observe how product and service information is communicated and a mutual understanding obtained;

c) presenting evidence of communication with potential and existing customers that indicates not only that product and service information has been received but has been acknowledged by the customer as being understood;

d) showing that any person who interfaces with the organization's customers are aware of how important it is that inaccurate information is not formally or informally transmitted to customers;

e) showing how unguarded comments by a representative of the organization do not create expectations that prove difficult to satisfy;

f) presenting evidence of frequent training and monitoring directed specifically at maintaining the quality of communication relative to product and service information.

Handling customer enquiries (8.2.1b)

What does this mean?

The requirement refers only to the customer communication elements of processes in which enquiries from customers are received.

Why is this necessary?

When handling a customer enquiry, customers need to be fed correct information and treated in a manner that maximizes the opportunity of a sale. If the personnel receiving a customer enquiry are uninformed or not competent to deal effectively with an enquiry, a customer may not receive the treatment intended by the organization and either go elsewhere or be misled. Both situations may result in lost business. As the person dealing with the enquiry could be the first contact the customer has with the organization, it is vital that they are competent to do the job.

How is this addressed?

Some typical customer interfaces are identified in Figure 34.1 by the two cartoon characters.

Enquiries from potential customers are the result of the effectiveness of the promotion strategy and the process that implements it. If this has been successful, customers will be contacting the organization to seek more information, clarify price, specification or delivery or request tenders, proposals or quotations

How is this demonstrated?

Demonstrating that there is effective communication with customers in relation to enquiries and that it is managed through an effective process may be accomplished by:

a) showing what provisions have been made to enable customers to enquire about the products and services the organization can and does supply;
b) showing that the channels and media available to the customer to submit enquiries are compatible with what typical customers will expect;
c) showing that the people responding to customer enquiries either verbally or in written material have the necessary product/service knowledge and communication skills;
d) inviting auditors to witness how staff process enquiries and interact with customers;
e) presenting evidence of how enquiries from customers have been processed and a mutually satisfactory outcome achieved;
f) presenting evidence of frequent training and monitoring directed specifically at maintaining the quality of communication in dealing with customer enquiries. A common approach with telephone sales is for customers to be informed through a recorded message that the conversation may be monitored for quality assurance purposes.

Handling customer contracts, including changes (8.2.1b)

What does this mean?

The requirement refers only to the customer communication elements of processes in which contracts are received, negotiated, accepted and changed. Contracts differ from orders

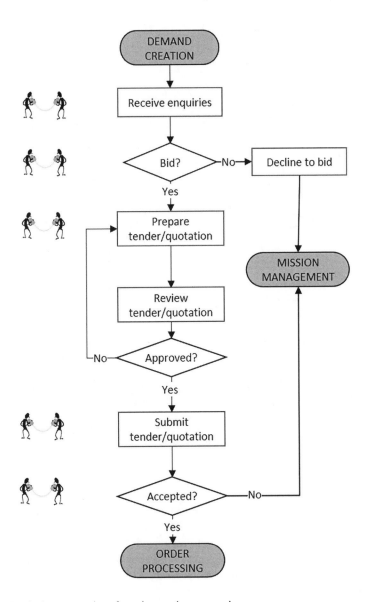

Figure 34.1 Typical customer interfaces in enquiry conversion

because with a contract the product, service and conditions for their supply can be varied whereas with an order, the product, service and conditions for their supply are non-negotiable see below.

Why is this necessary?

Customer enquiries may or may not result in contracts. However, when a contract is received, it is necessary to pass it through an effective process that will ensure both parties are in no doubt as to the expectations under the contract.

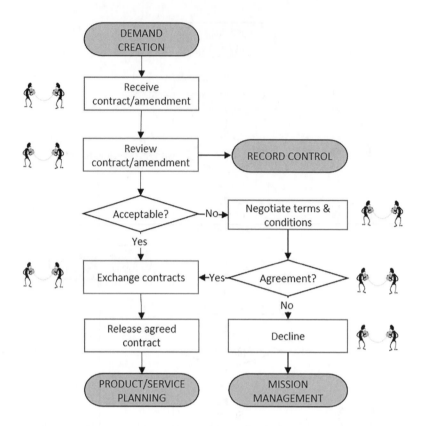

Figure 34.2 Typical customer interfaces in contract management

How is this addressed?

Typically, communication elements of handling contracts are primarily part of a contract management process. Some typical customer interfaces in this process are identified Figure 34.2 by the two cartoon characters.

When a contract is received, several activities need to be performed in addition to the determination and review of product/service requirements and in each of these activities, communication with the customer may be necessary to develop an understanding that will secure an effective relationship

How is this demonstrated?

Demonstrating that effective processes have been established for communicating with customers in relation to contract and changes thereto may be accomplished by:

a) showing what provisions have been made to enter into contracts with prospective customers for the supply of products and services;

b) showing that the staff assigned to negotiate contracts on behalf of the organization have the necessary legal knowledge and communication skills;

c) showing that changes to contracts pass through the same process and are thus subject to the same rigour as the initial contract;

d)   inviting auditors to witness how staff negotiate contracts and interact with customers;
e)   presenting evidence of how contracts from customers have been processed and a mutually satisfactory outcome achieved;
f)   presenting evidence of how changes to contracts have been processed and a mutually satisfactory outcome achieved whether the change is initiated by the customer or the organization;
g)   presenting evidence of frequent training and monitoring directed specifically at maintaining the quality of communication in relation to contracts.

Handling orders, including changes (8.2.1b)

What does this mean?

The requirement refers only to the customer communication elements of processes in which orders are received, accepted and changed.

Why is this necessary?

Customer enquiries may or may not result in orders. However, when an order is received, it is necessary to pass it through an effective process that will ensure both parties are in no doubt as to what to expect.

How is this addressed?

As stated above, with an order the product, service and conditions for their supply are none-negotiable. The customer is presented with a description of the product or service, a price and delivery or availability and can either accept these conditions or go elsewhere. If there are variations, they are usually pre-defined by the supplier thereby limiting any customization.

Typically, these communication elements are primarily part of an order processing process. Some typical customer interfaces in this process are identified Figure 34.3 by the two cartoon characters.

How is this demonstrated?

Demonstrating that effective processes have been established for communicating with customers in relation to orders and changes thereto may be accomplished by:

a)   showing what provisions have been made to enable customers to place orders for the supply of products and services through one or more channels and media;
b)   showing that the information provided conveys everything that a reasonable person would expect to know to make an informed decision such as description, conditions and limitations of use, price and delivery and any restrictions on changing orders;
c)   showing that the staff assigned to accept orders on behalf of the organization have the knowledge and communication skills;
d)   showing that there is provision for processing changes to orders before their fulfilment;
e)   inviting auditors to witness how orders are processed and how they interact with customers;
f)   presenting evidence of how orders from customers have been processed and a mutually satisfactory outcome achieved;

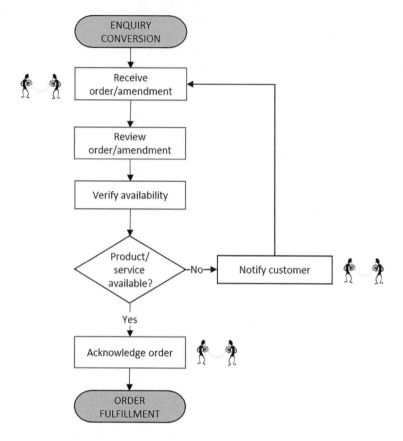

Figure 34.3 Typical customer interfaces in order processing

g) presenting evidence of how changes to orders have been processed and a mutually satisfactory outcome achieved whether the change is initiated by the customer or the organization (e.g. split delivery);

h) presenting evidence of frequent training and monitoring directed specifically at maintaining the quality of communication in relation to contracts.

Obtaining customer feedback (8.2.1c)

What does this mean?

The requirement refers only to the customer communication elements of processes where customer feedback is obtained.

Why is this necessary?

The views and perceptions of customers and other stakeholders constitute the feedback the organization needs to be self-regulating; otherwise, it is akin to driving in the dark without lights and correcting course only after hitting obstacles. One could judge performance only by monitoring sales, but this tends to be a lagging measure. It is much better to have foresight, observe trends and be proactive than receive complaints and be reactive.

When gathering feedback from customers it is vital that the way this is done does not materially affect the results, either by injecting bias or distorting the signals being received (see Box 34.2).

Box 34.2   Customer perception of quality

There was an old tailor in my town who was often heard instructing his partner, "Turn on the blue light. This guy wants a blue suit."

Contributed by Charles Scalies

How is this addressed?

Information can be gathered actively or passively. An example of active information gathering is where marketing staff conduct market surveys seeking to understand customer preferences, or testing perceptions about the organization and its products and services or sales staff conducting order completion surveys by phone or post. An example of passive information gathering is where customers are presented with the opportunity to rank preferences from a list or rate the organization and its products and services through a website or feedback cards. Typically, these communication elements are primarily part of the environment scanning process (see Chapter 12) and the customer satisfaction monitoring process.

How is this demonstrated?

Demonstrating that effective processes have been established for obtaining customer feedback may be accomplished by:

a) presenting evidence that provisions have been made to seek the views and perceptions of potential and existing customers through one or more channels and media;
b) presenting evidence that the staff assigned to obtain the views and perceptions of potential and existing customers have the necessary knowledge and communication skills;
c) inviting auditors to witness how views and perceptions of customers are obtained and how they interact with them;
d) presenting evidence of how views and perceptions of customers have been obtained and any bias and distortion avoided;
e) presenting evidence of frequent training and monitoring directed specifically at maintaining the quality of communication in relation to the acquisition of customer views and perceptions.

Communicating with customers in relation to customer complaints (8.2.1c)

What does this mean?

The requirement refers only to the customer communication elements of processes where customer complaints are received and resolved. A customer complaint is "an expression of dissatisfaction made to an organization, related to its products or services, or the complaints-handling process itself, where a response or resolution is explicitly or implicitly expected". (ISO 9000:2015). The requirements for handling customer complaints are addressed in Chapter 58.

Why is this necessary?

As stated in the preview to this chapter, registering a complaint is one of the six principal stages in the customer–supplier relationship, and therefore messing this up almost certainly will deter a customer from coming back. Making a good impression at this stage by rapidly recovering from a mistake will undoubtedly build customer loyalty and return business.

How is this addressed?

Complaints

There cannot be communication with the customer on complaints unless the organization has opened a channel for communication (e.g. an address and a person or department to contact). For the organization to act promptly, some key information will be needed and provision needs to be made to collect this information. Organizations with a website often provide online forms for their customers to complete which have drop-down menus that save the customer time in locating product and service details.

Not all customers will complain or even feedback dissatisfaction. In some cultures it is considered rude to complain and in others it's considered helpful; it all depends on how the complaint is expressed. In the service sector, customer-facing staff need to read the signs and take appropriate action on the spot. If an opportunity is missed to placate a customer who appears dissatisfied, it's likely they'll simply not return.

When complaints come in, provision needs to be made for prompt acknowledgment, unique reference numbers and a contact name as there is nothing more disconcerting for a customer than not knowing whether the complaint is being dealt with. The next issue for a customer is that when making contact in person, the organization's representative exhibits the appropriate interpersonal and language skills. Provision also needs to be made for escalating resolution of the complaints to a higher level when the front-line person is unable to pacify the customer. Further guidance may be found in ISO 10002 (see Table 4.1).

Disputes

When a customer is dissatisfied with an organization, they will boycott its products and services, and if enough customers are dissatisfied this will force down sales. Customers may take legal action if they have incurred unnecessary costs or hardship because of the organization's failure. Further guidance may be found in ISO 10003 (see Table 4.1).

How is this demonstrated?

Demonstrating that effective processes have been established for communicating with customers in relation to customer complaints may be accomplished by:

a)   presenting evidence that there is an open channel through which customers can register complaints either verbally or in writing at any time;

b) presenting evidence that customers are provided with sufficient information to enable them to determine what information to provide, their eligibility for replacement or refund and the steps in the process;

c) presenting evidence that provisions have been made to engage with customers who have been compelled to complain and to notify them of progress in its resolution;

d) presenting evidence that the staff assigned to capture and resolve customer complaints have the necessary knowledge and communication skills to engage with dissatisfied customers and are objective, open, impartial, sensitive and non-discriminatory;

e) inviting auditors to witness how customer complaints are handled and how they interact with customers through to their resolution;

f) presenting evidence of how customers have been treated following registration of a complaint and misunderstanding avoided;

g) presenting evidence of frequent training and monitoring directed specifically at maintaining the quality of communication in relation to the handling of customer complaints.

Communicating with customers in relation to customer property (8.2.1d)

What does this mean?

The requirement refers only to the customer communication elements of processes where customer property is handled and controlled. Customer property is any property owned or provided by the customer and can include:

- Intellectual property and personal data
- Any property supplied for incorporation into product to be supplied to customers
- Customer-owned tooling and returnable packaging also constitutes customer-supplied products
- Property being used on customer premises such as tools, software and equipment
- Property made available for the organization's use such as test and development facilities

The property being supplied may have been produced by a competitor, by the customer or even by your own organization under a different contract.

Box 34.3   Sony PlayStation account hacking

Sony has gone to considerable lengths to combat hackers since the PlayStation network was infamously compromised in 2011, but initially this was the scenario: Customer's account was hacked; Sony makes customer responsible; customer makes claim for losses on their bank which reclaims money from Sony; Sony then banishes customer from accessing games downloaded until the money is repaid, all of which did not acknowledge Sony's obligations for protecting customer property.

Documentation is not considered customer property because it is normally freely issued and ownership passes from customer to supplier on receipt. However, if the customer requires the documentation to be returned at the end of the contract, it should be treated as customer property. The requirements for handling customer property are addressed in Chapter 47.

Why is this necessary?

When a customer supplies property to an organization certain conditions will be attached regarding its use. Ownership of such property remains in the hands of the customer, and therefore the organization is under an obligation to take due care of it and use it in the agreed manner or as prescribed by law. It is therefore vital that the customer has confidence that the organization has a clear understanding of these obligations.

How is this addressed?

Communication with the customer regarding its supplied property may occur when:

*   taking an order (e.g. credit card details, identity, address);
*   negotiating a contract (e.g. agreeing to conditions of use);
*   taking receipt of product for use during the tenure of a contract;
*   reporting its damage, loss or malfunction;
*   returning property after the end of the contract.

In each of these transactions care needs to be taken to ensure that the customer understands the purpose of the communication and as applicable:

*   is reassured about what will happen to the property being handed over;
*   given the opportunity to ask questions;
*   is satisfied with the outcome before the transaction is completed.

How is this demonstrated?

Demonstrating that effective processes have been established for communicating with customers in relation to customer property may be accomplished by:

a)   presenting evidence that provisions have been made for assuring customers that their property will be protected while in the organization's possession;
b)   presenting evidence that exposes the way contact is made with the customer in the event of loss, damage, misuse or malfunction of their property;
c)   presenting written reports that have been submitted to customers concerning their property.

Establishing requirements for contingency action (8.2.1e)

What does this mean?

The requirement refers only to the customer communication elements of processes where requirements for contingency action are formulated. Typically, these communication elements are primarily part of disaster recovery, business continuity or risk management processes.

Why is this necessary?

ISO 9001 is primarily about giving customers an assurance of product and service quality, and therefore anything that may disrupt the provision of products and services following an agreement to supply is a matter of great concern to customers.

How is this addressed?

The requirement includes the phrase *when relevant* implying that there might be circumstances when communicating with customers about contingency actions is not relevant. Such communication might not be relevant if contained in the supplier's terms and conditions of sale is a disclaimer discharging the supplier from liability for indirect or consequential loss incurred by the customer due to matters beyond its control. If the customer has accepted these terms they have accepted the risk.

Box 34.4   A contingency plan that worked

Cantor Fitzgerald, a financial services company that had its offices on the 101st–105th floors of the North Tower of the World Trade Center in New York, is a prominent example of a successful implementation of a business contingency plan. In the space of two hours, the firm lost 658 of its 960 New York employees in September 11 attacks, as well as much of its office space and trading facilities. Despite these significant losses, the firm was able to resume business within a week.

(Wikipedia (7), 2016)

Where continuity of supply is vital for the customer, it is likely that special terms and conditions will be negotiated and specific requirements for contingency action agreed. In general, contingency actions are implemented in the event of a failure of existing planned actions or circumstances outside the organization's control arising that prevent planned actions from being executed.

Such circumstances include anything that may disrupt the provision of products and services that adversely affects the customer should they occur. They may therefore include natural disasters, major accidents, terrorist attack, criminal activity such as cyber-crime, theft of vital resources or information compromising continuity of supply and unlikely events such as losing key members of staff, key suppliers and access to information.

How is this demonstrated?

Demonstrating that effective processes have been established for communicating with customers in relation to specific requirements for contingency action may be accomplished by:

a) presenting evidence that a channel of communication is open for customers to express their requirement on contingency action (e.g. correspondence on the topic);

b) presenting evidence that there is consensus about:

    i   the identified risks to continuity of supply;

   ii   the nature of their impact on the customer;

  iii   the actions to be taken to recover from a disruption in supply;

  iv   the channels of communication to be opened on detection of an incident;

   v   what validation of the proposed contingency actions is to be undertaken and when;

  vi   what maintenance the organization is expected to undertake to be confident the planned contingency actions remain valid.

Bibliography

Wikipedia (7). (2016, March 26). Retrieved from Wikipedia: https://en.wikipedia.org/wiki/Cantor_Fitzgerald#September_11_attacks

ISO 9000:2015. (2015). *Quality Management Systems – Fundamentals and Vocabulary*. Geneva: ISO.

35 Requirements for products and services

Introduction

There will be requirements that apply to:

- product and service design;
- product manufacture, delivery, installation and servicing;
- service delivery;
- post-delivery activities and support.

A process for determining these requirements should be designed so that it takes as its input the identified need for a product or service and passes this through several stages where requirements from various sources are determined, balanced and confirmed as the definitive requirements that form the basis for development and production of a product or development and provision of a service.

In this chapter, we examine the four requirements of clause 8.2.2, namely:

- Defining product and service requirements (8.1a) and 8.2.2a)
- Defining applicable statutory and regulatory requirements (8.2.2b)
- Defining organizational requirements (8.2.2c)
- Meeting claims for products and services offered (8.2.2b)

Defining product and service requirements (8.2.2a)

What does this mean?

To determine such requirements means that the needs and expectations that are either stated (verbally, or in writing), implied or obligatory must be defined. Defined means stated or described exactly the nature, scope or meaning of (which can be done verbally or in writing).

The heading to clause 8.2.2 is "Determining the requirements related to products and services", which implies not only requirements for the products and services themselves, but also requirements for their fulfilment, delivery and post-delivery; conditions omitted from 8.2.2a) where it requires only that the requirements for the products and services be defined. This is assumed to be an oversight because in clause 8.2.3.1a) it requires a review before committing to supply products and services to a customer, to include requirements specified by the customer, including the requirements for delivery and post-delivery activities.

The products and services to be offered to customers will either be those which the organization intends to:

a)   design and supply to its own specifications (i.e. proprietary design);
b)   design and/or supply to a customer specification (i.e. custom design);
c)   supply but has not designed (i.e. a product manufacturer or service franchise).

Customers are not simply the person or organization that pay the invoice, but include users as well as makers or service providers that purchase products and services for embodiment or use in the product or service which they provide to their customers.

Although a customer may have provided detailed requirements for the product or service expressed in performance terms, it is unwise to believe that this is the only information needed. Some aspects may not have been considered by the customer because they are not specialists. Design input data may come from several sources, as shown in Figure 38.1.

Why is this necessary?

The purpose of the requirements is to ensure that the requirements for products and services are established before work on their design, procurement, production or delivery commences. This is one of the most important requirements of the standard. Most problems downstream can be traced either to a misunderstanding of the requirements of the market or the customer, or insufficient attention being paid to the resources required to meet these requirements. Get these two things right and you are halfway towards providing products and services that satisfy customer needs and expectations.

How is this addressed?

Proprietary products

The organization's requirements should be defined in market specifications or a design brief and include:

- A clear definition of the purpose of the product.
- The conditions of use including the operating environment.
- The functional and physical requirements.
- Intended transportation, transmission or other means for conveying the product to the customer in a specified condition.
- Intended customer support to maintain, service, assist or otherwise retain the product in a serviceable state and to recycle or dispose of the product or any equipment associated with the service.
- Intended in-service date.

Proprietary services

A way of determining service requirements is to imagine you are the customer of the service your organization is offering, ask the following questions and record how you imagine the customer would answer them:

1   What service do we intend to offer?
2   What are the customer's expectations relative to this service?

3   What is the sequence of actions the customer would take from the point of having a need to that need being satisfied?
4   What outputs would the customer expect at each stage through this process to determine if his/her expectations have been met?

By way of an example, Table 35.1 provides answers to these questions for the first two stages customer passes through in a fast food outlet in which 11 stages have been identified. A more comprehensive version is available on the companion website.

Table 35.1 Determining requirements for a fast food outlet

Question	Answers
What service do we provide?	We provide customers in a hurry with good wholesome food at competitive prices
What are the customer's expectations relative to this purpose?	E1 fast, E2 safe, E3 clean, E4 hygienic, E5 has a counter service, E6 represents value for money and E7 offers nutritious food.
What is the sequence of actions the customer would take from the point of having a need to that need being fulfilled? In the example there are 11 steps.	S1 Locate fast food outlet, S2 Observe conditions, S3 View menu, S4 Collect tray, S5 Select food, S6 Pay bill, S7 Choose cutlery and condiments, S8 Find table, S9 Consume food, S10 Dispose of waste, S11 Visit toilets/washroom, S12 Exit premises.
What outputs would the customer expect in order to determine if his or her expectations have been met at stage 1?	The customer expects to be able to locate the food outlet quickly (E1), and therefore signs advertising the outlet would need to be in the right location pointing to the outlet. This output becomes requirement (R1).
Pass the customer expectations through Step 1. Note that not all expectations will apply at each stage.	The customer would expect the signs to provide some indication of the type of outlet (E5). The signs would therefore need to display a clear and relevant message, and this output becomes requirement (R2).
	The customer would also expect that the route to the outlet is not hazardous, (E2) that he is not going to get his clothes dirty along this route (E3) and probably doesn't expect to encounter vermin (E4). The approach to the outlet therefore needs to be clean, safe and vermin free, and this output becomes requirement (R3)
What outputs would the customer expect in order to determine if his expectations have been met at stage 2?	The customer expects to be able to establish whether service will be quick (E1) and therefore queue length needs to be short. This output becomes requirement (R4). The customer also expects a counter service (E5) and the conditions to be safe (E2) and clean with a pleasant décor (E3). Therefore, an output would be a well-designed entry, serving and dining area. This is output becomes requirement (R5). On casting an eye around, customers might expect to see that it is no-smoking establishment, good lighting, directional signs to the toilets, and no vermin or insects (E4). Therefore, all areas would be maintained to high standards. This is output becomes requirement (R6).

Customer-specific products and services

Customers will convey their requirements in various forms. Many organizations do business through purchase orders or simply order over the telephone or by electronic or surface mail. Some customers prefer written contracts; others prefer a handshake or a verbal telephone agreement. However, a contract does not need to be written and signed by both parties to be a binding agreement. Any undertaking given by one party to another for the provision of products or services is a contract, whether written or not. The requirement for these provisions to be determined rather than documented places the onus on the organization to understand customer needs and expectations, not simply react to what the customer has transmitted. It is therefore necessary in all but simple transactions to enter into a dialogue with the customer to understand what is required. Through this dialogue, assisted by checklists that cover your product and service offerings, you can tease out of the customer all the requirements that relate to the product or service. Sometimes, the customer wants one of your products or services but in fact needs another but has failed to realize it. Customer wants are not needs unless the two coincide. It is not until you establish needs that you can be certain that you can satisfy the customer. There may be situations when you won't be able to satisfy the customer's needs because the customer simply does not have sufficient funds to pay you for what is necessary!

Many customer requirements will go beyond end product or service requirements. They will address delivery, quantity, warranty, payment, recycling, disposal and other legal obligations. With every product one provides a service. For instance, one may provide delivery to destination, invoices for payment, credit services (if they don't pay on delivery they are using your credit services), enquiry services, warranty services, etc., and the principal product may not be the only product either – there may be packaging, brochures, handbooks, specifications, etc. With services, there may also be products such as brochures, replacement parts and consumables, reports, certificates, etc.

In ensuring the contract requirements are adequately defined, you should establish where applicable that:

- there is a clear definition of the purpose of the product or service you are being contracted to supply;
- the conditions of use are clearly specified;
- the requirements are specified in terms of the features and characteristics that will make the product or service fit for its intended purpose;
- the quantity, price and delivery are specified;
- the contractual requirements are specified including warranty, payment conditions; acceptance conditions, customer supplied material, financial liability, legal matters; penalties, subcontracting, licences and design rights, recycling and disposal;
- the management requirements such as points of contact, programme plans, work breakdown structure, progress reporting, meetings, reviews, interfaces are specified;
- the quality assurance requirements such as quality management system standards and quality plans;
- reports, customer approvals and surveillance, product approval procedures and concessions are specified.

A specified requirement does not necessarily imply that it must be documented, but it is wise to have the requirements documented in case of a dispute later. The document also acts as a reminder as to what was agreed, and when either of the parties that made the agreement move on, it enables their successors to continue the relationship. This becomes very difficult if the agreements were not recorded, particularly if your customer representative moves on before you have submitted your first invoice. The document needs to carry an identity and if subject to change, an issue status. In the simple case this is the serial numbered order, and in more complicated transactions, it will be a multipage contract with official contract number, date and signatures of both parties.

How is this demonstrated?

Demonstrating that requirements for product and service to be offered to customers are defined may be accomplished by:

a) presenting evidence of a process for determining product and service requirements;
b) selecting a representative sample of products and services and presenting evidence of the requirements that were determined from an analysis of:

  i   market research data;
  ii information provided by the customer.

Defining applicable statutory and regulatory requirements (8.2.2b)

What does this mean?

Almost all products and services are governed by legal requirements that constrain or prohibit certain inherent characteristics of a product or service or the practices employed in their production and or delivery. Customers may specify certain legal requirements within orders and contracts but it should not be assumed that no others are applicable. Organizations have an obligation to comply with legal requirements whether or not they are invoked by their customers.

Many legal requirements apply to an organization and what it does to protect people and property such as those on working time, taxation, discrimination, remuneration, holidays, etc., that don't apply to products and services and others that do (e.g. electrical and mechanical safety, prohibited materials, food hygiene, product labelling and emissions).

Why is this necessary?

Whether or not a customer invokes the applicable legal requirements they will expect a supplier of products and services to be fully aware of them and be compliant without exception. Customers are not always aware of the legal requirements and may require a feature of a product or service that contravenes legal requirements. Were the organization to be ignorant of such requirements, it cannot be used in their defence, and therefore the onus is on the organization to determine the legal requirements that apply and not be compromised by customers unwilling to pay the price. These requirements are non-negotiable and must be complied with. Failure to comply with a legal requirement may result in prosecution.

How is this addressed?

Before you can identify the relevant legal requirements, you need to understand the context of the organization (see Chapter 12), and a profile of the intended markets and the products or services that the organization intends to offer to those markets to enable you to identify:

- The sector of the population the products or services will be designed to serve, for example, if designed for or inclusive of disabled people, legislation protecting the disabled will be relevant.

- The markets into which the product or service will be sold. There are national markets, common markets, single markets and unified markets, each with different regulations. It is only in a unified market that there is imposition of uniform product standards. There are currently six common markets and two proposed as of 2016. All these promote the four freedoms of people, goods, services and capital but each is at a different stage of development (see Chapter 1 under *ISO 9001 and the free movement of goods and services*).

- The countries of the world where the product will be used or the service delivered, for example, certain substances, materials, practices may be forbidden in some countries but allowed in others.

- The industrial sector into which the product or service will be sold. There are factors that are common to particular industries such as aerospace, automotive, oil and gas, food and drugs which drive the regulation of products or services in those sectors and as a result, a product that conforms to one sector's regulations may be nonconforming with another sector's regulations. If the organization maintains a register of requirements (ROR), this will include those legal requirements that apply to the type of organization it is. Among these will be legal requirements that apply to the type of products and services it offers to its customers and the type of processes it uses to produce those products and deliver those services. Beyond this, there may be additional legal requirements that only apply to a specific product, service or process.

There are legal requirements that don't apply to the product or service directly but which are related to the product or service through the processes that generate them. Although there may be no pollution or safety hazard from using the product, there may be pollution or safety hazards from making the product and, therefore, regulations that apply to production processes are indeed product related. However, these are not within the scope of this requirement of ISO 9001, but if noncompliance results in disruption to supply it may affect the degree to which an organization can consistently provide product or service that meet customer requirements. Therefore, a noncompliance constitutes a risk as defined in clause 6.1.1b, for example,

- A failure to observe government health and safety regulations could close a factory for a period and suspend an organization's ability to supply customers.

- Health and safety hazards could result in injury or illness and place key personnel out of action for a period and thus impact continuity of supply of product or services to customers.

- Environmental claims made by customers regarding conservation of natural resources, recycling, etc., may be compromised if environmental inspections of the organization show a disregard for such regulations.

- The unregulated discharge of waste gases, effluent and solids may result in public concern in the local community and enforce closure of the plant by the authorities and thus affect your customers.

- A failure to take adequate personnel safety precautions may put products at risk.

- A failure to dispose of hazardous materials safely and observe fire precautions could put the plant at risk and potentially affect customers.

- A failure to provide safe working conditions for personnel may result in public concern and local and national inquiries that may harm the reputation of the organization, leading to customer dissatisfaction.
- A failure to observe the codes of conduct of a professional body that regulates the professional services provided by an organization may result in disqualification of key staff and clients or their property being harmed or put at risk.

The requirement also applies to products that are purchased and are resold under the original manufacturer's label or rebadged under your label or incorporated into your product. There may be legal requirements that only apply to products you have purchased because of their particular form, function or material properties. Such regulations may not apply to your other products but they are part of what you have defined.

In the service sector, many services are delivered within an environment inclusive of customers (e.g. hospitality industry, healthcare, social work, education, public transport, sport and recreation), and therefore their protection is paramount and must be designed into the service and therefore health and safety regulations and certain environmental legislation does apply (e.g. fire regulations, hygiene, air quality, ventilation and temperature control).

The conduct of professional staff of organizations providing services, such as architects, lawyers, solicitors and medical practitioners, will be governed by their respective professional body or by law, and these regulations are an applicable legal requirement. If the legal requirement constrains the design, use or disposal of the product or design, use/delivery or termination of a service, in the context of ISO 9001 it is an applicable legal requirement. If the legal requirements constrain the processes by which products are produced they may be outside the scope of ISO 9001 (e.g. occupational health and safety regulations, environmental legislation), but measurement traceability, which is not an inherent characteristic of a product, may be a legal requirement that applies to the measurement process not the product but is within the scope of ISO 9001 see clause 7.1.5.

If you intend exporting the product, it would be prudent to determine the regulations that would apply before completing the design requirements. Failure to meet some of these requirements can result in refusal to grant an export or import licence.

How is this demonstrated?

Demonstrating that applicable statutory and regulatory requirements will be and have been defined may be accomplished by:

a) presenting evidence that a process has been established for this purpose and is, resourced and in operation;
b) presenting evidence of analysis carried out to determine statutory and regulatory requirements that are applicable to specific products and services to be provided;
c) showing that the sources of information used are legitimate (e.g. legal counsel, libraries, agencies, trade associations and government departments).

Defining organizational requirements (8.2.2c)

What does this mean?

In addition to the requirements specified by the customer or in the design brief and the regulations that apply, requirements may be imposed by the organization's policies that impinge on the particular products or services that are to be supplied. The product policy may impose certain style, appearance, reliability and maintainability requirements or prohibit use of certain technologies or materials. Other requirements may serve to aid production or distribution that are of no consequence to the customer but necessary for the efficient and effective production, storage and supply of the product. The organization's requirements may also include requirements for traceability, for example, product and service identification to enable fault diagnostics, recall, segregation, etc.

Why is this necessary?

The requirement is necessary in order that market needs, customer requirements and relevant organizational policies and objectives are deployed through the product and service offerings. A failure to identify such requirements and constraints at the requirement definition stage could lead to abortive design work or, if left undetected, lead to the supply of products or services that harm the organization's reputation and fail to satisfy customers. Often, an organization is faced with the task of balancing customer needs with those of other stakeholders. It may therefore be appropriate in some circumstances for the organization to decline to meet certain customer requirements because they conflict with the needs of certain stakeholders. On the other hand, it may cheat, as in the case the Volkswagen's emissions scandal (see Box 35.1).

Box 35.1   Volkswagen's emissions scandal

The company admitted that 11 million cars worldwide may have been fitted with a defeat device designed to cheat emissions tests. The device recognizes when the car is being tested and immediately cuts emissions to a level much lower than normal but which is unsustainable under normal driving conditions.

   The German company said it was putting aside a provisional €6.5 billion to "cover the necessary service measures and other efforts to win back the trust of our customers".

(The Guardian, 2015)

How is this addressed?

The organization's requirements should be defined in technical manuals that are used by designers, production and distribution staff. These will often apply to all the organization's products and services but will, however, need to be reviewed to identify the specific requirements that apply to particular products and services. Customers need to be advised if conformity to the organization's own requirements will impact their expectations (e.g. providing unexpected enhancements).

How is this demonstrated?

Demonstrating that product and service requirements considered necessary by the organizations have been determined may be accomplished by:

a)   presenting evidence of a process for determining product and service requirements;
b)   showing that there is stage in this process where requirements of the organization are assessed for relevance and where applicable included among the requirements for a specific product or service;

c)   selecting a representative sample of product and service requirements and showing which were considered necessary by the organizations in addition to customer and legal requirements;

d)   presenting evidence that any additional requirements differing from those of the customer were agreed with the customer.

Meeting claims for products and services offered (8.2.2b)

What does this mean?

When an organization makes descriptions of its products and services accessible to customers it is making certain claims about them which lead customers to believe they will be met. Failing to meet such expectations is a breach of trust. In some countries where goods and services are sold by description they are legally required to correspond with that description (e.g. UK Supply of Goods and Services Act 1982).

Claims for products and services include:

*   Claims referring to the features and characteristics of a product or service and the benefits they bring (i.e. what they can do or don't do).
*   Claims for their availability (i.e. how quickly they can be delivered, or days and times when a service can be accessed).
*   Claims for what the quoted price and warranty include.
*   Claims for where they were manufactured or the origin and nature of the components or ingredients.
*   Claims that imply protection against something that is unsafe or unhealthy.

Why is this necessary?

Any description of a product or service creates expectations that customers will trust to be satisfied. It is exemplary of a customer-focused organization to honour its promises.

How is this addressed?

Any description of products and services through advertising literature, proposals, tenders, displays, the media or verbally falls within the scope of this requirement. There will therefore be several stages in the marketing and sales processes where information relating to the product or service is accessible to customers.

Prior to product/service launch

Once development of a product or service is underway, the marketing team often wants to generate interest to attract advanced orders. There is a risk that the information available will change, and therefore great care must be taken to avoid misleading customers. Stating that the information is preliminary, that it may change and that the organization cannot accept liability for the consequence of its use is what many organizations do to mitigate this risk.

Another issue is myths, that is, phenomena that are believed to be true when there is no corroborated evidence proving it to be true (see Box 35.2).

Box 35.2   The blue light myth

Some dispensing opticians are making claims about the health benefits to be gained from filtering the blue light that is emitted from the display screens of computers, electronic notebooks, smartphones and other digital devices. They claim that the blue light can cause eyestrain, fatigue, headache, sleep, cell damage and age-related macular degeneration (AMD). They are consequently recommending lenses and filters that protect the eye from blue light which adds up to £70 to the cost. However, according to Dr John O'Hagan, who heads the Laser and Optical Radiation Dosimetry Group at Public Health England, the level of blue light emitted from such devices is <1% of the level that would cause eye damage and about one third of that of natural light (BBC Watchdog, 9 November 2016).

This is also confirmed by the Macular Society who states that these devices emit very little blue light, well below internationally established safety limits.

When selling products and services

Customers may issue invitations to tender for the supply of products or services which require new product or service development and in such cases the same precautions as stated above apply. Where existing products and services are offered for sale the claims made about them should be derived from information validated by actual use or design verification and validation. This can be accomplished by:

a) requiring customer-facing employees to become familiar with the features and characteristics of the product or service through communication with its producer or provider;

b) requiring any advertising material to be subject to review by the producer before being released.

The *producer* might be the chef in the restaurant, the author of the book, the designer of the product or service, the manager of the store, the provider of the service.

Care needs to be taken with products and services that are offered in different grades such that:

a) The grades offered match the grades available.

b) The price quoted is consistent with the grades offered and is unambiguous.

c) There is no ambiguity regarding discounted prices as to what is being offered for the price.

How is this demonstrated?

Demonstrating that the organization has confirmed it can meet the claims for the products and services it offers may be accomplished by:

a) presenting evidence of a process for the control of marketing and sales information by any means;

b)   presenting evidence of a process for familiarizing customer facing employees with the organizations products and services;

c)   selecting a representative sample of marketing and sales information and showing that the claims made:

    i   are consistent with what the organization can provide;

    ii   comply with relevant advertising regulations;

    iii   are justified based on objective evidence testifying the veracity of the information provided.

Bibliography

The Guardian. (2015, September 22). VW scandal: Chief Executive Martin Winterkorn Refuses to Quit. Retrieved from The Guardian: www.theguardian.com/business/2015/sep/22/vw-scandal-escalates-volkswagen-11m-vehicles-involved

36 Review of requirements for products and services

Introduction

A customer-focused organization will find out before they offer certain products and services to customers whether they have the capability to honour their obligations, and that the products and service offered will satisfy customer requirements. A company's capability is not increased by offering products and services to a specification it can't meet, or accepting contracts beyond its current level of capability. There may be penalty clauses in the contract or the nature of the work may be such that the organization's reputation could be irrevocably damaged as a result.

It is therefore prudent to take another look at all the various requirements, preferably by someone other than those who gathered the information to safeguard the organization's reputation and the customer interests. The review may be quite independent of any order or contract but may need to be repeated should an order or contract for the product be received.

In some situations, such as internet sales, a formal review is impractical for each order. Instead, the review can cover relevant product information such as catalogues or advertising material.

In this chapter, we examine the 10 requirements of clause 8.2.3 and 8.2.4, namely:

- Ensuring ability to meet the defined requirements (8.2.3.1)
- Reviewing requirements before committing to supply (8.2.3.1)
- Reviewing customer specified requirements before committing to supply (8.2.3.1a)
- Reviewing requirements necessary for intended use (8.2.3.1b)
- Reviewing requirements specified by the organization (8.2.3.1c)
- Reviewing statutory and regulatory requirements (8.2.3.1d)
- Reviewing and resolving requirements differing from those previously expressed (8.2.3.1e)
- Handling undocumented customer requirements (8.2.3.1)
- Retaining documented result of the review (8.2.3.2)
- Handling changes to requirements for products and services (8.2.4)

Ensuring ability before committing to supply (8.2.3.1)

What does this mean?

The period before the submission of a tender or acceptance of a contract or order is a time when neither side is under any commitment and presents an opportunity to take another look at the requirements before legal commitments are made. The organization needs to be able to honour the obligations it intends to enter into with its customers, and therefore checks need to be made to ensure that the necessary capacity and capability are available or will be available to discharge these obligations when required.

Often the contract for design and development and the contract for production or service delivery are two separate contracts. They may be placed on the same organization, but it is not unusual for the production contract to be awarded to an organization that did not design the product for cost reasons. This requirement is concerned with business capability rather than process capability and addresses the question, do we as a business have:

a)   the capability to make this product in the quantity required and deliver it in the condition required to the destination required over the period required and for the price to be paid? Or;

b)   the capability to deliver this service to the designated people or organizations in the designated locations under the stipulated conditions for the price to be paid?

Why is this necessary?

The purpose of the requirement review is to ensure that the requirements are complete, unambiguous and attainable by the organization. It is therefore necessary to conduct such reviews before a commitment to supply is made so that any errors or omissions can be corrected in time. There may not be opportunities to change the agreement after a contract has been signed without incurring penalties.

When the customer places an order or awards a contract, there is a commitment on both sides. The customer commits to pay an agreed sum in return for specified products and/or services. If the supplier is later found to be unable to deliver, they are in breach of the terms and conditions which has legal connotations and it may have adverse consequences for the customer.

Customers will not be pleased by organizations that have underestimated the cost, time and work required to meet their requirements and may insist that organizations honour their commitments – after all, an agreement is a promise and organizations that break their promises do not survive for long in the marketplace.

How is this addressed?

Provision should be made in the demand creation or sales process for a requirement review to take place before offers are made, tenders are submitted, contracts are signed or orders accepted.

To ensure this happens those personnel with responsibility for making offers, submitting tenders, signing contracts or accepting orders need to:

a)   understand the requirements;
b)   have access to information on the organization's capability and
c)   be competent to determine whether the organization can meet these requirements.

This may be role of a salesperson or a panel of experts depending on the complexity and significance of the commitment. This process is needed also for any amendments to the contract or order so that the organization takes the opportunity to review its capability with each change.

To ensure that the organization can meet the requirements for products and services a business risk assessment needs to be carried out that addresses the following questions:

- What new technical capabilities will be needed?
- Will we be able to develop the additional capabilities within the timescales permitted?
- Can we make this product in the quantities required or provide this service to the estimated number of users in the timescales required?
- In consideration of the timescales and quantities required, can we make this product or provide this service at a price that will provide an acceptable profit?
- Do we currently have the slack in our capacity to accommodate a programme of this size?
- Do we have the human, physical resources required and, if not, can we obtain the financial resources in time to acquire them?
- If we can obtain the additional resources, can we ensure they'll reach the level of competence/capability required in the timescales?
- Where applicable do we (or our partners) have the capability to transport this product to the required destination and protect it throughout the journey?
- Where applicable do we (or our partners) have the capability to deploy this service to the required locations?

Many organizations do not have staff waiting for the next contract so it is a common practice for companies to bid for work for which they do not have the necessary numbers of staff. However, they need to know how quickly they would be able to obtain the appropriate staff. If a contract requires specialist skills or technologies that are not already possessed, research will be necessary to determine the probability that these skills and technologies can be acquired before the contract is placed. No organization can expect to hire extraordinary people at short notice; in fact, all you can expect to be available are average people and you may well have no choice than to accept less-than-average people. With good management skills and a good working environment you may be able to get these average people to do extraordinary things, but it is not guaranteed!

Provisions will need to be made to ensure sales personnel promising a short delivery to win an order have confirmation that it won't place an impossible burden on the company. Sales personnel will need access to reliable data on the capability of the organization and its products and services, they should not exceed their authority and always obtain the agreement of those who will execute the contractual conditions before their acceptance.

In telephone sales transactions or transactions made by sales personnel alone, sales personnel will need access to current details of the products and services available, the delivery times, prices and procedures for varying the conditions.

How is this demonstrated?

Demonstrating the organization reviews requirements and assesses its ability to meet them before committing to supply products and services to a customer may be accomplished by:

a)   presenting evidence of a process for reviewing requirements before tenders are submitted and contracts or orders and any changes are accepted;

b) presenting evidence of a process for establishing the organization's capability to meet customer requirements and feeding this information to decision makers;

c) selecting a representative sample of orders and contracts and presenting evidence that:

    i the decision to offer products and services was based on the organization's ability to meet requirements;

    ii current information on the organization's capability was taken into account;

    iii this information was provided by those with responsibility of meeting such requirements.

Reviewing customer-specified requirements before committing to supply (8.2.3.1a)

What does this mean?

Customer specified requirements relating to products and services are those needs and expectations of customers that are defined in customer supplied documented information. Specified requirements relating to products and services may apply to product and service characteristics, but also to the processes by which they are to be produced, supplied or managed and may include:

- Characteristics that the product is required to exhibit (i.e. the inherent characteristics).
- Price and delivery requirements.
- Procurement requirements that constrain the source of certain components, materials or the conditions under which personnel may work.
- Management requirements related to the way the project will be managed, the product developed, produced and supplied.
- Post-delivery requirements such as installation, servicing, repair, customer support.
- Security requirements relating to the protection of information.
- Financial arrangements for the deposit of bonds, payment conditions, invoicing, etc.
- Commercial requirements such as intellectual property, proprietary rights, labelling, warranty, resale, copyright, etc.
- Licensing requirements relating to individuals permitted to provide a service such as a pilot's licence, driving licence, professional licence to practice.
- Personnel arrangements such as access to the organization's facilities by customer personnel and vice versa.

Why is this necessary?

The review referred to in this requirement is necessary to establish that the output of the requirement determination process is correct.

How is this addressed?

The information gathered as a result of determining the various product requirements should be consolidated in the form of a specification, contract or order and then subject to review. The personnel who should review these requirements depend on their complexity, and there are three situations that you need to consider:

a)   development of new product or service to satisfy identified market needs – new product/
    service development;
b)   sales against the organization's requirements – proprietary sales;
c)   sales against specific customer requirements – custom sales.

New product/service development

In setting out to develop a new product or service there may not be any customer orders –
the need for the product/service may have been identified as a result of market research and
from the data gathered a definitive requirement is developed. The requirement review is
performed to confirm that the requirements do reflect a product/service that will satisfy the
identified needs and expectations of customers. This review may be the same as the design
input review, but there are other outputs from market research such as the predicted quantities, the manner of distribution, packaging and promotion considerations. The review should
be carried out by those functions representing the customer, design and development, production, service delivery and in service support so that all views are considered.

Proprietary sales

In a proprietary sales situation, you may simply have a catalogue of products and services
advertising material and a sales office taking orders over the telephone or over the counter.
There are two aspects to the review of requirements. The first is the initial review of the
requirements and advertising material before they are made available for potential customers
to view and the second is where the sales person reviews the customer's request against the
catalogue to determine if the particular product is available and can be supplied in the quantity required. We could call these requirement review and transaction review. As a customer
may query particular features, access to the full product specification or a technical specialist
may be necessary to answer such queries.

Custom sales

In custom sales situation, the product or service is being produced or customized for a
specific customer and with several departments of the organization having an input to the
contract and its acceptability. These activities need coordinating so that you ensure all are
working with the same set of information. You will need to collect the contributions of those
involved and ensure that they are properly represented at meetings. Those who negotiate
contracts on behalf of the company carry a great responsibility.

One aspect of a contract often overlooked is the shipment of finished goods. You have
ascertained the delivery schedule and the place of delivery, but how do you intend to ship it
(by road, rail and ship or by air)? It makes a lot of difference to the costs. Also, delivery dates
often mean the date on which the shipment arrives, not the date it leaves. An appropriate lead
time for shipping to the required destination by the means agreed therefore needs to be built
into your schedules. If you are late, you may need to employ speedier means, but that will
incur a premium for which you may not be paid. Your financial staff will therefore need to
be involved in the requirement review.

Having agreed the requirements, you need to convey them to their point of implementation in sufficient time for resources to be acquired and put to work. Policy deployment and
Quality Function Deployment (QFD) are tools you can use for this purpose.

How is this demonstrated?

Demonstrating that requirements specified by the customer are being reviewed may be accomplished by:

a)   presenting evidence of a process for reviewing customer requirements;
b)   presenting evidence that these reviews include delivery and post-delivery activities;
c)   selecting a representative sample of tenders, contracts and orders and presenting evidence that the customer requirements have been reviewed consistent with the process.

Reviewing requirements necessary for intended use (8.2.3.1b)

What does this mean?

This requirement applies when a customer has specified requirements for products or services they are seeking to obtain and may or may not have specified their intended use. The products or services the customer requires may be those which the organization offers but which may have been designed for a different application than needed by the customer. Alternatively, the products or services the customer requires may not be those which the organization offers but which the organization may choose to provide and therefore needs to know the intended use.

Why is this necessary?

It is important to identify requirements necessary for intended use. Unless the conditions of use are made abundantly clear at the time of purchase the organization has an obligation to establish the conditions under which the products or services it supplies will be used to safeguard against any harm to property, people or the environment that may arise by improper use. For instance, after delivery of a product, a customer could inform you that your product does not function properly and you establish that it is being used in an environment that is outside its design specification. You would not have a viable case if the customer had informed you that it was going to be placed near high-voltage equipment and you took no action.

How is this addressed?

The customer is not likely to be an expert in your field. The customer may not know much about the inner workings of your product and service offerings and may therefore specify the requirements only in performance terms. In such cases, the onus is on the organization to determine the requirements that are necessary for the product or service to fulfil its intended use. For example, if a customer requires an electronic product to operate close to high-voltage equipment, the electronics will need to be screened to prevent harmful radiation from affecting its performance. The customer may not know that this is necessary but during your dialogue, you establish the conditions of use and as a result identify several other requirements that need to be met. These are *requirements not specified by the customer but necessary for known intended use*.

Careful examination of customer needs and expectations is needed to identify all the essential product requirements. A useful approach is to maintain a check list or datasheet of the products and services offered which indicates the key characteristics and the limitations, what they can't be used for and what your processes are not capable of but might be expected to be capable of. Of course, such data needs to be kept within reasonable bounds. It is therefore important to establish what the customer intends to use the product for, where and how they intend to use it and for how long they expect it to remain serviceable. With proprietary products, many of these aspects can be clarified in the product literature supplied with the goods or displayed close to the point of service delivery. With custom designed products and services, a dialogue with the customer is vital to understand exactly what the product will be used for through its design life.

How is this demonstrated?

Demonstrating that requirements not stated by the customer, but necessary for the customer's specified or intended use have been reviewed may be accomplished by:

a) presenting evidence of a process in which customer requirements are reviewed before committing to supply products and services to a customer. (This might be an enquiry conversion process or tendering process);

b) showing that there is a stage in this process where the intended use of the products or services is established in cases where conditions of use are not made abundantly clear;

c) selecting a representative sample of review records and presenting evidence that this review was carried out;

d) presenting an analysis of customer complaints showing that there were none where the cause was a misapplication of use.

Reviewing requirements specified by the organization (8.2.3.1c)

What does this mean?

Requirements specified by the organization were addressed in Chapter 35.

Why is this necessary?

When determining requirements related to products and services, both customer requirements and the organization's own requirements will be identified. Before committing to supply products and services to a customer, it is therefore necessary to review these requirements to identify any that may be in conflict. Sometimes organizations may require its products to have characteristics that aid production or distribution or utilize its own technologies and equipment that are of no consequence to the customer but necessary for the efficient and effective production, storage and supply of the product, for example, packaging of items primarily serves protection, but also serves transportation and display, and these arrangements can be frustrating to customers as they try to remove the packaging to get at the items they have purchased. From a service perspective, the organization may impose restrictions at the customer interface to safeguard staff, protect property and serve its long-term interests that might appear to customers as a barrier or time wasting, for example, the increasing security arrangements in banks, the form filling in insurance companies and the number of gates through which a person seeking an entry visa must pass before finally getting an entry visa.

How is this addressed?

Quite simply, one compares the organization's specified requirements to those of the customer and identifies any of the organization's specified requirements that conflict with those of the customer.

How is this demonstrated?

Demonstrating that requirements specified by the organization are reviewed before committing to supply products and services to a customer may be accomplished by:

a) presenting evidence of a process for reviewing requirements specified by the organization;
b) selecting a representative sample of accepted orders/contracts and presenting evidence showing that:

    i   the requirements specified by the organization were reviewed against those of the customer;
    ii conflicting requirements were identified.

Reviewing statutory and regulatory requirements (8.2.3.1d)

What does this mean?

Statutory and regulatory requirements will have been defined when specifying the product or service which the organization offers to potential customers. When a customer expresses an interest, it may do so in the form of needs and expectations which explicitly include legal requirements which may differ from those the organization has identified or imply different regulations apply, for example, the product or service may be been designed for a slightly different environment or country in which the customer is intending to use it.

Why is this necessary?

The organization has an obligation to determine customer needs and expectations so that they can establish that what is offered will meet those needs and expectations prior to entering into a commitment to supply.

How is this addressed?

Quite simply, one compares the statutory and regulatory requirements already determined by the organization as being applicable with those specified by the customer and in the region in which the customer will use the product or service and identify any conflicting and additional requirements.

How is this demonstrated?

Demonstrating that statutory and regulatory requirements applicable to the products and services have been reviewed may be accomplished by:

a) presenting evidence of reviews which show that a comparison has been made between the legal requirements offerings were designed to meet and those which the customer has specified;

b) presenting evidence that this review and a guide to its conduct is mandated through maintained documented information so that it happens as a routine.

Reviewing and resolving requirements differing from those previously expressed (8.2.3.1e)

What does this mean?

Previously expressed requirements are those that may have been included in an invitation to tender issued by the customer. Whether or not you have submitted a formal tender, any offer you make in response to a requirement is a kind of tender. Where a customer's needs are stated and you offer your product or service, you are implying that it responds to your customer's stated needs. You need to ensure that your *tender* is compatible with your customer's needs otherwise the customer may claim that you have sold a product or service that is not fit for purpose.

Why is this necessary?

In situations where the organization has responded to an invitation to tender for a contract, it is possible that the contract (when it arrives) may differ from the draft conditions against which the tender was submitted. It is therefore necessary to check whether any changes have been made that will affect the validity of the tender. Customers should indicate the changes that have been made, but they often don't.

How is this addressed?

If the product or service you are offering is in any way different from the formal contract or order requirements that has been received, you need to point this out to your customer and reach an agreement before you accept the order, as any changes may affect the quality, price or delivery of the product or service. Proceeding without checking or resolving the differences before work commences may create problems later.

How is this demonstrated?

Demonstrating you have reviewed and resolved contract or order requirements differing from those previously expressed prior to the organization's commitment to supply may be accomplished by:

a) presenting evidence of the review and of any changes identified that materially affect the quality, price or delivery;

b) presenting evidence of a satisfactory resolution of any differences prior to the organization accepting a commitment to supply;

c) presenting evidence that this review and a guide to its conduct is mandated through maintained documented information so that it happens as a routine.

Handling undocumented customer requirements (8.2.3.1)

What does this mean?

Customers often place orders by telephone or in face-to-face transactions where no paperwork passes from the customer to the organization. Confirmation of customer requirements is an expression of the organization's understanding of the obligations it has committed to honour.

Why is this necessary?

Confirmation is necessary because when two people talk, it is not uncommon to find that although they use the same words, they each interpret the words differently. Confirming an understanding will avoid problems later. Either party to the agreement could move jobs leaving their successors to interpret the agreement in a different way. If at some stage the customer appears to be requiring something different, you can point to your letter of confirmation.

How is this addressed?

The only way to implement this requirement is for the organization to send a written acknowledgement to the customer confirming the requirements that form the basis of the agreement. In this way, there should be no ambiguity, but if later the customer appears to be requiring something different, you can point to the letter of confirmation. If you normally use e-mail for correspondence, obtain an e-mail receipt that it has been read (not merely received as it could be overlooked) otherwise always send confirmation by post as e-mails can easily be inadvertently lost or deleted. Keep a copy of the e-mail and the letter and bring them under records control. Saving specific e-mails as text files in an appropriate directory on the server is better than simply keeping them on your local hard drive as a message.

How is this demonstrated?

Demonstrating you have confirmation of customer requirement in cases where no documented statement has been provided may be accomplished by:

a) presenting evidence that the organization has sent a written acknowledgement to the customer confirming the requirements that form the basis of the agreement;
b) proving that the confirmation was sent prior to acceptance of the requirements;
c) presenting a record of delivery of a confirmatory e-mail or letter when appropriate;
d) showing that the e-mail and the letter are archived under records control.

Retaining documented result of the review (8.2.3.2)

What does this mean?

The result of the review may be a decision but could include a list of actions to be executed to correct the requirements or amend the offer, or a list of concerns that need to be addressed. If the review is conducted with customer's representatives present, records of the review could include modifications, interpretations and correction of errors that may be held back until the first contract amendment. In such cases the review records act as a temporary extension to any contract

Why is this necessary?

Many problems that arise after acceptance of a contract or order are caused by poor understanding or poor definition of requirements. Access to records of the review will be necessary to recall accurately what took place to investigate the cause of problems. During the processing of orders and contracts, records of the requirement review indicate the stage in the process that has been reached and are also useful if the process is interrupted for any reason.

How is this addressed?

There should be some evidence that a person with the authority to do so has accepted each product and service requirement, order or contract. This may be by signature or by exchange of letters or e-mails. The decision may be executed by providing a requirement review field in a database that must be checked or signed and dated before the process may continue.

You should also maintain a register of all contracts or orders and in the register indicate which were accepted and which were declined. This is useful when assessing the effectiveness of the demand creation process. If you prescribe the criteria for accepting a contract, the signature of the contract or order together with this register can be adequate evidence of requirement review. If requirement reviews require the participation of several departments in the organization, their comments on the contract, minutes of meetings and any records of contract negotiations with the customer will represent the records of product/service requirement review. It is important, however, to be able to demonstrate that the requirement being executed was reviewed for adequacy, differences in the tender and for supplier capability, before work commenced. The minimum is a signature accepting an assignment to do work or supply goods but you must ensure that those giving consent know what they are consenting to. Criteria for accepting orders or contracts can be included in the appropriate procedures. It cannot be stressed too strongly the importance of these actions. Most problems are caused by the poor understanding or poor definition of requirements.

How is this demonstrated?

Demonstrating that the results of the review of contract or order requirements, have been retained may be accomplished by:

a) retrieving evidence showing that a person with the necessary authority has accepted the product or service requirement, order or contract. This may be by signature or by exchange of letters or e-mails.
b) retrieving evidence of:

    i precisely what was reviewed to confirm that all the relevant requirements and any subsequent changes were reviewed;

    ii the date of the review and the date of contract or order acceptance to confirm the review took place prior to undertaking a commitment to supply;

    iii who participated in the review to confirm those parties responsible for meeting the requirements were given a timely opportunity to judge their acceptability;

iv   the criteria against which the requirements were reviewed to confirm which factors affecting the organizations ability to meet them were considered and to confirm that those participating in the review were aware of what they were accepting in case of subsequent dispute;

v   any differences in the tender that were identified;

vi   the decisions taken and any actions to be undertaken before acceptance.

c)   retrieving a register of all contracts or orders that indicates which were accepted and which were declined with traceability to the records. This is useful when assessing the effectiveness of the process.

Handling changes to requirements for products and services (8.2.4)

What does this mean?

The process for controlling change is addressed in Chapter 33 under the heading "Control of planned changes". This requirement refers to changes that have been approved and applies to the stage in that process where documented information is amended and communicated.

Why is this necessary?

Changes are normally made to product or service requirements with the intention that they are implemented. If such information carriers were not to be changed, or the people who normally use this information were not aware of them, the previously issued requirements would continue in use and any benefit would be lost putting customer satisfaction and regulatory compliance at risk. It is therefore necessary for there to be a process for communicating changes in a way that users work with information that is valid for what they are doing.

How is this addressed?

Requirements for products and services may not only include the inherent product or service characteristics, but also any other requirements relating to their supply (quantities and timing), installation, commissioning, verification, operation, maintenance, disposal or withdrawal. Therefore, changes to requirements for products and services may vary widely in what they address and consequently activate different organizational processes, for example, a change to the quantity ordered or the hours a service is available may not affect the inherent characteristics of the product or service.

Product or service requirements may be changed by the customer, by the regulators or by the organization, and this may be made verbally or by changing the affected requirement documents.

We need to draw a distinction here between document control and configuration management. Document control is concerned with controlling the information carriers, whereas configuration management is concerned with controlling the information itself regardless of which documents carry it. A document carrying requirements may be changed but the change may not affect the product or service (e.g. changes that correct errors or remove ambiguity in the way the requirements are expressed).

If there is only one product or service specification and no related information, configuration management is like document control (see Chapter 32). When there are many specifications and related information, configuration management introduces a further dimension of controlling the compatibility between all the pieces of information. If a system parameter changes there may be a knock-on effect through the subsystems, equipment and components. The task requires all the items affected by the change to be identified and as each item will have a specification, this task will result in a list of affected specifications, user instructions, etc. These will constitute the *relevant* documented information. Although the list looks like a list of documents, it is really a list of items that are affected by the change. Identify what is affected, and you should be able to identify who should be informed.

How is this demonstrated?

Demonstrating that relevant documented information is amended and that relevant personnel are made aware of changed requirements may be accomplished by:

a) showing which processes would be triggered following notification of an approved change in product and service requirements from customers, regulators or internally;

b) presenting a description of the processes through which changes to product or service requirements are passed;

c) presenting evidence of the analysis carried out to determine, the documentation that needs to be amended and who has been assigned responsible for undertaking the amendments;

d) presenting evidence that all the documentation affected by the change has been amended;

e) presenting evidence that those affected by the amended documentation are aware of the change.

Implementation of the change is executed through another stage in the change control process.

37 Design and development planning

Introduction

Design is a creative process. It takes an idea, problem or a need and produces a solution in the form of either a conceptual model or a detail specification that is capable of being used to create a fully functioning product or service. Design is often a process which strives to set new levels of performance, new standards or creates new wants and as such can be a journey into the unknown. On such a journey, we can encounter obstacles we haven't predicted, which may cause us to change our course, but our objective remains constant having been set by the requirements addressed in Chapter 35.

Box 37.1 Design and development defined

Design and development is defined as a "set of processes that transform requirements for an object into more detailed requirements for that object" (ISO 9000:2015)

This definition was modelled on the former definition of a process which contained the word *transform* and was not amended when the definition of a process was changed and the word *transform* was changed to *use*. Requirements are not in fact transformed in the design process. They remain the same at the end as they were at the beginning, and what actually happens is that requirements are used in the preparation of more detailed set of requirements.

The two words *design* and *development* are sometimes used synonymously and sometimes used to define different stages of a process. Through a creative process a designer creates a concept from an understanding of customer needs, and through a translation process, the developer translates the concept into technical specifications for a product that can be manufactured or a service that can be delivered. Design is creative, development is constructive.

Design and development can be as simple as replacing a part in an existing product with one of a different specification or as complex as the design of an electricity generator that extracts power from the ocean. Design can be of hardware, software (or a mixture of both) and can be of new services or modified services.

In this chapter, we examine the 11 requirements of clause 8.3.1, namely:

- Establishing a design and development process (8.3.1)
- Considering the nature, duration and complexity of design and development (8.3.2a)
- Considering process stages (8.3.2b)
- Considering design and development verification and validation (8.3.2c)
- Considering design and development responsibilities and authorities (8.3.2d)
- Considering resource needs (8.3.2e)
- Controlling organizational interfaces (8.3.2f)
- Involving customers and users in the design and development process (8.3.2g)
- Considering requirement for provision of products and services (8.3.2h)
- Considering the level of external control (8.3.2i)
- Confirming verification and validation data capture (8.3.2j)

A historical perspective

In the 1994 version, the requirements were expressed in terms that only applied to products and not services. They were also based on the premise that design could be controlled by adherence to documented procedures; but that changed in the 2000 version when processes were required to be planned and documented for design and development. A further 12 requirements were added with more detail to what is to be included in design inputs, the reasons for conducting design reviews and a requirement for completing design validation prior to the delivery or implementation of the product wherever practicable. However, this requirement was removed in the 2015 version. The basic structure of the requirements has remained virtually the same, but the number of design and development requirements has almost doubled since 1987. Requirements for design review, verification and validation have now been subsumed under the heading of design control, which is ironic as in 1987, that was virtually the title of the whole section of the standard.

Establishing a design and development process (8.3.1)

What does this mean?

Before design commences there is either a requirement for a new or modified product or service or simply an idea or a problem to be solved that is intended to result in a new or modified product or service. The design and development process takes this input and creates an output that can be used to produce products and services that fulfil customer needs and expectations and meet applicable legal requirements.

The requirement defines the objective of the design and development process where it states "to ensure the subsequent provision of products and services". The word *ensure* means *to make certain*, and this requires the design and development process to be managed. Managing a design and development process involves keeping the design on course towards its objectives and as such will address all the factors that may prevent the design from achieving its objectives.

It is required that a design and development process is established, implemented and maintained as though such a process continually exists or that you have a design and development process waiting to be implemented. Each time design and development is undertaken it is through a different process because of different objectives, constraints, resources, etc.

A design and development process will have been established when it has been planned and resourced and therefore in organizations that are not continually designing new products and services, this event may not happen very often. One can therefore only implement and maintain the process that has been planned until the design is complete and not following its completion. Once a design has been completed, the process ceases to exist until it is activated once again. This means that a new process is established for each new or modified product or service, and this is what the word *appropriate* refers to in the requirement. Of course, a generic model process can be tailored for each new product or service design, but it would be unwise to claim this is the actual design and development process and is appropriate for any specific design.

A commonly held misconception is that diagrams such as those in Figure 37.1 and Figure 37.2 can be presented as evidence that processes have been established. These are just conceptual models for the sequence of activities in a process. The real sequence will differ and without resources, the process has not been established at all.

Control of design and development does not mean controlling the creativity of the designers – it means controlling the process through which new or modified designs are produced so that the resultant design is one that truly reflects customer needs. It therefore controls the inputs and the selection of elements to be used in the design such as components and technologies, and it controls the outputs but within these constraints the designer is free to innovate at will.

These requirements apply to the products and services that will be supplied to the organization's customers including any packaging. They are not intended to apply to tooling or anything that might come into contact with the product but is not shipped with the product or provided with the service. They also apply in cases where a design has been purchased and the related product manufactured or service delivered.

Why is this necessary?

Without control over the design and development process several possibilities may occur:

- Design will commence without an agreed requirement.
- Costs will escalate as designers pursue solutions that go beyond what the customer really needs.
- Costs will escalate as suggestions get incorporated into the design without due consideration of the impact on development time and cost.
- Designs will be released without adequate verification and validation.
- Designs will be expressed in terms that cannot be implemented economically in production or service delivery.

The bigger the project, the greater the risk that the design will overrun budget and timescale. Design control aims to keep the design effort on course so that the right design is released on time and within budget.

How is this addressed?

In a design and development process that is being managed effectively there may be up to 10 primary stages, but the order may vary depending on what is being designed and what is the best way of designing it. It should also be recognized that design is an iterative process where one or more stages will be repeated over and over until the best solution emerges:

1 Establish the customer needs.
2 Convert the customer needs into a definitive specification of the requirements.
3 Plan for meeting the requirements.
4 Organize resources and materials for meeting the requirements.
5 Conduct a feasibility study to discover whether accomplishment of the requirements is feasible.
6 Produce one or more design concepts which might fulfill the requirement.
7 Conduct a project definition study to discover which of the many possible solutions will be the most suitable.
8 Develop a specification for the proposed design solution that details all the features and characteristics of the product or service.
9 Produce a prototype or model of the proposed design.
10 Conduct extensive trials to discover whether the product, service or process that has been developed meets the design requirements and customer needs.
11 Feed data back into the design and repeat the process until the product or service is proven to be fit for the task.

Control of the design and development process requires the application of the same principles as any other process. The standard identifies the controls that need to be applied to each design, but other controls are needed to apply the process approach as defined by the requirements of clauses 4.4.1a) to 4.4.1h). Typical product and service design activity flow charts are illustrated in Figure 37.1 and Figure 37.2, respectively. (Note that the activity flow reflects traditional sequential design and not the more modern agile software development.)

The design process is a key process in enabling the organization to achieve its objectives. These objectives should include those that apply to the design process (see Chapter 16 on establishing quality objectives). Consequently, there need to be:

- Objectives for the design process.
- Measures for indicating achievement of these objectives.
- A defined sequence of sub-processes or activities that use the design inputs to create the optimum design outputs.
- Links with the resource management process so that human and physical resources are made available to the design process when required.
- Review stages for establishing that the process is achieving its objectives.
- Processes for improving the effectiveness of the design process.

How is this demonstrated?

Demonstrating that an appropriate design and development process has been established and is being implemented and maintained may be accomplished by:

a) presenting evidence of processes for design and development which are executed for each new or modified design;
b) selecting a representative sample of designs and producing evidence that the design process was managed according to the designated process description.

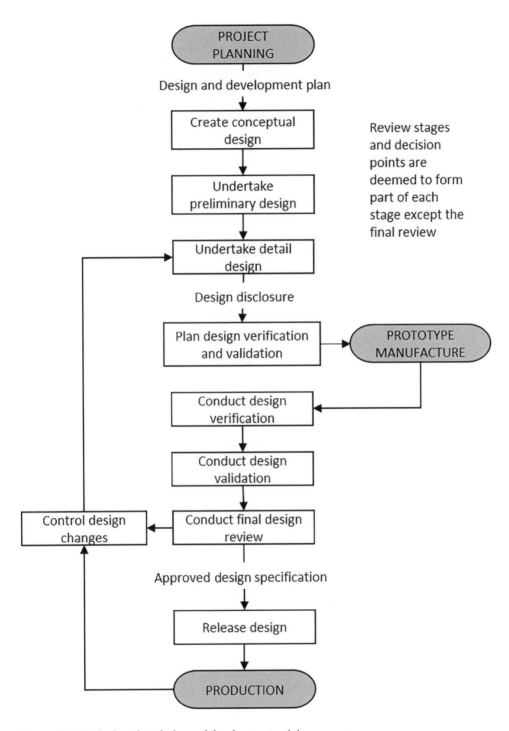

Figure 37.1 Typical product design and development activity sequence

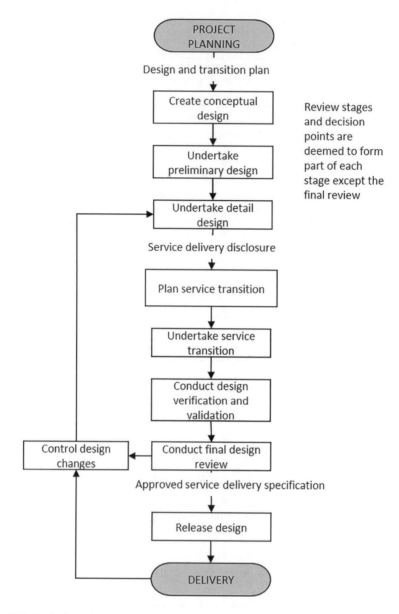

Figure 37.2 Typical service design and transition activity sequence

Considering the nature, duration and complexity of design and development activities (8.3.2a)

What does this mean?

The nature of the design and development activities refers to their characteristics in terms of the number and difficulty of the technical and financial decisions that will need to be made to produce a satisfactory design solution. Complexity may be reflected in the number of technical and organizational interfaces, which will influence the range and number of characteristics that need to be specified and the size of the design team that will be specifying them. If the time permitted to complete the design is specified, this will also have a bearing on the size of the team involved.

Why is this necessary?

The purpose of planning is to determine the provisions needed to achieve an objective. In most cases, these objectives include not only a requirement for a new or modified product or service but also requirements governing the costs and timescales (quality, cost and delivery or QCD). Remove these constraints and planning becomes less important, but there are few situations when cost and time are not the constraints. It is therefore necessary to work out in advance whether the objective can be achieved within the budget and timescale. One problem with design is that it is often a journey into the unknown, and the cost and time it will take cannot always be predicted. Without a best guess, some projects would not get underway, so planning is a vital first step to get the funding and second, to define the knowns and unknowns so that risks can be assessed and quantified.

How is this addressed?

A design and development plan should be prepared for each new design and for any modification of an existing design that radically changes the performance of the product or service. For modifications that marginally change performance, control of the changes required may be accomplished through the design change process.

Design and development plans need to identify the activities to be performed, by whom they will be performed and when they should commence and should be proven as meeting the design requirements. Therefore in drawing up a design and development plan you will need to cover the planning of design verification and validation activities. The plans should identify as a minimum:

- The design requirements.
- The design and development programme showing activities against time.
- The work packages and names of those who will execute them. (Work packages are the parcels of work that are to be handed out to either internal or external parties.)
- The work breakdown structure (WBS) showing the relationship between all the parcels of work.
- The reviews to be held for authorizing work to proceed from stage to stage.
- The resources in terms of finance, people and facilities.
- The risks to success and the plans to minimize them.
- The controls that will be exercised to keep the design on course.

Although developed for complex systems within the defence and space industries the WBS is a disciplined approach that provides several benefits:

a) Divides a product or service into its component parts, clarifying the relationship among them and the relationship of the tasks to be completed both to each other and to the end product or service.

b)   Facilitates effective planning and assignment of management and technical responsibilities.
c)   Aids status tracking of technical efforts, risks, resource allocations, expenditures, and cost/schedule/technical performance.
d)   Helps ensure that external providers are not unnecessarily constrained in meeting requirements for specific items.
e)   Provides a common basis for allowing consistency in understanding program cost and schedule performance.

Further details of the work breakdown structure practice may be found in (Mil Std 833C, 2011).

Planning for all phases at once can be difficult as information for subsequent phases will not be available until earlier phases have been completed. So, your design and development plans may consist of separate documents, one for each phase and each containing some detail of the plans you have made for subsequent phases.

Your design and development plans may also need to be subdivided into plans for special aspects of the design such as reliability plans, safety plans, electromagnetic compatibility plans and configuration management plans see ISO 10007. Although purchasing is dealt with in clause 7.4 of the standard, the requirements also apply to design activities.

How is this demonstrated?

Demonstrating that consideration has been given to the nature, duration and complexity of the design and development activities may be accomplished by:

a)   presenting design and development plans for the products and services that are under development or have recently completed development;
b)   where appropriate, presenting the work breakdown structure or equivalent evidence showing the work to be undertaken, the timescales over which it will be carried out and the identity of those authorized to undertake it.

Considering process stages (8.3.2b)

What does this mean?

A stage in design and development is a point at which the design reaches a phase of maturity. Before a design proceeds from one stage to another, a review normally takes place to decide whether the design has reached a sufficient degree of maturity to proceed to the next stage or to halt further development. This is sometimes referred to as *a stage-gate process*.

Why is this necessary?

Any endeavour is more easily accomplished when undertaken in small stages. By processing a design through several iterative stages, a more robust solution will emerge than if the design is attempted in one cycle, but it rather depends on what it is that is being developed. The stage gate's main aim is to keep control of costs whilst maintaining control over feature creep. It also prevents effort from being wasted by progressing a design solution that is not feasible or sufficiently developed.

How is this addressed?

One size does not fit all; there is no single, universal product/service development strategy that is right for all organizations at all times. Different strategies are used for new product/service development depending on the context of the organization and the risks it is prepared to take. Three strategies used by Hewlett Packard are described next (MacCormack, Crandall, & Henderson, 2012):

- Efficient development strategy – Well-defined stage-gated process with clear entry/exit criteria, explicit tasks and deliverables, and rigorous checkpoint review meetings; monitoring to plan.
- Agile development strategy – Evolutionary process based on frequent design-build-test iterations, milestone releases, and beta versions with actual customers; continually re-prioritizing features.
- Emergent development strategy – Lightweight process with fluid objectives; rapid exchange of information with potential customers to identify the customer value proposition.

Once an organization has defined the styles available, it should define how the choice between them should be made for individual projects based on the risks each presents to the organization. Although the specific industry and the organization's position in that industry will determine the broader environment for a project, the type of innovation being pursued – incremental, platform, or breakthrough – is a critical factor in choosing the appropriate style for development (MacCormack, Crandall, & Henderson, 2012).

Other strategies include:

- Lean product development
- Design for Six Sigma
- Flexible product development
- Quality function deployment. (ISO 16355–5 is being developed by ISO/TC 69 on QFD)
- Phase-gate model
- User-centred design

There are several common stages in a stage-gate design process; the names may vary but the intent remains the same.

- **Feasibility stage**: The stage during which studies are made of a proposed objective to determine whether practical solutions can be developed within time and cost constraints. This stage usually terminates with a design brief which is reviewed for adequacy and suitability and a decision made whether to progress the design to the next stage.
- **Conceptual design stage**: The stage during which ideas are conceived and theories tested. This stage usually terminates with a preferred concept in the form of a design requirement, which is reviewed for adequacy and suitability, and a decision made whether to progress the design to the next stage (see also Chapter 38 on design and development inputs).
- **Design definition stage**: The stage during which the architecture or layout takes form and the risks assessed and any uncertainty resolved. This stage usually terminates with

design requirement specifications for the components comprising the product, service or process which are reviewed for adequacy and suitability and a decision made whether to progress the design to the next stage.

- **Detail design stage**: The stage during which final detail characteristics are determined and methods of product reproduction or service delivery established. This stage usually terminates with a set of specifications for the construction of prototypes which are reviewed for adequacy and suitability and a decision made whether to progress the design to the next stage. Process design and development will also commence as soon as engineering drawings and tooling requirements are released.
- **Development or service transition stage**: The stage during which the prototype is proven using models or simulations and refined. This stage usually terminates with a set of approved specifications for the implementation of the design which are reviewed for adequacy and suitability and a decision made whether to approve the design for implementation or transition in the case of services. Process validation also occurs during this phase even though it is referred to under the heading Control of production and service provision in clause 8.5.1.

Any more detail will probably be a breakdown of each of these stages initially for the complete design and subsequently for each element of it. If dealing with a system you should break it down into subsystems, equipment, assemblies and so on. It is most important that you agree the system hierarchy and associated terminology early in the development programme otherwise you may well cause both technical and organizational problems at the interfaces.

How is this demonstrated?

Demonstrating that the required process stages and reviews have been considered may be accomplished by presenting a new product development plan which describes the chosen development strategy and the stages through which the design will pass from feasibility to release for production or service provision.

Considering design and development verification and validation (8.3.2c)

What does this mean?

Each design stage is a process that takes inputs from the previous process and delivers outputs to the next stage. Within each process are check points that feedback information into the process to produce a further iteration of the design depending on the chosen development strategy. The further along the design cycle, the more rigorous and complex the check points will need to be. The verification stages are those stages where design output of a stage is checked against the design input for that stage to ensure that the output is correct. The validation stages occur sequentially or in parallel or at the end of the process to confirm that the output is the right output by comparing it with the design brief or requirement.

Why is this necessary?

The checks necessary to select and confirm the design solution need to be built into the design process so that they take place when they will have the most beneficial effect on the design. Waiting until the design is complete before commencing verification or validation will probably result in extensive rework and abortive effort.

How is this addressed?

The stages of verification and validation should be identified in the design and development plan, but at each stage there may need to be supplementary plans to contain more detail of the specific activities to be performed. This may result in a need for a separate design verification plan.

The design verification plan should be constructed so that every design requirement is verified and the simplest way of confirming this is to produce a verification matrix of requirement against verification methods. Another matrix in a similar form is a Quality Function Deployment chart.

You need to cover all the requirements, those that can be verified by test, by inspection, by analysis, by simulation or demonstration or simply by validation of records. For those requirements to be verified by test, a test specification will need to be produced. The test specification should specify which characteristics are to be measured in terms of parameters, limits and the conditions under which they are to be measured.

The verification/validation plan may need to cover some or all the following details as appropriate:

- A definition of the design standard that is being verified/validated.
- The objectives of the plan (separate plans may be needed covering different aspects of the requirements).
- Definition of the specifications and procedures to be employed for determining that each requirement has been achieved.
- Definition of the stages in the development phase at which verification/validation can most economically be carried out.
- The identity of the various models that will be used to demonstrate achievement of design requirements. (Some models may be simple space models, others laboratory standard or production standard depending on the need.)
- Definition of the verification/validation activities that are to be performed to qualify or validate the design and those which need to be performed on every product in production or every service as a means of ensuring that the qualified design standard has been maintained.
- Definition of the resources needed to carry out the verification/validation activities.
- Definition of the timescales for the verification/validation activities in the sequence in which the activities are to be carried out.
- Identification of the location of the verification/validation activities.
- Identification of the organization responsible for conducting each of the verification/validation activities.
- Reference to the controls to be exercised over the verification/validation activities in terms of the procedures, specifications and records to be produced, the reviews to be conducted during the programme and the criteria for commencing, suspending and completing the verification/validation operations. (Provision should also be included for dealing with failures, their remedy, investigation and action on design modifications.)

As part of the verification/validation plan, you should also include an activity plan that lists all the planned activities in the sequence they are to be conducted and use this plan to progressively record completion and conformance. The activity plan should make provision for planned and actual dates for each activity and for recording comments such as recovery plans when the programme does not proceed exactly as planned. It is also good practice to conduct test reviews before and after each series of tests so that corrective measures can be taken before continuing with abortive tests

However simple the design, the planning of its verification and validation is vital to the future of the product or service. Lack of attention to detail can rebound months (or even years) later during its implementation.

How is this demonstrated?

Demonstrating that the required design and development verification and validation activities have been considered may be accomplished by presenting new product development verification and validation plans which identify the agreed verification and validation activities to be carried out.

Considering design and development responsibilities and authorities (8.3.2d)

What does this mean?

To cause the activities in the design and development plan to happen, they must be assigned to either a person or an organization. Once assigned and agreed by both parties, the assignee becomes responsible for delivering the required result. The authority delegated in each assignment conveys a right to the assignee to make decisions affecting the output. The assignee becomes the design authority for the items designed but this authority does not extend to changing the design requirement – this authority is vested in the organization that delegated or sponsored the design for the item.

Why is this necessary?

Responsibility for design activities needs to be defined so that there is no doubt as to who has the right to take which actions and decisions. Authority for design activities needs to be delegated so that those who are responsible for the output have the right to control their own output. Also, the authority responsible for the requirements at each level of the design needs defining so that there is a body to which requests for change can be routed. Without such a hierarchy, there would be anarchy resulting in a design that failed to fulfil its requirements.

How is this addressed?

Within the design and development plan the activities need to be assigned to a person, group or organization equipped with the resources to execute them. Initially the feasibility study may be performed by one person or one group but as the design takes shape, other personnel or other external organizations may be required.

One way of assigning responsibilities is to use the work package technique. With this approach, you specify on work statements the work required, what is included and what is excluded, the inputs to be used, the deliverables to be produced and the hours/person or days estimated to do the work. By obtaining the group's acceptance you have their commitment to the task and means of ensuring no gaps in or overlapping assignments.

If you subcontract any of the design activities, the supplier's plans need to be integrated with your plans and your plan should identify which activities are the supplier's responsibility. There needs to be a clause in subcontracts that prohibit subcontracting without your approval thereby enabling you to retain control.

How is this demonstrated?

Demonstrating that responsibilities and authorities involved in the design and development process have been considered may be accomplished by presenting a new product development plan or work statements which define how and to whom the work has been allocated and the responsibilities and authority of those undertaking it.

Considering resource needs (8.3.2e)

What does this mean?

The resources required for product and service design and development can vary enormously from a few hours of work/person in one department of an organization to several years worth of effort/person involving many organizations in different countries spread over a decade or more. Sometimes a joint venture company may be established to undertake the work.

Why is this necessary?

ISO 9001 duplicates the requirement for resources as shown in the Introduction to Part 7. Although it may appear to differ in that it requires internal and external resource needs to be determined, this simply duplicates 7.1.1(a) and (b). There will be the resources needed for the organization to function as a viable concern and in addition the resources needed to design and develop specific products and services.

How is this addressed?

A prerequisite for the determination of any resources is an objective and a programme of work required to achieve that objective. The work breakdown structure, or WBS, defined earlier allows for each item of work to be costed in terms of labour, materials and equipment either as internal costs or as externally provided costs.

The design and development process is an exploratory process (see Chapter 9). Unlike cybernetic processes, design can be a journey into the unknown as stated earlier, and therefore new knowledge and technologies may be necessary with consequences as to the consumption of resources.

How is this demonstrated?

Demonstrating that consideration has been given to the internal and external resource needed for the design and development of products and services may be accomplished by presenting evidence of resource budgets itemizing the work to be undertaken against each objective, the labour, materials, equipment/facilities and knowledge required and from where it will be obtained.

Evidence presented to demonstrate conformity with 4.4.1d, 7.1.1, 6.2.2b and 8.1c might mean that it is also deemed to satisfy conformity with 8.3.2e, but not vice versa, because there are other types of resources than those needed for the design and development of products and services.

Controlling organizational interfaces (8.3.2f)

What does this mean?

Interfaces need to be controlled where people involved in the design are in different locations or in different teams. Where people are at the same location and in the same team, there is likely to be less need for formal control due to the close contact they will have.

Control over interfaces between persons involved is as much about the control of technical interfaces as it is about the human interfaces. Where there are many different groups of people working on a design, they need to work together to produce an output that meets the overall requirement when all outputs are brought together. To achieve this each party needs to know how the design work has been allocated and to which requirements each party is working so that if there are problems, the right people can be brought together.

Why is this necessary?

If the interfaces between design groups are not properly managed, there are likely to be technical problems arising from groups changing interface requirements without communicating the changes to those affected. Political problems might arise from groups assuming the right to do work or make decisions that have been allocated to other groups. Cost overruns might arise from groups not communicating their difficulties when they are encountered. Control is largely by information, and it can often tend to be historical information by the time it reaches its destination. So it is important to control changes to the interfaces. If one small change goes unreported, it may cause months of delay correcting the error.

How is this addressed?

You should identify where work passes from one organization/team to another and the means used to convey the requirements such as specification, work instructions, work package descriptions or contracts. In multinational projects, it's important to agree a common language, time zone and the units of measure as such differences can have catastrophic consequences.

Often in design work, the product requirements are analysed to identify further requirements for constituent parts. These may be passed on to other groups as input requirements for them to produce a design solution. In doing so these groups may in fact generate further requirements in the form of development specifications to be passed to other groups and so on. Some of these transactions may be in-house but some might be subcontracted. Some organizations only possess certain design capabilities and subcontract most of the hardware, software or specialist service components to specialists such as the IT architecture and network design. In this way, they concentrate on the business they are good at and get the best specialist support through competitive tenders. These situations create organizational interfaces that require effective information control processes.

In managing the organizational interfaces, you will need to define:

- the customer and the supplier in the relationship;
- the product requirements that the supplier is to meet (the objectives and outputs);
- the work that the supplier is to carry out with the budget and time constraints;
- the responsibility and authority of this work (who does what, who approved what);
- the process used for conveying information and receiving feedback;
- the reporting and review requirements for monitoring the work;
- and conduct regular interface review meetings to check progress and resolve concerns and periodically review the interface control process for its effectiveness.

One mechanism of communicating technical interface information is to establish and promulgate a set of baseline requirements that are to be used at commencement of design for a particular phase. This baseline listing becomes a source of reference and if managed properly ensures that no designer is without the current design and interface information.

Interfaces should be reviewed along with other aspects of the design at regular design reviews scheduled prior to the completion of each phase or more often if warranted. Where several large organizations are working together to produce a design, an interface control board or similar body may need to be created to review and approve changes to technical interfaces. Interface control is especially difficult with complex projects. Once underway an organization like a large ship gains momentum and takes some time to stop.

How is this demonstrated?

Demonstrating that consideration has been given to the need to control interfaces between persons involved in the design and development process may be accomplished by:

a) presenting evidence of a work breakdown structure or equivalent showing the persons or organizations involved in the project and their respective locations and relationships;
b) presenting an assessment of the risks present due to the complexity of the relationships;
c) presenting evidence of interface agreements, service level agreements or contracts that define the technical specifications, reporting and change procedures intended to mitigate the risks to the technical interfaces.

Involving customers and users in the design and development process (8.3.2g)

What does this mean?

Whether the organization is in a business-to-business (B2B) relationship or a business-to-consumer (B2C) relationship, there will be stages in the design and development process where customer and end-user involvement should be encouraged. It may mean customer and end-user involvement in project and design reviews through each stage of the design and development in a B2B relationship or customer focus groups to surface or test new ideas in B2C relationship.

Why is this necessary?

The customer is of prime importance in the design and development process. The customer may have specified their involvement in the contract in which case face to face meetings provide an opportunity to confirm understanding of their specified requirements and get feedback on progress in developing the design concept. In cases where the customer has not specified the design requirements, it may be prudent to seek their involvement at key stages to confirm their needs have been fully understood. In a B2C relationship where the design requirements have been produced as a result of market research, there is less certainty than in a B2B situation, and testing the market is necessary before a commitment to production or service delivery is made.

How is this addressed?

Whether in a B2B or B2C relationship, time slots will need to be allocated in the design and development schedule when customer involvement is planned to occur. Information about these events needs to be specified in terms of:

- The objective the event is expected to achieve, for example, to get agreement on some aspect of the design, to report progress, to resolve a technical issue, to demonstrate concepts, prototypes or to witness development tests and simulations or to get authorization to proceed to the next stage.
- The measures which will be used to determine the success of the event.
- The information required to be sent to the customer or expected from the customer.
- The process by which the objective is expected to be achieved.
- The outputs expected, for example, agreement to proceed to the next stage, approval of specifications, test results etc.

In B2B relationships where the customer has brought together a user group, it's common for there to be a user group on the organization's project team that liaises with the external users at the concept stage to refine the human interfaces. This group may develop a human interface specification that captures the ergonomic features and characteristics that need to be built into the product or service.

On large developments, the customer is not one person but several, each representing a discipline and each with the potential to convey a different interpretation of the requirements that have been agreed in the contract. On occasions, the relationship between the customer's representative and the organization's opposite number can be a very close relationship where there is a potential for *mission creep* – a tweak here, an adjustment there – and unless the customer-facing staff are vigilant the work expands and the costs escalate. Although it's often said that *the customer is always right*, you need to manage your customer to make sure they don't try to correct the errors in their specification without considering the impact they have on the agreed completion date and where necessary, having to pay for them.

How is this demonstrated?

Demonstrating that consideration has been given to the need for involvement of customers and users in the design and development process may be accomplished by:

a)   presenting the design and development plan and pointing out the provisions made to involve the customer and users;

b) explaining the controls that are being or have been exercised to ensure the needs of customers and users have been accommodated;

c) explaining the controls that are being or have been exercised to ensure the results of customer and user involvement have been captured and used as appropriate.

Considering requirement for provision of products and services (8.3.2h)

What does this mean?

Box 37.2   New design planning requirement

In the 2008 version there was a requirement for appropriate information to be made available in the design outputs for (subsequent) purchasing, production and for service provision. This requirement remains in the 2015 version although worded differently. However, there now is a different requirement for the subsequent provision of products and services to be considered when planning the design and development of products and services. This has been interpreted as concurrent engineering activities for which provisions need to be made (time, resources, etc.) that need to be carried out in parallel with design and development.

Requirements for subsequent provision of products and services are those requirements that may influence design and development planning for which provision needs to be made, for example,

• Target dates for getting to market before the competition.
• Critical milestones for synchronizing with concurrent development of related products and services.
• Prototyping.
• Developing and validating processes for the manufacture of the product.
• Transition arrangements for commencement of service delivery.
• Documented information required to support the product or service through its life cycle that needs to be prepared concurrently with design and development.

Requirements that influence product or service features such as ease of assembly, manufacturability, testability, etc., should be included in the design input see Chapter 38.

Why is this necessary?

If we approach the design and development of a product or service without considering the other parties that depend on the design being complete, it's likely that the organization will beset with many problems and the time to get the product to market will slip. If all elements of a product's life cycle, are taken into careful consideration in the early design phases, errors and redesigns can be discovered early in the design process when the project is still flexible, and this will reduce the elapsed time required to bring a new product or service to market. This concept is referred to as concurrent engineering, and it is therefore necessary to plan for concurrent engineering activities from the outset.

How is this addressed?

Concurrent engineering

A technique used in the automotive industry that is designed to reveal design weaknesses early in the design and build the capability necessary for production is Advanced Product Quality Planning (APQP). Developed by Ford, GM and Chrysler in the 1990s it involves the use of cross-functional teams, a technique which has been proven to save time and money over the product life cycle. The major elements of such a programme are shown in Figure 37.3. More details can be found on the Internet.

Process development

There was a note in ISO 9001:2008 that suggests that the design and development requirements may be applied to the development of operations processes. The development of business processes has been addressed in Chapter 16, but the requirements of clause 8.3 could certainly be applied to the design and development of the processes for producing the product or delivering the service, although some tailoring would obviously be required.

Supporting documented information development

The information required for proceeding through design and development from one stage to another and also required for subsequent production and service delivery throughout its life cycle needs to be determined and its preparation planned for. These are often

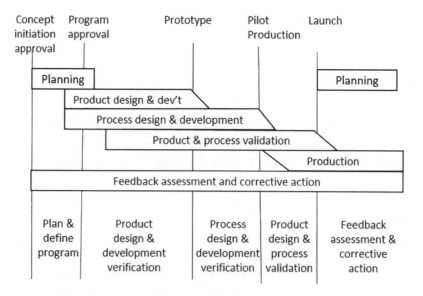

Figure 37.3 Product planning timing chart

referred to as program deliverables. The information that is output from one project phase is the input to the next project phase. Details of the deliverable documented information required should be included in the design and development plan or in separate Work Package Descriptions that align with the WBS (see 8.3.2a earlier). As an example only, Table 37.1 shows typical deliverables from each phase of the development of an automotive product that uses APQP. In addition, user manuals, maintenance manuals may be required.

How is this demonstrated?

Demonstrating that consideration has been given to requirements for subsequent provision of products and services may be accomplished by:

a)   selecting a representative sample of design and development plans;
b)   showing the provisions made for involving other disciplines in the design and development process;
c)   showing the provisions made to design the production and service delivery processes concurrently with products and services to reveal design weaknesses early;
d)   showing the provisions made to ensure synchronization of concurrent development of related products and services;
e)   showing the provisions made for preparing the documented information needed for each phase of the product or service life cycle.

Table 37.1 Advanced Product Quality Planning deliverables

Inputs	Project Phase	Outputs
Market Research Historical Warranty and Quality Information Team Experience Business Plan/Marketing Strategy Product/Process Benchmark Data Product/Process Assumptions Product Reliability Studies Customer Inputs	PLAN & DEFINE PROJECT	Design Goals Reliability and Quality Goals Preliminary Bill of Material Preliminary Process Flow Chart Preliminary Listing of Special Product and Process Characteristics Product Assurance Plan Management Support
Design Goals Reliability and Quality Goals Preliminary Bill of Material Preliminary Process Flow Chart Preliminary Listing of Special Product and Process Characteristics Product Assurance Plan Management Support	DESIGN & DEVELOP PRODUCT	Design Failure Mode and Effects Analysis (DFMEA) Design for Manufacturability and Assembly Design Verification Design Reviews Prototype Build – Control Plan Engineering Drawings (Including Math Data) Engineering Specifications Material Specifications Drawing and Specification Changes

(Continued)

Table 37.1 (Continued)

Inputs	Project Phase	Outputs
		New Equipment, Tooling and Facilities Requirements
		Special Product and Process Characteristics
		Monitoring and Measuring Equipment Requirements
		Team Feasibility Commitment and Management Support
Design Failure Mode and Effects Analysis (DFMEA)	DESIGN & DEVELOP PROCESS	Packaging Standards
		Product/Process Quality System Review
Design for Manufacturability and Assembly		Process Flow Chart
Design Verification		Floor Plan Layout
Design Reviews		Characteristics Matrix
Prototype Build – Control Plan		Process Failure Mode and Effects Analysis (PFMEA)
Engineering Drawings (Including Math Data)		Pre-Launch Control Plan
Engineering Specifications		Process Instructions
Material Specifications		Measurement Systems Analysis Plan
Drawing and Specification Changes		Preliminary Process Capability Study Plan
New Equipment, Tooling and Facilities Requirements		Packaging Specifications
Special Product and Process Characteristics		Management Support
Gages/Testing Equipment Requirements		
Team Feasibility Commitment and Management Support		
Packaging Standards	VALIDATE PRODUCT AND PROCESS	Production Trial Run
Product/Process Quality System Review		Measurement Systems Evaluation
Process Flow Chart		Preliminary Process Capability Study
Floor Plan Layout		Production Part Approval
Characteristics Matrix		Production Validation Testing
Process Failure Mode and Effects Analysis (PFMEA)		Packaging Evaluation
Pre-Launch Control Plan		Production Control Plan
Process Instructions		Quality Planning Sign-Off and Management Support
Measurement Systems Analysis Plan		
Preliminary Process Capability Study Plan		
Packaging Specifications		
Management Support		

Considering the level of external control (8.3.2i)

What does this mean?

Box 37.3   New organizational interface requirement

The 2008 version required the requirements specified by the customer to be determined but did not relate these to the planning of design and development. The 2015 version has identified one such customer requirement and now requires consideration to be given to the level of control expected by customers and other relevant interested parties when planning design and development.

In a B2B relationship the customer may stipulate in the contract key stage-gates or milestones where they and their representatives wish to exercise some degree of control over the progression of the design. Often these interventions are subject to negotiation as they can require a lot of preparation to provide the customer with sufficient information on which they can make decisions.

Why is this necessary?

With most customer-funded projects, the customer generally wants to exercise some control so as to obtain confidence that the design is proceeding in a direction that is acceptable to them. By intervening at key stages during design and development the customer is provided with the opportunity to correct misunderstandings in the requirements, share their expertise in the resolution of problems and grant approval for the design to proceed to the next stage with confidence.

How is this addressed?

The stage-gates or design reviews where the customer wishes to intervene will be identified in the design and development plan (see 8.3.2b earlier). The customer will usually stipulate the deliverables to be submitted prior to the event. Provision will need to be made in the plan, the schedule and the budget for these deliverables (see also 8.3.2h). Time needs to be allowed for follow-up action after the review, as the customer may stipulate that certain design work is paused until after the review in case changes are required.

Where the customer wishes to witness critical design verification and/or validation events, these also need to be designated in the design and development plan so that the event does not take place without them in attendance. Having to repeat a demonstration because someone forgot to inform the customer may be an expensive mistake you don't want to repeat.

Sometimes these milestones are linked to stage payments where funding is released on presentation of evidence that the design has reached a certain stage. These, too, should be designated in the design and development plan so that the design team is aware of the significance of the event and plans accordingly.

How is this demonstrated?

Demonstrating that consideration has been given to the level of control expected by customers and other relevant interested parties may be accomplished by:

a)   selecting a representative sample of products and presenting the design and development plans;
b)   presenting the contracts for these products and pointing out the controls that have been stipulated by the customer;
c)   showing the provisions made for enabling the customer and other interested parties to exercise the control they required over the design and development process.

Confirming verification and validation data capture (8.3.2j)

What does this mean?

This requirement is an extension to that in 8.3.2c for the organization to consider verification and validation activities. These activities will reveal whether design and development requirements have been met and therefore these results are to be classed as documented information or design verification and validation records.

Why is this necessary?

It is necessary to do this as part of design and development planning so that recording provisions are built into the process thereby preventing loss of valuable data that cannot be retrieved without repeating the activities and incurring additional costs.

How is this addressed?

Recording requirements should be considered when planning design verification and validation and provision made to capture the data. It may only be possible to capture evidence of some activities and results by manual observation and recording. It may also be more effective to capture other evidence on digital media automatically by the equipment being used. Therefore, when specifying the equipment, data recording requirements need to be addressed.

How is this demonstrated?

Demonstrating that consideration has been given to the documented information needed to demonstrate that design and development requirements have been met may be accomplished by presenting plans, specifications or procedures showing how verification and validation data are captured.

Bibliography

ISO 9000:2015. (2015). *Quality Management Systems: Fundamentals and Vocabulary*. Geneva: ISO.
MacCormack, A., Crandall, W., & Henderson, P. (2012). Do You Need a New Product-Development Strategy? *Research Technology Management* 55(1), 34–43.
Mil Std 833C. (2011). *Work Breakdown Structures for Defense Materiel Items*. Washington: US Department of Defense.

38 Design and development inputs

Introduction

Design and development can be viewed as a function of a department of specialists, but can also be viewed as a process with inputs and activities performed by cross-functional teams which together produce outputs of requite quality, and this is way it is viewed in ISO 9001. The process objective is to produce solutions to technical problems that, when implemented, produce products and services with features that reflect customer needs and expectations. The quality of the inputs to this process will therefore have a direct bearing on the quality of the outputs. We might almost say garbage in, garbage out, to quote an expression from the field of computer sciences, except that the design process is not wholly computerized and still relies extensively on the contribution of highly qualified specialists.

In this chapter, we examine the nine requirements of clause 8.3.3, namely:

- Determining requirements essential for the specific type of products and services (8.3.3)
- Considering functional and performance requirements (8.3.3a)
- Considering information from similar designs (8.3.3b)
- Considering statutory and regulatory requirements (8.3.3c)
- Considering standards or codes of practice (8.3.3d)
- Considering potential consequences of failure (8.3.3e)
- Ensuring the adequacy of inputs (8.3.3)
- Resolving conflicting inputs (8.3.3)
- Retaining documented information on inputs (8.3.3)

Determining requirements essential for the specific type of products and services (8.3.3)

What does this mean?

The design inputs are the requirements governing the design of the intended product or service. It may appear that this requirement duplicates those addressed in clause 8.2 of the standard (see Chapter 35), but those requirements may apply to different stages in the product or service life cycle. Of the requirements that apply to the product or service, those that cannot be satisfied by existing products and services qualify as design input requirements. They become the requirements essential for the specific type of products and services to be designed and developed.

Why is this necessary?

The design input requirements constitute the basis for the design without which there is no criteria to judge the acceptability of the design output.

How is this addressed?

Although a customer may have provided detailed requirements for the product or service expressed in performance terms, it is unwise to believe that this is the only information needed. Some aspects may not have been considered by the customer because they are not specialists. Design input data may come from several sources (see Figure 38.1).

Design inputs should reflect the customer's, regulator's and organization's needs and be produced or available before any design commences. The requirement specifications should include, as appropriate:

- The purpose of the product or service.
- The conditions under which it will be used, stored and transported.
- The skills and category of those who will use and maintain the product or service.
- The countries to which it will be sold and the related regulations governing sale and use of products.
- The special features and characteristics which the product or service is required to exhibit including any that need to be built-in to protect the product, protect the users, property or the natural environment in the event of failure, misuse or disposal (see 8.3.3e).
- The constraints in terms of timescale, operating environment, cost, size, weight or other factors.
- The standards with which the product or service needs to comply.
- The products or services with which it will directly and indirectly interface and their features and characteristics.
- The documentation required of the design output necessary to manufacture, procure, inspect, test, install, operate or maintain a product or service.

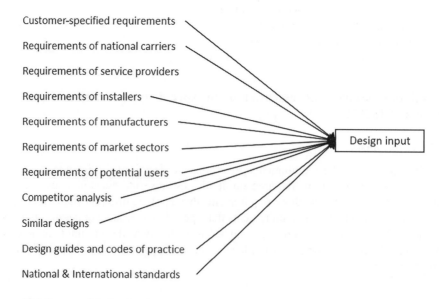

Figure 38.1 Source of design inputs

Organizations have a responsibility to establish their customer requirements and expectations. If you do not determine conditions that may be detrimental to the product and you supply the product as meeting the customer needs and it subsequently fails, the failure is your liability. If the customer did not provide reasonable opportunity for you to establish the requirements, the failure may be the customer's liability. If you think you may need some extra information to design a product that meets the customer needs, you must obtain it or declare your assumptions. A nil response is often taken as acceptance in full.

The result of competitive analysis should be used to ensure that product design requirements are not putting the product at a distinct disadvantage even before design commences.

One specific series of requirements that may not emerge from the forgoing are technical interface requirements. Some of these may need to be written around a particular supplier. However, within each development specification the technical interfaces between systems, subsystems, equipment, etc., should be specified so that when all these components are integrated they function properly.

The design output must reflect a product that is producible or a service that is deliverable. The design input requirements may have been specified by the customer and consequently not have considered your capability. The product of the design may therefore need to be producible within your current capability using your existing technologies.

The requirements should not contain any solutions at this stage to provide freedom and flexibility to the designers. If the design is to be subcontracted, this makes for fair competition and removes from you the responsibility for the solution. Where specifications contain solutions, the supplier is being given no choice and if there are delays and problems the supplier may have a legitimate claim to renegotiate the budget.

How is this demonstrated?

Demonstrating that requirements essential for the specific type of products and services to be designed and developed have been determined may be accomplished by;

a)   selecting a representative sample of products and services that the organization claims to have designed;
b)   presenting performance specifications for those selected products and services;
c)   presenting evidence showing how these requirements have been derived from stakeholder needs and expectations (e.g. a process description).

Considering functional and performance requirements (8.3.3a)

What does this mean?

Products and services possess distinguishing features that are referred to as characteristics in ISO 9000:2015. To design products and services that will satisfy the needs of those who are intended to use them these characteristics need to be specified in performance terms to allow the widest possible range of solutions to be developed. Quite why this requirement calls for both functional and performance requirement is not known because performance requirements are generally expressed in terms of physical and functional characteristics.

Functional requirements are those related to actions that product or service is required to perform or exhibit with or without external stimulus. Performance requirements relate to the results produced or behaviours exhibited under stated conditions. The intent of the requirement is that all characteristics that the product is required to exhibit should be included in the design input requirements and expressed in terms that are measurable.

Why is this necessary?

All the essential characteristics need to be stated otherwise the resultant design may not reflect a product or service that fulfils the conditions for intended use. Two products or services may possess the same performance characteristics but perform differently due to the arrangement of their component parts and the materials and processes used in their construction or delivery.

How is this addressed?

Considering that quality is defined as the "degree to which a set of inherent characteristics of an object fulfils requirements" (ISO 9000:2015), when determining design inputs, one should be defining the characteristics which the product or service needs to exhibit to fulfil customer needs and expectations.

A characteristic is a distinguishing feature which can be inherent or assigned, qualitative or quantitative and there are various classes of characteristic, such as the following:

a) Physical (e.g. size, appearance, mechanical, electrical, chemical or biological characteristics).
b) Sensory (e.g. related to smell, touch, taste, sight, hearing).
c) Behavioural (e.g. courtesy, honesty, veracity).
d) Temporal (e.g. punctuality, reliability, availability, continuity, maintainability, durability, flammability).
e) Ergonomic (e.g. physiological characteristic, or related to human safety).
f) Functional (e.g. speed, power, safety, portability).

Therefore, when determining the requirements essential for the specific types of products and services to be designed and developed one should be considering these. From the statement of product or service purpose, the conditions of use and the skills of those who will use the product or service, the most obvious characteristics can be derived and divided into different classes of characteristics as appropriate.

How is this demonstrated?

Demonstrating that consideration has been given to functional and performance requirements may be accomplished by:

a) presenting product and service design requirement specifications that define the key characteristics required;
b) presenting evidence showing how these characteristics have been derived from stakeholder needs and expectations.

Considering information from similar designs (8.3.3b)

What does this mean?

Most designs are a development of a product or service which was designed previously. It is rare for a design to be completely new. Even if the product concept is new, it may contain design solutions used previously. The history of these previous designs contains a wealth of information that may be applicable to the application that is currently being considered.

Why is this necessary?

Today's successes and failures are the result of yesterday's solutions; therefore, using information from previous designs may be advantageous, but only if lessons were learnt. If past design failures resulted in corrective action that eliminated the cause of the failure the use of such designs is not only wise but essential to sustain success and prevent failure. If previous design history is not utilized, problems may recur and the successes not recur.

How is this addressed?

In principle, the design history of a product should be archived and made available to future designers. Design history can be placed in a database or library that is accessible to future designers. A rather old way of doing this was for companies to create design manuals containing datasheets, fact sheets and general information sheets on design topics, which is a sort of design guide that captured experience. Companies should still be doing this but many will by now have converted to electronic storage media with the added advantage of a search engine. Information will also be available from trade associations, libraries and learned societies. Often professional journals, published literature and even newspapers can contain useful information for designers. In your model of the design process you need to install a research process that is initiated at some stage in the design of a product or service. The database or libraries need to structure the information in a way that it will return relevant data on previous designs. One advantage of submitting the design to a review by those not involved in it is that they bring their experience to the review and identify approaches that did not work in the past, or put forward more effective ways of doing such things in the future.

Within the design input requirements, such information would appear either as preferred solutions or non-preferred solutions, either directly or by reference to learned papers, standards, guide etc. See also Chapter 28 on organizational knowledge.

How is this demonstrated?

Demonstrating that consideration has been given to information derived from previous similar design and development activities may be accomplished by:

a) presenting process descriptions that show how inputs from archive material is researched and selected for inclusion in design input data;
b) selecting a representative sample of design input data and showing how it refers to or is traceable to data from previous designs.

Considering statutory and regulatory requirements (8.3.3c)

What does this mean?

The terms *statutory* and *regulatory requirements* are defined in Box 1.5. At the end product level, the applicable statutory and regulatory requirements are those addressed by clause

8.2.2a in Chapter 35. However, as the design unfolds, additional statutory and regulatory requirements may become applicable as specific solutions emerge. The only thing to add here is that the requirements referred to are those which directly affect product or service characteristics and will exclude those that are merely product or service related.

Why is this necessary?

It is far better to determine the legal requirements before and during design and development than wait until design verification when the product or service may fail or worse may enter service risking litigation.

How is this addressed?

The register of requirements referred to in Chapter 18 should include current statutory and regulatory requirements that are relevant to the organization and the markets in which it operates. Many of the statutory and regulatory requirements that apply to any type of product or service the organization will provide are likely to be included. However, as every product or service will be different to some degree, the process of determining design inputs should include a research activity that searches for statutory and regulatory requirements that may apply to a specific product or service due to its special characteristics.

A common approach is to list every possible statutory and regulatory requirement that could apply, leaving it to the designer to work out which are relevant. Much time could be saved if a test for relevance was carried out before the high-level design requirements are released.

How is this demonstrated?

Demonstrating that applicable statutory and regulatory requirements have been determined may be accomplished by presenting the evidence identified in Chapter 35 on clause 8.2.2b).

Considering standards or codes of practice (8.3.3d)

What does this mean?

In addition to the requirements identified there may be requirements that are dictated because of the organizational policies, national and international politics as was addressed under clause 8.2.2c in Chapter 35.

Why is this necessary?

The organization may wish to maintain a certain profile or reputation through its designs and therefore may impose requirements that may impact the design input requirements.

How is this addressed?

In addition to customer requirements industry practices, national standards, company standards and other inputs to the design requirements may need to be considered. An organization that has been operating for some time will have developed design guides or codes of practice that prescribe preferred technologies and design solutions and proscribe non-preferred design practices.

For manufactured products, consideration may need to be given to ease of assembly, manufacturability, testability and the constraints imposed by current technologies. It may therefore be appropriate to invoke codes of practice on design for assembly (DFA), design for manufacture (DFM) or design for X where X is a variable which can have one of many possible values such as manufacturability, power, variability, cost, yield, reliability, logistics, test, safety, etc.

How is this demonstrated?

Demonstrating that standards or codes of practice that the organization has committed to implement have been determined may be accomplished by:

a) presenting evidence identifying the organization's preferred standards and codes of practice;
b) selecting a representative sample of design requirements and showing that the preferred standards and codes of practice have been referenced when relevant.

Considering potential consequences of failure (8.3.3e)

What does this mean?

Although every effort will be made to ensure users can rely on a product or service until it becomes obsolete, but if it should either fail prematurely or fail to fulfil its purpose the user will not want the consequences to be detrimental to their wealth, health, safety, to the environment or other critical factors (e.g. ships can sink, aircraft can crash, food can kill, personal details can be stolen).

Here we are not referring to a specific design because design has not yet commenced, and therefore we need to distinguish between consequences that are due to the nature of a product or service and those that are due to the inherent reliability of its design (see Box 38.1).

Box 38.1   Distinguishing consequence due to the nature of the product from those due to the design of the product

A ship hits an iceberg and sinks because that's what ships do when they meet an immovable force. The consequences are that many passengers are lost – the cause of failure is that there were no means of escape. The solution is to fit sufficient lifeboats and stay away from icebergs. This is a consequence due to the nature of the product.

A car hits a stationary object and the airbag in front of the passenger seat fails to inflate. The consequence is that the passenger is injured – the cause of failure is a malfunction of the sensor. The solution is to add a redundant sensor. This is a consequence due to the inherent reliability of its design.

We are not only referring to failures resulting from normal usage, but those resulting from abnormal usage and accidental or deliberate damage or destruction. We are also not referring to the unintended consequences that may arise due to normal functioning of the product or service (e.g. the impact on the environment by burning fossil fuel), but it's fair to include the unintended side effects of various medicines.

Why is this necessary?

When failures do occur, they may have undesirable consequences unless provision has been made in the design to prevent them, to recover from them or to render the product or service safe.

How is this addressed?

The place to start with this requirement is to identify what is critical. Here are some examples:

- **Wealth**: If the product to be designed is a replaceable item such as a light bulb in a video projector, a fuse in a domestic power distribution unit or a battery in a mobile phone, one of the consequences of failure, due to the nature of the product, is that the user loses use of it, the consequence of which may be costly to the user.
- **Health**: If the service to be designed is a fast food restaurant, one of the consequences of failure, due to the nature of the service, may be food poisoning.
- **Safety**: If the product to be designed is a form of public transport, a failure may result in a crash and one of the consequences, due to the nature of the product, is that passengers may be trapped. Consider the *Challenger* disaster of 1986 where NASA had failed to take any precautions in the event of a catastrophic but possibly survivable accident. There was no equipment to arrest the craft's fall or to allow the astronauts to ditch it, nor even an emergency locating transmitter. The crew could do nothing but ride it down.
- **Environment**: If the product to be designed will be used in space, we know the hostile environment causes gases trapped within materials to be released and one of the consequences, due to the nature of the product, is that these gases can condense onto optical elements thereby reducing their performance.

In the previous examples, we have chosen wealth, health, safety and the environment as the critical factors but there may be others.

Remember we are not looking at how a product or service may fail – that comes later when addressing the robustness of its design. Think of the product or service as a black box. You don't need to know what's inside the box, only what the box is required to do, where it's required to do it and the type of people who are going be using it or come into contact with it.

Many of the requirements in national and international standards and codes of practice were conceived as a result of past failures to prevent recurrence and protect future users.

How is this demonstrated?

Demonstrating that potential consequences of failure due to the nature of the products and services have been determined may be accomplished by:

a) selecting a representative sample of design requirements;
b) presenting evidence identifying the potential consequences of failure and showing how they have been addressed through the design requirements.

Ensuring the adequacy of inputs (8.3.3)

What does this mean?

Adequacy in this context means that the design input requirements are a true reflection of the customer needs while providing freedom and flexibility to the designers. Ambiguities arise where statements imply one thing but the context implies another.

Why is this necessary?

The determination of design inputs results in information that needs to be reviewed prior to its release; otherwise, incorrect information may enter the design process.

How is this addressed?

It is advisable to hold an internal design review at this stage so that you may benefit from the experience of other staff in the organization. The review needs to be a systematic review, not a superficial glance. Design work will commence based on what is conveyed in the requirements or the brief, although you should ensure there is a process in place to change the information should it become necessary later. In fact, a change process should be agreed at the same time as agreement to the requirement is reached.

To detect incomplete requirements, you either need experts on tap or checklists to refer to. It is often easy to comment on what has been included but difficult to imagine what has been excluded. It is also important to remove subjective statements.

For those designs commissioned by the customer, it is also prudent to obtain customer agreement to the design requirements before commencing the design. In this way, you will establish whether you have correctly understood and translated customer needs.

How is this demonstrated?

Demonstrating that design and development inputs are adequate, complete, and unambiguous may be accomplished by:

a)   presenting evidence of a process for reviewing design input information and resolving issues arising from the review;
b)   presenting the criteria against which the adequacy of design and development inputs will be determined;
c)   selecting a representative sample of design inputs and retrieving evidence that all were reviewed against these criteria prior to release.

Resolving conflicting inputs (8.3.3)

What does this mean?

As the design input requirements come from many sources, as shown in Figure 38.1, there may be conflict between them. The conflicts may be due to differences in context or be unintentional, and some may arise because of ignorance or error. Sometimes there may be pressure from a particular interested party to influence a design in a particular way to get around legal requirements.

Why is this necessary?

The reader finds that in one document it requires X and in another it requires Y and wonders which is correct. There will be minor issues such as the same items shown on one diagram may be shown differently in another, or cross-references may conflict or terms may not be used consistent with the definitions provided. There may also be issues posing greater significance such as parameters and requirements specified differently in different source documents and recent changes in legislation in conflict with the organization's traditional approach to a situation. These inconsistencies create doubt as to what is required. Were the design inputs to pass to one person, this doubt could be resolved in the mind of that person, but when the information passes to several people or different groups, there is a risk of multiple interpretations which may result in design solutions that don't work when the parts are integrated into the whole resulting in project delays and abortive effort.

How is this addressed?

To detect conflicting requirements, you need to read statements and examine diagrams and tables very carefully. You need to check many aspects before being satisfied the information is fit for use. Any inconsistencies with either internally or externally generated information found should be recorded. Issues with internally generated information should be conveyed to the person responsible for it with a request for action. Issues with externally generated information and any unresolved issues with internally generated information should be brought to the conceptual design review panel for resolution (see Chapter 39).

Any changes to correct the errors should be self-evident so that all the information does not need to be reviewed again.

How is this demonstrated?

Demonstrating that conflicting design and development inputs are resolved may be accomplished by:

a)   presenting evidence of a process for reviewing design input information and resolving issues arising from the review;
b)   selecting a representative sample of design input review records where issues were identified;
c)   presenting evidence which shows that issues were resolved in accordance with the specified policies.

Retaining documented information on inputs (8.3.3)

What does this mean?

The documented information to be retained is records that describe the activities carried out to ensure the adequacy of design inputs and the results of any deliberations to resolve conflicting information. However, documented information on design and development inputs includes information that should also be maintained, which are the specifications that describe the design and development requirements for the products or services to be offered to customers.

Why is this necessary?

Documented information on design and development inputs needs to be maintained to ensure effective communication of requirements to those involved in the design and development process. Records from the process of determining design and development inputs need to be retained to support the resolution of any issues arising subsequently and to demonstrate probity in the execution of the design input determination process. Without the records, organizations are left to rely on knowledge stored in the minds of its employees, who either may not recall it accurately or may not be around when needed.

How is this addressed?

Having identified the design input requirements, regardless of their complexity, it's sensible to document them in a specification that when approved is brought under document control (see Chapter 32).

Records from the process of determining design and development inputs could be in the form of a checklist which verifies the completion of the required process stages and ensures the essential factors are addressed with cross reference to supporting information.

How is this demonstrated?

Demonstrating that documented information on design and development inputs is retained and maintained may be accomplished by:

a) presenting evidence of a process for documenting the agreed design and development inputs;
b) retrieving records from the archive for the selected samples which shows these requirements:

i   have been generated through the prescribed process;
ii   have been reviewed in accordance with the process requirements.

Bibliography

ISO 9000:2015. (2015). *Quality Management Systems – Fundamentals and Vocabulary*. Geneva: ISO.

39 Design and development controls

Introduction

Previous versions of ISO 9001 have included requirements on design verification, design validation and design reviews under separate headings. In the 2015 version, all these have been brought together under the heading of design and development controls. It is not uncommon for checking activities to be regarded as *controls*, but in fact, they alone control nothing without there being standards against which to measure performance and actions to adjust performance if the standards are not being met. In this new section, standards are indeed included through use of the phrase *results to be achieved are defined* and actions are included through use of the phrase *any necessary actions are taken on problems determined*.

Unlike previous versions, the standard now focuses on the design and development process but the controls are not limited to those required in clause 8.3.4. Clause 8.3.3 not only refers to requirements specific to product and service design but also requirements governing the design process such as "standards or codes of practice that the organization has committed to implement".

In this chapter, we examine the six requirements of clause 8.3.4, namely:

- Defining design and development process objectives (8.3.4a)
- Controlling design reviews (8.3.4b)
- Controlling design verification (8.3.4c)
- Controlling design validation (8.3.4d)
- Taking necessary actions (8.3.4e)
- Retaining documented information on design and development activities (8.3.4f)

Defining design and development process objectives (8.3.4a)

What does this mean?

The results to be achieved by the design and development process are those that the organization has deemed necessary to accomplish its goals. These are the process objectives and success measures which have been derived from stakeholder needs and expectations. The design process will more than likely be unique for each new product or service although each may be based on a particular design and development strategy (see 8.3.2b in Chapter 37).

Why is this necessary?

Without an objective or a standard against which performance can be evaluated there is no basis for proclaiming a process is under control.

How is this addressed?

The design and development process objectives may include as appropriate:

a)   Requirements the product or service is expected to meet (design input);
b)   Requirements constraining how these requirements are to be met (design budget);
c)   The milestones by which each stage of design is to be complete;
d)   The target cost to produce the product or supply the service. Sometimes market research will identify a price beyond which a customer is unwilling to pay, and there will be a profit margin that needs to be met to recover development costs which together produce a maximum production or service provision cost.

How is this demonstrated?

Demonstrating that results to be achieved have been defined may be accomplished by presenting a design and development plan that includes statements of the process objectives and the criteria for determining whether these objectives have been achieved.

Controlling design reviews (8.3.4b)

What does this mean?

ISO 9000:2015 defines a review as "determination of the suitability, adequacy or effectiveness of an object to achieve established objectives". The ISO 9000:2015 definition implies that a design review is an activity undertaken to determine the suitability, adequacy and effectiveness of a design to meet the design requirement. In this context, *suitability* means an appropriate design solution has been developed; *adequacy* means the design solution meets all the design requirements and *effectiveness* means it is the right design objective. Design reviews are therefore not document reviews.

The standard no longer specifically requires systematic reviews, but a note to the requirement states that they can be conducted separately or in any combination, as is suitable for the products and services of the organization which amounts to the same thing. Systematic reviews probe the design solution and delve into the detail to explore how requirements are fulfilled.

Why is this necessary?

A design represents a considerable investment by the organization. There is therefore a need for a formal mechanism for management and the customer (if the customer is sponsoring the design) to evaluate designs at major milestones to detect weaknesses before they proceed to the next stage. This is so that abortive work is avoided were these weaknesses to be detected at a later stage.

How is this addressed?

The purpose of the review is to determine whether the proposed design solution is the most practical and cost-effective way of meeting the requirements and should continue or should be changed before proceeding to the next phase. It should also determine whether the documentation for the next phase is adequate before further resources are committed. Design review is that part of the design control process which measures design performance, compares it with pre-defined requirements and provides feedback so that deficiencies may be corrected before the design is released to the next phase.

Suitable stages are at the transition between the various phases of design maturity in the design process (see Chapter 37 on 8.3.2b). Development commences with a detail design and proceeds through several iterations and may continue through several enhancements before the design becomes obsolete and a new design idea is conceived.

Review schedules

A schedule of design reviews should be established for each product, process or service being developed. In some cases, there will need to be only one design review. After completion of all design verification activities, but depending on the complexity of the design and the risks, you may need to review the design at some or all the following intervals:

Design requirement review: To establish that the design requirements can be met and reflect the needs of the customer before commencement of design.
Conceptual design review: To establish that the design concept fulfils the requirements before project definition commences.
Preliminary design review: To establish that all known risks have been addressed and development specifications have been produced for each sub-element of the product or service before detail design commences.
Critical design review: To establish that the detail design for each sub-element of the product or service complies with its development specification and that product specifications have been produced before manufacture of the prototypes.
Design validation readiness review: To establish the configuration of the baseline design and readiness before commencement of design validation.
Final design review: To establish that the design fulfils the requirements of its development specification before preparation for its production.

Design review input data

The input data for the review should be distributed and examined by the review team well in advance of the time when a decision on the design must be made. A design review is not a meeting. However, a meeting will often be necessary to reach a conclusion and to answer questions of the participants. Often analysis may need to be performed on the input data by the participants for them to determine whether the design solution is the most practical and cost-effective way of meeting the requirements.

Conducting a design review

Although design documents may have been through a vetting process, the purpose of the design review is not to review documents. By using a design review methodology, flaws in the design may be revealed before it becomes too costly to correct them. Design reviews also serve to discipline designers by requiring them to document the design logic and the process by which they reached their conclusions, particularly the options chosen and the reasons for rejecting others.

The experiences of previous designs provide a wealth of information of use to designers that can alert them to potential problems. In compiling this information, designers can feed off the experience of others, not only in the same organization, but also in different organizations and industries. By using technical data available from professional institutions, associations, research papers, etc., checklists can be compiled that aid the evaluation of designs.

Participants at design reviews

The chairman of the review team should be the authority responsible for placing the development requirement and should make the decision as to whether the design should proceed to the next phase based on the evidence substantiated by the review team.

The review team should have a collective competency greater than that of the designer of the design being reviewed and possess sufficient practical experience to provide advance warning of potential problems with implementing the design. For a design review to be effective, it must be conducted by someone other than the designer. The requirement for participants to include representatives of all functions concerned with the design stage means those who have an interest in the results. The representation at each review stage may well be different.

How is this demonstrated?

Demonstrating that design and development reviews evaluate the ability of the results of design and development to meet requirements may be accomplished by:

a)   presenting evidence of a process for conducting design and development reviews;
b)   Selecting a representative sample of designs and presenting evidence that they had been reviewed in accordance with the pre-defined process;
c)   Showing how the results of design and development were evaluated against the requirements.

Controlling design verification (8.3.4c)

What does this mean?

Defining design verification

Verification is defined as "confirmation, through the provision of objective evidence that specified requirements have been fulfilled" (ISO 9000:2015). There are two types of verification: verification performed during design on component parts or elements to verify conformance to specification and verification performed on the completed design to verify performance against the design input. Design is not complete until the criteria for accepting versions for production or service delivery have been established and this cannot be done until the design has been verified.

When verification is to be performed

The standard does not state when design verification is to be performed, but in clause 8.3.5 it requires design and development outputs to be adequate for the subsequent processes for the provision of products and services, and therefore design verification should take place before release of design outputs for production or service delivery (see Chapter 40 for further explanation). The stages of verification will therefore mirror the design review schedule but may include additional stages. Design verification needs to be performed when there is a verifiable design output but verification of the design after launch of the product into production may have unforeseen consequences. If this is at all likely the risks should be identified and addressed at the planning stage.

Why is this necessary?

Verification is fundamental to any process, and unless the design is verified, there will be no sound basis for declaring that the resultant design meets the requirements.

How is this addressed?

Product design verification

Design verification need only be as complex as the complexity of the design solution and the range of the design requirements. Without design verification products of production standard may contain features and characteristics that have not been proven. Verifying production acceptance criteria during design verification permits verification to be performed on production models to be limited to those features and characteristics that are subject to change due to the variability in manufacturing, either of raw materials or of assembly processes.

Verification may take the form of a document review, laboratory tests, alternative calculations, similarity analyses or tests and demonstrations on representative samples, prototypes, etc. In all these cases the purpose is to prove that the design is right (i.e. it meets the requirements).

Design verification process

During the design process, many assumptions may have been made and will require proving before commitment of resources to the replication of the design. Some of the requirements such as reliability and maintainability will be time-dependent. Others may not be verifiable without stressing the product beyond its design limits. With computer systems, the wide range of possible variables is so great that proving total compliance would take years. It is, however, necessary to subject a design to a series of tests and examinations to verify that all the requirements have been achieved and that features and characteristics will remain stable under actual operating conditions. The tests that confirm features and characteristics under nominal operating conditions are called verification tests whereas, those tests that confirm features and characteristics at the operating extremes are called qualification tests. These differ from other tests because they are designed to establish the design margins and prove the capability of the design (see the next section).

The product design verification process should provide for the following:

a) Verification requirements that define the features and characteristics that are to be verified and the method of verification.
b) Verification strategy that explains the rationale for choosing which requirements are to be verified by:

   i   Inspection – used to verify physical characteristics, size, weight, quantities.
   ii Test – used to verify functional characteristics (e.g. performance, reliability, durability or corrosion resistance).

iii   Analysis of records – used to verify properties of materials or material traceability, conformity with regulations such as Conformité Européene Marking (CE Marking).

iv   Calculation – used where verification of characteristics may only be possible by calculation. In such cases the design calculations should be checked either by being repeated by someone else or by performing the calculations by an alternative method. When used the margins of error permitted should be specified.

v   Simulation – used when the intended operating conditions cannot be replicated (e.g. zero gravity).

vi   Comparison with similar design – used to avoid unnecessary costs by comparing the design with a similar one that has been proven to meet the same requirements. Care must be taken when using this method that the requirements are the same and that evidence of compliance is available to demonstrate compliance with the requirements. Marginal differences in the environmental conditions and operating loads can cause the design to fail if it was operating at its design limit when used in the previous design.

vii   Practical demonstration – used to verify features such as evacuation timing on transport vehicles, flammability, maintainability and servicing.

c)   Verification plans see Considering design and development verification and validation (8.3.2c).

d)   Procedures that describe how the planned verification activities are to be carried out together with the tools and equipment to be used and the data to be recorded.

Development models

Many different types of models may be needed to aid product development, test theories, experiment with solutions, etc. However, when the design is complete, prototype models representative in all their physical and functional characteristics to the production models may need to be produced.

If design is proven on uncontrolled models, it is likely that there will be little traceability to the production models. Production models may therefore contain features and characteristics that have not been proven. The only verification that needs to be performed on production models is for those features and characteristics that are subject to change due to the variability in manufacturing, either of raw materials or of assembly processes.

When building prototypes, the same materials, locations, suppliers, tooling and processes should be used as will be used in actual production to minimize the variation.

Development tests will not yield valid results if obtained using uncontrolled measuring equipment; therefore, the requirements of clause 7.1.5 on measuring devices apply to the design process. Design is not complete until the criteria for accepting production versions have been established. Products need to be designed to be testable during production using the available production facilities. The proving of production acceptance criteria is therefore very much part of design verification.

Development tests

Where tests are needed to verify conformance with the design specification, development test specifications will be needed to specify the test parameters, limits and operating conditions.

For each development test specification, there should be a corresponding development test procedure that defines how the parameters will be measured using particular test equipment and considering any uncertainty of measurement. Test specifications should be prepared for each testable item. Two principal factors to consider are:

- Testable items sold as spare parts.
- Testable items the design and/or manufacture of which are subcontracted.

If you conduct trials on parts and materials to prove reliability or durability, these can be classed as verification tests.

Service design verification

Where the service involves facilities and equipment and produces a tangible output such as a meal in the restaurant, a repaired automobile, results of metallurgical, forensic or other scientific analysis, many of the elements of the product design verification process may equally apply. However, it is often not possible to verify service design other than by simulation without actually providing the service because of the human factors. Models can be produced to simulate the flow of information or people through service delivery and test the robustness of the design using different demand parameters to reveal if there will be bottlenecks and delays.

How is this demonstrated?

Demonstrating that verification activities have been conducted to ensure the design and development outputs meet the input requirements may be accomplished by:

a)   presenting evidence of a process for verifying that design outputs meet the corresponding design input requirements;
b)   selecting a representative sample of designs and presenting evidence that the product design verification process has been executed as planned including evidence that:

    i   all measuring instruments were within calibration during any tests;
    ii the measuring processes were capable and took account of the uncertainty of measurements;
    iii the samples used for design verification successfully passed all planned in-process and assembly inspections and tests prior to commencing verification tests;
    iv the configuration of the product in terms of its design standard, deviations, non-conformities and design changes was recorded prior to and after the tests;
    v   reviews were conducted before verification commenced to ensure that the product, the facilities, tools, documentation and personnel were in a state of operational readiness for verification;
    vi activities were conducted in accordance with the prescribed specifications, plans and procedures;
    vii results of all activities and the conditions under which they were obtained were recorded;
    viii deviations were recorded, remedial actions taken and the product subject to re-verification prior to continuing with verification;

ix   reviews were conducted following verification to confirm that sufficient objective evidence had been obtained to demonstrate that the verification requirements had been fulfilled.

Controlling design validation (8.3.4d)

What does this mean?

Validation is "confirmation through the provision of objective evidence that requirements for a specific intended use or application have been fulfilled" (ISO 9000:2015). Specified requirements are often an imperfect definition of needs and expectations, and therefore to overcome inadequacies in the way requirements can be specified, the resultant design needs to be validated against intended use or application.

Why is this necessary?

Merely requiring that the design output meets the design input would not produce a quality product or service unless the input requirements were a true reflection of the customer needs and this is not always possible.

Products may not be put to their design limits for some time after their launch into service, probably far beyond the warranty period. Customer complaints may appear years after product launch. When investigated, this may be traced back to a design fault which was not tested for during the validation programme. Such things as loading, evacuation, corrosion, insulation, resistance to wear, heat, chemicals, climatic conditions, etc., need to be validated as being of a standard commensurate with actual operating conditions with a margin sufficient to avoid costly repercussions in use.

How is this addressed?

Product design validation overview

Product design validation (also known as design qualification) is a process of evaluating a product design to establish that it fulfils the intended use requirements. It goes further than design verification in that validation tests and trials may stress the product of such a design beyond operating conditions to establish design margins of safety and performance. Product design validation can also be performed on mature designs to establish whether they will fulfil different user requirements to the original design input requirements. An example is where a component designed for one environment can be shown to possess a capability that would enable it to be used in a different environment. Multiple validations may therefore be performed to qualify a design for different applications.

Timing of design validation

There is no requirement stipulating when design validation should be performed. In the 2008 version, validation was to be completed prior to the delivery or implementation of the product or service wherever practicable. The problem is that it is often not possible to simulate intended use conditions so that they are sufficiently realistic. Any decision to launch into production or into operation involves risk. There are some characteristics such as safety and reliability that need to be demonstrated before launching into production; otherwise, unsafe or unreliable products might be put onto the market. One has only to scan the recall programmes accessible on the Internet to notice that many products are indeed launched into production with major faults. In practice, it depends on knowing what the risks are and therefore is a balance between risk and the impact any delay in production launch or going operational may have. It would therefore be prudent to conduct a risk assessment in such circumstances. However, it should be noted that there is no mean time between failures (MTBF) until you have a failure, so you need to keep on testing until you know anything meaningful about the product's reliability.

Qualification tests

Product design validation may take the form of qualification tests which stress the product up to and beyond design limits and include performance trials, reliability and maintainability trials where products are put on test for prolonged periods to simulate usage conditions.

As the cost of testing vast quantities of equipment would be too great and take too long, qualification tests particularly on hardware are usually performed on a small sample. The test levels are varied to take account of design assumptions, variations in production processes and the operating environment.

Another form of design validation is beta testing or public testing. These tests are conducted on software products where tens or hundreds of products are distributed to designated customer sites for trials under actual operating conditions to gather operational performance data before product launch. Sometimes, commercial pressures force termination of these trials and premature launch of products to beat the competition.

Product design acceptance tests

Following qualification tests, your customer may require a demonstration of performance to accept the design. These tests are called design acceptance tests. They usually consist of a series of functional and environmental tests taken from the qualification test specification supported by the results of the qualification tests. When it has been demonstrated that the design meets all the specified requirements, a design certificate can be issued. The design standard that is declared on this certificate is the standard against which all subsequent changes should be controlled and from which production versions should be produced

Demonstrations

Tests exercise the functional properties of the product. Demonstrations on the other hand, serve to exhibit usage characteristics such as access, maintainability including interchangeability, repairability and serviceability. Demonstrations can be used to prove safety features such as the crash tests filmed at high speed. When the film is played at normal speed, the crumpling of the steel and movement of the dummy against the air bag show up characteristics that prove whether the safety features behave as intended.

Product producibility trials

One of the most important characteristics that need to be demonstrated is producibility. Can we make the product economically in the quantities required? Does production yield a profit or do we need to produce 50 to yield 10 good ones? The demonstrations should establish whether the design is robust. Designers may be selecting components at the outer limits of their capability. A worst-case analysis should have been performed to verify that under worst-case conditions, i.e. when all the components fitted are at the extreme of their tolerance range, the product will perform to specification. Analysis may be costlier to carry out than a test and by assembling the product with components at their tolerance limits you may be able to demonstrate economically the robustness of the design.

Product approval

A product approval process is often required in large-scale production situations such as the automotive and domestic appliances sectors. When one considers the potential risk involved in assembling unapproved products into production models, it is hardly surprising that the customers impose such stringent requirements. The process provides assurance that the product meets all design criteria and is capable of production in the qualities required without unacceptable variation. It is intended to validate that products made from production materials, tools and processes meet the customer's engineering requirements and that the production process has the potential to produce product meeting these requirements during an actual production run at the quoted production rate. Until approval is granted, shipment of production product may not be authorized. If any of the processes change, then a new submission is required. Shipment of parts produced to the modified specifications or from modified processes would not be authorized until customer approval is granted.

Service design validation

Service design validation is often only possible after service transition during initial trials either with real customers or employees performing the role of customers.

How is this demonstrated?

Demonstrating that validation activities have been conducted to ensure that products and services resulting from the design and development process are capable of meeting the requirements for the specified application or intended use may be accomplished by:

a) presenting evidence of a process for validating product and service design and development;
b) Selecting a representative sample of designs and presenting evidence that the product design validation process has been executed as planned including evidence as indicated for design verification.

Taking necessary actions (8.3.4e)

What does this mean?

This requirement simply tells us that not only is it sufficient to determine actions resulting from design reviews, design verification and design validation activities but that these actions are to be taken.

In the 2008 version, the same requirement for taking action was stated for design reviews, design verification and design validation. In the 2015 version, this requirement is stated only once.

Why is this necessary?

The design is not complete until the actions resulting from all the reviews, verification and validation activities have been taken and produce the required result.

How is this addressed?

The results of the reviews, verification and validation activities need to include the actions to be taken to bring the result into conformity with the requirements. These actions need to be planned, scheduled and implemented as planned. Sometimes, the actions may involve nothing more that correcting documented information, but in a worst-case scenario they might involve repeating some or all the design verification and validation activities.

How is this demonstrated?

Demonstrating that any necessary actions are taken on problems determined during the reviews, or verification and validation activities may be accomplished by:

a) Presenting evidence of a process for design and development in which provision is shown to have been made:

  i   for feedback into the process after these activities; and
  ii   for verifying that the results are acceptable before the design proceeds to the next stage.

b) Selecting a representative sample of designs and presenting evidence showing that outstanding actions were resolved to the satisfaction of the design authority before the design proceeded to the next stage.

Retaining documented information on design and development activities (8.3.4f)

What does this mean?

The activities for which documented information is required to be retained are the results to be achieved, design reviews, design verification, design validation and the resulting actions taken. The requirement is limited to clause 8.3.4 as there are additional requirements specifying the documented information for design inputs, design outputs and design changes.

Why is this necessary?

Without evidence of activities carried out and the results of those activities it would be difficult to perform reviews and make decisions. Any decision to proceed either to the next stage of development or into production or operations needs to be based on fact and these records provide the facts.

How is this addressed?

Documented information on results to be achieved

The results to be achieved by the design and development process will usually be recorded in the design and development plan in the form of process objectives and success measures.

Documented information on design and development reviews

The results of the design review should be documented in a report rather than minutes of a meeting because they represent objective evidence that may be required later to determine compliance with requirements, investigate design problems and compare similar designs. Even when no problems are found, the records of the review provide a baseline that can be referred to when making subsequent changes.

The report should have the agreement of the full review team and should include:

- The criteria against which the design has been reviewed;
- A list of the documentation that describes the design being reviewed and any evidence presented which purports to demonstrate that the design meets the requirements;
- The decision on whether the design is to proceed to the next stage;
- The basis on which confidence has been placed in the design;
- A record of any uncompleted corrective actions from previous reviews;
- The recommendations and reasons for corrective action – if any;
- The members of the review team and their roles.

Documented information on design verification

The results of design verification comprise:

- The criteria used to determine acceptability;
- Data testifying the standard of the design being subject to verification;
- The verification methods;
- Data testifying the conditions, facilities and equipment used to conduct the verification;
- The measurements;
- Analysis of the differences between planned and achieved results.

In planning design verification, consideration needs to be given to the results, their format and content. The basic content is governed by the design specification but the data to be recorded before, during and after verification need to be prescribed. Some data may be generated electronically, and other data may be collected from observation. Often lots of different pieces of evidence need to be collected, collated and assembled into a dossier in a secure format. These factors need to be sorted out before commencing verification so that all the necessary information is gathered at the time. After verification, a report of the activities may also be necessary to explain the results, possible causes of any variation and recommendations for action for presentation at a design review.

Documented information on design validation

The results of validation are like those required for verification except that duration of testing and trials is important in quantifying the evidence. The results should not only indicate that the product meets intended use requirements but also satisfy market need. If by the time you have the validation results, the anticipated demand for the product has declined, it might not be prudent to launch into production.

As with design verification, consideration of the output, its format and content needs to be given early in the design phase so that the correct data is captured during validation trials.

Documented information on actions taken on problems found

Records of actions should include:

- The actions to be taken on the differences between the requirements and the results of design review, verification and validation;
- The actual actions taken;
- Confirmation that the actions taken resolve the original problem.

How is this demonstrated?

Demonstrating that documented information of design reviews, design verification, design activities and actions taken on problems detected is being retained may be accomplished by:

a) presenting evidence of a process for producing this documented information;
b) selecting a representative sample of designs and retrieving the relevant documented information that has been retained in accordance with the pre-defined process.

Bibliography

ISO 9000:2015. (2015). *Quality Management Systems – Fundamentals and Vocabulary*. Geneva: ISO.

40 Design and development outputs

Introduction

An output is the "result of a process" (ISO 9000:2015); therefore, anything produced by the design and development process is an output. This will include data, information, knowledge, understanding, models, assumptions, specimens, calculations and the rationale for the chosen solution. It is not simply the specifications or drawings, because should the design need to be changed, the designer may need to revisit the design data and re-examine the models from which the data was produced to modify parameters and assumptions. During design and development, a lot of information will be acquired and produced, for example, there will be information and artefacts that:

a) enter the design and development process, are used and returned to source leaving the knowledge behind;
b) are produced and used in producing other information but are only used for this purpose;
c) exit the process and are used for various purposes.

It is all this information and artefacts that constitutes the design and development outputs.
  In this chapter, we examine the five requirements of clause 8.3.5, namely:

- Ensuring outputs meet input requirements (8.3.5a)
- Ensuring outputs are adequate for subsequent processing (8.3.5b)
- Ensuring outputs reference monitoring and measuring requirements (8.3.5c)
- Ensuring products and services are fit for intended purpose (8.3.5d)
- Retaining documented information on design and development (8.3.5)

Ensuring outputs meet input requirements (8.3.5a)

What does this mean?

As stated in the introduction, design output is the result of the design and development process, and this will include all manner of things but not all results will have a corresponding requirement among the design and development inputs. What is of relevance here is only that information and models necessary to demonstrate that a product or service possessing the prescribed characteristics meets the design input requirements.
  It is interesting to note that the requirement omits validation. This is because design outputs are verified against design inputs, whereas the design is validated against the original

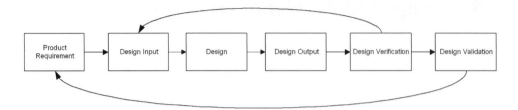

Figure 40.1 Relationship between design output and design validation

product requirement using a product or simulation that accurately reflects the design, thereby by-passing the design input and output as illustrated in Figure 40.1.

The requirement does not stipulate when the design outputs are to meet the input requirements. Design is an iterative process; therefore, preliminary design outputs will be used to produce prototypes, conduct trials and simulations. Such prototypes and beta models may be supplied to customers to acquire field data before the design is finalized.

Why is this necessary?

Clearly a design that does not reflect a product or service that meets the design requirement is nonconforming, but it does not mean the resultant product or service would be unfit for use. There may be cases where the design requirements prohibit the use of certain materials, methods or technologies but advances provide opportunities for using those materials, methods or technologies in ways that overcome the problem which the prohibition was intended to avoid.

How is this addressed?

The design review process should produce evidence that the design output meets the design input requirements. This evidence will be compiled from the results of the design verification and validation processes. Design verification is often an iterative process; therefore, the results may be cumulative. Unless the design output is expressed in a form that enables verification, it will not be possible to verify the design with any certainty. The characteristics of the resultant design should be directly or indirectly traceable to the design input requirements. A dimension may be stated in the design input which is easily verified when examining the design specifications, drawings, etc. In some cases, the input requirement may be stated in performance terms that are translated into several functions which, when energized, provide the required result. In other cases, a parameter may be specified above or below the design input requirement to allow for variation in production.

The characteristics of the product need to be expressed in measurable terms, and therefore the form, fit and function would need to be specified in units of measure with allowable tolerances or models and specimens to be capable of use as comparative references.

How is this demonstrated?

Demonstrating that design and development outputs meet the input requirements for design and development may be accomplished by:

a) presenting evidence of a process for determining that design outputs meet the input requirements;

b) presenting evidence that as features are determined, their compliance with the requirements is checked by analysis of design information;

c) presenting evidence that compliance with the requirements is checked by suitable methods in the case of products on representative models or in the case of services during service transition.

Ensuring outputs are adequate for subsequent processing (8.3.5b)

What does this mean?

Design and development is not undertaken for its own sake; the outputs are produced with the intention that they be used to construct a product or deliver a service that is representative of the design and which is fit for its intended purpose. This can be interpreted in two ways:

*   The design facilitates procurement, production and servicing of component parts, as well as assembly, installation and commissioning of the complete product or that facilitates the setup, resourcing and transition of a service. Tooling for production is part of the production process, but information within the design output is needed to enable tooling to be designed.

*   That transfer from development to production or service provision only takes place when the outputs are adequate, that is, actions outstanding or arising from development have been completed or are otherwise managed such that there is no adverse impact on the customer. (A requirement to this effect was in an early draft of ISO 9001 but was later removed.)

The instructions needed to produce, inspect, test, protect, preserve, install and maintain the product may be produced by the designers but are strictly outputs of the production and service provision planning processes.

Why is this necessary?

A design description alone will not result in its realization unless provision has been made in the design to enable its effective construction or delivery.

How is this addressed?

Product documentation

Products should be designed to facilitate procurement, manufacture, storage, transportation, installation and servicing and therefore additional characteristics to those required for end use may be necessary. Examples include, geometric tolerances, specific part numbers, part marking, assembly aids, error proofing, lifting points, transportation and storage protection. Techniques used to identify such design provisions are as follows:

*   Failure mode and effects analysis
*   Producibility analysis
*   Testability analysis
*   Maintainability analysis

In addition to the documents that serve product manufacture and installation, documents may be required for maintenance and operation. The product descriptions, handbooks, operating manuals, user guides and other documents which support the product or service in use are as much a part of the design as the other product requirements. Unlike the manufacturing data, the support documents may be published either generally or supplied with the product to the customer. The design of such documentation is critical to the success of the product as poorly constructed handbooks can be detrimental to sales.

Service documentation

Services should be designed to facilitate their resourcing and delivery and therefore additional characteristics to those required for end use may be necessary. For example, a retail outlet may require access for vehicles delivering stock in addition to customer access which may extend the area originally estimated for the service. The access requirements may prohibit delivery of the service in certain locations. Extensive storage areas for holding stock until required for use may be needed which limit the locations where the service may be delivered.

Other outputs of service design should include the information that service users require to access the service, select options, engage in the process and report problems.

How is this demonstrated?

Demonstrating that design and development outputs are adequate for the subsequent processes for the provision of products and services may be accomplished by:

a) selecting a representative sample of designs and retrieving evidence of analysis carried out to determine whether the outputs are adequate for the provision of products and services;

b) selecting a representative sample of products and services and searching records of production and services queries, problem reports and design changes for evidence of the inadequacy of design outputs for the provision of products and services.

Ensuring outputs reference monitoring and measuring requirements and acceptance criteria (8.3.5c)

What does this mean?

Monitoring and measurement requirements and acceptance criteria are the requirements that, if met, will deem the product acceptable. It means that characteristics should be specified in measurable terms with tolerances or limits. These limits should enable all production versions to perform to the product specification, providing such limits are well within the limits to which the design has been tested. It means that every requirement should be stated in such a way that it can be verified – that there is no doubt as to what will be acceptable and what will be unacceptable.

Why is this necessary?

It is not sufficient to specify the requirements a product or service should meet. It is also necessary to specify how conformity with those requirements is to be verified because of the wide variation in the quality of methods that can be used. Where product characteristics are specified in terms that are not measurable or are subjective, they lend themselves to misinterpretation and variation such that no two products produced from the same design will be the same and will exhibit inconsistent performance.

How is this addressed?

The design input requirements should have been expressed in a way that would allow several possible solutions. The design output requirements should therefore be expressed as all the inherent features and characteristics of the design that reflect a product that will satisfy these requirements. It should therefore fulfil the stated or implied needs (i.e. be fit for purpose).

A common method used to ensure characteristics are stated in terms of acceptance criteria is to define them by reference to product standards. These standards may be developed by the organization or may be of national or international status. Standards are employed to enable interchangeability, repeatability and to reduce variety.

Where there are common standards for certain features, these may be contained in a standards manual. Where this method is used, it is still necessary to refer to the standards in the relevant specifications to ensure that the producers are always given full criteria.

Some organizations omit common standards from their specifications. This makes it difficult to specify different standards or to subcontract the manufacture of the product or outsource the operation of a service without handing over proprietary information.

The requirements within the product and service specification need to be expressed in terms that can be verified. You should therefore avoid subjective terms such as *good-quality components*, *high reliability*, *commercial standard parts*, etc., as these requirements are not sufficiently definitive to be verified in a consistent manner.

Product specifications

Product specifications should specify requirements for the manufacture, assembly and installation of the product in a manner that provides acceptance criteria for inspection and test or other means of verification. They may be written or CAD-generated specifications, engineering drawings, diagrams, inspection and test specifications and schematics. With complex products, you may need a hierarchy of documents from system drawings showing the system installation to component drawings for piece-part manufacture. Where there are several documents that make up the product specification, there should be an overall listing that relates documents to one another.

Service specifications

Service specifications should provide a clear description of the way the service is to be provided, the criteria for its acceptability, the resources required, including the numbers and skills of the personnel required, the numbers and types of facilities and equipment necessary and the interfaces with other services and suppliers.

How is this demonstrated?

Demonstrating that design and development outputs include or reference monitoring and measuring requirements, and acceptance criteria may be accomplished by:

a)   presenting evidence of a process for specifying the product and service verification requirements among the design outputs;

b) selecting a representative sample of designs and presenting the product and service verification requirements that are used to confirm the product or service conforms to product and service requirements.

Ensuring products and services are fit for intended purpose (8.3.5d)

What does this mean?

Certain characteristics will be critical to the installation, operation or maintenance of the product or provision of the service. These are sometimes called critical to quality characteristics (CTQs). These can be divided into two types. Those characteristics that the product or service needs to exhibit to function correctly and those characteristics that are exhibited when the product or service is put together/provided, used or maintained incorrectly.

Why is this necessary?

Alerting assemblers, users and maintainers to CTQ characteristics increases their sensitivity, provides the awareness to plan preventive measures and thus reduces the probability of an incident or accident

How is this addressed?

Identifying CTQ characteristics

Some CTQ characteristics will be identified in the design input data, including those from statutory and regulatory requirements and will be independent of any design solution. Others may be identified during design and development from an analysis of the inherent characteristic of the chosen design solution.

Specifying CTQ characteristics

The design output data should identify by use of symbols or codes the CTQ characteristics. This will enable the manufacturers or service providers to determine the measures needed to ensure no variation from specification when the characteristics are initially produced and ensure no alteration of these characteristics during subsequent processing or operation.

Documented information should also indicate the warning notices required, where such notices should be placed and how they should be affixed. Examples that indicate improper function or potential danger are red lines on tachometers to indicate safe limits for engines, audible warnings that signal unsafe loading, for example, a stall warning on board an airliner or "No Step" notices on the flying surfaces of an aircraft wing or warnings on computers to indicate an incorrect command, etc. In some cases, it may be necessary to mark dimensions or other characteristics on drawings to indicate that they are critical and employ special procedures for dealing with any variations. In passenger vehicle component design, certain parts are regarded as safety critical because they carry load or need to behave in a certain manner under stress. Others are not critical because they carry virtually no load so there can be a greater tolerance on deviations from specification.

Failure mode and effects analysis and hazard analysis are techniques that aid the identification of characteristics crucial to the safe and proper functioning of the product.

When complying with legal requirements you may not have specified a dangerous substance in the product specification, but the characteristics you have specified in the product specification may be such that can only be produced by using a dangerous substance, and therefore it is not simply the product that may have to meet regulations, but also the materials used in making the product. These materials are a direct consequence of the chosen design solution; therefore, in general:

- The product must be safe during use, storage and disposal.
- The product should present minimum risk to the environment during production, storage, use and disposal.
- The materials used in the manufacture of the product must be safe during use, storage and disposal.
- The materials used in manufacture of the product should present minimum risk to the environment during use and disposal.

In service design, there may be similar CTQs, for example,

- Service availability in a telecommunications data centre
- Allergies in a restaurant
- Response time in an ambulance service

Box 40.1   Not paying attention to customers can land you in jail

A restaurant owner was jailed for six years for the manslaughter of a customer with a peanut allergy after he supplied him with a curry containing peanuts. The customer had told restaurant staff that his meal must be nut-free. The owner, who had a debt of £300,000, had swapped almond powder in recipes for cheaper groundnut mix, containing peanuts, despite warnings.

(The Guardian, 2016)

How is this demonstrated?

Demonstrating that design and development outputs specify the product and service characteristics that are essential for their intended purpose and their safe and proper provision may be accomplished by:

a) presenting evidence of a process for identifying and specifying critical to quality characteristics in the design output information;

b) selecting a representative sample of designs and presenting evidence that the CTQ characteristics are being identified and specified in accordance with the pre-defined process.

Retaining documented information on design and development outputs (8.3.5)

What does this mean?

The documentation information on design and development outputs is all the information generated during the design and development process that is needed to:

a)   demonstrate the design outputs meet the design inputs;
b)   justify the decisions made;
c)   enable the manufacture, purchasing, installation, commissioning, operation, maintenance, upgrade and disposal of the product;
d)   enable the transition, provision, maintenance, upgrade and termination of the service.

Why is this necessary?

Documented information emerging from the design and development process is used in many ways and at different times. Initially some of the information is needed to produce the product or transition the service but other information is needed by users, maintainers, even other designers when they come to modify the design. If that information is not retained it won't be possible to carry out these activities effectively and efficiently.

In this clause, there is a requirement only to retain documented information, but in clause 8.1 it requires certain documented information to be retained and maintained; however, this, too, omits to require information generated during design that is needed for operation, maintenance and disposal.

Box 40.2   Anything can happen on a building site

Something can't go where it was planned to go, the customer changes his mind and wants the control room on the other side of the building, the council wants the foundations strengthened or an additional load-bearing pier to be installed. No one went back to update the original plans, so a few years later after receiving a contract to update the control system, along comes the installer with a new control system and a set of cables with connectors he has had made of a length that conforms to the original plan and finds the old control system isn't where he thought it would be. This happens when the designer takes it for granted that the building conforms to the plans.

We see this sort of thing quite often with national infrastructure. The council workers don't know where the drains are, the telecoms people cut through the gas main because it's not marked on the plans, etc.

Design and development output information needs not only to be retained but also maintained for the poor guys who come along later and put a pick through a gas main that wasn't supposed to be there!

How is this addressed?

First, you need to establish what needs to be retained and what can be ditched. Clearly any information that meets the criteria in a) to d) earlier should be retained. The record of this type of information is sometimes referred to as a Master Record Index (MRI) or Configuration Index (CI). There may be a need to separate the CI by hardware, software and thus produce an HCI and SCI.

This leaves a lot of information that was used to create the information in the CI. One should consider whether to retain all versions of documents, all notes of meetings, all e-mails on the project, all design calculations and all design solutions that were rejected for one reason or another as the reason for rejection may be useful to other designers.

If the information is stored on digital media, there might be no urgency or pressure to delete any of it. Perhaps the most critical issue is the ability to retrieve the information when required. If there is so much information that identifying the correct file becomes difficult, some "weeding" certainly needs to be carried out and the information put into a structure that makes navigation easy and intuitive (see also Chapter 32).

How is this demonstrated?

Demonstrating that documented information on design and development outputs is retained may be accomplished by:

a) presenting evidence of a process for classifying and retaining design and development documented information to be retained and maintained;

b) selecting a representative sample of designs and presenting evidence that the relevant documented information can be retrieved with ease.

Bibliography

ISO 9000:2015. (2015). *Quality Management Systems – Fundamentals and Vocabulary*. Geneva: ISO.

The Guardian. (2016, May). Restaurant Owner Jailed for Six Years Over Death of Peanut Allergy Customer. Retrieved from The Guardian: www.theguardian.com/society/2016/may/23/restaurant-owner-mohammed-zaman-guilty-of-manslaughter-of-peanut-allergy-customer

41 Design and development changes

Introduction

The need to change the design of a product or service can arise at any time from the moment the design process is triggered to the time after the product or service has been superseded by a later design. There is an implication in the 2008 requirement that design changes being referred to are changes after design release, rather than changes of the design as it passes through the design process. In the 2015 version, the requirement applies to changes made during or after the design and development of products and services but the clue to its applicability lies in the words *to the extent necessary.* While a designer works on a design and perhaps goes through several iterations before finally settling on a solution, this is work in progress and not subject to change control. However, as soon as information is released to other designers for them to begin designing interfacing components or systems, it becomes necessary to control the changes because of the impact the changes may have on work already started. The controls exercised within in-house design do not have to be as rigid as those that will be needed were some of these designers to be external providers because of the contractual implications.

Design and development changes are changes to the features and characteristics of a product, service or process. Changes to design documents are not design changes unless the features and characteristics of the item are altered. Changes in the presentation of design information or to the system of measurement (imperial units to metric units) are not design or development changes. Therefore, not all design documentation changes are design changes. This is why design change control should be treated separately from document control. You may need to correct errors in the design documentation, and none of these may materially affect the product. The mechanisms you employ for such changes should be different from those you employ to make changes that do affect the design. By keeping the two types of change separate, you avoid bottlenecks in the design change loop and only present the design authorities with changes that require their expert judgement.

The other issue is between design changes and development changes as there are two quite different control processes. Design changes can occur at any stage in the design process from the stage at which the requirement is agreed to the final certification that the product, service or process is proven for production or service provision. Development changes can occur at any time in the life cycle of a design that extends until the product, service or process is superseded. Following design certification, changes to the product, service or process to incorporate design changes are generally classed as *modifications*.

Box 41.1   Revised design change control requirement – a bit of a mess!

There are some anomalies in the way the requirement has been expressed in the 2015 version. It requires the organization to *identify, review and control changes made*, implying that these actions are to be performed after changes have been made to the design. Perhaps this should have been *changes to be made*.

The 2008 version was clearer with "Design and development changes shall be identified and records maintained. The changes shall be reviewed, verified and validated, as appropriate, and approved before implementation." The phrase *before implementation* was assumed to mean that the design change should be approved before being implemented in production or service provision.

It seems that the words *verified and validated* in the 2008 version have been exchanged for *control* in the 2015 version, which is not the same thing. Verifying and validating a change is done on a changed design, whereas control of change will stop the change from being made in addition to verifying and validating the changed design.

In this chapter, we examine the four requirements of clause 8.3.6, namely:

- Identifying changes during and subsequent to design and development
- Reviewing changes during and subsequent to design and development
- Controlling changes during and subsequent to design and development
- Retaining information on design and development changes

Controlling changes during and subsequent to design and development (8.3.6)

What does this mean?

As indicated in Box 41.1 there are some anomalies with the wording of this requirement.

The key word is *control* because to control design changes, there must be a process that prevents undesirable changes being made and a process that causes desirable changes to be made. This means there is a process for requesting and authorizing a change and a process for making the change, which means cycling through the design process again, and this will involve review, verification, validation and approval before the change is implemented in production or service provision. Throughout the process the change will be identified; otherwise, it can't be controlled effectively.

Why is this necessary?

Once a baseline set of requirements has been agreed, it is necessary to control any changes to the baseline such that accepted changes are promptly made, and rejected changes are prevented from being made. Change control during the design process is a good method of controlling costs and timescales because once the design process has commenced every change will cost time and effort to address. This will cause delays whilst the necessary changes are implemented and provide an opportunity for additional errors to creep into the design. *If it's not broken, don't fix it*! is a good maxim to adopt during design. In other words, don't change the design unless it already fails to meet the requirements or you have discovered a requirement to be wrong. Designers are creative people who love to add the latest devices, the latest technologies, to stretch performance and to go on enhancing the design regardless of the timescales or costs. It is not uncommon to find good ideas being implemented before the ramifications of the change have been assessed. One reason for controlling design changes is to restrain the otherwise limitless creativity of designers to keep the design within the budget and timescale.

How is this addressed?

Changes to design before production is authorized

The imposition of change control is often a difficult concept for designers to accept. They would prefer change control to commence after they have completed their design rather than before they have started. They may argue that until they have finished there is no design to control. They would be mistaken. When a particular design solution is complete and has been found to meet the requirements at a design review, it should be brought under change control. Between design reviews the designers should be given complete freedom to derive solutions to the requirements. Between the design reviews there should be no change control on incomplete solutions. However, design work often proceeds in parallel and information is often shared between designers working on different subsystems or modules. In such cases a less formal change control process is needed to avoid abortive work, and one way of alerting designers is to use the notation *Advanced Information* or similar so that designers who use it are aware they are taking a risk it may change.

Proposing changes to the design

At each design review a design baseline should be established and recorded which identifies the design documentation that has been approved and change control procedures employed to deal with any changes. These change procedures should provide a means for formally requesting or proposing changes to the design and for promulgating design changes after their approval. A change proposal template is available on the companion website. You will need a central registry to collect all proposed changes and provide a means for screening those that are not appropriate either because they duplicate proposals already made or because they may not satisfy certain acceptance criteria. On receipt, the change proposals should be identified with a unique number that can be used on all related documentation that is subsequently produced. The change proposal needs to:

- identify the product of which the design is to be changed;
- state the nature of the proposed change;
- identify the principal requirements, specifications, drawings or other design documents which are affected by the change;
- state the reasons for the change either directly or by reference to failure reports, nonconformity reports, customer requests or other sources; and
- provide for the results of the evaluation, review and decision to be recorded.

Design change review and approval

Following the commencement of design, you will need to set up a change control board or panel comprising those personnel responsible for funding the design, administering the contract and accepting the product. All change proposals should be submitted to such a body for evaluation and subsequent approval or disapproval before the changes are implemented. By providing a two-tier system you can also submit all design documentation changes through such a body. They can filter the alterations from the modifications and the minor changes from the major changes.

The change proposals need to be evaluated to:

- validate the reason for change;
- determine whether the proposed change is feasible;
- judge whether the change is desirable;
- determine the effects on performance, costs and timescales;
- determine the impact of the change on other designs with which it interfaces and in which it is used;
- examine the documentation affected by the change and consequently programme their revision; and
- determine the stage at which the change should be embodied.

The evaluation may need to be carried out by a review team, by suppliers or by the original proposer. However, regardless of who carries out the evaluation, the results should be presented to the change control board for a decision.

During development, there are two decisions the board will need to make:

- whether to accept or reject the change; and
- when to implement the change in the design documentation.

If the board accepts the change, the changes to the design documentation can be processed through your document control procedures. With CAD systems, there is no reason why changes cannot be incorporated immediately following their approval. One does not need to accumulate design changes for incorporation into the design when design validation has been completed.

During production, the change control board will need to make four decisions:

- Whether to accept or reject the change;
- When to implement the change in the design documentation;
- When to implement the modification in new product; and
- What to do with existing product in production, in store and in service.

The decision to implement the modification will depend on when the design documentation will be changed and when new parts and modification instructions are available. The modification instructions can either be submitted to the change control board or through your document control procedures. The primary concern of the change control board is not so much the detail of the change, but its effects, its costs and the logistics of its embodiment. If the design change has been made for safety or environmental reasons, you may need to recall product to embody the modification. Your modification procedures need to provide for all such cases. With safety issues, there may be regulatory procedures that need to be implemented to notify customers, recall product, implement modifications and to release modified product back into service.

Design change verification and validation

Depending on the nature of the change, the verification may range from a review of calculations to a repeat of the full design verification programme. The changes may occur before the design has reached the validation phase and therefore not warrant any change to the validation programme. It is therefore necessary when evaluating a design change to determine the extent of any verification and validation that may need to be repeated. Some design changes warrant being treated as projects thereby recycling the full design process. Other changes may warrant verification on samples only or verification may be possible by an analysis of the differences with a proven design.

In some cases, the need for a design change may be recognized during production tests and to define the changes required you might wish to carry out trial modifications or experiments. Any changes to the product during production should be carried out under controlled conditions. To allow such activities as trial modifications and experiments to proceed you will need a means of controlling these events such that production items are not degraded or their status invalidated.

Changes to design after product enters service

As modifications are changes to products resulting from design changes, the identity of modifications needs to be visible on the product that has been modified. If the revision status of the product specification changes, you will need a means of determining whether the product should also be changed. Changes to the drawings or specifications that do not affect the form, fit or function of the product are usually called *alterations* and those which affect form, fit or function are *modifications*. Alterations should come under *document control*, whereas modifications should come under *configuration control*. You will therefore need a mechanism for relating the modification status of products to the corresponding drawings and specifications. Modification notation relates to the product, whereas revision notation relates to the documentation that describes the product. You will need a modification procedure that describes the notation to be used for hardware and software. Within the design documentation you will need to provide for the attachment of modification plates on which to denote the modification status of the product.

Prior to commencement of production, design changes do not require any modification documentation, the design changes being incorporated in prototypes by rework or rebuild. However, when product is in production, instructions will need to be provided so that the modification can be embodied in the product. These modification instructions should detail:

- The products that are affected by part number and serial number;
- The new parts that are required;
- The work to be carried out to remove obsolete items and fit new items or the work to be carried out to salvage existing items and render them suitable for modification;
- The markings to be applied to the product and its modification label;
- The tests and inspections to be performed to verify that the product is serviceable;
- The records to be produced as evidence that the modification has been embodied.

Modification instructions should be produced after approval for the change that has been granted and should be submitted to the change control board or design authority for approval before release. For further guidance see ISO 10007 on configuration management.

How is this demonstrated?

Demonstrating that changes to be made during or subsequent to the design of products and services may be accomplished by:

a)   presenting evidence of a design change process that provides for:

    i   proposed changes to be identified, described and reasons for change given;

    ii   the impact of the proposed change to be explained on the product and related products;

    iii proposed changes to be evaluated and either authorized or declined;

    iv authorized changes to be subject to same rigour of design control as the original design including verification and validation;

    v   changing the design of products that are already in production but not yet delivered;

    vi changing the design of products that are already in service.

b)   selecting a representative sample of products being designed and presenting evidence that design changes are being controlled according to the prescribed policies and procedures;

c)   selecting a representative sample of products in service and presenting evidence that changes to the design have been controlled according to the prescribed policies and procedures.

Retaining information on design and development changes (8.3.6)

What does this mean?

This requirement means that organizations retain the information used and produced in the design change process, including the initial idea for change, the process through which it passed, the results of that process, who authorized the change and what verification and validation were carried out that prove the change was effective. This may include, as applicable, correspondence on design changes with customers and subcontractors.

Why is this necessary?

Information on design changes needs to be retained to verify that the changes are being faithfully implemented and the change procedures have been followed correctly. The information is useful when investigating problems and demonstrating design changes are under control.

How is this addressed?

The documentation for design changes should comprise the change proposal, the results of the evaluation, the instructions for change and traceability in the changed documents to the source and nature of the change. You will therefore need:

•   **A change request** which contains the reason for change and the results of the evaluation – this is used to initiate the change and obtain approval before being implemented.

- **A change notice** that provides instructions defining what must be changed – this is issued following approval of the change as instructions to the owners of the various documents that are affected by the change. A change notice is probably unnecessary for process changes.
- **A change record** that describes what has been changed – this usually forms part of the document that has been changed and can be either in the form of a box at the side of the sheet (as with drawings) or in the form of a table on a separate sheet (as with specifications). For processes, the change record could be incorporated into the document that describes the process be it a specification, flow chart, or control plan.

Where the evaluation of the change requires further design work and possibly experimentation and testing, the results of such activities should be documented to form part of the change documentation.

How is this demonstrated?

Demonstrating that documented information on design and development changes has been retained may be accomplished by:

a) referring to the change record in the latest version of the approved design requirements, selecting a particular change and following the trail of information forward through the designated design change control process to the relevant design verification and validation records to establish that all the information that is required to be recorded can be retrieved from the archives;

b) referring to the product or service configuration records, selecting a particular change and retrieving the relevant design verification and validation records and following the trail of information back through the designated design change control process to the change record in the latest version of the approved design requirements to establish that all the information that is required to be recorded can be retrieved from the archives.

42 Control of externally provided processes, products and services

Introduction

Many organizations prefer to concentrate on the things they are good at and engage others to support them in their endeavours. This support takes many forms, but typically includes the supply of materials, components, semi-finished products, equipment, labour, professional services, business services and laboratory services. Whether embodied in the end product or simply used in its design, development, production or delivery, the quality of these supplies will affect the quality of the product or service provided to customers.

Maintaining the capability to carry out certain activities can be costly especially when they are not needed regularly or are not easily scalable. In seeking to reduce costs and mitigate risks, organizations seek specialists in activities that would normally be performed internally but can be performed more effectively by external providers. This is referred to as outsourcing and defined as "an arrangement where an external organization performs part of an organization's function or process" (ISO 9000:2015).

It is not intended that these requirements are applied to processes, products and services that in no way affect the quality of products and services offered to the organization's customers. The controls exercised over externally provided products and services need to be based on the risks they present to the organization's ability to satisfy its customers, whereas the controls exercised over external providers need to be based on the effectiveness of the controls applied by the external provider. This distinction is brought out in the requirements of clause 8.4.2. No organization wants problems in the supply chain, and therefore getting the balance right between these two controls is important. There is little solace in using external providers of outstanding repute for the supply of products and services of minor importance if a few external providers of products and services of major importance are those with a questionable reputation for quality.

In this chapter, we examine three requirements of clause 8, namely:

- Ensuring externally provided processes, products and services conform to requirements (8.4.1)
- Defining controls applied to products, services and their providers (8.4.2b, c and d)
- Defining controls applied to outsourced processes and their providers (8.1, 8.4.1 and 8.4.2a)

Ensuring externally provided processes, products and services conform to requirements (8.4.1)

What does this mean?

The requirement means that any processes, products and services acquired from external providers must meet the requirements specified by the organization and any requirements of its customers or statutes and regulations that apply to these products or services. This would include any requirements limiting the conditions or the source of supply. Any deviation from these requirements renders the process, product, or service nonconforming and subject to the controls of clause 8.7.

External providers are providers not part of the organization that provide products or services needed by the organization (see Chapter 11 for a further explanation of external providers).

Why is this necessary?

As a rule, the requirements of ISO 9001 do not, provide a reason why they are necessary. However, in this case it is clearly stated that the reason for controlling externally provided processes, products and services is to ensure they do not adversely affect the organization's ability to consistently deliver conforming products and services to its customers. This becomes the overall objective for every one of the processes used in managing externally provided processes, products and services.

How is this addressed?

Purchased products and services

Organizations ensure that externally provided processes, products and services conform to requirements by applying controls to external providers and the products and services they provide. Once the make or buy decision has been made, control of any procurement activity follows a common series of activities.

There are four key processes in the product and service procurement process, as illustrated in Figure 42.1, each of which should be managed effectively:

*   The specification process, which starts once the need has been identified and ends with a request to purchase or outsource.

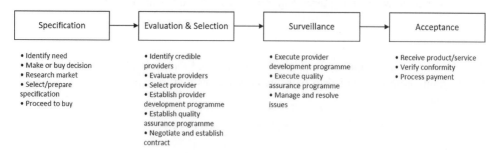

Figure 42.1 Key stages in product or service procurement

- The planning process, which starts following the request to purchase or outsource and ends with the placement of an order or contract on a chosen provider.
- The surveillance process, which starts following placement of order or contract and either ends on delivery of a product or continues for as long as the service is to be delivered.
- The acceptance process, which starts following an indication by the provider of readiness to deliver and either ends with the entry of products into the inventory and/or payment of invoice or continues for as long as the service is to be delivered.

Other than outsourcing, whatever you procure the processes will be very similar, although there will be variations for purchased services such as subcontract labour, computer maintenance, consultancy services, outsourced processes, etc. Where the procurement process is relatively simple, one route may suffice, but where the process varies, you may need separate routes to avoid all purchases, regardless of value and risk, going through the same process and incurring unnecessary costs and delay.

Outsourced processes

Outsourcing presents a special case in procurement as it's a device that enables an organization to retain its capability without owning the resources and incurring the cost of their maintenance.

At one time, an organization would develop all the processes it required and keep them in-house because it was believed it had better control over them, as illustrated in Box 42.1.

Box 42.1   Henry Ford on outsourcing

When Henry Ford built his first automotive plant in 1903, he hired a core of young, able men who believed in his vision and would make Ford Motor Company into one of the world's great industrial enterprises. By 1927, all steps in the manufacturing process, from refining raw materials to final assembly of the automobile, took place at the vast Rouge Plant, characterizing Henry Ford's idea of mass production. In time it would become the world's largest factory, making not only cars but the steel, glass, tires and other components that went into the cars.

(Benson Ford Research Center, 2013)

As trade became more competitive, organizations found that their none-core processes were absorbing a heavy overhead and required significant investment just to keep pace with advances in technology. They realized that if they were to make this investment, they would diminish the resources given to their core business and not make the advances they needed to either maintain or grow the business. As the cost of maintaining an in-house capability increases and the scale of its application decreases, it becomes increasingly more economical, including in terms of mitigating risks, to outsource these non-core processes to organizations for which they are their core processes. This will result in gaining a competitive advantage over the big players that attempt to do everything themselves.

Processes extend across functions, and therefore if it is decided to outsource a function, several parts of different processes may be involved, and each may interface with functions remaining in the organization making the interfaces more problematical.

Box 42.2   Outsourcing criteria

Outsourcing is a business model for the delivery of a product or service by a provider to a client as an alternative to the provision of those products and services within the client organization where:

*   The outsourcing process is based on a sourcing decision (make or buy)
*   Resources can be transferred to the provider
*   The provider is responsible for delivering outsourced services for an agreed period of time
*   The service can be transferred from an existing provider to another
*   The client is accountable for the outsources services, and the provider is responsible for performing them

(ISO 37500, 2014)

Outsourcing is an opportunity to add value, extend capability and/or mitigate risk. When a process or function is outsourced, the organization has chosen not to maintain that particular capability in-house. However, the outsourced function or process remains within the scope of the QMS because it retains control of that capability as indicated earlier.

ISO 37500 identifies four phases of an outsourcing life cycle model and divides each phase into several processes. Figure 42.2.summarizes the primary processes identified in ISO 37500 and the activities that would be carried out to deliver the process outputs.

ISO 37500 defined the purpose, trigger and principal outputs of each of these processes as follows:

Strategy analysis – the purpose of this process is to evaluate and initiate outsourcing opportunities and establish and maintain an outsourcing strategy that satisfies business goals. The trigger is when top management observes an outsourcing opportunity and the principal output is an outsourcing strategy and execution plan with milestones.

Initiation and selection – the purpose of this process is to specify the requirements for outsourcing and the outsourcing goals, to select adequate providers and to establish outsourcing agreements. The trigger is the positive decision to outsource a service, and the principal output is a successfully established outsourcing agreement.

Transition – the purpose of this process is to enable the provider to establish delivery capabilities that are set into full operation in the deliver value process. The trigger is the selection of a provider with the signing of the outsourcing agreement, and the principal output is established delivery capability.

Deliver value – the purpose of this process is to ensure that both client and provider realize and sustain the benefits of the outsourcing agreement through collaboration. The trigger is when the transition process has been completed, and the principal output is delivered service performance.

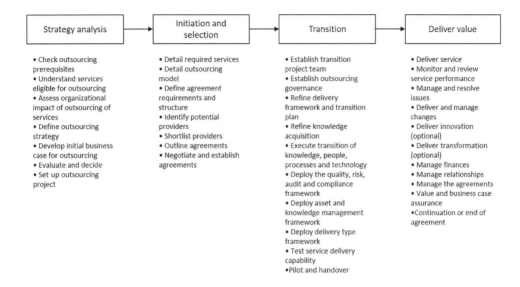

Figure 42.2 Key stages in outsourcing processes

How is this demonstrated?

Demonstrating that processes, products and services acquired from external providers conform to the requirements may be accomplished by:

a) selecting a representative sample of processes, products and services acquired from external providers;

b) retrieving the orders, contracts or service level agreements for these providers and presenting the requirements with which the selected processes, products and services are required to conform;

c) presenting evidence demonstrating that the selected processes, products and services conform to the specified requirements.

Defining controls applied to products, services and their providers (8.4.2b, c and d)

What does this mean?

There is one requirement for verification (8.4.2d) that picks on one aspect of control and four separate requirements governing what the organization does in defining procurement controls for processes, products and services.

Controls applied to products and services

Controls that are to be applied to products and services refer to the requirements to be met by the product or service and the verification to be performed to provide assurance that the product or service meets these requirements either before, during or after its production or delivery, whether on the provider's or on the organization's premises.

Controls applied to external providers

Controls that are to be applied to the provider refer to the requirements to be met by the provider to provide assurance that the provider's processes are capable of consistently delivering conforming product.

Potential impact of the externally provided processes, products and services

Externally sourced processes, products and services can have varying degrees of impact on the processes of the organization and the products and services it provides to its customers. A product or service with a critical impact would warrant stringent control over every stage of its acquisition, whereas a process, product or service with negligible impact may warrant no more than a simple check to verify receipt of the right product. Outsourcing a process with a critical impact on customers necessitates careful management of the risks.

Effectiveness of controls applied by the external provider

Rather than impose the same degree of control over external providers regardless of need, the standard quite sensibly recognizes that the effectiveness of the controls exercised by the supplier should be considered when establishing what control s are needed.

Why is this necessary?

All organizations have external providers of one form or another to provide products and services to their customers. Some of them directly or indirectly affect the product or service being supplied to the organization's customers and others may have no impact at all.

As it would not be prudent to exercise no control over external providers, it would also be counterproductive to impose rigorous controls over every provider and their products and services regardless of the risks and their consequences. A balance should be made based on risk to the processes in which the externally provided product or service is to be used and the final product or service into which the externally provided product or service may be installed.

How is this addressed?

Where the provider can demonstrate an effective degree of control commensurate with the risk, minimal intervention may be required by the organization. However, if the risks are too great to rely entirely on the provider's controls, additional measures will need to be taken.

Selecting the degree of control

You need some means of verifying that the provider has met the requirements of your order, and the more unusual and complex the requirements, the more control will be required. The degree of control you need to exercise over your providers and their products and services depends on:

a)   the confidence the organization has in the ability of its external provider to meet its requirements for quality and delivery;

b)   the impact of externally provided processes, products and services on the organization's ability to meets its customer requirements;

c)   the effectiveness of the controls applied by the external provider.

If you have high confidence in a particular provider, you can concentrate on the areas where failure is more likely. If you have no confidence, you will need to exercise rigorous control until you gain sufficient confidence to relax the controls. The fact that a provider has gained ISO 9001 registration for the products and services you require should increase your confidence, but if you have no previous history of their performance it does not mean they will be any better than the provider you have used for years which is not registered with ISO 9001. Your purchasing process needs to provide the criteria for selecting the appropriate degree of control and for selecting the activities you need to perform.

In determining the degree of control to be exercised, there are several options available depending on the risk. Imposing ISO 9001 on your providers is not necessarily the best solution in all situations.

QUALITY ASSURANCE BY PRODUCT VERIFICATION ON RECEIPT USING STANDARD RESOURCES

Where the quality of the product can be verified on receipt using your normal inspection and test techniques, this may be the least costly of methods and usually applies where achievement of the requirements is measurable by examination of the end product in your facilities. You would not normally impose ISO 9001 on such suppliers until you had confidence to remove verification on receipt. Sampling inspection on receipt should be used when statistical data is unavailable to you or you don't have the confidence for permitting ship-to-line.

This option is often chosen for providers of proprietary product where your choices are often limited because you have no privileges. If your confidence in a provider is low, you can increase the level of verification and if high you can dispense with receipt verification and rely on in-process controls to alert you to any deterioration in provider performance.

QUALITY ASSURANCE BY PRODUCT VERIFICATION ON RECEIPT USING ADDITIONAL RESOURCES

Where the quality of the product can be verified on receipt providing you acquire additional equipment or facilities this option may be economic if there is high utilization of the equipment. You would not normally impose ISO 9001 on such providers until you had confidence to remove verification on receipt or if the burden of receipt verification because too onerous and the next level was a not practical for some reason.

QUALITY ASSURANCE BY PRODUCT VERIFICATION ON SUPPLIER'S PREMISES

Where the quality of the product can be verified by witnessing final acceptance of the product on the provider's premises, this method may be an economic compromise if you don't possess the necessary equipment or skill to carry out product verification. This method should yield as much confidence in the product as the previous methods.

You do, however, need to recognize that your presence on the provider's premises may affect the results. They may omit tests that are problematical or your presence may cause them to be particularly diligent, a stance that may not be maintained when you are not present. This is a level where ISO 9001 applies but may not be necessary if full confidence can be gained by being on site.

QUALITY ASSURANCE BY PRODUCT VERIFICATION BY THIRD PARTY

Where safety is of paramount importance the verification of the product could be contracted to a third party such as a part evaluation laboratory. This can be very costly and is usually only applied with highly complex products. The provider should be given the option of seeking ISO 9001 certification or paying for the independent product verification.

QUALITY ASSURANCE BY CONTROL OF PRODUCTION, INSTALLATION AND SERVICING

Where the product characteristics are such that they cannot be verified by examination of the end product alone and can only be verified by having complete confidence in the provider's manufacturing, installation and servicing processes you can require certain manufacturing and verification documents to be submitted for approval and carry out periodic audit and surveillance activities either directly or using third parties (clause 8.4.3). This is a level where ISO 9001 can be invoked in contracts supplemented by customer specific requirements as necessary.

Quality assurance by control of design

Where the product characteristics are such that they cannot be verified by control of production alone and can only be verified by having complete confidence in the provider's design processes you can require certain design and verification documents to be submitted for approval and carry out periodic audit and surveillance activities either directly or using third parties (clause 8.4.3). This is a level where ISO 9001 can be invoked in contracts supplemented by customer-specific requirements as necessary.

To relate the degree of inspection to the importance of the item, you should categorize purchases as follows:

*   If the subsequent discovery of nonconformity before use will not cause design, production, installation or operational problems of any nature, a simple identity, carton quantity and damage check may suffice. An example of this would be mechanical fasteners.
*   If the subsequent discovery of nonconformity before use will cause minor design, production, installation or operational problems, you should examine the features and characteristics of the item on a sampling basis. An example of this would be electrical, electronic or mechanical components.
*   If the subsequent discovery of nonconformity before use will cause major design, production, installation or operational problems then you should subject the item to a complete test to verify compliance with all prescribed requirements. An example of this would be an electronic unit.

These criteria would need to be varied depending on whether the items being supplied were in batches or separate. However, these are the kinds of decisions you need to take to apply practical receipt verification procedures.

Externally sourced products intended for incorporation

Products from external providers that are intended for incorporation into the organization's own products may include materials, components and assemblies that are designed and produced by the provider or designed by the organization but produced by the provider and these may be delivered from anywhere in the world along a supply chain (e.g. automobile and aircraft assembly).

Externally sourced products provided directly to the customer

Products from external providers that are provided directly to the customer may include materials, finished products and spares in situations where they will be assembled and/or commissioned on site (e.g. civil engineering projects).

Externally sourced services intended for incorporation

Services from external providers that are intended for incorporation into the organization's own products are those that contribute to the design or production of the product and may include design services, measurement services, non-destructive testing services, etc. These services would be classed as outsourced processes if undertaken by the same provider for all projects.

Externally sourced services provided directly to the customer

Services from external providers that are provided directly to the customer may include call centres, billing, mailing, training, fault diagnostics and resolution, and these may be designed by the organization and delivered by a service provider or designed and delivered by a service provider (e.g. a training organization using freelance instructors to deliver its training courses).

Determining the effectiveness of controls applied by the external provider

If the provider is one used previously, you should have retained data on their performance, and therefore an analysis and evaluation of this data should indicate whether the controls applied by them are effective. If the provider is one that has not been used previously, an on-site audit may be necessary as part of the selection process. During the audit, you need to examine the evidence they can provide in demonstrating their controls are effective.

How is this demonstrated?

Demonstrating that appropriate controls have been determined for externally provided products and services may be accomplished by:

a) presenting evidence of a process for determining control of externally provided products and services;
b) selecting a representative sample of externally provided products and services and presenting the plans made for controlling external providers and resulting outputs;
c) presenting the results of an assessment of risks these providers and their products and services present to the organization's ability to meets its customer requirements;
d) presenting the results of an assessment of the controls the providers apply to meet the organization's requirements for quality and delivery;
e) showing how the plans made will mitigate the risks and any weaknesses in the provider's ability to meet the requirements.

Defining controls applied to outsourced processes and their providers (8.1, 8.4.1 and 8.4.2a)

What does this mean?

The standard requires the organization to ensure that externally provided processes remain within the control of its QMS. When an organization chooses to outsource (either permanently or temporarily) a process that affects product conformity with requirements, it cannot simply ignore this process, nor exclude it from the QMS. What this means is that in addition to identifying the outsourced processes in the system description, you need to describe how you manage these processes, and this will differ from the way ordinary purchases are managed.

Box 42.3   Outsourced processes become services

When an organization outsources a function or a process to an external provider, that function or process becomes a service provided by an external body. However, the service is provided under fundamentally different conditions to other services the organization may transact. It is for this reason that the standard refers to externally provided processes, products and services rather than externally provided products and services.

In purchasing products and services to the provider's own specification, the organization is not outsourcing processes or subcontracting. It is simply buying products and services. An outsourced process is one which historically has been operated internally that is executed by another organization on behalf of the parent organization. As part of the outsourcing agreement, resources including people, equipment and knowledge may therefore be transferred to the service provider.

Why is it necessary?

With an outsourced process, the organization is using the labour and the facilities of another organization because it is believed to be commercially advantageous but is still accountable for the process and therefore needs to have control over it.

How is this addressed?

The organization needs to have control over all the processes required for it to achieve its objectives otherwise it can't be confident that it will satisfy its customers. Outsourcing therefore needs to be a carefully chosen strategy based on sound business decisions (see ISO 37500 for more detail).

Selecting the degree of control

Two scenarios to consider are:

a)   The organization designs the process and requests external providers to bid to take on the operation of the process using its own resources. The organization would work with the provider in developing process capability and exercise approval over any changes including substitute staff, materials, equipment, facilities, etc. This strategy is suitable when the organization needs close control over the process because of the many interfaces it has with other processes within the organization.

b) The organization sets the process objective and measures of success and requests external providers to bid to design and operate a process using its own resources. The organization would work with the provider in developing process capability but only exercise approval over changes affecting performance. This strategy is suitable when the organization doesn't need close control over the process because its interface is limited to a single input and output.

Outsourcing governance process

To have confidence that the outsourced process is being run as effectively as if it were in-house, there needs to be a governance process in place. The governance process is defined in ISO 37500 and depicted in Figure 42.3.

Quality assurance process

As part of the transition process (see Figure 42.3) the outsourced processes should be subject to the same provisions as made for in-house processes (see Chapter 16), and therefore during this phase a quality assurance programme that addresses the following activities as necessary should be performed:

a) Acquire the documented information that declares the external provider's plans for providing a service that meets the requirements of the service agreement;

b) Produce a plan that defines how an assurance of service quality will be obtained;

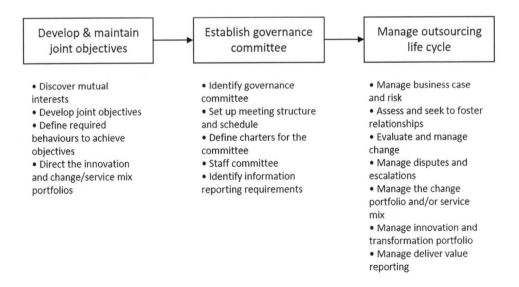

Develop & maintain joint objectives	Establish governance committee	Manage outsourcing life cycle
• Discover mutual interests • Develop joint objectives • Define required behaviours to achieve objectives • Direct the innovation and change/service mix portfolios	• Identify governance committee • Set up meeting structure and schedule • Define charters for the committee • Staff committee • Identify information reporting requirements	• Manage business case and risk • Assess and seek to foster relationships • Evaluate and manage change • Manage disputes and escalations • Manage the change portfolio and/or service mix • Manage innovation and transformation portfolio • Manage deliver value reporting

Figure 42.3 Outsourcing governance process

c) Organize the resources to implement the plans for assuring the quality of service;

d) Establish whether the providers proposed service possesses characteristics that will satisfy the organization's requirements;

e) Assess the provider's operations and determine where and what the quality risks are;

f) Establish whether the provider's plans make adequate provision for the control, elimination or reduction of the identified risks;

g) Assess the extent to which the provider's plans are being implemented and risks contained;

h) Establish whether the service being supplied has the prescribed characteristics.

Following completion of service transition activities and commencement of routine service provision, the following activities should be undertaken against a schedule that takes due account of the risks and performance trends:

a) Review the impact of planned changes and reported problems on previously assessed arrangements and take appropriate action;

b) Continue to assess the extent to which the provider's plans are being implemented and risks contained;

c) Establish whether the service being supplied has the prescribed characteristics.

How is this demonstrated?

Demonstrating that the controls to be applied to externally provided processes have been determined may be accomplished by:

a) presenting evidence of a process for controlling outsourced processes;

b) selecting a representative sample of outsourced processes and presenting evidence that they are being controlled in accordance with the pre-defined process.

Bibliography

Benson Ford Research Center. (2013). The Innovator and Ford Motor Company. Retrieved from The Henry Ford: www.thehenryford.org/exhibits/hf/The_Innovator_and_Ford_Motor_Company.asp

ISO 37500. (2014). *Guidance on Outsourcing*. Geneva: International Organization for Standardization.

ISO 9000:2015. (2015). *Quality Management Systems: Fundamentals and Vocabulary*. Geneva: ISO.

43 Evaluation, selection and monitoring of external providers

Introduction

The evaluation and selection of external providers is a key stage in the procurement planning process. In some markets, there are so many capable providers choosing the right one is not difficult, but in others the market may be awash with rouge traders, counterfeit goods or simply be an immature market where there are few, if any, shining stars.

In this chapter, we examine the three requirements of clause 8.4.1 paragraph 3, namely:

- Evaluation, selection and re-evaluation of external providers including preliminary assessment, pre-qualification, qualification, tendering and contract negotiation process
- Monitoring of external providers, including verification of products and services
- Retaining evidence of evaluation, selection and monitoring

Evaluation, selection and re-evaluation of external providers (8.4.1)

What does this mean?

In searching for an external provider, you need to be confident that they can provide the product or service you require. This means that the decision to select an external provider should be based on knowledge about that provider's capability to meet your requirements. The decision should be based on facts gathered as a result of an evaluation against criteria that you have established.

Why is this necessary?

It would be foolish to select an external provider without first verifying that it could meet your requirements in some way or other. Failure to check out the provider and its products and services may result in late delivery of the wrong product or provision of the wrong service. It may also mean that you might not know immediately that the product or service does not meet your requirements and discover much later that it seriously affects commitments you've made to your customer.

How is this addressed?

Classification process

The process for selection of external providers varies depending on the nature of the products and services to be procured. The more complex the product or service, the more complex the process. You either purchase products and services to your specification (custom) or to the provider's specification (proprietary). For example, you would normally procure stationery, fasteners or materials to the provider's specification but procure an oil platform, radar system or road bridge to your specification. There are grey areas where proprietary products or services can be tailored to suit your needs and custom made products or services that primarily consist of proprietary products or services configured to suit your needs.

Selection process

There is no generic model for the selection process – each industry seems to have developed a process to match its own needs. However, we can treat the process as several stages some of which do not apply to simple purchases as shown in Table 43.1. At each stage the number of potential providers is whittled down to end with the selection of what is hoped to be the most suitable that meets the requirements. With *custom* procurement, this procurement cycle may be exercised several times. Provider capability will differ in each phase. Some providers have good design capability but lack the capacity for quantity production; others have good research capability but lack development capability.

You need to develop an evaluation and selection process, and in certain cases this may result in several closely related procedures for use when certain conditions apply. Do not try to force every purchase or contract through the same selection process. Having purchasing policies that require three quotations for every purchase regardless of past performance of the current supplier is placing price before quality. Provide flexibility so that process complexity matches the risks anticipated. Going out to tender for a few standard nuts and bolts would seem uneconomical. Likewise, placing an order for £1 million of equipment based solely on the results of a third-party ISO 9001 certification would seem reckless.

Preliminary provider assessment

The purpose of the preliminary provider assessment is to select a credible provider and not necessarily to select a provider for a specific purchase. There are millions of providers in the world, some of which would be happy to relieve you of your wealth given half a chance, and others that take pride in their service to customers and are a pleasure to have as partners. You need a process for gathering intelligence on potential providers and for eliminating unsuitable ones so that the buyers do not need to go through the whole process from scratch with

Table 43.1 External provider evaluation and selection stages

Stage	Purpose	Proprietary	Tailored	Custom
Preliminary provider assessment	To select credible providers	✓	✓	✓
Pre-qualification of providers	To select capable bidders		✓	✓
Qualification of providers	To qualify capable bidders			✓
Request for quotation (RFQ)	To obtain prices for products or services	✓	✓	
Invitation to tender (ITT)	To establish what bidders can offer			✓
Tender or quote evaluation	To select a provider	✓	✓	✓
Contract or service agreement negotiation	To agree terms and conditions	✓	✓	✓

each purchase. The first step is to establish the type of products and services you require to support your business and then search for providers that claim to provide such products and services. In making your choice, look at what the provider says it will do and what it has done in the past. Some of the checks needed to establish the credibility of providers are time consuming and would delay the selection process if undertaken only when you have a specific purchase in mind. You will need to develop your own criteria, but typically unsuitable providers may be those that:

- are unlikely to deliver what you want in the quantities you may require;
- are unable to meet your potential delivery requirements;
- cannot provide the after-sales support needed;
- are unethical such as use of child labour or produce or sell counterfeit goods;
- do not comply with the health and safety standards of your industry;
- do not comply with the relevant environmental regulations;
- do not have a system to assure the quality of products or services;
- are not committed to continuous improvement; and
- are financially unstable.

The provider assessment will therefore need to be in several parts:

- **Technical assessment** – to establish that products, processes or services are what the provider claims them to be.
- **Quality management system assessment** – to establish whether the provider can demonstrate an ability to consistently provide products and services that meet the organization's requirements.
- **Financial assessment** – to check the credit rating, insurance risk, stability, etc.
- **Ethical assessment** – to check probity, conformance with professional standards and codes.

These assessments do not need to be carried out on the provider's premises. Much of the data needed can be accumulated from a questionnaires and searches through directories and registers of companies. You can also choose to rely on assessments carried out by accredited third parties to provide the necessary level of confidence. (There is no single directory of all registrations.)

Classification of external providers

The assessments may yield providers over a wide range and you may find it beneficial to classify providers as follows:

Class A: ISO 9001 certified and demonstrated capability – This is the class of those certified providers with which you have done business for a long time and gathered historical evidence that proves their capability.

Class B: Demonstrated capability – This is the class of those providers you have done business with for a long time and warrant continued patronage on the basis that it's better to deal with those providers you know than those you don't. They may not even be contemplating ISO 9001 certification, but you get a good product, a good service and no hassle.

Class C: ISO 9001 certified and no demonstrated capability – This is the class of those certified providers with whom you have done no business. This may appear a contradiction because ISO 9001 certification is obtained based on demonstrated capability, but you have not established their capability to meet your requirements.

Class D: Capable with additional assurance – This is the class of first-time providers with which you have not done sufficient business to put in Class B and where you may need to impose ISO 9001 requirements or similar to gain the confidence you need.

Class E: Unacceptable performance that can be neutralized – This class is for those cases where you may be able to compensate for poor performance if they are sole providers of the product or service.

Class F: No demonstrated capability – This is the class of those providers you have not used before and therefore have no historical data.

Class G: Demonstrated unacceptable performance – This is the class of those providers that have clearly demonstrated that their products and services are unacceptable and it is uneconomic to compensate for their deficiencies.

Pre-qualification of providers

Pre-qualification is a process for selecting providers for known future work. The product or service design will have proceeded to a stage where an outline specification of the essential parameters has been developed. You know roughly what you want, but not in detail. Pre-qualification is undertaken to select those providers that can demonstrate that they have the capability to meet your specific requirements on quality, quantity, price and delivery. A provider may have the capability to meet quality, quantity and price requirements but not have the capacity available when you need the product or service. One that has the capacity may not offer the best price, and one that meets the other criteria may not be able to supply product or service in the quantity you require.

A list of potential bidders can be generated by searching for providers that match given input criteria specific to the particular procurement. However, the evidence you gathered to place providers on your supplier database may now be obsolete. Their capability may have changed, and therefore you need a sorting process for specific purchases. If candidates are selected that have not been assessed, an assessment should be carried out before proceeding any further.

Once the list is generated a request for quotation (RFQ) or invitation to tender (ITT) can be issued depending on what is required. RFQs are normally used where price only is required. This enables you to disqualify bidders offering a price well outside your budget. ITTs are normally used to seek a line-by-line response to technical, commercial and managerial requirements. At this stage, you may select a few potential providers and require each to demonstrate its capability. You know what they do but you need to know if they have the capability of producing a product or providing a service with specific characteristics and can control its quality.

When choosing a bidder, you also need to be confident that continuity of supply can be assured. One of the benefits of ISO 9001 certification is that it should demonstrate that the provider has the capability to consistently supply certain types of products and services. However, it is not a guarantee that the provider has the capability to meet your specific requirements. Providers that have not gained ISO 9001 certification may be just as good. If the product or service you require can only be obtained from a non-registered contractor, using an ISO 9001 registered provider should enable you to reduce your controls, so by using a non-ISO 9001 registered provider you will need to compensate by performing more quality assurance activities yourself, or employing a third party to do so. You may also make a preliminary visit to each potential bidder but would not send out an evaluation team until the qualification stage.

Qualification of providers

Of those potential bidders that are capable, some may be more capable than others. Qualification is a stage executed to compile a short list of bidders following prequalification. A detail specification is available at this stage, and production standard models may be required to qualify the design. Some customers may require a demonstration of process capability to grant production part approval.

During this stage of procurement, a series of meetings may be necessary. A pre-bid meeting may be held on the customer's premises to enable the customer to clarify the requirements with the bidders. A mid-bid meeting or pre-award assessment may be held on the provider's premises at which the customer's supplier evaluation team carries out a capability assessment of representative products or services on site. This assessment may cover:

- an evaluation of the product;
- an audit of design and production plans to establish that, if followed, they will result in compliant product;
- an audit of operations to verify that the approved plans are being followed;
- an audit of processes to verify their capability; and
- an inspection and test of product (on-site or off-site) or service on site to verify that it meets the specification.

The result of provider qualification is a list of capable providers that will be invited to bid for specific work.

ISO 9001 certification was supposed to reduce the amount of supplier assessments by customers and it has in certain sectors. However, the ISO 9001 certification, although focused on a specific scope of registration, is often not precise enough to give confidence to customers for specific purchases.

The evaluation may qualify two or three suppliers for a specific purchase. The tendering process will yield only one winner but the other providers are equally suitable and should not be disqualified because they may be needed if the chosen provider fails to deliver.

Invitation to tender

Once the bidders have been selected, an ITT needs to be prepared to provide a fixed baseline against which unbiased competitive bids may be made. The technical, commercial and managerial requirements should be finalized and subject to review and approval prior to release. It is important that all functions with responsibilities in the procurement process review the tender documentation. The ITT will form the basis of any subsequent contract.

The requirements you pass to your bidders need to include as appropriate:

- The tender conditions, date, format, content, etc.
- The terms and conditions of the subsequent contract.

- A specification of the product, service or process that you require that transmits all the relevant requirements of the main contract (see Chapter 44).
- A specification of how the requirements are to be demonstrated (Chapter 44).
- A statement of work which you require the provider to perform. It might be design, development, management or verification work and will include a list of required deliverables such as project plans, quality plans, production plans, drawings, test data, etc. (You need to be clear as to the interfaces, both organizationally and technically.)
- A specification of the requirements which will give you an assurance of quality. This might be a simple reference to ISO 9001, but as this standard does not give you any rights or flow down your customer's requirements, you will probably need to amplify the requirements (see Chapter 44).

In the tendering phase, each of the potential providers is in competition, so observe the basic rule that what you give one must be given to all! It is at this stage that your provider conducts the tender review defined in clause 8.2.3 of ISO 9001.

Tender or quote evaluation

On the due date when the tenders should have been received, record those that have been submitted and discard any submitted after the deadline. Conduct an evaluation to determine the winner – the provider that can meet all your requirements (including confidence) for the lowest price. The evaluation phase should involve all your staff that were involved with the specification of requirements. You need to develop scoring criteria so that the result is based on objective evidence of compliance.

The standard does not require that you only purchase from *approved providers*. It does not prohibit you from selecting providers that do not fully meet your requirements. There will be some providers that provide a product or service with the right functions, but quality, price and delivery or availability may not be satisfactory. If the demonstrated capability is lacking in some respects you can adjust your controls to compensate for the deficiencies – obviously not a preferred option – but you may have no choice. Deming advocated in the fourth of his 14 points that organizations should "move toward a single provider for any one item, on a long-term relationship of loyalty and trust".

In some cases, your choice may be limited to a single source because no other provider may market what you need. On other occasions, you may be spoilt for choice. With some proprietary products, you can select particular options to tailor the product or service to your requirements. It remains a proprietary product or service because the provider has not changed anything just for you. Most products and services you will purchase from providers are likely to be from catalogues. The designer may have already selected the item and quoted the part number in the specification. Quite often you are buying from a distributor rather than the manufacturer and so need to ensure that both the manufacturer and the distributor will meet your requirements.

Contract negotiation

After selecting a winner, you may need to enter contract negotiations to draw up a formal subcontract. It is most important that none of the requirements are changed without the provider being informed and given the opportunity to adjust the quotation. It is at this stage that your provider conducts the requirement review defined in clause 8.2.3.1e) of ISO 9001. It is pointless negotiating the price of products and services that do not meet your needs. You will just be buying a heap of trouble! Driving down the price may also result in the provider selling their services to the highest bidder later and leaving you high and dry!

Criteria for periodic evaluation

For one-off purchases, periodic re-evaluation would not be necessary. Where a commitment is made from both parties to supply products and services continually until terminated, some means of re-evaluation is necessary as a safeguard against deteriorating standards.

The re-evaluation may be based on provider performance, duration of supply, quantity, risk or changes in requirements and conducted in addition to any product verification that may be carried out. Providers are no different from customers in that their performance varies over time. People, organizations and technologies change and affect the quality of the service obtained from providers. The increasing trend for customers to develop partnerships with providers has led to supplier development programmes where customers work with providers to develop their capability to improve process capability, delivery schedules or reduce avoidable costs. These programmes replace re-evaluations because they are ongoing and any deterioration in standards is quickly detected. In addition, the effect of recent mergers, acquisitions and affiliations on the effectiveness of the quality system should be verified.

How is this demonstrated?

Demonstrating that external providers are evaluated, selected and re-evaluated based on their ability to meet requirements may be accomplished by:

a) presenting evidence of a process for evaluating, selecting and re-evaluating external providers and explaining the typical stages of the process;
b) showing how this process differs depending on whether:

   i   the products are intended for incorporation into the organization's own products and services;
   ii   the products are provided directly to the customers by external providers;
   iii a process, or part of a process, is provided by an external provider

c) presenting evidence that in determining the activities and decisions for specific applications of this process consideration has been given to the risks that may be encountered;
d) presenting the criteria used to evaluate, select and re-evaluate external providers and showing how its application will assess their ability to meet requirements for processes, products and services;
e) selecting a representative sample of external providers of processes, products and services and retrieving evidence that their evaluation and selection was undertaken in accordance with the declared processes.

Monitoring the performance of external providers (8.4.1)

What does this mean?

Monitoring the performance of external providers is the activity of periodically determining whether providers are meeting the organization's requirements. Criteria for monitoring are the principles or standards to be applied to determine the nature, frequency and by whom the monitoring is to be undertaken.

Why is this necessary?

Monitoring the performance of external providers serves to ensure that the output from the procurement process continually meets the organization's requirements. When you purchase items as individuals, it is a natural act to inspect what you have purchased before you use it. To neglect to do this may result in you forfeiting your rights to return it later if found defective or nonconforming. When we purchase items on behalf of our employers, we may not be as tenacious, so the company should enforce its own monitoring policy as a way of protecting itself from the mistakes of its external providers.

How is this addressed?

When carrying out surveillance of external providers, you will need a plan which indicates what you intend to do and when you intend to do it. You will also need to agree the plan with your provider. If you intend witnessing certain tests, the provider will need to give you advanced warning of commencing such tests so that you may attend.

The resources allocated for monitoring, measurement, analysis and evaluation need to be appropriate to the potential risks to the organization and its customers. It is therefore necessary to focus on those providers that indicate the greatest risk to the organization's performance.

Order value

One group of external providers at risk is those that provide the highest value of products and services. High-order value implies significant investment by the organization because such decisions are not taken lightly. Should the provider fail, the organization may not be able to recover sufficiently to avoid dissatisfying its customers.

The supplier database should identify the value of orders with suppliers, and from an examination of these data, a Pareto analysis may reveal the proportion of providers that receive the highest value of orders. The performance of those in the top 20% in the order value list will obviously have more effect on the organization than the performance of the other 80% and may warrant closer attention. If you establish the effort spent on developing these providers as opposed to the others, it may reveal that the priorities are wrong and need adjustment.

Order quantity

Another group at risk is providers that process the greatest number of orders. If there is a systemic fault in their processes, many deliveries may contain the same fault. It is possible that some of the high-order value providers are the same as the high-order quantity providers such as those supplying consumables. The performance of the top 20% in the order quantity list may affect all your products, especially if the product is a raw material, fasteners, adhesives or any item that forms the basis of the product's physical nature.

Quality risk

The third group at risk comprises suppliers that supply products or services that a failure mode analysis has shown are mission critical regardless of value or quantity. You may only need a few of these and their cost may be trivial, but their failure may result in immediate customer dissatisfaction. The FMEA should show the likelihood of failure, and therefore the Pareto analysis could reveal the top 20% of products and services that are critical to the organization in terms of quality.

Delivery risk

The fourth group at risk is those providers that must meet delivery targets. Some items are on a long lead time with plenty of slack, others are ordered when stocks are low and others are ordered against a schedule that is designed to place product on the production line Just-In-Time (JIT) to be used. It is the latter that are most critical, although a JIT scheme does not have to be in place for delivery to be critical. The top 20% of these suppliers deserve special attention, regardless of value, quantity or product or service quality risk. A late delivery may have ramifications throughout the supply chain.

Providers per item

The fifth group of providers is not necessarily a group at risk. Many organizations insist on having more than one qualified provider for a given item or service just in case a supplier under-performs. As Deming points out: *A second source for protection in case of ill luck puts one vendor out of business temporarily or forever, is a costly policy. There is lower inventory and a lower total investment with a single supplier than with two.* Remember you don't have to be ordering from different providers at the same time for a second source to be a costly policy. There are the costs associated with the evaluation and approval, which are double for maintaining a viable second source. An analysis of purchased items by provider will reveal how many items are sourced from more than one provider. Those items sourced from the most number of providers are therefore candidates for a provider reduction programme.

Costs

Cost is also a factor, but often is only measured when there is a target for providers to reduce costs year on year. An analysis of these providers may reveal the top 20% that miss the target by the greatest amount.

Once the top 20% have been identified in each group, further analysis should be carried out to establish how each of these providers perform on quality, cost and delivery; the amount of effort spent in developing these providers; and the degree to which these suppliers respond to requests for action.

A common method of assessing providers was to send out questionnaires that gathered data about them, but they often add little value. A measure of how many of your suppliers have ISO 9001 certification does not reveal anything of value because it does not indicate their performance. Analysis of provider data should only be performed to obtain facts from which decisions are to be made to develop the provider or terminate supply.

How is this demonstrated?

Demonstrating that the monitoring of external providers is based on their ability to provide processes or products and services in accordance with requirements may be accomplished by:

a)   presenting evidence of a process for the monitoring of external providers;
b)   showing how this process differs depending on whether:

i   the products are intended for incorporation into the organization's own products and services;

ii the products are provided directly to the customers by external providers;

iii a process, or part of a process, is provided by an external provider.

c) showing how the monitoring criteria are determined and applied;

d) showing how the monitoring criteria differs depending on the potential impact of the externally provided processes, products and services on the intended results of the QMS;

e) presenting evidence that the monitoring being undertaken is consistent with the planned arrangements.

Retaining evidence of evaluation, selection and monitoring (8.4.1)

What does this mean?

Documented information of the evaluation, selection and monitoring of external providers are documents containing information detailing the activities carried out, the criteria and methods used and the results achieved.

Why is this necessary?

Records of provider evaluation are necessary to select providers based on facts rather than opinion. They are also necessary for comparisons between competing providers as a mere listing provides little information on which to judge acceptability. Information on the methods and criteria used is useful when conflicting information is provided, for example, evidence from monitoring performance may reveal problems but a formal evaluation may tell a different story. Access to information on who performed the evaluation and the methods and criteria used may show that the scope of evaluation to be too narrow or biased. This may help improve evaluation methods and evaluator selection to prevent future problems. However, the amount of information that is maintained and retained is at the organization's discretion.

How is this addressed?

Types of documented information

Documented information on the evaluation, selection and monitoring of external providers can be classified in four groups:

•   Initial evaluation for provider selection;
•   Listing of providers
•   Provider performance monitoring; and
•   Re-evaluation to confirm approval status.

Initial evaluation records

The initial evaluation records would include the evaluation criteria, the method used, the results obtained and the conclusions. They may also include information relevant to the provider such as provider history, advertising literature, catalogues and approvals. These records may not contain actions and recommendations because the evaluation may have been carried out under a competitive tender. The actions come later, when re-evaluations are performed and continued supply is decided.

Listing providers

Prior to computerization, organizations maintained a list of approved suppliers or vendors, but this only included those that were deemed acceptable. It is important that you record those providers that should not be used due to previously demonstrated poor performance so that you don't repeat the mistakes of the past. Assessing providers is a costly operation, and therefore it's prudent to list all providers that have been evaluated to avoid duplicating the effort.

A computerized database of suppliers or providers that captures details of the provider, their approval rating, the products and services that have been evaluated, when and by whom the provider was evaluated with details of orders placed and performance history enables greater flexibility and decision-making.

Provider monitoring records

You should monitor the performance of all your providers and classify each according to prescribed guidelines. Provider performance will be evident from audit reports, surveillance visit reports and receipt inspections carried out by you or the third party if one has been employed. You need to examine these documents for evidence that the provider's QMS is controlling the quality of the products and services supplied. You can determine the effectiveness of these controls by periodic review of the provider's performance – what some firms call *vendor rating*. By collecting data on the performance of providers over a long period you can measure their effectiveness and rate them on a suitable scale. In such cases, you should measure at least three characteristics: quality, delivery and cost.

Re-evaluation records

Re-evaluation records would include all the same information as the initial evaluation but in addition contain follow-up actions and recommendations, the provider's response and evidence that any problems have been resolved.

How is this demonstrated?

Demonstrating that adequate documented information of provider evaluation, selection, monitoring has been retained may be accomplished by:

a) presenting evidence of a process for collecting and retaining evidence of the evaluation, selection, monitoring and re-evaluation of external providers;
b) selecting a representative sample of external providers and retrieving documented information that indicates:

  i   which providers have been evaluated or are awaiting evaluation;
  ii the evaluation criteria and methods;

   iii which evaluated providers were approved and the scope of their approval;

   iv the evidence that the decisions were made using the results of the evaluations conducted;

   v the results of monitoring performance;

   vi the actions taken as a result of monitoring.

44 Information for external providers

Introduction

The quality of communication between an organization and its external providers plays an important role in ensuring providers supply exactly what the organization needs. It is often the case that problems with supplies can be traced to a weakness in communication between the two parties and that these weaknesses arise as a result of one party or the other making invalid assumptions. Key to effective communication is the information that is transmitted, but as explained in Chapter 31, the quality of communication depends on how *meaning* is conveyed within the documented information. As is explained in the section on communication processes in Chapter 34, there is much more to communication than sending messages. Therefore, where it states *the organization shall communicate to external providers its requirements*, it means that the organization should obtain feedback from the external provider that the requirements have been received, are understood and will be implemented.

Information for external providers is what the 1994 edition referred to as *purchasing data* and the 2008 edition referred to as *purchasing information*. There is no change in the intent, but there are some changes in the detail. The information to be communicated to external providers may be different for each stage in selection process as indicated in Table 43.1.

In this chapter, we examine the seven requirements of clause 8.4.3, namely:

- Communicating product and service requirements (8.4.3a)
- Communicating approval requirements (8.4.3b)
- Communicating personnel competence conditions (8.4.3c)
- Communicating requirements for interaction (8.4.3d)
- Communicating performance control and monitoring conditions (8.4.3e)
- Communicating intent to perform verification on provider's premises (8.4.3f)
- Ensuring adequacy of specified requirements (8.4.3)

Communicating processes, product and service requirements (8.4.3a)

What does this mean?

Communicating process, product and service requirements means conveying in terms that will be understood by the potential provider, requirements for the processes, products and services to be provided including those pertaining to their quality, quantity, delivery or availability.

Why is this necessary?

The provider needs to know what the organization requires before it can satisfy the need, and although the standard does not specifically require the information to be recorded, you need to document procurement requirements so that you have a record of what was ordered. This can then be used when the products or services have been delivered and the respective invoices arrive to confirm that what has been received was what was ordered. The absence of such a record may leave you unable to resolve to your satisfaction unwanted product or unsatisfactory service.

How is this addressed?

The essential procurement information must be communicated to providers so that they know what you require, but it may take different forms depending on whether you are procuring proprietary products or services, subcontracting manufacture or service delivery, commissioning bespoke services or outsourcing processes.

Product identification

The product or service identification should be sufficiently precise as to avoid confusion with other similar products or services. The supplier may produce several versions of the same product and denote the difference by suffixes to the main part number. To ensure you receive the product you require you need to carefully consult the literature provided and specify the product in the same manner as specified in the literature or as otherwise advised by the supplier.

Purchasing specifications

If you are procuring the services of a provider to design and/or manufacture a product or design and/or deliver a service, you will need specifications which detail all the features and characteristics which the product or service is to exhibit. The reference number and version of the specifications need to be specified in the event that they change after placement of contract. This is also a safeguard against the repetition of problems with previous supplies.

These specifications should also specify how the requirements are to be verified so that you have confidence in any certificates of conformity that are supplied. Products required for particular applications need to be qualified for such applications and so the procurement documents will need to specify what qualification tests are required.

Service level agreements

If you are outsourcing a process, you'll need a service level agreement (SLA). This needs to include:

- The nature of the services required and their outcomes;
- The service levels in terms of quality and quantity for the key characteristics;
- The metrics and KPIs relative to the impact on the client organization;
- The critical elements of the process, including the trigger conditions, inputs, sequence of activities, outputs and resources;
- The boundary between the organization and the service provider;

- The provisions to be put in place to mitigate known risks;
- The reporting requirements.

A more comprehensive list is provided in ISO 37500 Annex F.

Statutory and regulatory requirements

In addition to specific product and service requirements, the regulations that apply in the country of manufacture and the country of sale need to be identified. This may result in two different sets of requirements. The regulations may be tougher in the country of sale than in the country of production. A process will therefore be needed that will enable you to keep track of changes in import and export regulations and identify all the relevant requirements you need to invoke in purchaser orders and contracts of service level agreements.

How is this demonstrated?

Demonstrating that the organization has communicated requirements for the processes, products and services to be provided may be accomplished by:

a) selecting a representative sample of external providers of processes, products and services;
b) retrieving the orders, contracts or service level agreements for these providers and showing that the specifications for the processes, products and services were included;
c) presenting feedback from the external providers that the requirements have been understood and will be implemented.

Communicating approval requirements (8.4.3b)

What does this mean?

Approval by the organization is a decision made after it has been demonstrated that products and services, methods, processes or equipment meet the requirements and perform the function or produce the results required.

Why is this necessary?

To have confidence in an external provider's capability to provide conforming products and services of the quality required, it is sometimes necessary to validate their products and services, methods, processes or equipment before authorizing production or service delivery to commence. This enables problems to be detected and resolved on site before proceeding with production or service delivery and avoid stoppages when the products or services are used.

How is this addressed?

In general, the approval requirements may include as applicable:

- The application form to be submitted by the provider when applying for approval.
- The procedures to be followed.

- The specifications defining the product or service characteristics.
- Production or service delivery process descriptions.
- Process capability studies.
- Risk assessments identifying the risks, their probability of occurrence and the provisions made to mitigate them by design.
- Hazards analysis and resulting warning notices and precautions to be taken in handling, storage and use.
- Verification and validation reports testifying conformity with the declared specifications.
- The conditions under which approval is granted and by which product may be released or services delivered, including advertising restrictions.
- The information that is to be supplied with every delivery.

How is this demonstrated?

Demonstrating that approval requirements have been communicated to external providers may be accomplished by:

a) selecting a representative sample of external providers of processes, products and services where their approval by the organization is deemed necessary;
b) retrieving the orders, contracts or service level agreements for these providers and showing that the conditions communicated to providers included requirements for approval of products and services, methods, processes or equipment;
c) presenting feedback from the external providers that the requirements have been understood and will be implemented.

Communicating personnel competence conditions (8.4.3c)

What does this mean?

When procuring services, the quality of the output often depends far more on the competence of the people supplying the service than the materials or equipment they may be using. If it is expected that the people supplying the service be qualified to a particular level or have demonstrated their competence in a particular field, occupation or activity to a professional body, this needs to be specified. Processes that are outsourced will be expected to be carried out by personnel at least as competent as when the process was carried out in house.

Why is this necessary?

Products can often be tested to prove they satisfy requirements regardless of the qualifications or competence of the people who designed and built them. With services, it is somewhat different as a service is often experienced as it is being delivered, and therefore if the people delivering it are not qualified or competent they could do serious harm to the organization's personnel, property or reputation. There may be situations where the provider's staff interface directly with your customer, and you will need to have confidence that such staff will not say or do anything that will harm your relationship with your customer.

How is this addressed?

If there are specific qualifications or levels of competence required these should be stated in the procurement information, for example, gas fitters should be registered with Gas Safe Register, teachers should hold a nationally recognized teaching certificate, welders should hold a nationally recognized certificate in welding and medics should hold a nationally recognized medical qualification and be registered with the appropriate professional body.

For outsourced processes the qualifications and competences required of the personnel engaged in the process should be no different than if the process was being carried out in-house. Where providers' staff interface directly with your customers, you will probably need to vet them to be confident they exhibit the appropriate inter personal skills and specialist knowledge and this will need to be written into the information provided.

There is an expectation when engaging professional services that their staff will be appropriately qualified and behave in a professional manner. Sometimes the uniform is accepted as proof the wearer is competent, but we can be easily fooled. As a way of protecting your interests you might consider including a general clause in the terms and conditions of procurement to the effect that the client expects that personnel engaged by the provider are qualified to relevant national standards and are competent to carry out the work detailed in the order/contract to the standards specified.

How is this demonstrated?

Demonstrating that your organization has communicated its requirements for competence and qualification of persons to external providers may be accomplished by:

a) selecting a representative sample of external providers of processes, products and services where requirements for competence and qualifications are deemed necessary;
b) retrieving the orders, contracts or service level agreements for these providers and showing that the conditions communicated to providers included requirements for competence and qualifications;
c) presenting feedback from the external providers that the requirements have been understood and will be implemented.

Communicating requirements for interaction (8.4.3d)

Box 44.1   Revised requirement on QMS requirements

In the 2008 edition purchasing information was required to include quality management system requirements. This requirement does not appear in the 2015 edition, and in its place are three separate requirements:

8.4.2b: define the controls that it intends to apply to an external provider (these are the QMS requirements)

8.4.2b: define the controls that it intends to apply to the resulting output (these are the product and service monitoring, verification and approval requirements)

8.4.3d: to communicate requirement to external providers for their interactions with the organization (these are the organizational interface requirements)

What does this mean?

Interaction between an organization and its external providers occurs at various stages throughout the duration of a partnership, contract or in executing an order. The standards from which ISO 9001 was derived in the 1980s contained clauses requiring the contractor to interface with the customer's representative for various reasons and this requirement is a modern equivalent (see Box 44–1).

Why is this necessary?

There needs to be interaction between an organization and its external providers for both parties to be certain as to what the other party requires, to plan, avoid surprises, progress issues, gain confidence and maintain a mutually beneficial relationship.

How is this addressed?

Either within the standard terms and conditions for procurement, the contract, order or separate QMS requirements the organization transmits information to the provider and requires, for example,

- Access to certain records;
- Access to witness certain events or verify product or service;
- Acknowledgment to the receipt of information;
- Attendance at certain meetings;
- Collaboration in certain investigations;
- Investigation and reporting of certain problems;
- Notification of any measurement capability that exceeds the known state of the art;
- Notification of any opportunities for saving costs, reducing delivery timescales or improving quality that are within the scope of the agreement;
- Notification of any potential or actual situation that may adversely affect the quality, delivery or cost of the products and services to be supplied;
- Notification of certain events;
- Notification of corrective actions taken;
- Notification of damage, loss or malfunction in property provided by the organization
- Notification of nonconformities in products from subcontractors designated by the organization;
- Submission of certain reports and requests for concession.

How is this demonstrated?

Demonstrating that the organization has communicated to external providers its requirements for their interactions with the organization may be accomplished by:

a) selecting a representative sample of external providers of processes, products and services where specific requirements for interaction are deemed necessary;

b) retrieving the orders, contracts or service level agreements for these providers and showing that the conditions communicated to providers included requirements for interaction;

c) presenting feedback from the external provider that the requirements have been understood and will be implemented.

Communicating performance control and monitoring conditions (8.4.3e)

What does this mean?

Box 44.2   New requirement for control and monitoring

In the 2008 version the actions of the organization relative to the external provider's performance was implied within the requirement to select suppliers based on their ability to supply product in accordance with the organization's requirements. This new requirement implies performance data would be available from external providers on which to base such decisions.

Basically, this means getting agreement from your external providers to the methods you'll use to control and monitor their performance. Control in this context means the ability to regulate progress and deliveries, etc., during the tenure of the contract/agreement.

As is stated in clause 8.4.1, external providers are required to be selected on their ability to provide processes or products and services in accordance with requirements. Knowledge of a provider's ability is gained either from past performance or a planned evaluation as explained in the section "Classification of external providers" in Chapter 42. Past and current performance enables organizations to make decisions.

Why is this necessary?

There are three reasons:

- The organization may need to control the progression of work carried out by the provider to grant stage approval or to align with their schedules and avoid abortive work.
- In defining the controls, the organization intends to apply to an external provider, clause 8.2.2c (2) requires the effectiveness of the controls applied by the external provider to be taken into consideration. The organization won't be able to do this unless it is monitoring the external provider's performance.
- Where an organization maintains a vendor rating system, it will assign a rating to external providers based on data collected about their performance. This will be used in the selection process and in continuing a mutually beneficial relationship. It is therefore prudent to tell your external providers of the ratings being assigned and the methods employed to calculate them.

How is this addressed?

You need to determine what data you need to collect about your external providers and whether you intend to take a hands-off approach or a hands-on approach. With a simple purchase, you place an order and if your provider is customer focused you'll receive:

- an acknowledgement of the order;
- notification that the consignment has been prepared for delivery;
- notification that the consignment has been dispatched;
- the goods within the notified time.

These are four stages that may be monitored which you don't normally need to require your provider to notify you about. A hands-off approach is one you'll take when you have considerable confidence in a provider's capability but you may still want to monitor achievement of key milestones.

A hands-on approach is one you'll take with more complex purchases where you need to examine the output at key milestones before permitting the provider to proceed or to authorize transition from development into production, the release of product or the start or stoppage of a service.

How is this demonstrated?

Demonstrating that the organization has communicated to its external providers the requirements for the control and performance monitoring it will apply may be accomplished by:

a) selecting a representative sample of external providers of processes, products and services where specific requirements for control and performance monitoring by the organization are deemed necessary;
b) retrieving the orders, contracts or service level agreements for these providers and showing that the conditions communicated to providers included requirements for control and performance monitoring by the organization;
c) presenting feedback from the external provider that the requirements have been understood and that they will cooperate.

Communicating intent to perform verification on provider's premises (8.4.3f)

What does this mean?

If you choose a verification method other than receipt inspection that involves a visit to the provider's premises, the provider has a right to know, and the proper vehicle for doing this is through the procurement information such as a contract, order or service level agreement.

Why is this necessary?

The provider needs to know if you or your customers intend to enter its premises to verify product before shipment or very service while it is being delivered so that they may make the necessary arrangements and establish that the proposed methods are acceptable to them.

How is this addressed?

Verification by the organization

The acceptance methods need to be specified at the tendering stage so that the provider can make provision in the quotation to support any of your activities on site. When you visit a provider, you enter its premises only with their permission. The product remains their property until you have paid for it, and therefore you need to be very careful how you behave. The contract or order is likely to only give access rights to products and areas related to your contract and not to other products or areas. You cannot dictate the methods the provider should use unless they are specified in the contract, as will likely be the case with outsourced processes. It is the results in which you should be interested, not the particular practices, unless you have evidence to demonstrate that the steps they are taking will adversely affect the results.

Verification by your customer

In cases where your customer requires access to your providers to verify the quality of products and services, you will need to transmit this requirement to your provider in the procurement information and obtain agreement. Where a firm's business is wholly that of contracting to customer requirements, a clause giving their customers certain rights will be written into their standard trading conditions. If this is an unusual occurrence, you need to identify the need early in the contract and ensure it is passed on to those responsible for preparing subcontracts. You may also wish to impose on your customer a requirement that you are given advanced notice of any such visits so that you may arrange an escort. Unless you know your customer's representative very well, it is unwise to allow unaccompanied visits to your providers. You may, for instance, have changed, for good reasons, the requirements that were imposed on you as the main contractor when you prepared the subcontract and in ignorance your customer could inadvertently state that these altered requirements are unnecessary. They may be unnecessary for their purposes but not yours.

Customer visits are to gain confidence and not to accept product or service. The same rules apply to you when you visit your external providers. The final decision is the one made on receipt or sometime later when the product is integrated with your equipment and you can test it thoroughly in its operating environment or equivalent.

How is this demonstrated?

Demonstrating that your organization has communicated its requirements for verification or validation activities that it, or its customer, intends to perform at the external provider's premises may be accomplished by:

a) selecting a representative sample of external providers of processes, products and services where on site activities were deemed necessary;
b) retrieving the orders, contracts or service level agreements for these providers and showing that the conditions communicated to providers included requirements for verification or validation activities that it, or its customer intended to perform on the provider's premises;
c) presenting feedback from the external providers that the requirements have been understood and will be implemented.

Ensuring adequacy of specified requirements (8.4.3)

What does this mean?

The adequacy of information for external providers is judged by the extent to which it accurately reflects the requirements of the organization for the products concerned.

Communication of such requirements to the supplier can be verbal or through documentation and processed by post or electronically.

Why is this necessary?

The acceptance of an order or contract by an external provider places the organization under an obligation to accept product or service that meets the stated requirements. It is therefore important that such information is deemed adequate before being released to the supplier.

How is this addressed?

Prior to orders being placed, the procurement information should be checked to verify that it is fit for its purpose. The extent to which you carry out this activity should be based on risk, and if you choose not to review and approve all purchasing information, your procedures should provide the rationale for your decision. In some cases, orders are produced using a computer and transmitted to the external provider directly without any evidence that the order has been reviewed or approved. The purchase order does not have to be the only procurement document. If you enter procurement data onto a database, a simple code used on a purchase order can provide traceability to the approved documents.

You can control the adequacy of the procurement data in at least four ways:

- Provide the criteria for staff to operate under self-control.
- Classify orders depending on risk and only review and approve those that present a certain risk.
- Select those orders that need to be checked on a sample basis.
- Check everything they do (this is Theory X management and not recommended).

A situation where staff operate under self-control would be in the case of telephone orders where there is little documentary evidence that a transaction has taken place. There may be an entry on a computer database showing that an order has been placed with a particular provider. To ensure the adequacy of purchasing requirements, in such circumstances you would train the buyers in the use of the database and route purchase requisitions only to trained buyers for processing.

How is this demonstrated?

Demonstrating the adequacy of specified requirements prior to their communication to external providers may be accomplished by:

a) presenting evidence of a process for transmitting procurement information to external providers;
b) presenting evidence of the methods employed to ensure the adequacy of information before communication to the chosen provider;
c) selecting a representative sample of orders, contracts and service level agreements and retrieving evidence which confirms that the procurement information was transmitted to providers in accordance with the prescribed methods by the authorized people at the stages identified in the process description;
d) retrieving records pertaining to these samples which demonstrate that no issues with information adequacy were reported following its receipt by the provider.

45 Control of production and service provision

Introduction

The requirements referred to in this section of the standard refer to the controls to be exercised over two of the principal processes: production and service delivery. These are the processes that are cycled repeatedly to replicate products and services to the same standard every time. The design process is a journey into the unknown, whereas the production and service delivery process is a journey along a proven path with what is expected to be a predictable outcome. The design process requires control to keep it on course towards an objective, the production and service delivery processes require control to maintain output to a prescribed standard.

Production and service provision are controlled in two different ways. Product quality can be controlled by controlling the product that emerges from the producing processes or by controlling the processes through which the product passes. Process control relies on control of the factors that influence the results of the process, whereas product control relies on verification of the product as it emerges from the process. If you concentrate on the process using the results of the product verification and eliminate the root cause of variation, you will gradually reduce rework until all output products are of consistent quality. It will therefore be possible to reduce dependence on output verification (the third of Deming's 14 points).

Controlling a service is more like controlling a process than controlling a product. At each stage, there may be interaction with a customer and more opportunities to enhance customer satisfaction or to disappoint. The outcome is the accumulation of satisfaction or disappointment through the process, and therefore with services, particularly customer-facing services, there are fewer opportunities to hide mistakes and rework.

The headings in clause 8.5 of ISO 9001 are a little confusing. The title of the first section, *Control of production and service provision*, implies that the subsequent sections serve some other purpose than the control of production and service provision when in fact they are wholly part of it. It is also confusing that release of product is mentioned in 8.5.1 but then addressed in 8.6 and not 8.5. However, the logic behind this is that these two clauses address requirements that are applicable to both *production and service provision* and *externally provided processes, products and services*. It's yet another example of how requirements cannot be taken in isolation or that clause headings cannot be interpreted as being all inclusive.

Implementing production and service provision under controlled conditions means that there are:

a) Known inputs (i.e. we know what inputs to expect).
b) Known outputs (i.e. we know what outputs to expect).
c) Known customers for the outputs.

d) Known sequence and interaction of sub-processes (the sequence in which activities are to be carried out and which processes we interact with to obtain inputs and deliver outputs).

e) Known methods (for preparing, carrying out, checking and releasing work).

f) Known criteria (for how well the work is to be carried out and the results to be achieved).

g) Known and available resources.

h) Known responsibilities and authorities.

i) Known capability (i.e. we know what the process will produce in terms of its quantity, quality and how long on average it will take).

j) Known influences on the process that must be managed to minimize risks to its capability.

All of these basically come from looking at clause 4.4a) to g), thus promoting the use of the process approach as per clause 5.1.1d).

Top management is also required in clause 5.1.1d) to promote the use of risk-based thinking which means that the degree of control exercised over production and service provision needs to be proportionate to the potential impact of identified risks on conformity of products and services.

In this chapter, we examine seven requirements of clause 8.5.1, namely:

*   Availability of documented information (8.5.1)
*   Availability and use of suitable monitoring and measuring resources (8.5.1e)
*   Monitoring and measurement activities (8.5.1c)
*   Use of suitable infrastructure and environment (8.5.1d)
*   Competence and qualification of personnel (8.5.1e)
*   Validation of processes (8.5.1f)
*   Actions to prevent human error (8.5.1g)

Release, delivery and post-delivery (8.5.1h) are addressed in Chapter 50.

The information, infrastructure, environment, monitoring, measurements and competent people are determined are provided by other processes, and therefore the focus here is on making sure that they are available to and used in production and service delivery processes.

Availability of documented information (8.5.1a)

What does this mean?

The documented information to be available is intended to define what is to be produced or provided, what requirements it has to meet and what work is required to be carried out. These requirements are expressed in terms of

a) the characteristics of the products to be produced;

b) the characteristics of the services to be provided;

c) the activities to be performed;

d) the results to be achieved.

The organization is given the option to choose which are applicable because documented information on each of these may not be necessary if the personnel possess the knowledge through training or experience, for example, the competent plumber does not need to have available the information defining basic pipe jointing and routing criteria or the sequence of activities in installing a central heating system, as this will be learnt during training. However, there should be a checklist to prompt the plumber to make sure he has not forgotten anything of importance.

The information is the input to the production or service delivery process, usually coming out of the planning process, but may be direct from customers. It may take the form of definitive specifications, service level agreements, drawings, layouts or any information that specifies the physical and functional characteristics that the product or service is required to exhibit.

Why is this necessary?

Without information specifying the product to be produced or the service to be provided, there is no sound basis on which work can be controlled. The people doing the work need to know what it is they are required to so and when it is required to be completed.

How is this addressed?

To ensure the right information is available, a communication channel needs to be opened between the operations planning process and the production and service delivery processes. Along this channel needs to pass all the information required to produce the product and deliver the service. At a minimum, there should be documented information defining the work to be carried out, when it's to be completed and for whom it's being provided so the provider knows who the customer is and can therefore access any relevant information. Although competent personnel don't need to be told how to do their job, they are not clairvoyant!

Provision also needs to be made for transmitting changes to this information in such a manner that the recipients can readily determine the correct information to use, what the changes are and why they have been made so that the user is more aware of the results to be achieved.

How is this demonstrated?

Demonstrating that documented information that defines the characteristics of the products and services is available may be accomplished by:

a)   presenting evidence of a process for making information defining product and service characteristics available to production and service personnel;
b)   selecting a representative sample of people engaged in production and service provision and presenting evidence that the information they need to create product and service characteristics is available to them in accordance with the pre-defined process.

Availability and use of suitable monitoring and measuring resources (8.5.1e)

What does this mean?

Measurement is one of the key factors needed to control processes. This means providing the equipment needed to measure product features and monitor process performance and providing adequate training and instruction for this equipment to be used as intended.

Why is this necessary?

Product and service quality can only be determined if the equipment needed to measure and monitor characteristics are available and used as intended.

How is this addressed?

When designing the process for producing product or delivering service, you should have provided stages at which product or service features are verified. You may also need to install monitoring devices that indicate when the standard operating conditions have been achieved and whether they are being maintained. The equipment used to perform measurements needs to be available where the measurements are to be performed. The monitoring devices need to be accessible to process operators for information on the performance of the process to be obtained. The monitoring equipment may be located in inaccessible places providing the signals are transmitted to the operators controlling the process.

How is this demonstrated?

Demonstrating that suitable monitoring and measuring resources are available and used may be accomplished by:

a) presenting evidence of a process for making the planned monitoring and measuring resources available in production and service delivery processes;
b) presenting evidence of a process for training personnel in the use of the monitoring and measuring resources;
c) selecting a representative sample of people engaged in production and service provision and presenting evidence that:

    i   the monitoring and measuring resources specified in the documented information they have been provided with is available for their use;
    ii they have been trained in its use.

Monitoring and measurement activities (8.5.1c)

What does this mean?

Monitoring and measurement is the means by which product and process characteristics are determined, and this requirement means carrying out the planned monitoring and measurement activities. The specifications define the target values and the process description or plan defines when measurements should be taken to ascertain whether the targets have been met.

Why is this necessary?

To control product quality, the achieved characteristics need to be measured and the process operating conditions need to be monitored. All controls need a verification stage and a feedback loop. You cannot control production or service delivery processes without performing verification.

How is this addressed?

Either the planning documents, checklists or training will designate the stages where monitoring and measurement activities are to be carried out and may include monitoring facilities as well as a verification of process parameters and process outputs.

To carry out the planned monitoring and measurement at appropriate stages:

- The monitoring provisions need to be in place routing data from sensors to those responsible for monitoring and measurement at the designated stages (7.1.5 and 8.5.1b).
- The measurement equipment needs to be available where measurements are to take place (7.1.5 and 8.5.1b).
- The physical and psychological environment needs to be suitable for ensuring valid and reliable results (7.1.4 and 8.5.1d).
- The people need to be competent in the operation of the equipment and interpretation of the data (7.2 and 8.5.1e).
- The acceptance criteria need to be available (8.5.1a).

Of all these conditions, the psychological environment is crucial because the people must be motivated to carry out the planned monitoring and measurement and not be put under pressure to skip those stages in the plan because of failures in other processes or approaching deadlines.

How is this demonstrated?

Demonstrating that monitoring and measurement activities are being implemented at appropriate stages may be accomplished by:

a) presenting evidence of the plans for carrying out monitoring and measurement of processes and process outputs in production and service delivery;
b) selecting a representative sample of planned monitoring and measurement activities and providing evidence that they have been carried out at the designated stages using the acceptance criteria provided.

Use of suitable infrastructure and environment (8.5.1d)

What does this mean?

This means using the infrastructure and environment that has been provided as per clauses 7.1.3 and 7.1.4 rather than carrying out production and service delivery in facilities and an environment which has not been designated for this purpose.

Why is this necessary?

Process outputs cannot be achieved unless the physical resources that are essential to perform the work are used. In any other state, people would be employed to compensate for the inadequacies of the infrastructure and environment – a state that can be sustained in some circumstances but not for long without degrading the quality of the work.

How is this addressed?

The infrastructure and environment needed will have been determined in planning the processes. To ensure that personnel engaged in production and service delivery processes know which equipment and facilities to use, they can be:

a)   Identified in job orders, work instructions or other process documentation or;
b)   Designated areas can be provided to produce specific products or delivery of specific services where all the infrastructure and physical environment is available.

How is this demonstrated?

Demonstrating that infrastructure and environment suitable for the operation of processes is used may be accomplished by:

a)   presenting documented information which identifies any special infrastructure and environment that is to be use in production and service delivery processes;
b)   selecting a representative sample of production and service delivery activities and providing evidence that they have been carried out using the designated infrastructure and environment.

Competence and qualification of personnel (8.5.1f)

What does this mean?

This means that people performing production and service activities are to possess the ability to achieve intended results. The reason for adding *any required qualifications* is that people may have the ability to achieve results but not have any formal qualifications. In certain occupations, there may be legal or contractual requirements for accredited qualifications. This is one such situation where this requirement is applicable. There may also be situations where persons undergoing training are appointed and in such cases this requirement will apply to their supervisors.

Why is this necessary?

By not appointing competent people to perform production and service activities, you run the risk of mistakes being made and nonconforming outputs being produced or provided. Some services are regulated by law (e.g. healthcare services, legal services and certain engineering services) and unqualified people are prohibited from practicing; therefore, by not appointing appropriately qualified people you run the risk of being prosecuted.

How is this addressed?

The determination and assessment of competence is addressed in Chapter 29 and the provision of people in addressed in Chapter 24.

How is this demonstrated?

Demonstrating that the competent persons possessing required qualifications have been appointed may be accomplished by:

a) presenting evidence of a process for appointing personnel to undertake work in production and service delivery processes;

b) presenting evidence that the appointments are made based on competence and any necessary qualifications;

c) selecting a representative sample of personnel and presenting evidence that their appointments were made in accordance with the designated process.

Validation of processes (8.5.1g)

What does this mean?

Many processes do not present any difficulty in the verification of the resultant output before delivery regardless of the tools, personnel, facilities or other means used to carry out the process. However, there are some processes where the output is totally dependent on the competence of personnel, the capability of the equipment and the facilities and, what is more, where its performance cannot be fully verified by examination of the output at any stage before delivery. For communication purposes, these processes are commonly referred to as *special processes*.

Among such processes are welding, soldering, adhesive bonding, casting, forging, forming, heat treatment, protective treatments and inspection and non-destructive test techniques, environmental tests and mechanical stress tests. The standard only requires process validation where, as a consequence of not being able to verify the output, deficiencies become apparent only after the product is in use.

In service industries, there are many cases where there is no opportunity to verify the output before its delivery such as with medical, financial and legal services. The service is provided instantaneously, and it's only after the service has been rendered that it's possible to verify whether it has been done properly.

Why is this necessary?

If any of these factors on which the performance of a process depends is less than adequate, deficiencies may not become apparent until long after the product or service is installed, or used. Normally, product characteristics are verified before release, but when this is not possible without destroying the product, the process needs to be qualified as capable of only producing conforming product or delivering conforming service.

How is this addressed?

To limit the potential for deficiencies to escape detection before the product is released or service is used, measures should be taken that ensure the suitability of all equipment, personnel, facilities and prevent varying conditions, activities or operations. A thorough assessment of the processes should be conducted to determine their capability to maintain or detect the conditions needed to consistently produce conforming product or service. The limits of capability need to be determined and the processes applied only within these limits.

You should produce and maintain a list of processes that have been validated as well as a list of the personnel who are qualified to operate them. In this way, you can easily identify an unqualified process, an unauthorized person or an obsolete list if you have neglected to maintain it.

Where process capability relies on the competence of personnel, personnel operating such processes need to be appropriately educated and trained and undergo examination of their competency. If subcontracting manufacturing processes you need to ensure that the supplier only employs qualified personnel and has qualified process equipment and facilities. In production, you need to ensure that only those personnel, equipment, materials and facilities that were qualified are employed in the process; otherwise, you will invalidate the qualification and inject uncertainty into the results.

Where there is less reliance on personnel but more on the consistency of materials, environment and processing equipment, the particular conditions need to be specified. Where necessary restrictions should be placed on the use of alternative materials, equipment and variations in the environment. Operating instructions should be used that define the setup, operation and shutdown conditions and the sequence of activities required to produce consistent results. The resultant product needs to be thoroughly tested using such techniques that will enable the performance characteristics to be measured. This may involve destructive tests to measure tensile and compressive strength, purity, porosity, adhesion, electrical properties, etc. In production, samples should be taken at set frequencies and the tests repeated.

The records of qualified personnel using special processes should be governed by the training requirements. Regarding the equipment, you will need to identify the equipment and facilities required within the process specifications and maintain records of the equipment. This data may be needed to trace the source of any problems with product that was produced using this equipment. To take corrective action you will also need to know the configuration of the process plant at the time of processing the product. If only one piece of equipment is involved, these records will give you this information, but if the process plant consists of many items of equipment which are periodically changed during maintenance, you will need to know which equipment was in use when the fault was likely to have been generated.

How is this demonstrated?

Demonstrating that special processes are subject to validation and periodic re-validation may be accomplished by:

a)   presenting evidence of the analysis undertaken to determine which production and service delivery processes qualify as *special processes*;

b)   selecting a representative sample of *special processes* and presenting evidence of validation of their capability to produce consistent results;

c)   presenting evidence of planned and unplanned changes to production and service delivery processes and showing that following the changes re-validation was carried out on the special process affected by the change.

Actions to prevent human error (8.5.1g)

What does this mean?

Actions to prevent

There is an implication in the requirement that all human errors can be prevented, but despite error proofing by design, everyone can make errors no matter how well trained and motivated they are Deming used The Red Bead Experiment to illustrate that differences between people arise almost entirely from action of the system in which they work and not from the people themselves (Deming, 1982). The source of the errors is therefore more likely to be in the system rather than the person. As the system is subject to influences outside the control of management, any action taken to address the risk of human error should be proportionate to the potential impact on the conformity of products and services (see clause 6.1.1).

What is human error?

The term *human error* is defined by ISO as: "human action or inaction that can produce an unintended result" (ISO/IEC 2382:2015). The terms *mistake* and *error* are accepted alternatives; thus, ISO doesn't draw any distinction between them or whether the actions were intended or unintended unlike other authorities. There is little more in the standard to help categorize human errors. However, James Reason, a professor of psychology at the University of Manchester has become somewhat of an authority about human error, and his research was used as a basis for the UK Health and Safety Executive's guide to *Reducing error and influencing behaviour* (HSG48, 1999). This chapter draws on this guide which, although focused primarily on health and safety, is equally relevant to product and service quality.

HSG48 refers to human failure and draws a distinction between errors and violations. A human error is defined as "an action or decision which was not intended, which involved a deviation from an accepted standard, and which led to an undesirable outcome". A violation is defined as "a deliberate deviation from a rule or procedure" (HSG48, 1999). The guide describes how the various types of human failure are related, and these are shown in Figure 45.1 and described next.

> ***Slips and lapses*** are unintentional and occur in very familiar tasks which we can carry out without much need for conscious attention. These tasks are called *skill-based* and are very vulnerable to errors if our attention is diverted, even momentarily. Slips and lapses are the errors which are made by even the most experienced, well-trained and highly-motivated people. They are described as *actions-not-as-planned.*
>
> ***Slips*** are failures in carrying out the actions of a task.

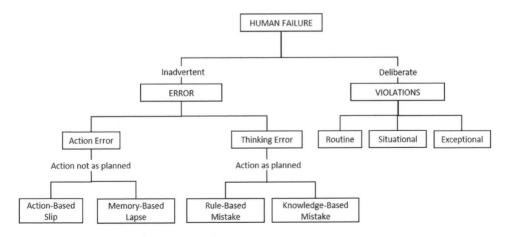

Figure 45.1 Types of human failure (Adapted from HSG48, 1999)

Lapses cause us to forget to carry out an action, to lose our place in a task or even to forget what we had intended to do.

Mistakes are an unintentional consequence of an intentional action and are a more complex type of human error where we do the wrong thing believing it to be right. The failure involves our mental processes which control how we plan, assess information, make intentions and judge consequences. Two types of mistakes exist: rule-based and knowledge-based.

Rule-based mistakes occur when our behaviour is based on remembered rules or familiar procedures. We have a strong tendency to use familiar rules or solutions even when these are not the most convenient or efficient.

Knowledge-based mistakes occur when in unfamiliar circumstances we have to revert to consciously making goals, developing plans and procedures. Planning or problem solving needs us to reason from first principles or use analogies. Misdiagnoses and miscalculations can result when we use knowledge-based reasoning.

Violations are any deliberate deviations from rules, procedures, instructions and regulations and are therefore intentional. Most violations are motivated by a desire to carry out the job despite the prevailing constraints, goals and expectations. Very rarely are they wilful acts of sabotage or vandalism. Violations are divided into three categories:

Routine violations are where breaking the rule or procedure has become a normal way of working within the work group.

Situational violations are where breaking the rule is due to pressures from the job such as being under time pressure, insufficient staff for the workload, the right equipment not being available or even extreme weather conditions. It may be very difficult to comply with the rule in a particular situation, or staff may think that the rule is unsafe under the circumstances.

Exceptional violations rarely happen and only then when something has gone wrong. To solve a new problem, you feel you need to break a rule even though you are aware that you will be *taking a risk*. You believe, falsely, that the benefits outweigh the risks.

To which humans is it referring

Although humans throughout the organization will make errors no matter what level their competence, in the context of *control of production and service provision* human error relates to the people engaged in these processes doing something which produces an undesirable result. These people will include not only the people operating the machines, making the product or the front office staff delivering the service but also those who are supervising them and the back-office staff supporting them. However, the controls put in place in the processes that feed production and service delivery processes should have been designed to prevent human errors in these processes.

The ISO definition is broad enough to place slips, lapses, mistakes and violations within the category of human error.

The causes of human error that can result in incidents, accidents and near misses can equally result in nonconforming product or service.

Box 45.1   Human factors in service delivery

In 1972 a British European Airways Hawker Siddeley Trident airliner came down in a field near Staines, UK, with the loss of 118 lives. The crew had failed to maintain the correct air speed three minutes after taking off from London Heathrow airport. In the subsequent air accident investigation, it was found that the crash took place against the background of a pilots' strike. The pilot had a pre-existing heart condition and had suffered a potentially distressing arterial event two hours before the crash which was likely to have been caused by an altercation between the captain and his co-pilot before take-off as they argued about the strike. It was also found that there was a lack of crew training on how to manage pilot incapacitation, and the crew wrongly disabled the stall recovery system due to being unaware of the stall protection systems.

There were other causes but the accident illustrates the importance of avoiding stress and distractions before undertaking a safety critical operation (Wikipedia (8) 2017).

Why is this necessary?

Modern production processes have reduced opportunities for human error primarily by use of robotics, but there remain many other opportunities for error in the service industry where the services are primarily provided by people. Also, the consequences of human error can vary considerably between industries. A human error in the manufacture of a component may soon be detected before the component is released for delivery to a customer, but a human error in the operation of a nuclear power plant, an oil production platform, a surgical operation or in the transport of people by rail, sea or air may cost many lives.

How is this addressed?

Human error is a conclusion that is determined with hindsight, a social evaluation of behaviour after the fact, and therefore actions to prevent human error tend to be actions derived from experience. In the worst case, they are often the result of an inquiry where the recommendation is: *This must never be allowed to happen again.*

Box 45.2   Changing conditions, not people

We cannot change the human condition, but we can change the conditions under which humans work.

(Reason, 2000)

Determining the risk of human errors and violations

The standard requires actions for preventing human error to be implemented but before we can do this, we need to have determined what actions are necessary and before that, what the risks are. The risks associated with the type of work undertaken by the organization and actions to address them should have been identified when "planning for the QMS" (as addressed by 6.1.1), but further risks may be present that are associated with the particular configuration of the production and service delivery processes, the working environment, infrastructure and the equipment being used.

One way of identifying risks is to ask what could go wrong in the process but as Reason points out, "when we think about boiling an egg and at what stages and in how many ways this relatively simple operation can be bungled, there would be a very long list of possibilities making it appear highly unlikely that we could ever adequately chart the varieties of human error." However, it turns out that human error is neither as abundant nor as varied as its vast potential might suggest. Not only are errors much rarer than correct action – they also tend to take a surprisingly limited number of forms (Reason, 1991).

Reason applies the following three question algorithms to distinguish between different kinds of intentional behaviour:

a)   Were the actions directed by some prior intention?
b)   Did the actions proceed as planned?
c)   Did they achieve their desired end?

In any situation, if all the questions can be answered in the affirmative, there's no human error, but a negative result provides an indication of the nature of the error.

These forms of human error were identified earlier, and now we turn our attention to some typical risks under each category:

SLIPS

- Performing an action too soon in a procedure or leaving it too late.
- Omitting a step or series of steps from a task.
- Carrying out an action with too much or too little strength (e.g. over-torqueing a bolt).
- Performing the action in the wrong direction (e.g. turning a control knob to the right rather than the left, or moving a switch up rather than down).
- Doing the right thing but on the wrong object (e.g. switching the wrong switch).
- Carrying out the wrong check but on the right item (e.g. checking a dial but for the wrong value).

LAPSES

- Distracted by sudden noise, a disturbance in the area or something that attracts our gaze away from what we are doing.
- Interrupted by a phone call, a co-worker, a supervisor.
- Preoccupied by anxieties in personal life or a work situation or mentally planning something unrelated to the work in hand.

MISTAKES

- Rules are unclear leading to users making assumptions.
- Procedures over-complicated.
- Legitimate information not readily available so worker reverts to memory.

Box 45.3   Performance-influencing factors

Performance-influencing factors (PIFs) are the characteristics of the job, the individual and the organization that influence human performance. Optimizing PIFs will reduce the likelihood of all types of human failure. (NB: This list is not exhaustive.)

Job factors

- Clarity of signs, signals, instructions and other information
- System/equipment interface (labelling, alarms, error avoidance/ tolerance)
- Difficulty/complexity of task
- Mismatch between job requirements and people's capabilities
- Routine or unusual
- Divided attention
- Procedures inadequate or inappropriate
- Preparation for task (e.g. permits, risk assessments, checking)
- Time available/required
- Tools appropriate for task
- Communication, with colleagues, supervision, contractor, other
- Working environment (noise, heat, space, lighting, ventilation)

Person factors

- Physical capability and condition
- Fatigue (acute from temporary situation, or chronic)
- Stress/morale
- Work overload/underload/ability to cope
- Competence to deal with circumstances
- Motivation vs. other priorities

Organization factors

- Work pressures (e.g. production vs. safety)
- Level and nature of supervision/leadership
- Communication
- Manning levels
- Peer pressure
- Clarity of roles and responsibilities
- Consequences of failure to follow rules/procedures
- Effectiveness of organizational learning (learning from experiences)
- Organizational or safety culture (e.g. everyone breaks the rule)

Source (HSE, 2016)

ROUTINE VIOLATIONS

- The desire to cut corners to save time and energy.
- The perception that the rules are too restrictive.
- The belief that the rules no longer apply.
- Lack of enforcement of the rule.
- New workers starting a job where routine violations are the norm.
- Not realizing that this is not the correct way of working.

Causes of human error

People perform differently in different situations. Some principle factors that will affect a person's performance are listed in Box 45.4: mismatches between job requirements and people's capabilities provide the potential for human error.

Actions to reduce human error and violations

Errors can be prevented by designing products, services and processes that make it hard for people to do the wrong thing and easy for people to do the right thing. Many of the actions to prevent human error and violations can be taken in product and service design as well as in planning and the creation of procedures, instructions, forms and regulations, using such techniques as poka-yoke or error proofing; for example,

- Design forms so that data entry is intuitive, devoid of ambiguity and easy to check.
- Design jobs to avoid the need for tasks which involve very complex decisions, diagnoses or calculations (e.g. by writing procedures for rare events requiring decisions and actions).
- Check that job aids such as procedures and instructions are clear, concise, available, up to date and accepted by users.
- Consider using photographs or videos to demonstrate critical tasks.
- Make rules and procedures relevant, practical and where possible easily understood by people at a seventh grade level using the Felsch–Kincaid readability tests (a feature offered in Microsoft Office).

If the risk cannot be eliminated by design, features can be provided so that the worker is alerted before an error is about to be made (e.g. the pop-up dialogue in a computer programme before a file is closed or deleted) or when an error has been made (an audible alarm) so that action can be taken to prevent a nonconformity. For those errors that cannot be prevented by design, other measures need to be taken. It is important to involve the workforce in both determining the actions to take and implementing those actions. HSG48 provides some useful tips in this area:

Human error

- Reduce the stresses which increase the frequency of errors;
- Make certain that arrangements for training are effective;
- Ensure proper supervision particularly for inexperienced staff, or for tasks where there is a need for independent checking;
- Consider the possibility of human error when undertaking risk assessments;
- Think about the different causes of human errors during incident investigations to introduce measures to reduce the risk of a repeat incident; and
- Monitor that measures taken to reduce error are effective.

Violations

- Take steps to increase the chances of violations being detected (e.g. by routine monitoring);
- Think about whether there are any unnecessary rules;
- Explain the reasons behind certain rules or procedures and their relevance;
- Improve design factors that affect the likelihood of corner cutting;
- Involve the workforce in drawing up rules to try to increase acceptance;
- Improve the working environment;
- Provide appropriate supervision;
- Establish a quality-first culture;
- Provide more training for abnormal and emergency situations;
- During risk assessments think about the possibility of violations; and
- Try to reduce the time pressure on staff to act quickly in novel situations.

Box 45.4   Risk treatment in rail transport

Driving involves spending long periods of time in the cab of a train. Drivers can be susceptible to fatigue and loss of alertness which could increase the probability of a human error. Shift rosters were redesigned to reduce the disruptions of circadian rhythms, and changes to the cab environment were designed to improve alertness.

How is this demonstrated?

Demonstrating that actions to prevent human error are being implemented may be accomplished by:

a) presenting evidence of an analysis of risks in production and service delivery, which identifies the actions to be taken to reduce the likelihood of human error;
b) selecting a representative sample of planned actions and presenting evidence that they have been implemented;
c) retrieving records of nonconformities and customer complaints and showing that none were deemed to be caused by human error or violation of rules;
d) retrieving records indicating the achieved result of production and service delivery processes and showing that no performance failures were deemed to be caused by human error or violation of rules.

Bibliography

Deming, W. E. (1982). *Out of the Crisis*. Cambridge, MA: The MIT Press.

HSE. (2016, August). Performance Influencing Factors (PIFs). Retrieved from Health and Safety Executive: www.hse.gov.uk/humanfactors/topics/pifs.pdf

HSG48. (1999). *Reducing Error and Influencing Behaviour*. London: HSE.

ISO/IEC 2382:2015. (2015). *Information Technology – Vocabulary*. Geneva: International Organization of Standardization.

Reason, J. (1991). *Human Error*. Cambridge: Cambridge University Press.

Reason, J. (2000, March 18). Human error: models and management. *British Medical Journal*, 320(7237), 768–770.

Wikipedia (8). (2017, April 2). *British European Airways Flight 548*. Retrieved from Wikipedia: https://en.wikipedia.org/wiki/British_European_Airways_Flight_548

46 Identification and traceability

Introduction

The requirement for product identification and traceability was included in the first edition of ISO 9001. The change in the use of the word *product* has caused some minor alterations in the 2015 edition, as it is replaced by the word *output*, but other than that it has not changed over the years.

Output identity is vital in many situations to prevent inadvertent mixing, to enable re-ordering, to match outputs with documents that describe them and to do that basic of all human activities – *communicate*. Often the names we give to things to describe them is sufficient for us to know immediately what the other person is talking about and to locate an item, but sometimes we need more definitive information, especially when all the things we are looking at look alike. We need to be sure not only that we have located the right item but, that it's ready for use so we don't inadvertently use it, post it, ship it, etc. In addition, if we encounter a faulty item we might want to find all others with the same fault so that we can contain any consequences that might arise from their use.

In this chapter, we examine the three requirements of clause 8.5.2, namely:

- Identifying process outputs
- Identifying the status of process outputs
- Maintaining traceability

Identifying process outputs (8.5.2)

What does this mean?

The requirements for the identification of process outputs are intended to enable outputs with one set of characteristics to be distinguishable from those with another set of characteristics. Where such situations cannot prevail, process outputs don't require an identity.

First, it only applies to outputs during production and service provision that require a distinguishable identity, and second it only applies to outputs planned for incorporation into deliverable products and services. This includes conforming and non-conforming outputs which should be identified accordingly (see next section). Unintended outputs such as waste, noise, pollution are excluded.

In the example of a fast food outlet referred to in Chapter 35, signs advertising the outlet in the right location pointing to the outlet were classed as an output a customer would look for. Although a process did place these signs at the designated location, these outputs were outputs of the sign production process, not outputs of the service delivery process and therefore not the type of output to which this requirement applies.

Why is this necessary?

Without codes, numbers, labels, names and other forms of identification we cannot adequately describe the product or service to anyone else or be certain we are looking at the right product. The output must be identified in one way or another otherwise it cannot be matched to its specification.

How is this addressed?

Output identity during production

Separate identity is necessary where it is not inherently obvious. If outputs are so dissimilar that inadvertent mixing would be unlikely to occur, a means of physically identifying the products is probably unnecessary. *Inherently obvious* in this context means that the physical differences are large enough to be visible to the untrained eye. Functional differences, therefore, no matter how significant as well as slight differences in physical characteristics such as colour, size, weight and appearance would constitute an appropriate situation for documented identification procedures.

Identifying product and its components should start at the design stage when the product is conceived. The design should be given a unique identity, a name or a number and that should be used on all related information. When the product emerges into production, the product should carry the same number or name; in addition it should carry a serial number or other identification to enable product features to be recorded against specific products. If verification is on a go or no-go basis, product does not need to be serialized. If measurements are recorded some means should be found of identifying the measurements with the product measured. Serial numbers, batch numbers and date codes are suitable means for achieving this. This identity should be carried on all records related to the product.

Apart from the name or number given to a product, you need to identify the version and the modification state so that you can relate the revision of the drawings and specifications to the product they represent. Products should either carry a label or markings with this type of information in an accessible position or bear a unique code number that is traceable to such information.

You may not possess any documents that describe purchased product. The only identity may be marked on the product itself or its container. Where there are no markings, information from the supplier's invoice or other such documents should be transferred to a label and attached to the product or the container. Information needs to be traceable to the products it represents.

The method of identification depends on the type, size, quantity or fragility of the product. You can mark the product directly (provided the surface is not visible to the end user unless of course identity is part of the brand name) tie a label to it or the container in which it is placed. You can also use records remote from the product providing they bear a unique identity that is traceable to the product.

There are, of course, situations where attaching an identity to a product would be impractical such as for liquids or items too small but the output nevertheless has an identity that is conveyed through the packaging and associated information. In the food industry, the biscuits on the conveyor might not carry an identity but the box into which they are packed does as does the instruction that ordered the biscuits to be produced. Thus, identifying a product by suitable means might require the product to be labelled, or might require the container to be labelled.

Marking products has its limitations because it may damage the product, be removed or deteriorate during subsequent processing. If applied directly to the product, the location and nature of identification should be specified in the product drawings or referenced process specifications. If applied to labels which are permanently secured to the product, the identification needs to be visible when the product is installed to facilitate checks without its removal. If the identity is built into the forging or casting, it is important that it is legible after machining operations. One situation which can be particularly irritating to customers is placing identification data on the back of equipment and then expecting the customer to state this identity when dealing with a service call, thus causing delay while the customer dives under the desk to locate the serial number and drops the telephone in the panic!

Output identity during service provision

Services are somewhat different. Many are not identified other than by the nature of what the organization does by generic categories such as investment, mortgage, financial planning services of banks. Where there are differences for instance in interest rates, the *products* are given different names such as instant access account, 90-day account and so on. Process outputs are often in the form of documented information which at the least should carry a distinguishing title and date.

How is this demonstrated?

Demonstrating that suitable means to identify process outputs have been used may be accomplished by:

a)   presenting evidence of a product identification process and identification conventions;
b)   selecting a representative sample of process outputs and producing evidence that they are identified in accordance with the prescribed conventions;
c)   searching the database of recorded nonconformities, including customer complaints to show that none were traceable to an output lacking identity or having an unsuitable identity.

Identifying the status of process outputs (8.5.2)

What does this mean?

Output status with respect to monitoring and measurement means an indication as to whether the output conforms or does not conform to specified requirements. Thus, identifying output status enables verified outputs to be distinguishable from unverified outputs and conforming outputs to be distinguishable from nonconforming outputs.

Why is this necessary?

Measurement does not change an output but does change our knowledge of it. Therefore, it is necessary to identify which outputs conform and which do not so that inadvertent mixing, processing or delivery is prevented.

How is this addressed?

Identifying status in production

The most common method of denoting output status is to attach labels either to the item or to containers holding it. Green labels are often used for acceptable good and red labels for reject goods. Labels should remain affixed until the product is either packed or installed. Labels should be attached in a way that prevents their detachment during handling. If labels need to be removed during further processing, the details should be transferred to inspection records so that later the status of the components in an assembly can be checked through the records. At dispatch, product status should be visible. Any product without status identification should be quarantined until re-verified and found conforming. Once a product has passed through the demand fulfilment process and is in use, it requires no status identity unless it is returned to the production process for repair or other action.

It should be possible when walking through a machine shop, for example, to identify which items are awaiting verification, which have been verified and found conforming and which have been rejected. If, by chance, an item was to become separated from its parent batch, it should still be possible to return the item to the location from whence it came. A machine shop is where this type of identification is essential – it is where mix-ups can occur. In other places, where mix-ups are unlikely, verification status identification does not need to be so explicit.

Identifying output status is not just a matter of tying a label on the item. The status should be denoted by an authorized signature, stamp, mark or other identity which is applied by the person making the accept or reject decision and which is secure from mis-use. Signatures are acceptable as a means of denoting verification status on paper records but are not suitable for computerized records. Secure passwords and write-only protection must be provided to specific individuals. Signatures in a workshop environment are susceptible to deterioration and illegibility that is why numbered inspection stamps with unique markings evolved. The ink used has to survive the environment, and if the labels are to be attached to the product for life, it is more usual to apply an imprint stamp on soft metal or bar code.

In some situations, the location of an item can constitute adequate identification of item status. However, these locations need to be designated as Awaiting Inspection, Accepted Product or Reject Product or other such labels as appropriate to avoid the inadvertent placement of items in the wrong location. The location of an item in the normal production flow is not a suitable designation unless an automated transfer route is provided.

If you use inspection stamps, you will need a register to allocate stamps to particular individuals and to indicate which stamps have been withdrawn. When a person hands in his stamp, it is good practice to avoid using the same number for 12 months or so to prevent mistaken identity in any subsequent investigations.

Identifying status in service provision

At any stage of providing a service where there are items awaiting to be checked there may be items that fail the checks that need to be reworked or destroyed and the same provisions as described for production would apply. When a service or part of a service is out of action, users need to be notified. Public conveniences are closed while being cleaned and a sign placed across the entry; in hotels rooms are denoted *not ready* on the accommodation register until the housekeeping staff have vacated when the status is changed. If there is a possibility that a service that is not ready can be inadvertently be perceived as serviceable provisions should be made to mitigate this risk.

Services that rely on products should carry a label or a notice to denote their serviceability when accessed. A bank cash machine is one example where a notice is displayed when the machine is out of service. In some cases, customers may need to be informed by letter or telephone.

With software, the verification status can be denoted in the software as a comment or on records testifying its conformance with requirements.

With documentation, you can either denote verification status by an approval signature on the document or by a reference number, date and issue status that is traceable to records containing the approval signatures.

How is this demonstrated?

Demonstrating that the status of outputs is identified with respect to monitoring and measurement throughout production and service provision may be accomplished by:

a)   presenting evidence of a process for identifying verification status;
b)   selecting a representative sample of items at various stages of production and service provision and showing that their verification status can be determined.

Maintaining traceability (8.5.2)

What does this mean?

Traceability is a process characteristic. It provides the ability to trace something through a process to a point along its course either forwards or backwards through the process and determine as necessary; its origin, its history and the conditions to which it was subjected. Traceability may be a requirement of the customer, legislation or statutes or simply a requirement of the organization to conduct investigations when events do not proceed as planned. Traceability therefore does not need to be a customer requirement for this requirement of ISO 9001 to apply.

Why is this necessary?

One needs traceability to find the root cause of problems. If records cannot be found which detail what happened to an item, then nothing can be done to prevent its recurrence. Traceability is key to enabling corrective action, product recall and defending product liability claims.

In situations of safety or national security, it is necessary to be able to locate all products of a batch in which a defective product has been found to eliminate them before there is a disaster. It is also very important in the aerospace, automobile, medical devices and food and drugs industries – in fact, any industry where human life may be at risk due to a defective product being in circulation.

Traceability is also important to control processes. You may need to know which products have been through which processes and on what date if a problem is found sometime later. The same is true of test and measuring equipment. If, on being calibrated, a piece of test equipment is found to be out of calibration, it is important to track down all the equipment which has been validated using that piece of measuring equipment. This in fact is a requirement of clause 7.1.5.2, but no requirement for traceability is specified.

How is this addressed?

Providing traceability can be an onerous task. Some applications require items to be traced back to the original ingot from which they were produced. Traceability is achieved by coding items and their records such that you can trace an item back to the records at any time in its life. The chain can be easily lost if an item goes outside your control. For example, if you provide an item on loan to another organization for investigation and it is returned sometime later, without a certified record of what was done to it and under what conditions it was stored and/or used, you have no confidence in its integrity. Traceability is only helpful when the chain remains unbroken. It can also be costly to maintain. The system of traceability that you maintain should be carefully thought out so that it is economic. There is little point in maintaining an elaborate traceability system for the once in a lifetime event when you need it, unless your very survival or society's survival depends on it. However, if there is a field failure, to prevent recurrence, you will need to trace the component back through the supply chain to establish which operation on which component was not performed correctly simply to rule out any suggestion that other products might be affected.

The conventions you use to identify product and batches need to be specified in the product specifications and the stage at which product is marked specified in the relevant procedures or plans. Often such markings are automatically applied during processing, as is the case with printed circuits, mouldings, ceramics, castings, products, etc. Process setting up procedures should specify how the marking equipment or tools are to be set up.

If you do release a batch of product prior to verification being performed and one out of the batch is subsequently found to be nonconforming, you will need to retrieve all others from the same batch. This may not be as simple as it seems. To retrieve a component which has subsequently been assembled into a printed circuit board, which has itself been fitted into a unit along with several other assemblies, not only would you need a good traceability system, but also one that is constantly in operation.

It would be considered prudent to prohibit the premature release of product if you did not have an adequate traceability system in place. If nonconformity will be detected by the end product tests, allowing production to commence without the receipt tests being available may be a risk worth taking. However, if you lose the means of determining conformity by premature release, don't release the product until you have verified it as acceptable.

How is this demonstrated?

Demonstrating that the identification of outputs is controlled when traceability is a requirement and that any documented information necessary is retained may be accomplished by:

a) presenting evidence of policies stipulating what needs to be traceable about an item and from what source it needs to be traceable;
b) presenting codes of practice that prescribe the conventions used to provide traceability;
c) selecting a representative sample of items that have been designated as traceable and using the established codes following the trail through the records to show that an unbroken chain exists to the intended source.

47 Property belonging to external providers

Introduction

Property belonging to external providers is any property owned or provided by a customer or supplier for use or embodiment during the course of a contract. Typical examples include:

- Your organization is building a house, vehicle, or other structure for the external provider into which they want certain items incorporated before handing over or delivery of the product to them.
- Your external provider makes available their facilities for your organization to use in developing or testing products on their premises.
- Your external provider makes available their equipment for your organization to use in developing or testing products on your premises.
- Your organization is contracted to run a service from your external provider's premises utilizing some of their equipment.
- Your external provider provides clothing, security devices and protective equipment whilst on their premises.
- An external provider leaves his property with your organization for repair.
- An external provider provides credit card data or other personal data in order that your organization may process payment or register them as a member or patient.

Documentation is not considered external provider's property because it is normally freely issued and ownership passes from provider to receiver on receipt. However, if the external provider requires the documentation to be returned or destroyed at the end of the contract, it should be treated as externally provided property.

In this chapter, we examine the five requirements of clause 8.5.3, namely:

- Care of property belonging to external providers
- Identifying property belonging to external providers
- Verifying property belonging to external providers
- Protecting property belonging to external providers
- Reporting problems to external providers

Care of property belonging to external providers (8.5.3)

What does this mean?

Exercising care means that precautions are taken to retain the property in the same condition as it was when received and to prevent loss and damage from whatever cause.

The property being supplied may have been produced by a competitor, by the customer, by your supplier on loan, or even by your own organization under a different contract. It may include intellectual property, personal data, products to be incorporated into products that are to be supplied to your customers, returnable packaging, customer- or supplier-owned tooling, software, equipment, development facilities on customer premises.

Why is this necessary?

External providers will expect any property they supply for your organization's use to be treated with due care and protected from loss or damage.

How is this addressed?

For externally provided property that is used on your own premises, you should maintain a register containing the following details as appropriate:

- Name of property, part numbers, serial numbers and other identifying features.
- Name of external provider and source of property if different.
- Delivery note reference, date of delivery.
- Receipt inspection requirements.
- Condition on receipt including reference to any rejection note.
- Storage conditions and place of storage.
- Maintenance specification if maintenance is required.
- Current location and name of custodian.
- Date of return to external provider or embodiment into supplies.
- Part number and serial number of product embodying the externally provided property.
- Dispatch note reference of assembly containing the property.

These details will help you keep track of the property whether on embodiment loan or contract loan and will be useful during customer audits or in the event of a problem with the item either before or after dispatch of the associated assembly.

There might also be a need for a definition of responsibilities – a table showing which of the three parties (customer, supplier and your organization) is responsible for the acquisition, verification, repair, return or externally provided property and investigation of defects, etc., and what the associated financial liabilities are.

If the property requires any maintenance you should be provided with a maintenance specification and the appropriate equipment to do the job. Maintenance may include both preventive and corrective maintenance but you should clarify with your external provider which it is. You may have the means for preventive maintenance such as lubrication and calibration but not for repairs. Always establish your obligations in the contract regarding customer-supplied property, because you could take on commitments for which you are not contractually covered if something should go wrong. You need to establish who will supply the spares and re-certify the equipment following repair.

How is this demonstrated?

Demonstrating that the organization is exercising care with property belonging to external providers may be accomplished by presenting evidence of a process for registering externally provided property which shows how responsibility for its care is transferred into the organization and information relating to its storage, use, maintenance, and return is retained.

Identifying property belonging to external providers (8.5.3)

What does this mean?

Identifying externally provided property means attaching labels or other means of identification that denote its owner.

Why is this necessary?

If externally provided property carries an identity that distinguishes it from other property, it will prevent inadvertent disposal or unauthorized use.

How is this addressed?

Externally provided property may carry suitable identification but if not, labels, containers or other markings may be necessary to distinguish it from organization owned property. As externally provided property may have been supplied by the organization originally as in the case of a repair service, labels indicating the owner should suffice. In a vehicle service area, for instance, a label is attached to the car keys rather than labelling the car itself.

When deciding the type of marking, consideration needs to be given to the conditions of use. Markings may need to be permanent to be durable under the anticipated conditions of use. It would be wise to seek guidance from the external provider if you are in any doubt as to where to place the marking or how to apply it. Bar coding is often the most practical solution as it can contain the external provider's identity, date of supply, contract and limitations of use.

How is this demonstrated?

Demonstrating that the organization has identified the external provider's property may be accomplished by:

a)   presenting evidence of a process for identifying externally provided property so that it's distinguishable from organization owned property;

b)   selecting from the register of externally provided property a representative sample and tracking it to the place where it is located.

Verifying property belonging to external providers (8.5.3)

What does this mean?

External providers may supply property purchased from other suppliers for installation in an assembly purchased from your organization.

Why is this necessary?

Items need to be verified before incorporation into the organization's product, regardless of their source, first, to establish the condition of the item on receipt in the event that it is damaged, defective or is incomplete and second, to verify that it is fit for the intended purpose before use. If you fail to inspect the property on receipt you may find difficulty in convincing your external provider later that the damage was not your fault.

How is this addressed?

When property is received from an external provider it should be processed in the same way as purchased product so that it is registered and subject to receipt inspection. The inspection you carry out may be limited if you do not possess the necessary equipment or specification, but you should reach an agreement with the external provider as to the extent of any receipt inspection before the property arrives. You also need to match any delivery note with the property because the external provider may have inadvertently sent you the wrong property. Unless you know what you are doing, it is unwise to activate any mechanical or electrical property without proper instructions from the external provider.

How is this demonstrated?

Demonstrating that external provider's property provided is being verified by the organization may be accomplished by:

a) presenting evidence of a process for verifying the integrity of externally provided property before use;

b) selecting a representative sample of externally provided property from the register and producing evidence that it has been verified prior to its readiness for use.

Protecting property belonging to external providers (8.5.3)

What does this mean?

Protection means safeguarding against loss, damage, deterioration and misuse.

Why is this necessary?

As the property will either be returned to the external provider on completion of contract or will be incorporated into your products, it is necessary to protect the property from conditions that may adversely affect its quality.

How is this addressed?

Where the externally provided property is in the form of property that could be inadvertently degraded, it should be segregated from other products to avoid mixing, inadvertent use, damage or loss. Depending on the size and quantity of the items and the frequency with which your external provider supplies such property you may require a special storage area. Wherever the items are stored you should maintain a register of such items, preferably separate from the store (e.g. in inventory control or the project office). The authorization for releasing external provider property from stores may need to be different for inventory control reasons. You also need to ensure that such property is insured. You will not need a corresponding purchase order and they may not therefore be registered as stock or capital assets. If you receive externally provided property very infrequently, you will need a simple process that is only activated when necessary rather than being built into the inventory control process. Under such circumstances it is easy to lose this property and forget they are someone else's property. You need to alert staff to take extra care especially if they are high-value items that cannot readily be replaced.

How is this demonstrated?

Demonstrating that external provider's property provided is being protected by the organization may be accomplished by:

a)   presenting evidence of a process for protecting externally provided property while under the organization's control;

b)   selecting a representative sample of externally provided property from the register and producing evidence that it is being protected in accordance with the provider's requirements.

Reporting problems to external providers (8.5.3)

What does this mean?

While externally provided property is on your premises, it may be damaged, develop a fault or be lost. Also, when using externally provided property on the provider's premises, events may occur that result in damage or failure to the property.

Why is this necessary?

It is necessary to record and report any damage, loss or failure to the provider so that as owners they may decide the action that is required. Normally, the organization does not have responsibility to alter, replace or repair externally provided property unless authorized to do so under the terms of the contract.

How is this addressed?

The external provider is responsible for the property they supply wherever it came from in the first place. It is therefore very important that you establish the condition of the property before you store it or use it. If you detect that the property is damaged, defective or is incomplete, you should place it in a quarantine area and report the condition to the external provider. Even if the property is needed urgently and can still be used, you should obtain the agreement of its external provider before using inferior property; otherwise, you may be held liable for the consequences.

   You could use your own reject note or nonconformity report format to notify the external provider of defective property but these are not appropriate if the property is lost. You also need a response to the problem and so a form that combined both a statement of the problem and of the solution would be more appropriate.

How is this demonstrated?

Demonstrating that problems with external provider's property are reported and records retained may be accomplished by:

a) presenting evidence of a process for recording and reporting problems with externally provided property;

b) scanning the register of externally provided property and selecting items that have been recorded as lost, damaged or otherwise found to be unsuitable for use;

c) presenting evidence that the provider had been promptly informed about the items selected and that records detailing the occurrence have been retained.

48 Preservation of process outputs

Introduction

During production and service provision, each process will produce outputs that are invariably destined as inputs to another process. As this transition is often not instantaneous, during the interim period, the outputs may be vulnerable to conditions that potentially cause loss of integrity.

Although an output may be waiting verification the lapsed time may be sufficient for its integrity to be degraded due to contact with people or objects, airborne particles, microbes and vapours. So whether the item was or was not conforming, the mere passage of time in the wrong environment can adversely affect its conformity.

The range of environments and preservation methods are too great to address in this chapter, but some of the more common factors and methods of dealing with them are covered.

In this chapter, we examine the requirement of clause 8.5.4 for preservation of process outputs and look at it from several perspectives.

Preservation (8.5.4)

What does this mean?

These requirements are concerned with conformity control i.e. preserving conforming outputs so that they don't lose their identity or deteriorate when being handled, moved, stored, transported or further processed. They apply equally to service operations that involve the supply of product such as in the hospitality industry.

Why is this necessary?

As considerable effort will have gone into producing a conforming output, it is necessary to protect it from adverse conditions that could change its physical and functional characteristics. In some cases, preservation is needed immediately the characteristics have been generated (e.g. surface finish). In other cases, preservation is only needed when the item leaves the controlled environment (e.g. food, chemicals and electronic goods). Preservation processes need to be controlled in order that items remain in their original condition until required for use.

How is this addressed?

Determination of preservation requirements commences during the design phase or the operations planning phase by assessing the risks to output quality during production and service provision.

The preservation processes should be designed to prolong the life of the item by inhibiting the effect of natural elements and human contact. Whereas the conditions in the workplace can be measured, those outside the workplace can only be predicted.

Having identified there is a risk to output quality you may need to prepare instructions for the handling, storage, packing, preservation and delivery of particular items. In addition to issuing the instructions you will need to reference them in the appropriate work instructions in order that they are implemented when necessary. Whatever the method, you will need traceability from the identification of need to implementation of the provisions and from there to the records of achievement.

Identification

Packages for export may require different markings than those for the home market. Those for certain countries may need to comply with particular laws. Unless your customer has specified labelling requirements, markings should be applied both to primary and secondary packaging as well as to the output itself. Markings should also be made with materials that will survive the conditions of storage and transportation. Protection can be given to the markings while in storage and in transit but this cannot be guaranteed while items are in use. Markings applied to items therefore need to be resistant to cleaning processes both in the workplace and in use. Markings on packaging are therefore essential to warn handlers of any dangers or precautions they must observe. Limited Life Items should be identified to indicate their shelf life. The expiry date should be visible on the container and provisions should be made for such items to be removed from stock when their indicated life has expired. Whereas a well-equipped laboratory can determine the difference between different items and materials, the consumer needs a simple practical method of identification, and labelled packets often provide a reliable and economic alternative. For items that start to deteriorate when the packaging seal is broken, the supplier's responsibility extends beyond delivery to the point of use. In such cases markings need to be applied to the containers to warn the consumers of the risks.

Handling

Handling provisions serve two purposes both related to safety. Protection of the item from the individual and protection of the individual handling the item. This latter condition is concerned with safety and addressed through other provisions; however, the two cannot, and should not, be separated, and handling procedures should address both aspects.

Handling items can take various forms depending on the hazard you are trying to prevent from happening. In some cases, notices on the item will suffice, such as "LIFT HERE" or "THIS WAY UP" or the notices on batteries warning of acid. In other cases, you will need to provide special containers or equipment. There follows a short list of handling provisions that your procedures may need to address:

- Lifting equipment
- Pallets and containers
- Conveyors and stackers
- Design features for enabling handling of items
- Handling of electrostatic-sensitive devices
- Handling hazardous materials
- Handling fragile materials

Contamination control

In certain industries such as pharmaceuticals, semiconductors, space systems, biotechnology and decorative finishes, contamination by dust, fibres, pollutants, microbes, vapours, foreign objects and material is detrimental to product quality. These contaminants are generated by people, process, facilities and equipment. The air in a typical office building contains from 500,000 to 1,000,000 particles (0.5 microns or larger) per cubic foot of air. Airborne particles can prevent the production of blemish-free surfaces, can damage moving surfaces, degrade food and drugs and can cause deterioration in performance, for example, film contaminants of only 10 nm (nanometres) can drastically reduce coating adhesion on a semiconductor wafer. *Contamination control* is the generic term for all activities aiming to control the existence, growth and proliferation of contamination in certain areas. *Cleanrooms* are areas designated for activities requiring levels of contamination control. Air cleanliness is designated by the concentration of particles and classified as in ISO 1464–1:2015.

Risk assessments should be carried out to establish the maximum particle size that can be permitted and a corresponding economic level of contamination control. Regulations governing the operation of a cleanroom should be developed and enforced.

Segregation

Segregation is vital in many industries where products can only be positively identified by their containers. It is also important to prevent possible mixing or exposure to adverse conditions or cross-contamination. Examples where segregation makes sense are:

- Toxic materials
- Flammable materials
- Limited-life items
- Explosives

Segregation is not only limited to the product but also to the containers and tools used with the product. Particles left in containers and on tools – no matter how small – can cause blemishes in paint and other finishes as well as violate health and safety regulations. If there are such risks in your manufacturing process, then procedures need to be put in place that will prevent product mixing.

Segregation may also be necessary in the packaging of products, not only to prevent visible damage, but also to prevent electrical damage as with electrostatic-sensitive devices. Segregation may be the only way of providing adequate product identity as is the case with fasteners. In other cases, items may need to be stored in sealed containers to retard decay, corrosion and/or contamination.

Packaging

Packaging is the material and containers that protect items from damage, interference, contamination and deterioration. There are two basic layers of packaging: primary packaging, which designates the layer of packaging in immediate contact with the item, and secondary packaging, which designates the layer of packaging in immediate contact with the external environment. It is imperative that primary packaging keep the item absolutely sealed off from its environment.

Packaging design should be governed by the requirements of clause 8.3, although if you only select existing designs of packaging these requirements would not be applicable but they do need to be assessed as being fit for purpose.

Packing processes should be designed to protect an item from damage and deterioration under the conditions that can be expected during its storage and transportation. You will need a means of identifying the packaging and marking requirements for particular items and of determining processes for the design of suitable packaging including the preservation and marking requirements. Depending on your processes you may need to devise packages for various storage and transportation conditions, preservation methods for various types of product and marking requirements for types of product associated with their destination.

Where applicable, preservation processes should require that items be cleaned before being packed and preservative applied.

Unless your customer has specified packaging requirements, there are several national standards that can be used to select the appropriate packaging, marking and preservation requirements for your products. Your procedures should make provision for the selection to be made by competent personnel at the planning stage and for the requirements thus selected to be specified in the packing instructions to ensure their implementation.

Packing instructions should not only provide for protecting the product but also for including any accompanying documentation such as:

* Assembly and installation instructions
* Licence and copyright notices
* Certificates of conformity
* Packing list identifying the contents of the container
* Export documents
* Warranty cards

The packing instructions are likely to be one of the last instructions you provide and probably the last operation you will perform for a consignment. This also presents the last opportunity for you to make mistakes! They may be your last mistakes but they will be the first your customer sees. The error you made on component assembly probably won't be found, but the slightest error in the packaging, the marking, the enclosures will almost certainly be found; therefore, this process needs careful control. It may not be considered so skilled a process but all the same it is vital to your image

Protection

Protection of items can take different forms therefore its necessary to firstly determine what hazard you need to protect the product from. There is protection from hazards in the external environment and this is the function of packaging (see earlier). There is protection from hazards due to unauthorized access, and this is the function of security. Security measures may be enacted through the design of primary packaging (e.g. the copyright protection afforded commercial DVDs, medicine bottles fitted with childproof caps and the tamper-evident band attached below a screw cap on bottles).

There may hazards in the workplace before conforming items are placed into storage because of their fragility or susceptibility to contamination for which they need protection. Indeed, even before items are subject to verification precautions may need to be taken to protect surfaces, edges and liquids.

Storage

To preserve the quality of items that have passed receipt verification, they should be transferred to stockrooms in which they are secure from damage and deterioration. You need secure storage areas for several reasons:

- For preventing personnel from entering the stockrooms and removing items without authorization.
- For preventing items from losing their identity because once the identity is lost it is often difficult, if not impossible to restore complete identification without testing material or other properties.
- For preventing vermin damaging the stock.
- For preventing climatic elements causing stock to deteriorate.

Although loss of product may not be considered a quality matter, it is if the product is externally provided property or if it prevents you from meeting your customer requirements. Delivery on time is a quality characteristic of the service you provide to your customer, and therefore secure storage is essential.

To address these requirements, you will need to identify and specify the storage areas that have been established to protect items pending use or delivery. Although it need be only a brief specification, the requirements to be maintained by each storage area should be specified based on the type of product, the conditions required preserving its quality, its location and environment. Items that require storage at certain temperatures should be stored in areas that maintain such temperatures. If the environment in the area in which the room is located is either uncontrolled or at a significantly higher or lower temperature, an environmentally controlled storage area will be required.

All items have a limit beyond which deterioration may occur and therefore temperature, humidity, pressure, air quality, radiation, vibration, etc., may need to be controlled. At some stage, usually during design or operations planning, the storage conditions need to be defined and displayed. In many cases, dry conditions at room temperature are all that is necessary but problems may occur when items requiring non-standard conditions are acquired. You will need a means of ensuring that such items are afforded the necessary protection and your storage procedures need to address this aspect. It is for this reason that it is wiser to store items in their original packaging until required for use. If packets need to be opened to verify identity, etc., the packaging design is already noncompliant.

Any area where items are stored should have been designated for that purpose in order that the necessary controls can be employed. If you store items in undesignated areas, then there is a chance that the necessary controls will not be applied. Designation can be accomplished by placing notices and markers around the area to indicate the boundaries where the controls apply.

Each time the storage controller retrieves an item for issue, there is an opportunity to check the condition of stock. However, some items may have a slow turnover in certain storage areas (e.g. where spares are held pending use). It is also necessary to plan and carry out regular checks of the overall condition of the stockroom for damage to the fabric of the building. Rainwater may be leaking onto packaging and go undetected until that item is removed for use.

Some items such as electrolytic capacitors and two part adhesives may deteriorate when dormant. Others such as rubber materials, adhesive tape and chemicals deteriorate with the passage of time regardless of use. These are often referred to as *shelf-life items* or *limited life items*. Dormant electronic assemblies can deteriorate in storage, and in the unlikely event that items would remain in storage for more than one year, provision should be made for their periodic inspection or retest.

The assessment interval will depend on the type of building, the stock turnover, the environment in which the stock is located and the number of people allowed access. The interval may vary from storage area to storage area and should be reviewed and adjusted as appropriate following the results of the assessment.

Transmission

Transmission in this context means the conveyance of a product by electronic means across the Internet. When this means of delivery is chosen, there needs to be agreement with the customer on:

a)   Pre-transmission arrangements such as notification of intended delivery
b)   The level of protected against corruption in transit (e.g. password protected)
c)   The security protocols (authenticated, encrypted, etc.)
d)   The access settings to the customer server
e)   Any size or format constraints
f)   Post-transmission arrangements such as delivery receipt/acknowledgement

Transportation

The methods of transportation of product between locations within the organization and to the customer needs to be determined at an early stage because arrangements will vary depending on the size, weight, security, speed, cost, destinations, customs and other factors. Special rigs, containers and handling equipment may be needed to load items onto or into the vehicle and off at its destination. A logistic plan may be needed to work out in detail how the product will be moved from A to B. The carriers chosen are external providers and, unless relying on national postal services, need to be subject to the same provisions as those addressed in clause 8.4.

In choosing the most suitable form of transport a risk assessment should be carried out, weighing up the advantages and disadvantages of different modes of transport.

How is this demonstrated?

Demonstrating that outputs are being preserved to ensure conformity during production and service provision may be accomplished by:

a)   presenting evidence of an analysis to determine the risks items are exposed to during production and service provision that may adversely affect conformity;
b)   presenting evidence of the provisions made to mitigate the identified risks;
c)   presenting evidence that the provisions made are justified;
d)   selecting a representative sample of process outputs and showing that they are being protected as planned.

49 Control of changes

Introduction

When products enter production or services commence delivery, they have passed through a transition phase where the bugs (special cause variation) have been ironed out and the only variation present (common cause variation) is within acceptable limits. The intent is that these processes consistently produce outputs of the quality required thereafter until formally changed.

The requirements for controlling changes in the 2008 version were limited to changes in design, whereas the 2015 version includes several additional requirements that address change as illustrated in Figure 32.2.

In this chapter, we examine the two requirements of clause 8.5.6, namely:

- Control of changes (8.5.6)
- Retaining information on planned changes (8.5.6)

Control of changes (8.5.6)

What does this mean?

A change is a situation where performance of any of the process parameters increases or decreases to a level outside the previously accepted tolerances. Natural variation within the accepted tolerances is not classed as change in this context. Different types of changes may arise: planned and authorized, planned and unauthorized, unplanned and unauthorized.

Planned and authorized changes

A planned and authorized change may be made to the design of a product or service which requires changes to be made to the fulfilment processes. Also, planned and authorized changes may be made to the fulfilment processes to make them more efficient or effective.

Planned and unauthorized changes

A planned and unauthorized change is one which has passed through all the stages except authorization and is released for execution. This happens when it is decided that the benefits of proceeding without authorization exceed the risks of delay.

Unplanned change

An unplanned change may be made by unknown causes which are detected during a regular review of process performance.

Reviewing changes

Reviewing a change in this context means determining the suitability, adequacy or effectiveness of a change to achieve established objectives, and therefore the review occurs before and after the change has been made. The action taken to determine what might be affected by the change before its execution is an evaluation.

Controlling changes

Controlling changes means causing desirable changes to be made and preventing undesirable changes from being made. An unauthorized change may be both desirable and undesirable depending on the risks.

Why is this necessary?

Were change to be allowed to be made in production and service delivery processes without some degree of control, it's highly likely that performance would be adversely affected. The consequences will vary, but unless the effects are studied before changes are made, there is no telling how detrimental to performance the changes might be. If changes are detected in performance for which there is no obvious explanation an investigation is necessary to understand what happened and contain the impact.

As a QMS is a systemic perspective of the organization, it follows that a change in any of its elements will influence other elements as everything is connected within a functioning whole.

How is this addressed?

Once processes are stable and producing outputs of acceptable quality a set of baseline conditions need to be established (sometimes referred to as standard operating conditions or procedures [SOPs]; [conditions may include procedures but not be limited to procedures]), and staff need to be discouraged from tinkering with the processes. Key process parameters need to be routinely monitored and process owners alerted to unplanned changes in performance. Any problems need to be recorded, reported and subject to investigation to establish the cause and action only taken after a proper evaluation of the impact.

Controlling changes

Planned changes to products and services need to be routed through the design change process (see Chapter 41), the output of which should be a plan for making the change that includes:

*   An announcement advising the date the change is planned to become effective.
*   The name of the person or organization authorizing the change.
*   A timeline for the activities to be carried out to undertake and validate the change.

- Instructions on what action to take on products currently in production, in stock and in service.
- Instructions for changing infrastructure, equipment, processes and other resources including training of personnel.

Everything affected by the change should be scheduled for change and careful consideration given to:

- How and when current production or service delivery will cease if necessary.
- The extent to which current production or service delivery can continue during the change.
- The sequence in which changes are made.
- How people are notified of the changes.
- Who will be involved in managing and undertaking the change.
- How items of infrastructure are removed and new items installed.
- How the processes will be re-built and restarted.
- How process capability will be verified or revalidated as applicable.
- When the effects of the change will be reviewed.
- What form recommencement of production or service delivery will take.
- Who will be notified.

Depending on the nature and extent of the change, information on the change may be as simple as a verbal instruction from the process owner or a dossier of documented information of considerable size which requires a dedicated task force to implement.

Reviewing changes

After the plan for change has been implemented a review should be scheduled by the change authority to find answers to the following questions:

a)   Were all the actions undertaken directed by the plan?
b)   Did the actions proceed as planned?
c)   Did they achieve their desired objective?

Those responsible for the processes affected and the key personnel involved in executing the change should participate in the review.

The cause of deviations should be established and either a plan for corrective action authorized and implemented or a revised plan authorized and implemented.

How is this demonstrated?

Demonstrating that changes for production or service provision are reviewed and controlled may be accomplished by:

a)   presenting evidence of process performance and showing that any unplanned change was subject to evaluation;
b)   presenting evidence of a process for evaluating changes to production and service delivery processes;

c) presenting evidence of the planning undertaken to execute changes;
d) selecting a representative sample of planned changes made to production and service delivery and presenting evidence that they were executed in accordance with the plan.

Retaining information on changes (8.5.6)

What does this mean?

The documented information to be retained is the results of the review that is carried out after the change has been made. The documented information of design changes is required to be retained through clause 8.3.6 (see Chapter 41).

Why is this necessary?

Every change has a purpose or an objective, and therefore it would be irresponsible not to establish if the change that has been executed actually achieved the intended objective. It would also be desirable to be able to provide proof to any appropriate authority.

How is this addressed?

A review of the change should generate the following information:

a) Authorization to proceed with the change.
b) Confirmation that the change was executed as planned.
c) Confirmation that the objectives for change were achieved.
d) Actions arising from the review.

After completion of actions arising the record of actions should be updated
   This information should be retained in the event it is required in subsequent analysis of performance and in providing assurance to customers or their representatives.

How is this demonstrated?

Demonstrating that documented information describing the results of the review and follow-up actions may be accomplished by:

a) presenting evidence of a process for capturing documented information on change reviews in production and service provision;
b) presenting a record of changes reviews that have been conducted;
c) selecting a representative sample of change reviews and presenting evidence that documented information can be retrieved which contains:

   i   the results of the review;
   ii the status of actions arising from the review;
   iii the name of the person authorizing the change.

50 Release, delivery and post-delivery of products and services

Introduction

Requirements for release, delivery and post-delivery of products and services have hitherto been less prominent in the standard. Release requirement were previously hidden among the final inspection requirements of the 1987 version and among product monitoring and measurement requirements in the 2008 version. The new structure imposed by Annex SL has resulted in product and service verification being moved from monitoring and measurement into operations (which is a logical move), but it implies that the monitoring measurement requirements of clause 9 are associated with the system and not control of products and services. What might not be so logical is to find product and service verification requirements under the title *Release of products and services* because it addresses product and service verification at appropriate stages, not simply their release implying an overlap with the requirement in clause 8.5.1 for the implementation of monitoring and measurement activities at appropriate stages to verify that acceptance criteria for products and services have been met.

Delivery requirements in the 1987 version were limited to protecting and identifying the product, and there were no requirements concerning post-delivery. In the 2008 version determination of delivery and post-delivery requirements were included among customer-related processes with their implementation addressed among the general requirements on production and service provision and product monitoring and measurement.

The 2015 version does bring some clarity to these requirements by placing release and post-delivery under separate headings, but delivery is not prominent. The order of these requirements in the 2015 version is also not logical with implementation of delivery and post-delivery requirements in clause 8.5.1, post-delivery in clause 8.5.5, followed by release requirements in clause 8.6. For clarity and convenience, these topics are addressed together in one chapter. The term *service delivery* is often interchangeable with *service provision*, making the phrase *product and service delivery* have a different meaning. The term *service fulfilment* has been used instead to indicate the final stage of providing a service that is equivalent to the delivery of a product.

However, it emphasizes the problem the authors of ISO 9001 have in presenting requirements in an order relative to a life cycle or an order relative to their objective or subject matter or whether they are generic or specific in a group. The standard still contains a mix of styles.

In this chapter, we examine the five requirements of clauses 8.5 and 8.6 that address the release, delivery and post-delivery of products and services, namely:

* Verifying product and service requirements have been met (8.6)
* Authorizing release of products and services (8.5.1h and 8.6)

- Retaining evidence of conformity (8.6)
- Product delivery and service fulfilment (8.5.1h)
- Post-delivery activities (8.5.1h and 8.5.5)

Verifying product and service requirements have been met (8.6)

What does this mean?

Activities that monitor and measure product and service are often referred to as inspection, test or verification activities. Appropriate stages mean the stages at which:

- the achieved characteristics are accessible for measurement;
- an economic means of measurement can be performed;
- the correction of error is less costly than if the error is detected at later stages.

It may be possible to verify some characteristics on the final product just prior to shipment, but it is costly to correct errors at this late stage resulting in delayed shipment. It is always more economical to verify product at the earliest opportunity.

The *planned arrangements* in this case are the plans made for verifying product and service in terms of what is to be verified, who is to verify it, when is it to be verified, how is to be verified, where is it to be verified and what criteria is to be used to judge conformity.

Product and service requirements are all the requirements for the product or service including customer, regulatory and the organization's requirements (see Chapter 35). Some of these may be met by inherent design features, others will be met in production, installation or in service.

The forms of verification that are used in product and service development should also be governed by these requirements as a means of ensuring that the product or service on which design verification is carried out conforms with the prescribed requirements. If the product or service is non-compliant, it may invalidate the results of design verification. Product and service verification also applies to any measuring and monitoring devices that you design and manufacture to ensure that they are capable of verifying the acceptability of product or service as required.

Why is this necessary?

One verifies products and services to establish that they meet requirements. If one could be certain that a product or a service would be correct without it being verified, verification would be unnecessary. However, most processes possess inherent variation due to common causes – variations that affect all values of process output and appear random. Although a process may be under statistical control, a special event could disturb performance and without checks on the output, its detection may go unnoticed. One can only check for those events we think might happen which is why our confidence in the system is shaken when we discover a condition with a cause we had not predicted.

How is this addressed?

As product and services requirements may include characteristics that are achieved by design, production, installation or service delivery, a high-level verification matrix is needed to provide traceability from requirement to the means of verification. This will undoubtedly lead to there being a few characteristics that need to be verified only once by design verification with many of the others being verified in production or service delivery. Characteristics that do not vary only need to be checked once. For example, a car designed with four wheels could not possibly be made with two, three or five wheels when put into production, but a body panel with screw inserts could emerge from the process without the inserts or a patient may emerge from hospital without surplus screw inserts being removed!

Having established that characteristics vary, the stage at which they need to be verified should be determined. This leaves three possibilities: on receipt, in-process or on completion. Receipt verification was addressed in Chapter 42. Here we address in-process and finished product verification.

In-process verification

In-process verification is carried out to verify those features and characteristics that would not be accessible to verification by further processing or assembly. When producing a product that consists of several parts, sub-assemblies, assemblies, units, equipment and subsystems, each part, sub-assembly, etc., needs to be subject to final verification but may also require in-process verification for the reasons given earlier. Your control plans should define all the in-process verification stages that are required for each part, sub-assembly, assembly, etc. In establishing where to carry out the verification, a flow diagram may help. The verification needs to occur after a specified feature has been produced and before it becomes inaccessible for measurement. This doesn't mean that you should check features as soon as they are achieved. There may be natural breaks in the process where the product passes from one stage to another or stages at which several features can be verified at once. If product passes from the responsibility of one person to another, there should be a stage verification at the interface to protect the producer even if the features achieved are accessible later. Your verification plans should:

- identify the product to be verified;
- define the specification and acceptance criteria to be used and the issue status which applies;
- define what is to be verified at each stage. (Is it all work between stages or only certain operations? The parameters to be verified should include those that are known to be varied by the manufacturing processes. Those that remain constant from product to product need verifying once only usually during design proving.);
- define the verification aids and test equipment to be used. (There may be jigs, fixtures, gauges and other aids needed for verification. Standard measuring equipment would not need to be specified because your verification staff should be trained to select the right tools for the job. Any special measuring devices should be identified.);
- define the environment for the measurements to be made if critical to the measurements to be made;
- identify the organization that is to perform the verification;
- make provision for the results of the verification to be recorded.

Finished product verification

Finished product verification is in fact the last verification of the product that you will perform before dispatch but it may not be the last verification before delivery if your contract includes installation. There are three definitions of finished product verification:

- The verification carried out on completion of the product; afterwards the product may be routed to stores rather than to a customer.
- The last verification carried out before dispatch; afterwards you may install the product and carry out further work.
- The last verification that you as a supplier carry out on the product before ownership passes to your customer; this is the stage when the product is accepted and consequently the term *product acceptance* is more appropriate and tends to convey the purpose of the verification rather than the stage at which it is performed.

There are two aspects to finished product verification. One is checking what has gone before and the other is accepting the product.

Final verification and test checks should detect whether:

- All previous verification activities have been performed.
- The product bears the correct identification, part numbers, serial numbers, modification status, etc.
- The as-built configuration is the same as the revision status of all the parts, subassemblies, assemblies, etc., specified by the design standard. A configuration record containing this data would avoid argument later as to whether certain specification changes were embodied in the product.
- All recorded nonconformities have been resolved and verified.
- All concession applications have been approved.
- All verification results have been collected.
- Any result outside the stated limits is either subject to an approved concession, an approved specification change or a retest that shows conformance with the requirements.
- All documentation to be delivered with the product has been produced and conforms to the prescribed standards.

Whenever a product or service is supplied, produced or repaired, rebuilt, modified or otherwise changed, it should be subject to verification that it conforms to the prescribed requirements and any deficiencies corrected before being released for use.

How is this demonstrated?

Demonstrating that planned arrangements for verifying that product and service requirements have been met have been implemented may be accomplished by:

a) presenting evidence of processes for verifying products and services at appropriate stages through production and service provision;

b)   selecting a representative sample of products and services that have not yet reached verification and presenting evidence of planned verification that is consistent with the pre-defined processes;

c)   selecting a representative sample of products and services that are at a stage after a planned verification stage and presenting evidence that the planned verification was carried out.

Authorizing release of products and services (8.5.1h and 8.6)

What does this mean?

The *planned arrangements* in this case are the plans made for producing the product or providing the service as addressed in Chapter 33 under operations planning.

   This requirement can impose unnecessary constraints if taken literally. Many activities in planned arrangements are performed to give early warning of nonconformities. This is to avoid the losses that can be incurred if failure occurs in later tests and inspections. The earlier you confirm conformance, the less costly any rework will be. One should therefore not hold shipment if later activities have verified the parameters, whether earlier activities have been performed. It is uneconomic for you to omit the earlier activities, but if you do, and the later activities can demonstrate that the end product meets the requirements, it is also uneconomic to go back and perform those activities that have not been completed. Your planned arrangements could cover installation and maintenance activities which are carried out after dispatch and so it would be unreasonable to insist that these activities were completed before dispatch or to insist on separate plans just to sanitize a point. A less ambiguous way of saying the same thing is to require no product to be dispatched until objective evidence has been produced to demonstrate that it meets the product requirements and that authorization for its release has been given.

Why is this necessary?

Having decided on the provisions needed to produce product that meets the needs and expectations of customers, regulators and the organization itself, it would be foolish to permit release of product before confirming that all that was agreed to be done has been done. However, circumstances may arise where nonconforming product has been produced and instead of shipping such product without informing the intended recipient, an organization committed to quality would seek permission to do so.

How is this addressed?

You need four things before you can release product whether it be to the stores, to the customer, to the site for installation or anywhere else:

1   Sight of the product.
2   Sight of the requirement with which the product is to conform including its packaging, labelling and other product related requirements.
3   Sight of the objective evidence that purports to demonstrate that the specific product meets the requirement.
4   Sight of an authorized signatory or the stamp of an approved stamp-holder who has checked that the specific product, the evidence and the requirement are in complete accord.

Once the evidence has been verified, the authorized person can make the release decision and endorse the appropriate record indicating readiness for release. Should there be any discrepancies, they should be validated and if proven valid, the nonconforming product process should be initiated.

If planned arrangements cannot be achieved, a concession might be obtained from the recipient to permit release of product that did not fully meet the requirements. The recipient could be the owner of the process receiving product for processing or the external customer receiving product in response to an order.

How is this demonstrated?

Demonstrating that products and services are not released to the customer until the planned arrangements have been satisfactorily completed or otherwise approved may be accomplished by:

a)   presenting evidence of a process for releasing product or service;
b)   selecting a representative sample of delivery receipts and presenting evidence that:

i   all the planned activities had been carried out and satisfactorily completed;
ii   any activities not carried out were either not necessary because subsequent events were confirmed correct or were approved by a relevant authority.

Retaining evidence of conformity (8.6)

What does this mean?

Evidence of conformity is the information recorded during product verification that shows the product to have exhibited the characteristics required.

Why is this necessary?

At a point in the process, product will be presented for delivery to the next stage in the process or to a customer. At such stages a decision is made whether to release product and this decision needs to be made based on facts substantiated by objective evidence.

How is this addressed?

This requires that you produce something like an acceptance plan which contains, as appropriate, some or all the following:

*   Identity of the product to be verified.
*   Definition of the specification and acceptance criteria to be used and the issue status that applies.
*   Definition of the verification aids and measuring devices to be used.
*   Definition of the environment for the measurements to be made.
*   Provision for the results of verification to be recorded – these need to be presented in a form that correlates with the specified requirements.

Having carried out these verification activities, it should be possible for you to declare that the product has been verified and objective evidence produced that will demonstrate that it meets the specified requirements. Any concessions given against requirements should also be identified. If you can't make such a declaration, you haven't done enough verification. Whether your customer requires a certificate from you testifying that you have met the requirements, you should be able to produce one. The requirement for a certificate of conformance should not alter your processes, your quality controls or your procedures. Your QMS should give you the kind of evidence you need to assure your customers that your product meets their requirements without having to do anything special.

Your verification records should be of two forms: one which indicates what verification activities have been carried out and the other which indicates the results of such verification. They may be merged into one record but when parameters need to be recorded it is often cleaner to separate the progress record from the technical record. Your procedures, quality plan or product specifications should also indicate what measurements should be recorded.

Don't assume that because a parameter is shown in a specification that an inspector or tester will record the result. A result can be a figure, a pass or fail or just a tick. Be specific in what you want recorded because you may get a surprise when gathering the data for analysis. If data collection is computerized, you shouldn't have the same problems but beware, too much data is probably worse than too little! In choosing the method of recording measurements, you also need to consider whether you will have sufficient data to minimize recovery action in the event of the measuring device subsequently being found to be out of calibration (see 7.1.5.2). As a rule, only gather that data you need to determine whether the product meets the requirements or whether the process is capable of producing a product that meets the requirements. You need to be selective so that you can spot the out of tolerance condition or variation in the measurement system. Sometimes, plotting the results as a histogram might indicate abnormalities in the results that are symptomatic of measurement errors. The acceptance criterion is therefore not simply specified upper and lower limits but evidence that results are located in a normal distribution that is centred on the nominal condition.

All verification records should define the acceptance criteria, the limits between which the product is acceptable and beyond which the product is unacceptable and therefore nonconforming.

How is this demonstrated?

Demonstrating that evidence of conformity has been retained on the release of products and services may be accomplished by:

a)   presenting evidence of a process for capturing documented information on product release activities;
b)   selecting a representative sample of delivery receipts and presenting evidence that details:

    i   the identity of the products released;
    ii   the criteria used for their acceptance;
    iii   the verification carried out to determine conformity;
    iv   the identity of the person who authorized release.

Product delivery and service fulfilment (8.5.1h)

What does this mean?

Delivery is an activity that serves the shipment or transmission of product to the customer and is one part of the distribution process. Delivery may include preparation for delivery such as packing, notification, transportation, customs, arrival at destination and unpacking on customer premises. In the consumer goods market, there may be intermediaries such as agents, wholesalers, retailers, resellers, etc., which exist to distribute product to the customer. In the service delivery process this means the fulfilment of the service and may include transmission of information and payment mechanisms.

Why is this necessary?

The process of moving goods from producer to customer is an important process in the QMS. Although good product design, economic production and effective promotion are vital for success, these are useless if the customer cannot access the product and take ownership without hassle. It is necessary to control delivery activities because conforming product may be degraded by the way it is protected during transit. It may also be delayed by the way it is transported. You may be under an obligation to supply product by certain dates or within so many days of order and therefore control of the delivery process is vital to honour these obligations

How is this addressed?

A distribution process needs to be designed which responds to customer demand and distributes products from storage locations to customers in a way that ensures they arrive in good condition.

The distribution or marketing channel promotes the physical flow of goods and services along with ownership title, from producer to consumer or business user. Often the logistics for moving goods to outlets where consumers can purchase them is a separate business but nevertheless starts out in the demand creation process when determining the distribution strategy. There are several different distribution channels depending on the type of products and services and the market into which they are to be sold, for example,

- Products may be sold directly to consumers and business users or through retailers via agents or warehouses.
- Services may be sold directly to consumers and business users or through agents.

Delivery takes place between each of the parties in the distribution chain and for each party there are several aspects to the delivery process:

- Preparation of product such as cleaning and preservation
- Packing of product
- User information
- Product certification

- Labelling and transit information
- Handling
- Customer notification
- Transportation
- Tracking

Preparation and packing of product is addressed under preservation in Chapter 48 as the methods also apply to internal processing. However, within the delivery process there will be specific packing stages that are different in nature to internal packing stages.

Sometimes, delivery is made electronically. The product may be a software package or a document stored in electronic form. Protection of the product is still required but takes a different form. You need to protect the product against loss and corruption during transmission or downloading.

When shipping consumer goods it is necessary to include user information such as operating instructions, handbooks, warranty and return instructions.

Customers may require product certificates testifying the fulfilment of contracts or order requirement. Customs may require certain legal information on the outside of the package otherwise the consignment will be held at the port of entry and customers will be none too pleased.

The type of transport employed is a key factor in getting shipments to customers on time.

On-time delivery

To guarantee shipment on time, you either need to maintain an adequate inventory of finished goods, for shipment on demand or utilize only predictable processes and obtain sufficient advanced order information from your customer. Without sufficient lead-time on orders you will be unlikely to meet the target. There will be matters outside your control and matters over which you need complete control. It is the latter that you can do something about and take corrective action should the target not be achieved.

First, you need to estimate the production cycle time during the production trial runs in the product and process validation phase, assess risk areas and build in appropriate contingencies. An assessment of your supplier's previous delivery performance will also enable you to predict their future performance. When new processes become stabilized over long periods and the frequency of improvement reduces as more and more problems are resolved, you will be able to reduce lead-time.

Your planning and delivery procedures need to record estimated and actual delivery dates and require the data to be collected and analysed through delivery performance monitoring. When targets are not met, you should investigate the cause under the corrective action procedures and formulate corrective action plans. Where the cause is found to be a failure of the customer to supply some vital information or equipment, it would be prudent not to wait for the periodic analysis but react promptly.

Customer notification

A means for notifying the customer of pending delivery is often necessary. Your organization might be linked with the customer electronically so that demands are transmitted from the customer to trigger the delivery process. However, the customer may need to change quantities and delivery dates due to variations in production. This does not mean that the changes will always be to shorten delivery times, but on occasion the delivery times may need to be extended due to problems on the assembly line or because of problems with other suppliers. The customer may not have made provision to store your product so needs to be able to urgently inform you to hold or advance deliveries. If the customer reduced the quantity required from that previously demanded, you could be left with surplus product and consequently need protection through the contract for such eventualities.

How is this demonstrated?

Demonstrating that the organization has planned and has control of product delivery activities may be accomplished by:

a) presenting evidence of delivery requirements the organization is committed to satisfy;
b) presenting evidence of a delivery process that has been designed to deliver products that meet customer requirements;
c) selecting a sample of delivery activities and presenting evidence that:

  i   the processes are being carried out in accordance with the prescribed policies and procedures;
  ii the processes are delivering conforming product to the correct destination;
  iii customer confirms receipt of consignment in good condition.

Post-delivery activities (8.5.1h and 8.5.5)

What does this mean?

Post-delivery activities are those performed after ownership passes from the organization to the customer. These may be post-installation if an installation service was included in the contract. They include return material or merchandise authorization (RMA) or return goods authorization (RGA) where a customer returns a product to receive a refund, replacement, or repair during the product's warranty period.

Why is this necessary?

Control of post-delivery activities is just as important as pre-delivery if not more so as the customer may be losing use of the product and want prompt resolution to the problems encountered. Post-delivery performance is often the principal reason why customers remain loyal or choose a competitor. Even if a product does give trouble, a sympathetic, prompt and courteous post-delivery service can restore confidence.

How is this addressed?

The wide range of post-delivery services makes a detailed analysis impractical in this book. However, some simple measures can be taken that would apply to all types of post-delivery activities:

•   Define the nature and purpose of the post-delivery service.
•   Define post-delivery policies that cover such matters as handling complaints, offering replacement product, service staff conduct, end-of-life disposal considering the duration over which the organization is committed to provide the service.

- Establish conditions of post-delivery services, including as applicable:

  - Customer feedback
  - Technical support
  - Warranty claims
  - Servicing
  - Repair
  - Product recall
  - Hazard alerts
  - Recycling and disposal

- Specify objectives and measures for service features such as response time, resolution time.
- Review the statutory and regulatory requirements that may apply following delivery (e.g. consumer protection, disposal, product recall) and ensure the post-delivery process takes these into account.
- Communicate the policies and objectives and ensure their understanding by those involved.
- Define the stages in the process needed to achieve these objectives.
- Identify the information needs and ensure control of this information.
- Identify and provide the resources to deliver the service.
- Plan verification stages to verify achievement of stage outputs.
- Provide communication channels for feeding intelligence into production and service design processes.
- Make provision for alerting customers to hazards associated with its products that it was unaware of at the time of delivery.
- Determine methods for measuring process performance.
- Measure process performance against objectives.
- Determine the capability of the process and make changes to improved performance.
- Determine process effectiveness and pursue continual improvement.

Although the ISO 9001 requirement for retention applies only to records, you may also need to retain tools, jigs, fixtures, and test software – in fact anything that is needed to diagnose failures, repair or reproduce equipment to honour your long-term commitments.

How is this demonstrated?

Demonstrating that the organization has determined and has control over post-delivery activities may be accomplished by:

a) presenting evidence of post-delivery requirements the organization is committed to satisfy to its customers;

b) presenting evidence of post-delivery processes that have been designed to deliver outputs that meet the post-delivery requirements;

c) selecting a sample of post-delivery activities and presenting evidence that:

  i   the processes are being carried out in accordance with the prescribed policies and procedures;
  ii the processes are delivering outputs that fulfil the process objectives;
  iii customers are satisfied with the service being provided.

51   Control of nonconforming outputs

Introduction

Nonconformities are caused by factors that should not be present in a process. There will always be variation, but variation is not nonconformity. Nonconformity arises when the variation exceeds the permitted limits. The factors that cause nonconformity on one occasion will (unless removed) cause nonconformity again and again. As the objective of any process must be to produce conforming output, it therefore follows that it is necessary to control nonconforming output.

During the execution of a process work is undertaken and objects produced. These objects only become outputs (*a result of a process*) when put out of the process. At any other stage, it's work-in-progress. A process will produce many results, but those outputs that are intended to be subject to the requirements of clause 8.7 are the outputs of operational processes which are destined:

a)   for delivery to a customer or;
b)   for incorporation into a product or service that is destined for delivery to a customer or;
c)   which materially affects conformity of products and services destined for delivery to a customer.

They are intended to apply to operations processes and not management processes. The requirements are also not intended to be applied to tools and equipment used in production or service delivery.

In the 2008 version, these requirements were included under *Measuring and monitoring* rather than *Product realization* because control of nonconforming product could not be excluded from the QMS as could other requirements of Section 7. This condition no longer applies.

Nonconforming output control represents a feedback loop within a process for handling outputs that failed to pass verification checks. This is more pertinent to production processes than to service delivery processes where there is often no intermediate step between producing an output and delivering it to a customer (e.g. a teacher delivers a lesson to the students directly and not to some intermediary who checks it before it's passed on to the student). These are special processes and are addressed in Chapter 45 under 8.5.1f). There are some services where outputs can be checked before delivery such as those that provide documented information.

The change in terminology in the 2015 version from *product* to *outputs, products and services* makes the sentence structure clumsy in some cases, and therefore when the distinction between output, product or service does not need to be made the word *item* will be used instead.

In this chapter, we examine four requirements of clause 8.7, namely:

- Preventing unintended use (8.7.1)
- Taking appropriate action (8.7.1)
- Verifying corrected nonconformities (8.7.1)
- Retaining information of actions taken on nonconforming process outputs (8.7.2)

Preventing unintended use (8.7.1)

What does this mean?

Outputs that do not conform to their requirements are those that have been examined against pre-defined requirements and judged to be at variance with those requirements either at a verification stage or a subsequent stage. The requirements are not limited to customer requirements (see also Chapter 24), and therefore a nonconforming output is one that fails to meet one or more of the following:

- the specified customer requirements.
- the applicable regulatory requirements.
- the intended usage requirements.
- the stated or implied needs.
- the organization's own requirements.
- the customer expectations.

There are three different types of nonconformity as shown in Box 51.1. The requirements of clause 8.7 apply only to actual and suspect nonconformities.

Box 51.1   Types of nonconformity

Potential nonconformity is when conditions are present that may cause a nonconformity if no action is taken. For example, (1) a process that will produce nonconforming output if not arrested or (2) a design that produces an output that will become nonconforming under certain conditions of use.

Actual nonconformity is a verified non-fulfilment of a requirement. For example, (1) when measuring an output's characteristic, the observed value is different from the required value or (2) when subjecting an output to the specified operating conditions, it fails to function as required.

Suspect nonconformity is when there is the possibility that conditions could have caused nonconformity. For example, (1) an output from a batch in which some have been found nonconforming; (2) an output that possesses some of the same characteristics as the nonconforming output; or (3) an output that has been accepted using equipment subsequently found out of calibration; or (4) output is mishandled but shows no obvious signs of damage; or (5) an output with no indication of verification status.

Why is this necessary?

Nonconforming outputs need to be prevented from use or delivery to maintain an uninterrupted flow through the production and service delivery processes. Process owners expect process inputs to conform to requirements therefore were nonconforming outputs to be released they would create problems downstream leading to process and product failures closer to the point of delivery to customers. If they weren't detected before delivery the organization would be put in the embarrassing position of knowingly supplying nonconforming product or service.

How is this addressed?

The only sure way of preventing inadvertent use of nonconforming items is to destroy them, but that may be a little drastic in some cases. It may be possible to eliminate the nonconformity by repair, completion of processing or rework. A more practical way of preventing the inadvertent use or installation of a nonconforming item is to identify the item as nonconforming and then place it in an area where access to it is controlled. These two aspects are covered further below.

Items suspected of being nonconforming should be treated in the same manner as items that have been deemed nonconforming and quarantined until dispositioned. However, until nonconformity can be proven, the documentation of the nonconformity merely reveals the reason for the item being suspect.

Identifying nonconforming outputs

The most common method is to apply labels to the item that are distinguishable from other labels. It is preferable to use red labels for nonconforming items and green labels for conforming items. In this way, you can determine their status at a distance and reduce the chance of confusion. You can use segregation as a means of identifying nonconforming items but if there is the possibility of mixing or confusion then this means alone should not be used.

The labels should identify the item by name and reference number, specification and version if necessary and either a statement of the nonconformity or a reference to the nonconformity report containing full details of its condition. Finally, the person or organization testifying the nonconformity should be identified either by name or code.

Controlling nonconforming outputs

To control a nonconforming item, you need to:

a) know when it became nonconforming;
b) know who decided it was nonconforming;
c) know of its condition;
d) know where it is located;
e) know that it is unable to be used.

On detection of a nonconformity, details of the item and the nonconformity should be recorded to address (a), (b) and (c) earlier. A nonconformity report template is available on the companion website.

Segregating a nonconforming item (or separating good from bad) places it in an area with restricted access and addresses (d) earlier. Such areas are called quarantine areas or quarantine stores. Items should remain in quarantine until disposal instructions have been issued. The store should be clearly marked and a register maintained of all items that enter and exit the store to track their movements. Where items are too large to be moved into a quarantine store or area, measures should be taken to signal to others that the item is not available for use. Cordons or floor markings can achieve this. With services the simplest method is to render the service unavailable or inaccessible.

How is this demonstrated?

Demonstrating that outputs that do not conform to their requirements are identified and controlled to prevent their unintended use or delivery may be accomplished by:

a)   presenting evidence of a process for controlling the use of nonconforming items;
b)   selecting a representative sample of nonconforming items and showing that they have been identified and controlled in accordance with the pre-defined process;
c)   retrieving records of nonconforming items and presenting evidence that shows the items were not used or delivered.

Taking appropriate action (8.7.1)

What does this mean?

Some terms within the requirement warrant explanation.

Appropriate action

Appropriate action means taking an action that is not detrimental to conformity of the product or service. In some cases, it may be cost effective to eliminate the nonconformity, and in others it may not be cost effective and therefore such action would be inappropriate and the item would be disposed of.

Correction

Correction is an action taken to remove the detected nonconformity, not to be confused with corrective action which removes the cause of the nonconformity to stop it happening again.

Concession

Where the nonconformity may have little or no effect on the form, fit or function of the product or service concerned it may be offered to the acceptance authority for acceptance as is (i.e. without correction). If accepted, the acceptance authority grants permission (a concession) to use or release a product or service that does not conform to specified requirements.

Segregation

Segregation is an action taken to remove a nonconforming item from the process flow while a decision is made on what to do with it. This is like *custody.*

Containment action

Containment action is an action taken to limit the extent of a nonconformity by removing an immediate cause thus allowing normal operations to continue until product, service or process design change removes the root cause. This is like *first aid*.

Nonconformities detected during or after delivery

These are nonconformities detected by the organization during or after delivery of a product or service. With capable processes, samples are taken for testing and a batch may be released based on historic data that there is only a slight risk that the sample would fail the test. In the event the test results are negative, the items in transit or that have arrived at their destination are deemed to contain nonconforming items.

Why is this necessary?

A blanket approach to all deviations from requirements would not be economically viable, and over the last 50 years or so the range of options identified within the requirement has emerged as a more pragmatic way of dealing with this issue.

How is this addressed?

Determining appropriate action

Determining the action to take depends on whether:

a)   it's an isolated incident;
b)   the item is one of a batch of identical items;
c)   the item is one of a batch of similar items possessing the same characteristic that is nonconforming and;
d)   items of the same specification are in transit to customers, have already been delivered or are in use.

Box 51.2   Examples of correction

- While observing the performance of a process you notice that the values are drifting towards the upper limits. You adjust the process and bring it back under control. You have corrected the process and avoided an occurrence of product non-conformity but not a future occurrence; you have merely delayed its occurrence
- You are delivering a service and the customer points out an error which you correct immediately.
- Sometimes a product may be inadvertently submitted for verification before all operations have been completed and in such cases correction involves resuming normal operations to complete the item.
- A product can be made to conform by continuation of processing; this type of correction is called rework.
- An unknown state can be corrected by carrying out verification and declaring the product conforming or nonconforming.

You will also need to consider when the nonconformity was detected. A nonconformity may be detected:

a) prior to release and not implicate any other items that have been produced or;
b) following its release by a subsequent user of the item within the organization or;
c) prior to release and implicate items already in use such as when subsequent analysis reveals inaccurate measurements or when verification methods or acceptance criteria change.

Such items may not have failed in service because it has not been used in a manner needed to cause failure but if part of the same batch or lot contains a common cause of nonconformity all the items are suspected. Action taken because of latent nonconformity may involve product recall, product alerts or the issue of instructions for correction.

When making the decision your action therefore needs to address:

•   action on the nonconforming item to remove the nonconformity;
•   a search for other similar items which may be nonconforming (i.e. suspect product);
•   action to recall product containing suspect nonconforming product.

If a process is under statistical process control, don't ascribe the nonconformity to a special cause and adjust the process (see Chapter 58).

Correction

When an item is found to be nonconforming, there are three decisions you need to make based on the following questions:

•   Can the item be made to conform?
•   If the item cannot be made to conform, is it fit for use?
•   If the item is not fit for use, can it be made fit for use?

The authority for making these decisions will vary depending on the answer to the first question. If, regardless of the severity of the nonconformity, the item can be made to conform simply by rework or completing operations, these decisions can be taken by operators or inspectors, providing rework is economical. Decisions on scrap, rework and completion would be made by the fund-providing authority rather than the design authority. If the product cannot be made to conform by using existing specifications, decisions requiring a change or a waiver of a specification should be made by the authority responsible for drawing up or selecting the specification.

It may be sensible to engage investigators to review the options to be considered and propose actions for the authorities to consider. In your procedures, you should identify the various bodies that need to be consulted for each type of specification. Departures from customer requirements will require customer approval; departures from design requirements will require design approval; departures from process requirements will require process engineering approval, etc. The key lies in identifying who devised or selected the requirement in the first place.

To deal with corrections, you will need a method of recording the decision and assigning the responsibility for the action to be taken. These documents also need to stipulate the verification requirements to be implemented following any correction. When deciding on correction, you may need to consider whether the result will be visible to the customer on the exterior of the product. Rework or repairs that may not be visible when a part is fitted into the final assembly might be visible when these same parts are sold as service spares. To prevent on-the-spot decisions being at variance each time, you could:

- identify in the specifications those products that are supplied for service applications, i.e. for servicing, maintenance and repair;
- provide the means for making rework invisible where there are cost savings over scrapping the item;
- stipulate on the specifications the approved rework techniques.

Requesting concessions

If the requirements cannot be achieved at all then this is not a situation for a concession but a case for a change in requirement. If you know in advance of producing the product or service that it will not conform to the requirements, you can then request a deviation from the requirements. This is often referred to as a *production permit* or *deviation*. Concessions apply after the product has been produced. Production permits or deviations apply before it has been produced. Both are requests that should be made to the acceptance authority for the product. The relevant authority is the authority that specified the requirement that has not been met. This authority could therefore be the customer, the regulator or the designer.

All specifications are but a substitute for knowledge of fitness for use. Any departure from such specification should be referred to the specification authors for a judgment. To determine whether a nonconforming product could be used, an analysis of the conditions needs to be made by qualified personnel. There are two ways of doing this. Either you refer all such nonconformities to the relevant authority or the authority appoints representatives who can make these decisions within prescribed limits. A traditional method is to classify nonconformities, assign authority for accepting concessions for each level and define the limits of their authority. These levels could be as follows:

- **Critical Nonconformity**. A departure from the requirements which renders the product or service unfit for use.
- **Major Nonconformity**. A departure from the requirements included in the contract or customer specification.
- **Minor Nonconformity**. A departure from the requirements not included in the contract or customer specification.

The only cases where you need to request concessions from your customer are when you have deviated from one of the customer requirements and cannot make the product conform. When you repair a product, provided it meets all the customer requirements, there is generally no need to seek a concession from your customer. Although it is generally believed that nonconformities indicate an out-of-control situation, if you detect and rectify them before release of the product, you have quality under control and have no need to report nonconformities to your customer. However, if the frequency of nonconformity exceeds process capability targets, the process has become unstable and requires corrective action.

In informing your customer when nonconforming product has been shipped you obviously need to do this immediately when you are certain that there is a nonconformity. If you are investigating a suspect nonconformity it only becomes a matter for reporting to your customer when the nonconformity remains suspect after you have concluded your investigations. Alerting your customer every time, you think there is a problem will destroy confidence in your organization. Customers appreciate zeal, but not paranoia!

Production permits or deviations are generally permitted for specific batches or a defined time. This is to allow time for corrective action to be taken. It is therefore necessary to keep a log of the items and quantities produced that are subject to the production permit or authorized deviation. It is also necessary to ensure that when the batch or date when the corrective action becomes effective arrives, the production permit or deviation is withdrawn. Flags should be inserted into production schedules alerting planners to batches that are subject to authorized concession or production permit and when the date or batch beyond which authorization is invalid arrives, the flags are removed.

When delivery subject to authorized concession or production permit commences, the packaging should be duly annotated.

Regrading

Not in every case need a nonconforming item be either scrapped, corrected or accepted on a concession. One alternative might be to regrade it so that it may be used in other applications. In some cases, products are offered in several models, types or other designations but are basically of the same design. Those which meet the higher specification are graded as such and those which fail may meet a lower specification and can be regraded. The grading should be reflected in the product identity so that there is no confusion.

Regrading can be accomplished by assigning a new identity to the product. Scrapping an item should not be taken lightly – it could be an item of high value. Scrapping may be an economical decision with low-cost items, whereas the scrapping of high-value items may require prior authorization as salvage action may provide a possibility of yielding spares for alternative applications.

Suspension of production

In some cases, the nonconformity may be so severe that it is necessary to suspend production until containment action has been put in place. It may necessitate reverting to 100% inspection until temporary measures have been proven effective. Those given this responsibility and authority should be communicated. For continuous production lines, its common practice to give this authority to the person on the spot, running the line because in the time taken to alert supervision, vast quantities of product may have been produced and have to be destroyed.

Suspension of service

Unlike products, nonconforming services are usually rendered unavailable for use by notices such as *Out of Order* or by announcements such as *Normal service will be resumed as soon as possible*. Products are often capable of operation with nonconformities, whereas services tend to be withdrawn or suspended once the nonconformity has been detected, however trivial the fault. With services, such as the supply of utilities, cash dispensing and any other service vulnerable to interruption due to the weather, vandalism or terrorism contingency plans need to be in place to make the situation as painless as possible because suspending a service may cause serious problems for certain groups of people, particularly the vulnerable. Those given this responsibility and authority should be communicated.

Action on nonconformities detected after release

Nonconformities detected after release indicate that the controls in place are not effective and should give cause for concern. Details should be recorded and an investigation conducted to establish why the planned verification did not detect the problem. Action should then be taken to improve the verification methods by changing procedures, acceptance criteria, equipment or retraining personnel.

When a nonconformity is detected by verification personnel in an item where items of the same type are in use, an analysis is needed to establish whether the nonconformity would previously have escaped detection. If not, there is no cause for alarm, but if something has now changed to bring the nonconformity to light, an evaluation of the consequences needs to be conducted. It may only be a matter of time before the user detects the same nonconformity.

The procedures should cover:

- the rationale for notifying the customer and determining the appropriate course of action;
- the method of receiving and identifying returned product;
- the method of logging reports of nonconformities from customers and other users;
- the process of responding to customer requests for assistance;
- the process of dispatching service personnel to the customer's premises;
- a process for investigating the nature of the nonconformity;
- a process for replacing, or repairing nonconforming product and restoring customer equipment into service;
- a process for assessing all products in service that are nonconforming, determining and implementing recall action if necessary.

If you need to recall product that is suspected as being defective you will need to devise a recall plan, specify responsibilities and timescales and put the plan into effect. Product recall is a *correction*, not a *corrective action*, because it does not prevent a recurrence of the initial problem.

It is, of course, a matter for the organization's management to decide the appropriate action. However, it would not be conducive to strong customer relations if you were to neglect to inform the customer of any nonconformity with customer specified or legal requirements.

How is this demonstrated?

Demonstrating that appropriate action is taken based on the nature of the nonconformity and its effect on the conformity of products and services may be accomplished by:

a) presenting evidence of a process for determining the action to take on nonconforming outputs detected before, during and after their delivery;
b) presenting the criteria used in deciding the actions to be taken;
c) selecting a representative sample of nonconforming items and presenting evidence that:

  i production and service delivery was suspended while containment actions were put in place;

ii   the customer was promptly informed about nonconformities in products or services that they had received and notifies of the action being taken;

iii   authorization was obtained from the customer before delivering a product that had been deemed nonconforming with their requirements;

iv   the actions taken proceeded as planned.

Verifying corrected nonconformities (8.7.1)

What does this mean?

Any action taken to correct the nonconformity will change it and therefore it needs to be subject to re-verification. This may involve verification against different requirements to the original requirements.

Why is this necessary?

If a nonconforming item is accepted without correction, no re-verification is necessary, but if the item is changed the previous verification is no longer valid.

How is this addressed?

Any item that has had work done to it should be re-verified prior to it being released to ensure the work has been carried out as planned and has not affected features that were previously found conforming. There may be cases where the amount of re-verification is limited and this should be stated as part of the recovery action plan. However, after rework or repair the re-verification should verify that the product meets the original requirement; otherwise, it is not the same item and must be identified differently.

The verification records should indicate the original rejection, the disposition and the results of the re-verification in order that there is traceability of the decisions that were made.

How is this demonstrated?

Demonstrating that requirements are verified when nonconforming outputs are corrected may be accomplished by:

a)   presenting evidence of a policy for verifying nonconforming outputs following their correction;

b)   selecting a representative sample of nonconforming items and presenting evidence that verification has been conducted in accordance with the prescribed policy.

Retaining information of actions taken on nonconforming process outputs (8.7.2)

What does this mean?

Documented information that describes the nonconformity are the documented details of the output (its identity), the specific deviations from requirements (what it is and what it should have been), the conditions under which the nonconformity was detected (the environmental or operating conditions – what was happening at the time when the nonconformity was detected), the time and date of detection, the name of the person detecting it and the actions taken with reference to any instructions, revised requirements and decisions.

Why is this necessary?

Records of nonconformities are needed for presentation to the authorities responsible for deciding on the action to be taken and for subsequent analysis. Without such records, decisions may be made on opinion resulting in the means for identifying opportunities for improvement being absent.

How is this addressed?

There are several ways in which you can document the presence of a nonconformity.
   You can record the condition:

*   on a label attached to the item;
*   on a form unique to the item such as a nonconformity report;
*   of functional failures on a failure report and physical errors on a defect report;
*   in a logbook for the item such as an inspection history record or snag sheet;
*   in a logbook for the workshop or area.

The detail you record depends on the severity of the nonconformity and to whom it needs to be communicated. In some cases, a patrol inspector or quality engineer can deal with minor snags daily as can an itinerant designer. Where the problem is severe and the necessary action complicated, a panel of experts may need to meet. The description of the nonconformity and the actions taken to deal with it are often combined on one form separate to the corrective action because the corrective action may not be determined concurrently with the other action for various reasons. It is important when documenting the nonconformity that you record as many details as you can because they may be valuable to any subsequent investigation to help diagnose the cause and prevent its recurrence.
   To assist decision makers provision should be made to record the impact of the nonconformity on both the customer and the organization. This provision enables personnel to demonstrate compliance with clause 7.3d (see Chapter 30).

How is this demonstrated?

Demonstrating that documented information that described the nonconformity and the actions taken and concessions obtained is being retained may be accomplished by:

a)   presenting evidence of a policy to retain documented information of nonconformities and what information is to be recorded;
b)   selecting a representative sample of records and presenting evidence that they:

   i   describe the nonconformities;
   ii   identify the authority deciding the action to be taken;
   iii   describe the action taken and that it accords with the plan;
   iv   identify any concessions obtained;
   v   describe the actions taken.

Key messages from Part 8

Chapter 33   Operational planning and control

1   If a product possesses the wrong features or executes those features using unreliable materials or processes, it won't sustain a market, but if a service delivers the wrong advice, a wrong decision or the wrong product, it may not sustain a business.
2   The processes needed to meet the requirements for the provision of products and services are those needed to specify, develop, produce and supply the products or services required.
3   The process design requirements should address the risks and opportunities identified from the analysis of internal and external issues that affect the QMS.
4   When planning these processes, the risks arising from inherent design weaknesses need to be identified and addressed so that the required output quality is assured.
5   It's necessary to express criteria for the acceptance of products, services and processes in terms that their achievement can be verified with objective evidence.

Chapter 34   Customer communication

6   The quality of communication with customers is directly proportional to organizational success.
7   Information provided in any form by an organization about its products and services needs to create expectations that can be satisfied without recourse to extraordinary action.
8   Personnel who encounter the organization's customers may make a lasting impression so it's vital they share its vision and values because its entire reputation is in their hands.
9   Anything that may disrupt the provision of products and services following an agreement to supply will be a matter of great concern to customers and therefore of interest to them

Chapters 35 and 36   Requirements for products and services

10   Where requirements for products and services are not properly established and understood by both customer and supplier before work commences, situations are more likely to arise downstream that invariably incur delays and additional costs.
11   With every product one provides a service and the quality of this service needs just as much care and attention as the quality of the product.

12 Organizations have an obligation to comply with legal requirements whether or not they are invoked by their customers.

13 Where the requirements of customers differ from those of other stakeholders, the customer requirements must take precedence unless a waiver can be negotiated or the opportunity to supply can be declined.

14 Customer-focused organizations neither make claims about their products and services they cannot meet in full, nor do they deceive their customers by claiming their products and services provide benefits that are unreal.

Chapter 37–41   Design and development

15 Design can be a journey into the unknown, and on such a journey, we can encounter obstacles we haven't predicted, which may cause us to change our course but our objective remains constant having been set by the requirements of our customers.

16 Design is not a repeatable process because once a design has been completed the process ceases to exist and a new process is established for each new or modified product or service.

17 Control of design and development does not mean controlling the creativity of the designers but controlling the process through which designs are developed so that the resultant output is one that truly reflects customer requirements.

18 Today's successes and failures are the result of yesterday's solutions; therefore, using information from previous designs may be advantageous but only if lessons were learnt.

19 Product and service features need to be proved as being of a standard commensurate with actual operating conditions with a margin sufficient to avoid costly repercussions in use.

20 Change control during the design process controls not only the design but also costs and timescales because once the design process has commenced every change will cost time and effort to address.

Chapter 42–44   Externally provided processes, products and services

21 Externally sourced processes, products and services can have varying degrees of impact on the processes of the organization and its products and services, and therefore, the controls applied need to be proportionate to the level of risk.

22 When external providers are selected based on their capability to meet requirements deduced from facts gathered as a result of an evaluation against soundly based criteria, the likelihood of a successful relationship is assured.

23 The effort spent on developing external providers should be proportional to the consequence of their failure on meeting the organizations objectives.

Chapter 45–49   Production and service provision

24 The production and service delivery process is a journey along a proven path with what is expected to be a predictable outcome.

25 Process control comes about by operators knowing what results to achieve, by knowing what causes results to vary and by being able to correct performance when necessary.

26 If effort is concentrated on the process using the results of output verification and the root cause of variation is eliminated, rework will gradually reduce until all outputs are of consistent quality.

27 There is a perception that human errors are preventable, but despite mistaking proofing by design, everyone can make errors no matter how well trained and motivated they are.

28 Determination of preservation requirements commences during the design phase or the manufacturing or service planning phase by assessing the risks to product or service quality during its production or delivery.

Chapter 50   Release, delivery and post-delivery of products and services

29 If one could be certain that a product or a service would be correct without it being verified, verification during or after its generation would be unnecessary.

30 It would be considered prudent to prohibit the premature release of product if you did not have an adequate traceability system in place

31 Unless otherwise deemed unnecessary, verification needs to occur after a specified feature has been produced and before it becomes inaccessible for measurement.

32 Mistakes made at the packing stage may be the last mistakes made but they will be the first the customer sees.

Chapter 51   Control of nonconforming outputs

33 Nonconformities are caused by factors that should not be present in a process.

34 The factors that cause nonconformity on one occasion will (unless removed) cause nonconformity again and again.

35 In service delivery processes, there is often no intermediate step between producing an output and delivering it to a customer so the process has to be deemed capable before going live.

36 Actions taken on nonconforming items detected before delivery need to be proportionate to the effort that went into creating them.

37 Actions taken on nonconforming items after delivery need to be proportionate to the external consequences and consistent with customer expectations.

38 The only cases where you need to request concessions from your customer are when you have deviated from one of the customer requirements, cannot make the product or service conform but can demonstrate it's fit for its intended use.

Part 9

Performance evaluation

Introduction to Part 9

The 2015 version puts more emphasis on results than previous versions, and therefore Section 9 of the standard contains a collection of requirements on monitoring, measurement, analysis and review which are aimed at answering three questions:

- What are we doing? – performance
- How well are we doing it? – performance evaluation
- What are we going to do about the results? – management review

The standard is structured to reflect the PDCA cycle, and Section 9 is intended the contain requirements for checking or studying performance. Although there are such requirements in other sections, a close examination will show that they mostly require the checks determined from Section 9 requirements to be implemented, but there are no cross-references to Section 9 (which is most odd considering its importance). PDCA is a fractal (a repeating pattern that displays at every scale); therefore, measurement, analysis and evaluation are sub-processes within each process. However, to strip all the verification requirements from other sections and plonk them in Section 9 would have left those sections incoherent. The requirements in Section 9 are therefore more generic than the checking requirements in others sections and apply to all processes.

Measurement, analysis, evaluation and improvement processes are vital to the achievement of quality. We measure parameters using soundly based methods and devices of known integrity to produce data about an object. We monitor those parameters to observe how they change over time. In analysis, we interpret the data objectively and put them in a form which conveys meaning. We evaluate the results of analysis to assess their significance, compare them with norms and draw conclusions as to whether the results are good or bad, valid or invalid. Finally, we review the conclusions reached from the evaluation, decide what we are going to do about them and commission a coherent plan of action.

Management review, which was positioned as part of the section on management responsibility in the 2008 version, now joins other requirements that relate to performance evaluation.

This part of the Handbook comprises five chapters (Chapters 52–56) that address the following clauses of ISO 9001:2015:

- Clause 9.1.1 on monitoring, measurement is addressed by Chapter 52
- Clause 9.1.2 on customer satisfaction is addressed by Chapter 53
- Clause 9.1.3 on analysis and evaluation is addressed by Chapter 54

- Clause 9.2 on internal; audit of addressed by Chapter 55
- Clause 9.3 on management review is addressed by Chapter 56

The placement of the requirements within Section 9 is not consistent with the clause headings, thus reducing ease of use in auditing. Instead of making clause 9.1.1 monitoring and measurement and clause 9.1.2 analysis and evaluation, clause 9.1.1 is a general heading under which there are requirements for both monitoring and measurement and analysis and evaluation and in addition a separate clause on analysis and evaluation. Also, customer satisfaction is separated for special attention and is addressed three times; indirectly in clause 9.1.1, directly in clause 9.1.2 and again directly in clause 9.1.3. However, for consistency with the standard, the chapters of the Handbook mirror the standard.

52 Monitoring, measurement, analysis and evaluation

Introduction

To have confidence that the intended results are being achieved, before we measure we need standards, targets and requirements we can use to judge the results of measurement. The target value is therefore vital, but arbitrary values de-motivate personnel. Targets should always be focused on purpose so that through the chain of measures from corporate objectives to component dimensions there is a soundly based relationship between targets, measures, objectives and the purpose of the organization, process, product or service.

Measurement is a doing process. It's a "process to determine a value" (ISO 9000:2015), and the value determined is generally the value of a quantity of something for decision-making. What is being measured doesn't have to be a quantity. It can be entirely qualitative. It also doesn't mean that measurement removes uncertainty. Many decisions are made with full knowledge of uncertainty. Even on a production line where sampling techniques are used in place of 100% inspection, there is a degree of uncertainty that the results obtained from a sample are not representative of the population from which the sample is taken. Therefore, we can say that any information that reduces uncertainty counts as a measurement.

Monitoring is a doing process. Monitoring is necessary to observe variation in the measurements over time and what those variations mean to the stability of processes, and questions whether it is natural variation or something unusual. By keeping an eye on performance, we can be prepared to act when there are unexpected changes.

The measurement process

Measurement begins with a definition of the measure, the quantity that is to be measured, and it always involves a comparison of the measure with some known quantity of the same kind. If the measure is not accessible for direct comparison, it is converted or *transduced* into an analogous measurement signal. As measurement always involves some interaction between the measure and the observer or observing instrument, there is always an exchange of energy, which, although in everyday applications is negligible, can become considerable in some types of measurement and thereby limit accuracy.

Any measuring requirement for a quantity requires the measurement process to be capable of accurately measuring the quantity with consistency. For this to happen, the factors that affect the result need to be identified and a process designed that considers the variations in these factors and delivers a result that can be relied on as being accurate within defined limits, for example, measuring a characteristic that is produced from a process that is unstable may enable you to decide whether the item you measured is or is not of the standard required, but it won't tell you anything about the next item that is produced.

To measure something, we need the following:

- Sensor (a detecting device that can be human with or without measuring instruments).
- Converter if the sensor is not human (a device for converting the signal from the sensor into a form that the human senses can detect).
- Transmitter where the measurement is done remotely (a device for transmitting the signal to a receiver for analysis)

Sensors need to be accurate, precise, reliable and economic. Sensors that tell lies are of use only to those who wish to deceive. It is too easy to look at a clock, a speedometer, a thermometer or any other instrument and take it for granted that it is telling you the truth. We often put more credence into the readings we get from instruments than we do from our own sensors but both can be equally inaccurate.

In this chapter, we examine the five requirements of clause 9.1.1, namely:

- Determining what needs to be monitored and measured (9.1.1a)
- Determining monitoring, measurement, analysis and evaluation methods (9.1.1b)
- Determining when the monitoring and measuring are to be performed (9.1.1c)
- Determining when results are to be analysed and evaluated (9.1.1d)
- Retaining documented information as evidence of results (9.1.1)

There is a requirement to evaluate the effectiveness of the QMS, but it is repeated in clause 9.1.3c), and therefore the topic will be addressed along with other analysis and evaluation requirements in Chapter 54.

Determining what needs to be monitored and measured (9.1.1a)

What does this mean?

Box 52.1   Revised requirements on monitoring and measurement

In the 2008 version the organization was required to plan and implement the monitoring, measurement, analysis and improvement processes needed to demonstrate conformity to product requirements. In the 2015 version the organization is required to determine what needs to be monitored and measured. Reference to conformity to product requirements has gone, and therefore monitoring and measurement is not intended to be limited to production and service delivery and that it extends to the monitoring and measurement of other aspects of performance.

In the 2008 version there were also specific requirements to measure QMS processes and product characteristics. In the 2015 version there is no requirement to measure anything specifically. It is now at the discretion of the organization what it needs to measure. However, analysis and evaluation of conformity of products and services and the performance and effectiveness of the QMS (a requirement in clause 9.1.3) cannot be undertaken without measuring characteristics of products, services and processes.

We shouldn't monitor and measure things without a reason but the statistician David Moore, the 1998 president of the American Statistical Association, goes so far as to say: "If you don't know what to measure, measure anyway. You'll learn what to measure." (Hubbard, 2010). However, the requirement is stated in the context of managing quality and therefore we have a reason for measuring, the difficulty, if there is one, comes in knowing how to measure.

Why is this necessary?

We measure things to provide information for making decisions, to quantify uncertainty so we put our effort into measuring things we are uncertain about as it would be wasteful to measure things we were certain about. We need to know what to measure so that the information produced is of value to the decisions we intent to make.

How is this addressed?

Understanding the value of information tells us what to measure and how much effort we should put into it. Therefore, if we have a capable process that has been producing the outputs required in the quantity and quality required and doing it economically for a considerable time we wouldn't put a lot of effort into monitoring and measuring it. We'd draw the conclusion that there is sufficient monitoring and measurement going on and leave well alone. But this process was not always like this. In the beginning, we knew nothing about its capability. We had no idea what was being produced, how the quantity produced varied, how the quality of its outputs varied, how much resource it consumed, and what was causing the variation and the waste. The more uncertainty you have the more data you need to reduce uncertainty significantly.

To determine what you need to monitor and measure Hubbard advises that we:

1. Determine a decision problem and the relevant uncertainties (i.e. what decision do we need to make and what are the variables about which we are uncertain that affect that decision?);
2. Define what you know now so that your uncertainty can be quantified and the risks can be determined;
3. Compute the value of additional information as it reduces the risk in the decision and allows us to identify what to monitor or measure.

Box 52.2   The value of additional information

Sometimes managers indicate they need certain information to make decisions and wheels are set in motion to produce reports on a regular basis. However, managers' interests change as priorities change and therefore, reports which at one time were of great value to them are no longer relevant and are rarely read but have become a burden on the staff producing them. Once the managers realize the reports have no bearing on the decisions they take they lose their value.

Monitoring and measurement to inform decisions on conformity

There are two decisions to be made in this regard:

a)   Is the process capable of producing outputs of the required quality?
b)   Is the process producing outputs of the right quality?

The monitoring and measurement data generated should serve to inform each of these decisions.

We can't measure quality directly. We can measure mass, length and time and derivatives of these but there is no unit of quality. However, we can measure the degree of conformity with requirements which customers accept as a substitute for what they look for when making judgements about quality. By tradition, most of the monitoring and measurement undertaken relative to quality has been in production processes. A lot of the uncertainty about what to measure is removed by specifications which define the characteristics the process, product or service is required to possess. If the specifications have been developed from an understanding of customer needs, what is being measured will correlate with what customers require and the resultant process, product or service will be of the desired quality. Specifications are not perfect so in some circumstances there may be latent issues using this method. A cause-and-effect analysis enables the variables that can affect conformity with the specified characteristics to be identified which itself identifies what to monitor and measure.

Monitoring and measurement to inform other decisions

Several clauses directly or indirectly require monitoring and/or measurement, and these are identified in Table 52.1. In two cases, there is a requirement to evaluate rather than measure or monitor, but without data quantifying the value of an object there is nothing to evaluate. In two other cases, there is a requirement to evaluate the effectiveness of an object rather than

Table 52.1 Monitoring and measurement requirements

Clause	Monitoring and/or measurement requirement	Addressed in
4.1	Monitor information about (the identified) external and internal issues	Chapter 12
4.2	Monitor information about (the relevant) interested parties	Chapter 13
4.4.1	Determine the monitoring and measurement needed to ensure the effective operation and control of the processes; (needed by the QMS)	Chapter 16
6.1.2	Evaluate the effectiveness of actions taken to address risks and opportunities	Chapter 21
6.2.1	Monitor quality objectives	Chapter 22
7.2	Evaluate the effectiveness of the actions taken to acquire the necessary competence	Chapter 29
8.4.1	Apply criteria for the monitoring of performance of external providers	Chapter 43
8.4.2	(Determine) the effectiveness of the controls applied by the external provider	Chapter 42
9.1.1	Evaluate the performance and the effectiveness of the quality management system)	Chapter 54
9.1.2	Monitor customers' perceptions	Chapter 53
9.1.3	(Measure) conformity of products and services	See below
9.3.2c)	(Monitor) process performance	See below
9.3.2d)	(Monitor) the adequacy of resources	See below
10.1.2c	Review the effectiveness of any corrective action taken	Chapter 58

measure or monitor it but without the analysis and evaluation of data, obtained from monitoring and/or measurement, decisions on the object's effectiveness won't be soundly based. The same argument is valid for three other entries in the table where *monitor and measure* are in brackets. In these three cases information is either needed for management review or expected from the analysis and evaluation of data.

These are the subjects for monitoring and measurement specified in ISO 9001, but don't be fooled by the brevity of the list because it hides a raft of subjects through the term *criteria* and the general way in which the requirements are expressed. In each of the cases you need to apply the three steps of Hubbard's decision-making process noted earlier.

Removing the barriers to measurement

One of the problems we face with determining what to measure is a belief that some things are immeasurable so it's not worth putting them on the list of things to measure. Even though we should have information on X to make Y decision, we delude ourselves into believing that it's not possible to measure it so we don't bother. We'll address how to measure things in the next section, but unless we choose to measure something, we won't put any effort into figuring out how to do it.

Quality is an intangible until we specify what we mean by quality and express it in terms of characteristics we can measure. We measure quality because it's of value to our customers. We know our customers will detect whether the products and services they received are of insufficient quality and so we set about measuring product and service quality. We can apply the same approach to anything by answering some key questions adapted from those in *How to Measure Anything* Chapter 3 (Hubbard, 2010):

a)   Would it matter if we didn't have a value for X?
b)   If it does matter it must be detectable by a customer or another stakeholder.
c)   If it's detectable, it can be detected as an amount or a range of possible amounts.
d)   If it can be detected as an amount or a range of possible amounts, it can be measured.

How is this demonstrated?

Demonstrating that what needs to be monitored and measured has been determined may be accomplished by:

a)   presenting evidence which identifies the strategic and operational decisions that need to be made in the management of product and service quality;
b)   presenting evidence of a process for determining what is to be monitored and measured to inform these decisions;
c)   presenting evidence of the results of the analyses which defines what is it be monitored and measured relative to the identified decisions.

Determining monitoring, measurement, analysis and evaluation methods (9.1.1b)

What does this mean?

The methods for monitoring, measurement, analysis and evaluation are simply systematic ways of carrying out these activities from the first to the last step which can be explained and performed.

Why is this necessary?

If people responsible for carrying out these important activities were to decide what methods are appropriate, it might produce variable results, as different methods vary in their effectiveness. There will be cases where the choice of method can be left to the user's discretion, often because it can't be pre-determined, and other cases where the choice needs to be limited to ensure valid results.

How is this addressed?

Throughout the standard there are several clauses requiring monitoring, measurement, analysis or evaluation. Some require only one of these activities and clause 9.1.3 invokes the others but all four are related. Clause 4.4.1c) requires methods needed to ensure the effective operation and control of processes to be determined including monitoring and measurements methods so the requirement in clause 9.1.1b) duplicates part of this requirement.

When considering measurement, there are four rules we need to adopt which have been adapted from Frank Price's Three Rules of Quality (Price, 1984):

1   No measurement without recording.
2   No recording without analysis.
3   No analysis without evaluation.
4   No evaluation with action.

There are some simple lessons in these rules. If you are going put the effort into measuring a variable and not record the result, you'll regret it later when you come to analyse the data and find you have none so must take the measurement all over again. After putting in the effort to give meaning to the data and not use the information produced to draw conclusions from it, why do it? And, after doing all this and not act, why measure in the first place?

Price also asks four simple questions which help us determine whether we have control of product quality. They have been adjusted to apply equally to the control of service quality.

1   CAN we make/provide it OK? is a decision we make before we go into production or the service goes live. We need to know if the processes are capable of producing a product or delivering a service that conforms to its specification; otherwise, we'll end up with a pile of product we can't sell and a bunch of unhappy customers. If we can't make or provide it OK, we are either using the wrong processes or resources or we have the wrong design for the technology we are using.
2   ARE we making/providing it OK? is a decision we should periodically make because of the inherent variation in the people, materials and the processing environment.
3   HAVE we made/provided it OK? is a decision we make based on objective evidence. It's the final check before we deliver the product or we present the bill to the customer.
4   CAN we make/provide it better? There's always room for improvement, fewer errors, greater efficiency, better design.

Determining monitoring and measuring methods

There are various forms of monitoring and measuring and different technologies that can be used, but before selecting a method, you need to know:

- what you are monitoring or measuring;
- why you are monitoring or measuring it;

- how the data will be captured;
- how will the data be collected;
- the quality of data you need;
- to whom the data will be transmitted.

Measuring product conformity (8.5.1c)

As we have already stated, we can't measure quality directly as there are no units of quality, and so we provide specifications which substitute for a direct measurement of product quality. These specifications will define the characteristics which the product needs to possess and invariably they are expressed on quantitative terms.

Conformity with qualitative and quantitative characteristics may be measured by visual or metrological inspection methods. For functional characteristics, acceptance test specifications usually define the test methods. In some cases, conformity cannot be measured using metrology and workmanship standards are defined on which to base the acceptability of characteristics created by human manipulation of materials by hand or with the aid of hand tools. For characteristics that are not visible to the human eye non-destructive testing techniques (NDT) are employed.

Measuring service conformity (8.5.1c)

The service specification will define the service required in terms of the outcomes, and the outputs that are deemed necessary to produce those outcomes. Service conformity refers to the extent to which the outputs conform to the requirements. Service quality refers to the extent to which the service outcomes meet customer needs and expectations as is addressed under customer satisfaction in Chapter 53.

Providing a service is equivalent to running a process and therefore measuring service conformity is accomplished using methods such as inspection and audit. Outputs are inspected either 100% or on a sample basis depending on confidence in the provider and risk to the intended outcome. Audits examine the process used to produce the output to determine whether the steps taken were consistent with the policies, regulations and practices stipulated in the service specification.

With professional services where accuracy is vital, as in a forensic science lab, proficiency testing is undertaken to test for examiner bias. In some cases, the test is conducted *non-blind* so the participants are aware that their accuracy is being evaluated. In other cases, the tests are conducted blind to remove the influence of measurement. In addition, a blind verification or audit is conducted by another qualified examiner during routine casework using previously examined evidence to discover whether the first examiner made any errors.

Monitoring processes (8.5.1c)

Process monitoring is carried out by capturing data on process parameters such as throughput and yield which is computed from the measurement of outputs. Parameters are often not measured directly but through analysis of data. This topic is addressed in Chapter 54.

Monitoring the adequacy of resources (9.3.2d)

Each result requires resources, and even when the required results are being achieved there may be better ways or more efficient ways of achieving them. Therefore, targets may be set for efficiency, and these, too, need to be monitored and measured. One method is to monitor resource utilization. Another is to conduct benchmarking against other processes to find the best practice.

Resources are required to be determined, provided and maintained in clause 7.1, and there are several ways in which the adequacy of the resources can be monitored by:

a) process reviews carried out by the process owner which reveal whether the resources were adequate;

b) independent process audits which reveal whether the planned resources were provided and maintained;

c) analysis of project and process data for evidence that the adequacy of resources was a contributory factor in causing a nonconformity.

If the resource estimates were correct, the resources spent on running a process should be within the estimate but often things don't go to plan. There may be bottlenecks in the flow through the process when input rate exceeds the output rate. The resources needed to resolve problems encountered may exceed the contingency included in the estimates. Flow rate and delays may therefore be parameters that need to be monitored. Methods will need to be devised to collect, analyse and evaluate the information and determine opportunities for improvement.

How is this demonstrated?

Demonstrating that the methods for monitoring, measurement, analysis and evaluation have been determined may be accomplished by:

a) presenting evidence of a process for planning monitoring, measurement, analysis and evaluation activities relative to products, services and processes;

b) selecting a representative sample of products, services and processes and presenting evidence of the monitoring, measurement, analysis and evaluation methods that have been determined.

Determining when the monitoring and measuring are to be performed (9.1.1c)

What does this mean?

Monitoring is often planned to commence after a series of activities commence and continue until deliberately stopped. Measurement, on the other hand, is planned to take place when certain conditions are met and it's this judgement that is being referred to in this requirement.

Why is this necessary?

The timing of monitoring and measurement is an important factor in the validity of the results. Pick the wrong time, and the data produced may be unreliable, invalid and not useful for further analysis and evaluation.

How is this addressed?

Monitoring

Where monitoring equipment is built into plant and facilities monitoring can commence as soon as these facilities are commissioned and handed over for operational use. Where work has been undertaken to determine the status of something such as customer satisfaction, the internal and external environment or progress on a project, and it's important to know when there are changes, monitoring should commence once that piece of work has been completed. In this way, the organization is alerted to changes and has the opportunity to act on them.

Measurement

PRODUCT MEASUREMENT

Measurement of product characteristics should be carried out after a specific characteristic has been produced and before it becomes inaccessible for measurement unless there are good economic reasons for not doing so. If measurement is not taken as a characteristic is produced, it's necessary to determine whether it's economical to continue. If variation in a characteristic will affect other characteristics the process may need to be stopped if the characteristic is not within tolerance limits. If variation in a characteristic will not affect other characteristics there may be natural breaks in the process where the item passes from one stage to another or stages at which several characteristics can be verified at once.

If product passes from the responsibility of one person to another, there should be a stage verification at the interface to protect the producer even if the characteristics achieved are accessible later.

When producing a product that consists of several parts, sub-assemblies, assemblies, units, equipment and subsystems, each part, sub-assembly, etc., needs to be subject to final verification but may also require in-process verification for the reasons given earlier. Your control plans should define all the in-process verification stages that are required for each part, sub-assembly, assembly, etc.

SERVICE MEASUREMENT

With services that produce an output such as advice produced from legal services or a finding as in forensic science services, measurement may take place at key stages verifying the work is ready to pass onto the next stage or verify the output meets the defined criteria.

With services where the customer is a participant in the production of the service offering such as in a hospital or restaurant, there will be a diversity of customers using the service each having different perceptions of what constitutes a quality service. From the provider's viewpoint, service conformity measurement should be instantaneous, that is, the provider needs to decide whether an action has yielded the right result and therefore should measure conformity immediately the service is provided and not afterwards because it will too late. The margin for self-correction is limited but there are certain characteristics where interaction with the customer is possible to confirm responses or resolve uncertainties, for example, dentists depend on their patient alerting them to a breach of pain threshold, a hairdresser constantly checks the style is what the client wants, a tailor will invite the client to several fittings before the garment is complete, a subject will sit for an artist several times before the painting is complete. However, these interactions are limited to certain variables due to heterogeneity. The characteristics that are not customer dependent have to be got right first time.

Analysis and evaluation methods

These are addressed in Chapter 54.

How is this demonstrated?

Demonstrating that when the monitoring and measuring is to be performed has been determined may be accomplished by:

a) presenting evidence which identifies the strategic and operational decisions that need to be made in the management of product and service quality;
b) presenting evidence of a process for determining when monitoring and measurement is to be performed to inform these decisions;
c) presenting evidence of the results of the analyses which defines when monitoring and measurement is to be performed relative to the identified decisions.

Determining when results are to be analysed and evaluated. (9.1.1d)

What does this mean?

The results of monitoring and measurement are used for different purposes, and therefore when their analysis and evaluation is carried out depends on there being sufficient data to make a decision.

Why is this necessary?

The timing of analysis and evaluation of data can be critical to process control as results are needed to decide whether a process should stop or continue so speed of response is critical here. In other cases, the data are being used as a lagging measure to judge effectiveness, so the scale of data is critical here.

How is this addressed?

Those designing a process need to build data logging, analysis and evaluation into the design so that operators analyse and evaluate the results of measurement to make timely decisions that enable them to control the process. On a longer timescale the data may be collected offline and subject to different analysis and evaluation to detect outliers and anomalies, examine trends and investigate the causes of variation. The performance record needs to be long enough to have statistical significance.

How is this demonstrated?

Demonstrating that monitoring and measurement result are analysed and evaluated at appropriate times may be accomplished by:

a) presenting evidence that the analysis and evaluation of measurement results is determined as part of process design;

b) selecting a representative sample of processes and showing that provision has been made for undertaking analysis and evaluation at a stage consistent with the priority given the decisions they serve.

Retaining documented information as evidence of results (9.1.1)

What does this mean?

Appropriate documented information of results refers to the form it takes and not whether results are documented.

A statement in ISO Guidance on the requirements for documented information of ISO 9001:2015 (ISO/TC 176/SC2/N1276, 2015) implies that all the documented information required by the standard (see Table 32.1) is the objective evidence required to demonstrate the QMS is effective. The documented information that contains evidence of activities performed and results achieved is certainly relevant, but some items in the table are not relevant.

Why is this necessary?

Unless documentation information of monitoring and measurement is retained, decisions will lack the information necessary to be evidence based. However, there is no requirement to retain this information for any specific period, but it would make sense to retain it for audit and review purposes; otherwise, the decisions cannot be seen to have been evidence based.

How is this addressed?

Throughout ISO 9001 there are requirements for certain documented information to be retained. This is evidence of results which include:

- Documented information to the extent necessary to have confidence that the processes are being carried out as planned (clause 4.4)
- Evidence of fitness for purpose of monitoring and measuring resources (clause 7.1.5.1)
- Evidence of the basis used for calibration of the monitoring and measurement resources (when no international or national standards exist) (clause 7.1.5.2)
- Evidence of competence of person(s) doing work under the control of the organization that affects the performance and effectiveness of the QMS (clause 7.2)
- Results of the review and new requirements for the products and services (clause 8.2.3)
- Records needed to demonstrate that design and development requirements have been met (clause 8.3.2)
- Records of the activities of design and development controls (clause 8.3.4)
- Records of design and development outputs (clause 8.3.5)
- Design and development changes, including the results of the review and the authorization of the changes and necessary actions (clause 8.3.6)
- Records of the evaluation, selection, monitoring of performance and re-evaluation of external providers and any and actions arising from these activities (clause 8.4.1)

How is this demonstrated?

Demonstrating that appropriate documented information as evidence of monitoring and measurement results are being retained may be accomplished by:

a)   presenting evidence of a process for retention of monitoring and measurement results;
b)   selecting a representative sample of monitoring and measurement activities and retrieving documented information from the archive that contains the results of that monitoring and measurement.

Bibliography

Hubbard, D. W. (2010). *How to Measure Anything: Finding the Values of Intangibles in Business.* Hoboken, NJ: John Wiley.

ISO 9000:2015. (2015). *Quality Management Systems: Fundamentals and Vocabulary.* Geneva: ISO.

ISO/TC176/SC2/N1276. (2015). Guidance on the Requirements for Documented Information. Retrieved from ISO/TC 176 Home Page: http://isotc.iso.org/livelink/livelink/fetch/2000/2122/-8835176/-8835848/8835872/8835883/Documented_Information.docx.

Price, F. (1984). *Right First Time.* Aldershot, England: Wildwood House.

53 Customer satisfaction

Introduction

Customer satisfaction is the motive behind ISO 9001; therefore, it becomes of paramount importance that the organization has up-to-date intelligence on the perceptions of current customers about its products and services. Any organization that is not monitoring customer perceptions or is ignoring them will ultimately fail for customers are the lifeblood of every organization. It is important also to use a factual approach to collecting and analysing such data. This will influence tactical and perhaps strategic decisions about product features (quality of design) and production/delivery processes (quality of conformance).

Customer satisfaction should be one of the key performance indicators, and data on customer perceptions will serve to validate not only the business outputs but also the assumptions made about customer requirements that were inputs to the process which determines requirement related to products and services.

In this chapter, we examine the two requirements of clause 9.1.2, namely:

*   Methods for obtaining customer perceptions (9.1.2)
*   Methods for monitoring and reviewing customer perceptions (9.1.2)

Monitoring customer perceptions (9.1.2)

What does this mean?

By combining definitions of the terms *customer satisfaction* and *requirement*, ISO 9000:2015 defines customer satisfaction as the "customer's perception of the degree to which the customer's stated or implied needs or expectations have been fulfilled". To satisfy customers, you should therefore go beyond the stated requirements.

Customers are people who differ in their perceptions as to whether the transaction has been satisfactory. The term *perception* is used because satisfaction is a subjective and human condition, unlike acceptance, which is based on objective evidence. Customers may accept a product but not be wholly satisfied with it or the service they have received. Whether you have done your utmost to please the customer, if the customer's perception is that you have not met their expectations, they will not be satisfied. You could do the same for two customers and find that one is ecstatic about the products and the services you provide and the other is dissatisfied.

As the ISO 9000:2015 definition of a customer is "an organization or person that could or does receive a product or service that is intended for or required by this person or organization" and includes consumer, client, end user, retailer, beneficiary and purchaser, the search for perceptions needs to go beyond the immediate customer to which your products and services are provided.

Information relating to customer perception is any meaningful data from which a judgement can be made about customer satisfaction and would include compliments, complaints, sales statistics, survey results, etc. The requirement refers to the monitoring of customer perception rather than the measurement of customer satisfaction. One difference is that monitoring involves systematic checks on a periodic or continuous basis, whereas measurement may be a one-off event.

Why is this necessary?

The primary purpose of the management system is to enable the organization to achieve its objectives, one of which will be the creation and retention of satisfied customers. It therefore becomes axiomatic that customer satisfaction needs to be monitored.

How is this addressed?

ISO 10004 provides guidance for monitoring and measuring customer satisfaction and should be consulted. There are several ways of monitoring information relating to customer perceptions ranging from unsolicited information to customer focus meetings.

Customer surveys

The most important part of a customer survey is to ask the right questions and Hill, Self and Roche suggest that the survey will provide a measure of satisfaction only if the questionnaire covers those things the customer was looking for in the first place (Hill, Self, & Roche, 2002). There is an eight-step process:

1   Identify the customer's requirements (needs and expectations).
2   Determine a representative sample of customers.
3   Determine which type of survey is appropriate.
4   Design the questionnaire.
5   Conduct the survey.
6   Analyse the data.
7   Present the results.
8   Take action on the results.

Several types of surveys can be used:

* Personal interview
* Telephone interview
* Self-completion questionnaires

DESIGN OF THE QUESTIONNAIRE

It should be noted that self-completion questionnaires by themselves are not an effective means of gathering customer opinion. It is much better to talk face to face with your customer using an interview checklist. Think for a moment how a big customer like Apple or

Wal-Mart would react to thousands of questionnaires from their suppliers. They would either set up a special department just to deal with the questionnaires or set a policy that directs staff not to respond to supplier questionnaires. Economics alone will dictate the course of action that customers will take.

The personal form of survey is conducted through interview such as a customer service person approaching a customer with a questionnaire while the customer is on the organization's premises. This may apply to hotels, airports, entertainment venues and large restaurants. With this method, there is the opportunity for dialogue and capturing impromptu remarks that hide deep-rooted feelings about the organization.

The self-completion form relies on responses to questionnaires and seeks to establish customer opinion on several topics ranging from specific products and services to general perceptions about the organization. The questionnaires can be sent to customers in a mail shot, included with a shipment or filled in before a customer departs (as with hotels and training courses). These questionnaires are somewhat biased because they only gather information on the topics perceived as important to the organization.

DETERMINING SAMPLE SIZE

An important question is how big a sample do you take to gather statistically significant data? With services, it is the size of the population that use the service in terms of location, period and user attributes that is relevant (e.g. all patients over the age of 65 admitted to hospitals in Wales in 2016). Within the sample there will be non-responses and false responses, so there will be a difference between the actual and the target population that may affect the validity of the sample as being representative of the population being studied. Representiveness is a critical characteristic of the sampling method. As Lisch points out, describing a sample as representative should keep the research question and related variables in mind (Lisch, 2014).

DETERMINING THE RELEVANT VARIABLES

Customer perceptions of a product or service differ in all kind of ways. There is no standard customer so when monitoring perceptions, we need to look at the demographics of a given population and this contains the variables. For example, is it relevant if the customers who use a service are male or female, are of certain age groups, of certain ethnic origin, religion, or how far they should travel to use the service? Is it relevant whether they use the online portal or visit the provider in person? Is it relevant whether the customer is a new user or a regular user and is the reason for using the service relevant? Are they returning because their previous transaction didn't go quite as expected? If the method doesn't account for the *failure demand* the statistics will be skewed. Many variables need to be considered so that the method of measuring service quality is representative.

Repeat orders

The number of repeat orders (e.g. 75% of orders are from existing customers) is one measure of whether customers are loyal, but this is not possible for all organizations, particularly those that deal with consumers and do not capture their names (justification for having loyalty cards). Another measure is the period over which customers remain loyal (e.g. 20% of our customers have been with us for more than 10 years). A marked change in this ratio could indicate increasing success or impending disaster.

Competition

Monitoring what the competition is up to is an indicator of your success or failure. Do they follow your lead or are you always trying to catch up? Monitoring the movement of customers to and from your competitors is an indicator of whether your customers are being satisfied.

Referrals

When you win new customers find out why they chose your organization in preference to others. Find out how they discovered your products and services. It may be from advertising or may be your existing customers referred them to you.

Demand

Monitoring the demand for your products and services relative to the predicted demand is also an indicator of success or failure to satisfy customers. It could also be an indicator of the effectiveness of your sales promotion programme, therefore analysis is needed to establish which it is. However, beware of *failure demand*. Seddon writes about this in connection with public services (Seddon, 2014).

Effects of product transition

When you launch a new product or service, do you retain your existing customers or do they take the opportunity to go elsewhere?

User surveys

If your immediate customer is not the end user, you could seek the opinion of users to learn of their experiences with your products and get a better understanding of their needs and expectations. These surveys might yield opportunities for future uses, modifications or applications of your products. Depending on the type of product you supply you might need to do a pilot survey to test its validity before running it on the user population.

Lost business analysis

An analysis of lost business may provide an insight into your tendering practices, although it is often difficult to get reliable data. A potential customer to which you have submitted your tender or quotation may cooperate but they are often not interested as they don't perceive it of being any benefit to them; in other words, you are seen as wasting their time. However, it depends on how you approach it and what your relationship has been with that customer. Certainly, a questionnaire through the post or by e-mail is unlikely to be answered. Lost business analysis needs a personal touch.

Delivery feedback

If the arrangement with your customer is for scheduled deliveries over a long period, you might receive immediate feedback without requesting it. Otherwise, you set up a delivery receipt mechanism that provides the recipient to comment on their experience with the delivery. Some organizations include a customer feedback card with the delivery, and online sales often have a feedback link that customers can follow to record their experience.

The problem with delivery feedback is that many satisfied customers won't bother to register feedback so you can't measure customer satisfaction by how many feedback returns you receive. You can only assess those you do receive and ignore the proportion of the potential number of returns. As with all customer surveys, a 3% response should be regarded as good.

Dealer reports

If you ship product to a dealer, you will have a different relationship than you have with the dealer's customer and may get useful feedback.

Warranty claims

Warranty claims are tangible therefore they can be assessed objectively. They provide an immediate indicator of customer satisfaction particularly if the product represents a substantial investment for the customer. On the other hand, if the product is a throw-away item, you may only receive warranty claims from disgruntled customers; others might not consider the effort worthwhile.

Focus meetings

A personal form of obtaining information on customer satisfaction is to arrange to meet with your customer. Seek opinions from the people within the customer's organization such as from marketing, design, purchasing, quality assurance and manufacturing departments, etc. Target key product features as well as delivery or availability, price and relationships. This form is probably only suitable in B2B relationships.

Complaints

As stated previously, if your processes have not been designed to alert you to customer dissatisfaction, you may be under the illusion that your customers must be satisfied. The process for handling customer complaints is addressed under "Customer feedback" in Chapter 58. Here the topic is monitoring, and therefore you should be looking at the overall number of complaints, the upward or downward trend and the distribution of complaints by type of customer, location and nature of complaint. Coding conventions could be used to assign complaints to various categories covering the product (or parts thereof) packaging, labelling, advertising, warranty, support, etc. Any complaint, no matter how trivial, is indicative of a dissatisfied customer. The monitoring methods need to take account of formal complaints submitted in writing by the customer and verbal complaints given in conversation via telephone or meeting. Everyone who encounters customers should have a method of capturing customer feedback and communicating it reliably to a place for analysis.

Compliments

Compliments are harder to monitor because they can vary from a passing remark during a sales transaction to a formal letter. Again, all personnel who encounter customers should have a non-intrusive method for conveying to the customer that the compliment is appreciated and will be passed on to the staff involved.

How is this demonstrated?

Demonstrating that customer's perceptions of the degree to which their needs and expectations have been fulfilled are being monitored may be accomplished by:

a) presenting evidence of a process for monitoring customer perceptions;
b) selecting a representative sample of customers and presenting evidence that customer perceptions are being monitored as planned.

Determining methods for monitoring and reviewing customer perceptions (9.1.2)

What does this mean?

This requirement seeks to take the data generated by the monitoring process addressed earlier and through analysis produce meaningful information on whether customers are in fact satisfied with the products and services offered by the organization

Why is this necessary?

Customer satisfaction is not something one can monitor directly by installing a sensor. One has to collect and analyse data to draw conclusions.

How is this addressed?

We can now look at those ways by which data can be collected relative to the different techniques of monitoring customer satisfaction:

*   Repeat orders: These data can be collected from the order processing process.
*   Competition: These data are more subjective and result from market research.
*   Referrals: These data can be captured from sales personnel during the transaction or later follow-up calls.
*   Demand: These data can be collected from sales trends.
*   Effects of product transition: These data can be collected from sales trends following new product launch.
*   Surveys: These data can be collected from survey reports.
*   Focus meetings: These data can be collected from the meeting reports.
*   Complaints: These data can be collected from complaints recorded by customers or by staff on speaking with customers.
*   Compliments: These data can be collected from written compliments sent in by customers or by staff on speaking with customers.

As indicated earlier there are several sources of data, several ways in which they can be collected and several functions involved. Provisions need to be made for transmitting the data from the processes where it can be captured to the place where it is to be analysed. It is evident that sales and marketing personnel are involved and as information on customer perceptions is vital for these functions to manage their own operations effectively, it may be appropriate to locate the analysis process within one of these departments. In some organizations, customer support groups are formed to provide the post-sales interface with customer and in such cases, they would probably perform the analysis.

The customer perception monitoring process

The integrity of your process for determining the degree of customer satisfaction is paramount; otherwise, you could be fooling yourselves into believing all is well when it is far from reality. The process therefore needs to be free from bias, prejudice and political influence.

In defining the process, you will need to:

- determine the sources from which information is to be gathered;
- determine the method of data collection – the forms, questionnaires and interview checklists to be used;
- determine the frequency of data collection;
- devise a method for synthesizing the data for analysis;
- analyse trends;
- determine the methods to be used for computing the customer satisfaction index;
- establish the records to be created and maintained;
- identify the reports to be issued and to whom they should be issued; and
- determine the actions and decisions to be taken and those responsible for them.

Pareto analysis can be used to identify the key areas on which action is necessary. For example, it may turn out that 80% of the sales come from repeat orders indicating a decline in the number of new customers. Also 80% of the complaints may be from one market sector that generates only 20% of the sales – which is an indication that 80% of customers may be satisfied. Alternatively, 80% of the compliments may come from 20% of the customers but as they represent 80% of the sales it may prove very significant. The important factor is to look for relationships that indicate major opportunities and not insignificant opportunities for improvement. Use the results to derive the business plans, product development and process development plans for current and future products and services.

Frequency of measurement

Frequency needs to be adjusted following changes in products and services and major changes in organization structure such as mergers, downsizing, plant closures, etc. Changes in fashion and public opinion should also not be discounted. Repeating the survey after the launch of new technology, new legislation or changes in world economics affecting the industry may also affect customer perception and consequently satisfaction.

Trends

To determine trends in customer perception you will need to make regular measurements and plot the results preferably by particular attributes or variables. The factors will need to include quality characteristics of the product or service as well as delivery performance and price. The surveys could be linked to your improvement programmes so that following a change, and allowing sufficient time for the effect to be observed by the customer, customer feedback data could be secured to indicate the effect of the improvement.

Customer dissatisfaction will be noticeable from the number and nature of customer complaints collected and analysed as part of your corrective action procedures. This data provides objective documentation or evidence and again can be reduced to indices to indicate trends.

By targeting the final customer using data provided by intermediate customers, you will be able to secure data from the users, but it may not be very reliable. A nil return will not indicate complete satisfaction so you will need to decide whether the feedback is significant enough to warrant attention. Using statistics to make decisions in this case may not be a viable approach because you will not possess all the facts!

Customer satisfaction index

A customer satisfaction index that is derived from data from an independent source would indeed be more objective. The Swedish Customer Satisfaction Barometer (SCSB) was the first national satisfaction index established in 1989 and developed for the Swedish economy by Claes Fornell at the University of Michigan. Using the Swedish model as a basis, Claes Fornell went on to develop the American Customer Satisfaction Index in 1994. The ACSI measures the satisfaction of U.S. household consumers with the quality of products and services offered by both foreign and domestic firms with significant share in U.S. markets (ACSI, 2016). The European Customer Satisfaction Index (ECSI) is a variation of the ACS I and differs from the ACSI by distinguishing service quality and product quality and excludes the incidence of complaint behaviour as a consequence of satisfaction (see Figure 53.1).

The basic ACSI technology is now used in more than 16 countries, including the United States, Great Britain, Brazil, India, South Africa, Singapore, Sweden and South Korea. In the UK, The Institute of Customer Service is an independent, professional membership body for

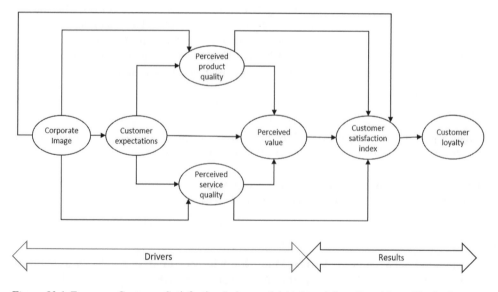

Figure 53.1 European Customer Satisfaction Index model (Adapted from Ronald van Haaften)

Chapter 53 *Customer satisfaction* 755

customer service that was founded in 1996. UKCSI rates customer satisfaction at national, sector and organizational level across 13 sectors of the economy – individually rating many of the leading organizations in each sector (UKICS, 2016).

Ronald van Haaften compares the various models in his online book and concludes that although the objective of all customer satisfaction models is to provide results that are relevant, reliable and valid and have predictive financial capability, nevertheless they have some obvious distinctions in model structure and variable selection so that their results cannot be compared with each other (Haaften, 2016).

How is this demonstrated?

Demonstrating that the organization has determined methods for monitoring and reviewing customer perceptions may be accomplished by presenting evidence that data obtained from monitoring customer perception have been analysed and reviewed.

Bibliography

ACSI. (2016, May 10). About the American Customer Satisfaction Index. Retrieved from American Customer Satisfaction Index: www.theacsi.org/about-acsi

Haaften, R. v. (2016, May 10). Abstract: Customer Satisfaction Models. Retrieved from Rovaka: www.van-haaften.nl/customer-satisfaction/customer-satisfaction-models/101-abstract-customer-satisfaction-models

Hill, N., Self, B., & Ross, G. (2002). *Customer Satisfaction Measurement for ISO 9000:2000*. Oxford: Butterworth Heinemann.

Lisch, R. (2014). *Measuring Service Performance*. Abingdon: Routledge.

Seddon, J. (2014). *The Whitehall Effect*. Axminster: Triarchy Press.

UKICS. (2016, May 9). UK Customer Satisfaction Index. Retrieved from The Institute of Customer Service: www.instituteofcustomerservice.com/research-insight/uk-customer-satisfaction-index

54 Analysis and evaluation

Introduction

The monitoring and measurement processes produce data, and these need to be turned into information and the information studied to figure out if what is being produced or provided is okay and if not what to do about it. This is the purpose of the analysis and evaluation processes.

Analysis

Analysis is a thinking process. It is objective. Analysing the results of monitoring and measurement will enable us to understand what they mean and whether they are in line with what we expect. It interprets and explains the measurements putting them in a form that reveals understanding of what the data are telling us.

Steps in the analysis process include:

- Collect the data from monitoring and measurement activities.
- Sort, classify, summarize, calculate, correlate, present, chart or otherwise translate the original data into meaningful information for the evaluators.
- Transmit the assimilated information to the evaluators.

Evaluation

Evaluation is judgemental and reaches a conclusion after assessing the results of the analysis. Evaluation of the results of analysis will therefore tell us their significance relative to the object of the original measurement and will draw conclusions as to whether the results are good or bad, valid or invalid, whether what we are doing is efficient and effective and therefore acceptable or not. It may even tell us whether the target is the right target to aim at.

Steps in the evaluation process include

- Verify the validity of the information.
- Evaluate the economical and statistical significance of the information.
- Draw conclusions from the results and
- Transmit to the decision makers.

Statistical methods

It is not the purpose of this book to cover statistical methods even though they are central to the control of quality and therefore readers are referred to the *Handbook of Engineering Statistics* which is available for free online (NIST, 2013).

In this chapter, we examine the seven requirements of clause 9.1.3, namely:

* Evaluating product and service conformity (9.1.3a)
* Evaluating the performance and effectiveness of the QMS (9.1.3c)
* Evaluating implementation of planning (9.1.3d)
* Evaluating effectiveness of actions to address risks and opportunities (9.1.3e)
* Evaluating performance of external provider (9.1.3f)
* Evaluating the need for improvements to the QMS (9.1.3g)

There is a requirement to evaluate customer perceptions in clause 9.1.3b), but it is repeated in clause 9.1.2, and therefore the topic is addressed in Chapter 53. There is also an omission because process performance is to be considered for inclusion among the inputs into management review but there is no requirement to use the results of the analysis of data and information arising from monitoring processes to evaluate the performance of processes. An additional section is therefore included in this chapter to address the evaluation of process performance.

Evaluating product and service conformity (9.1.3a)

What does this mean?

The data from monitoring and measuring product and service characteristics are used to determine conformity to specification that is required to be analysed and evaluated. The data collected from customer feedback are also included where the cause of the complaint is product or service nonconformity.

Monitoring product and service conformity differs from monitoring product and service quality by the criteria used. Conformity is judged by the extent to which a specification is met and normally determined by the producer whereas, quality is judged by the extent to which needs and expectations are met and determined by the customer or their representative. With services, customers often don't possess a specification to judge conformity, they recognize a quality service when they experience it.

Why is this necessary?

Data need to be collected, analysed and evaluated for all products and services to determine whether the objectives for conformity and consistency are being achieved.

How is this addressed?

Measuring product conformity

Data on conformity and nonconformity can be collected from the product verification points in each process. Collect what data are available to start with because analysis is an iterative process. After analysing the data, you may discover you want additional data which requires data entry requirement to be changed. It is also important to collect data on the size of the population from which they were taken. The data for critical to quality characteristics and from customer complaints, returns and repairs should also be included.

The information that is generally produced from an analysis of product conformity data are the overall ratio of conformity to nonconformity by product or service type, by characteristic, by customer, by time, by location, etc. What you are looking for are trends to indicate specific types of variation that are symptomatic of certain types of problems. Conformity might vary at different times, at different locations, on different machines or different service operatives. From the nonconformity data, a Pareto analysis of the nonconformities will reveal the most common types of nonconformity and investigation of the causes will reveal the most common cause of nonconformity.

Measuring service conformity

In the service specification, the required characteristics may be expressed qualitatively (e.g. competence and courtesy), but these are not variables we can measure directly. Lisch identified two types of variables, those which hide from direct measurement, which he refers to as *latent variables*, and those which are accessible by direct measurement, which he refers to as *manifest variables*. For example, competence (the latent variable) of a customer service hotline could be measured in terms of the number of calls until an issue is finally settled (the manifest variable) but it won't be sufficient to limit the translation from latent to manifest to a single variable (Lisch, 2014).

With products where conformity can be measured precisely, it is often taken for granted, the only argument is about whether the product possesses the features the customer is looking for. With service conformity, the customer is often a participant in the delivery of the service and is judging conformity at every stage through the process. To measure service conformity, you have to perform the role of customer so that you witness what they will witness.

A common latent variable in the service sector is responsiveness and an equally common manifest variable is the number of rings until a phone call is answered. State-of-the-art telephone systems can even record the performance and provide related analyses. But the same state-of-the-art telephone systems also request you to go through various levels before you get to the desk of someone who may be able to respond and even though the call was answered in two rings (meeting the target) it could take a further 20 minutes before you reach a real person. We therefore need to define exactly what is being measured before devising a method of measuring it.

How is this demonstrated?

Demonstrating that the results of an analysis of monitoring and measurement are used to evaluate conformity of products and services may be accomplished by: (a) presenting evidence of a process for collecting analysing and evaluating the results of product and service monitoring and measurement; (b) selecting a representative sample of products and services and presenting evidence that the results of monitoring and measurement are being analysed and evaluated using the prescribed methods and that the results are used to determine the action to be taken.

Evaluating process performance (9.3.2c)

Process analysis

If you monitor the difference between the measured value and the required value of a characteristic and plot the results on a horizontal timescale in the order the products were produced, you would notice that there is variation over time. There will be a natural scattering of the measured values about some central tendency value. This scattering about a central value is known as a distribution. The distribution can be characterized by:

- location (typical value);
- spread (span of values from smallest to largest);
- shape (the pattern of variation; whether it is symmetrical, skewed, etc.).

A process is deemed stable if it runs in a consistent and predictable manner. This means that the average process value is constant and the variability is controlled. If the variation is uncontrolled, then either the process average is changing or the process variation is changing or both.

If we can demonstrate that the process is stabilized about a constant location, with a constant variance and a known stable shape, then we have a process that is both predictable and controllable. This is required before you can set up control charts or conduct experiments. There are two type of variation, that which is attributable to an assignable or special cause and that which is attributable to a common cause.

First, you need to know if you can make the product or deliver the service in compliance with the agreed specification. For this you need to know if the process is capable of yielding conforming product. Statistical process control (SPC) techniques will give you this information. Second you need to know if the product or service produced by the process actually meets the requirements. SPC techniques will also provide this information. However, having obtained the results you need the ability to change the process in order that all product or service remains within specified limits and this requires either real-time or offline process monitoring to detect and correct variance. To verify process capability, you periodically rerun the analysis by measuring output product characteristics and establishing that the results demonstrate that the process remains capable.

A process may be designed to deliver outputs that meet specification therefore a measure of performance is the ratio of conforming output to total output. If the ratio is less than 1 the process is not capable. Most processes fall into this category because some defective output is often produced, but it is possible to design processes so that they only produce conforming output. This does not mean perfect output but output that is within the limits defined for the process. The target yield for a process may be 97% implying that 100% is not feasible; therefore, a yield of 98% is good and a yield of 96% bad – it depends what standards have been set.

This type of measurement requires effective data collection, transmission and analysis points so that information is routed to analysts to determine performance and for results to be routed to decision makers for action. With a process such as order processing, in addition to each order being checked the process should be monitored to establish it is meeting the defined objectives for processing time, customer communication etc. and that there is no situation developing that may jeopardize achievement of the order processing objectives. Therefore, every process will have at least two verification stages – one for verifying output quality and another for verifying process performance against objectives.

The object of process capability studies is to compute the indices and then take action to reduce common cause variation by preventive maintenance, error proofing, operator training, revision to procedures and instructions, etc. The inherent limitations of attribute data prevent their use for preliminary statistical studies because specification values are not measured. Attribute data have only two values, (conforming or nonconforming, pass or fail, go or no-go, present or absent) but they can be counted, analysed and the results plotted to show variation. Measurement can be based on the fraction defective such as parts per million (PPM). Whereas variables data follow a distribution curve, attribute data vary in steps because you can't count a fraction. There will either be zero errors or a finite number of errors. Process capability studies can only commence once the process is under control (i.e. the results are predictable and exhibit a normal distribution). Under control does not necessarily mean that the process only delivers conforming product. A process that is under control is illustrated by the bell curve on the left in Figure 54.1. A capable process is illustrated by the bell curve on the right in Figure 54.1 where all results are within specification limits.

Special cause

The cause of variations in the location, spread and shape of a distribution is considered special or assignable because the cause can be assigned to a specific or special condition that does not apply to other events. They are causes that are not always present. Wrong material, inaccurate measuring device, worn-out tool, sick employee, weather conditions, accident, stage omitted are all one-off events that cannot be predicted. When they occur, they make the shape, spread or location of the average change as shown in Figure 54.2. The process is not predictable while special cause variation is present. Eliminating the special causes is part

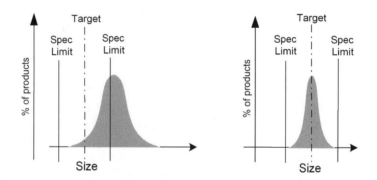

Figure 54.1 Difference between a process that is under control (left) and a capable process (right)

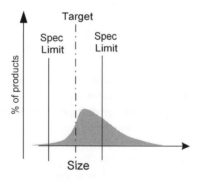

Figure 54.2 Variation in the shape from assignable cause

of quality control, and many of these problems can be detected before they result in nonconforming product through preparatory measures and routine checks.

Once all the special causes of variation have been eliminated, the shape and spread of the distribution and the location of the average become stable, the process is under control – the results are predictable. However, it may not be producing conforming product. You may be able to predict that the process could produce one defective product in every 10 produced. There may still be considerable variation but it is random. A stable process is one with no indication of a special cause of variation and can be said to be in statistical control. Special cause variation is not random – it is unpredictable. It occurs because something has happened that should not have happened, so you should search for the cause immediately and eliminate it. The person running the process should be responsible for removing special causes unless these causes originate in another area when the source should be isolated and eliminated.

Common cause

Once the special cause of variation has been removed, the variation present is left to chance, it is random or what is referred to as common cause. This does not mean that no action should be taken but to treat each deviation from the average as a special cause will only lead to more problems. The random variation is caused by factors that are inherent in the system. The operators have done all they can to remove the special causes, the rest are down to management. This variation could be caused by poor design, working environment, equipment maintenance or inadequacy of information. Some of these events may be common to all processes, all machines, all materials of a particular type, all work performed in a particular location or environment, or all work performed using a particular method.

If the average value is the target value you will get a distribution similar to the curve as in Figure 54.3, and this shows that the process is under control. If the average value is outside the upper specification limit you will get a distribution similar to the bell curve as on the left in Figure 54.1. The goal is to get the average value on the target value with a spread of variation within the upper and lower limits as shown by the bell curve in Figure 54.3, but

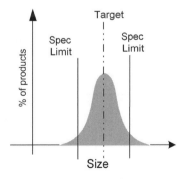

Figure 54.3 Symmetrical about a mean on target

this process still produces nonconforming product, as depicted by the edge of the bell curve going outside the spec limits.

If every time you plot the results of a batch of product the distribution remains the same, the process is under control regardless of the average being on the target value.

The factors causing these variations are referred to as *common causes* and these need to be managed effectively.

By removing special causes, the process settles down and although nonconformities remain, performance becomes more predictable as shown in Figure 54.4. Further improvement will not happen until the common causes are reduced and this requires action by management. However, the action management takes should not be to look for a scapegoat – the person whom they believe caused the error, but to look for the root cause – the inherent weakness in the system that causes this variation.

Common cause variation is random, and therefore adjusting a process on detection of a common cause will destabilize the process. The cause must be removed, not the process adjusted. When dealing with either common cause or special cause problems, the search for the root cause will indicate whether the cause is random and likely to occur again or a one-off event. If it is random, only action on the system will eliminate it. If it is a one-off event, no action on the system will prevent its recurrence, it just has to be fixed. Imposing rules will not prevent a nonconformity caused by a worn-out tool that someone forgot to replace.

With a stable process, the spread of common cause variation will be within certain limits. These limits are not the specification limits but are limits of natural variability of the process. These limits can be calculated and are referred to as the upper and lower control limits (UCL and LCL, respectively). The control limits may be outside the upper and lower specification limits to start with, but as common causes are eliminated, they close in and eventually the spread of variation is all within the specification limits. Any variation outside the control limits will be rare and will signal the need for corrective action. This is illustrated in Figure 54.5.

Keeping the process under control is process control. Keeping the process within the limits of the customer specification is quality control. The action needed to make the transition from process control to quality control is an improvement action.

Figure 54.4 Inducing stability in processes

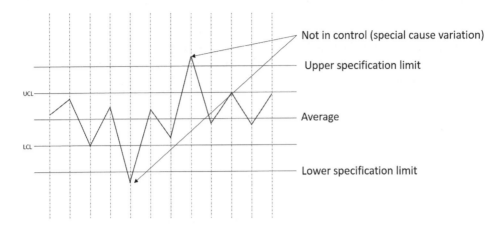

Values of a characteristic measured on items
from the same batch or taken on the same
service at different times

Figure 54.5 Control limits and specification limits for a specific characteristic

Evaluating the performance and effectiveness of the QMS (9.1.3c)

What does this mean?

The performance of the QMS (what it does) is determined from monitoring, measurement and analysis, whereas the effectiveness of the QMS (the extent to which what it does is what it is intended to do) is determined from evaluation of the results of the analysis. There is an anomaly in that in addition to QMS effectiveness the management review serves to determine its suitability, adequacy and alignment with the strategic direction of the organization. The variables that constitute these parameters also require monitoring, measurement and evaluation but appear to be excluded from clause 9.1.1. The implication is that system effectiveness is not considered conditional on the QMS conforming to ISO 9001 or being suitably aligned with the strategic direction.

Why is this necessary?

There is a requirement in clause 5.1.1g) for top management to ensure the QMS achieves its intended results and the requirement to evaluate the performance and effectiveness of the QMS serves to provide evidence as to whether top management have achieved this objective.

How is this addressed?

Effectiveness is defined as the "extent to which planned activities are realized and planned results achieved" (ISO 9000:2015) therefore what is required is:

a) an evaluation of results to establish the extent to which objectives have been achieved (an objective is a result to be achieved which means it's an intended result and therefore must be a planned result)

b)   an examination of plans to establish the extent to which the plans have been realized (i.e. caused to happen or carried out which means they have been implemented)

We can scan through ISO 9001 to find all the clauses that require results or evidence to be retained (see Table 32.1), collect all the evidence and place it in the following groups relative to the critical to quality characteristics (CTQs):

*   Products
*   Services
*   Processes
*   Resources
*   External providers
*   Customer satisfaction

For each CTQ the measured results can be tabulated or presented graphically as line or bar charts to depict the trends. Evaluating these data should indicate whether the planned objectives are being achieved and whether the planned activities have been realized.

Clause 9.3.2 provides the following list of items that are required inputs to the management review for providing information on the performance and effectiveness of the QMS:

1   Customer satisfaction and feedback from relevant interested parties.
2   The extent to which quality objectives have been met.
3   Process performance and conformity of products and services.
4   Nonconformities and corrective actions.
5   Monitoring and measurement results.
6   Audit results.
7   The performance of external providers.

The following items which are omitted from the earlier list also need to be addressed:

*   Monitor information about (the relevant) interested parties.
*   The effectiveness of actions taken to acquire the necessary competence.
*   The effectiveness of actions taken to address risks and opportunities.
*   Adequacy of resources.
*   Opportunities for improvement.
*   The effectiveness of leadership.

There may be a small number of factors on which the performance of the QMS and hence the organization depends, and these above all others should be monitored. For example, in a telephone data centre, processing thousands of transactions each day, system availability is paramount. In a fire department or an ambulance service, response time is paramount. In an air traffic control centre the number of near misses is paramount because the centre exists to maintain aircraft separation in the air. Analysis of the data that the system generates should reveal whether the targets are being achieved.

It is also important to establish whether the system provides useful data with which to manage the business. This can be done by providing evidence showing how business decisions have been made. Those made without using available data from the QMS show either that poor data is being produced or management is unaware of its value. One of eight quality management principles is evidenced-based decision-making which implies decisions should be made using data collected, analysed and evaluated by the defined QMS processes.

How is this demonstrated?

Demonstrating that the performance and effectiveness of the QMS is being evaluated may be accomplished by:

a) presenting evidence of a process for collecting data from monitoring, measurement and audits and showing how the data collected are sufficient for analysing and evaluating the performance of the QMS;

b) presenting the results of the performance evaluation and showing how the effectiveness of the QMS has been determined.

Evaluating implementation of planning (9.1.3d)

What does this mean?

We plan in order to achieve objectives and for no other reason and in ISO 9001 there are several things that are specifically required to be planned (see Box 54.1). The combination of evidence that the processes are being carried out as planned (clause 4.4.2b) and evidence of the implementation of the audit programme (clause 9.2.2f) is what is to be analysed so that the results can be used to evaluate if planning has been implemented effectively.

When planning is implemented the planned actions are carried out. When planning is implemented effectively the objectives the planning is intended to achieve are achieved.

Box 54.1   Planning required by ISO 9001

Clauses which invoke or infer a requirement for planning:

- Processes needed for the QMS (4.4.1)
- Changes to the QMS (5.3e)
- Planning for the QMS (6.1.1)
- Actions to address risks and opportunities (6.1.2)
- Achievement of quality objectives (6.2.2)
- Processes to provide products and services (8.1)
- Design and development (8.3.2)
- Control of externally provided products and services (8.4.2b)
- Verification of products and services (8.6)
- Internal audit programme (9.2.2)
- Management reviews (9.3.2)

Why is this necessary?

Search the Internet for why plans fail and one will find several articles giving the top 5, 7 or 10 reasons why strategic plans fail. Among the reasons will be found: writing the plan and putting it on the shelf, it never gets implemented and failure to execute. Although these all refer to strategic plans, all plans have one thing in common: their success depends on people doing what they say they will do and the intended result being achieved.

How is this addressed?

As the internal audit programme is supposed to provide information on whether the QMS is effectively implemented and maintained, and all the plans referred to in Box 54.1 are part of the QMS, it follows that this is the primary source of the information to be evaluated. The audit programme will cover a wider scope than planning and indeed not all planning will have been implemented prior to an audit. It may therefore be necessary to conduct separate planning audits so that the results of the evaluation are timely. There is little point in reporting that planning was not effectively implemented months after the end of a project. The results of the evaluation should be released at a time when they can be acted upon.

The reasons for the disparity between what was planned and the result achieved should be investigated as part of the evaluation. The sponsors of the plan will need a report that provides more than an affirmative or negative answer, they need to know of the aspects that were deficient and their underlying cause so they can do something about them.

How is this demonstrated?

Demonstrating that the results of monitoring and measurement are used to evaluate if planning has been implemented effectively may be accomplished by:

a)   presenting evidence that the effectiveness of planning is being evaluated;
b)   presenting evidence that the results of analysing monitoring and measurement results have been used in the evaluation.

Evaluating effectiveness of actions to address risks and opportunities (9.1.3e)

What does this mean?

From the way the requirement is expressed we can deduce that it is not requiring an evaluation of the process for identifying risks and opportunities so, for example, if the results of monitoring and measuring revealed an issue that had not been previously identified, that issue would be classified as an opportunity for improvement.

If plans have been made for evaluating the effectiveness of actions taken to address risks and opportunities as required by clause 6.1.2b (2), this requirement means that those plans are to be implemented.

Why is this necessary?

If effort has been spent to control risks and enhance opportunities, it is necessary to discover whether it was worthwhile so as to inform future decisions. See also Chapter 21 on clause 6.2.1b (2).

How is this addressed?

The plan for evaluating the effectiveness of actions to address risks and opportunities simply needs to be implemented.

How is this demonstrated?

Demonstrating that the effectiveness of the actions taken to address risk and opportunity has been evaluated may be accomplished by:

a)   presenting evidence of a plan for evaluating the effectiveness of the actions taken to address risk and opportunity;
b)   selecting a representative sample of risks and opportunities from the plan and presenting evidence that the plan has been effectively implemented.

Evaluating performance of external providers (9.1.3f)

What does this mean?

The data this requirement refers to is that which should have been produced and retained from addressing clause 8.4.1 (see Chapter 43).

Why is this necessary?

See Chapter 43.

How is this addressed?

See Chapter 43.

How is this demonstrated?

See Chapter 43.

Evaluating the need for improvements to the QMS (9.1.3g)

What does this mean?

Clause 9.1.3c) requires the performance and effectiveness of the QMS to be evaluated and this requirement goes one step further by indirectly requiring recommendations to be deduced from the results. It's a case of answering three questions instead of two:

•   What does the QMS do?
•   How well does it do it?
•   Can it do it better?

Why is this necessary?

If one is going to put effort into evaluating the performance and effectiveness of the QMS, it would seem sensible to go the extra mile and identify where improvements can be made.

How is this addressed?

To evaluate the need for improvement to the QMS, having completed an evaluation of its performance and effectiveness, there are two further steps to take before confirming a need for improvement. The first is that of identifying opportunities for improvement and the second is establishing whether the improvement is likely to happen if we mobilize for it.

Identifying opportunities for improvement

Observing that planned results have not been achieved may indicate an opportunity for improvement. However, Deming cited two mistakes frequently made in attempts to improve results, both being costly. The first was to react to an outcome as if it came from a special cause, when it came from a common and is thus random cause of variation. The second was to treat an outcome as if it came from common causes of variation, when it came from a special cause.

Juran writes on improvement thus "Putting out fires is not improvement of the process – Neither is discovery and removal of a special cause detected by a point out of control. This only puts the process back to where it should have been in the first place" (Deming, 1982). This we call restoring the status quo. If eliminating special causes is not improvement but maintaining the status quo, it means that an action to correct a nonconformity is not improvement and although it is deemed to be improvement by ISO in the note to clause 10.1 it must be an error. So, using the ISO 9000:2015 definition of improvement (*activity to enhance performance*) it means there two areas where improvement is desirable – the reduction of common cause variation and the raising of standards.

An examination of the results should find opportunities for reducing common cause variation by discovering the root cause and proposing corrective action to eliminate it.

It is then necessary to analyse the results of monitoring changes in the needs and expectations of stakeholders and the internal and external issues and assess whether the results the QMS produces need to change and if so you will have found opportunities for improving:

a)   alignment between the results of the QMS and stakeholder needs;
b)   the suitability of the QMS for the environment in which it operates.

Having discovered an opportunity for improvement, before proposing that change is needed, a study should be undertaken to establish that a change is both desirable and feasible. Even though you may have objective evidence of a disparity between actual performance and expected performance, it does not follow that every proposal for improving performance will be granted approval as it will be dependent on management priorities. Everyone wants better results, but some will bring greater benefits than others and cost less and take less time than others and if your proposal doesn't capture the attention of management it won't go any further.

Breakthrough in attitudes

The first step is to achieve a breakthrough in attitudes. To convince others of the need for and the urgency of improvement requires you quantify the benefits of improvement. Use the language of money – the cost savings, the increase in productivity, reduction of warranty costs, increase in sales etc. Another ploy is to sensitize the manager. Managers become desensitized to variances. Some people are motivated by firefighting. It turns attention to the special causes of variation and away from the common causes but there is often more to be gained in the long term by reducing common causes than special causes.

A 90% yield from a process may have become the norm such that no alarm bells ring unless the yield falls below 90%. An opportunity to raise normal performance to 95% may attract management attention but not unless it is feasible. Getting buy-in to raise standards that have been stable for years need a compelling case to be made.

A compelling case for change

The second step therefore is to conduct a feasibility study and develop a compelling case for change; to move from what to change to how to change. The change may well save money, but how will it be done? Some improvements may only be dependent one thing being changed, whereas others may require lots of things to be changed. One way of doing this is to apply the Pareto principle[1]:

1   Make list of all that stands between you and making the change.
2   Arrange the list in order of importance.
3   Identify the vital few to be dealt with individually.
4   Identify the trivial many to be dealt with as a group.

Although exact numbers will vary, it may be found, for instance, that 10% of nonconformities account for 80% of customer complaints; therefore, dealing with the 10% on an individual basis may bring about a significant change in performance. The same relationship may be found for maintenance costs and that an analysis of the cost of maintenance by cause may find that 10% of the causes resulted in 80% of all maintenance costs.

Having identified the vital few, the next step is to discover the root cause of the variation and this requires the use of problem solving tools (see "Common problem-solving tools" in Chapter 58). In some cases, the organization may possess the knowledge to do this but may have to bring in diagnostic experts or use diagnostic laboratories.

Once management has accepted that improvement is both desirable and feasible work can begin on planning the change (see Chapter 23).

How is this demonstrated?

Demonstrating that the results of analysis of monitoring and measurement are being used to evaluate the need for improvements to the QMS may be accomplished by:

a)   presenting evidence of the analysis undertaken to evaluate the performance and effectiveness of the QMS and showing that opportunities for improvement are identified;
b)   presenting evidence showing the link between the results on monitoring and measurement and the identified opportunities for improvement on the QMS;
c)   selecting a representative sample of identified opportunities and presenting evidence of the work done to convince management of the need for improvement and authorization to proceed.

Note

1   Dr Juran attributed the separation of the vital few from the trivial many to Vilfredo Pareto and henceforth this has been referred to as the Pareto principle.

Bibliography

Deming, W. E. (1982). *Out of the Crisis*. Cambridge, MA: The MIT Press.

ISO 9000:2015. (2015). *Quality Management Systems – Fundamentals and Vocabulary*. Geneva: ISO.

NIST. (2013). Retrieved from *Handbook of Engineering Statistics*: www.itl.nist.gov/div898/handbook/index.htm

55 Internal audit

Introduction

Both customers and managers have a need for an assurance of quality as they are not able to oversee operations for themselves. They need to place trust in the functions and processes of the organization, thus avoiding constant intervention. All organizations should have control over their own operations and put in place arrangements that will act as a safeguard against deterioration in performance. This is the role of the internal audit.

The audit concept is explained in Box 55.1, but in simple terms its purpose is to establish, by an unbiased means, factual information on quality performance. Quality audits are therefore the measurement component of the QMS.

Box 55.1   What is an audit?

ISO 9000:2015 defines an audit as a "systematic, independent and documented process for obtaining audit evidence and evaluating it objectively to determine the extent to which agreed criteria are fulfilled".

Independent means freedom from responsibility for the activity being audited.
Audit evidence means records, statements of fact or other information, which are relevant to the audit criteria and verifiable.
Agreed criteria means set of policies, procedures or requirements used as a reference against which objective evidence is compared.

The concept of internal audit is common in the financial sector. It provides assurance of financial probity by establishing that:

a) the plan for controlling the organization's finances is sound and will, if followed, ensure all financial transactions can be accounted for and;
b) the plan is being followed as prescribed.

A similar approach applies to internal quality audits but goes one step further by establishing that:

a) the plan for managing the quality of the organization's products and services is sound and will, if followed, ensure they meet customer and regulatory requirements;
b) the plan is being followed as prescribed and;
c) that the agreed criteria are being fulfilled.

As the requirement is placed under the heading "Internal audit", the certification body audit cannot be a substitute even though it is performed with the same purpose in mind. There are internal and external audits. External audits may be performed by customers or regulators or by independent third parties. Internal audits may be performed by dedicated auditors from different departments or by personnel engaged by the manager responsible for the results of a process.

In this chapter, we examine seven requirements of clause 9.2, namely:

- Auditing objectives (9.2.1)
- Planning the audit programme (9.2.2a)
- Defining audit criteria and scope (9.2.2b)
- Ensuring audit objectivity and impartiality (9.2.2c)
- Defining audit methods (9.2.1)
- Reporting audit results to management (9.2.2d)
- Undertaking correction and corrective action (9.2.2e)
- Retaining evidence of the implementation (9.2.2f)

Audit objectives (9.2.1)

What does this mean?

Clause 9.2.1 specifies the objectives of the audit programme which are to provide information on whether the QMS:

a)   conforms to the organization's own requirements for its QMS;
b)   conforms to the requirements of ISO 9001;
c)   is effectively implemented and maintained.

Box 55.2   Revised audit requirement

The 2008 version required internal audits to determine whether the QMS conforms to the planned arrangements (with a cross-reference here to planning of product realization) and to the QMS requirements established by the organization. This had several implications (a) that the QMS requirements established by the organization were somehow different to or excluded the planned arrangements for product realization (b) that the QMS had to conform to the product realization plan or (c) that audits should verify whether the QMS is capable of enabling the organization to implement the plans for product realization. The 2015 version corrects this ambiguity by removing reference to product realization (or operations) because planning of product realization is a process within the QMS, not outside it.

The phrase *QMS requirements established by the organization* was also ambiguous as it could be interpreted as requirements *for the QMS* or requirements *of the QMS*.

The organization's own requirements

The organization's own requirements for its QMS are those requirements that have been established as a result of understanding the context of the organization, understanding the needs of interested parties as explained in Chapter 13 and shown in Figure 22.2. (see also

Chapter 17). These requirements include any requirements of stakeholders that will influence the performance of the QMS.

The requirements of ISO 9001

The requirements of ISO 9001 are those specified in the standard by the verb *shall* but they also include customer specific QMS requirements additional to those in ISO 9001 where ISO 9001 requires conformity with customer requirements.

Effective implementation

Effective implementation means that the processes are being run as intended and delivering the expected outputs. This means that the policies, procedures and other directives are being implemented as intended. A process may be run as intended but not achieve the desired results indicating a design weakness in the process that may be caused by a policy or procedure being inadequate or unsuitable.

Effectively maintained

Effectively maintained means that the processes continue to remain capable despite changes in the quantity, condition or nature of the human, physical and financial resources and changes in the requirements for the outputs. It means that people are maintaining their competence, infrastructure standards are being maintained and required staff levels are being maintained.

Why is this necessary?

Audit objectives need to be derived from the organization's own requirements and conformity with ISO 9001 will be one of those requirements.

How is this addressed?

If every internal audit has an objective that serves to provide information on whether the QMS conforms to (a) the organization's own requirements for its QMS, or (b) the requirements of ISO 9001 or (c) is being effectively implemented and maintained, the requirement is met. The way the audit objective is expressed may not reflect these words but serve the same purpose, for example, an audit having the objective of determining whether an outsourced process was being managed effectively is serving (c). Similarly, if an organization requires the QMS to be flexible and an audit which examines the QMS with the objective of finding evidence of inflexible practices would be serving (a).

How is this demonstrated?

Demonstrating that internal audits are being conducted at planned intervals to provide information on whether the QMS conforms to the requirements of ISO 9001 and the organization's own requirements may be accomplished by:

a) presenting evidence of the requirements against which the QMS is required to conform;
b) presenting evidence of a process for planning and conducting internal audits;

c)   presenting evidence that audit objectives are consistent with providing evidence of conformity with the organization's own QMS requirements, ISO 9001 requirements and its effective implementation.

Planning the audit programmes (9.2.2a)

What does this mean?

This clause requires an audit programme or programmes to be planned and for the audit frequency, audit methods, audit responsibilities and requirements for planning and reporting to be determined, taking into consideration several factors.

The audit programme is defined in ISO 9000:2015 as: "a set of one or more audits planned for a specific time frame and directed towards a specific purpose". However, the programme is not simply a calendar showing the frequency of audits but is required to include methods, responsibilities, planning requirements and reporting. It would therefore appear that the definition falls short of what ISO 9001 require of an audit programme (see Box 55.3).

The frequency is the interval over which the audit is to be repeated and can be daily, weekly, monthly, quarterly, annually or other intervals. The methods are how the audit is to be planned, conducted, reported and completed. The planning and reporting requirements are those pertaining to individual audits.

Box 55.3   Revised audit programme requirement

The 2008 version required a set of audits to be planned, referred to as an audit programme, and it also required the audit criteria, scope, frequency and methods to be defined but didn't prescribe where it had to be defined. However, a procedure was required to define the responsibilities and requirements for planning and conducting audits, establishing records and reporting results.

In the 2015 version, the requirement for a procedure has been removed and therefore audit frequency, methods, responsibilities, planning requirements and reporting is now to be included in the audit programme and the audit criteria and scope defined as part of each audit. The implication here is that the audit programme is now expected to be more that a calendar of audits and to be a collection of information.

The term *planned intervals* implies a specific length of time between audits regardless of need whether resulting from a risk assessment or simply a status check. However, an audit programme that is derived from risk assessments or status checks following which an audit interval is determined would be a preferable interpretation.

Why is this necessary?

Without defined methods of auditing it is likely that each auditor will choose a different way of performing the audit – some methods will be effective and some not so effective. To run an effective management system, auditing should aim for best practice and by defining and refining auditing methods best practice is established for the benefit of the organization.

How is this addressed?

ISO 19011 provides guidance on planning and conducting internal audits, but it does not describe how to do the audit so you have a free choice as to the methodologies you adopt see "Defining the audit method" further on in this chapter.

The first audit can be any time after the decision to conform to ISO 9001 has been made or the organization's own requirements specified. It can therefore include audits carried out to determine the gap between the way quality is managed currently and the way ISO 9001:2015 or the organization requires quality to be managed and conducted at specific intervals thereafter. However, having performed an initial audit against the requirements of ISO 9001 and/ or the organization's own requirements, one should only need to repeat this audit when the system changes or when the requirements change.

Audit programme

The audit programme covers a range of audits which collectively determine the effectiveness of the QMS. There is only one audit programme for a QMS but there may be a need for different types of audit programmes depending on whether the audits are of the QMS, contracts, projects, processes, products or services. It would therefore be expected that all audits in an audit programme would serve the same purpose.

An audit of one requirement of a policy, standard, process, procedure, contract etc. in one area only will not be conclusive evidence of compliance if the same requirements are also applicable to other areas. Where operations are under different managers but performing similar functions, you cannot rely on the evidence from only one area – management style, commitment and priorities will differ. To ensure that an audit programme is comprehensive you will need to draw up a matrix showing the areas or processes or products, etc., to be audited and the dates when the audits are to be carried out. Supporting each audit programme an analysis of process maturity and importance should be performed and the key aspects to be audited identified. The programme should also include shift working so that auditors need to be very flexible. One audit per year covering 10% of the QMS in 10% of the organization is hardly comprehensive. However, there are cases where such an approach is valid. If sufficient confidence has been acquired after conducting a comprehensive series of audits over some time, the audit programme can be adjusted so that it targets only those areas where change is most likely, thus auditing more stable areas less frequently.

The QMS will contain many provisions, not all of which may be verified on each audit. This may either be due to time constraints or work for which the provisions apply not being scheduled. It is therefore necessary to record those aspects that have or have not been audited and devise the programme so that over a one- to three-year cycle all provisions are audited in all areas at least once.

Audit frequency and importance

Within the audit programme would be a calendar chart showing where and when the audits will take place There is little point in conducting in-depth audits on processes that add little value. There is also little point auditing processes that have only just commenced operation. You need objective evidence of conformity and effectiveness and that may take some time to collect.

This requirement focuses on the criteria for choosing the areas, activities, processes, etc., to audit and following initial audits to verify that the system is functioning as planned, subsequent audits should be scheduled depending on the maturity and importance of the aspects to be audited. Maturity in this context means its relative state of development. A newly developed process might require more frequent auditing than a mature, well used process that has been proven effective.

The frequency might need to be suspended because of organizational changes until they have had time to bed in or reduced where the results of previous audits have revealed a higher than average performance in an area (such as zero nonconformities on more than three occasions). However, where the results indicate a lower-than-average performance (such as a much higher-than-average number of nonconformities) the frequency of audits would be increased.

On the importance of the process, you need to establish to whom is it important – to the customer, the managing director, the public or your immediate superior? You also need to establish the importance of the activity relative to the effect of nonconformity with the audit criteria. Importance also applies to what may appear minor decisions in the planning or design phase but if the decisions are incorrect it could result in major problems downstream. If not detected, getting the units of measure wrong can have severe consequences particularly if the customer specified dimensions in metric units and the purchase order has them specified in imperial units. Rather than check the figures or the units of measure, audits should verify that the appropriate controls are in place to detect such errors were they to arise before it is too late.

The importance of the activities will determine whether the audit is scheduled once a month, once a year or left for three years. Any longer, and the activity might be considered to have no value in the organization.

There is no requirement to document the results of any analysis carried out to determine the importance of the areas to be audited but it might useful to do so that the next time you prepare the audit programme you can look back and review the reasons for the decisions that were taken and determine whether they remain valid or should change.

Audit responsibilities

Depending on the complexity of the audit those involved will perform several different roles for which they discharge certain responsibilities.

- If there is an audit team, the role of the team leader needs to be defined including the responsibility for planning the audit, selecting the team, coordinating the activities of the team and ensuring the audit objectives are achieved.
- The role of the auditors need to be defined including the responsibility for gathering the evidence in a way that does not alienate the auditee and ensuring findings are confirmed and recorded.
- If there are observers or advisors involved their role needs to be defined.

Planning the audit

The detail plan for each audit may include dates if it is to cover several days but the main substance of the plan will be what is to be audited, against which requirements and by whom. At the detail level, the specific requirements to be checked should be identified based on risks, past performance and when it was last checked. Overall plans are best presented as programme charts and detail plans as checklists. Audit planning should not be taken lightly. Audits require effort from auditees as well as the auditor so a well-planned audit designed to quickly discover pertinent facts is far better than a rambling audit that jumps from area to area looking at this or that without any obvious direction.

Although checklists may be considered a plan, in the context of an audit they should be considered only as an aid in preparing the auditor to follow trails that may lead to the discovery of pertinent facts. However, there is little point in drawing up a checklist then putting it aside. Its rightful place is after the audit to verify that there is evidence indicating:

* those activities that were compliant;
* those activities that were not compliant;
* those activities that were not checked and;
* those activities where there were opportunities for improvement.

Audits of practice against procedure or policy should be recorded as they are observed and you can either do this in note form to be written up later or directly on to observation forms especially designed for the purpose. Some auditors prefer to fill in the forms after the audit and others during the audit. The weakness with the former approach is that there may be some dispute as to the facts if presented sometime later. It is therefore safer to get the auditee's endorsement to the facts at the time they are observed. In other types of audits there may not be an auditee present. Audits of process documentation against policy can be carried out at a desk. One can check whether the documents of the QMS address the relevant clauses of the standard at a desk without walking around the site, but you can't check whether the system is documented unless you examine the operations in practice as there may be activities that make the system work that are not documented. Further guidance is provided in ISO 19011.

Different types of audit

There are different types of audits, each being defined by the objective it seeks to achieve. In all the following cases, the audits will reveal (a) areas of conformity which should be presented as opportunities for celebration and (b) areas of nonconformity which should be presented as opportunities for improvement.

SYSTEM AUDIT

A system audit takes as its objective the intended results of the QMS (see Chapter 12) and establishes that:

a) the planned arrangements will, if implemented enable the organization to deliver results consistent with its purpose and strategic direction;
b) the planned arrangements are being implemented effectively;
c) the results being achieved are consistent with those intended.

There are two ways for conducting the system implementation audit:

* planning a series of audits that will cover the entire system in one cycle or;
* analysing the results of process audits, product audits and service audits and determining effectiveness by correlation.

PROCESS AUDIT

A process audit takes as its objective a result that the organization either desires or is required to achieve and seeks to establish that the processes that are intended to deliver this result are being managed effectively (Hoyle & Thompson, 2009). The audit examines whether:

- the activities are being performed as planned;
- the resources are being utilized effectively;
- the desired/required results are being achieved.

The audits could be performed by personnel external to the process, such as internal auditors or managers or be performed by the process owners.

PRODUCT AUDIT

A product audit takes as its objective the product requirement and establishes whether a product conforms to its specification. The various activities specified by the planned arrangements should result in an output that conforms in full with the specified requirements. However, there is variation in all processes and although the processes may be deemed capable, incidents can occur that escape detection. The product audit determines whether the controls in place are effective.

SERVICE AUDITS

Providing a service is a process therefore a service audit can be treated as a process audit.

PROJECT AUDITS

A project audit takes as its objective an invitation to tender, a contract, or an undertaking that requires the development of new or modified products, processes and services. Project audits are often referred to as horizontal audits as they cross functional boundaries.

Audit programme review

As the programme is executed, periodic reviews should be undertaken using data from a variety of sources to confirm that the audit programme is effective. The data should be analysed to determine whether the maturity and importance of the areas being audited remains unchanged and if necessary the programme modified to take account of changes in maturity and importance. This might result in altering the frequency of audits, the depth of the audits and areas covered.

   This requirement can be met in one of several ways:

- system audits;
- project audits followed by process audits;
- process audits conducted by personnel external to the process;
- process audits conducted by personnel operating the process and
- product and service audits.

How is this demonstrated?

Demonstrating that internal audits are being conducted at planned intervals to provide information on whether the QMS is effectively implemented and maintained may be accomplished by:

a) presenting evidence of a process for planning and conducting internal audits;

b) presenting evidence that these audits are conducted at planned intervals;

c) presenting evidence from the audits that the QMS is being effectively implemented and maintained.

Defining audit criteria and scope (9.2.2b)

What does this mean?

The audit criteria are the standards for the performance being audited. They may include policies, procedures, regulations or requirements. Examinations without such a standard are surveys not audits. The scope of the audit is a definition of what the audit is to cover – the boundary conditions including the areas, locations, shifts, processes, departments, etc. In defining these aspects, they may be described through procedures, standards, forms and guides.

Why is this necessary?

Without audit criteria, there is no basis for deciding whether performance is good or bad, acceptable or unacceptable.

How is this addressed?

For each audit the auditor should as a matter of routine always define the standard against which the audit is to be performed and the scope of the audit. The standards should also be those with which the auditee has agreed to conform unless the audit is being carried out to establish whether there is an opportunity to raise standards. In such cases, there will be no correction and corrective action required.

The audit criteria and scope should be explained to the auditee at the beginning of the audit and agreement sought so that there is no dispute later.

How is this demonstrated?

Demonstrating that the audit criteria and scope for each audit has been defined may be accomplished by presenting evidence of audit plans and reports that specify the criteria and scope of the audit.

Ensuring audit objectivity and impartiality (9.2.2c)

What does this mean?

This requirement means that auditors and their association with the work being audited and the people doing the work should not influence their judgement. The requirement suggests that anyone auditing their own work may be influenced such that they overlook, hide or ignore facts pertinent to the audit.

Why is this necessary?

If personnel have personally produced a product, they are more likely to be biased and oblivious to any deficiencies than someone totally unconnected with the product. They may be so familiar with the product that they are blind to its full strengths and weaknesses. A second pair of eyes often catches the errors overlooked by the first pair of eyes. However, auditors are human and if there is a personal relationship between the auditor and the auditee, the judgement of the auditor may be prejudiced. Depending on the nature of any problems found, the auditor being a friend, relation or confidant of the auditee, may be reluctant to or may be persuaded not to disclose the full facts if the findings indicate a serious deficiency. Even a customer may fail to exercise objectivity when it is found that the cause of problem is the inadequacy of the customer requirement!

How is this addressed?

Apart from an auditor not auditing their own work, any other competent person with experience in the area to be audited could be selected as an auditor.

The requirement for objectivity and impartiality does not mean that one must rule out supervisors, managers, friends, relations or internal customers as auditors. These conditions do not necessarily mean such a person cannot be objective and impartial – there is simply an inherent risk. This risk is overcome by the selection being made on a person's character and track record. It would be foolish to limit the selection of auditors to those who are totally independent because in some small organizations there may be no one who fits this criterion. The difficulty arises in demonstrating after the audit, that the selected auditor exercised objectivity and impartiality. In organizations that observe a set of shared values, where honesty and trust are prevalent and frequently reinforced, it should not be necessary to demonstrate that the selected auditors meet this criterion. For other organizations, a solution is for the auditors to be selected based on having no responsibility for the work audited and no personal relationship with any of the auditees concerned.

By being divorced from the audited activities, the auditor is unaware of the pressures, the excuses, the informal instructions handed down and can examine operations objectively without bias and without fear of reprisals. To ensure their objectivity and impartiality, auditors need not be placed in separate organizations. Although it is quite common for quality auditors to reside in a quality department, it is by no means essential. There are several solutions to retaining impartiality:

- Auditors can be from the same department as the activities being audited, provided they do not perform the activities being audited.
- Separate independent quality audit departments could be set up staffed with trained auditors.
- Audits could be carried out by competent personnel at any level provided they do not perform the activities being audited.

A competent person in this regard is one that can satisfy certain acceptance criteria. A common trap that many fall into is to put people through an ISO 9001 training course as if by magic, they become qualified auditors. It is true that such people have been exposed to the requirements of ISO 9001, but they could have had equal exposure simply by reading the standard, and many of these one- or two-day courses are no more than tutored reading. An in-depth understanding is what is required, and this comes from application followed by examination. Training courses that offer audit simulation provide a more effective learning environment. Slide shows teach little, if anything. Examinations that test memory are not effective. An examination that explains a situation or presents objective evidence and asks the auditor to study the evidence, draw conclusions and explain his or her rationale is more conducive to yielding competent internal auditors. However, a more effective method is where the auditor is observed asking questions of people at their place of work, assessing the answers and evidence presented, drawing conclusions and writing a report of their findings.

How is this demonstrated?

Demonstrating that there is objectivity and impartiality in the internal audit process may be accomplished by:

a) presenting criteria for the selection of auditors and showing steps taken to ensure auditor objectivity and impartiality;
b) selecting a representative sample of audits and establishing that the auditors selected met the prescribed criteria.

Defining the audit method

What does this mean?

The audit method is the way the evidence will be gathered.

Why is this necessary?

The integrity of the audit result will depend on the method used to gathering the evidence as well as the competence of the auditor. With internal audits, it will be a mixture of checking documented information and interviewing people or observing activities, but how an auditor proceeds to carry out the audit can differ significantly. When analysing audits for trends, knowing which audit method was used is necessary to be able to compare the results.

How is this addressed?

ISO 19011 Appendix B identifies several auditing methods some involving human interaction and others with no human interaction and there is useful information in the sections which address these methods:

- Conducting interviews on site and off site
- Completing checklists
- Conducting document reviews
- Sampling
- Observation of work performance

What is not addressed in ISO 19011 are different ways of gathering evidence, and two of these are addressed next.

Requirements-driven audit

The requirements-driven method starts with a table of requirements and a description of the evidence that should exist which demonstrates conformity with these requirements. There are four steps to this audit:

1   Gather all the requirements with which the QMS is required to conform (e.g. ISO 9001), legislation applying to products and services, customer-specific requirements.
2   Prepare an exposition table, an extract of which is shown in Table 55.1.
3   Appoint someone else to verify that the exposition is complete and the evidence indicated in the table is valid as a response to the requirement. This is a desk audit, checking that documentation describes arrangements that satisfy the requirement.
4   Perform an implementation audit to verify that the intent that is expressed in the exposition is translated into action.

Table 55.1 Extract from an ISO 9001 exposition

Req No	Clause	Subject	Evidence of Compliance
	4	**Context of the organization**	
	4.1	**Understanding the context of the organization**	
1.	4.1	Determining external and internal issues that are relevant to TSL's purpose and its strategic direction that affect the results of the QMS	a) TSL's strategic planning process description b) TSL's mission and vision statement c) TSL's environmental scanning process description d) System maps and analysis reports e) Strategic issues register
2.	4.1	Monitoring and reviewing information about these external and internal issues	a) TSL's environmental scanning process description b) Quarterly situation analysis reports c) Quarterly board meeting reports d) Quality programme review reports
3.	4.2a	Determining interested parties that are relevant to the QMS	a) TSL's strategic planning process description b) TSL's environmental scanning process description c) TSL's stakeholder requirements register
4.	4.2b	Determining the requirements of these interested parties that are relevant to the QMS	a) QMS development process description b) The applicability column in TSL's stakeholder requirements register
5.	4.3	Determining the boundary and applicability of the QMS	QMS development process description describes how the scope of the QMS is determined
6.	4.3	The scope of the organization's QMS shall be available and be maintained as documented information	The boundary of the QMS is shown on the external system map. The applicability of the QMS is shown on the internal system map. Both maps are included in the QMS description
7.	4.4	Establishing and maintaining a QMS	QMS development process description describes how the QMS is established and maintained
8.	4.4	Continual improvement of the QMS	Continual improvement of the QMS is evidenced by the performance trends against KPIs contained in quarterly QMS performance reports

To build this table, examine each requirement and translate it into a subject statement, then explain where in the system description can evidence of compliance be found. This will only apply to requirements that can be met through design verification. Conformity with some requirements such as the requirement for the quality policy to be communicated, understood and applied cannot be verified except by interview or by analysis of records during an implementation audit.

Through hyperlinks you can link to the location in the system description or other documents that demonstrates compliance (i.e. documents that express intentions and documents that record activities carried out or results achieved). As there are over 300 requirements, this will be a large and complicated matrix, so one solution is to compile the matrix as a series of layers. The first layer would list the requirements as a series of clauses against the core business processes. For each process a list of applicable requirements can then be identified and cross-referenced to the process stage where it is implemented. A report can be printed and the assumptions, conclusions and recommendations added. Sometimes requirements are expressed in such a way that a single response is almost impossible and the only way of collecting evidence of conformity with some requirements is to interview people. For example, with the requirement for a quality policy we can translate them into 12 separate requirements that could elicit a different response if you asked each of the questions in Table 55.2.

Table 55.2 Sample conformity audit

#	Clause	Requirement	Response
1	5.2.1	Is there a quality policy?	Yes
2	5.2.1	Has it been established, implemented and maintained by top management?	No because the quality manager produced the statement and top management were simply presented with it to approve
3	5.2.1a)	It is appropriate to the purpose and context of the organization?	No evidence found as there is no published statement of the organization's purpose or mission
4	5.2.1a)	Does the policy support the strategic direction of the organization?	No evidence found as there is no published statement of the organization's strategic direction
5	5.2.1b)	Does it provide a framework for establishing and reviewing quality objectives?	No because the statements are motherhood statements not definable values or principles
6	5.2.1c)	Does it include a commitment to satisfy applicable requirements?	Yes
7	5.2.1d)	Does it include a commitment to continual improvement of the quality management system?	Yes
8	5.2.2.a)	Is it available as documented information?	Yes – there is a framed version in strategic locations
9	5.2.2.a)	Is it maintained as documented information?	Yes – it was last reviewed in July 2015
10	5.2.2b)	Is it communicated, understood and applied within the organization?	Not effective, as only 5 out of 20 understood how it applied to what they do
11	5.2.2b)	Is it applied within the organization?	Not effective, as evidence found that plans for improving the QMS were put on hold on several occasions over the last three years
12	5.2.2c)	Is it available to relevant interested parties?	Yes – it is displayed on the organization's website

Once a requirements-driven conformity audit has been completed and it has been established that the system has been designed to meet all requirements of ISO 9001 or the organization's own requirements, it only needs to be checked again if the requirements of the standard change or the system design changes. The system design is represented by the QMS description, and although you can require any changes to pass through a formal change control process, some changes may not have been passed through that process because they were deemed temporary or not perceived as changes to the QMS by the persons concerned. Such changes should be picked up on *implementation audits*.

Result-driven conformity audit

If we begin with the end in mind, we would start an audit by wanting to know if the system was effective, then establish that it conforms to the requirements. One of the problems with the requirements-driven approach is that the end is conformity and not effectiveness. A result-driven conformity audit starts with what the organization is achieving and finishes with checking for conformity with the requirements

1   First, look at what results are being achieved relative to the organization's quality objectives.
2   Establish whether these results are consistent with the intent of the requirements, that is, consistently conforming product, satisfied customers, compliance with relevant statutes and regulations and continual improvement.
3   Discover what processes are delivering these results.
4   Determine whether these processes are being managed effectively.
5   Lastly check that what is being done demonstrates conformity with the requirements.

In many cases, it may not be possible to get beyond Step 1 simply because the organization cannot demonstrate what it is achieving as there is no simple way of showing performance. Second, it might not be possible to separate the objectives from the masses of data the organization presents and therefore not get beyond Step 2.

However, once a result-driven audit has been completed and all deficiencies resolved (which might take a considerable time relative to other audits) subsequent audits will be a fraction of the time. If there has been no change in the organization's objectives and no adverse change in its performance when Step 2 is completed, the system remains compliant and therefore there is no justification for checking conformity with the requirements. If there are changes that will affect the integrity of the system, an analysis needs to be carried out to establish the impact of these changes and repeat Steps 1–5 on the processes affected by these changes. A video on management system auditing is available on the companion website.

How is this demonstrated?

Demonstrating that the audits methods have been planned may be accomplished by:

a)   presenting evidence of an audit planning process in which the audit methods are selected
b)   selecting a representative sample of audit plans and showing that the audit methods have been described

Reporting audit results to management (9.2.2d)

What does this mean?

If the objective of the audit is to determine whether the QMS conforms to certain requirements the results of the audit disclose the degree of conformity. If the objective of the audit is to determine whether the QMS is effectively implemented and maintained the results of the audit disclose the degree of implementation and maintenance. The degree of conformity or implementation and maintenance is often measured in terms of the number of nonconformities or a percentage of conformity but this will present an imbalanced result as it takes no account of the significance of the nonconformities.

The relevant management is the managers responsible for the processes, project, products, services or departments that have been audited.

Why is this necessary?

Top management is accountable for the effectiveness of the QMS and therefore they need information on its effectiveness so they can discharge their responsibilities. In a very small business, the manager can audit the QMS and resolve any problems, but in larger organizations, the task is delegated, and it is therefore incumbent on the auditor to report back to management the results of the audit and for them to decide on the action to be taken.

How is this addressed?

Reporting the audit

AUDIT SUMMARY REPORT

The audit summary report summarizes the audit in terms of:

- What was audited when;
- Who was involved;
- Why the audit was conducted;
- What was found;
- What was or will be done about it;
- What impact this might have had on the organization if nothing was done;

The summary is intended for top management as they don't need the detail. If this is a separate report to a more detailed report, it might get their attention. One way to deter top management is to place a thick report on their desk.

AUDIT DETAIL REPORT

Records of an audit simply define what was recorded, whereas the results of an audit go much further. There follows a comprehensive list of topics that should be addressed in the audit report:

1 General location of the audit (site, plant, etc.).
2 Audit objectives (what the audit intends to achieve).
3 Audit criteria (the basis for determining success).
4 Audit strategy (basic approach to conducting the audit or summary of the audit process, the key issues to be examined such as critical aspects, previous audit findings, organizational changes, trends observed on previous audits).
5 Audit plan (showing the processes that were audited with dates and times and those not audited (but within scope).
6 Processes or parts thereof not covered but within the audit scope and the reasons for exclusion.

7 Annotated audit checklists (showing what was checked and not checked in each process).
8 Audit team (names and position where appropriate).
9 Personnel interviewed on each process with dates.
10 Audit findings reports, including the agreed proposals for remedial and corrective action.
11 Obstacles encountered that affect the conclusions.
12 Any unresolved issues between the auditors and the auditees.
13 Audit conclusions (whether audit objectives had been achieved relative to the success criteria).
14 Recommendations for improvement.
15 Follow-up action plan; and
16 Report distribution list.

AUDIT FINDINGS

It is usual for the nonconformities detected to be reported on a form which allows for the action to be taken by the auditee to be recorded. An Audit findings report is shown on the companion website. This form simply conveys findings not the conclusions of the audit (i.e. that would be addressed in a summary report). The form does not need to have provision for corrective action unlike so many audit reports simply because this may be addressed by a separate corrective action form. What is stated on so many audit reports as corrective action is action to correct errors because it invariably does not get to the root cause. Many such forms contain actions to correct documentation that should have been correct. The authors of such reports obviously believed that by correcting the documents they were preventing recurrence but all they were doing was making good. Most auditors do not explore the reason why the documents were incorrect. Also, the audit findings report should have provision for an impact assessment. Very few audit reports do this, but clause 7.3d) clearly requires personnel to be aware of the implication of any nonconformities.

How is this demonstrated?

Demonstrating that the results of the audits are reported to relevant management may be accomplished by:

a) presenting evidence of the internal audit process in which the reporting arrangements are specified;
b) selecting a representative sample of audit reports and showing that they were presented to the manager responsible for the activity/process/area audited.

Undertaking correction and corrective action (9.2.2e)

What does this mean?

Correction of a nonconformity is the "action (taken) to eliminate a detected nonconformity" and corrective action is the "action taken to eliminate the cause of a nonconformity and to prevent recurrence" (ISO 9000:2015). Action without undue delay means that those responsible for taking these two separate actions are expected to act before the detected problem adversely impacts subsequent results. An appropriate correction is one that eliminates the nonconformity. An appropriate corrective action is one that reduces the likelihood of recurrence to an acceptable level because in some instances, it may be impossible to eliminate the cause of a nonconformity (see Chapter 54).

Why is this necessary?

There is no point in conducting audits and finding problems if those responsible for them do not intend to take action to prevent such problems adversely affecting results.

How is this addressed?

Appropriateness of actions

It often arises that problems detected by internal audits are perceived by management to have no impact on results, so it delays taking any action. Some managers may judge taking no action as an appropriate action due to the insignificance of the detected nonconformity. The solution therefore is for auditors to justify the action needed based on risk. However, there is a risk that any action taken may not remove the cause of the nonconformity; it may be palliative in consideration of forthcoming changes and therefore there may be a probability of a problem recurring at some future time. As stated in clause 6.1.1: "Actions taken to address risks and opportunities shall be proportionate to the potential impact on the conformity of products and services." Action can only be taken on known circumstances. When those circumstances change, it's not possible to predict with certainty what might happen.

Responsibility for action

For those results over which the auditee has control, the auditee should take responsibility for correcting the nonconformity but, depending on its cause may not have responsibility for preventing its recurrence. Unless the auditee is someone with responsibility for correcting the nonconformity and taking the corrective action, the auditee's manager should determine the actions required. If the action required is outside that manager's responsibility, the manager and not the auditor should seek out the appropriate authority and secure a proposal.

Acting without undue delay

Management's responsibility for acting on the findings of internal audits without undue delay should be stated in the relevant documents e.g. policy manual, job descriptions or procedures.

Target dates should be agreed for all actions and the dates should be met as evidence of commitment. Third-party auditors will search your records for this evidence so you will need to impress on your managers the importance of honouring their commitments. The target dates also should match the magnitude of the deficiencies. Small deficiencies which can be corrected in minutes should be dealt with at the time of the audit otherwise they will linger on as sores and show a lack of discipline. Others which may take 10 to 15 minutes should be dealt with within a day or so. Big problems may need months to resolve and require an orchestrated programme to be implemented. The actions in all cases when implemented should remove the problem (i.e. restore conformity). An action should not be limited to generating another form or procedure because it can be rejected by another manager thereby leaving the deficiency unresolved.

To ensure actions are implemented without undue delay, the auditor needs to be sure that a failure to act will in fact affect performance of the process or the system. Management will not implement actions that have no effect on performance even if the action required restores conformity with ISO 9001. It is therefore sensible for the auditor to explain the impact of the detected nonconformity within the audit report – possibly by using a classification convention from critical to minor. But it would be more effective if the potential impact was stated, and this requires the auditor to have a greater knowledge of the requirements be they in a contract, standard, policy, procedure or work instruction and the consequences of failing to implement them.

Actions to verify that the agreed actions have been taken and to verify that the original nonconformity has been eliminated are addressed in Chapter 58.

When all the agreed nonconformities have been eliminated, the audit report can be closed. The audit remains incomplete until all actions have been verified as being completed. Should any action not be carried out by the agreed date, the auditor needs to make a judgement as to whether it is reasonable to set a new date or to escalate the slippage to higher management. For minor problems, when there are more urgent priorities facing the managers, setting a new date may be prudent. However, you should not do this more than once. Not meeting the agreed completion date is indicative either of a lack of commitment or poor estimation of time, and both indicate that there may well be a deeper-rooted problem to be resolved.

How is this demonstrated?

Demonstrating that appropriate correction and corrective actions are being taken without undue delay may be accomplished by:

a)   selecting a representative sample of audit reports and showing that those responsible for the resolution of nonconformities:

  i   have been alerted as to the risks of not taking action;
  ii   have agreed a date by which the nonconformity will be corrected;
  iii have agreed a date by which action to eliminate the cause of the nonconformity will be completed.

a)   presenting evidence which justifies why no action to eliminate the nonconformity or the cause of the nonconformity will be taken;
b)   presenting evidence that the agreed actions have been taken by the agreed dates;
c)   presenting evidence justifying why the dates were not met;
d)   presenting evidence that the outstanding actions have been escalated to higher management for action.

Retaining evidence of the implementation (9.2.2f)

What does this mean?

The records of the audit would be records of the audit programme for the organization, audit plans for individual audits, the audit reports and records of follow-up audits. Records of the results of the audit should be included in these records. These are explained further next.

Why is this necessary?

Without adequate records of audits there is no sound basis on which to review performance of the audit process, the audit programme, individual audits and individual auditors, neither will there be objective evidence on which to base decisions for changing any of these.

How is this addressed?

Presenting a pile of audit records is one thing and easily accomplished; demonstrating you have carried out an effective audit and have an effective audit programme is quite another.

The audit programme as a record

The audit programme should be retained from year to year as a record of:

- the programme objectives;
- when audits were carried out;
- the factors affecting the timing and depth of the audit;
- the processes, functions, departments and locations audited.

These records might be useful in establishing the last time an area was audited and the reasons for the audit at that time. This means that the programme should be maintained showing planned and executed audits. It is not much use if all it shows is what you planned to audit as plans can be abandoned.

   In addition, the results of the audit programme reviews should be recorded and these records maintained as evidence that you are managing the audit programme rather than simply doing the same thing over and over regardless of the impact it is having.

Specific audit records

For a specific audit the following records should be maintained:

- audit plans (if not included in the audit reports);
- audit checklists (if not included in the audit reports);
- audit reports (summary and detailed reports);
- audit findings reports (if not included in the audit reports);
- remedial and corrective action reports (if not contained in the audit findings report).
- auditor records.

Auditor records

Records should also be maintained for individual auditors including:

- Audit experience (type and number of audits conducted, position on audit team, duration, dates, etc.).
- Auditor competence assessment.
- Auditor development (training, coaching, etc.).

How is this demonstrated?

Demonstrating that evidence of the implementation of the audit programme and the audit results has been retained may be accomplished by:

a) presenting evidence of an audit process and the organization's requirements for retaining documented information on internal audit programmes and audit results;
b) presenting the audit programmes for the current year and previous three years;
c) selecting a representative sample of audits from the audit programme and establishing that the information retained is that which is required to be retained.

Bibliography

Hoyle, D., & Thompson, J. (2009). ISO 9000 Auditor Questions Second Edition. Monmouth: Transition Support Ltd.

ISO 9000:2015. (2015). *Quality Management Systems – Fundamentals and Vocabulary*. Geneva: ISO.

56 Management review

Introduction

Top management have set the quality policy and agreed the quality objectives and means for their achievement (i.e. the quality management system). Performance monitoring, measurement and evaluation will provide information that enables top management to understand whether the organization is:

a) doing what they planned it should do or has been diverted from its purpose;
b) achieving what they wanted it to achieve or is missing the target by a long way;
c) achieving the results in the best way or is wasting time and other resources;
d) pursuing the right objectives relevant to its purpose and strategic direction or is resistant to change.

Presented with this information top management will be better informed to make decisions and set priorities for action that will influence the future direction of the organization relative to product and service quality.

It follows therefore that periodic checks are needed to establish that the QMS continues to fulfil its purpose. Failure to do so will inevitably result in deterioration of standards and performance as inherent weaknesses in the system build up to eventually cause catastrophic failure (i.e. decline in customer satisfaction and orders, lost markets and business). It may be argued that this won't happen because people won't let it happen – they will take action. If these conditions persist, what will emerge is not a managed system but an unmanaged system that is unpredictable, unreliable with erratic performance. A return to the days before the QMS was established.

Top management should never be complacent about the organization's performance. Even maintaining the status quo to maintain market position, keep customers and retain capability requires improvement. If the management review restricts its agenda to examining audit results, customer complaints and nonconformities month after month without a commitment to improvement, the results will not get any better – in fact they will more than likely get worse.

In this chapter, we examine the five requirements of clause 9.3, namely:

*   Organizing the management review (9.3.1)
*   Scheduling management reviews (9.3.1)
*   Carrying out management reviews (9.3.1)
*   Deciding what actions to take (9.3.2)
*   Retaining evidence of the results of management reviews (9.3.3)

Organizing the management review (9.3.1)

What does this mean?

Top management

Top management are the people who direct and control the organization at the highest level, so in a small organization they will be the managing director and in a large organization the senior management team headed by the chief executive officer or chief operating officer depending on the scope of the QMS.

Review purpose

The purpose of the management review is clearly stated in the requirement but it might appear that the three terms *adequacy*, *suitability* and *effectiveness* are three different ways of expressing the same thing. However, according to ISO they are three different concepts, as shown in Box 56.1.

Box 56.1   Definition of management review terms

Suitability – the extent to which the MS "fits" and is right for the organization's purpose, its operations, culture, and business systems

Adequacy – the extent to which the MS is sufficient in meeting the applicable requirements; and

Effectiveness – the extent to which planned activities are realized and planned results achieved.

(ISO/TMB/JTCG N 360, 2013)

ISO defines the term *review* as "determination of the suitability, adequacy or effectiveness of an object to achieve established objectives" (ISO 9000:2015), and therefore by requiring the QMS to be reviewed it was unnecessary to state the reason because by doing so the requirement becomes tautological, for example, top management shall review (*determine the suitability*) of the QMS at planned intervals to ensure its continuing suitability. It might have been better had the term review being defined simply as *another look at something*. However, there are some other anomalies:

a)   An effective QMS can be one that is unsuitable or inadequate.
b)   An adequate QMS is one that conforms to ISO 9001 which requires top management to ensure the QMS is suitable, adequate and effective therefore an adequate system is a suitable and effective QMS;
c)   If the planned activities are realized and the planned results are achieved but the plan was made against the wrong objectives the QMS would be ineffective, but it would also be inadequate and unsuitable because it would not conform to ISO 9001.

The QMS will evolve as the organization evolves, the two are inseparable and if the QMS doesn't respond to changing needs of society, of customers, of regulators and of other stakeholders the QMS and by association, the organization is not effective. There can be no situation where the organization is effective and the QMS is not effective and vice versa. However, there may be situations where specific elements of the system are unsuitable, inadequate and ineffective and need to change but that doesn't mean the whole system is unsuitable, inadequate and ineffective. These deficiencies simply reduce the degree of suitability, adequacy and effectiveness by a few percentage points.

Alignment with the strategic direction of the organization

There is an implication in the requirement that a QMS can be suitable, adequate and effective and yet not aligned with the strategic direction of the organization. From the definition of suitability in Box 56.1 a suitable system would be aligned with the strategic direction of the organization, if it wasn't it would be unsuitable. It would also appear that a system that was resource hungry in an economic climate that is depressed would also be unsuitable, implying that a system that is aligned with the strategic direction of the organization needs to be an efficient system as it's inconceivable that an inefficient system would be so aligned. The only use of the term *efficiency* in ISO 9001 is used in connection with the process approach where it states: "Understanding and managing interrelated processes as a system contributes to the organization's effectiveness and efficiency in achieving its intended results." It is therefore expected that the QMS be efficient so could it be that the addition of this requirement is code for *requiring* the QMS to be efficient?

Why is this necessary?

There is a need for top management to look at the results of evaluating data the system generates and determine:

a) whether the system and by association, the organization they established is doing the job they wanted it to do;
b) whether it will do the job it needs to do in the future and if not;
c) what needs to change.

Financial performance is reviewed regularly and a statement of accounts is produced every year. There are significant benefits to be gained if quality performance is treated in the same way because it is quality performance that causes the financial performance. Underperformance in any area will be reflected in the financial results.

How is this addressed?

Deciding who conducts the review

The requirement emphasizes that top management conduct the review – not the management representative or quality manager. Exactly who the description fits depend on a) the scope of the QMS and b) who directs and controls the organization at the highest level the organization(s) that is/are within that scope. Some organizations may have many divisions, and each may have a QMS, so the divisional manager and his or her direct reports fit the description of top management.

   In many ISO 9001–registered organizations, the management review is a chore, an event held once each year, on a Friday afternoon before a national holiday – perhaps a cynical view but nonetheless often true. To provide evidence of its commitment to conducting management reviews, top management would need to demonstrate that it planned for the reviews, studied the input material, considered the recommendations provided, discussed various options and decided what action should be taken.

Getting commitment

Top management will not regard the management review as important unless they believe it is essential for running the business. The way to do this is to treat it as a business performance review. This is simpler than it may appear. If the quality policy is now accepted as corporate policy and the quality objectives are accepted as corporate objectives, any review of the QMS becomes a performance review and is no different from any other executive reviews. The problem with management reviews that were focused on the documented quality system was that they allowed discussion on the means for achieving objectives to take place in other management meetings, leaving the management review to a review of errors, mistakes and documentation that few people were interested in. The QMS is the means for achieving quality objectives; therefore, it makes sense to review the means when reviewing the ends so that actions are linked to results and commitment secured for all related changes in one transaction.

One of the reasons that the management review may not work is when it is considered something separate from management's job, separate from running the business, a chore to be carried out just to satisfy the standard. This is partially due to perceptions about quality. If managers perceive quality to be about the big issues like new product or service development, major investment programmes for improving plant, for computerizing business processes, etc., then the management review will take on a new meaning. If, on the other hand, it is about reviewing nonconformities, customer complaints and internal audit records, it will not attract a great deal of attention.

How is this demonstrated?

Demonstrating that top management intends to review the QMS for its continuing suitability, adequacy, effectiveness and alignment with the strategic direction of the organization may be accomplished by:

a)   presenting evidence of a policy which declares that reviews of the QMS are to be conducted by people who direct and control the organization at the highest level;

b)   presenting evidence that the objectives of the review have been defined by top management and showing that they are consistent with the requirements of the standard.

Scheduling management reviews (9.3.1)

What does this mean?

Planned intervals mean that the time between the management reviews should be determined in advance (e.g. annual, quarterly or monthly reviews). The plan can be changed to reflect circumstances but should always be looking forward.

Why is this necessary?

By requiring reviews at planned intervals, it indicates that some forethought is needed so that performance is reviewed on a regular basis thus enabling timely decisions to be made.

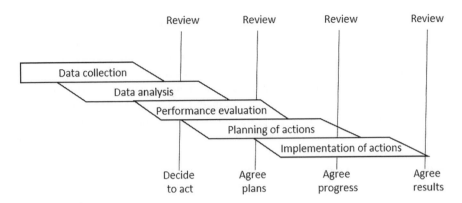

Figure 56.1 Review cycle

How is this addressed?

Duration of review

The review should be in five stages as shown in Figure 56.1. The meeting takes place to review and discuss the results of analysis and evaluation and decide on a course of action. Subsequent reviews may approve action plans, progress and the results of implementing actions. When you have a real understanding of the intentions of the review you will realize that its objectives cannot be accomplished entirely by a single meeting. If raw data are presented to top management their eyes will glaze over and they'll become bored very quickly. Therefore, all the number crunching is done well before a meeting of the top management. It could take months to collect, analyse and evaluate some aspects of performance, and only when conclusions have been reached and recommendations made should the information be presented to top management. This is not to say that interim reports should not be provided to top management at intermediate reviews as they will want to know how the work is progressing.

Number of reviews

The standard does not require only one review. In some organizations, it would not be practical to cover the complete system in one review. It is often necessary to consolidate results from lower levels and feed into intermediate reviews so that departmental reviews feed results into divisional reviews that feed results into corporate reviews. The fact that the lower-level reviews are not performed by top management is immaterial provided the results of these reviews are submitted to top management as part of the system review. From a scheduling perspective, there may need to be a review schedule that is synchronized with lower level reviews. It is also not necessary to separate reviews based on ends and means. A review of financial performance is often separated from technical performance, and both are separated from management system reviews. This situation arises in cases where the management system is perceived as procedures and practices. In organizations that separate their performance reviews from their management system reviews, one should question whether they are gaining any business benefit or in fact whether they have understood the purpose of the management system.

If there are management reviews by different levels of management, it would probably be better to call the review of the QMS by top management the management system review.

Timeline for reviews

A simple bar chart or table indicating the timing of management reviews over a given period will meet this requirement. The bar of the chart should indicate the three stages that culminate in a meeting to make decisions. The frequency of management reviews should be matched to the evidence that demonstrates the effectiveness of the system. Initially the reviews should be frequent say monthly, until it is established that the system is suitable, adequate and effective. Thereafter the frequency of reviews can be modified. If performance has already reached a satisfactory level and no deterioration appears within the next three months, extend the period between reviews to six months. If no deterioration appears in six months extend the period to 12 months. It is unwise to go beyond 12 months without a review because something is bound to change that will affect the system. A review should be held shortly after a reorganization, the launch of a new product or service, breaking into a new market, securing new customers, etc., to establish if performance has changed. After new technology is planned, a review should be held before and afterwards to measure the effects of the change.

Another method is to schedule reviews at set intervals with a changing agenda. The agenda would be driven by decisions that need to be made in a developing situation. Decisions would be put to management for review more frequently than decisions that arise less often e.g. it may be agreed that for a given period outsourced processes be reviewed monthly until they become stable, but the review of external issues could be done on an annual basis to coincide with the frequency of data collection.

How is this demonstrated?

Demonstrating that management reviews are being carried out at planned intervals may be accomplished by:

a)   presenting evidence of a schedule for management reviews;
b)   selecting a representative sample of records of management reviews and showing that they were carried out according to the schedule.

Planning and carrying out management reviews (9.3.2)

What does this mean?

The requirement means that management reviews should not be done on the spare of the moment as there are many important issues that need to be addressed. Having collected data, analysed and evaluated the results they need to be brought to the attention of top management. Some of the decisions will affect the whole system which only top management can make. In reviewing the QMS top management is reviewing the way the organization creates and retains its customers and as customers provide the revenue, it's important for the review to be highly productive.

Why is this necessary?

The requirement for planning management reviews is based on the principle that if unplanned reviews are carried out it is likely that the information needed to make decisions won't be accessible or in a form suitable for top management to decide a course of action. When reviews are planned, they lead to a more productive use of time and the achievement of objectives.

How is this addressed?

Gathering the inputs

The standard identifies six main categories of information and seven sub-categories of information about performance of the QMS which is generated from different processes. In broad terms, there are three questions that need answers:

- whether performance is in line with objectives;
- whether there are better ways of achieving these objectives;
- whether the objectives are relevant to the needs of stakeholders.

The processes to obtain the answers are represented in Figure 56.2. The inputs to the management review cover a wide range of subjects – so wide, in fact, that it's likely that opportunities for improving products, services, processes and external providers will be reviewed and action taken at lower levels than top management. In such cases reports of such reviews would be provided to top management as evidence that the arrangement they set up are working. If there needs to be a realignment of the QMS to respond to changes in stakeholder needs this should be decided upon by top management.

It should be expected that the inputs to the management review in clause 9.3.2 would match the outputs from analysis and evaluation of data of clause 9.1.3 and that these match the requirements within the standard but there is a mismatch as shown in Table 56.1. Unlike

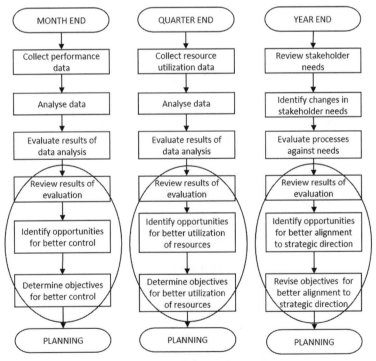

The information produced by the activities that are
encircled is presented for management review.

Figure 56.2 Information sources for management review

Table 56.1 Alignment between clauses 9.1.3, 9.3.2 and documentation requirements

Analysis and evaluation outputs (9.1.3)	Management review input (9.3.2)	Documentation requirements
a) conformity of products and services;	c3) conformity of products and services;	8.2.3.2a) products and services review 8.3.4b) design reviews results 8.3.4c) design verification results 8.3.4d) design validation results
b) the degree of customer satisfaction;	c1) customer satisfaction and feedback from relevant interested parties;	
c) the performance and effectiveness of the quality management system;	c) information on the performance and effectiveness of the quality management system (not identified elsewhere in this column)	9.1.1 results of performance and the effectiveness of the QMS
	c2) the extent to which quality objectives have been met;	
	c3) process performance	4.4.2 and 8.1e)1 providing confidence that the processes are being carried out as planned 8.3.4a) on the results to be achieved by the design and development process
	c4) nonconformities and corrective actions;	10.2.2 as evidence of the nature of the nonconformities and any subsequent actions taken
	c5) monitoring and measurement results;	8.6a) evidence of conformity of products and services with the acceptance criteria
d) if planning has been implemented effectively	c6) audit results;	9.2.2 audit programme and audit results
e) the effectiveness of actions taken to address risks and opportunities;	e) the effectiveness of actions taken to address risks and opportunities (see 6.1);	
f) the performance of external providers;	c7) the performance of external providers;	8.4.1 results of the evaluations and monitoring the performance of the external providers
g) the need for improvements to the quality management system.	f) opportunities for improvement.	
	a) the status of actions from previous management reviews	
	b) changes in external and internal issues that are relevant to the quality management system;	
Implied in d)	d) the adequacy of resources;	7.6. evidence of fitness for purpose of monitoring and measurement resources 7.6 the basis used for calibration or verification

the 2008 version there is no requirement to specifically measure characteristics of the product or service but it's implied by clause 8.6, in fact the onus is on the organization to determine what to monitor and measure, therefore for decisions to be soundly based, they need to be based on the analysis and evaluation of data, obtained from monitoring and/or measurement (see Chapter 5).

The status of actions from previous management reviews

As with all reviews, decisions will be made to act on the results of the review and progress on these actions needs to be monitored and redirection given when necessary.

Current performance on follow-up actions from earlier management reviews should address not only whether they are open or closed but how effective they have been and how long they remain outstanding as a measure of planning effectiveness. This is shown in Figure 56.1 by the four reviews.

Improvement opportunities relative to actions from management reviews may cover:

• the prioritization of actions;
• the reclassification of problems relative to current business needs;
• the need to re-design the management review process.

Changes in external and internal issues that are relevant to the quality management system

Changes in the issues that were determined to understand the context of the organization and set priorities for action will be of immediate concern to top management because they may affect the strategic direction in which the organization is moving. But as top management are accountable for the effectiveness of the QMS (clause 5.1.1a), they also need to understand how these issues affect the QMS. Clause 9.1.3g) requires the results of analysis and evaluation to be used to evaluate the need for improvement, and this is where the results of PESTLE and SWOT are reported, although only those aspects that are relevant to the QMS would be addressed by the management review.

Information on the performance and effectiveness of the QMS

Several parameters are identified in clause 9.3.2 as being indicative of the effectiveness of the QMS, and it's significant that trends are required to be presented for review and for the most part, these trends can be represented graphically. In this way top management are more able to see any problem in context. However, unchanging performance is not necessarily indicative that nothing is wrong. The evaluation should look at the bigger picture to see whether opportunities for improvement exist.

One of the items in the list of 9.3.2c) is monitoring and measurement results, which seems like it's included to cover all the monitoring and measurement that has not been specifically mentioned in 9.3.2c). They may indeed be other parameters the organization has chosen to monitor or measure to determine the effectiveness of the QMS.

In clause 9.3.2c)6 it requires audit results and not internal audit results implying it could include first, second and third party audit results. However, second-party audits, supplier audits or audits of external providers (whichever term you prefer) may be addressed by 9.3.2c)7 depending on whether conformity with QMS requirements is considered a performance parameter.

The audit results will have been subject to analysis and evaluation and therefore the information presented for review would include an overview of the audit programme showing what has and has not been audited, an overview of the extent to which the QMS conforms to regulatory requirements including ISO 9001, any specific policies of management and the extent to which the prescribed requirements, policies, plans and procedures have been implemented. The audits should give top management confidence in the veracity of the performance data being reported.

Information on the adequacy of resources is also required in 9.3.2d) but was omitted from the list in 9.1.3. It is assumed that the adequacy of resources is evaluated when evaluating implementation of planning referred to in 9.1.3d).

The effectiveness of actions taken to address risks and opportunities

This will be a difficult topic to address because having identified a risk and then taken the risk (whether mitigated or not), if no undesirable outcome occurred, there will be a temptation to believe the action was effective. The undesirable circumstances may or may not have arisen so that the provisions could be tested. With some risk mitigation measures, threats can be monitored (e.g. number of attacks by computer hackers).

With opportunities, there are opportunities that arise and must be seized to take advantage of them and therefore there may be tangible consequences of missing such opportunities that can be reported.

Opportunities for improvement in the QMS

Opportunities for improvement may be discovered during the evaluation of processes and suppliers that may lead to improvements in the QMS, and this can be reported to management review for information. The opportunities for improvement of interest to top management are those that will bring about a breakthrough in performance of the system and are likely to arise where the objectives of the primary processes of demand creation, demand fulfilment, resource management and mission management are changed.

How is this demonstrated?

Demonstrating that the management review is planned and carried out taking into consideration pre-defined inputs may be accomplished by:

a)   presenting evidence of a process for planning management reviews;
b)   explaining the criteria for selecting and validating input data;
c)   showing the steps that are taken to ensure that input data to the review has been through planned analysis and evaluation stages and is in a form suitable for presentation to top management.

Deciding what actions to take (9.3.3)

What does this mean?

The review is of current performance and hence there will be some parameters where objectives or targets have not been accomplished thus providing opportunities for improvement. There will also be some areas where the status quo is not good enough for the growth of the organization or to meet new challenges.

Improving the effectiveness of the management system is not about tinkering with documentation but enhancing the capability of the system so that the organization fulfils its objectives more effectively. The QMS comprises processes; therefore, the effectiveness of these too must be improved. Improvement of product related to customer requirements means not only improving the degree of conformity of existing product but enhancing product features so that they meet changing customer needs and expectations.

Why is this necessary?

The outputs of the management review should bring about beneficial change in performance. The performance of products is directly related to the effectiveness of the processes that produce them and the performance of these processes is directly related to the effectiveness of the system that connects them. Without adequate resources, no improvement would be possible.

How is this addressed?

There will be reports about new marketing opportunities, reports about new legislation, new standards, the competition and benchmarking studies. All these may provide opportunities for improvement.

The implication of this requirement is that the review should result in decisions being made to improve products, processes and the system in terms of the actions required. In this context improvement means improvement by better control (doing things better), improvement by better utilization of resources (doing things smarter) and improvement by innovation (doing new things).

Actions related to improvement of the system should improve the capability of the system to achieve the organization's objectives by undertaking action to improve the interaction between the processes.

Actions related to improvement of processes should focus on making beneficial changes in methods, techniques, relationships, measurements, behaviours, capacity, competency etc. A quick fix to overcome a problem is neither a process change nor a process improvement because it only acts on a particular problem. If the fix not only acts on the present problem but will also prevent its recurrence, it can be claimed to be a system improvement. This may result in changes to documentation but this should not be the sole purpose behind the change – it is the performance that should be improved.

Actions related to improvement of products should improve:

*   The quality of design. The extent to which the design reflects a product that satisfies customer needs.
*   The quality of conformity. The extent to which the product or service conforms to the design.
*   The quality of use. The extent to which the user can secure continuity of use of the product or service and low cost of ownership from the product.

Such actions may result in providing different product and service features or better-designed product and service features as well as improved reliability, maintainability, durability and performance. Product improvements may also arise from better packaging, better user instructions, clearer labelling, warning notices, handling provisions, etc.

Actions related to resource needs are associated with the resource planning process that should be part of the management system. Such actions may serve to improve the utilization of existing resources (efficiency) or serve to improve the allocation of resources (effectiveness). It is always a balance between time, effort and materials. If the effort cannot be provided the time has to expand accordingly.

How is this demonstrated?

Demonstrating that management review outputs include decisions on improvement may be accomplished by:

a)   presenting evidence of a process for conducting management review and showing that the outputs of the review are intended to include decisions and actions related to opportunities for improvement.
b)   selecting a representative sample of records from management reviews and presenting evidence of decisions that were made and actions taken to seize opportunities for improvement in products, processes, resources and the system.

Retaining evidence of the results of management reviews (9.3.3)

What does this mean?

The results of management reviews are what they produce, and these will be tangible things such as the decisions taken but will also be intangible results such as confidence. However, the results won't be understood unless placed in the right context and therefore the circumstances under which the review was conducts also need to be documented and retained.

Why is this necessary?

Documented information from management reviews needs to be retained for several reasons:

*   To convey the actions from the review to those who are to take them.
*   To convey the decisions and conclusions as a means of employee motivation.
*   To enable comparisons to be made at later reviews when determining progress.
*   To define the basis on which the decisions have been made.
*   To demonstrate system performance to stakeholders.

How is this addressed?

The information from management reviews that needs to documents includes the following:

*   The date and location of review.
*   Contributors to the review and the role they played.
*   The criteria against which the QMS is being judged for suitability, adequacy and effectiveness.
*   The results of analysis and evaluation submitted, testifying the current performance of the QMS.

- Any changes in the internal and external environment (identified risks, opportunities and potential stakeholder needs).
- Conclusions (is the QMS effective and capable of coping with the projected changes in the environment and if not why not?)
- Actions and decisions (what will stay the same and what will change).
- Responsibilities and timescales for undertaking the agreed actions (who will do it and by when is it required to be completed).

How is this demonstrated?

Demonstrating that evidence of the results of management reviews is being retained may be accomplished by:

a)   presenting evidence of a process for capturing the results of management review
b)   selecting a representative sample of management reviews and retrieving documented information describing the results of the review.

Bibliography

ISO 9000:2015. (2015). *Quality Management Systems – Fundamentals and Vocabulary*. Geneva: ISO.
ISO/TMB/JTCG N 360. (2013, December 3). Concept Document to Support Annex SL. Retrieved from ISO Standards Development: http://isotc.iso.org/livelink/livelink?func=ll&objId=16347818&objAction=browse&viewType=1

Key messages from Part 9

Chapter 52 Monitoring, measurement, analysis and evaluation

1 You know nothing about an object until you can measure it, but you must measure it accurately and precisely.
2 To have confidence that the intended results are being achieved, before we measure we need standards, targets and requirements we can use to judge the results of measurement.
3 Measurement begins with a definition of the measure, the quantity that is to be measured, and it always involves a comparison of the measure with some known quantity of the same kind.
4 Quality is an intangible until we specify what we mean by quality and express it in terms of characteristics we can measure.
5 There should be no measurement without recording, no recording without analysis, no analysis without evaluation and no evaluation with action even if it is to keep going.
6 Pick the wrong time for measuring or monitoring a quantity and the data produced may be unreliable, invalid and useless for further analysis and evaluation.

Chapter 53 Customer satisfaction

7 Data on customer perceptions will serve to validate not only the business outputs but also the assumptions made about customer requirements.
8 The most important part of a customer survey is to ask the right questions, and these need to be derived from what customers expect from your products not what you want from them.
9 There is no standard customer so we need to look at the demographics of a given population to find the variables and determine their relevance.
10 The integrity of the process for determining the degree of customer satisfaction is paramount; otherwise, you could be fooling yourselves into believing all is well when it is far from reality.

Chapter 54 Analysis and evaluation

11 Analysis is a thinking process – it's objective, whereas evaluation is judgemental and reaches a conclusion after assessing the results of the analysis.
12 Conformity is judged by the extent to which a specification is met and normally determined by the producer whereas, quality is judged by the extent to which needs and expectations are met and determined by the customer or their representative.

13 When analysing product conformity, look for trends to indicate specific types of variation that are symptomatic of certain types of problems.

14 When analysing service conformity, variables that cannot be measured directly need to be translated into variables that can be measured directly but it won't be sufficient to limit the translation to a single variable.

15 If we can demonstrate that the process is stabilized about a constant location, with a constant variance and a known stable shape, then we have a process that is both predictable and controllable.

16 The cause of variations in the location, spread and shape of a distribution is considered special or assignable because the cause can be assigned to a specific or special condition that does not apply to other events.

17 Common cause variation is random and therefore adjusting a process on detection of a common cause will destabilize the process.

18 If effort has been spent to control risks and enhance opportunities, it is necessary to discover whether it was worthwhile so as to inform future decisions.

Chapter 55   Internal audit

19 Both customers and managers have a need for an assurance of quality as they are not able to oversee operations for themselves – this is the role on the internal audit.

20 If we begin with the end in mind, we would start an audit by wanting to know if the system was achieving its intended results, and only afterwards establish if it conformed to ISO 9001.

21 To run an effective management system, auditing should aim for best practice and by defining and refining auditing methods best practice is established for the benefit of the organization.

22 Auditing is a skill, and proficiency in auditing is more effectively demonstrated when the auditor is observed asking questions of people at their place of work, assessing the answers and evidence presented, drawing conclusions and presenting evidence substantiating their findings.

Chapter 56   Management review

23 When presented with evidence of whether the QMS they established is doing the job they wanted it to do, will it do the job it needs to do in the future and, if not, what needs to change, top management will be better informed to make decisions and set priorities for action relative to product and service quality.

24 Top management will not regard the management review as important unless they believe it is essential for running the business.

25 The QMS is the means for setting and achieving quality objectives; therefore, it makes sense to review the means when reviewing the ends so that actions are linked to results and commitment secured for all related changes in one transaction.

26 Improving the effectiveness of the management system is not about tinkering with documentation but enhancing the capability of the system so that the organization fulfils its objectives more effectively.

Part 10

Improvement

Introduction to Part 10

The focus of improvement

ISO 9000:2015 defines improvement as: "activity to enhance performance" and it results from a commitment to do things better or do different things. Although there is a note in clause 10 stating that examples of improvement can include correction, it must be stated that correction is not an improvement because it simply restores the status quo (i.e. the current level of performance) – it's removal of special cause variation (i.e. nonconformities) such as putting out fires and is thus an element of quality control. Enhancing performance is increasing the level of performance beyond the current level and correcting errors won't enhance performance unless action is taken to prevent them recurring. If we want to increase the level of performance, we either should reduce common cause variation or pursue different objectives, and this means we should act on the system. So improvement is not concerned with correcting errors but with doing things better and doing different things.

Another anomaly is that improvement is not an activity but an objective or outcome. We can undertake activities with the object of improving a situation, but unless and until the situation improves there is no improvement. The action we took may not yield improvement at all, and therefore it is only with hindsight that we can say that the activities we undertook enhanced performance. Therefore, improvement is a result of activities that bring about enhanced performance.

There is a second dimension to improvement – it is the rate of change. We could improve *gradually* or by a *step change*. Gradual change is also referred to as *incremental improvement, continual improvement or kaizen*. *Step change* is also referred to as *breakthrough* or a *quantum leap*. Gradual change arises out of refining the existing methods, modifying processes to yield more and more by consuming less and less. Breakthroughs often require innovation, new methods, techniques, technologies and new processes.

The improvement process

Improving the level of performance and effectiveness can be accomplished by the following 10 steps adapted from those advocated by Dr Juran.

1 Determine the objective to be achieved (e.g. to counter threats or seize opportunities in the external environment, to increase competitiveness, to improve efficiency or effectiveness). (This is addressed by clause 6.2.)

2   Determine the policies needed for improvement (i.e. the broad guidelines to enable management to cause or stimulate the improvement). (This is addressed by clause 6.3.)
3   Conduct a feasibility study. This should discover whether accomplishment of the objective is feasible and propose several strategies or conceptual solutions for consideration. If feasible, approval to proceed should be secured. (This stage is addressed in clause 9.1.3 under evaluating the need for improvement.)
4   Produce plans for the improvement that specifies how the objective will be achieved. (This is addressed by clauses 6.2, 6.3, 8.1 and 8.3.)
5   Organize the resources to implement the plan. This includes organizing project teams and developing new knowledge when necessary. (This is addressed by clauses 5.3, 7.1.1 and 7 1.6.)
6   Carry out research, analysis and design to define a possible solution and credible alternatives. (For products and services, this is addressed by clause 8.3.4 and also for processes but only if clause 8.3 is applied to the design and development of processes.)
7   Model and develop the best solution and carry out tests to prove it fulfils the objective. (For products and services, this is addressed by clause 8.3.4 and also for processes but only if clause 8.3 is applied to the design and development of processes.)
8   Identify and overcome any resistance to changing the level of performance required. (This is addressed by clause 5.1.1.)
9   Implement the change (i.e. put new products into production and new services into operation). (This is addressed by clause 8.5.)
10 Put in place the controls to hold the new level of performance. (This is addressed by clauses 4.4 and 8.5.)

Structure of requirements

This part of the standard is intended to align with the Act part of the PDCA cycle. After checking whether what we did was correct, it's not unreasonable to assume that the next stage would be to Act on what we discovered from the checks. This would normally involve putting right what was wrong and it might appear from clause 0.3.2 of ISO 9001 that we are on the right track. However, PDCA is not a linear flow but a cyclic flow. It's a cycle for learning and improvement of a product, service or process, not a cycle for controlling a product, service or process. Therefore, the Act part of the cycle in this context is concerned with deciding what action if any should be taken having been presented with the results of an evaluation into the performance of products, services, processes, external providers as well as of the whole QMS. These results will provide knowledge about existing products, services, processes and external providers as well as those that are new or have been subject to change and will therefore help you decide:

a)   that no change is necessary at the present time or;
b)   that the changes we made brought about the desired improvement so can be adopted or;
c)   that the changes we made didn't bring about the desired improvement so must be abandoned or;
d)   that the objective for change remains valid but the plan for achieving it needs to be revised or;
e)   the performance evaluation revealed further need for improvement.

There are several decision-making processes, but some decisions are made to decide whether a process should stop or continue. In other cases, decisions are made to decide whether there should be change in the level of performance and effectiveness and this is the purpose of product, service, process and system reviews.

However, management review is in clause 9 on performance evaluation when it may have been more appropriate if it had been included in clause 10. Also in clause 9 is a requirement to evaluate the need for improvement to the QMS and therefore opportunities for improvement will have already been identified before we reach clause 10.

However, we should bear in mind that the order of requirements is not significant and therefore whatever the subject of the performance evaluation (the Check part of the PDCA cycle) proposed improvements need to be prioritized before authorization of planning.

Dispersion of improvement requirements

There several requirements on improvement in the standard which can be grouped together in terms of their purpose.

Commitment to improvement

1 Top management to promote improvement (5.1.1).
2 Top management to declare a commitment to continual improvement of the QMS (5.2.1).
3 Top management to assign the responsibility and authority for reporting on opportunities for improvement (5.3).
4 To provide the resources needed for continual improvement of the QMS (7.1.1).

The intent of improvements

1 To determine the risks and opportunities that need to be addressed to achieve improvement (6.1.1d).
2 Use the results of improvements in processes, products and services to enhance organizational knowledge (7.1.6).

Determining opportunities for improvements

1 Take into consideration opportunities for improvement when planning the management review (9.3.2).
2 Use the results of analysis to evaluate the need for improvements to the QMS (9.1.3).
3 Determine and select opportunities for improving the performance and effectiveness of the QMS (10.1).
4 Consider results of analysis, evaluation and management review to determine opportunities for continual improvement (10.3).

Action to bring about improvement

1 Include in the outputs of the management review decisions and actions related to opportunities for improvement (9.3.3).
2 Implement (improvement) actions to meet customer requirements and enhance customer satisfaction (10.1).
3 Continually improve the suitability, adequacy and effectiveness of the QMS (10.2).

Had the word *improvement* been mentioned in clause 6.3, it would have closed the PDCA loop, as shown in Figure 5.4, and signalled that improvement is not a separate action but a result, the achievement of which needs to be planned and implemented. In the PDCA cycle the A means *Act* but the action is not carrying out the improvement but deciding what action to take. The next stage is planning the change and only when implementing the change will improvement be brought about.

This part of the Handbook comprises three chapters (Chapters 57–59) that address the following clauses of ISO 9001:2015:

* Clause 10.1 – Determining and selecting opportunities for improvement
* Clause 10.2 – Nonconformity and corrective action
* Clause 10.3 – Continual improvement of the QMS

Bibliography

Juran, J. M. (1964). *Managerial Breakthrough*. New York: McGraw-Hill.

57 Determining and selecting opportunities for improvement

Introduction

The concept of improvement

The road to improvement starts when we are motivated to monitor and measure the intended outputs of a process because we want to know whether we are doing it right, is our output good or bad. By collecting and analysing all that data over several iterations of the process we will observe the level at which we are performing, and by evaluating these results we will learn whether we are meeting the performance targets. It is now that we realize how big or small the gap in our performance is and can identify opportunities for improvement. By measuring the unintended outputs of the process, we get an idea whether we are doing it in the best way and notice the inefficiencies and thereby identify further opportunities for improvement. Lastly we want to know if we are aiming at the right target and on looking again at what our customers need we discover some misalignment and thereby identify more opportunities for improvement. These three processes are illustrated in Figure 56.2.

In many cases, organizations have focused on improving the work processes, believing that there would be an improvement in business outputs, but often such efforts barely have any effect. It is not until we stand back that the system comes into view. A focus on work processes and not business processes is the primary reason why ISO 9001, TQM and other quality initiatives fail. They resulted in sub-optimization – not optimization of organizational performance. If the business objectives are functionally oriented, they tend to drive a function-oriented organization rather than a process-oriented organization. When process-oriented objectives, measures and targets, focused on the needs and expectations of external stakeholders are established, the functions will come into line and organizational performance can be optimized.

References to improvement

In the introduction to Part 10 we listed the 13 references to improvement according to their purpose. Following we list the requirements for improvement in clause order so as point out the duplication:

1 The organization is to improve the processes and the QMS (clause 4.41h).
2 The organization is required to continually improve a QMS in accordance with the requirements of ISO 9001 (clause 4.4.1).
3 Top management is to demonstrate leadership and commitment with respect to the QMS by promoting improvement (5.1.1).

4 The quality policy is required to include a commitment to continual improvement of the QMS (5.2.1).
5 Top management is to assign the responsibility and authority for reporting opportunities for improvement (5.3c).
6 When planning for the QMS to determine the risks and opportunities that need to be addressed to achieve improvement (6.1.1d).
7 Resources needed for the continual improvement of the QMS are required to be determined and provided (7.1.1).
8 Persons doing work under the organization's control are to be aware of the benefits of improved performance (7.3).
9 The results of analysis are to be used to evaluate the need for improvements to the QMS (9.1.3).
10 Management review inputs are to take into consideration opportunities for improvement (9.3.2f).
11 Management review outputs are to include decisions and actions related to opportunities for improvement (9.3.3a).
12 The organization is to determine and select opportunities for improvement and implement any necessary actions to meet customer requirements and enhance customer satisfaction (10.1).
13 The organization is required to continually improve the suitability, adequacy and effectiveness of the QMS (10.3).

Although this might appear as a gross duplication of requirements, there are only three which may be regarded as duplicates as each of the others address a different aspect of improvement. Requirements (1) and (2) express the same requirement but in slightly different ways. From an auditing perspective evidence of conformity with (2) would satisfy (1) but not vice versa because the improvement may not be continual. Requirement (2) indirectly refers to (13) so can be discounted as a separate requirement. Requirements (12) and (13) appear similar in intent but the difference is brought out by the separation of improvements in products and services from improvement in the QMS in the subsequent clauses. Improving products and services without improving the QMS is addressed in the introduction to Chapter 59.

In this chapter, we examine the four requirements of clause 10.1, namely:

- Determining and selecting opportunities for improvement (10.1)
- Improving products and services (10.1a)
- Correcting, preventing or reducing undesirable effects (10.1b)
- Improving quality management system results (10.1c)

The first requirement is a general requirement, meaning that opportunities for improvement are not intended to be limited to the three other specific areas for improvement.

Determining and selecting opportunities for improvement (10.1)

What does this mean?

The way the requirement is expressed implies that at any time, at any stage and at any level the organization is to determine and select opportunities for improving customer satisfaction.

It may appear tautological to place *meet customer requirements* and *enhance customer satisfaction* in the same sentence, but one may satisfy customers by meeting most of their requirements and enhance customer satisfaction by meeting *all* customer requirements.

The note to the requirement refers to different ways of bringing about improvement:

Corrective action is intended to bring about improvement by removing the cause of an existing systemic problem and preventing its recurrence and is therefore reactive. It's reducing common cause variation and is different from removing a nonconformity by correcting the error because it may recur.

Continual improvement is recurring activity to enhance performance and is therefore incremental or gradual in nature. The activity occurs over a period, but with intervals of interruption (unlike *continuous* which indicates occurrence without interruption).

Breakthrough is an improvement to a level of performance that has hitherto not been achieved in the organization and is therefore a step change.

Innovation is an improvement that comes about by application of knowledge that is new to the world or new to a particular field of study. It can produce an incremental improvement, a step change improvement or a transformation. Any improvement may be accomplished either by utilizing existing knowledge or creating new knowledge (innovation).

Transformation is an improvement that comes about by changing an organization from one form to another, for example, changing it from a reactive to a proactive organization, from an organization with uncertain values to one with shared values, from a small local business to a global business. It therefore affects everyone and everything in the organization. Deming's 14 points for management were developed as a basis for *transforming* American industry (Deming, 1982). Some organizations made the transition, but many more hang on to obsolete styles of management.

Why is this necessary?

When organizations cease changing they become fossils. Even maintaining the status quo requires organizations to change because all around them is changing. There are fast lanes, slow lanes and stop lanes. As Deming wisely remarks, "Learning is not compulsory, it's voluntary, improvement is not compulsory, it's voluntary, but to survive we must learn" (Voehl, 1995).

How is this addressed?

The quality policy includes a commitment to continual improvement of the QMS which implies a commitment to improving anything that will ultimately enhance customer satisfaction. Deploying this policy to all levels should result in the creation of a culture in which everyone is motivated to look for opportunities for improvement – this deals with the *determine opportunities for improvement* part of the requirement. Many of these opportunities may arise as a result of the planned analysis of monitoring and measurement results as we discussed in Chapter 54, but it's unwise to limit the source of such opportunities to the formal processes as some may arise out of a casual observation when a person is engaged on doing something else. A suggestions box or form such as an opportunities for improvement (OFI) form could be devised for this purpose and a means provided for capturing the suggestions in the formal improvement processes.

The identified opportunities need to pass through an evaluation process to select those that will potentially improve the organization's ability to meet customer requirements and enhance customer satisfaction. Every improvement has to improve the performance of the system as a whole; otherwise, it's wasted effort (see Box 57.1).

Box 57.1   Improving the whole and not the parts taken separately

If we have a system of improvement that is directed at improving the parts taken separately you can be absolutely sure that the performance of the whole will not be improved. We don't improve the quality of a part unless by doing so we improve the quality of system of which it forms a part.

(Ackoff, 1994)

This evaluation process is also addressed in Chapter 54. At this stage, it may be possible to classify the potential improvement as a corrective action, continual improvement, breakthrough, innovation or transformation simply by the way the objective is expressed, but it's not necessary to classify potential improvements at all. Perhaps after the changes have been made and found to be successful, you may be able to say it's an innovation, a breakthrough or an incremental change.

The third part of this requirement is for necessary actions to be taken. This requires a planning process which is addressed in Chapter 23 and the linkage shown in Figure P6.1. Each improvement plan will differ. The corrective action could be planned in a sentence on a form, the continual improvement follows the PDCA cycle, with each iteration bringing about a gradual change. Planning a breakthrough in science, technology, human relations, communications, etc., will be treated as a project and may require research and development. Innovation can be part of any type of improvement but creating innovative technologies, products and services may require a breakthrough in knowledge and then development planning. Transformations are quite different because its companywide and often involves a reorientation of attitudes and behaviours. It mobilizes the workforce around a new set of goals and strategies and is high risk.

How is this demonstrated?

Demonstrating that opportunities for improvement are being determined and selected to enhance customer satisfaction may be accomplished by:

a) presenting evidence of a process or processes to determine and select improvements which enhance customer satisfaction;
b) presenting evidence that the improvement policy is inducing staff to identify improvement opportunities and that these are captured by the improvement processes;
c) selecting a representative sample of improvement initiatives and presenting evidence that each one was:

   i   evaluated for its impact on system effectiveness;
   ii selected using soundly based criteria;
   iii routed through an appropriate planning process;

iv   implemented as planned and the resultant change measured, analysed and evaluated;

v   reviewed and a decision taken to adopt it, abandon it or rethink it.

Improving products and services (10.1a)

What does this mean?

In general, this clause requires improvement in product and service conformity to meet current needs and improvement in product and service features to meet future needs.

Why is this necessary?

Improving product and service conformity is necessary to meet customer requirements and enhance customer satisfaction. Customer needs and expectations change, and therefore it's necessary to predict how they will change and initiate improvement to products and services either by modifying existing products and services or developing new products and services.

How is this addressed?

The result of monitoring and/or measuring product and service characteristics and the analysis and evaluation of these results revealed opportunities for improving conformity as we discussed in Chapter 54 relative to clause 9.1.3a). However, as indicated in clause 4.2 there is a requirement to also monitor the needs and expectations of interested parties and this will include the future needs and expectations of customers. As was shown in Table 56.1, there is provision in the inputs to management review for the future needs and expectations of customers to be considered.

Improvement in product and service conformity is generally addressed through the corrective action process. This is a problem-solving process which identifies the root cause of variation and closes the gap. (see Chapter 58 for further details). Improvement in product and service features is addressed through the design and development process which is addressed in Chapter 37.

How is this demonstrated?

Demonstrating that improvements are being made to product and service conformity as well as to address future needs and expectations may be accomplished by presenting evidence that the improvement processes identify opportunities for improving product and service conformity as well as improving product and service features to meet future needs.

Correcting, preventing or reducing undesirable effects (10.1b)

What does this mean?

The way in which the requirement is expressed implies that undesirable effects are those effects which affect customer satisfaction rather than other aspects of an organization's performance.

The requirement draws attention to one tenet of quality management – that of prevention, detection and correction – but the requirement to prevent undesirable effects appears to imply some degree of prediction as to what might cause undesirable effects and could be indirectly invoking risk assessment. Any improvement will therefore make the undesired effects less likely or have less severe consequences. The inclusion of the word *prevent* here seems to resolve an anomaly in clause 6.1.1 where the risks and opportunities concerned were only those to be addressed when planning *for the* QMS and not when planning the QMS or its products and services. On the other hand, it may be referring to preventing (recurrence of) nonconformities (i.e. corrective action).

An error that has an immediate undesirable effect is likely to be subject to a customer complaint and will be dealt with using the processes that address clause 8.2.1c) (see Chapter 34). However, the error may be of a type where the boiling frog metaphor applies (see Box 57.2). Customers probably won't complain if the errors are imperceptible but increase over time – eventually they take their business elsewhere without giving notice.

Box 57.2   The boiling frog metaphor

The premise is that if a frog is put suddenly into boiling water, it will jump out, but if it is put in cold water which is then brought to a boil very slowly, it will not perceive the danger and will be cooked to death. Although not based on fact (because a frog put in boiling water will die), the story is often used as a metaphor for the inability or unwillingness of people to react to or be aware of threats that rise gradually.

The requirement to reduce undesirable effects is addressing the current situation and the action necessary may cover the whole range of improvement types as defined earlier.

Why is this necessary?

One reason for improvement is survival and an organization won't survive unless it tackles the problems that either could potentially cause or are currently causing an undesirable effect on customer satisfaction.

How is this addressed?

Correcting undesirable effects as defined above, is addressed through control of nonconformity and corrective action (see Chapter 58).

Preventing undesirable effects is addressed by undertaking risk assessment on products, services and processes see:

- Considering potential consequences of failure (8.3.3e) in Chapter 38
- Controlling design verification (8.3.4c) in Chapter 39
- Process risk assessment in Chapter 16

Reducing undesirable effects is addressed by pursuing opportunities for improvement identified from analysis and evaluation of monitoring and measurement as explained in Chapter 54.

How is this demonstrated?

Demonstrating that opportunities include correcting, preventing or reducing undesirable effects may be accomplished by presenting evidence of opportunities for improvements that have been addressed by corrective action, process, product and service risk assessments.

Improving quality management system results (10.1c)

What does this mean?

This requirement duplicates that in clause 9.1.3g) which requires the results of analysis to be used to evaluate the need for improvements to the quality management system. It also duplicates clause 9.3.3 which requires management review outputs to include decisions and actions related to any need for changes to the quality management system (to ensure its continuing suitability, adequacy and effectiveness).

Bibliography

Ackoff, R. L. (1994). Beyond Continuous Improvement. Retrieved from www.youtube.com/watch?v=OqEeIG8aPPk.

Deming, W. E. (1982). *Out of the Crisis*. Cambridge, MA: The MIT Press.

Voehl, F. (1995). *Deming the Way We Knew Him*. New York: Taylor & Francis.

58 Nonconformity and corrective action

The best-laid plans of mice and men oft go awry
Translated from the Scots poem "Tae a Moose" by Robert Burns (1785)

Introduction

Every system needs a self-correction mechanism. When we undertake planning, we are predicting that the plans we make will ensure success because we think we know the right things to do and how to do things right. However, we have assumed that we have addressed the significant risks, created the conditions that will motivate our people to achieve the goals and provided them with all the resources needed for them to deliver the required performance. But in reality, we can't know the future and can't control every variable.

Within ISO 9000:2015 there is a term used for when plans go awry, a term for fixing a problem, a term for restoring normality and a term for preventing plans going awry again, and these are explained further next.

Nonconformity

A nonconformity is the non-fulfilment of a need or expectation that is stated, generally implied or obligatory. Therefore, the term can apply over a range of situations from a failure to carry out a task an individual is expected to perform, to a failure to meet the needs and expectations of customers and other stakeholders and all situations in between.

This definition prompts questions about applicability and scope because Annex SL has only one clause on nonconformity whereas ISO 9001 has two. Clause 10.2 addresses both the control of nonconformity (clause 10.2.1a) and action to prevent its recurrence (clause 10.2.1b onwards). Clause 8.7 only deals with the control of nonconforming process outputs in production and service provision but includes delivered product and service. Clause 10.2.1a) is unspecific and appears to have the same intent as clause 8.7. To resolve the conundrum, we draw on two other ISO documents.

- ISO TS 9002 explains that the intent of clause 8.7 is "to prevent nonconforming outputs from progressing to the next stage or to the customer" (ISO/TC 176/SC 2/N1338, 2015) but the clause also applies to nonconforming products and services detected after delivery of products, during or after the provision of services. However, apart from a brief mention of correcting nonconformity it makes no reference to the similarity between the requirements of 8.7 and 10.2.1a).

- The supporting guide to Annex SL states that "the intent of the clause on Nonconformity and corrective action (clause 10.2) is to specify the requirements for responding when the MSS and MS (including operational) requirements are not satisfied" (ISO/TMB/JTCG N 360, 2013).

The conclusion we draw on the question of applicability is that, clause 10.2.1a) has the same intent as clause 8.7 but 10.2.1a) applies to all nonconformities and 8.7 only applies to process outputs that are destined for customers in one way or another as explained in Chapter 51.

Furthermore, as the requirements on internal audit no longer require follow-up action the requirement to review the effectiveness of any corrective action taken in clause 10.2.1 also applies to actions taken on the results of internal audits. We therefore cannot conclude 10.2 only applies to systemic failures.

On the question of scope, clause 8.7 specifically addresses actions required to deal with nonconforming outputs which are not included in 10.2.1a), but unlike clause 8.7, clause 10.2.1a) addresses the consequences of nonconformity. To sum up, it's a bit of a mess but as we have stated several times, the order in which the requirements are stated is not significant and therefore one should apply a pick and mix approach and a dash of common sense.

Correction

Correction is the term used to describe the action of removing an actual or suspect nonconformity in a product before its acceptance. Correction does not stop the nonconformity recurring. As correction is applied before a product is completed, actions intended to restore, recover or remedy the situation are inappropriate as a conforming condition has not been reached. Examples are:

- While observing the performance of a process you notice that the values are drifting towards the upper limits. You adjust the process and bring it back under control. You have corrected the process and avoided an occurrence of product nonconformity but not a future occurrence because you have merely delayed its occurrence.
- You are delivering a service and the customer points out an error which you correct immediately.
- Sometimes a product may be inadvertently submitted for verification before all operations have been completed and in such cases correction involves normal operations to complete the item.
- A product can be made to conform by continuation of processing; this type of correction is called rework.
- An unknown state can be corrected by carrying out verification and declaring the product conforming or nonconforming.

Remedial action

Remedial action is the term used to describe the action of removing an actual nonconformity in a product that was previously deemed conforming. Examples are:

- If a conforming product has been damaged or in some other way becomes nonconforming, action to remove the detected nonconformity is a remedial action.

- An automated cash machine in a bank has malfunctioned and the supervisor investigates, locates the problem and resets the operating conditions.
- Repair is a remedial action that restores an item to an acceptable condition but unlike rework, it may involve changing the product so that it differs from the specification but fulfils the intended use.

Recovery action

Recovery action is the action of restoring the status quo which may involve seeking out products with the same characteristics as those found nonconforming and/or, taking an action to deal with the consequences of nonconformity and restore a situation to normal. The latter may be significant, as was the case with the Deepwater Horizon drilling rig explosion of 2010. It is not a corrective action if it removes a nonconformity and not the cause of the nonconformity. Recall action is a recovery action even though the nonconformity might not have been exhibited.

Corrective action

Corrective action is the term used to describe the pattern of activities that traces the symptoms of an actual or suspect nonconformity to its cause, produces solutions for preventing its recurrence, implements the change and monitors that the change has been successful. Such an action prevents the recurrence of the nonconformity. A problem must exist for you to take corrective action. Corrective action uses root cause analysis to discover and eliminate the actual causes. Containment action removes an immediate cause, thus allowing production to continue until product or process design change removes the root cause. Examples are:

- The root cause of nonconformity is found to be a deficiency in design practices. The action to introduce new design practices is a corrective action.
- If the conditions for nonconformity are present because they have been detected in other similar products but have not yet resulted in failure, the action of preventing failure is a corrective action. This argument is based on the premise that a nonconformity is a non-fulfilment of a requirement, therefore a product that has not exhibited failure but possess the potential for failure is a nonconforming product not a potentially nonconforming product (i.e. we know the nonconformity exists, it's just a matter of time before it causes a failure).
- If a process is already running and will produce nonconforming output either now or in the future, it is not capable. Any action undertaken to eliminate an existing flaw in the process before a nonconforming product is produced is a corrective action. However, by taking corrective action on the process your action becomes a risk treatment on the product.

The steps in the corrective action process are identified in Table 58.1 with cross-reference to the sub-clauses of 10.2.1. Clause 10.2.1f) is superfluous as corrective actions change the QMS anyway – that is its purpose, so no further changes are necessary. If it is believed that further changes are necessary, there is probably a misunderstanding not only of what a corrective action is, but also of what a QMS is.

The responsibilities vary through the corrective action process as shown in Table 58.2.

Table 58.1 Steps in the corrective action process

	Action	Clause 10.2.1 requirement
1.	Collect the nonconformity data and classify	Reviewing and analysing nonconformities (10.2.1b)
2.	Conduct Pareto analysis to identify the vital few and trivial many and evaluate the need for action	
3.	Organize a diagnostic team	Determining the cause and extent of nonconformity (10.2.1b)
4.	Postulate causes and test theories	
5.	Determine the root cause of nonconformity	
6.	Determine if similar nonconformities exist or could exist elsewhere	
7.	Determine the effects of nonconformity and the need for action	Implementing any action needed (10.2.1c)
8.	Determine the action needed to prevent nonconformity recurring	
9.	Organize an implementation team	
10.	Create or choose the conditions which will ensure effective implementation	
11.	Implement the agreed action	
12.	Record the nature of the nonconformities	Retain documented information (10.2.2)
13.	Record the results of Pareto analysis	
14.	Record the causes of nonconformity	
15.	Record the criteria for determining severity or priority	
16.	Record the proposed actions to be taken	
17.	Record the actions actually taken	
18.	Record the results of actions taken	
19.	Assess the actions taken and determine if they were those required to be taken	Reviewing corrective actions (10.2.1.d)
20.	Determine whether the actions were performed in the best way	
21.	Determine whether the nonconformity has recurred	
22.	If nonconformity has recurred repeat steps 1–22	
23.	Review risk assessments and opportunities and update if necessary	Update risks and opportunities determined during planning (10.2.1e)

Table 58.2 Corrective action responsibilities

Action	Responsibility
Reviewing variations	Process owner
Dealing with the nonconformity	Process owner
Seeking for suspect product	Process owner
Handling the consequences	Rapid reaction force if necessary
Determining the cause	Diagnostic team
Evaluating the action needed	Diagnostic team
Implementing the corrective actions	Implementation team
Reviewing corrective actions	Diagnostic team

In this chapter, we examine six requirements of clause 10.2, namely:

*   Reacting to nonconformity (10.2.1a)
*   Evaluating the need for action to eliminate the cause (10.2.1b)
*   Implementing actions that are appropriate to the effects (10.2.1c)
*   Updating risks and opportunities during planning (10.2.1e)
*   Retaining evidence of the nature of nonconformity and actions taken (10.2.2a)
*   Reviewing effectiveness of corrective actions and retaining evidence of results (10.2.1d and 10.2.2b)

Reacting to nonconformity (10.2.1a)

What does this mean?

Reacting to a nonconformity simply means that having recognized there is a nonconformity, you can't ignore it and are required to take the following action: First of all to control it, which means not allowing it to cause problems elsewhere before it is corrected. Second to correct it, which means putting right that which is wrong and third, deal with the consequences, which means dealing with the fall-out or the effect of the problem in terms of its impact on the organization and its stakeholders. The requirement acknowledges that there may be circumstances when all three actions may not be applicable, for example, an error in a document detected before its release may not require further control and may not have consequences of any significance, but an error in legislation that has become law may require an amendment which may take months to pass through the legislature with serious consequence for those who fall foul of the law or their victims in the meantime.

Nonconformity with any time bound characteristic cannot be corrected, for example, if a waiting time target is not met in a hospital Accident and Emergency (A&E) department, there is nothing that can be done for those patients affected as the clock cannot be turned back. If an event planned for a particular date does not take place, the only action that can be taken is to set a new date which may mean changing the requirement.

Why is this necessary?

The inclusion of a requirement to control as well as correct a nonconformity imposes a constraint, without which you'd be free to take your time over correcting the nonconformity without considering its effects. It's not uncommon for management to know of problems and allow them to go unresolved while they deal with other issues which they consider to be of higher priority. By requiring action to control nonconformity, it puts management under an obligation to contain the effects of a nonconformity if action to correct it cannot be taken immediately or if the effects have already occurred (e.g. food poisoning caused by nonconformity with hygiene regulations).

How is this addressed?

How one reacts to nonconformity is an indication of professionalism and leadership.
    Before managers will take action, they need to know:

*   What is the problem or potential problem?
*   Has the problem been confirmed?

- What are the consequences of doing nothing (i.e. what effect is it having)?
- What is the preferred solution?
- How much will the solution cost?
- How much will the solution save?
- What are the alternatives and their relative costs?
- If I need to act, how long have I got before the effects damage the business?

Confirming nonconformity

Whatever you do, don't act on suspicion, always confirm that a problem exists or that there is a certain chance that a problem will exist if the current trend continues. Validate symptoms before proclaiming action!

A nonconformity has occurred therefore only when a non-fulfilment of a requirement has been confirmed. This may be at a planned verification stage when an object is submitted for review, approval or acceptance. At other times a nonconformity may be suspected as explained in Box 51.1. Depending on the grounds for suspicion it may be prudent to take action to confirm the nonconformity but only in conjunction with those responsible for the object. If a person has not finished a task and you suspect the output will be nonconforming, you either wait to see if the person takes the necessary action before releasing the output or you intervene diplomatically. Not everyone will perform a task in the same way. Some people work methodically, while others may appear to work erratically (depending on your perspective) but achieve the same objective. However, where health and safety are likely to be compromised, intervention is mandated because in most jurisdictions, every citizen has a duty of care.

Customer complaints and returned products

The customer can be mistaken, and customer complaints therefore need to be validated as genuine nonconformities before entering the corrective action process. Parts returned from dealers, customer manufacturing plants, etc., might not be nonconforming. Parts may be obsolete, surplus to requirements, have suffered damage in handling or have been used in trials, etc. Products may have failed under warranty and not be logged as a complaint but nonetheless they are nonconforming. See Box 58.1 for an example from the service sector.

Whatever the reason for return, you need to record all returns and perform an analysis to reveal opportunities for corrective action when appropriate. Prior to expending effort on investigations, you should establish your liability and then investigate the cause of any non-conformities for which the organization is liable.

Controlling nonconformity

Measures for controlling nonconforming outputs from production and service delivery processes are addressed in Chapter 51. The primary method for products is segregation to prevent inadvertent use or delivery coupled with containment action to limit the extent of a nonconformity by removing an immediate cause thus allowing normal operations to continue.

Controlling other incidents of nonconformity may require a different treatment. The objective of control is to maintain conditions within pre-set limits. The immediate priority is recognizing when action needs to be taken to contain the nonconformity from spreading.

Box 58.1   Customer preference

Nonconformity

In the service sector a dish returned to the kitchen in a restaurant may conform to the description on the menu but the menu may not have stated that there was cheese in the dish. The customer preferred a dish without cheese but neglected to say so when ordering. In this case the object of nonconformity is the menu, not the dish.

Correction

The correction is straightforward; offer an alternative dish.

Containment action

To remove an immediate cause and allow continued use of the menu a sticker stating *contains cheese* could be placed next to the item on the menu.

Corrective action

To prevent recurrence of the nonconformity, review and revise where necessary all menus ensuring that the descriptions are accurate and indicate essential ingredients. Review and revise criteria for dish descriptions in the menu design guide and re-educate the menu designer.

It's crucial to recognize when they are one-off events or are a symptom of an underlying cause that is capable of producing multiple nonconformities. It is equally important not to treat a one-off event as if it is capable of producing multiple nonconformities as it can create unnecessary panic.

Correcting nonconformity

The correction of process outputs in production and service delivery is addressed in Chapter 51. The methods used for correction of other outputs will vary depending on the nature of the output and the process which generates it. Often work is simply returned to the producer for action with comments as to the deficiencies. In some cases, a nonconformity cannot be corrected as is the case with time-based characteristics and effort must be put in to eliminating the cause or changing the requirement if the process simply isn't capable of meeting it.

Consequences of nonconformity

The consequences of a nonconformity may be trivial or catastrophic. Some consequences are suggested in the severity classification in Table 58.3 which is used in risk assessments. How one deals with these consequences is for the organization to decide, but they need to be filtered through the needs and expectations of the stakeholders as revealed in the stakeholder analysis (see Chapter 13) to ensure alignment with their values and balancing of their

Table 58.3 Suggested severity classification in the automotive sector

Effect Criteria	Severity of Effect	Ranking
Hazardous-without warning	May endanger machine or assembly operator. Very high severity ranking when a potential failure mode affects safe vehicle operation and/or involves noncompliance with government regulation. Failure will occur without warning.	10
Hazardous-with warning	May endanger machine or assembly operator. Very high severity ranking when a potential failure mode affects safe vehicle operation and/or involves noncompliance with government regulation. Failure will occur with warning.	9
Very high	Major disruption to production line. May have to scrap 100% of product. Vehicle/item inoperable, loss of primary function. Customer very dissatisfied.	8
High	Minor disruption to production line. Product may have to be sorted and a portion (less than 100%) scrapped. Vehicle operable, but at a reduced level of performance. Customer dissatisfied.	7
Moderate	Minor disruption to production line. A portion (less than 100%) of the product may have to be scrapped (no sorting). Vehicle/item operable, but some comfort/convenience item(s) inoperable. Customer experiences discomfort.	6
Low	Minor disruption to production line. May have to rework 100% of product. Vehicle/item operable, but some comfort/convenience item(s) operable at reduced level of performance. Customer experiences some dissatisfaction.	5
Very Low	Minor disruption to production line. The product may have to be sorted and a portion (less than 100%) reworked. Fit and finish/squeak and rattle item does not conform. Defect noticed by most customers.	4
Minor	Minor disruption to production line. A portion (less than 100%) of the product may have to be reworked on-line but out-of-station. Fit and finish/squeak and rattle item does not conform. Defect noticed by average customers.	3
Very Minor	Minor disruption to production line. A portion (less than 100%) of the product may have to be reworked on-line but in-station. Fit and finish/squeak and rattle item does not conform. Defect noticed by discriminating customers.	2
None	No effect.	1

needs. For example, after receiving a shipment of defective product which set production back weeks, a supplier that developed a contingency plan in conjunction with the customer will receive more brownie points than presenting an apology with a box of chocolates! (See Chapter 34 on clause 8.2.1e.)

There have been some horrific consequences of nonconformity in recent decades as indicated in Table 58.4 where we examine the 1980s. This was a decade during which there were seven major accidents where human error was a contributory factor. When examining the combined legacy of the incidents, it is striking that there is nothing revolutionary about the corrective actions and that in an industry that has been historically more risk conscious than the others – the aerospace industry – it took a second catastrophe for the culture to eventually change.

Table 58.4 Major incidents in the 1980s

Incident	Lessons learnt
Union Carbide Bhopal, India (Chemical processing) 1984	No safety measures can prevent an accident if there is no safety culture that governs the behaviour of management and employees; the application of the principles of intrinsically safe design and learning from accidents through knowledge transfer.
Space Shuttle Challenger (Aerospace) 1986	The same flawed decision-making process that had resulted in the Challenger accident was responsible for Columbia's destruction 17 years later.
Chernobyl (Nuclear industry) 1986	Tell the truth, evacuate, closely monitor radiation levels in food, comply with safety rules and plan ahead.
King's Cross Fire (Transport sector) 1987	Smoking on underground trains was banned.
Herald of Free Enterprise (Transport sector) 1987	Indicators fitted displaying the state of the bow doors on the bridge, watertight ramps fitted to the bow sections of the front of ships.
Clapham Junction (Transport sector) 1988	Testing mandated on rail signalling work and the hours of work of employees involved in safety critical work was limited.
Piper Alpha (Offshore) 1988	Having both production and safety overseen by the same agency was a conflict of interest and led to the adoption of the Offshore Installations (Safety Case) Regulations 1992.

It is not expected that organizations would have a set of procedures to follow in such circumstances as each one will be different but there may be contingency plans for specific scenarios such as disaster recovery plans.

How is this demonstrated?

Demonstrating that the organization reacts and takes appropriate action when a nonconformity occurs may be accomplished by presenting evidence of a process for confirming, controlling and correcting nonconformities and, dealing with the consequences.

Evaluating the need for action to eliminate the cause (10.2.1b)

What does this mean?

The requirement implies that on every occasion a nonconformity occurs we should evaluate the need for action to eliminate its cause. However, further on in clause 10.2.1, there is recognition that corrective actions should be appropriate to the effects of the nonconformities encountered thereby allowing us discretion as to when such action is necessary. There will be occasions when action to eliminate the cause of a nonconformity is necessary because of its severity such as if life, property or the environment is threatened, and there will also be occasions when the severity is very minor and it's not practical to investigate every single one but periodically take another look at these minor nonconformities to see if eliminating their cause is economically viable.

The reference to customer complaints is that every customer complaint is a nonconformity with some requirement. They may not all be product requirements. Some may relate to product delivery, to the attitude of staff or to false claims in advertising literature. Any complaint implies that a requirement (expectation, obligation or implied need) has not been met even if that requirement had not been determined previously. We should accept that we could have overlooked something. Just because it was not written in the contract does not mean that the customer is wrong.

Why is this necessary?

Were we to simply deal with problems as they arose we would inevitably find ourselves spending all day sorting out problems, giving us no time to see where we are going and prepare for the future. This is a situation which some organizations find themselves in and unless they invest in removing the cause of those problems they will eventually cease to function.

How is this addressed?

After correction and containment of a nonconformity we need to take another look at it and consider the following five actions:

1   Review the nonconformity when first detected and ensure it is classified and recorded.
2   Rank the nonconformities to prioritize action.
3   Establish by use of problem-solving tools if the more significant nonconformities can be prevented from happening again.
4   Reduce the frequency of occurrence in the population of nonconformities.
5   Establish by analysis if the nonconformity had been predicted in the planning phase and why the preventive action measures were not effective. This is addressed under "Updating risks and opportunities during planning (10.2.1e)".

Classifying nonconformities

At the time a nonconformity occurs, an initial judgement should be made as to its severity so that an investigation into its cause is initiated in a timely manner. The severity classification of the type used for risk assessment should be used (see Table 58.3) to prioritize the investigation.

If the nonconformity has been detected in a process that is under statistical process control beware of the two common mistakes which both Shewhart and Deming observed were frequently made in process control (Deming, 1982):

1   Ascribing a variation to a special cause when it belongs to the system (from common causes)
2   Ascribing a variation to the system (a common cause) when the cause was special

Over adjustment or tampering is a common example of mistake 1 and never doing anything to try to find a special cause is an example of mistake 2.

Sorting the vital few from the trivial many

If the nonconformity has been classified as not warranting immediate corrective action the data should be collected and a Pareto analysis undertaken to find the vital few nonconformities out of the total population that provide the bulk of improvement potential When dealing with nonconformity, the question we need to ask is: *What are the few sources of nonconformities that comprise the bulk of all nonconformities?* If we can find these noncon-formities and eliminate their cause, we will reduce variation significantly.

The first step is to sort the nonconformities by characteristic, process, product, and service. Then rank the nonconformities in order of occurrence so that the nonconformity having the most occurrences would appear at the top of the list. The result might be that for a particular product or process, a few types of nonconformity would account for the greatest proportion of nonconformities.

Another way of ranking the nonconformities is by seriousness. Not all nonconformities will have the same effect on product or service quality. Some may be critical and others may be insignificant. By classification of nonconformities in terms of criticality a list of those most serious nonconformities can be revealed using the Pareto analysis. Even though the frequency of occurrence of a particular nonconformity may be high, it may not affect any characteristic that impacts customer requirements. This is not to say the cause should not be eliminated but there may be other more significant problems to eliminate first.

When parts are rejected after their delivery it is indicative that your processes are not under control. Rejected parts analysis should be focused on determining the reason why the process failed to detect nonconformity. There could be some weakness in the process that if not corrected, further nonconforming parts might be shipped.

Determining the cause

To eliminate the cause of nonconformity the cause needs to be known, and therefore the first step is to conduct an analysis of the symptoms to determine their cause. Nonconformities are caused by factors that should not be present in a process and so when nonconformities arise there has clearly been some deficiency in the planning that needs to be investigated. Some preliminary questions need to be answered:

a) Was the type of nonconformity predicted when risk assessment was conducted during planning?
b) If it had, why didn't the existing controls prevent it occurring? The assessment may have underestimated the likelihood or the consequences or it could be that the recommended actions for controlling the risk were not implemented. But it could also be the case that the recommended actions didn't work;
c) (If applicable) why did the process capability studies conclude the process was capable when clearly it wasn't? It could be of course that the rejected parts are simply within the variation that was expected from a capable process. However, if the number of parts rejected already exceeds this limit, process improvement action will be necessary.

Simply asking why an event occurred might reveal a cause but don't accept the first reason given because there is usually a reason why this previous event occurred. Toyota discovered that asking *why* successively five times would invariably discover the root cause. There may be more or less than five steps to the root cause but it is critical to stop only when you can't go any further. The following example illustrates the technique:

A trainer arrives to conduct a training course to discover that the materials have not been delivered from head office as he expected them to be.

1   Why were the materials not delivered? – Answer: Because the administrators thought the trainer would bring his own materials.

2 Why did they think the trainer would bring his own materials? – Answer: Because they had not been informed otherwise.
3 Why weren't the administrators given the correct information? – Answer: Because the office manager had not communicated the agreed division of responsibility when setting up training courses.
4 Why had the office manager not communicated the agreed division of responsibility? Answer: Because the office manager had put other matters before internal communication in his order of priorities.
5 Why had the office manager not got his priorities right? Answer: Because he was not yet competent.
6 Why was the office manager not yet competent? Answer: Because the top management had made the appointment in haste.
7 Why had top management made the appointment in haste? Answer: Because they were not effective leaders.

Therefore, a lack of leadership (the second quality management principle) is the root cause. It took seven questions to get there but if we had stopped at question 3, and assumed that giving the administrators the correct instructions would prevent recurrence of the problem, we would be wrong. It might well prevent recurrence of the specific problem with a particular office manager but not similar problems with other managers. If the office manager forgot to issue the instructions, it indicates that he did not complete the process that commenced when the division of responsibility was agreed. This is quite typical. A meeting is held and agreements reached and when everyone departs they get on with what they were doing before the meeting, not realizing that a process has been initiated that needs to continue and be completed outside the meeting. If the staff were competent, they would complete the process before moving on.

Another technique is the cause and effect diagram (also known as the Ishikawa diagram or fishbone diagram). This is a graphical method of showing the relationship between cause and effect. Each type of nonconformity (an effect) would be analysed to postulate the causes so, for example, the question would be put to a diagnostic team – What could cause too much solder on a joint? The team would come up with several possibilities. Each one would be tested either by experiment or further examination of the soldering process and a root cause will be established.

Some nonconformities appear random but often have a special cause. To detect these causes, statistical analysis may need to be carried out. The causes of such nonconformities are generally due to noncompliance with (or inadequate) working methods and standards rather than the methods or standard being in error. Other nonconformities have a clearly defined special or unique cause that must be corrected before the process can continue. Common cause problems generally require the changing of unsatisfactory designs or working methods. They may well be significant or even catastrophic. These rapidly result in unsatisfied customers and loss of profits. To investigate the cause of nonconformities you will need to:

a) identify the requirements which have not been achieved;
b) collect data on nonconforming items, the quantity, frequency and distribution;
c) identify when, where and under what conditions the nonconformities occurred;
d) identify who was carrying out what operations at the time.

Nonconformity reduction

Previously it was suggested that action be taken on the vital few nonconformities that dominated the population. If this plan is successful, these nonconformities will no longer appear
in the list the next time the analysis is repeated. As the vital few nonconformities are tackled,
the frequency of occurrence will begin to decline until there are no nonconformities left to
deal with. This is nonconformity reduction and can be applied to specific products, services
or processes. If you were to aggregate the nonconformities for all products, services and
processes you would observe that it is quite possible to take corrective action continuously
and still not reduce the number of nonconformities – no matter how hard you try, you cannot
seem to reduce the number. This is because the objectives and targets keep changing. They
rarely remain constant long enough to make valid comparisons from year to year. There is
always some new process, practice or technology being introduced that triggers the learning
cycle all over again.

Common problem-solving tools

There are many tools you can use to help you determine the root cause of problems. These
are known as disciplined problem-solving methods. A common method in the automotive
industry is known as 8D, meaning eight disciplined methods. Originally conceived by the
Ford TOPS (Team Oriented Problem Solving) programme in 1987, it was upgraded and
renamed Prevent Recurrence in 1992.

* D1 – Establish the team
* D2 – Describe the problem
* D3 – Develop an interim containment action
* D4 – Define or verify root cause
* D5 – Choose or verify permanent corrective action
* D6 – Implement or validate permanent corrective action
* D7 – Prevent recurrence
* D8 – Recognize the team

   The notion of a permanent corrective action implies that there is a temporary corrective
action. Corrective action either removes the root cause or it doesn't. If the *permanent corrective action* were effective, it would also deal with the issues raised under D7 like modifying
specifications but other interpretations indicate that this would form part of D6. However,
it is not too important what the steps are called provided all the steps are completed. Whilst
8D has a certain meaning, disciplined methods are simply those proven methods that employ
fundamental principles to reveal information. There are two different approaches to problem-solving. The first is used when data are available as is the case when dealing with nonconformities. The second approach is when not all the data needed are available.
   The seven quality tools in common use are as follows:

1   Pareto diagrams – used to classify problems according to cause and phenomenon.
2   Cause-and-effect diagrams – used to analyse the characteristics of a process or situation.
3   Histograms – used to reveal the variation of characteristics or frequency distribution
    obtained from measurement.
4   Control charts – used to detect abnormal trends around control limits.

5   Scatter diagrams – used to illustrate the association between two pieces of corresponding data.
6   Graphs – used to display data for comparative purposes.
7   Check-sheets – used to tabulate results through routine checks of a situation.

The further seven quality tools for use when not all data are available are as follows:

8   Relations diagram – used to clarify interrelations in a complex situation.
9   Affinity diagram – used to pull ideas from a group of people and group them into natural relationships.
10   Tree diagram – used to show the interrelations among goals and measures.
11   Matrix diagram – used to clarify the relations between two different factors (e.g. QFD).
12   Matrix data-analysis diagram – used when the matrix chart does not provide information in sufficient detail.
13   Process decision program chart – used in operations research.
14   Arrow diagram – used to show steps necessary to implement a plan (e.g. PERT).

The source of causes is not unlimited. Nonconformities are caused by one or more deficiencies in:

a)   communication,
b)   information,
c)   personnel training and motivation,
d)   materials,
e)   tools and equipment,
f)   the operating environment.

Each of these is probably caused by not applying one or more of the eight quality management principles.

Once you have identified the root cause of the nonconformity you can propose corrective action to prevent its recurrence if it is economical to do so. A cost–benefit analysis may be needed to establish if the benefits to be gained from its elimination outweigh the costs that would be incurred to eliminate it.

Searching for other similar nonconformities

When a nonconformity is detected there may be the possibility that it's a symptom of a condition that could exist elsewhere but has yet to be detected (see Box 51.1). Once the cause is known a search should be made for other situations where the cause of nonconformity is or has been present. If the cause was rooted in certain practices, equipment, behaviours, materials, environmental conditions, etc., a search for products or their component parts and services that have been subject to those conditions should be made and the suspect products and services identified.

There may also be situations where work is planned to be subject to the conditions that have already led to a nonconformity and unless the planned work is halted, nonconformities may result. Should any such situations be identified consideration should be given to alerting those concerned so that nonconformity may be avoided.

How is this demonstrated?

Demonstrating that nonconformities are reviewed and analysed to evaluate the need for action to eliminate their cause may be accomplished by:

a) presenting evidence of a process for determining the action needed eliminate nonconformities;
b) selecting a representative sample of nonconformities and presenting evidence that:
   i   they have been reviewed to classify and prioritize diagnostic effort so it's proportional to the risks presented;
   ii diagnosis has been carried out to determine the possible cause(s);
   iii the need for action to eliminate the causes has been evaluated.

Implementing actions that are appropriate to the effects (10.2.1c)

What does this mean?

The action needed is the action that will eliminate the cause of the nonconformity and therefore prevent its recurrence. The action needed to fix the nonconformity was addressed earlier. Also, included in this section is the requirement for corrective actions to be appropriate to the effects of the nonconformities encountered. This means that where the system has failed the system should be fixed so the nonconformity does not recur and this was the case with the incidents listed in Table 58.4 but nonconformities that had zero consequences will require no corrective action as it's more cost effective just to fix them when they occur.

Why is this necessary?

Getting at the root of the problem is crucial to corrective action. There are countless corrective action procedures being implemented that do not get close to eliminating the cause of nonconformity. They focus on the immediate cause and not the root cause. Action on the immediate cause is only a palliative – which is a temporary measure. Fixing the immediate cause will result in another nonconformity eventually appearing somewhere else. It is also important to minimize the time taken to isolate and eliminated the root cause to minimize the impact on the customer of further deliveries. While you are contemplating the cause of the nonconformity, batches of nonconforming product might be on their way to already dissatisfied customers, or patients may be stuck in A&E for hours and hours or perhaps another ambulance doesn't arrive within the allotted time and too late the save a person's life.

How is this addressed?

Having determined the root causes of the nonconformity and the scale of what else is affected by them, consideration needs to be given to devising a plan that will eliminate these causes in a way that does not itself cause other problems. An analysis may therefore need to be carried out to determine how and when the interventions can be made without causing unnecessary disruption (see Chapter 23).

There needs to be a plan of corrective action that details the actions to be taken to eliminate the cause, who will be responsible for carrying out these actions and the date by which a specified reduction in nonconformity is to be achieved. Note that a corrective action is not complete until it has been proven effective therefore putting in place new procedures, or staff training or error-proofing devices is not indicative of *job done* as it may turn out that they are not effective and the nonconformity recurs.

If the nonconformity has severe consequences, and the diagnosis reveals that it resulted from an unanticipated interaction of multiple failures in a complex system – it's a serious system failure and will require multiple corrective actions. There may have been a cascade of failures because in a system all the elements are interconnected and therefore the failure of one element may lead to the failure of others. Even if there is a single cause at the end of the chain, it should not be assumed that removing that single cause of failure will prevent a recurrence if the other links in the chain remain vulnerable. All links may need to be strengthened. Case studies of relevance are *Apollo 13* (1970), Three Mile Island (1979), and ValuJet Flight 592 (1996).

Your management system needs to accommodate various corrective action strategies, from simple intradepartmental analysis with solutions that affect only one area, procedure, process, product or service, to projects that involve many departments, occasionally including suppliers and customers. Your corrective action procedures need to address these situations in order that when the time comes you are adequately equipped to respond.

Before we leave this topic, let us not overlook the most obvious corrective action: that of changing the requirement. It does not always follow that the requirement is correct but it rather depends firstly on when the nonconformity is detected and with whose requirement the product or service is nonconforming.

- If the nonconformity arises from a process the capability of which has been proven, it is highly likely that the nonconformity is due to special cause variation induced by a change in an otherwise stable parameter. However, one should not overlook the possibility that the standards/limits/targets imposed might be far tighter than necessary.
- If the nonconformity arises in a process, the capability of which has not been formally determined, such as a design process, a purchasing process or recruitment process, etc., and it is not an isolated case, the requirement might be too stringent and a relaxation might be the most sensible solution.
- If the nonconformity arises from a failure to meet a customer requirement after product or service approval, it is likely that no change to the requirement would be permitted, however;
- If arising during development and again it is not an isolated case, there might be a valid argument for changing the requirement. Requirements are sometimes ambiguous, inconsistent or simply not achievable. Designers sometimes make assumptions and impose limits that are far tighter than needed.

How is this demonstrated?

Demonstrating that the actions implemented are appropriate to the effects of nonconformity may be accomplished by:

a) presenting evidence of a process for implementing corrective actions that are appropriate to the effects of nonconformity;

b) selecting a representative sample of nonconformities and presenting evidence:

  i   of a plan for implementing corrective action with dates and responsibilities;
  ii that consideration has been given to the interdependencies of such actions and the provisions necessary to minimize further problems;
  iii that the planned actions have been carried out.

Updating risks and opportunities during planning (10.2.1e)

What does this mean?

When nonconformities occur, it provides data for validating any estimate of likelihood and consequences that have previously been made. Also, the corrective actions that are implemented will change the basis on which the original determination of risks and opportunities was carried out and therefore it may need to be updated.

Why is this necessary?

For the risk assessments to be useful they need to reflect the current configuration of the QMS.

How is this addressed?

Statistics on nonconformities will be collected, analysed and evaluated on a regular basis (see Chapter 54), and these data will enable estimates made when undertaking risk assessment to be validated. If the data show the estimates to be wildly inaccurate, they should be revised. This should happen independently of whether corrective action is taken as it's likely that the estimates will change again afterwards.

Whatever changes are needed to the QMS to prevent a recurrence of nonconformities, any risk assessments undertaken originally will need to be undertaken on the changes so that estimates of likelihood and consequences can be revised and controls appropriate to the changes put in place as part of the corrective action.

How is this demonstrated?

Demonstrating that risks and opportunities determined during planning are updated when necessary after nonconformities are detected may be accomplished by:

a)   presenting evidence of a process for revising risk assessments following an analysis of nonconformity data and when changes are made to the QMS as a result of corrective actions;
b)   selecting a representative sample of nonconformity data and presenting evidence that the incidence of nonconformity is consistent with the estimates used in risk assessments;
c)   selecting a representative sample of risk assessments and presenting evidence that they have been updated following changes to the QMS as a result of corrective actions.

Retaining evidence of the nature of nonconformity and actions taken (10.2.2a)

What does this mean?

The nature of the nonconformities is the information that places the nonconformity in context, for example, describes what is nonconforming, what requirement has not been met, the conditions under which it was detected, etc. The subsequent actions refer to the action taken to correct and contain the nonconformity, if any, and those taken to deal with the consequences and prevent recurrence.

Why is this necessary?

It's necessary to retain some documented information on nonconformity so that it can be reliably communicated to those concerned with correction, control and diagnosis of nonconformity and, so that actions and decisions can be reviewed at a later stage and the logic of the analysis and subsequent actions can be checked. This is more important where nonconformities have severe consequences and involve independent enquiries.

How is this addressed?

By treating the detection and correction of the nonconformity and the elimination of its cause as three separate activities requiring different information we can identify several things that are worthy of consideration.

Information describing the nature of the nonconformity

Certain basic information needs to be retained such as:

a)   The identity of the object that is nonconforming.
b)   The specific requirement with which the object does not conform.
c)   The date and time it was detected.
d)   The person who detected it and their location.
e)   The prevailing environmental conditions at the time, when relevant.
f)   What was happening at the time the nonconformity was detected, when relevant.

When capturing this formation, it's important to consider the needs of two groups of people: those who will need the information to correct, control and deal with the consequence and those who will need the information for subsequent analysis, evaluation and improvement. Once the moment has passed it may be impossible to rewind to the instant when the nonconformity occurred and run through it again to refresh the memory.

Information describing actions taken to deal with the nonconformity

The actions taken to deal with the nonconformity will vary and may include:

a)   Actions to halt the generation of further nonconformities (e.g. containment action, suspension of production or service delivery).
b)   Actions to prevent inadvertent use or delivery (e.g. segregation, quarantine).
c)   Actions to correct the nonconformity (e.g. rework, repair, return to supplier, scrap, concession).
d)   Actions to locate other related items that are actually or potentially nonconforming.
e)   Actions to deal with the immediate impact (e.g. delays, product recall, clean up, warranty, compensation, prosecution).

The description of the nonconformity and the actions taken to deal with it are often combined on one form separate to the corrective action because the corrective action may not be determined concurrently with the other action for various reasons. A single form is not always appropriate and in some cases a freeform layout might be more suitable for serious nonconformities. An example is included on the companion website.

Information describing actions taken to eliminate the cause of nonconformity

The actions taken to discover the cause of the nonconformity and ensure it does not recur or occur elsewhere may include the:

a)   results of Pareto analysis;
b)   likely causes and the root cause;
c)   criteria for determining severity or priority;
d)   tests conducted to validate the root cause;
e)   actions proposed to eliminate the root cause;
f)   actions proposed to render interacting processes, products and service less vulnerable to contributing to nonconformity
g)   actions taken;
h)   results of the actions taken.

A sample corrective action form is included on the companion website which might be suitable for most circumstances, but for multiple corrective actions it's probably better to use a free-form report to record the results of an investigation into the effectiveness of corrective actions.

For it to be possible to verify the actions taken, records need to exist to provide traceability. For example, if your corrective action report (e.g. CAR023) indicates that procedure XYZ requires a change, a reference to the document change request (e.g. DCR134) initiating a change to procedure XYZ will provide the necessary link. The Change request can reference the corrective action report as the reason for change. If you don't use formal change requests, the amendment instructions can cross reference the corrective action report. Alternatively, if your procedures carry a change record, the reason for change can be added. There are several methods to choose from, but whatever the method you will need some means of tracking the implementation of corrective actions. This use of forms illustrates one of the many advantages of form serial numbers.

How is this demonstrated?

Demonstrating that documented information is being retained as evidence of the nature of the nonconformities and any subsequent actions taken may be accomplished by:

a)   presenting evidence of a process for capturing information addressing the nature of the nonconformity and subsequent actions taken;
b)   selecting a representative sample of nonconformities and retrieving documented information that described:

   i   the nature of the nonconformity;
   ii   the actions taken to deal with the nonconformity including its consequences;
   iii the actions taken to eliminate its causes.

Reviewing effectiveness of corrective actions and retaining evidence of results (10.2.1d and 10.2.2b)

What does this mean?

We have combined two clauses here because the review of effectiveness generates the evidence to be retained.

The review of the effectiveness of corrective actions means establishing that the actions have been effective in eliminating the cause of the nonconformity. As the requirement for follow-up audit has not been included in clause 9.2, the action taken after an audit to verify that agreed actions from the audit have been completed as planned is contained in the action referred to in clause 10.2.1d).

Why is this necessary?

Every process should include verification and review stages not only to confirm that the required actions have been taken but also that the desired results have been achieved. It is only after a reasonable time has elapsed without a recurrence of a particular nonconformity that you can be sure that the corrective action has been effective.

How is this addressed?

This requirement implies four separate actions:

- A review to establish what actions were taken.
- An assessment to determine whether the actions were those required to be taken.
- An evaluation of whether the actions were performed in the best possible way.
- An investigation to determine whether the nonconformity has recurred.

The effectiveness of some actions can be verified at the time they are taken but quite often the effectiveness can only be checked after a considerable lapse of time. Remember it took an analysis to detect the nonconformity therefore it may take further analysis to detect that the nonconformity has been eliminated. In such cases the corrective action report should indicate when the checks for effectiveness are to be carried out and provision made for indicating that the corrective action has or has not been effective.

Some corrective actions may be multidimensional in that they may require training, changes to procedures, changes to specifications, changes in the organization, changes to equipment and processes – in fact so many changes that the corrective action becomes more like an improvement programme. Checking the effectiveness becomes a test of the system carried out over many months. Removing the old controls completely and committing yourselves to an untested solution may be disastrous therefore it is often prudent to leave the existing controls in place if possible until your solution has been proven to be effective.

The nonconformity data should be collected and quantified using one of the seven quality tools, preferably the Pareto analysis. You can then devise a plan to reduce the 20% of causes that account for 80% of the nonconformities.

When all the agreed nonconformities have been eliminated, the corrective action report can be closed. The action remains incomplete until all actions have been verified as being completed. Should any action not be carried out by the agreed date, a judgement needs to be made as to whether it is reasonable to set a new date or to escalate the slippage to higher management. For minor problems, when there are more urgent priorities facing the managers, setting a new date may be prudent. However, you should not do this more than once. Not meeting an agreed completion date is indicative either of a lack of commitment or poor estimation of time, and both indicate that there may well be a deeper-rooted problem to be resolved.

You will need to monitor the reduction of nonconformity therefore the appropriate data collection measures need to be in place to gather the data at a rate commensurate with the operations schedule. Monthly analysis may be too infrequent. Analysis by shift may be more appropriate but take care not to degrade other processes by your actions.

Follow-up action is necessary to verify that the agreed action has been taken and verify that the original nonconformity has been eliminated. Follow-up action may be carried out immediately after the planned completion date for the actions or at some other agreed time. However, unless verification is carried out relatively close to the agreed completion date, it will not be possible to ascertain if the actions were carried out without undue delay as is required for nonconformities detected during internal audits.

How is this demonstrated?

Demonstrating that the effectiveness of any corrective action taken has been reviewed and records retained may be accomplished by:

a)   presenting evidence of a process for reviewing the effectiveness of corrective actions and recording the results;
b)   selecting a representative sample of corrective actions and presenting evidence:

   i   that the planned actions have been implemented;
   ii   of the result being achieved following implementation of the actions;
   iii that the effectiveness of the actions has been reviewed and the needs for any further action determined.

Bibliography

Deming, W. E. (1982). *Out of the Crisis*. Cambridge, MA: The MIT Press.
ISO 9000:2015. (2015). *Quality Management Systems – Fundamentals and Vocabulary*. Geneva: ISO.
ISO/TC 176/SC 2/N1338. (2015). *Quality Management Systems – Guidelines for the Application of ISO 9001:2015*. Geneva: International Organization for Standardization.
ISO/TMB/JTCG N 360. (2013, December 3). Concept Document to Support Annex SL. Retrieved from ISO Standards Development: http://isotc.iso.org/livelink/livelink?func=ll&objId=16347818&objAction=browse&viewType=1.

59 Continual improvement of the QMS

Introduction

The concept of continual improvement

The idea of continual improvement of the system emerged from the work done by Deming in Japan in the early 1950s (Deming, 1982). This was embedded in points 1 and 5 of his 14 Points for Management as follows:

Point 1. "Create constancy of purpose towards improvement of product and service, with the aim of becoming competitive and to stay in business and to provide jobs."

Point 5 "Improve constantly and forever the system of production and service to improve quality and productivity and thus constantly decrease costs."

Continual improvement is defined by ISO as a "recurring activity to enhance performance" whereas improvement is defined as an "activity to enhance performance" (ISO 9000:2015). It is also defined as gradual change or incremental change. Some might say it's changing bit by bit, but that can give the wrong impression. The activity does not have to be applied to the same characteristic to qualify as continual as we might do when using PDCA. The principle is applied to the organization as a whole; therefore, whether a change is accomplished through corrective action, breakthrough, innovation or transformation is not pertinent to qualifying an improvement as continual. It's the recurring nature of improvement within the organization regardless of what is being improved or the magnitude of improvement that is relevant.

We may think we can remove one problem after another in a never-ending cycle but when discontinuous changes occur such as a breakthrough in technology, an economic crisis, a natural disaster or a change in leadership, improvements that have been accumulated over months or years, may be swept away in an instant. But all may not be lost because, as Ackoff remarked, continual improvement is not nearly as important as discontinuous improvement. Creativity is a discontinuity (Ackoff, 1994).

References to continual improvement

There are four references to continual improvement within the requirements of the standard:

1   The organization is required to continually improve a QMS in accordance with the requirements of ISO 9001 (4.4.1).

2   The quality policy is required to include a commitment to continual improvement of the QMS (5.2.1).
3   Resources needed for the continual improvement of the QMS are required to be determined and provided (7.1.1).
4   The organization is required to continually improve the suitability, adequacy and effectiveness of the QMS (10.3).

The first of these requirements indirectly refers to the last requirement, thus adding no additional requirement to be audited.

Structure of requirements

In Annex SL clause 10.1 is "Nonconformity and corrective action" and 10.2 is "Continual improvement"; therefore, the new clause *10.1 General* is a discipline-specific clause for ISO 9001. Examples of improvement given in the note to clause 10.1 include continual improvement, and both clauses require improvement in the effectiveness of the QMS, so there does not appear to be any justification for clause 10.3; however, it could not be removed as it's a mandatory Annex SL clause. TC 176 did attempt to merge the discipline-specific requirements with those on continual improvement and strike out the word *continual* in the heading, but it was not permitted so we are left with a compromise that is a trifle ambiguous. However, there are three differences between clauses 10.1 and 10.3 relative to continual improvement worthy of note:

a)   Continual improvement of the QMS is a requirement in clause 10.2 but not in clause 10.1.
b)   Improvement in the suitability, adequacy and effectiveness of the QMS is required in clause 10.3 but is limited to improving the performance and effectiveness of QMS in clause 10.1.
c)   A further analysis process for identifying opportunities for continual improvement is required in addition to those already required by clauses 9.1.3, 9.3.3 and 10.1.

The difference between improvement in products and services and improvement in the QMS

In the 2008 version, there was a requirement to continually improve the effectiveness of the QMS but no requirement to improve products and services which there now is in the 2015 version. In fact, a published interpretation for the 2008 version (RFI 025) stated that the realization of a new product to improve an old one could be one of the results of the management review but this is not considered to be a continual improvement by RFI 024 which states that "the improvement addressed in ISO 9001:2008 Clause 8.5.1 does not include product improvement." This is contradictory to Deming's first point (see earlier).

It is true that improvement in products and services is not improvement in the QMS, because if by executing a process of the QMS, products or services are improved that is not a change to the QMS because it was doing what it was designed to do. However, if the QMS is changed to improve its suitability, adequacy or effectiveness and as a result products and services improve either by design change or reduction of variation this makes product and service improvement a consequence of continual improvement in the QMS. The 2015 version does now require both improvement in the QMS and improvement of products and services, but it does not require continual improvement of both, which means that either RFI 025 is not valid for the 2015 version or it is valid and ISO 9001 is still in conflict with Deming's original intentions stated above.

In this chapter, we examine the two requirements of clause 10.3, namely:

- Improving the suitability adequacy and effectiveness of the QMS
- Determining additional opportunities for improvement

Continually improving the suitability adequacy and effectiveness of the QMS (10.3)

What does this mean?

Continual improvement of the QMS is specifically focused on the attributes of suitability, adequacy and effectiveness and therefore using the definitions in Box 56.1:

- Continually improving suitability means gradually improving the extent to which the QMS "fits" and is right for the organization's purpose.
- Continually improving adequacy means gradually improving the extent to which the QMS is sufficient in meeting the applicable requirements.
- Continually improving effectiveness means gradually improving the extent to which planned activities are realized and planned results achieved.

The implication here is that changes that don't produce an observable improvement in the suitability, adequacy and effectiveness of the QMS won't qualify as continual improvements to the QMS, although they may qualify as changes to some other part of the management system.

Why is this necessary?

The reason for requiring continual improvement of the QMS rather than simply its improvement reflects the nature of the QMS. As a system, interaction between the elements is continual and once interaction ceases the system ceases to exist (see Chapter 8 for further explanation). Within the system there are processes with feedback loops that trigger change, either to correct an error or to prevent that error recurring or to revisit the objectives and question their alignment with the needs and expectation of the interested parties. Each of these changes brings about an improvement; some of them will be implemented through minor changes to policies and practices and others may take on the form major projects involving new technologies. These changes in themselves may not be recurring on the aspects they affect but taken as a group, they will represent a recurring activity to enhance performance of the QMS – somewhere within the system, an activity will be taking place with the intent that it enhances its performance.

How is this addressed?

A representation of continual improvement is shown in Figure 59.1. There are periods of relatively stable performance followed by improvement to a new level of performance. The intervals are shown in years, but depending on what is being improved it may be achieved in much shorter intervals. The duration of the improvement phase is generally as long as it takes to run through the PDCA cycle until the improvement objective is achieved and stable performance at a new level is obtained.

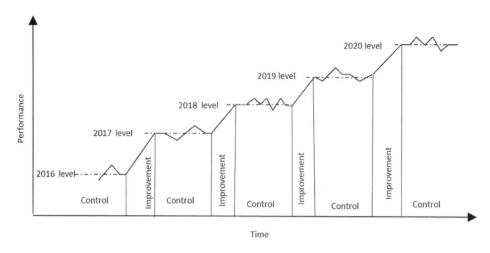

Figure 59.1 Continual improvement

If we are planning on improving the suitability adequacy and effectiveness of the QMS we need some units of measure and to measure these attributes before and after changes are made to the QMS. Before determining and selecting improvements we also need data, and these will come from the analysis and evaluation processes addressed in Chapter 54. The following tables have been created for illustration purpose only and omit both the methods of measurement and the root causes.

Measuring and improving QMS suitability

The phrase *extent to which* implies a measure of suitability; therefore, we need to work out a way to measure the extent to which the QMS *fits* and is right for the organization's purpose. There are four questions that will help us do this:

1   What are the observable consequences of a QMS that is right for the organization's purpose?
2   How would we detect if the suitability was better or worse than before we made improvements?
3   What improvement actions would gradually increase QMS suitability?
4   What is the measure of suitability before and after improvement?

We are not referring to whether a tool is fit for its purpose but whether the organization is fit for its purpose, and therefore, the type of answers you should be looking for will be more like those in Table 59.1.

Measuring and improving QMS adequacy

Next we need to work out a way to measure the extent to which the QMS is adequate in meeting the applicable requirements. These will be QMS requirements such as ISO 9001, any customer specific QMS requirements, applicable regulations and organizational require-ments. Using the same approach as we did with QMS suitability there are another four ques-tions for which we require answers as indicated in Table 59.2.

Table 59.1 Measuring and improving QMS suitability

Observable Consequences	Measure of suitability	Continual improvement action	Suitability Before	Suitability After
The products and services being offered are consistent with the strategic direction	Ratio of aligned offerings to total offerings	Gradually withdraw non-aligned products and services	85%	95%
The products and services being offered closely match customer needs	Change in sales revenue	Increase frequency of customer focus meetings from biannually to annually	−2%	+1%
The time to market is as good as our competitors	Average time from approval of concept to launch	Gradually introduce standardization in design to minimize design content of a project	Between 18 and 24 months on average	Between 12 and 18 months on average
After observing changes in the external environment corresponding changes to the QMS are made on time	On time completion of major QMS changes	Appoint project managers to manage major QMS changes	Between 6 and 18 months late on average	Between 0 and 6 months late on average

Table 59.2 Measuring and improving QMS adequacy

Observable Consequences	Measure of adequacy	Continual improvement action	Adequacy Before	Adequacy After
Conformity with ISO 9001 requirements	Detected Nonconformity	Undertake corrective action	65	25
Conformity with organization's own requirements	Detected Nonconformity	Undertake corrective action	60	34
Conformity with applicable regulations	Detected Nonconformity	Undertake corrective action	24	10

As an organization is not homogeneous, there are infinite ways in which conformity with a requirement can be tested, and therefore it's not practical to calculate a conformity ratio. However, improvement can be measured by the change in the number of nonconformities providing the method of measurement and the person judging conformity remains the same.

Measuring and improving QMS effectiveness

In this case, we need to work out a way to measure the extent to which planned activities are realized and planned results achieved. The inputs to management review listed in clause 9.3.2c provide a basis for measurement as shown in Table 59.3.

Coordinating continual improvement initiatives

In many cases, organizations have focused on improving the work processes believing that as a result there would be an improvement in business outputs but often such efforts barely

Table 59.3 Measuring and improving QMS effectiveness

Observable consequences	Measure of effectiveness	Continual improvement action	Effectiveness	
			Before	After
Customers claim they gain the benefits they expected from our products and services	Customer loyalty	Redesign customer survey	90%	92%
Feedback from other stakeholders is favourable	Employee turnover	Introduce exit interviews	20%	10%
	Supplier loyalty	Introduce supplier development programmes	80%	85%
	Community support	Introduce open days	25%	50%
Improvement initiatives are delivering results	Quality objectives achieved on time and within budget	Reprioritize initiatives	40%	65%
Conformity of products	First-time pass rate	Various	93%	94%
Conformity of services	First-time pass rate	Various	85%	90%
Processes perform as intended	Process yield	Various	95%	96%
	Resource usage	Various	125%	105%
	Adverse effects	Various	£150K	£95K
Quality problems don't recur	Ratio of old problems to new problems	Introduce root cause analysis training	55%	45%
External providers deliver on time and in conformity with requirements	First-time pass rate	Introduce supplier development programmes	75%	78%
The quality problems we experience are not ones we could have prevented	Ratio of preventable to unpreventable problems	Introduce risk management training	50%	10%
Internal audits find quality problems that are preventing us from achieving our goals	Ratio of significant findings to trivial audit findings	Refresh auditor training	20%	35%

had any effect. Unless a change in a work processes improves system suitability, adequacy or effectiveness there is little point in making it. It's worth repeating the quote from Ackoff on continual improvement in Chapter 5: *Until managers take into account the systemic nature of their organizations most of their efforts to improve their performance are doomed to failure.*

The danger inherent in the tables is that there is a tendency to treat each improvement independent of the others,, which is why it's necessary to coordinate these improvements in a way that represent a coherent programme. Consider the interdependences in the following sequence:

1   A programme for reducing variation reduces rework time and waste;
2   A reduction in waste leads to less storage space required;
3   A reduction in storage space might lead to a reduction in insurance premiums, rent, etc.
4   A reduction in insurance and rent might enable a reduction in price;
5   A reduction in price might draw in more customers.

Is it for this reason that when top management assign responsibility and authority for reporting on opportunities for improvement, the person they appoint should have knowledge of systems thinking so they can spot the improvement opportunities that would not yield improvement in the system or be counterproductive if implemented in conjunction with others.

How is this demonstrated?

Demonstrating that the suitability, adequacy and effectiveness of the QMS are being continually improved may be accomplished by:

a) presenting evidence of a process for evaluating the suitability, adequacy and effectiveness of the QMS and identifying opportunities for improvement;
b) presenting the results of assessments made over the last three years by showing how QMS suitability, adequacy and effectiveness have been measured;
c) presenting evidence that improvement in any elements of the QMS has not been taken in isolation and that enhancement in QMS suitability, adequacy or effectiveness has been a recurring feature.

Determining additional opportunities for continual improvement (10.3)

What does this mean?

The organization is required to consider the results of analysis and evaluation, and the outputs from management review, to determine if there are needs or opportunities to be addressed as part of continual improvement but this is a requirement that is stated elsewhere in the standard and here is the reasoning:

a) The results of analysis are required by clause 9.1.3 to be used to evaluate the need for improvements to the QMS.
b) The inputs to the management review are required by clause 9.3.3 to take into consideration opportunities for improvement thereby making the output from clause 9.1.3 an input to clause 9.3.2.
c) The outputs of the management review are required by clause 9.3.3 to include decisions and actions related to opportunities for improvement and therefore decisions will have been made on the outputs of the analysis and evaluation of measuring and monitoring results.

The final requirement in clause 10.3 therefore implies that the opportunities for improvement determined by the management review for action are to be subject to a further review to determine if any of them can be addressed as part of continual improvement. The fact these improvement opportunities are a result of a management review that recurs at planned intervals means they have been addressed as part of continual improvement. Thus, there is no additional requirement to meet.

Bibliography

Ackoff, R. L. (1994). Beyond Continuous Improvement. Retrieved from http://www.youtube.com/watch?v=OqEeIG8aPPk.

Deming, W. E. (1982). *Out of the Crisis*. Cambridge, MA: The MIT Press.

ISO 9000:2015. (2015). *Quality Management Systems – Fundamentals and Vocabulary*. Geneva: ISO.

Key messages from Part 10

Chapter 57   Determining and selecting opportunities for improvement

1 We cannot improve anything unless we know its present condition, and this requires measurement and analysis to tell us whether improvement is both desirable and feasible.
2 If we want to increase the level of performance, we should either reduce common cause variation or pursue different objectives and this requires that we act on the system.
3 We can undertake activities with the object of improving a situation but unless and until the situation improves there is no improvement.
4 Seeking individual opportunities for improvement without considering their impact on the system leads to sub-optimization – not optimization of organizational performance.
5 Every improvement must improve the performance of the system as a whole; otherwise, it's wasted effort.
6 Customers probably won't complain if the errors are imperceptible but increase over time – eventually they take their business elsewhere without giving notice.

Chapter 58   Nonconformity and corrective action

7 Every system needs a self-correction mechanism.
8 It is not a corrective action if it removes a nonconformity and not the cause of the nonconformity.
9 Validate symptoms before proclaiming action.
10 Beware of ascribing a variation to a special cause when it belongs to the system and ascribing a variation to the system when the cause was special.
11 Simply asking why an event occurred might reveal a cause, but don't accept the first reason given because there is usually a reason why this previous event occurred.
12 A diagnosis that reveals that nonconformity resulted from an unanticipated interaction of multiple failures in a complex system is a serious system failure and will require multiple corrective actions.
13 When nonconformities occur it's important to know whether their frequency and consequences had been predicted during system, process, product and service planning and for risk assessments to be validated or revised accordingly.
14 It is only after a reasonable time has elapsed without a recurrence of a specific nonconformity that you can be sure that the corrective action has been effective.

Chapter 59   Continual improvement of the QMS

15 When organizations cease changing they become fossils. Even maintaining the status quo requires organizations to change because all around them is changing.

16 It's the recurring nature of improvement within the organization regardless of what is being improved or the magnitude of improvement that qualifies improvement as being continual improvement.

17 When discontinuous changes occur such as a breakthrough in technology, an economic crisis, a natural disaster or a change in leadership, improvements that have been accumulated over months or years, may be swept away in an instant.

18 Improvement in products and services is not an improvement in the QMS, because if by executing a process of the QMS, products or services are improved, the QMS is doing what it was designed to do.

19 Improvement in the suitability, adequacy and effectiveness of the QMS requires appropriate units of measure and measurement to be undertaken before and after changes are made to the QMS as evidence improvement has occurred and is recurring periodically.

Appendix A
Common acronyms

The following acronyms used in this Handbook and are normally explained on first use.

ANSI	American National Standards Institute
APQP	Advanced Product Quality Planning
AQPC	American Quality and Productivity Center
B2B	Business to business
B2C	Business to consumer
BSI	British Standards Institution
CAD	Computer-aided design
CCF	Common cause failure
CIPD	Chartered Institute of Personnel Development
CM	Configuration management
CSA	Canadian Standards Association
CSF	Critical success factor
CTQ	Critical to quality
DFA	Design for assembly
DFM	Design for manufacture
DIS	Draft international standard
EMS	Environmental management system
FDIS	Final draft international standard
FMEA	Failure mode and effects analysis
GM	General Motors
HACCP	Hazard Analysis and Critical Control Points
HCI	Hardware configuration index
IAF	International Accreditation Forum
ISO	International Organization for Standardization
ISP	Internet service provider
ITT	Invitation to tender
JIT	Just-in-time
KPI	Key performance indicator
LCL	Lower control limit
MR	Management representative
MRI	Master record index
MS	Management system
MSS	Management system standard
MTBF	Mean time between failures

NATO	North Atlantic Treaty Organization
NGO	Non-governmental organization
NSB	National standards body
OEM	Original equipment manufacturer
OFI	Opportunity for improvement
OHSMS	Occupational Health and Safety Management System
PAS	Publically Available Specification
PERT	Program Evaluation Review Technique
PESTLE	Political, Economic, Social, Technological, Legal and Environmental
PFMEA	Process failure mode and effects analysis
QCD	Quality, cost and delivery
QFD	Quality function deployment
QMS	Quality management system
RFQ	Request for quotation
RGA	Returned good authorization
RMA	Returned material authorization
ROR	Register of requirements
SCI	Software configuration index
SLA	Service level agreement
SMS	Safety management system
SOP	Standard operating procedures (or conditions)
SPC	Statistical process control
SWOT	Strengths weaknesses, threats and opportunities
TC	Technical Committee
TQM	Total quality management
UCL	Upper control limit
USP	Unique selling point
WBS	Work breakdown structure

Appendix B
Glossary of terms

This list includes some of the terms and common words used in this book that have acquired a special meaning in the field of quality management. The definitions of terms included in ISO 9000:2015 are excluded from this list but where the term is included in ISO 9000 but defined differently in this book, the corresponding clause reference is given.

Acceptance authority. An organization with the right to decide on the acceptability of something, typically products, services, designs, projects or proposals for changing a design or project. Also referred to as design authority or project authority.

Acceptance criteria. The standard against which a comparison is made to judge conformance.

Accreditation. A process by which organizations are authorized to conduct certification of conformity to prescribed standards.

Action and its derivatives:

- Action – the doing of something
- Counteraction – action in opposition
- Interaction – reciprocal action or influence
- Reaction – resisting action
- Transaction – that which is done

Activity. An element of work that produces an output required by a process. Activities comprise tasks or operations.

Adequacy. The extent to which something is sufficient in meeting the applicable requirements.

Applicable. In the context of documents, applicable means capable of being applied to the activities to be undertaken. In the context of activities applicable means where it applies, for example, if there is a requirement for all electronic circuits to be grounded and the product in question contains no electronic circuits, the requirement cannot be applied – it is therefore not applicable.

Applicability. It is a technical issue, unlike appropriateness which can be subjective.

Appropriate. Means suitable for its purpose or to the circumstances and required knowledge of this purpose or circumstances. Without criteria, an auditor is left to decide what is or is not appropriate based on personal experience.

Approved. Something that has been confirmed as meeting the requirements.

Assessment. The act of determining the extent of compliance with requirements.

Assurance. Evidence (verbal or written) that gives confidence that something will or will not happen or has or has not happened.

Attribute data.   Qualitative data that can be counted for recording and analysis, for example, presence or absence of a required characteristic, number of failures in a production run, number of people eating in the cafeteria on a given day, etc.

Audit.   An examination of results to verify their accuracy by someone other than the person responsible for producing them. (See also ISO 9000:2015 clause 3.13.1.)

Authority.   The right to take actions and make decisions.

Authorized.   A permit to do something or use something that may not necessarily be approved.

Benchmarking.   A technique for measuring an organization's products, services and operations against those of its competitors resulting in a search for best practice that will lead to superior performance.

Bias.   In a measurement system, this is the difference between the observed average of the measurements and the reference value.

Business management system.   The set of interacting and managed processes that function together to deliver business results.

Business objectives.   Objectives the business needs to achieve to accomplish its mission. These are usually derived from an analysis of stakeholder needs and expectations.

Business plan.   Provisions made to fulfil the organization's mission, and vision and apply its values in terms of the strategy, objectives, measures, targets and enabling processes.

Business process.   A process that is designed to deliver outputs that satisfy business objectives.

Calibrate.   To standardize the quantities of a measuring instrument.

Certification body.   An organization that is authorized to certify organizations. The body may be accredited or non-accredited.

Certification.   A process by which a product, process, person or organization is deemed to meet specified requirements.

Class.   A group of entities having at least one attribute in common or a group of entities having the same generic purpose but different functional use.

Clause of the standard.   A numbered paragraph or subsection of the standard containing one or more related requirements such as 7.2.2. Note: each item in a list is also a clause.

Codes.   A systematically arranged and comprehensive collection of rules, regulations or principles.

Commitment.   An obligation a person or an organization undertakes to fulfil (i.e. doing what you say will do).

Common cause variation.   Random variation caused by factors that are inherent in the system.

Competence.   The ability to demonstrate the use of education, skills and behaviours to achieve the results required for the job.

Competence-based assessment.   A technique for collecting sufficient evidence that individuals can perform or behave to the specified standards in a specific role (Shirley Fletcher).

Competent.   An assessment decision that confirms a person has achieved the prescribed standard of competence.

Concession.   Permission granted by an acceptance authority to supply product or service that does not meet the prescribed requirements. (See also ISO 9000:2015 clause 3.12.5.)

Concurrent engineering.   See also simultaneous engineering.

Configuration control.   Systematic evaluation, coordination, approval or disapproval of all changes to the baseline configuration (NASA SP 6001).

Configuration management.   A discipline applying technical and administrative direction and surveillance to the identity, documentation, control and recording of the functional and physical characteristics of a product taking into account system interfaces (DEF STAN 05–57).

Conformity assessment.   Any activity concerned with determining directly or indirectly that relevant requirements are fulfilled (ISO/IEC Guide 2).

Conformity control.   Ensuring that products remain conforming once they have been certified as conforming.

Contract loan.   An item of customer-supplied property provided for use in connection with a contract that is subsequently returned to the customer.

Contract.   An agreement formally executed by both customer and supplier (enforceable by law) which requires performance of services or delivery of products at a cost to the customer in accordance with stated terms and conditions. Also agreed requirements between an organization and a customer transmitted by any means.

Contractual requirements.   Requirements specified in a contract.

Control charts.   A graphical comparison of process performance data to computed control limits drawn as limit lines on the chart.

Control procedure.   A procedure that controls product or information as it passes through a process.

Control.   The act of preventing or regulating change in parameters, situations or conditions.

Controlled conditions.   Arrangements that provide control over all factors that influence the result.

Core competence.   A specific set of capabilities including knowledge, skills, behaviours and technology that generate performance differentials.

Corrective action.   Action planned or taken to stop something from recurring. (See also ISO 9000:2015 clause 3.12.2.)

Corrective and preventive action (CAPA).   Action taken to remove a nonconformity and prevent its recurrence. Also interpreted as Corrective Action and Preventive Action where corrective action is taken to mean action to correct the nonconformity contrary to the ISO 9000 definition of the term.

Corrective maintenance.   Maintenance carried out after a failure has occurred that is intended to restore an item to a state in which it can perform its required function.

Critical success factors (CSFs).   Those factors on which the achievement of specified objectives depend.

Critical to quality (CTQ).   Key measurable product, service or process characteristics that must meet the agreed performance standards to satisfy the customer.

Cross-functional team.   See multidisciplinary team.

Customer complaints.   Any adverse report (verbal or written) received by an organization from a customer.

Customer feedback.   Any comment on the organization's performance provided by a customer.

Data.   Information that is organized in a form suitable for manual or computer analysis.

Define and document.   To state in written form, the precise meaning, nature or characteristics of something.

Demand creation process.   A key business process that penetrates new markets and exploits existing markets with products and a promotional strategy that influences decision makers and attracts potential customers to the organization.

Demand fulfilment process.   A key business process that converts customer requirements into products and services in a manner that satisfies all stakeholders.

Deming's 14 points of management

1   Create constancy of purpose
2   Adopt the new philosophy
3   Cease dependence on inspection

4 End the practice of awarding business based on price tag
5 Improve constantly and forever the system of production and service
6 Institute training on the job
7 Institute leadership
8 Drive out fear
9 Break down barriers between departments
10 Eliminate slogans, exhortations and targets
11 Eliminate quotas and management by objectives and by numbers
12 Remove barriers that rob the hourly worker of his right to pride of workmanship
13 Institute a vigorous program of education and self-improvement
14 Put everybody in the company to work to accomplish the transformation

Demonstrate.   To prove by reasoning, objective evidence, experiment or practical application.

Department.   A unit of an organization that may perform one or more functions. Units of organization regardless of their names are also referred to as functions (see Function).

Design and development.   Design creates the conceptual solution and development transforms the solution into a fully working model (See also ISO 9000:2015 3.4.8.)

Design review.   A formal documented and systematic critical study of a design by people other than the designer.

Design.   A process of originating a conceptual solution to a requirement and expressing it in a form from which a product may be produced or a service delivered.

Disposition.   The act or manner of disposing of something.

Documented procedures.   Procedures that are formally laid down in a reproducible medium such as paper or magnetic disk.

Effectiveness.   The extent to which planned activities are realized and planned results achieved.

Embodiment loan.   An item of customer-supplied property provided for incorporation into product that is subsequently supplied back to the customer or a party designated by the customer.

Enhanced customer satisfaction.   An outcome where satisfaction is derived from meeting all stated, generally implied or obligatory needs and expectations.

Ensure.   To make certain that something will happen.

Establish and maintain.   To set up an entity on a permanent basis and retain or restore it in a state in which it can fulfil its purpose or required function.

Evaluation.   To ascertain the relative goodness, quality or usefulness of an entity with respect to a specific purpose.

Externally provided product.   Hardware, software, documentation or information owned by an external provider such as a customer or supplier which is provided to an organization for use in connection with a contract or transaction.

Failure demand.   A demand that arises only because the service was not delivered properly to begin with.

Failure mode effects analysis (FMEA).   A technique for identifying potential failure modes and assessing existing and planned provisions to detect, contain or eliminate the occurrence of failure. (See also Risk assessment.)

Follow-up audit.   An audit carried out following and as a direct consequence of a previous audit to determine whether agreed actions have been taken and are effective.

Force majeure.   An event, circumstance or effect that cannot be reasonably anticipated or controlled.

Function. In the organizational sense, a function is a special or major activity (often unique in the organization) which is needed for the organization to fulfil its purpose and mission. Examples of functions are design, procurement, personnel, manufacture, marketing, maintenance, etc.

Hazard. Anything that may cause harm to people, product, property or the natural environment.

Hazard analysis. The process of collecting and evaluating information on hazards and conditions leading to their presence to decide which are significant for product safety and therefore should be addressed in the HACCP plan (ISO 15161).

Hazard Analysis and Critical Control Points (HACCP). A technique used particularly in the food industry for the identification of hazards and control of risks. The CCP is a step at which control can be applied and is essential to prevent or eliminate a food safety hazard or reduce it to an acceptable level (ISO 15161).

International Accreditation Forum. The world association of conformity assessment accreditation bodies in the fields of management systems, products, services, personnel and other similar programmes of conformity assessment.

Identification. The act of identifying an entity (i.e. giving it a set of characteristics by which it is recognizable as a member of a group).

Implement. To carry out a directive.

Implementation audit. An audit carried out to establish whether actual practices conform to the documented quality system. Note: Also referred to as a conformance audit or compliance audit.

Inspection. The examination of an entity to determine whether it conforms to prescribed requirements. (See also ISO 9000:2015 clause 3.11.7.)

Installation. The process by which an entity is fitted into larger entity.

Intellectual property. Creations of the mind: inventions, literary and artistic works and symbols, names, images and designs used in commerce. Intellectual property is divided into two categories: industrial property and copyright.

Interested party. Person or group having an interest in the performance or success of an organization which normally includes: customers, owners, employees, contractors, suppliers, investors, unions, partners or society (see also ISO 9000:2015 clause 3.2.3). Interested parties can be benevolent or malevolent and the latter group might include terrorists, criminals and competitors whose only interest is to harm the organization (See also Stakeholder.)

Just-In-Time. A method of lean production where the demand comes from the end of the process through to the beginning so that the only parts that are delivered are those that are needed at the time they are needed.

Kaizen. Continuing improvement in personal life, home life, social life and working life. When applied to the workplace it means continuing improvement involving everyone – managers and workers alike (Masaaki Imai).

Key performance indicators (KPI). The quantifiable characteristics that indicate the extent by which an objective is being achieved. (See also stakeholder success measures.)

Lagging measures. Measures that indicate an aspect of performance long after the conditions that created it have changed (e.g. profit and return on capital).

Leading measures. Measures that indicate an aspect of performance while the conditions that created it still prevail (e.g. response time, conformity).

Linearity. In a measurement system, this is the difference in the bias values through the expected operating range of the measuring equipment.

Manage work. To manage work means to plan, organize and control the resources (personnel, financial and material) and the tasks required to achieve the objective for which the work is needed.

Management representative. The person management appoints to act on their behalf to manage the quality management system.

Management system. A systemic view of an organization from the perspective of how it manages its performance. (See also ISO 9000:2015 3.5.3.)

Mass production. A method of production that is supply driven based on sales forecasts rather than firm orders. It produces large amounts of standardized products on parallel production lines that stretch from raw materials to finished product (vertical integration). In mass production, the job comes to the worker who passes it on to the next worker to perform the next operation on the line.

Measures. The characteristics by which performance is judged. They are the characteristics that need to be controlled in order than an objective will be achieved. They are the response to the question "What will we look for to reveal whether the objective has been achieved?"

Measurement. The act of measuring. It is a process of associating numbers with physical quantities and phenomena.

Measurement capability. The ability of a measuring system (device, person and environment) to measure true values to the accuracy and precision required.

Measurement process. Activities, measuring devices, personnel, operating environment and the measurement system to determine the value of a quantity.

Measurement system. The units of measure and the process by which standards for these units of measure are developed and maintained.

Measurement uncertainty. The variation observed when repeated measurements of the same parameter on the same specimen are taken with the same device.

Mission. An expression of the purpose of an organization, why it exists, what it is being mobilized to accomplish in the long term.

Mission management. A key business process that determines the direction of the business, continually confirms that the business is proceeding in the right direction and makes course corrections to keep the business focused on its mission.

Modifications. Entities altered or reworked to incorporate design changes.

Monitoring. To check periodically and systematically. It does not imply that any action will be taken.

Motivation. An inner mental state that prompts a direction, intensity and persistence in behaviour.

Multidisciplinary team. A team comprising representatives from various functions or departments in an organization, formed to execute a project on behalf of that organization.

Nature of change. The intrinsic characteristics of the change (what has changed and why).

Objective evidence. Information that can be proven true based on facts obtained through observation, measurement, test or other means. (See also ISO 9000:2015 clause 3.8.3.)

Objective. A result to be achieved usually by a given time.

Obsolete documents. Documents that are no longer required for operational use. They may be useful as historic documents.

Operating procedure. A procedure that describes how specific tasks are to be performed (might be called a work instruction).

Organizational goals. Where the organization desires to be in markets, in innovation, in social and environmental matters, in competition and in financial health.

Organizational interfaces. The boundary at which organizations meet and affect each other expressed by the passage of information, people, equipment, materials and the agreement to operational conditions.

Performance indicators.   Quantifiable measures of performance related to specific objectives. They respond to the question "What would we expect to see happening if this objective had been achieved?" (see also measures).

Political, Economic, Social and Technological Analysis.   A tool used in scanning the environment for changes affecting an organization's success in fulfilling its mission.

Plan.   Provisions made to achieve an objective.

Planned arrangements.   All the arrangements made by the organization to achieve the customer's requirements. They include the documented policies, objectives, plans, specifications and processes and the documents derived from such requirements.

Planned maintenance.   The maintenance carried out with forethought as to what is to be checked, adjusted, replaced, etc.

Poka-yoke.   Japanese term that means "mistake proofing", a concept introduced by Shigeo Shingo to Toyota in 1961. It is a device that prevents incorrect parts from being made or assembled, or prevents correct parts being assembled incorrectly. Previously the term *baka-yoke* was used but as this means "fool proofing" and is rather offensive so it was discontinued. Even mistake proofing has evolved into "error proofing" to avoid the personal implications. Error proofing is one of the two pillars of the Toyota Production System (TPS).

Policy.   A guide to thinking, action and decision.

Positively identified.   An identification given to an entity for a specific purpose which is both unique and readily visible.

Predictive maintenance.   Work scheduled to monitor machine condition, predict pending failure and make repairs on an as-needed basis.

Prevent.   To stop something from occurring by a deliberate planned action.

Preventive action.   Action proposed or taken to stop something from occurring (See also ISO 9000:2015 clause 3.12.1.)

Preventive maintenance.   Maintenance carried out at pre-determined intervals to reduce the probability of failure or performance degradation (e.g. replacing oil filters at defined intervals). Also referred to as planned maintenance.

Procedure.   A sequence of steps to execute a routine activity. (See also ISO 9000:2015 clause 3.4.5.)

Process.   A series of activities that use resources to produce a result. An effective process would be one in which the activities use resources to achieve a prescribed objective. The activities may be interrelated, interdependent and may interact. (See also ISO 9000:2015 clause 3.4.1.)

Process approach.   An approach to managing work in which the activities and resources (including behaviours) function together in such a relationship as to produce results consistent with the process objectives.

Process capability.   The inherent ability of a process to reproduce its results consistently during multiple cycles of operation.

Process description.   A set of information that describe the characteristics of a process in terms of its purpose, objectives, measures, design features, inputs, activities, resources, behaviours, outputs, constraints, measurements and reviews.

Process management.   The planning, operation and control of interrelated and interacting activities to produce a desired result.

Process measures.   Measures used to judge the performance of processes. They are generally a response to the question "What will we look for to reveal whether the process objectives have been met?"

Process parameters. Those variables, boundaries or constants of a process that restrict or determine the results.

Production. The creation of products.

Prototype. A model of a design that is both physically and functionally representative of the design standard for production and used to verify and validate the design.

Purchaser. One who buys from another.

Qualification test. Determination by a series of tests and examinations of a product, and its related documents and processes, that the product meets all the specified performance capability requirements under operational conditions.

Quality characteristics. Any characteristic of a product or service that is needed to satisfy customer needs or achieve fitness for use.

Quality control. A process for maintaining standards of quality that prevents and corrects change in such standards so that the resultant output meets customer needs and expectations. (See also ISO 9000:2015 clause 3.3.7.)

Quality function deployment (QFD). A technique to deploy customer requirements (the true quality characteristics) into design characteristics (the substitute characteristics) and deploy them into subsystems, components, materials and production processes. The result is a grid or matrix that shows how and where customer requirements are met.

Quality management system requirements. Requirements pertaining to the design, development, operation, maintenance and improvement of quality management systems.

Quality management system. A systemic view of an organization from the perspective of how it creates and retains customers. (See also ISO 9000:2015 3.4.3.)

Quality objectives. An objective that primarily benefits the customer by its achievement.

Quality of conformance. The extent to which the product or service conforms to the design standard. The design has to be faithfully reproduced in the product or service.

Quality of design. The extent to which the design reflects a product or service that satisfies customer needs and expectations. All the necessary characteristics should be designed into the product or service at the outset.

Quality of use. Extent by which the user is able to secure continuity of use from the product or service. Products need to have a low cost of ownership, be safe and reliable, need to be maintainable in use and need to be easy to use.

Quality planning. Provisions made to achieve the needs and expectations of organization's stakeholders and prevent failure.

Quality plans. Plans produced to define how specified quality requirements will be achieved, controlled, assured and managed for specific contracts or projects.

Quality problems. The difference between the achieved quality and the required quality.

Quality requirements. Those requirements which pertain to the features and characteristics of a product or service which are required to be fulfilled to satisfy a given need.

Quarantine area. A secure space provided for containing product pending a decision on its disposal.

Reductionism. A way of using logic and causal thinking to separate the individual parts of what is being studied and draw conclusions about a group based on the analysis of its constituent parts. It is not always possible to predict the behaviour of systems as any changes can lead to unintended consequences. Reductionism tends to ignore the influence that individual parts have on each other.

Registrar. See Certification body.

Registration. A process of recording details of organizations of assessed capability that have satisfied prescribed standards.

Regulator.   A legal body authorized to enforce compliance with the laws and statutes of a national government.

Regulatory requirements.   Requirements established by law pertaining to products or services.

Remedial action.   Action proposed or taken to remove a nonconformity in a product previously deemed conforming (see also Corrective and preventive action).

Repeatability.   In a measurement system, this is the variation in measurements obtained by one appraiser using one measuring equipment to measure an identical characteristic on the same part.

Representative sample.   A sample of product or service that possesses all the characteristics of the batch from which it was taken.

Reproducibility.   In a measurement system, this is the variation in the average of the measurements made by different appraisers using the same measuring instrument when measuring an identical characteristic on the same part.

Resources.   Anything physical or mental that can be used to obtain something else one needs or desires. Resources include time, personnel, skill, machines, materials, money, plant, facilities, space, information, knowledge, etc. Resources are used by processes resulting in some being reusable and others changed, lost or depleted by the process.

Resource management.   A key business process that specifies, acquires and maintains the resources required by the business to fulfil the mission and disposes of any resources that are no longer required.

Responsibility.   An area in which one is entitled to act on one's own accord or able to respond by having caused an event.

Review.   Another look at something.

Rework.   Continuation of work on a product to make it conform to the specified requirements without additional procedures or techniques.

Risk.   The likelihood of something happening that could have a negative effect. (See also ISO 9000:2015 3.7.9.)

Risk assessment.   A study performed to quantify potential risks associated with an event or situation. It identifies hazards or failure modes, their effect on people, product, property or natural environment, the probability of their occurrence and detection and the severity of their effect to identify provisions taken or needed to eliminate, control or reduce the root cause. (See also FMEA, HACCP.)

Risk management.   The process whereby organizations methodically address the risks attaching to their activities with the goal of achieving sustained benefit within each activity and across the portfolio of all activities (see IRM 2002).

Shall.   A provision that is binding.

Should.   A provision that is optional.

Simultaneous engineering.   A method of reducing the time taken to achieve objectives by developing the resources needed to support and sustain the production of a product in parallel with the development of the product itself. It involves customers, suppliers and each of the organization's functions working together to achieve common objectives.

Six Sigma.   Six standard deviations.

SMART objectives.   A technique for testing objectives as follows:

S Specific.   Objectives should be *specific* actions completed while executing a strategy or delivering an output. They should be derived from the mission and relevant to the process or task to which they are being applied. Objectives should be specified to a level of detail that those involved in their implementation fully understand what is required for their completion – not vague or ambiguous and defining precisely what is required.

M Measurable. Objectives should be *measurable* actions that have a specific end condition. Objectives should be expressed in terms that can be measured using available technology. When setting objectives, you need to know how achievement will be indicated, the conditions or performance levels that will indicate success.

A Achievable. Objectives should be *achievable* with resources that can be made available – they should be achievable by average people applying average effort.

R Realistic. Objectives should be *realistic* in the context of the current climate and the current and projected workload. Account needs to be taken of the demands from elsewhere that could jeopardize achievement of the objective.

T Timely. Objectives should be *time-phased* actions that have a specific start and completion date. Time-phased objectives facilitate periodic review of progress and tracking of revisions. The specific date or time does not need to be expressed in the objective unless it is relevant – in other cases the timing for all objectives might be constrained by their inclusion in the 2005 business plan, implying all the objectives will be achieved in 2005. The business plan for 2005–2008 implies all objectives will be achieved by 2008.

Special cause variation. A cause of variation that can be assigned to a specific or special condition that does not apply to other events (e.g. weather, power failure, tool breakage, etc.).

Specified requirements. Requirements prescribed by the customer and agreed by the organization or requirements prescribed by the organization that are perceived as satisfying a market need. Such requirements may or may not be documented.

Stability. In a measurement system, this is the total variation in the measurements obtained with a measurement system on the same part when measuring a single characteristic over a period.

Stakeholder. The individuals and constituencies that contribute, either voluntarily or involuntarily, to an organization's wealth-creating capacity and activities, and that are therefore its potential beneficiaries and/or risk bearers (Post, Preston and Sachs). They are the response to the question "Who are we working for?" (See also Interested party.)

Stakeholder measures. Measures used to judge the performance of an organization. They are generally a response to the question, "What measures will the stakeholders use to reveal whether their needs and expectations have been met?" (See also Key performance indicators.)

Statistical control. A condition of a process in which there is no indication of a special cause of variation.

Status. The relative condition, maturity or quality of something.

Strategy. The broad priorities adopted by an organization in recognition of its operating environment in pursuit of its mission (Paul Niven).

Subcontractor. A person or company that enters into a subcontract and assumes some of the obligations of the prime contractor.

Suitability. The extent to which an object "fits" and is right for the organization's purpose, its operations, culture, and business systems.

Strengths, weaknesses, opportunities and threats analysis. A tool for determining the capability of an organization to achieve prescribed objectives.

Systems approach. An approach to managing an organization that recognizes its performance results from the interaction of interrelated elements and cannot be predicted by analysing each element taken separately.

System audit. An audit carried out to establish whether the quality system conforms to a prescribed standard in both its design and its implementation.

System effectiveness. The ability of a system to achieve its stated purpose and objectives.

Systems thinking. The understanding of dynamic relationships between interacting variables or process outputs within the context of a larger whole. A scientific field of knowledge for understanding change and complexity through the study of dynamic cause and effect over time. (Kambiz E. Maani)

Targets. The level of performance to be achieved (e.g. standard, specification, requirement, budget, quota, plan).

Task. The smallest component of work. A group of tasks comprise an activity.

Technical interfaces. The physical and functional boundary between products or services.

Tender. A written offer to supply products or services at a stated cost.

Theory X. A label given to a belief that workers inherently dislike and avoid work and must be driven to it (Douglas McGregor, 1906–1964).

Total quality management. A management philosophy and company practices that aim to harness the human and material resources of an organization in the most effective way to achieve the objectives of the organization (BS 7850: 1992).

Traceability. The ability to trace the history, application, use and location of an individual article or its characteristics through recorded identification numbers. (See also ISO 9000:2015 3.6.13.)

Validation. A process for establishing whether an entity will fulfil the purpose for which it has been selected or designed. (See also ISO 9000:2015 3.8.13.)

Values. The fundamental principles that guide the organization in accomplishing its goals. They are what it stands for, such as integrity, excellence, innovation, inclusion, reliability, responsibility, equality, fairness, confidentiality, safety of personnel and property, etc. These values characterize the culture in the organization.

Verification activities. A special investigation, test, inspection, demonstration, analysis or comparison of data to verify that a product or service or process complies with prescribed requirements.

Verification requirements. Requirements for establishing conformance of a product or service with specified requirements by certain methods and techniques.

Verification. The act of establishing the truth or correctness of a fact, theory, statement or condition. (See also ISO 9000:2015 clause 3.8.12.)

Vision. An expression of the aspirations of an organization; what success will look like as it fulfils its mission.

Waiver. See Concession.

Work breakdown structure. A structure in which elements of work for a particular project are placed in a hierarchy.

Work instructions. Instructions that prescribe work to be executed, who will do it, when it is to start and be complete and, if necessary, how it is to be carried out.

Work packages. An assembly of related work elements.

Workmanship criteria. Standards on which to base the acceptability of characteristics created by human manipulation of materials by hand or with the aid of hand tools.

Zero defects. The performance standard achieved when every task is performed right first time with no errors being detected downstream.

Bibliography

Note: A version of this bibliography with active hyperlinks is available on the companion web site

Ackoff, R. L. (1971, July). Towards a System of Systems Concepts. *Management Science*, 17(11).

Ackoff, R. L. (1994). Beyond Continuous Improvement. Retrieved from http://www.youtube.com/watch?v=OqEeIG8aPPk

Ackoff, R. L. (1998). A Systemic View of Transformational Leadership. *Systemic Practice and Action Research*, 11(1), 23–36.

Ackoff, R. L. (1999). *Ackoff's Best: His Classic Writings on Management: Chapter 1*. New York: John Wiley & Sons, Inc.

ACSI. (2016, May 10). About the American Customer Satisfaction Index. Retrieved from American Customer Satisfaction Index: http://www.theacsi.org/about-acsi

AIAG. (2016, November). Potential Failure Modes and Effects Analysis Reference Manual. Retrieved from LEHIGH University: https://www.lehigh.edu/~intribos/Resources/SAE_FMEA.pdf

APQC. (2015, October 23). American Productivity & Quality Center. Retrieved from APQC Process Classification Framework (PCF): https://www.apqc.org/knowledge-base/documents/apqc-process-classification-framework-pcf-cross-industry-excel-version-70

Barron, R. A., & Greenberg, J. (1990). *Behaviour in Organizations*. Needhan Heights MA, USA: Allyn and Bacon.

Bateman, M. (2006). *Tolley's Practical Risk Assessment Handbook*. Abingdon, Oxon: Routledge.

BBC. (2011, December 2). Delivering Quality First. Retrieved from BBC: http://www.bbc.co.uk/aboutthebbc/insidethebbc/howwework/reports/deliveringqualityfirst.html

BBC. (2014, November 14). Captain Eric 'Winkle' Brown. Retrieved from Desert Island Discs: http://www.bbc.co.uk/programmes/b04nvgq1

BBC. (2016, November 9). Watchdog. Retrieved from BBC One: http://www.bbc.co.uk/programmes/b082wkj6

BD. (2016, January 12). Business Dictionary. Retrieved January 12, 2016, from http://www.business-dictionary.com

Benson Ford Research Center. (2013). The Innovator and Ford Motor Company. Retrieved from The Henry Ford: https://www.thehenryford.org/exhibits/hf/The_Innovator_and_Ford_Motor_Company.asp

Benson Ford Research Center. (2017, April). Henry Ford Quotations. Retrieved from Benson Ford Research Center: https://www.thehenryford.org/collections-and-research/digital-resources/popular-topics/henry-ford-quotes/

Billington, R. (1988). *Living Philosophy – An Introduction to Modern Thought*. Abingdon, Oxon: Routledge.

Boone, L. E., & Kurtz, D. L. (2013). *Contemporary Marketing*, 16 Edition. Boston: CENGAGE Learning Custom Publishing.

Bosch. (2017, April). Bosch, R (1885). Retrieved from The Finance Director: http://www.the-finance-director.com/contractors/business-process-outsourcing/bosch-communication-center/bosch-communication-center3.html

Box, G. E. (1976). Science and Statistics. *Journal of the American Statistical Association*, 71(356), 791–799.

Box, G. E. (1986). *Empirical Model-Building and Response Surfaces*. New Jersey: John Wiley & Sons.

British Library Preservation Advisory Centre. (2013). Managing the Library Archive Environment. Retrieved from http://www.bl.uk/aboutus/stratpolprog/collectioncare/publications/booklets/managing_library_archive_environment.pdf

Bryson, J. (2004). *Creating and Implementing Your Strategic Plan – A Workbook for Public and Non-Profit*. San Francisco: Jossey Bass.

CAA. (2016, June 19). Airport data 2015. Retrieved from Civil Aviation Authority: http://www.caa.co.uk/Data-and-analysis/UK-aviation-market/Airports/Datasets/UK-Airport-data/Airport-data-2015/

Canon. (2008, July). Canon Sustainability Report 2008. Retrieved from Canon: http://www.canon.com/csr/report/pdf/sustainability2008e.pdf

Carroll, L. (1865). *Alice's Adventures in Wonderland*. London: Macmillan & Co.

Carter, R. C., Martin, J. N., Mayblin , B., & Munday, M. (1983). *Systems, Management and Change: A Graphic Guide*. London: Paul Chapman Publishing Ltd.

Checkland, P. B. (1981). *Systems Thinking, Systems Practice*. Chichester: John Wiley & Sons.

Chong, J. (2012, December). Human Capital vs. Human Resources – A Word and a World of Difference. Retrieved from Dignity of Work: https://thedignityofwork.com/2012/12/04/human-capital-vs-human-resources-a-word-and-a-world-of-difference/#comments

Choppin, J. (1997). *Total Quality Through People*. Leighton Buzzard: Rushmere Wyne, England.

CIPD. (2016, August). Competence and Competency Frameworks. Retrieved from CIPD: http://www.cipd.co.uk/hr-resources/factsheets/competence-competency-frameworks.aspx

CIPD. (2015, October 12). Absence Survey. Retrieved from Chartered Institute for Personnel Development: http://www.cipd.co.uk/pressoffice/press-releases/absence-survey-121015.aspx

Covey, S. R. (1992). *Principle-Centered Leadership*. London: Simon & Shuster UK Ltd.

Crosby, P. B. (1986). *Quality Without Tears*. New York: McGraw-Hill.

Daily Telegraph. (2007, December 22). 'Doing a Ratner' and other Famous Gaffes. Retrieved from Daily Telegraph: http://www.telegraph.co.uk/news/uknews/1573380/Doing-a-Ratner-and-other-famous-gaffes.html

Daily Telegraph. (2012, July 30). Apple Design Chief: 'Our Goal Isn't to Make Money'. Retrieved from Daily Telegraph: www.telegraph.co.uk/technology/apple/9438662/Apple-design-chief-Our-goal-isnt-to-make-money.html

Dalkir, K. (2011). *Knowledge Management in Theory and Practice*, Second Edition. Cambridge, MA: MIT Press.

de Geus, A. (1999). *The Living Company Growth, Learning and Longevity in Business*. London: Nicholas Brealey Publishing.

Deming, E. W. (1994). *The New Economics for Industry, Government and Education* – Second Edition. MIT Press.

Deming, W. E. (1982). *Out of the Crisis*. Cambridge, MA: MIT Press.

Dennis, P. (2006). *Getting the Right Things Done – A leaders guide to planning and execution*. Cambridge. MA, USA: The Lean Enterprise Institute.

Drucker, P. F. (1974). *Management, Tasks, Responsibilities, Practices*. Oxford: Butterworth-Heinemann.

Drucker, P. F. (1999). *Management Challenges for the 21st Century*. Oxford: Butterworth Heinemann.

Edwards, W. (2013, Winter). Thinking About Processes, Transformations and Control. *Journal of the United Kingdom System Society "Systemist"*, 34(3), 100.

Feigenbaum, A. V. (1961). *Total Quality Control – Engineering and Management*, Second Edition. New York: McGraw-Hill.

Field, A. (2008, February 29). Customer Focused Leadership. Retrieved from Harvard Business Review: https://hbr.org/2008/02/leadership-that-focuses-on-the-1.html

Fletcher, S. (2000). Competence-based assessment techniques. London: Kogan Page.

FMC. (1990). *Worldwide Quality Q-101*. Dearborn, MI: The Ford Motor Company.

Ford, H., & Crowther, S. (1922). *My Life and Work*. New York: Doubleday Page & Company.

Fox, W. A. (1986). *Always Good Ships – Histories of Newport News Shipbuilding*. Virginia Beach: Donning Company Publishers.

Fraser, P. K. (2015, September, October and November). Business Process Principles. (P. Harding, Ed.) SAQI e-Quality Edge (193, 194 & 195).

Frost, A. (2016, July 28). Introducing Organizational Knowledge. Retrieved from KMT (Knowledge Management Tools): http://www.knowledge-management-tools.net/introducing-organizational-knowledge.html

Ghaleiw, M. A. (2015). Delivering Outstanding Performance through Highly Engaging Quality Culture. 2015 ASQ MENA Regional Quality Conference. Dubai: ASQ. Retrieved April 17, 2016, from http://www.asqmea.org/download/Highly-Engaging-Quality-Culture-Paper-Rev-3.pdf

The Guardian. (2015, September 22). VW Scandal: Chief Executive Martin Winterkorn Refuses to Quit. Retrieved from The Guardian: https://www.theguardian.com/business/2015/sep/22/vw-scandal-escalates-volkswagen-11m-vehicles-involved

The Guardian. (2016, May). Restaurant Owner Jailed for Six Years Over Death of Peanut Allergy Customer. Retrieved from The Guardian: https://www.theguardian.com/society/2016/may/23/restaurant-owner-mohammed-zaman-guilty-of-manslaughter-of-peanut-allergy-customer

Gharajedaghi, J. (2011). *Systems Thinking: Managing Chao and Complexity – A Platform for Designing Business Architectures*. Burlington, MA: Elsevier.

Haaften, R. v. (2016, May 10). Abstract – Customer Satisfaction Models. Retrieved from Rovaka: http://www.van-haaften.nl/customer-satisfaction/customer-satisfaction-models/101-abstract-customer-satisfaction-models

Hammer, M., & Champy, J. (1993). *Reengineering the Corporation*. New York: Harper Business.

Hill, N., Self, B., & Roche, G. (2002). *Customer Satisfaction Measurement for ISO 9000:2000*. Oxford: Butterworth Heinemann.

Hillson, D. ((2) 2016, June 20). Managing Risk in Practice – Workshop. Retrieved from https://www.youtube.com/watch?v=fVIqy5IoyS4

Hillson, D. ((1) 2016, June 23). Managing Risk in Projects. Retrieved from https://www.youtube.com/watch?v=GO2rpxjbi_A

Hodgetts, R. M. (1979). *Management, Theory, Process and Practice*. Philadelphia: W. B. Saunders Company.

Horstman, M. (2016). *The Effective Manager*. Hoboken, NJ: John Wiley & Sons.

Hoyle, D., & Thompson, J. (2009). *ISO 9000 Auditor Questions*, Second Edition. Monmouth: Transition Support Ltd.

HSE. (2016, August). Performance Influencing Factors (PIFs). Retrieved from Health and Safety Executive: http://www.hse.gov.uk/humanfactors/topics/pifs.pdf

HSG48. (1999). *Reducing Error and Influencing Behaviour*. London: HSE.

Hubbard, D. W. (2009). *The Failure of Risk Management: Why it's Broken and How to Fix It*. Hoboken, NJ: John Wiley & Son.

Hubbard, D. W. (2010). *How to Measure Anything: Finding the Values of Intangibles in Business*. Hoboken, NJ: John Wiley.

Hunsaker, P. L., & Alessandra, A. J. (1986). *The Art of Managing People*. New York: Simon & Shuster, Inc.

Hunsaker, P. L., & Alessandra, A. J. (2009). *The New Art of Managing People*. New York: Simon & Schuster, Inc.

Imai, M. (1986). *KAIZEN – The Key to Japanese Competitive Success*. New York: McGraw-Hill.

IMechE. (2015, November 2). Global Food – Waste Not, Want Not. Retrieved November 2, 2015, from Institution of Mechanical Engineers: http://www.imeche.org/knowledge/themes/environment/global-food

Ishikawa, K. (1985). *What Is Total Quality Control: The Japanese Way*. (D. J. Lu, Trans.) Englewood Cliffs, NJ: Prentice-Hall Inc.

ISO 10018. (2012). *Quality Management – Guidelines on People Involvement and Competence*. Geneva: International Organization of Standardization.

ISO. (2009). Selection and Use of the ISO 9000 Family of Standards. Retrieved from International Organization for Standardization: https://www.iso.org/files/live/sites/isoorg/files/archive/pdf/en/iso_9000_selection_and_use-2009.pdf

ISO. (2015). Quality Management Principles. Retrieved from International Organization for Standardization: https://www.iso.org/files/live/sites/isoorg/files/archive/pdf/en/pub100080.pdf

ISO 26800. (2011). *Ergonomics — General Approach, Principles and Concepts*. Geneva: International Organisation for Standardization.

ISO 27500. (2016). *The Human-Centred Organization – Rationale and General Principles*. Geneva: International Organization of Standardization.

ISO 31000. (2009). *Risk Management – Principles and Guidelines*. Geneva: International Organization of Standardization.

ISO 37500. (2014). *Guidance on Outsourcing*. Geneva: International Organization for Standardization.

ISO 8402. (1986). *Quality Vocabulary. International Terms*. Geneva: International Organization of Standardization.

ISO 9000:2015. (2015). *Quality Management Systems – Fundamentals and Vocabulary*. Geneva: ISO.

ISO Glossary. (2016). Guidance on Selected Words Used in the ISO 9000 Family of Standards. Retrieved from ISO: http://www.iso.org/iso/03_terminology_used_in_iso_9000_family.pdf

ISO Guide 73. (2009). *Risk Management – Vocabulary*. Geneva: International Organization of Standardization.

ISO/IEC. (2015). *ISO/IEC Directives, Part 1 — Consolidated ISO Supplement — Procedures Specific to ISO*. Geneva: ISO General Secretariat.

ISO/IEC 2382:2015. (2015). *Information Technology – Vocabulary*. Geneva: International Organization of Standardization.

ISO/IEC 27002. (2013). *Information Technology. Security Techniques. Code of Practice for Information Security Controls*. Geneva: International Organization for Standardization.

ISO/TC176/SC2/N1017. (2011). ISO 9000 User Survey Report. Retrieved from ISO survey seeks feedback from ISO 9001 users: http://www.iso.org/tc176/sc2/ISO9000UserSurvey

ISO/TC176/SC2/N1276. (2015). Guidance on the Requirements for Documented Information. Retrieved from ISO/TC 176 Home Page: http://isotc.iso.org/livelink/livelink/fetch/2000/2122/-8835176/-8835848/8835872/8835883/Documented_Information.docx

ISO/TC176/SC2/N1282. (2015, August). ISO 9001 Summary of changes. Retrieved 2016, from ISO/TC 176 Home Page: http://isotc.iso.org/livelink/livelink/fetch/2000/2122/-8835176/-8835848/8835872/8835883/ISO9001Revision.pptx

ISO/TC176/SC2/N1289. (2015). Retrieved from ISO TC/176/SC2 Home Page: https://www.iso.org/files/live/sites/isoorg/files/archive/pdf/en/iso9001-2015-process-appr.pdf

ISO/TC176/SC2/N1338. (2015). Quality Management Systems – Guidelines for the Application of ISO 9001:2015. Geneva: International Organization for Standardization.

ISO/TC176/SC2/N544R3. (2008). ISO 9000 – Quality Management. Retrieved from ISO: http://www.iso.org/iso/04_concept_and_use_of_the_process_approach_for_management_systems.pdf

ISO/TMB/JTCG N359. (2013, December 03). JTCG Frequently Asked Questions in support of Annex SL. Retrieved 2016, from Annex SL Guidance documents: http://isotc.iso.org/livelink/livelink?func=ll&objId=16347818&objAction=browse&viewType=1

ISO/TMB/JTCG N360. (2013, December 3). Concept Document to Support Annex SL. Retrieved from Annex SL Guidance documents: http://isotc.iso.org/livelink/livelink?func=ll&objId=16347818&objAction=browse&viewType=1

ISO/TS 19150. (2012). *Geographic Information – Ontology – Part 1: Framework*. Geneva: International Organization of Standardization.

ISO-SD. (2016, September 15). How Does ISO Develop Standards? Retrieved from ISO: http://www.iso.org/iso/home/standards_development.htm

Jackson, M. C. (2003). *Systems Thinking: Creative Holism for Managers*. Chichester: John Wiley & Son.

Jarvis, A., & MacNee, C. (2011). Improved Customer Satisfaction – Key Result of ISO 9000 User Survey. Retrieved from ISO: http://www.iso.org/iso/home/news_index/news_archive/news.htm?refid=Ref1543

Jeston, J., & Nelis, J. (2008). *Management by Process: A Road Map to Sustainable Business Process Management*. Oxford: Butterworth-Heinemann.

Jobs, S. (1990). Steve Jobs on Joseph Juran and Quality (ASQ, Interviewer). Retrieved November 9, 2015, from https://www.youtube.com/watch?v=XbkMcvnNq3g&feature=youtu.be

Juran Foundation, Inc. (1995). *A History of Managing for Quality*. Milwaukee: ASQC Quality Press.

Juran, J. M. (1964). *Managerial Breakthrough*. New York: McGraw-Hill.

Juran, J. M. (1974). *Quality Control Handbook*, Third Edition. New York: McGraw-Hill.

Juran, J. M. (1992). *Juran on Quality by Design*. New York: The Free Press, Division of Macmillan Inc.

Kenniston, K. (1976, February 19). The 11 Year Olds of Today Are the Computer Terminals of Tomorrow. New York Times.

Kline, N. (1999). *Time to Think: Listening to Ignite the Human Mind*. London: Octopus Publishing Group.

Knobbe, C. A. (2016, June 18). Cataract Surgery Complications. Retrieved from All about Vision: http://www.allaboutvision.com/conditions/cataract-complications.htm

Kohn, L. T., Corrigan, J. M., & Donaldson, M. S. (1999). To Err Is Human – Building a Safer Health System. Institute of Medicine. Retrieved July 2016, from http://www.nap.edu/download/9728

Lisch, R. (2014). *Measuring Service Performance*. Abingdon: Routledge.

MacCormack, A., Crandall, W., & Henderson, P. (2012). Do You Need a New Product-Development Strategy? *Research Technology Management*, 55(1), 34–43.

Martin, R. L. (2010, January-February). The Age of Customer Capitalism. Retrieved from Harvard Business Review: https://hbr.org/2010/01/the-age-of-customer-capitalism

Meadows, D. M. (2008). *Thinking in Systems – A Primer*. White River Junction, Vermont: Chelsea Green Publishing.

Mil Std 833C. (2011). *Work Breakdown Structures for Defense Materiel Items*. Washington: US Department of Defense.

Miller, R. (2016, April 11). What Is Customer-Focused Leadership. Retrieved from The Training Bank: http://www.thetrainingbank.com/what-is-customer-focused-leadership/

Moen, R. D., & Norman, C. L. (2010, November). Circling Back – Clearing Up Myths About the Deming Cycle and Seeing How It Keeps Evolving. *Quality Progress*, 22–28.

Morgan, G. (1997). *Images of Organization*. London: SAGE Publications.

MWD. (2015, November). Retrieved from Merriam-Webster Dictionary: http://www.merriam-webster.com/dictionary/risk

Neave, H. R. (1990). *The Deming Dimension*. Knoxville: SPC Press.

New York Times. (1996, January 3). Arthur Rudolph, 89, Developer of Rocket in First Apollo Flight. Retrieved from New York Times Archives: http://www.nytimes.com/1996/01/03/us/arthur-rudolph-89-developer-of-rocket-in-first-apollo-flight.html

NIST. (2013). Retrieved from Handbook of Engineering Statistics: http://www.itl.nist.gov/div898/handbook/index.htm

Nonaka, I. (1994). A dynamic theory of organizational knowledge creation. *Organizational Science*, 5(1), 14–37.

OED. (2013). Retrieved from Oxford English Dictionary: oed.com

Ofcom. (2006, August 17). Broadband Migrations: Enabling Consumer Choice. Retrieved 2015, from Ofcom: https://www.ofcom.org.uk/consultations-and-statements/category-2/migration

ONS. (2017, March 9). Sickness Absence in the Labour Market. Retrieved from Office of National Statistics: https://www.ons.gov.uk/employmentandlabourmarket/peopleinwork/labourproductivity/articles/sicknessabsenceinthelabourmarket/2016#main-pointsh

Polanyi, M. (2009). *The Tacit Dimension*. Chicago: University of Chicago Press.

Post, J. E., Preston, L. E., & Sachs, S. (2002). *Redefining the Corporation: Stakeholder Management and Organizational Wealth*. Redwood, CA: Stanford University Press.

Price, F. (1984). *Right First Time*. Aldershot, UK: Wildwood House.

Reason, J. (1991). *Human Error*. Cambridge: Cambridge University Press.

Reason, J. (2000, March 18). Human Error: Models and Management. *British Medical Journal*, 320(7237), 768–770.

Reichfeld, F. (2001). *The Loyalty Effect*. Harvard Business School Press.

Rittenberg, L., & Martens, F. (2012). Understanding and Communicating Risk Appetite. Retrieved from Committee of Sponsoring Organizations of the Treadway Commission: https://www.coso.org/Documents/ERM-Understanding-and-Communicating-Risk-Appetite.pdf

Robbins, H., & Finley, M. (1998). *Why Change Doesn't Work*. London: Orion Business Books.

Rollinson, D. (2008). *Organisational Behaviour and Analysis*. Harlow, Essex: Pearson Education Limited.

Rothwell, W. J. (2007). Beyond Rules of Engagement. Retrieved from Dale Carnegie Training: http://www.dalecarnegie.co.uk/assets/1/7/Beyond_Employment_Engagement.pdf

Rummler, G. A., & Branche, A. P. (1995). *Improving Performance: How to Manage the White Space on the Organization Chart*. San Francisco: Jossey-Bass Inc.

Russell, B. (1998). *The Problems of Philosophy*. Oxford: Oxford University Press.

Saltzer, J. H., & Schroeder, M. D. (1975). *The Protection of Information in Computer Systems*. Cambridge, MA: MIT.

Seddon, J. (2000). *The Case Against ISO 9000*. Oxford: Oak Tree Press.

Seddon, J. (2014). *The Whitehall Effect*. Axminster: Triarchy Press.

Senge, P. M. (1990). *The Fifth Discipline – The Art and Practice of the Learning Organization*. Random House.

Sherwood, D. (2002). *Seeing the Forest for the Trees – A Manager's Guide to Applying Systems Thinking*. London: Nicholas Brealey Publishing.

Shewhart, W. A. (1939 (Republished 1986)). *Statistical Methods from the Viewpoint of Quality Control*. New York: Dover Publications.

Small, J. E. (1997). *ISO 9000 for Executives*. Sunnyvale, CA: Lanchester Press Inc.

Smith, A. (1776). An Inquiry in the Nature and Causes of the Wealth of Nations.

Stacey, R. D. (2010). *Complexity and Organization Reality*. Abington, Oxford, UK: Routledge.

TAG. (2016, October). Retrieved from Threat Analysis Group: http://www.threatanalysis.com/2010/05/03/threat-vulnerability-risk-commonly-mixed-up-terms/

Taleb, N. N., Goldstein, D. G., & Spitznagel, M. W. (2009, October). The Six Mistakes Executives Make in Risk Management. Retrieved from Harvard Business Review: https://hbr.org/2009/10/the-six-mistakes-executives-make-in-risk-management

Taylor, F. W. (1911). *The Principles of Scientific Management*. New York: W W Norden & Company.

Toyota. (2010, May 12). Toyota Auris Hybrid Production: Quality First and Foremost. Retrieved from Toyota Media Site: http://media.toyota.co.uk/2010/05/toyota-auris-hybrid-production-quality-first-and-foremost/

Toyota. (2015, February 26). No More Than One Cup's Worth of Water Can Fit into a Single Cup. To Hold More Water, You Need More Cups. Retrieved from Toyota-Global: http://www.toyota-global.com/company/toyota_traditions/philosophy/jan_feb_2007.html

Toyota-Global. (2015, February 24). Guiding Principles at Toyota. Retrieved from Toyota-Global: http://www.toyota-global.com/company/vision_philosophy/guiding_principles.html

UKICS. (2016, May 9). UK Customer Satisfaction Index. Retrieved from The Institute of Customer Service: https://www.instituteofcustomerservice.com/research-insight/uk-customer-satisfaction-index

Voehl, F. (1995). *Deming the Way We Knew Him*. Taylor & Francis.

West Midlands. (2015, February 15). A Positive Approach to Risk & Personalisation. Retrieved from Think local act personal: https:// www.thinklocalactpersonal.org.uk/_assets/Resources/Personalisation/TLAP/Risk_personalisation_framework_West_Midlands.pdf

Wikipedia (1). (2016, September). Philip B Crosby. Retrieved from Wikipedia: https://en.wikipedia.org/wiki/Philip_B._Crosby

Wikipedia (2). (2015, November 4). Venetian Arsenal. Retrieved from Wikipedia: https://en.wikipedia.org/wiki/Venetian_Arsenal#Mass_production

Wikipedia (3). (2015, January 23). Necessity and Sufficiency. Retrieved from Wikipedia: https://en.wikipedia.org/wiki/Necessity_and_sufficiency

Wikipedia (4). (2015, February 15). Enron Scandal. Retrieved from Wikipedia: http://en.wikipedia.org/wiki/Enron_scandal#Causes_of_downfall

Wikipedia (5). (2015, March 31). Data Storage Device. Retrieved from Wikipedia Free Encyclopedia: https://en.wikipedia.org/wiki/Data_storage_device

Wikipedia (6). (2015, March 31). Cloud Storage. Retrieved from Wikipedia: https://en.wikipedia.org/wiki/Cloud_storage

Wikipedia (7). (2016, March 26). Cantor Fitzgerald. Retrieved from Wikipedia: https://en.wikipedia.org/wiki/Cantor_Fitzgerald#September_11_attacks

Wikipedia (8). (2017, April 2). British European Airways Flight 548. Retrieved from Wikipedia: https://en.wikipedia.org/wiki/British_European_Airways_Flight_548

World Health Organization. (2009). What Is Human Factors and Why Is It Important to Patient Safety? Retrieved from http://www.who.int/patientsafety/education/curriculum/who_mc_topic-2.pdf

Zeithami, V. A., Parasuraman , A., & Barry, L. L. (1990). *Delivering Quality Service: Balancing Customer Perceptions and Expectations*. New York: The Free Press.

Index

acceptance criteria 39, 486, 526–31, 612, 624–5, 708–14, 850
acceptance tests 259, 616
accreditation 850
accredited certification 8, 17
accuracy and precision 413–15
Ackoff, Russell 81, 108, 112, 273, 280, 318–19, 379, 468, 839, 844
action: to address risks 340; to address risks and opportunities 325, 339, 373, 766, 800; to bring about improvement 59, 809; to prevent human error 44, 678–84
analysis and evaluation methods 739, 744
analysis of data 296, 757
Annex SL: and conflicting improvement clauses 840; and the definition of risk 149–56; on implementation sequence 24; the intent and impact of 22–4; and multiple management systems 41; on nonconformity 818–19; and system elements 118
applicability: of ISO 9001 11, 20, 205; of the QMS 203, 782
approval: of changes 504; by customer 527, 617, 750; of documented information 493; of ISO 9001 6; of parts 594, 653; of product 617; requirements 663; signature 690
approved providers 654
archiving 434, 492
as-built configuration 711
assessing competence 442, 446–9
attribute data 759–60, 851
audit: checklists 786–9; of conformity 206, 783–4; criteria 771–4, 779–85; findings 46, 786–9, 844; follow-up 788, 837, 853; frequency and importance 774–5; impartiality 779–81; internal 17, 45–6, 280, 312, 766, 771–90, 805, 819, 844; method 774, 781; objectives 247, 772–6, 785–6; planning 777, 784; programme 57, 481, 766–74; programme review 778; report 785–9; responsibilities 774, 776; scope 779
auditor: competence 247, 789; records 789; selection of 780–1

awareness 262: of contribution 459; of customer requirements 306, 317; of implications 460; of quality objectives 458; of quality policy 457; of vision 217

back-up data storage 502
balancing: needs 188, 558, 825; objectives 140–1, 292; processes 133–4
behaviour: of management 258, 272, 826; of systems 70, 110, 119, 241
behavioural: approach 403; styles 385, 404
behaviour – customer 192, 193; intentional 682
behaviours – appropriate 317, 441, 516
benchmarking 184, 217, 345, 742, 801, 851
best practice 487, 742, 774
Billington, Ray 429–30
Boone & Kurtz 192
boundary: system 114, 116–17, 129, 200–3
breakthrough 813; in attitudes 768; and control 133, 365; and improvement 73, 193; in performance 297, 800
Bryson, John 106, 183
BS 5750 52, 84, 305
business: capability 563; environment 175; objectives 225, 349, 811, 851; performance 38, 61, 794; plan 353, 593, 753, 851
business excellence model 7
business goals 44, 135, 268
business process 26, 42, 67, 134–42, 220–36, 267–9, 313, 343, 522, 851
business process management (BPM) 7, 269
business management system 227, 264, 851

calibrating the estimators 333
calibration: of measuring equipment 420–3; recording the basis for 422; records 419–23; standards 425; status 423–5
cause and effect 69, 73, 119, 134, 283, 345, 399, 738, 829–30
certificate of conformance 714
certification body 8–9, 45, 512, 772, 851
certification to ISO 9001 17, 19, 53

change: authority 505, 706; a compelling case for 769; control 433, 505, 630, 631–6; control board 633–5; notice 636; proposal 534, 632–5; record 507, 636, 836; request 494, 504–7, 635, 836

changes: to design 630–4; to the QMS 26, 311, 365–71, 843

checklists 331, 554, 605, 611, 675, 753, 777, 781–9

Choppin, Jon 277

classification of external providers 649–51

classifying nonconformities 827–8

climate, economic 303, 380, 793

codes of practice 487, 492, 602–3

command and control 38, 485

commitment 261; management 26, 46, 215; with respect to the QMS 261–81

common cause failure 332, 342

common cause variation 704, 759, 761–2, 851

communicating: with customers 537–50; with external providers 661–70; the importance of effective quality management 26, 255, 273–5; quality objectives 360; quality policy 300; requirements 661, 663, 665; responsibility and authority 306, 308–10

communication 464–77; barriers to 465; internal 464, 468–70; oral and written 475–7; process 464–6; skills 467

comparative references 417, 622

competence 439–56; assessment methods 447–9; and documentation 489; levels of 450, 665; and the professions 443; and qualification of personnel 676

competence-based assessment 448, 851

complexity: of audits 776; of design and development 580; model of 130; organizational 183, 205

compliance risk 153, 331

concessions: requesting 725

concurrent engineering 591–2, 851

confidence, levels of 8

configuration control 503, 634, 851

configuration management 54, 532, 573–4, 851

configuration record 636, 711

conformity assessment 7, 17, 66, 85, 205, 852

conformity – demonstrate 268, 480, 736

consequences: of failure 28, 462, 603–4; of nonconformity 820, 824–5

consideration replaces prescription 21

constraints: and demands 103, 105; as objectives 106, 353

containment action 723, 726, 820, 824, 830

contamination control 700

contingency planning 426

continual improvement 4, 29, 34, 73, 217, 290, 296, 813, 839–45, 847

contract loan 693, 852

contract negotiation 572, 654

control charts 759, 852

controlled conditions 671, 852

controlled documents, types of 483

controlled experiment 196

controlling: external documents 511; organizational interfaces 588–9

control model 132

control of: changes 367, 504, 533, 704–7; externally provided processes 208, 637–48, 671, 731; nonconforming outputs 29, 719–29, 732, 818, 824; production and service provision 28, 671–85, 731

control plans 710, 743

core competence 118, 140, 852

corporate objectives 316, 735, 794

corporate policy 207, 484, 794

correcting nonconformity 824

corrective action: and nonconformity 818–38; process 820–1; report 789, 836; undertaking 786

corrective and preventive action 852

corrective maintenance 387, 392–4, 693, 852

Covey, Steven 277, 404

criminals 167, 191, 329, 854

critical success factors 222, 247, 852

critical to quality characteristics (CTQ) 359, 460, 626–7, 764, 852

Crosby, Philip B 94

culture 63, 65, 165, 214, 255, 283, 291, 300, 385, 402–8, 442, 466, 472–3, 485, 813

customer: communication 27, 537–50, 730; compliments 1, 748, 751–3; contracts 540; dissatisfaction 98, 189, 221, 557, 657, 751, 754; enquiries 540; feedback 28, 544, 718, 751, 852; needs and expectations 7, 188, 354, 565, 712, 815; orders 543; property 205, 208, 547–8; returned product 727, 823; surveys 412, 748

customer complaints 17, 62, 317, 545–7, 615, 751, 823, 852

customer first 66, 188, 283, 292, 293

customer focus 43, 70, 255, 283–8, 317–18, 320, 355, 559, 562, 731

customer requirements 9, 11, 38, 43, 70, 94, 102–7, 192, 247, 321, 452, 520, 554–71, 813

customer satisfaction 11, 19, 34, 189, 286, 390, 747–55, 798, 804, 853

customer satisfaction index 754

customer support 469, 522, 753

data collection 317, 753, 759

Davenport, Thomas 143

delivery performance 356, 716–17, 753

delivery process 715–18

delivery risk 657

demand creation process 227–32, 523, 852

demand fulfilment process 226, 227, 519, 521–3, 852

Deming, Edwards 3, 68, 77–80, 89, 94, 96, 157, 244, 273, 277, 292, 356, 403, 654, 657, 671, 679, 768, 813, 827, 839–41, 852

demonstrations, design validation 616

department: objectives 141, 144, 354; policy 485

design: authority 586, 618, 724, 850; baseline 632; calculations 613, 629; process 578, 583, 608, 612–13, 631, 731; requirements 578–610; review 589, 605, 609–11, 617–20, 632, 853; standard 585, 616, 711

design and development: process 522, 576–8; responsibilities 586; verification 584

development models 613

development tests 613

deviations from requirements 723–8

digital media: access to 499; changing 506–7; inadvertent changes to 497; loss of 501; retrieval of 499; storage of 501; version control 508

disaster recovery 234, 392, 548

disposition 853; of nonconformities 509; of records 510

distribution channels 192, 715

document control 433, 487–513, 573, 630–4

documented information: approval 493; availability 672; distribution of 498, 512; needed for the QMS 482; retrieval of 499; review 493

documented procedures 31, 45, 478–9, 483, 576, 853

documenting the QMS 30, 49, 209

Drucker, Peter 161, 177, 247, 248–9, 332, 439

effectiveness of the system 46, 59, 796

effectiveness of training 452

efficiency – improving 1, 72, 75, 145, 349, 354, 741, 793, 807

embodiment loan 693, 853

employee: engagement 76, 277–80, 300, 457–63; motivation 355, 802; satisfaction 64, 134, 194, 287, 403

empowerment 75, 396

end-to-end processes 144

engagement of people principle 71–2, 396

enquiry conversion process 541, 568

Enron's values 355

environmental: issues 492; management system 22, 41, 116, 199

environmental policy 462

environment – controlled 400, 698

environment – natural 116, 153, 160, 182, 399, 598

equality issues 492

ergonomics 222, 397, 401

error proofing 132, 461, 623, 684, 759, 833, 856

ethical assessment 651

EU directives 13

evidence-based decision making 73, 76

evidence of conformity 27, 421, 500, 508, 713–14, 783

exclusions 201, 205

exposition 206, 782

external providers: classification of 651; control of 642; evaluation of 767, 649–60; information for 661–70; level of control 595; property of 692

facility maintenance process 395

factual approach 73, 747

failure mode and effects analysis (FMEA) 250, 394, 400–1, 525, 593, 657, 853

failure severity ranking 825

feedback loops 119, 121, 132, 138, 239–41, 719, 841

Feigenbaum, Armand 110, 112

final inspection requirements 708

financial assessment 651

financial performance 283, 793, 795

financial resources 136, 216, 375, 377, 564

financial risk 153

finished product verification 711

flow charts 45, 128, 214, 370, 524, 578

focus meetings 748, 751, 843

Ford, Henry 141, 154, 295, 639

functional approach 138–43, 269

functional requirements 599

gap analysis 214

goals: business 135, 220, 268, 443; creating line of sight from 442; personal 103

Hammer and Champy 138

hard copy media, changing 506

hazards analysis 401, 664

health and safety 556–7, 679, 823

Hertzberg, Frederick 403, 408

Hillson, David 150–1, 155, 330

human capital 135, 153, 217, 228, 355, 379, 381

human error 28, 44, 404, 474, 678–85, 732, 825

human factors 39, 397–406, 614, 681

human resources 26, 377, 379

Hunsaker and Alessandra 385, 404, 476

identifying nonconforming outputs 721

Imai, Masaaki 77, 78, 81, 93, 295

improvement, process 67, 248, 315, 801

improving: products and services 367, 812, 815; QMS results 817

improving QMS adequacy, suitability & effectiveness 841–5

information: control 539, 588; security 22, 41, 497, 517; from similar designs 600

infrastructure 387–95, 514, 529, 675–6; drawings 124; IT 331; maintenance 120

innovation 33, 59, 94, 96, 100, 279, 338, 583, 801, 813–14
in-process controls 119, 643
in-process verification 710, 743
inspection stamps 689
insurance copies 499, 502
integrating actions 343
integrated QMS 173, 268
integrating QMS requirements 49, 267, 316, 320
integration: barriers to 269; requirement for 268
intellectual property 547, 693, 854
interaction of processes 126, 242
interested party/parties 24, 34, 63, 74, 159–67, 170, 188–98, 251, 854
interface control 589
internal communication 464–70
internal customer 164, 538
International Accreditation Forum (IAF) 13, 854
inventory 394, 485, 528, 657, 716
investors 102, 162, 165, 194, 229, 469
invitation to tender 570, 650, 653
Ishikawa, Kaoru 132, 280, 295, 829
ISO 9000 family of standards xx, xxii, 5, 9, 48, 53–60, 84
ISO 9001 certification xx, 3–4, 9, 13, 19, 38, 49, 66, 207, 644, 651–7
ISO 14001 22–3, 45, 66, 206
ISO standards development process 6
issues relevant to the QMS 185, 266, 325, 329
issuing authority 490

Jeston and Nelis 143, 314, 316
job descriptions 310–19, 470–5, 787
Jobs, Steve 3
Juran 94–8, 133, 243, 365, 369, 768–9, 807
just-in-time 69, 308, 657, 854

Kahneman, Daniel 333
kaizen 93, 295, 807, 854
key performance indicators (KPI) 64, 243–5, 263, 265, 276–7, 747
knowledge, organizational 375, 428–38, 441, 515, 809
knowledge management 26, 375, 428

lagging measures 244, 854
leadership: defining 257; demonstrating 258, 284, 520
leading measures 244, 854
learning: a flow diagram for 77–81; fostering 38; on the job 432, 451; from mistakes 279, 297; opportunity 258, 320; organization 138; scores 453
learning from mistakes 297
learning is not compulsory 813
legibility 501, 689
limited life items 699–703
lost business analysis 750

maintenance of equipment 392, 394, 417
maintenance process 394, 522
management commitment 26, 215, 261
management principles *see* quality management principles
management representative 32, 305–6, 313, 793
management review 17, 29, 46, 318, 365–7, 471, 733, 791–803
management system 108–25
management system standards 23
marketing: function 139–40; process 192–6
marketing and quality 3, 10, 27, 34, 119
Maslow 403
mass production 139, 297, 855
measurement: method 246, 356, 423, 843; of product 743; uncertainty 414, 423, 855
measuring: conformity with procedures 46; performance 351; product conformity 741, 757; resources 207, 409–27; risk 333; service conformity 741, 758
measuring QMS suitability, adequacy & effectiveness 841–5
measuring quality 98, 527, 738–41
mental models 73, 109, 114
metaphors: bicycle 260–6; boiling frog 816; and paradigms 115
methods for operation and control of processes 248
Mil-Q-9858 52, 84
mission critical 656
mission management process 227–9, 388, 519
mission statement 297–8, 363, 431
mistake proofing 856
modifications 581, 585, 630–4, 855
monitoring and measurement activities 252, 413–16, 674–5
monitoring processes 741, 757
motivation: definition of 855; employee 262, 354–5, 802; factors 403; process 403; theories; worker 398, 403–8
mutually beneficial relationships 190

nature of the product 28, 603–4, 649
nonconforming product 35, 719–29, 762, 818–20
nonconformities: classifying 827; confirming 823; consequences of 819–20, 824–6; correcting 824; and corrective action 29; and defect 93; detected during or after delivery 723; levels of 725; reduction 830; report 835; types of 720; verifying corrected 728

objective evidence: authorising release 712; to claim conformity 478, 516; definition of 855; relative to quality assurance 771–89; relative to results 237
objectives *see* business, objectives; corporate objectives; department, objectives; process, objectives; product, objectives; quality objectives; strategic, objectives

obsolete documents 499, 855
on-the-job training 281, 435
on-time delivery 64, 191, 356, 716
operating procedure 483–8
operational: policies 485; processes 31, 213, 363, 521–3
operations planning process 673
opportunities for improvement 29, 61, 316–17, 729, 764, 768, 777, 800–9, 811–17
order processing process 544, 752
organization: goals 129, 257, 855; purpose 76, 177, 354, 366
organizational: interfaces 588–9, 855; knowledge 375, 377, 428–38, 515; roles 118, 305–19
outsourced processes 639–41, 646–8
outsourcing: data storage 502; defined 637; and purchasing 28; risks 342
outsourcing agreement 640, 646

packaging 547, 699–702, 726, 801
paradigms xxii, xxiii, 115, 169, 271, 432
Pareto 656–7, 753, 758, 769, 821, 827, 830
performance: indicators 64, 221, 243, 276–7; requirements 599
personnel acquisition process 382–5
personnel development 449, 455
PESTLE analysis 181–2
plan-do-check-act (PDCA) 77–82, 85, 138, 248, 435, 733, 808–10, 839–41
planned arrangements 264, 658, 709, 712–13, 772, 777–8, 856
planned maintenance 393, 856
Polanyi, Michael 428
policy 289–305
poor quality 92–8, 216, 259, 292, 519
post-delivery 28, 522, 551, 565, 708–9, 717–18, 732
potential nonconformity 720
predictive maintenance 393, 856
pre-qualification of providers 650, 652
preservation: of legibility 501; of process outputs 698–703
pressure groups 159, 165, 167, 182, 195, 329, 341
preventive action 24, 29, 148, 156, 479, 827, 856
preventive maintenance 393–4, 418, 693, 759, 856
previous designs 601, 610, 731
problem solving tools 830
procedures *versus* processes 129
process: analysis 230, 758–63; audit 777–8, 742; characteristics 524, 593, 674; criteria 246, 530; descriptions 485; design 74, 525, 584; development 522, 592; efficiency 72; identification 225; inputs 234; interactions 240–2; models 131–2; objectives 353; parameters 741, 857; records 531; stages 582; validation 584, 677; verification 710

process approach 19, 26, 45–6, 70–6, 126, 138–46, 169, 219, 232, 269–71, 578, 672, 793, 856
process capability 353, 389, 594, 653, 759–63, 856
process control 530
processes: chain of 134, 141, 524; needed for the QMS 212, 219–50; *versus* procedures 129; *versus* systems 130; types of 133
processing environment 396–408, 515, 740
process management 315, 856
procurement process 123, 308, 522, 532, 638
producibility trials 616
product: acceptance 711, 419; approval 554, 617; audit 777–8; characteristics 94, 624, 644, 677, 743; documentation 623; identification 662, 686–90; liability 483, 509, 690; measurement 743; objectives 353; recall 426, 690, 718, 724, 727; requirements 565–7, 589, 736; specifications 610, 625, 691, 714; verification 643–4, 710–11
production: permit 725–6; process 13, 400, 415, 522, 616–17, 623, 681; schedule 394, 485, 726
production part approval 594, 653
productivity 62, 110, 194–5, 287, 351, 396–407, 428, 441, 768, 839
profit 1, 67, 188, 224, 244, 255, 258, 259, 287, 294–6, 320, 338, 353
project: audit 778; management 342; plan 226, 444, 532–3; review 434, 535; schedules 535
protection 701; documented information 496, 565; of property belonging to external providers 695; during transmission 716; against unauthorised denial of use 496
prototype 196, 462, 578, 584, 593–4, 612–13, 622, 857
provider reduction programme 657
psychological factors 44, 396–7, 401–3
purchasing specifications 662
purpose and direction 25–6, 173, 180–5, 227

qualification of providers 650–3
qualification tests 612, 616, 662
qualified personnel 441, 678, 725
quality: attainment levels of 100; characteristics 410, 527, 857; cost and delivery 8, 213, 308, 349–50, 581, 657; of design, conformity and use 801, 857; planning 363, 592–4; problems 99–100, 468, 844, 857; requirements 269, 857; risk 648, 656; tools 830–1; use of term in ISO 9001 100
quality assurance 12, 38, 52–3, 274, 533, 554, 643–4, 647; plans 533; process 647
quality control 274, 352, 761–2, 807, 857
quality first 258–63, 292–7, 685
Quality function deployment (QFD) 566, 583, 831, 857

quality improvement 66, 81, 274

quality issues 462, 509

quality management principles 19, 58, 64, 68-77, 83, 259, 295–6

quality management system (QMS): applicability 204; changes to 26, 365–72, 784, 843; commitment to 261; continual improvement of 839–45; demonstrating leadership of 258; description 214, 782, 784; documenting the 30; planning 325–7, 432, 511, 816; requirements 8, 26, 42, 210, 221, 267–9, 320, 461–3, 772; structure 121; types of 112

quality manager 47, 307

quality manual 30–1, 49, 100, 140, 201, 214, 478, 485

quality objectives 26, 75, 857; auditing 784; awareness of 458–9; deriving 265–7, 329; establishing 347–63, 374; framework for 298–300; planning to achieve 363–4, 374; reviewing achievement of 794

quality plans 100, 361, 532–4, 654

quality policy 33; availability of 303; communicating the 300; establishing 265–7; expressing 293; and leadership 255–8; maintenance of 302

quarantine area 696, 721, 857

records: of competence 454, 479; disposition of 510; retention of 509–10; validation of 585

recovery action 728, 820

reduction: of nonconformity 94, 830–2, 838; of providers 657; in risk 342–3, 648; in targets 141, 188; of variation 5, 768, 840; in waste 844

reductionism 232

reference documents 488

reference material 410, 417

regulatory requirements 9, 104, 555–7, 569–70, 601–2, 663, 858

release of products and services 708–18

reliability: characteristics 600–3; improvement in 801; plans 532, 582; service 527; studies 593; tests 416; verification of 612–16

remedial action 819, 858

reporting: audit results 784; problems to external providers 696–7; QMS performance 316

requirements: applying ISO 9001 205; assurance 60; communicating 661–5; customer 9, 520, 562–74; functional and performance 599; of interest parties 188–98; management 58, 557–9, 772; organization's own 267, 357, 568, 772; product and service 551–74; QMS 267, 772–4, 842

resistance to change 172, 215, 369

resolving differing requirements 570; conflicting inputs 605

resource: management process 227–8, 376, 389, 578; needs 377, 587, 802

responsibility and authority 306–7; for acting on audit findings 787; assigning and communicating 308–11; for ensuring conformity 311–13; for ensuring process performance 313–16; for ensuring QMS integrity 318–19; for promoting customer focus 317–18; for quality 317; for reporting QMS performance 316–17

returned product 823

revision conventions 506

rework 819–20, 835, 844, 858

risk: analysis 325, 333–9; appetite 338, 340; avoiding 341; component testing 345; financial 153; intrinsic and extrinsic 327; matrix 336; mitigation 148, 426, 509, 800; operational 153; positive and negative 149, 151–2; reduction 342–3; taking 151–2, 332, 384, 461; tolerance 338–9; transferring 342; types of 152; use of term in ISO 9001 156

risk assessment: contamination control 700; in design 616, 664; in mitigating environment effects 400–2; in planning audits 774; in plant and facilities maintenance 392; in preventing human error 684; in process planning 461, 524; in requirements review 563; in transportation 703

risk-based thinking (RBT) 22, 26, 39, 156–7, 269–71, 320, 461, 481–2, 672

risk management 148–9, 269, 332–3, 340, 345, 532, 844, 858

risks and opportunities 154–6; addressing 325–46; evaluating the effectiveness of actions to address 346, 766; requirements in ISO 9001 for managing 157; updating during planning 834

roles: clarity of 443; communication of 309; definition of 307; management 281

root cause analysis 820

Russell, Bertrand 148, 430

safety management system (SMS) 15, 23, 116

sales process 54, 230, 522, 525, 559

sample size 749

sampling 132, 643–4, 735, 749, 781

scope: of ISO 9001 8, 199, 349–50, 557; of the QMS 199–209, 252, 257, 280, 307, 350, 397, 482, 640; of registration 199–200, 653

Seddon, John 37–8, 750

segregation: of process outputs 700, 721–2; of product 823, 835

selection of external providers 649–60

self-control, conditions for 132, 309

Senge, Peter 119, 133, 136

sequence of processes 238–40

service: audits 777–8; delivery process 519, 522–3, 671, 719, 732; design validation 617; design verification 614; documentation 624; measurement 743; specifications 625, 741

service level agreement (SLA) 589, 662–73
service quality characteristics 527
service quality measures 413
service specifications 625
servicing 485, 623, 644, 718
six sigma 7, 138, 583, 858
SMART objectives 858
Smith, Adam 139
social factors 404–7
society: defined 165; impact on 160; needs of
   159, 194–5, 792; role of business in 161, 177;
   role of regulations/regulators in 104, 162, 165;
   satisfying 102, 283
special cause variation 760–1, 859
special processes 677–8, 719
stakeholder measures 244, 859
stakeholders 159–67, 188–98, 859
statistical: control 414–15, 515, 709, 761, 859;
   inferences 345; methods 756; studies 759
statistical process control (SPC) 724, 759, 827
statutory requirements 9
storage of documented information 500–3
storage of product 702–3
strategic: direction 46, 59, 178, 192; objectives
   225, 338, 353; planning 180–4, 220, 266, 332;
   risk 152
supplier assessment *see* external providers
supply chain 5, 38, 150, 194, 285, 637, 645,
   657, 691
surveillance: of external providers 656; process
   639
suspension of service or production 191, 726, 835
SWOT analysis 181–4, 213–14, 247, 331, 363
system: audit 120, 774–7, 860; effectiveness
   763, 860; map 121–2, 137, 181–3, 203–4,
   213, 782; maturity 75, 83; model 212, 370;
   purpose or function 117; structure 120
systems approach 24, 69, 70, 72–3, 108, 860
systems thinking 845, 860

targets, setting 244, 246
target value 246, 350, 411, 674, 735, 761–2
Taylor, Frederick Winslow 3, 52, 110, 139
technical interfaces 588–9, 599, 860
terminology: agreement of 584; changes in 32–5;
   consistency of 22, 62, 500; understanding
   xxiii, 87–170
terms of reference 310, 337
test reviews 586
Theory X 670, 860
third party audit 8–9, 42, 396
top management: definition of 87, 257;
   requirements of 255
total quality management (TQM) 319, 811, 860
Toyota 78, 291–5, 828, 856
traceability: maintaining 690

training courses 452–3, 780
training needs 439
Tversky, Amos 333

uncertainty: Bertrand Russell quotes 148, 430; and
   calibration 333; meaning of 149; measurement
   414; state of 150, 152; types of 150
undesirable effects, reducing 815–17
units of measure 334, 422, 588, 622, 776,
   842, 855
user surveys 750

validation: of processes 208, 677
values: aligning 221; changing 7;
   communicating 31; compromising 105;
   and culture 466; defined 178; Enron's 355;
   guiding 59, 258; and paradigms 115; personal
   384; refreshing 217; relative to context 214;
   shared 71, 260, 780; testing 259; transforming
   813; translating 289–98; in the work
   environment 404
variation: due to common cause 415, 704, 709,
   761; due to special cause 414, 760, 833; in the
   environment 400; in measurement processes
   415; natural in systems 357, 530; neglecting
   39; in production 622
verification: matrix 585, 709; of measuring
   equipment 420; plan 585, 710; process
   612–14; on provider's premises 668; on
   receipt 643; requirements 612, 625, 708, 733,
   860; status 417, 689–90
verification and validation: data capture 596;
   design change 634; stages of 585
version control 503, 508, 512
vision: achieving the 292, 305; aligning 221,
   298; and change 369; communicating 31,
   71, 285, 296; creating 255, 260; defined
   178, 860; expressing 178–9; Henry Ford's
   639; objectives and 347, 350; process for
   developing 135, 227; refreshing 217; sharing
   255, 431, 730

waiver *see* concessions
warranty: claims 315, 718, 751; services 522
work breakdown structure (WBS) 544, 581–2,
   587, 860
work environment 194, 396–408
working standards 425
work instructions 470, 488, 676, 699, 890
workmanship: criteria 527, 860; examples of
   528; standards 741
work package 581, 586, 593, 860
work processes 128, 135, 221, 226, 314, 443,
   521, 811, 844

zero defects 93, 860